International Conference on Aerospace and Mechanical Engineering (AeroMech17)

Recent Advancements and Innovations in Aerospace & Mechanical Engineering

IOP Conference Series: Materials Science and Engineering
Volume 370

Batu Ferringhi, Malaysia
21 – 22 November 2017

ISBN: 978-1-5108-6400-9
ISSN: 1757-8981

Printed from e-media with permission by:

Curran Associates, Inc.
57 Morehouse Lane
Red Hook, NY 12571

Some format issues inherent in the e-media version may also appear in this print version.

Printed by Curran Associates, Inc. (2018)

For permission requests, please contact the Institute of Physics
at the address below.

Institute of Physics
Dirac House, Temple Back
Bristol BS1 6BE UK

Phone: 44 1 17 929 7481
Fax: 44 1 17 920 0979

techtracking@iop.org

Additional copies of this publication are available from:

Curran Associates, Inc.
57 Morehouse Lane
Red Hook, NY 12571 USA
Phone: 845-758-0400
Fax: 845-758-2633
Email: curran@proceedings.com
Web: www.proceedings.com

Table of contents

Volume 370

International Conference on Aerospace and Mechanical Engineering (AeroMech17)

21–22 November 2017, Batu Ferringhi, Penang, Malaysia

Accepted papers received: 22 May 2018
Published online: 1 June 2018

Preface

Papers

Preface

As chairman and on behalf of the organising and scientific committee, I am delighted to welcome the delegates from all over the world to The 1st International Conference of Aerospace and Mechanical Engineering, AeroMech'17 on 21 till 22 November 2017 held at Parkroyal Hotel, Penang, Malaysia.

Organised in a beautiful place located at the north of Peninsular Malaysia, AeroMech'17 is an international podium to present the state of art research & development in Aerospace and Mechanical Engineering, while exchanging ideas, thoughts and information between the academia and industry.

This year, our theme is Recent Advancements and Innovations in Aerospace & Mechanical Engineering. Being the first international conference organized by School of Aerospace Engineering, we please to have accepted 67 papers out of the 94 submission. Even with our acceptance rate of 70%, Aeromech'17 managed to attract almost 200 authors' participation from all over the world.

We are also delighted to have two distinguish keynote speakers, Dr. Chan Weoi Leong, research scientist at Temasek Laboratories, National University of Singapore (NUS) and Mr. Izmir Yamin, Founder and CEO of Independence–X Aerospace Sdn. Bhd to address our conference aspiration.

Dr. Chan will deliver a talk on the Recent Development and Challenges on Insect-Inspired Flapping Wing Micro-Air-Vehicle, while Mr Izmir will be presenting on The Development of Flying Mechanism for Martian Surface.

I hope all delegates will participate actively in the conference by sharing their scope of interest within the aerospace and mechanical engineering areas. My gratitude and best wishes to the Dean of School Aerospace Engineering, the entire conference committee team and reviewers who have made the conference a success.

Thank you.

Associate Professor Ir. Dr. Parvathy Rajendran, CEng
Conference Chair
AeroMech'17 -
1st International Conference of Aerospace & Mechanical Engineering

IOP Conf. Series: Materials Science and Engineering **370** (2018) 011002 doi:10.1088/1757-899X/370/1/011002

Peer review statement

All papers published in this volume of *IOP Conference Series: Materials Science and Engineering* have been peer reviewed through processes administered by the proceedings Editors. Reviews were conducted by expert referees to the professional and scientific standards expected of a proceedings journal published by IOP Publishing.

IOP Conf. Series: Materials Science and Engineering **370** (2018) 012001 doi:10.1088/1757-899X/370/1/012001

A Computational Study to Investigate the Effect of Altitude on Deteriorated Engine Performance

Koh, W. C.[1], Mazlan, N. M.[2], Rajendran, P.[3], Ismail, M. A.[4]

[1,2,3] School of Aerospace Engineering, Engineering Campus, Universiti Sains Malaysia, 14300 Nibong Tebal, Pulau Pinang, MALAYSIA
[4] School of Mechanical Engineering, Engineering Campus, Universiti Sains Malaysia, 14300 Nibong Tebal, Pulau Pinang, MALAYSIA

Corresponding email: nmusfirah@usm.my

Abstract. This study presents an investigation on the effect of operational altitudes on the performance of the deteriorated engine. A two-spool high bypass ratio turbofan engine is used as the test subject for this study. The engine is modelled in Gas Turbine Simulation Program (GSP) based on an existing engine model from literature. Real flight data were used for the validation. Deterioration rate of 0.1% per day is applied for all turbofan components engine. The simulation is performed by varying the altitude from sea level until 9000m. Results obtained show reduction in air mass flow rate and engine thrust as altitude increases. The reduction in air mass flow rate is due to the lower air density at higher altitude hence reduces amount of engine thrust. At 1000m to 4000m, thrust specific fuel consumption (TSFC) of the engine is improved compared to sea level. However depleted in TSFC is shown when the aircraft flies at altitude higher than 4000m. At this altitude, the effect of air density is dominant. As a result, the engine is required to burn more fuel to provide a higher thrust to sustain the aircraft speed. More fuel is consumed hence depletion in TSFC is obtained.

1. Introduction

Commercial aviation is expected to grow significantly in the future. Demand in reliable and fast global transportation is increasing and has already set countless new standards for the aviation industry, at the same time, creating new possibilities and challenges in the field of science and technology. With the level of technology advancements today, the manufacturing of turbofan engines are usually of high quality and performance. However, once an engine is no longer a clean engine, deterioration of its performances occur. The phenomenon of deterioration is inevitable since every used engine will deteriorate over time. For aero gas turbines, the main mechanisms of deterioration are contributions of mechanical wear, thermal distress as well as abrasion, corrosion and erosion [1]. Deterioration will effect upon many aircraft operational parameters, which will impede the operations of aircrafts. In most cases, if the deterioration is beyond allowable limitations, it will cause losses in performance and economy to airline operators. Hence, in general, deterioration will cause more expenditure and is harmful to the environment.

This research was carried out to understand the effect of altitude on the overall performance specifically air mass flow rate, thrust and thrust specific fuel consumption (TSFC) of a deteriorated engine, aiming at helping aircraft operators saving expenditure on maintenance and fuel. An engine model within a simulation software was constructed and verified, as well as applying and performing the right conditions and configurations to simulate the research interest situations for the engine.

2. Literature Review

Most of the major propulsion systems for civil aircraft in service today are using gas turbine type of engines. Turbofan engines being the most widely used engine variant for short-to-medium and long range applications. Turbofan engines offers relatively good fuel efficiency and low noise levels compare to conventional turbojet engines [1]. In addition, turbofan engine has an efficient operation Mach number up to 0.85, which is an advantage compared to a turboprop engine [2]. Turbofan engines also has a medium to high bypass ratio direct drive turbofan engine, usually with two or three spool designs. These engines covers a conventional configuration with a large single-stage fan, a multi-stage low pressure compressor, a multi-stage high pressure compressor, an annular combustor, a multi-stage high pressure turbine and a multi-stage low pressure turbine as shown in Figure 1.

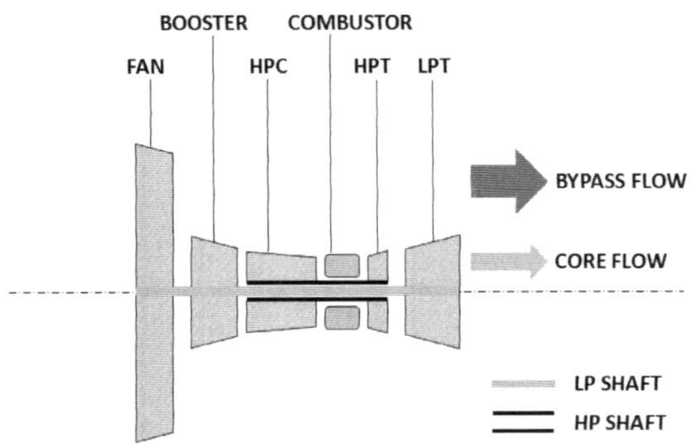

Figure 1. Typical two-shaft turbofan engine configuration [1]

Key parameters which describe the engine performance for an aircraft gas turbine engine is the net thrust (FN) and the specific fuel consumption (SFC) or sometimes thrust specific fuel consumption (TSFC). SFC is usually influenced by factors such as thermal efficiency, propulsive efficiency and combustion efficiency [3]. There are also three main design parameters of a turbofan engine. They are the turbine entry temperature (TET), overall pressure ratio (OPR) and bypass ratio (BPR). A change in these three parameters will significantly affect the engine thermal and propulsive performance.

The maximum TET in an aero engine combustors is usually limited by the mechanical integrity of the combustion chamber and the turbine parts which are exposed to the highest gas temperatures in the whole engine. Apart from the materials available used for manufacturing, these highly thermal stressed engine parts can be applied with active cooling to ensure efficient operation. Hence, an engine which allows higher TET will normally yield greater thermal performance[4]. The overall pressure ratio (OPR) represents the relationship of the total pressure at the compressor exit and the total pressure at the engine inlet. This heavily depends on the number of compressors and the individual compressor design ie: the number of stages. Maximum overall pressure ratios in aero engines are usually limited by the maximum permissible engine weight and the operation ranges of the combustor and the turbines. The engine bypass ratio (BPR) is the ratio between air mass flow rate of air which bypasses the core of the engine to the air mass flow rate passing through the core engine. The air which passes through the core will be involved in the combustion process. Maximum engine bypass ratio is aero engines are usually limited by the size of the fan diameter or by the decrease in size of the core engine diameter. Extremely large fan will increase the aircraft total drag disproportionally as well as the weight of the fan section. Also, a larger fan requires a higher shaft speed. On the other hand, decreasing the size of the core engine is limited by compressor stages pressure ratio and the size of the combustion chamber. More detailed elaboration of the correlations between engine overall pressure ratio (OPR), bypass ratio (BPR) and the fan pressure ratio (FPR) with detailed analyses and diagrams can be found in Walsh, P.P. and P. Fletcher [3] and in Bräunling, W.J [4].

2.1. Gas Turbine Engine Deterioration

For a mechanical turbomachinery such as a gas turbine engine, substantial wear and tear over its service life is inevitable. Gas turbines are subject to gradual deterioration of engine performance with increasing time of operation [5]. In industrial gas turbines 70 – 85% of overall performance deterioration is estimated which causing by deposition [6]. Large particles causing the erosion can be removed from the fluid through proper filtration. However the remaining large fraction of small particles that causing deposition is difficult to be removed [7]. Unlike industrial gas turbines, deposition and erosion in aircraft engine superimpose each other thus making quantification of the individual effects from in-service data is a challenging task. Sallee [8] and Kramer and Smith [9] evaluated in-service data of the Pratt & Whitney JT9D and General Electrics CF-6 turbofan engines and found that one of the main factors causing performance deterioration of high-pressure compressor is surface finish degradation. Among compressor components, front stages being more affected than rear stages, and stators being more affected that rotors with the deposition of mass on stators was approximately twice as high as that on rotor [10]. Rapid roughness was estimated to build up within the first 1000 flight cycles and remain approximately constant afterward.

Deterioration in aircraft can be categorized as (1) On-wing recoverable performance deterioration which can be recovered by on-wing maintenance such as compressor washing, (2) Off-wing recoverable performance deterioration which can be recovered by off-wing maintenance such as disassembly and replacing or refurbish the damaged parts, and finally (3) Permanent performance deterioration which cannot be recovered at economically justifiable expenses. This is usually happen due to natural ageing which constitutes to an unavoidable process.

Engine degradation monitoring or Engine Health Monitoring (EHM) is an essential instrument for gas turbine operation. It is used to determine engine performance and also to allow a prognosis for future performance trends prediction and estimation [11]. Typically, the minimum data that is recorded for performance analysis consists of pressure and temperatures of the engine gas path as well as shaft rotational speed and fuel flow. In addition to those parameters, vibration data, engine oil temperature and pressure is also used for analysis. Apart from obtaining the parameters through sensors, visual inspection is also done either using direct visual such as the borescope or indirect visual methods such as non-destructive testing (NDT) [12]. The measurement of engine exhaust gas temperature (EGT) is a common measurement used to determine actual engine, which in turn allows the calculation of exhaust gas temperature margin (EGTM). This method is used by many major airline operators and has proven to give reliable results on the engine health. It is usually measured at locations after the high pressure turbine exit or at the first stages of the low pressure turbine.

2.2. Previous study on the Effect of Engine Deterioration on Engine Performance

Unlike industrial stationary gas turbine, the impact of engine component deterioration on civil aero engines application has not been widely investigated or available in open literature. Igie et al [13] studied the effect of different level aero-engine compressor fouling on engine performance particularly at short and long-haul missions. Two different aircrafts with different two-spool engines were used. They observed increment in turbine entry temperature (TET) for both aircraft engines in order to maintain the same level of thrust as their clean condition. The highest TET is observed during take-off and climb when thrust setting is the highest.

Kellersman et al [14] performed numerical study in comparing the performance of a jet engine compressor front stage with Blade Integrated Disk (BLISK) geometry during a flight operation interval due to deterioration effects. BLISK is an arrangement of blades in which each stage cannot be changed during maintenance. Unlike conventional compressor front stage, deterioration effects will grow stronger in BLISK until part reaches its life limit. Results obtained from the numerical study shown increases in specific fuel consumption (SFC) and exhaust gas temperature (EGT), both having an influence on the On-Wing-Time of a jet engine and hence the maintenance intervals. However for the conventional blade rows, the blade can be arranged, so the coupled effects can be minimized and the performance can be restored. For a BLISK, the arrangement cannot be changed leading to mistuned stage hence drop engine performance and efficiency.

Jorgenson et al [15] utilized a computer tool in understanding the low pressure compressor and engine performance during an engine roll back event caused by ice accretion. Ice accretion can occurs in the fan low pressure compression system when the aircraft operates at high altitude in which ice water content is high. In general, air static temperature typically increases in fan and low pressure compression system. As ice crystals are ingested into the systems, a portion of the ice crystals melt due to the warmer air. This allows the ice-water mixture to stick to the metal surfaces of the compressor components. This will result to the blockage on stationary components such as the stator vanes hence resulted into the deterioration in performance of the compressor and consequently reduces engine thrust.

An experimental and analytical study was performed to investigate the influence of compressor deterioration on engine dynamic behaviour and transient stall-margin [16]. The analytical study was performed by modifying a two-dimensional, linear, compressible state-space analysis developed by Fuelner et al [17]. Unsteady pressure fluctuation forcing in the clade passages to account engine transients and deterioration were included in the latter model. Whilst the experiments were performed on large commercial aircraft engines in both undeteriorated and deteriorated states. Results showed losses in stall-margin of about 5% due to deterioration. Measurement on transient stall margin has yield to 12% losses for undeteriorated engine and 7% for deteriorated engine. This is because during acceleration transients the operating lines depart from the steady-state level toward the stall line, thus reducing instantaneous stall-margin.

All studies mentioned above observed deterioration effect of certain components on the engine performance. As far as author's knowledge, no study has been performed to investigate deterioration of all components in the engine on engine performance. Therefore this study is the first to analyse the effect of altitudes on engine performance when all components is assumed deteriorated.

3. Methodology

3.1. Modelling CFM56-3 Engine Design Point

This study used a two-spool high bypass turbofan type of engine, CFM56-3 which modelled in Gas Turbine Simulation Program (GSP). The configuration of the engine is shown in Figure 2. GSP is used to model design point and off-design condition for the engine. Within this study, design point of the engine is referred as clean engine where deterioration is absent. The engine is modelled based on the general configuration and characteristics obtained from previous study conducted by Radiaura and Rubio [44] who modelled the engine based on the CFM56-3 Correlation Test Report data [45].

The Correlation Test Report is a proprietary report, owned by the maintenance organization of TAP Portugal in collaboration with CFM International. Since the full report cannot be obtained, a partial informative report is found within the research of Ridaura and Julio Antonio Rubio [44]. From their study, performance data of a CFM56-3 on a test bed as well as their methodology on verifying the CFM56-3 model is obtained. Although the software they used to evaluate the engine performance is differed with the one used in this study, the calculation procedure however is similar and was applied to all engine performance simulation software. In addition, the results they achieved was verified through Kurzke's method [43] and results has been found to be satisfying. Engine parameters used in modelling the engine design point is shown in Table 1. This parameters were adopted in our engine design point model. The verification results are shown in Results and Discussion section.

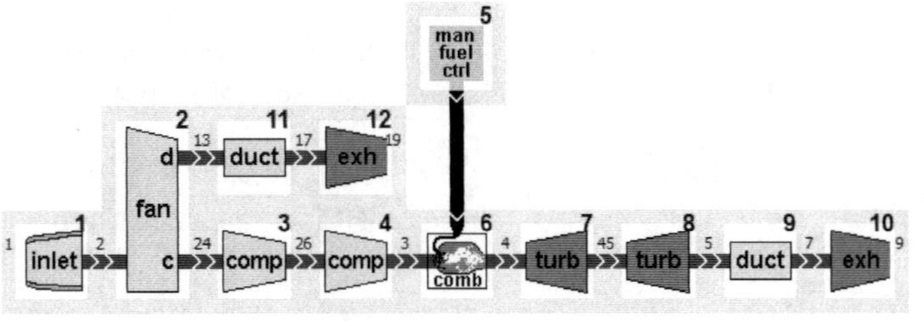

Figure 2. GSP11 CFM56-3 configuration

Table 1. Input parameters for CFM56-3 modelling

Input Parameter	Value
Ambient Temperature, T1 (K)	288.15
Ambient Pressure, P1 (kPa)	101.325
Intake Pressure Ratio	0.99
Inner Fan Pressure Ratio	1
Outer Fan Pressure Ratio	1.714
Intermediate Compressor (Booster) Pressure Ratio	2.237
High Pressure Compressor Pressure Ratio	10.672
Design Bypass Ratio	5.294
Burner Exit Temperature (K)	1636.99
Burner Design Efficiency	0.9995
Fuel Heating Value (LHV) (MK/kg)	43.38

To evaluate the effect of operational altitudes on performance of deteriorated engine, using the settings in the first case study as the base, the ambient conditions is manually varied using the altitude, with intervals of 1000m starting from sea level altitude to the average cruising altitude of similar engine aircraft of 9000m. The pressure and temperature of the respective altitudes are all based on the International Standard Atmosphere. Table 2 shows the respective pressure and altitude used for each altitude.

Table 2. Pressure and temperature at respective altitudes

Altitude (m)	Pressure (bar)	Temperature (K)
Sea Level	1.01	288.15
1000	0.89	281.65
2000	0.79	275.15
3000	0.70	268.65
4000	0.62	262.15
5000	0.54	255.65
6000	0.47	249.15
7000	0.41	242.65
8000	0.36	236.15
9000	0.31	229.65

4. Results and Discussion

4.1. Verification of CFM56-3 Gas Turbine Engine Model Design Point

Table 3 shows percentage difference between the engine simulation performed in GSP 11 and Correlation Test Data for several engine parameters. It is shown that the percentage difference between GSP11 and Correlation Test Data for all engine parameters are below 10% except for the bypass duct pressure (PT17) which is slightly higher than 10%. The results obtained are considered acceptable and the engine model is verified and reliable to be used for further case analysis study.

Table 3. CFM56-3 design point verification results

Performance Parameter	GSP11 Results	Correlation Test Data	Deviation (%)
IPC Temperature, TT26 (K)	387.18	372.70	3.88
HPC Pressure, PT3 (kPa)	2599.43	2371.01	9.63
HPC Temperature, TT3 (K)	787.17	786.19	0.12
Bypass Duct Pressure, PT17 (kPa)	193.80	167.59	15.64
Thrust Specific Fuel Consumption, TSFC (kg/kNs)	0.0109	0.0110	0.91
Exhaust Gas Temperature, EGT (K)	1168.05	1064.65	9.71

4.2. Off-Design Conditions Model Verification

The off-design conditions for CFM56-3 engine modelled in GSP11 is validated by comparing the engine performance with a real flight data of a CFM56-7B24 provided by a local airline. Within this study, the engine used for the validation will be referred as Engine X. Since there is limited data available from the local airline operator, the only comparable parameter for off-design verification is the takeoff exhaust gas temperature (EGT). The CFM56-3 model in GSP11 is set to simulate at the same ambient conditions of Engine X at an ambient pressure of 0.95bar and ambient temperature of 284.25K.

Error! Reference source not found. shows the EGT of Engine X for its first 30 days since new (clean engine). It shows that the engine has operated for almost 200 flights in 30 days. The number of daily flights of Engine X (the number of times the engine being used every day) is different each day for the period of 30 days hence varies the EGT. Additionally, the engine flies at different locations every day thus inconsistency in air humidity, ground temperature, weather etc will influence the fluctuation of EGT. However for simplification an average value of EGT is calculated by considering that the aircraft flies only once a day.

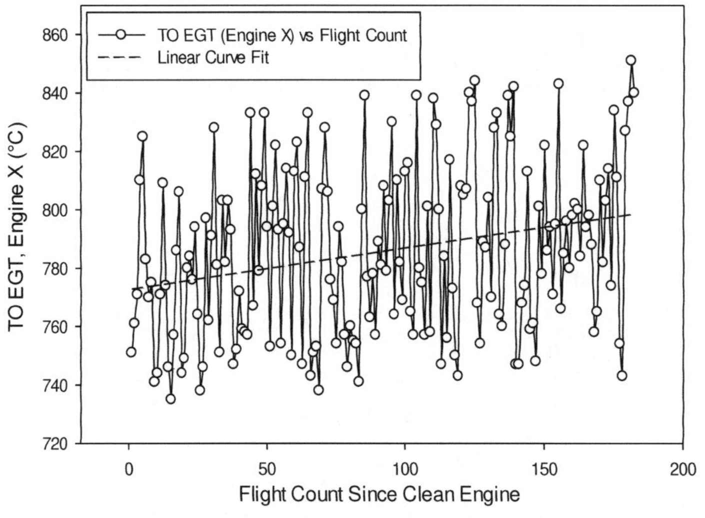

Figure 3. Takeoff exhaust gas temperature against flight count since clean engine [Source: Local airline operator]

The average value of EGT per day is presented as straight line as shown in Figure 4. This value will be used as comparable data with EGT value obtained from simulation.

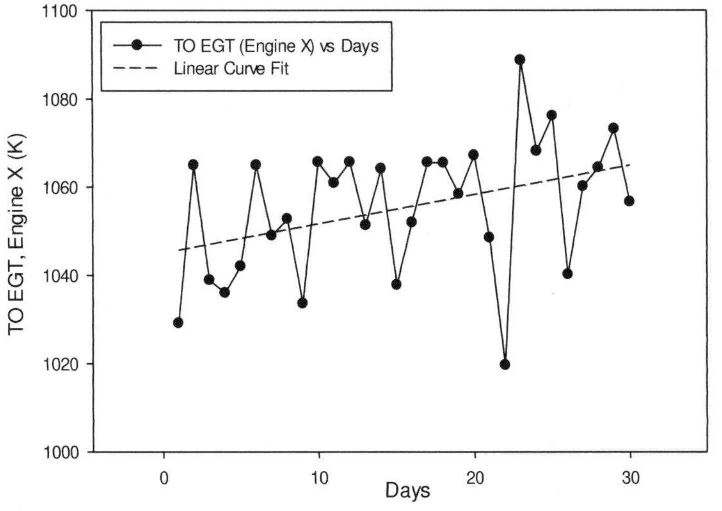

Figure 4. Average takeoff exhaust gas temperature against days since clean engine

The GSP11 modelled engine is applied with deterioration effect where the fan, booster (low pressure compressor), high pressure compressor, high pressure turbine and low pressure turbine are all assumed to have a deterioration rate of 0.1% per day. The results are shown in Figure 5.

It can be noticed that EGT changes when 0.1% deterioration effect was applied and the value increases with the number of operation days. Similar trend is observed as Engine X although absolute values are different. This is due to the fact that Engine X (CFM56-7B24) and CFM56-3 are not the same engine thus those deviations are expected. However considering that both engines are being in the same family, the difference in absolute value is not the main issue, the trend in results obtained are considered more representative than the absolute value. As far as the trend is concerned, the trend similarity between the engine model and real data obtained from local airline shown in the graph is validated and can be used for further analysis.

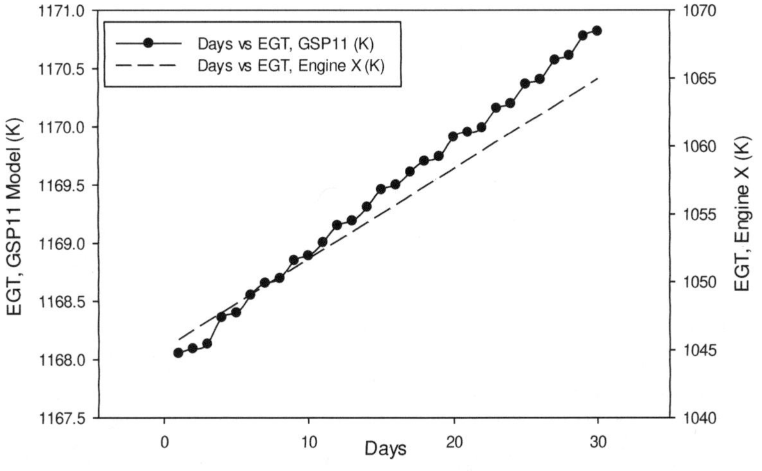

Figure 5. Exhaust gas temperature against day between GSP11 model and Engine X

4.3. The Effect of Altitude on Air Mass Flow Rate, W1

In this study, the effect of altitude on air mass flow rate of the deteriorated engine is observed (Figure 6). Very small changes in air mass flow rate is observed for all 30 days of engine operation is noticed. Furthermore, the deterioration trend is similar for all altitudes. In comparison to sea level, as the altitude increases, the air mass flow rate decreases. This is because as altitude increases, the air density decreases, hence reducing the amount of air goes into the engine.

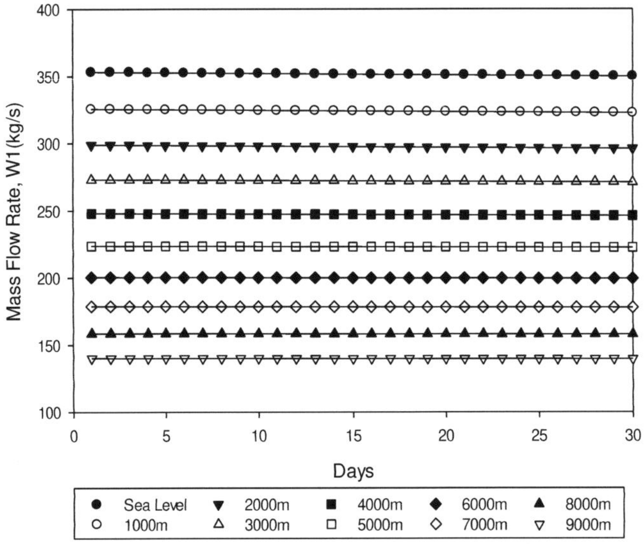

Figure 6. Effect of operation altitude on air mass flow rate

4.4. Effect of altitude on engine thrust, FN

The effect of altitude on engine thrust is depicted in Figure 7. It can be seen that the trend of the graph is similar with the effect of altitude on air mass flow rate. As for turbofan engine, amount of thrust generated by the engine comprises of a combination of thrust generated from the fan and thrust generated from the core engine. Thrust generated from the fan contributes about 80% of total thrust. Therefore decreases in the amount of air bypassing the fan will contribute to the reduction in total thrust of the engine. As for air mass flow rate, as the altitude increases, the air density decreases hence reducing the amount of air entering the fan. For that reason, the amount thrust generated by the engine reduces. As number of operation day increases, reduction in engine thrust is observed and it is consistent with all altitudes.

4.5. Altitude Effect on Thrust Specific Fuel Consumption, TSFC

Figure 8 shows the effect of altitude on the thrust specific fuel consumption at altitude ranging from 1000m to 9000m. Sea level depicted design point of the engine model. It can be seen that at 1000m to 4000m, TSFC of the engine is improved (low TSFC) compared to design point. However TSFC of the engine started to increase (depleted TSFC) when the aircraft flies at altitude higher than 4000m. TSFC of the engine is worst at altitude of 9000m. It is known that at high altitude air density decreases thus provide lower drag for the aircraft. The lower drag causing the engine to consume less fuel in providing enough thrust hence lower TSFC is observed. Meanwhile after 4000m (the altitude becomes much higher), pressure and density drops much quicker than temperature so that the effect of air density is much dominant. Reduction in pressure ratio causes the fuel flow decreases. This is required to maintain the proper fuel-air ratio in the engine. At this condition, thrust of the aircraft reduces. However, since TET of the engine is set to be constant in the beginning. As a result, for the engine to maintain the constant turbine entry temperature (TET) at higher altitudes, the turbofan engine needs to burn more fuel to provide a higher thrust to sustain the aircraft speed, hence after 4000m, the turbofan engine burns more fuel with a higher TSFC.

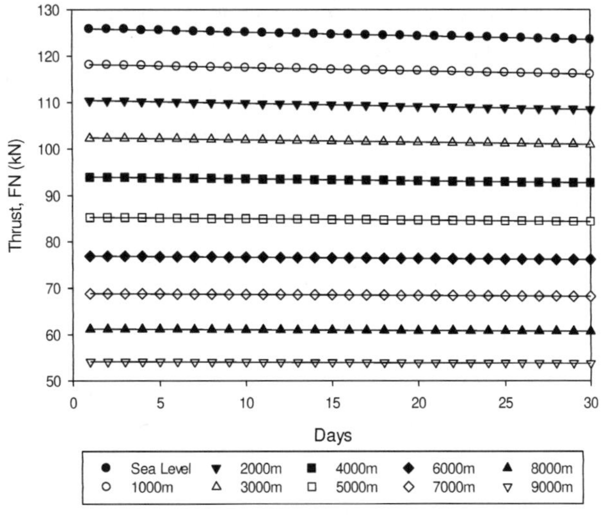

Figure 7. Effect of operation altitude on thrust

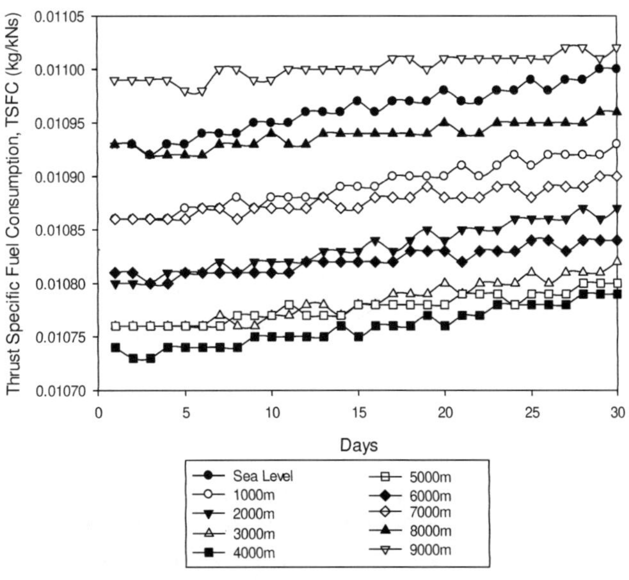

Figure 8. Effect of operation altitude on thrust specific fuel consumption

5. Conclusion

This study is performed to examine the changes in engine performance when all components in the engine is deteriorated. A two-spool high bypass turbofan engine was modelled in Gas Turbine Simulation Program (GSP) based on engine parameters obtain in literature. The model was validated with real flight data obtained by local airlines.

The effect of altitude on engine performance particularly air mass flow rate, thrust, and thrust specific fuel consumption was analysed by ranging the altitude from 1000m to 9000 m. The simulation was performed for 30 days on operation. As altitude increases, reduction in air mass flow rate is observed. This reduction contributes to the reduction in engine thrust due to small amount of air bypassing the fan. TSFC of the engine improved when the aircraft flies from 1000m to 4000m. However TSFC depleted when aircraft flies more than 4000m. The effect of air density is more dominant at higher altitudes. Therefore in order to maintain the same turbine entry temperature and velocity required by the aircraft, the engine has to burn more fuel thus increases TSFC.

Acknowledgement
This research and publication was supported by Universiti Sains Malaysia Grant No. 304/PAERO/ 60315002.

References

[1] Venediger B 2013 *Civil aircraft trajectory analyses-impact of engine degradation on fuel burn and emissions*
[2] Grieb H S 2004 Design of Turboprop Engines. Basel-Boston-Berlin
[3] Walsh P P and Fletcher P 2004 *Gas turbine performance: John Wiley & Sons*
[4] Bräunling W J 2015 *Aircraft engines: foundations, aero-thermodynamics, ideal and real circuit processes, thermal turbomachinery, components, emissions and systems:* Springer-Verlag
[5] Döring F, Staudacher S, Koch C and Weißschuh M 2017 *Journal of Turbomachinery* **139** 051003
[6] Diakunchak I S 1991 *Performance deterioration in industrial gas turbines: American Society of Mechanical Engineers*
[7] Kurz R and Brun K 2012 *Journal of engineering for gas turbines and power* **134** 032401
[8] Sallee G 1978 *Performance deterioration based on existing (historical) data; JT9D jet engine diagnostics program*
[9] Kramer W and Smith J 1978 *Long-term CF6 engine performance deterioration: Evaluation of engine S/N 451-380*
[10] Tarabrin A, Schurovsky V, Bodrov A and Stalder J-P 1998 *Influence of Axial Compressor Fouling on Gas Turbine Unit Perfomance Based on Different Schemes and With Different Initial Parameters.* In: ASME 1998 International Gas Turbine and Aeroengine Congress and Exhibition: American Society of Mechanical Engineers pp V004T11A6-VT11A6
[11] Roth B A, Doel D L and Cissell J J 2005 *Probabilistic matching of turbofan engine performance models to test data*
[12] Li Y 2002 *Proceedings of the Institution of Mechanical Engineers, Part A: Journal of Power and Energy* **216** 363
[13] Igie U, Goiricelaya M, Nalianda D and Minervino O 2016 *Proceedings of the Institution of Mechanical Engineers, Part G: Journal of Aerospace Engineering* **230** 1312
[14] Kellersmann A, Reitz G and Friedrichs J 2014 *Procedia CIRP* **22** 249
[15] Jorgenson P C, Veres J P and Jones S M 2014 Modeling the Deterioration of Engine and Low Pressure Compressor Performance during a Roll Back Event due to Ice Accretion
[16] Spakovszky Z, Gertz J, Sharma O, Paduano J, Epstein A and Greitzer E 1999 *ASME J. Turbomach* **122** 477
[17] Feulner M R, Hendricks G J and Paduano J D 1996 *Transactions-American Society Of Mechanical Engineers Journal Of Turbomachinery* **118** 1

IOP Conf. Series: Materials Science and Engineering **370** (2018) 012002 doi:10.1088/1757-899X/370/1/012002

Evaluation of Electromagnetic Near-Field Measurement Technique as Non-Destructive Testing for Composite Structures

Alaa Raad Hussein, Thar M. Badri Albarody* and Puteri Sri Melor Bt Megat Yusoff
Department of Mechanical Engineering, Universiti Technologi Petronas, Malaysia

*Corresponding author: dher.albarody@utp.edu.my

Abstract. Nowadays there is no viable non-destructive method that could detect flaws in complex composite products. Such a method could provide unique tools to allow engineers to minimize time consumption and cost during the evaluation of various product parameters without disturbing production. The latest research and development on propagation waves introduce micro, radio and millimetre waves as new potential non-destructive test methods for evaluation of mechanical flaws and prediction of failure in a product during production. This paper focuses on recent developments, usage, classification of electromagnetic waves under the range of radio frequency, millimetre and micro-waves. In addition, this paper reviews the application of propagation wave and proposed a new health monitoring technique based on Doppler Effect for vibration measurement in complex composite structures. Doppler Effect is influenced by dynamic behaviour of the composite structures and both are effect by flaws occurred inside the structure. Composite manufacturers, especially Aerospace industry are demanding these methods comprehensively inspect and evaluate the damages and defects in their products.

1. Introduction

Non-destructive testing (NDT) or so called non-destructive and evaluation (NDE) is an inspection technique to detect and locate flaws on the surface or through thickness of an object being tested, also this method is cost effective, high quality results, in-line diagnose [1] and speedy testing while required less operators such as labors or technologists [2]. So to satisfy these requirements, selection of suitable NDT instruments subjected to a number of criteria such as; types of materials forming composite that have different physical properties varies according to the weight fraction of composite constituent [3]; shape, size and types of flaws need to be detected; cost of the technique used for inspection [2, 4-6].

In the last decades composite materials are developing very rapidly and require sophisticated NDT system. NDT systems are able to identify the flaws in composite products such as fiber breakage, inter-phy delamination, crack propagation and overheating without induce stress or damages [7, 8]. Therefore, composite industries are demanding a new NDT system that could predict flaws during production and failure during operation as part as structural health monitoring (SHM) of the critical parts. In addition, it can reduce (up to 30%) of the total cost when compared to that of traditional

NDE [9]. This can be achieved by online monitoring which allows efficient solution to interrogate a wide range of the surface of composite structure [10].

Visual inspection is the most common, cheap and easy NDT [2] propagate wave is the most suitable method to visualize internal damages. Ultrasonic test, eddy current X-ray inspection, radiography, shearography, magnetic particle test [2, 4, 11-13] and terahertz spectroscopy could asses the crack initiation and are able to provide data and properties of a structure in advance [14]. Nevertheless, these methods are limited inspection capability, since, it cannot penetrate deeply inside complex shapes to investigate the defects as well as it represents a time consuming and costly especially for composite parts [15].

On the other hand, laser Doppler vibrometers is a suitable method for vibration measurement and provides good information on the bond joint in a composite structure but could not be employ to predict defects inside composites [14, 16, 17]. Microwave, radio frequency interferometry, millimeter wave and lamb wave [18] are very useful inspections methods and health monitoring systems [19] in providing remote inspection on surface and deep layers of structures [17-19].

This paper reviews current existing NDT methods based on their capacity to evaluate the mechanical defects in composite structures. It also proposes a new technique for inspection and monitoring defects in complex composites after production and during service using a near-field measurement technique based electromagnetic waves.

2. Waves Propagation Inspection Techniques

Near-field measurement technique based waves propagation attracts huge attention from NDT specialists due its deconvolution, reliable and repeatability, such inspection possess high sensitivity to damages providing a feasible solution that might minimize the total test time [20].

Commonly, micro, radio, millimeter waves have a good propagation in a wide range of dielectric composite materials [21]. Nevertheless, other composite materials could inspected by associated with short wavelength [22, 23] or ultrasonic and lamb waves, taking into account wave attenuation according to the properties of the material itself, such as: damping in material, wave dissipation, energy required to transfer the wave and so on [24, 25]. Generally, guided waves that are mentioned herein can be classified into five types namely ultrasonic, lamb, RF, millimeter and micro-waves, all are candidate of NDT. Table 1 shows the advantages and disadvantages and other properties of the mentioned waves.

Table 1. Guided wave properties.

Guided wave	Properties
Millimeter wave [19, 26]	It has the ability to penetrate in all the types of material such as metallic, ceramic and etc., the range of the frequency is start from 10 GHz up to 300 GHz or exceed this limit. It is safe for human being when they expose to.
Ultrasonic wave [18, 27]	Limited to material investigation and provide a minor failure of composite that can be on-line obtained from this method. Cannot interact with the internal structure of high porosity material. The range of frequency is start from 300 MHz to 30 GHz
Radio frequency [28, 29]	It is kind of electromagnetic wave and consider as low cost for tracking and identification for detect crack growth, strain and corrosion
Microwave [29, 30]	This wave excites thermography and sensitive to crack detection. Low attenuation associated with it propagation in dust and smoke, also it is suitable for critical application for dielectric materials and ceramic The range of frequency is from 300MHz to 30 GHz and it reach 300 GHz when associated with millimeter wave.
Lamb wave [31, 32]	It is a type of ultrasonic wave, its propagate in isotropic material. Plate thickness is playing important role of it velocity and propagation (thin plate) because this waves considered as dispersive waves, as well as it can identify the damages may occur in composites

As shown in Table 1, millimeter and micro-wave are the most suitable for defect evaluation due to the unique properties of penetrate most types of materials whether its conductive, dielectric, metallic or ceramic materials. Furthermore, millimeter wave can test objects behind barriers and provides information about the vibration amplitude for the operating objects [17, 33].

3. Electromagnetic Near-Field Health Monitoring Techniques

Vibration is one of the major problems that may affect the global integrity and reliability of composite structures and can lead to a catastrophic defects in the materials [34]. So vibration monitoring during operation plays an important role in terms of in-situ structural health monitoring [30]. There are different types methods to monitor and measure vibration, each one depends on the shape and design of the object [35]. A conventional method for measuring vibration was used by Giurgiutiu [36] is shown in Figure 1, the obtained results is suitable for SHM specially for vibration amplitude and mode shape. For NDT vibration can be calculated via scanning laser vibrometers which used by Ooijevaar et al. [37]. They considered laser vibrometer technique is a fast measurement method comparing with accelerometers which provides more confidence for the vibration analysis.

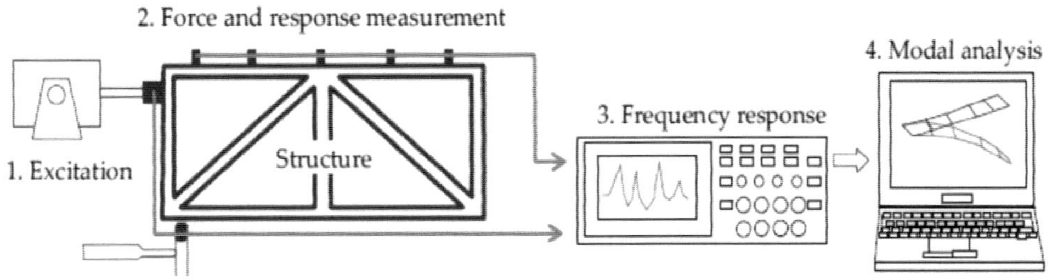

Figure 1. Conventional schematic for model analysis [37]

Continuous wave (CW) which is one of the methods employed for vibration detect (as shown in Figure 2), is contain of two antennas connected to a software, one of them for transmitting the signal and the other for receiving the scattered wave [35]. The scattered wave is the information wave of the phase, frequency and amplitude of the vibration. This wave is proportional to periodically motion of the object. CW method of vibration test is fast and does not need to be connected with sensors to the monitored item. The received signal will be analyzed by fast Fourier transform (FFT) to simulate the mode shape as in the frequency modulation continuous wave (FMCW) approach for vibration measurement and tracking moving objects [26, 38].

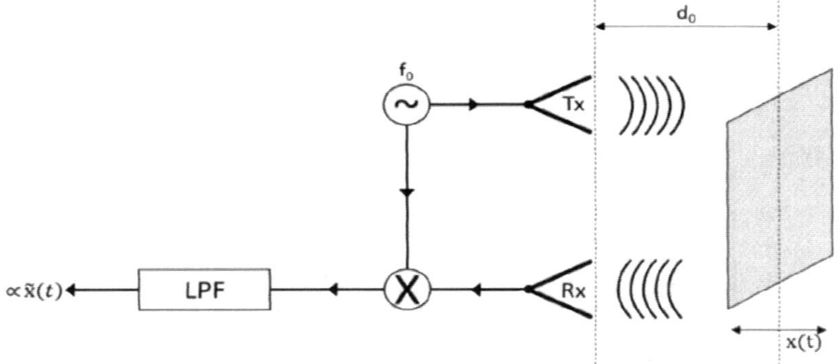

Figure 2. Continuous wave for vibration detect [35]

In contactless vibration measurement method for composite materials, a phase modulation is the key for analysis of vibration as well as wave length associated with frequency domain can control the phase modulation [39]. Consequently, Doppler radar sensor is widely used for vibration measurement using this method in a nonlinear mode. The measurement method basically depends on two waves (transmitting and backward), the backward waves received by antenna and go through specific process, the material being tested is a flat plate cover with aluminium foil to enhance better reflectivity [40].

4. Remote Defect Inspection

This study proposes a system where a complex composite structure may be exposed to electromagnetic waves under the range of RF, millimeter or micro-waves. These waves may either be reflected or transmitted through the breadth of product carrying instantaneous stress state information. Such dynamic information can be interpreted to estimate defects and flaws occurrence and evolution under operating conditions.

The system is basically consisting of two antennas one for transmitting and the other for receiving the wave (see Figure 3). The receiving antenna is placed behind the product that is designed to allow for penetration of electromechanical wave. The antennas are connected to a network analyzer [28]. The reflected waves will utilize to detect the vibration amplitude to identify flaws inside a complex composite under operational condition. The transmitted waves on the other side will pass through the product in shifted phase where the phase differences may carry defect information.

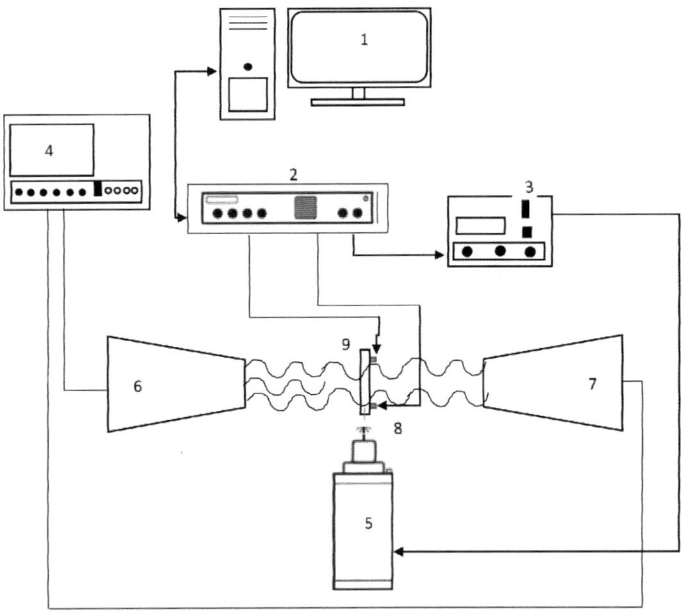

No.	Description
1	Computer
2	DAQ
3	Power amplifier
4	Vertical network analyzer
5	Shaker
6	Transmitting antenna
7	Receiving antenna
8	Accelerometer
9	Vibrating object

Figure 3. Vibration and flaw measurement (system conducted in this study)

5. Conclusion

Non-destructive testing was reviewed in terms of classification and properties for each advance method used recently to test composite materials that are developed for aerospace applications. Vibration measurement is one of parameters that plays an important role for the life cycle of a structure. Until to this study, the wave propagation in a material is not fully utilize especially in prediction of failure, crack propagation and delamination of complex composite structures. This paper proposes a comprehensive technique that utilized in operation and/or inspection of the SHM.

Acknowledgment
Authors would like to thank mechanical engineering department, universiti technologi PETRONAS. for its esteemed support.

References

[1] G.P. Malfense Fierro 2017, M. Meo, *Ultrasonics*, 74 30-47.
[2] P. Cawley 2001, *Proceedings of the Institution of Mechanical Engineers, Part L: Journal of Materials: Design and Applications*, 215 213-223.
[3] I. Scott, C. Scala 1982, *NDT international*, 15 75-86.
[4] A. Katunin, K. Dragan, M. Dziendzikowski 2015, *Composite Structures*, 127 1-9.
[5] T.M.B. Albarody, H.H. Al-Kayiem 2015, *Journal of Mechanical Science and Technology*, 28 4877-4891 Springer.
[6] T.M.B. Albarody, H.H. Al-Kayiem, W. Faris 2016, *Journal of Mechanical Science and Technology*, 30 77-87.
[7] M. Carboni, A. Gianneo, M. Giglio 2015, *Ultrasonics*, 60 51-64.
[8] S. Gholizadeh 2016, *Procedia Structural Integrity*, 1 50-57.
[9] M.H. Sherafat, N. Quaegebeur, L. Lessard, P. Hubert, P. Masson 2014, *EWSHM - 7th European Workshop on Structural Health Monitoring*, Nantes, France.
[10] V. Memmolo, F. Ricci, N.D. Boffa, L. Maio, E. Monaco 2016, *Procedia Engineering*, 167 48-55.
[11] D. Zhou, M. Pan, Y. He, B. Du 2017, *Measurement*, 105 136-145.
[12] B. Salski, W. Gwarek, P. Korpas, S. Reszewicz, A.Y.B. Chong, P. Theodorakeas, I. Hatziioannidis, V. Kappatos, C. Selcuk, T.-H. Gan, M. Koui, M. Iwanowski, B. Zielinski 2015, *Composite Structures*, 122 104-112.
[13] S. Kharkovsky, R. Zoughi 2007, *IEEE Instrumentation & Measurement Magazine*, 10 26-38.
[14] P. Kudela, T. Wandowski, P. Malinowski, W. Ostachowicz 2016, *Optics and Lasers in Engineering*.
[15] V. Giurgiutiu, C. Soutis 2010, *Encyclopedia of aerospace engineering*.
[16] M.H. Sherafat, R. Guitel, N. Quaegebeur, P. Hubert, L. Lessard, P. Masson 2016, *Materials & Design*, 90 787-794.
[17] J. Moll, K. Bechtel, B. Hils, V. Krozer 2014, *EWSHM - 7th European Workshop on Structural Health Monitoring*, Nantes, France.
[18] T. Clarke 2009, Imperial Collage London, 182.
[19] S. Kim, C. Nguyen 2014, pp. 90620E-90620E-90624.
[20] T. Cicero 2009, Imperial College London.
[21] W. Węglewski, K. Bochenek, M. Basista, T. Schubert, U. Jehring, J. Litniewski, S. Mackiewicz 2013, *Computational Materials Science*, 77 19-30.
[22] M. Bilodeau, N. Quaegebeur, P. Masson 2017, *NDT & E International*, 88 33-41.
[23] D.A. Cook, Y.H. Berthelot 2001, *NDT & E International*, 34 483-492.
[24] T. Wandowski, P. Malinowski, J. Moll, M. Radzienski, W. Ostachowicz 2015, *SPIE Smart Structures and Materials and Nondestructive Evaluation and Health Monitoring, International Society for Optics and Photonics*, pp. 94380Q-94380Q-94314.
[25] N. Guo, R.C. Qiu, S.S. Mo, K. Takahashi 2007, *EURASIP journal on Wireless Communications and Networking*, 48-48.
[26] M. scholz, S. rediske, A. nuber, H. friedmann, J. moll, P. arnold, V. krozer, P. kraemer, R. salman, D. pozdniakov 2016.
[27] V. Memmolo, F. Ricci, L. Maio, N. Boffa, E. Monaco 2016, *SPIE Smart Structures and Materials and Nondestructive Evaluation and Health Monitoring, International Society for Optics and Photonics*, pp. 980504-980504-980512.
[28] C. Zhang, L. Ma 2016, *Scientific Reports*, 6.
[29] J. Zhang, G.Y. Tian, A.B. Zhao 2017, *NDT & E International*, 86 89-99.

[30] J. moll, v. krozer 2016, *8th European Workshop on Structural Health Monitoring*, pp. 1-6.
[31] M. Mańka, M. rosiek, A. Martowicz, T. Stepinski, T. Uhl 2016, *Mechanical Systems and Signal Processing*, 78 71-83.
[32] P. Ochôa, V. Infante, J.M. Silva, R.M. Groves 2015 *Composites Part B: Engineering*, 80 291-298.
[33] S. Kim 2006, development of millimeter wave integrated-circut interferometric sensors for industrial sensing applications, Texas A & M University.
[34] A.R. Hussein, N.A.A. Jail, A.R. Abu Talib 2011, *Journal of Applied Sciences*, 11 348-353.
[35] A. Raffo, S. Costanzo, G. Di Massa 2017, *Sensors*, 17 115.
[36] V. Giurgiutiu 2014, high frequency vibration SHM with PAWS modal sensor - the electromechanical impedance method. 509-572.
[37] T.H. Ooijevaar, R. Loendersloot, L.L. Warnet, A. de Boer, R. Akkerman 2010, *Composite Structures*, 92 2007-2015.
[38] W. Sediono 2013 Proceedings of 2013 *IEEE International Conference on Teaching, Assessment and Learning for Engineering* (TALE), pp. 378-381.
[39] C. Gu, T. Inoue, C. Li 2013, *IEEE Microwave and Wireless Components Letters*, 23 566-568.
[40] S. Kim, C. Nguyen 2003, *IEEE Transactions on Microwave Theory and Techniques*, 51 1724-1728.

Background-Oriented Schlieren (BOS) for Scramjet Inlet-isolator Investigation

Azam Che Idris [1,*], Mohd Rashdan Saad[1], Kin Hing Lo[2] and Konstantinos Kontis[2]

[1]Faculty of Engineering, National Defence University of Malaysia, Kem Sungai Besi, Kuala Lumpur 57000, Malaysia

[2]School of Engineering, University of Glasgow, James Watt South Building, Glasgow G12 8QQ, Scotland;

Corresponding email: *azam.che@alumni.manchester.ac.uk

Abstract. Background-oriented Schlieren (BOS) technique is a recently invented non-intrusive flow diagnostic method which has yet to be fully explored in its capabilities. In this paper, BOS technique has been applied for investigating the general flow field characteristics inside a generic scramjet inlet-isolator with Mach 5 flow. The difficulty in finding the delicate balance between measurement sensitivity and measurement area image focusing has been demonstrated. The differences between direct cross-correlation (DCC) and Fast Fourier Transform (FFT) raw data processing algorithm have also been demonstrated. As an exploratory study of BOS capability, this paper found that BOS is simple yet robust enough to be used to visualize complex flow in a scramjet inlet in hypersonic flow. However, in this case its quantitative data can be strongly affected by 3-dimensionality thus obscuring the density value with significant errors.

1. Introduction

The Background-Oriented Schlieren (BOS) technique, which has been gaining its popularity today, is particularly suitable for calculating the density increase across an oblique shock. The calculated density gradient can be used to estimate the overall performance of a scramjet inlet using the concept shown by Idris *et al.* [1]. Compared to their method where average isolator flow properties were inferred from the pressure map on sidewall, two-dimensional BOS technique would result in spanwise-integrated density field for the whole isolator [2]. Concurrently, a three-dimensional BOS which requires simple modifications from two-dimensional method could provide truly tomographic reconstruction of the isolator density-field along the spanwise direction [3][4]. Hence, the calculated isolator exit Mach number and subsequently the inlet-isolator performances by using BOS method will be more reliable and accurate.

The method was first presented by Dalziel *et al.* in late 1990s as the "synthetic schlieren" [5 – 7]. They demonstrated that the displacements of light-ray path from random dots set up on the background of a changing density volume can be processed by using cross-correlation software "borrowed" from Particle Image Velocimetry (PIV) methods. The results are in the form of light-ray path-integrated density-gradient field, which explains why they associated it with the classical schlieren technique. At about the same time, Meier independently introduced and patented the concept of BOS in Germany [8].

This relatively new method has very simple setup, which requires only high resolution camera (typically Digital Single-Lens Reflex (DSLR) camera) and a suitable background image, thus making it adaptable for many applications. For example, Richard and Raffel visualised compressible vortices from spinning helicopter rotor blade using BOS [9]. They took images of a hovering helicopter with the camera positioned at an angle such that the tip of the rotor blade was set against background view of a grass field. The vortices from the blade tips were shown to be interacting with its engine exhaust plume.

Elsinga *et al.* compared this method with another quantitative schlieren technique called Calibrated Colour Schlieren (CCS) in analysing a two-dimensional wedge-plate model in flow Mach number of 1.94 [2][10]. BOS and CCS were shown to be effective in visualising shock and expansion wave. The resultant flow field properties calculated using BOS were close to the analytical value predicted from oblique shock relations and Pandtl-Meyer expansion theory for that particular geometry.

Venkatakrishnan and Meier applied this method to predict the flow around axisymmetric cone-cylinder geometry in a freestream Mach number of 2 [3]. Filtered back projection algorithm introduced in the article allows the authors to obtain slices of two-dimensional density-field around the model. The calculated density matched the published data of cone tables [11]. There are many other exciting and interesting examples of recent applications of BOS in flow-diagnostics, however, there has been no attempt yet to apply this method for scramjet inlet-isolator investigations.

2. Basic BOS Theory

The theory for BOS is relatively simple and can be described using figure 1 below. In the figure, x-axis is parallel to the freestream and starts at the model tip at half-symmetry plane, y-axis is parallel with the vertical plane and z-axis is along the line of sight. Random patterns of suitable shapes and sizes are printed onto single image and fixed on the background plane. The background plane is positioned at a distance (Z_D) from the middle plane of density gradient volume in investigation. The lens of the image capture device is located at length (Z_B) from the background plane. The notation (Z_I) in the figure denotes the focal length of the lens, where the light rays from the background plane converged onto the camera chip. From the figure, consider that as the control volume changes in density, the light ray from a particular point on the background plane will shift by (Δy) on the camera chip. This displacement can be calculated by using the PIV cross-correlation software using inputs of two still images of "wind-off" and "wind-on". The wind-off is the initial condition where the density field is known and is usually the wind-tunnel condition without flow. The wind-on, as the name implies, refers to the condition where the density field changes due to wind tunnel flow. This displacement is due to the light path being deflected vertically by angle ε_y. The deflection angle is typically very small and can be obtained by:

$$\sin(\varepsilon_y) \approx \tan(\varepsilon_y) \approx \varepsilon_y = \frac{\Delta y'}{Z_D} \tag{1}$$

Here, $\Delta y'$ denotes the virtual displacement of the point on background image and can be related to Δy displacement detected by the camera by:

$$\frac{\Delta y'}{Z_D} = \frac{\Delta y}{Z_I} \tag{2}$$

The deflection angle ε_y can be related to change in refractive index $\frac{\delta n}{\delta y}$ of the volume by:

$$\varepsilon_y = \frac{1}{n_0} \int_{Z_D - \Delta Z_D}^{Z_D + \Delta Z_D} \frac{\delta n}{\delta y} dz \tag{3}$$

Here, n_0 denotes the initial refractive index of the flow, "wind-off" conditions prior to change in density field. The change in refractive index $\frac{\delta n}{\delta y}$ can be related to density gradient $\frac{\delta \rho}{\delta y}$ by partially-deriving the Gladstone-Dale equation:

$$\frac{\delta(n-1)}{\delta y} = \frac{\delta(k\rho)}{\delta y} \tag{4}$$

$$\frac{\delta n}{\delta y} = k\frac{\delta \rho}{\delta y} \tag{5}$$

Here, k is the Gladstone-Dale constant and its value depends on type of gas used.

Equation (1) to (5) can be similarly applied to consider the light path deflection on x-axis. Thus, the final density gradient at any point in the flow field is a product of $\frac{\delta \rho}{\delta x}$ and $\frac{\delta \rho}{\delta y}$.

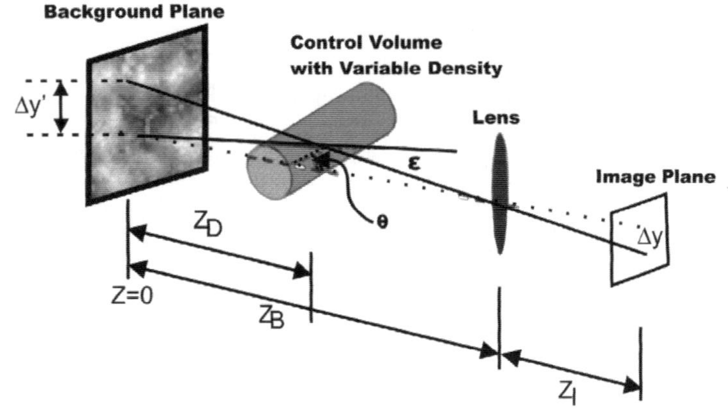

Figure 1. Deflection of light path from background image due to change in density gradient.

3. Cross-correlation Theory

The background images of wind-off and wind-on can be discretized into a finite number of interrogation windows for cross-correlation analysis. The windows must be of the same size, W, and contains enough shape elements for analysis. Since a window could contain many elements that have different displacement magnitudes, it is best to have the smallest window size possible to have a good representative of average particles displacement (see figure 2). The cross-correlation function can be calculated by using equation shown below [12]:

$$\emptyset(\Delta x, \Delta y) = \frac{\sum_{x,y}^{W}(I_a(x,y) - \bar{I}_a)(I_b(x+\Delta x, y+\Delta y) - \bar{I}_b)}{(\sum_{x,y}^{W}(I_a(x,y) - \bar{I}_a)^2 \sum_{x,y}^{W}(I_b(x,y) - \bar{I}_b)^2)^{0.5}} \tag{6}$$

Here, I_a denotes the pixel intensities at location (x,y) within the interrogation window in the wind-off image. Similarly, I_b denotes the pixel intensities at location (x,y) within the interrogation window in the wind-on image. $\bar{I}_a$ and $\bar{I}_b$ are the average I_a and I_b for the whole window size. $\emptyset(\Delta x, \Delta y)$ is the correlation function in terms of displacement in x-y axis, Δx and Δy. The function reaches peak value at the coincident points of elements match between wind-off and wind-on images.

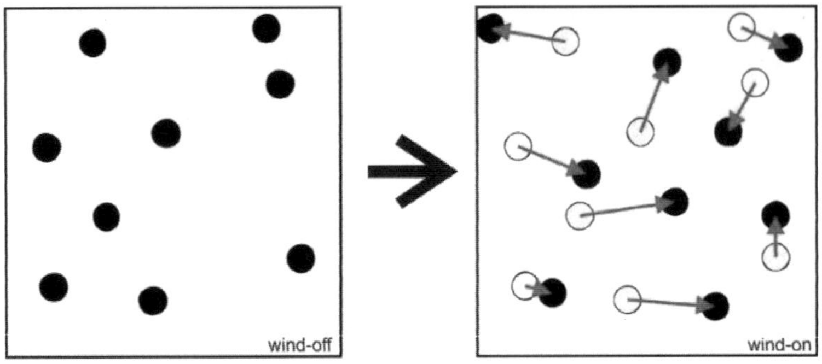

Figure 2. Movement of elements in an interrogation window for cross-correlation analysis

Depending on the interrogation window size, the computing requirements could be really expensive. Thus, Fast Fourier Transform (FFT) approximation is typically used to produce each displacement vector [13]. However, Pust had shown that Direct Cross Correlation (DCC), which directly solves the equation without approximation, is significantly better in producing realistic results in comparison to FFT [13].

4. Experimental Setup

The BOS setup for current scramjet inlet-isolator model was applied onto scramjet inlet model similar to Idris *et al.* [1]. The scramjet inlet was subjected to dry air flow with static pressure of 1220 Pa, static temperature of 62 K, total pressure of 645 kPa, total temperature of 370 K and Mach number of 5.

Due to the sensitivity of BOS system the model should be located halfway between the background and camera lens (i.e $Z_D = 0.5Z_B$) [3][14]. Hargather and Settles demonstrated that as the model is gradually moved towards the background, the sensitivity of system drops significantly [14]. The problem with this requirement is that with such small isolator height of only 6.8 mm, a sufficient compromise of camera focusing could not be found. If the lens is focused on the background image, then isolator section would appear heavily blurred, rendering the resultant image useless. On the other hand, if the focusing is made onto the middle plane of the isolator section, then the patterns on the background image would be too out of focus to be detected by cross-correlations software. Compromise was made by applying the background image onto one sidewall, while the other sidewall provided the optical access (see figure 3). We believe that sensitivity was not significantly compromised since the shockwaves inside the isolator section are strong and can be detected easily.

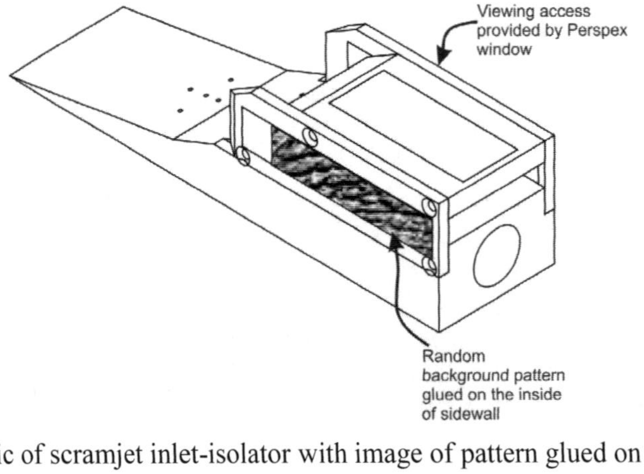

Figure 3. Schematic of scramjet inlet-isolator with image of pattern glued on the inside of sidewall

The setup for BOS experiments is shown in figure 4. The camera used was Canon DSLR EOS-600D with lens of focal length 135 mm. The camera was positioned as such the lens touched the test-section window. This was done in order to increase the ratio of Z_D/Z_B and thus increasing the sensitivity in measurement. The camera was set to capture a wind-off image after the test section was vacuumed right before the tunnel starting. The camera was also set to capture a series of image at 4 fps during the tunnel run, and the best image was chosen for wind-on. The two images chosen were then processed together to obtain the desired density gradient field.

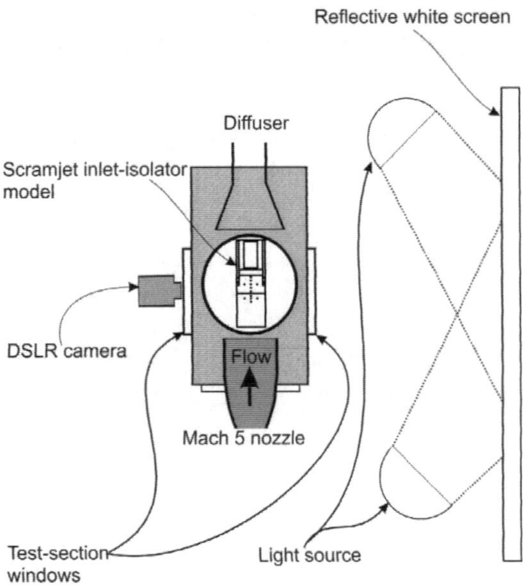

Figure 4. Top view schematic of BOS experimental setup

Two red-head lamps with power of 650 W each were directed towards a large white screen. The reflective lights from the screen gave uniform illumination to the background image printed on sidewall. The experiments were conducted in total darkness except for the two light sources to obtain the best contrast for easy detection of particles deflection.

The pattern for the background image was made using open source general PIV post-processing software, PIVMat© [15] written in MATLAB© environment. The software was set to draw 700,000 black particles of diameter 0.15 mm each in random pattern on an A4 sized image. Effectively, there were 11.22 particles/mm^2 and the filled surface ratio was 0.198. The image was then printed on projector transparency sheet (A4 size) to allow for background illumination of the particles. A small rectangular strip of size 55 × 6.8 mm^2 was cut and pasted onto the inner face of sidewall as shown in figure 4.

5. Results and Discussion

The raw pre-processed images of wind-off and wind-on conditions are shown in figure 5. Any changes in particles cluster shapes and locations cannot be detected by naked eye. The model also experienced slight clockwise rotation in figure 5(b) due to heavy downwards force on the compression ramps. Even though the rotation was very subtle and cannot be observed with naked eyes, it is still of significant magnitude in comparison to each particle deflection value and it can influence their final value as calculated by cross-correlation software. Thus, it must first be corrected prior to processing. The model movement was corrected using 'cp2tform' function in MATLAB 7.10.0 (R2010a), where control points were selected on both figures and the algorithm automatically applied rotation and translation on figure 5(b), so that the control points on both figures agree with each other.

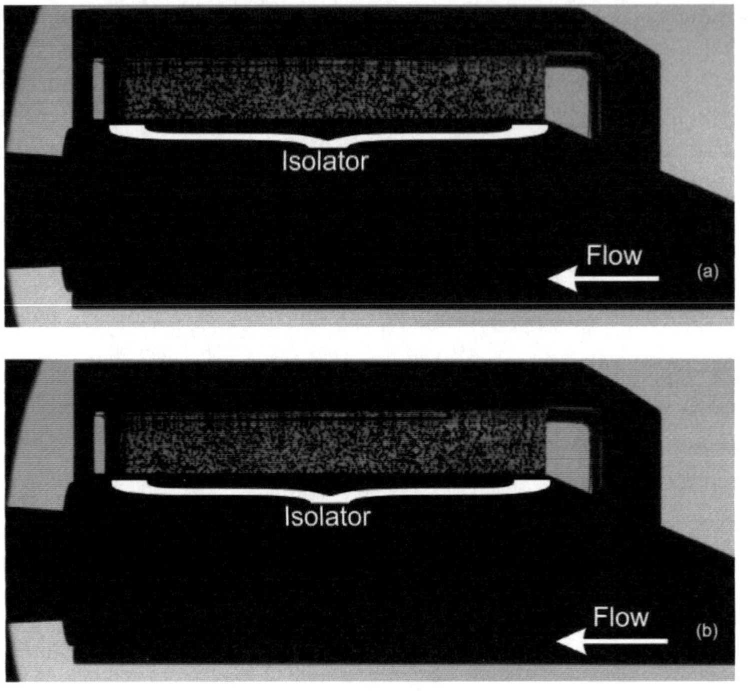

Figure 5. Pre-processed images of (**a**) wind-off condition; (**b**) wind-on

Both figures were processed by using open-source PIVLab© cross-correlation software written by Thielicke and Stamhuis [16]. The software's capability to cross-correlate PIV data has been demonstrated in papers by Mahanti *et al.*[17], Mirsepassi and Dunn-Rankin [18] and Ryerson and Schwenk [19]. The software was chosen since because of its robustness despite being freely available. Commercial software for PIV processing typically has more features but tends to be very expensive.

The small isolator size prohibited the use of small interrogation window size. This is due to requirements of sufficient particle elements in each interrogation window. Using smaller particles size and higher particles density in the background image might circumvent this issue, but they were limited by the printing capabilities. Thus, to overcome this problem, interrogation window of the size 192×192 pixel^2 was chosen, with overlap value of 96%. Spatial resolution was improved significantly with high overlap value, but it comes with large computing cost. If using DCC scheme, rendering all displacement vectors took about 7 hours on a Dell M6300 mobile workstation with Core2Duo Extreme X7900 (2.8 Ghz) processor and 4 GB of RAM. The images in figure 6 were also processed using FFT (Fast Fourier Transform) cross-correlation scheme, and the results came at much faster of less than 5 minutes on the same laptop. Comparison of the density gradient using standard colour schlieren (taken from Idris *et al.* [1]), DCC and FFT are given in the Fig. 6 below.

From figure 6(a), we can observe that a large flow separation at shoulder produced a very strong separation shock that propagates downstream in the form of a shock-train. The shock-train structure is very complex and identifying each shock location was not straight forward. There are at least two more separations that can be detected, one at just downstream of cowl-tip and another one at further downstream. Each separation would produce two shockwaves of their own, a separation shock and a re-attachment shock. All background shock-waves reflected as they propagated downstream and interacted with each other creating an overall complex system of shocks.

Figure 6(b) and (c) shows that all part of the shock-train can be visualized using both the DCC and the FFT calculation scheme of BOS. Even better, the BOS method detected two weak shocks that are barely visible in the colour schlieren image, at the end of the isolator section. However, since BOS detected only density gradient, and by definition a shockwave is a density gradient; region without a shockwave should only have colour 'dark blue' as per the legend in figure 6. This is not the case, such as that in figure 6(b) and 6(c) show readings of density gradient between successive shockwaves. The density gradient field calculated by the FFT method (see figure 6(c)) suffer less error than the one calculated by DCC (see figure 6(b)). This is because the FFT ignored small movement vectors inside each interrogation window in order to decrease its computational time thus making its signal to noise ratio more favourable.

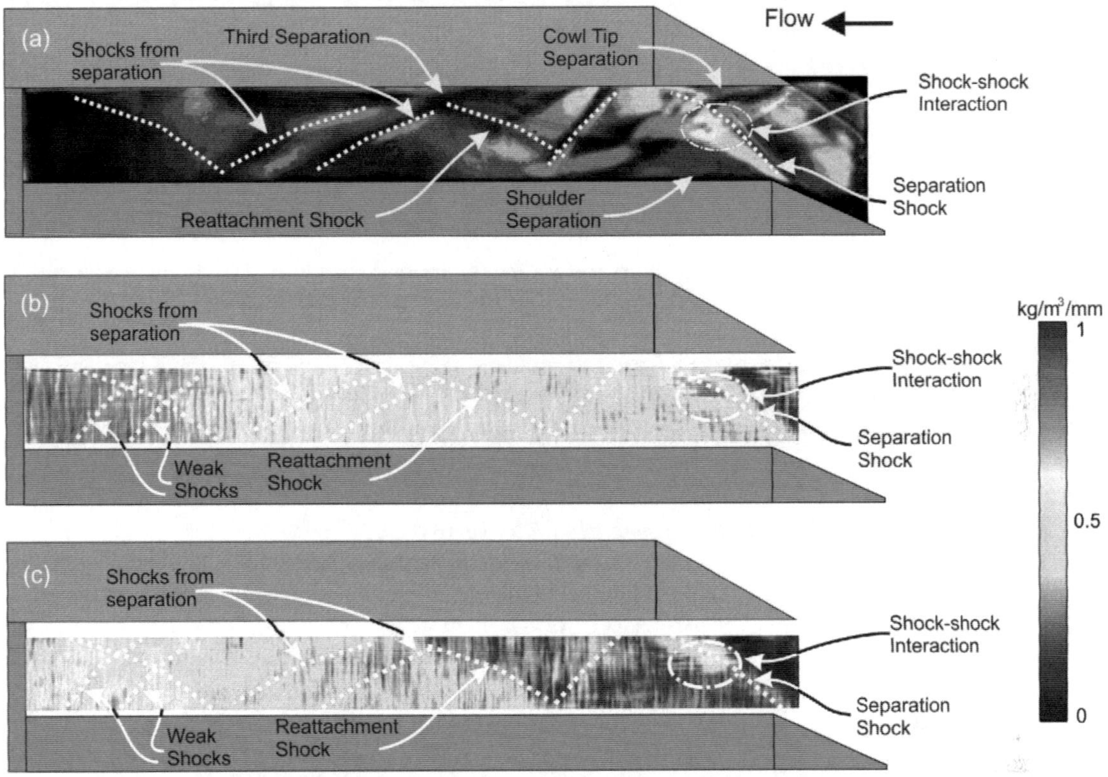

Figure 6. Schlieren images of baseline scramjet inlet-isolator model; (**a**) Colour schlieren (taken with permission from Idris *et al.*); (**b**) schlieren image rendered from density gradient calculated using Direct Cross Correlation (DCC) of BOS data; (**c**) schlieren image rendered from density gradient calculated using Fast-Fourier Transform Cross Correlation (FFT) of BOS data

The density of the flow exiting the isolator section can be calculated easily by integrating the density field and inserting the boundary condition. The boundary condition is the value of density at the start of isolator section and this is a known value of 0.29 kg/m^3[1]. If we do a line-integration at the middle of the field, from upstream towards the exit of the isolator, the density at the exit is 1.52 kg/m^3 for DCC and 1.01 kg/m^3 for FFT. Both values are off by a significant margin from the actual density calculated using equation of state with known value of pressure and temperature, which is about 0.39 kg/m^3 [1]. We believe that the error is not from the BOS setup per-se, instead, we argue that using the BOS method is not suitable for the current case. This is because, shock-train in a scramjet isolator is typically very complex and any schlieren based measurement, be it traditional, colour or even BOS will be presented with a big problem due to the smearing of shock waves as viewed by an observer. As explained by Raffel, the BOS is a line-of-sight integration technique thus if the flow is not fully 2-dimensional then the density-gradient will be distorted [20]. The flow inside the isolator suffered from 3-dimensionality effect close to the sidewall region [1]. Thus the shock waves appeared to be thicker than their theoretical size and their width was wrongly accounted into when the density gradient field was integrated.

6. Conclusion

An exploratory study of applying BOS method to characterize a scramjet inlet-isolator has been performed. The setup could not be simpler, requiring only a DSLR camera and suitable background image. The processing software needed for rendering the density-gradient field is open-source and easily available. Using FFT scheme of calculating cross-correlation is more beneficial in term of computational time and noise rejection. However, using schlieren-based density-measurement for scramjet inlet-isolator investigation would not produce an accurate reading of density since the shockwaves would appeared smeared to the observer.

References

[1] Idris A C, Saad M R, Zare-Behtash H and Kontis K 2014 *Sensors* **14(4)** 6606
[2] Elsinga G, Van Oudheusden B, Scarano F and Watt, D 2004 *Experiments in Fluids* **36** 309
[3] Venkatakrishnan L and Meier G 2004 *Experiments in Fluids* **37** 237
[4] Goldhahn E, Alhaj O, Herbst F and Seume J 2009 *Imaging Measurement Methods for Flow Analysis* 135
[5] Dalziel S, Hughes G O and Sutherland B R 2000 *Fluids* **28** 322
[6] Dalziel S B, Hughes G O, and Sutherland B R 1998 *Proc. 8th Int. Symp. on Flow Visualization*
[7] Dalziel S B 2000 *Proc. of the 5th Intern. symp. on stratified flows*
[8] Meier G 1999 *Deutsche Patentanmeldung* 856
[9] Richard H and Raffel M 2001 *Measurement Science and Technology* **12** 1576
[10] Elsinga G, Van Oudheusden B, Scarano F and Watt D 2003 *Proceedings of PSFVIP-4*
[11] Sims J L 1964 *Tables for supersonic flow around right circular cones at zero angle of attack* NASA SP-3004
[12] Willert C and Gharib M 1991 *Experiments in fluids* **10** 181
[13] Pust O 2000 *Proceedings of the 10th International Symposium on Applications of Laser Techniques to Fluid Mechanics* **27**.
[14] Hargather M J and Settles G S 2011 *HVAC Research* **17** 771
[15] Moisy F, Rabaud M and Pinsolle E 2008 *ISFV13-13th International Symposium on Flow Visualization, and FLUVISU12-12th French Congress on Visualization in Fluid Mechanics Paper No. 326*
[16] Thielicke W and Stamhuis E J 2014 *Journal of Open Research Software* **2(1):**e30
[17] Mahanti P, Keebaugh M and Weiss N 2014 *Proceedings of the COMSOL Conference*
[18] Mirsepassi A and Dunn-Rankin D 2012 *16th Int Symp on Applications of Laser Techniques to Fluid Mechanics*
[19] Ryerson W and Schwenk K 2012 *Journal of Experimental Zoology* 127–140
[20] Raffel M 2015 *Exp Fluids* **56**(60)

A practical approach in porous medium combustion for domestic application: A review

A. K. Ismail[1], N.H. Ibrahim[1], K. A. Shamsuddin[1], M.Z. Abdullah[2], M.Zubair[3]

[1]Universiti Kuala Lumpur – Malaysian Spanish Institute 09000 Kulim Hi-Tech Park, Kedah, Malaysia.
[2]School of Aerospace Engineering, Universiti Sains Malaysia, 14300 Nibong Tebal Penang, Malaysia.
[3]Department of Aeronautical & Automobile Engineering, Manipal Institute of Technology, Karnataka, India.

*Corresponding author: ahmadkamal@unikl.edu.my

Abstract. Combustion in porous media has been widely studied. Many application involving the combustion of porous media has been reported in various way with most consider on numerical works and industrial application. Besides, recent application of porous medium combustion for domestic is the topic of interest among researchers. In this paper, a review was conducted on the combustion of porous media in term of practical application for domestic consumers. Details on the type of fuel used including bio fuel and their system have been search thoroughly. Most of the system have utilized compressed air system to provide lean combustion in domestic application. Some self-aspirating system of porous medium burner was also reported. The application of new technology such as cogeneration by using thermoelectric cells in tandem with porous medium combustion is also revised according to recent work which have already been published. Besides, the recent advances which include coating of porous material is also considered at the end of this paper.

1. Porous media as a medium of combustion

Porous medium can be made from metal, ceramics, plastic or any materials. However, only high temperature resistance materials such as metal and ceramic are suitable to be used as medium of combustion. The selection between porous metal and ceramics materials is based on the application of the burner. Metallic materials have less thermal stability and high thermal inertia. Thus, metal porous materials were less use for porous medium combustion. The porous ceramic are frequently used than metals in porous burners based on the temperature stability produced during combustion with comparatively acceptable emission [1]. The work on combustion using ceramic porous media has been reviewed by Wood and Harris [2]. Recently, several studies have established the applications of porous ceramic inert media in combustion [3-5].

2. Types of fuels used in porous medium combustion

2.1 Gas combustion

In porous medium combustion, gas fuel such as methane, butane, propane and LPG are often used in many applications reported. LPG gas is well known in the domestic gas burner applications, while butane or propane is suitable to be used for portable burners. Combustion by using LPG in porous medium burner have been carried out by many researchers [6-9]. Most LPG combustion in porous medium concentrate on the application for domestic use such as porous medium gas stove [7, 8, 10]. A work by Cerri [11] on porous medium combustion using methane has been conducted on two ceramic foam burners with and without catalytic. It was found that HC concentrations always remained low on both cases. Gao [12] has published a few work on methane and propane gas combustion with latest work by using two layer porous burner with different foam.[13, 14]. There are many others recent works [15-21] using methane as fuel in porous medium combustion describing methane as the most allowable gas source to be produced and used in many ways especially with porous medium burner. Toledo [22] used butane fuel and wood pellets to produce syngas through rich and ultra-rich combustion in an inert porous medium . Propane has also been used [23] to produce hydrogen by using combustion in porous medium. The used of propane for porous medium combustion has been widely tested [24-28].

2.2 Liquid combustion

The combustion of liquid fuel in porous burner is less reported. The use of kerosene fuel in porous burner is tested to suit with domestic burner application normally in rural area where LPG is not available. The problem of using liquid fuel in porous burner is that the liquid fuel need to be vaporized before entering the porous medium reaction zone to produce flame such as in gas combustion. A study of kerosene combustion in porous medium has been done experimentally by Vijaykant [29] using carbon foam with SiC coated. The heat recirculation inside the foam pre vaporizes the liquid kerosene fuel before combustion. It was found that combustion of kerosene in porous medium reduced the emission by improving the vaporization and air fuel premixing. Aviation grade kerosene has been used by Periasamy [30, 31] to study the performance and emission in porous medium combustion. The kerosene was injected by using an air-blast atomizer to produce stable flame. From this study, it has been shown that the NOx produced was not dominated by parameter such as equivalence ratio and spray nozzle when porous medium is used. Latest study on kerosene non-spray combustion in porous medium can be found from work prepared by Wongwatcharaphon [32, 33]. Pastore [34, 35] has provided a study on syngas production through rich combustion of liquid fuel such as n-heptane, diesel oil, kerosene and rapeseed-oil methyl ester (RME) bio-diesel. Combustion of other liquid fuels such as heptane can be found from work performed by Kaplan [36] using a spray nozzle for fuel impingement in porous medium combustion zone. Kamal [37] has studied on the orientation of the spray nozzle in porous medium with swirled air supply. Porous medium which has been inserted at the burner exit has produced droplet evaporation while enlarging the recirculation zone before the fuel injector improving mixing and combustion efficiency. Trimis [38] produced a novel cool flame vaporizer for liquid fuel with porous medium burner. Krittacom [39] investigated the evaporation of diesel fuel and its combustion behavior. The fuel drop technique was used. The porous burner (PB) used pebbles with and without porous emitter (PE) where higher temperature profile has been found for PB with PE. More works can be found from literatures provided by Jugjai [40, 41] without using spray atomization for liquid porous medium combustion. Heavy oil or crude oil combustion in porous medium has been studied by Cinar [42].

2.3 Bio fuel combustion

A few recent work have covered on the use of vegetable cooking oil–kerosene fuel blends for porous burners. The burners are designed for domestic application which involved heat generation for cooking as well as power generation by using thermoelectric and thermophotovoltaic cells [43-45]. In this study, comparison has also being done. The test includes different mixtures blend composition of kerosene-vegetable cooking oil (VCO) such as: 100 kerosene, 95/5 KVCO, 90/10 KVCO, 80/20 KVCO, 75/25 KVCO and 50/50 KVCO. The power output was found to be increased in lean region with lower emission. However, these values started to increase beyond the stoichiometric ratio. Renewable energy source such as bio fuel become significant nowadays. By taking advantages of porous medium combustion, the use of bio fuel can be more efficient. The type of bio fuel that is currently tested in porous burner is vegetable oil [46]. There are bio fuel in liquid form such as blended kerosene fuel with palm oil or bio seed fuel. BIOFLAM project [47, 48] used renewable biodiesel which is also known as a Fatty Acid Methyl Ester (FAME). In this project, a specially designed ceramic premixed liquid fuel porous medium burner was used with cool flame vaporization technique to produce high power modulation with low emission level. In another work, vegetable (rapeseed) oil as a bio fuel has been tested with porous medium combustion [49]. A prototype integrated vaporizer system with a PIM burner based on the flow velocity flame stabilization technique was built and investigated. A stable combustion has been observed with the absence of CO emissions recorded. Canola methyl ester (CME) and soy methyl ester (SME) produced from vegetable were among the bio fuel tested in porous medium burner [50]. It has been found that the axial flame temperature produced by CME and SME were similar to commercial Jet-A fuel combustion in porous medium burner. However, some clogs were found due to solid particle decomposition stuck in porous medium and on the metal walls of the injector by combustion of CME and SME. Biogas is another type of renewable energy that can be used with porous medium burner. Due to the low heating value of biogas, the Reciprocal Flow Porous Burner (RFPB) has been used by [51] to increase the efficiency with low emission. A stable combustion was found for both study of methane and biogas combustion, when RFPB is applied, which is also more efficient having low emissions. Some interesting studies related to the used of biogas in porous medium combustion can be found from the work done by Gao and Keramiotis [52, 53].

2.4 Exhaust gas emission from porous medium combustion

The incomplete combustion of fossil fuel which is also known as hydrocarbon (HC) fuel produced unburned hydrocarbon (UHC) and carbon monoxide (CO). Another product of combustion which is dangerous is the nitrogen oxide gas or NOx. This type of gas will only occur at very high temperature where the nitrogen content in the atmospheric air oxidized in combustion flame. Whereas, high temperature combustion crack down the CO into CO_2 in atmospheric air as well as promoting the UHC to be completely burned. Therefore a trade of between CO and NOx in HC fuels existed. The strategy of using porous medium to cater with the trade off issue between CO and NOx has been found by Trimis [54]. By using porous medium, the combustion of HC fuel will be completely burned inside the porous media without leaving any UHC. High temperature combustion flame has been controlled by the use of porous medium and the nitrogen gas oxidation will not exist in the surrounding air avoiding the formation of NOx. A numerical study has been done by Zhou [55] to investigate the NO and CO emission reduction in porous medium . Due to very high temperature inside porous medium solid material, the heat transfer mechanism to the thermal load will be largely due to radiation mode which will not oxidize the nitrogen gas in the air. This phenomena can be found from studies on porous medium radiant burner where CO and NOx were both in relatively low and in acceptable range [36]. By manipulating the pressure and temperature of the inlet mixture into porous medium combustion system, the CO can be eliminated and NOx content was remarkably reduced as studied by Bakry [56].

3. Recent application of porous medium combustion devices

Numerous application of porous medium in combustion devices has been found. Some recent works on major application has been selected to describe the essential of porous medium combustion technology. Water purification is one of the examples of using porous medium combustion technique [57] [58]. A new technology of water purification of organic inclusions or solutes by using a filtration combustion reactor with alternating flow direction is investigated. An acetone aqueous solution served as a model liquid. Non-catalytic porous burner also can be used in syngas production as presented by Pastore [34]. A two-layer inert porous medium combustor was used and different fuels such as n-heptane, diesel, kerosene and bio-diesel were successfully reformed to syngas in a Zirconia foam burner with conversion efficiency over 60%. Micro heat-recirculating combustor using porous media plate, has been designed and tested by Cao [59]. The micro heat-recirculating combustor has a very important application value in developing the micro gas turbine power generation system. The micro porous media combustor possesses some merits such as wider combustion operation ranges, higher combustion efficiency and lower heat loss ratio and should be an ideal micro combustor for developing the micro gas turbine power generation system [60]. Stability of lean combustion in mini-scale porous media combustor with heat recuperation was studied by Xu [24]. In miniaturization of burners, it is very difficult to organize stable self-sustained combustion. A mini-scale porous media combustor with heat recuperation was set up to study the stability of lean combustion and its emission. The diameter of the porous media was only 20mm and the burner was about 140mm in length.

3.1 Domestic porous medium burner
Some attempts have been found to produce domestic porous medium burner especially on gas stove. A concept of Porous Radiant Re-circulated Burner (PRRB) has been introduced by Jugjai [61]. The PRRB re-circulated the exhaust gas captured by porous medium to preheat the incoming secondary air supply. Muthukumar [7] presented a detail study on LPG gas stove using porous medium technology in line with Pantangi [62]. However, the design of the stove needs improvement taking into account the application of porous medium burner without compressed air. A practical self-aspirating porous medium burner has been developed and tested by Yoksenakul [63] using LPG. This porous medium burner has been designed according to the allowable Peclet number calculated from Ergun equation on pressure losses. The burner has demonstrated stabilized flame with surface and submerged combustion flame without flashback. The use of packed bed with sphere balls has reduced the pressure drop for this naturally aspirated porous medium burner. However, the results of CO emission are higher compared to conventional burner (CB) with a small improvement in thermal efficiency. Another domestic application of porous medium burner is the heat production for household application [64]. This burner is capable for use in water heating by adopting customized heat exchanger.The use of porous ceramic foam in premixed porous medium LPG stove can be found from the work performed by Mujeebu [10, 65, 66]. Two different foams have been used to create preheater and reaction zone by using Alumina and Porcelain types of ceramic. The thickness of the foam has been studied to find the best thickness which provides better flame combustion, emission and flame stability. Porous radiant burner (PRB) was study by Pantangi [8]. The PRB was developed for domestic cooking stove by using LPG. PRB consist of SiC reaction layer and alumina balls preheating layer. Performance test has been done by using the water boiling test according to BIS (Bureau of Indian Standard) to compare the efficiency between PRB and conventional burner (CB) of cooking stove. It has been found that the PRB has higher efficiency of about 3% over CB. Lower CO and NOx is also detected for PRB compared with CB.Since most domestic burners operate by self-aspirating air supply system, the application of this type of burner with porous medium combustion technique continuously investigated. Laphirattanakul demonstrated on porous medium combustion with self-entrainment system [67]. The system concentrated on naturally aspirated air supply system which include primary and secondary natural draft. Different flame images have been compared at different thermal power output to investigate the combustion stability and propagation. This work has shown detail explanation on the direction of heat from flame propagation which gives advantages for the design of naturally aspirated porous burners which can also extend to the combustion model in numerical analysis.

3.2 Application of Thermoelectric and Thermo Photovoltaic for cogeneration
An experimental study has been carried out [68] on the use of a power-generated system with TE cells to generate electrical power for household use. Champier [69] applied the TE cells to their biomass cook stove for power cogeneration in rural areas. A portable power source using butane as fuel for heat power conversion by TE cell has been done by Posthill [70]. SiC coated porous medium foam has been used for thermoelectric power conversion by Mueller [71]. A few works on patent burners involving TE generator can be found from Cedar and Hu [72, 73]. An attempt on combining the TE and TPV to enhance the thermal efficiency with higher power generation has been done by Sark [74]

IOP Conf. Series: Materials Science and Engineering **370** (2018) 012004 doi:10.1088/1757-899X/370/1/012004

while Yang [75] proposed novel design of combining TE, TPV and hot water (HW) for solar power generation. Latest work conducted by Singh [76] provide study on small scale burner with self-aspirating system to be used for power generation. This system is said to produce continuous electrical power output for a long period by stabilizing the combustion process. For this study, thermoelectric devices were used for power generation with also considered on the environment condition which includes ambient temperature. From the effort published, it is found that the power generated from small scale heat source or domestic burner is crucial nowadays especially to be used for outdoor and at rural areas with the absence of electricity. Detail layout and photographic views of the porous medium burner with cogeneration by Ismail et.al [77] described the latest application of porous medium burner for heating and power generation system.

4. Effect of coating to the porous medium combustion

The used of coating techniques on porous medium for combustion was less reported. Materials such as SiC, Ni and Cr were selected due to the catalytic capabilities. These catalytically active materials were usually employed to improve the properties of combustion process. It is known that noble metals such as gold and platinum provide high catalytic activity, and are very good medium for combustion reactions. However these noble metals and its compounds are expensive, and it may experience sublimation when use at high temperatures. SiC which is a type of ceramic is also known to have thermal stability and chemical inertness as well as can be used as catalyst material. Therefore, these transitional metals (Ni and Cr), and SiC have been chosen and the coating technique was by using dip-coating method onto an inert Al_2O_3 porous substrate [78, 79].

For coating techniques, the plain substrate was initially prepared. Then, the sintered plain substrate is dipped in the coating bath (coating slurry) and re-sintered to obtain a uniform coating. A liquid deposition method was used to coat the SiC, Ni, and Cr-based coating slurry on the pre developed plain substrate. This process is also known as the dip-coating technique and is ideally suited for coating porous substrates having highly reticulated structure. In this process the dry plain substrate is first dipped into the coating slurry and after being withdrawn from this solution, results in a wet, dense and uniform coating on the surface of the substrate. As a conclusion, SiC coated Alumina have very less effect on the structure even after used for combustion.

In Fig.1.(i) results prove that during coating process, the coated substrate was not subjected to or experienced very little oxidation and phase changes during the sintering process using the vacuum furnace. The results after combustion in Fig. 1.(ii) shows only the occurrence of moissanite-6H (ICDD: 01-072-0018) phase on the SiC-coated substrate. Other results from the work by Ismail [78] shows that Ni-coated substrate undergoes an obvious oxidation reaction. Moreover, the XRD result for Cr-coated porous substrate also show an oxidized coating layer indicated by the presence of chromium oxide phase after the combustion test. As a conclusion, SiC coated Alumina has very less effect on the structure even after used for combustion.

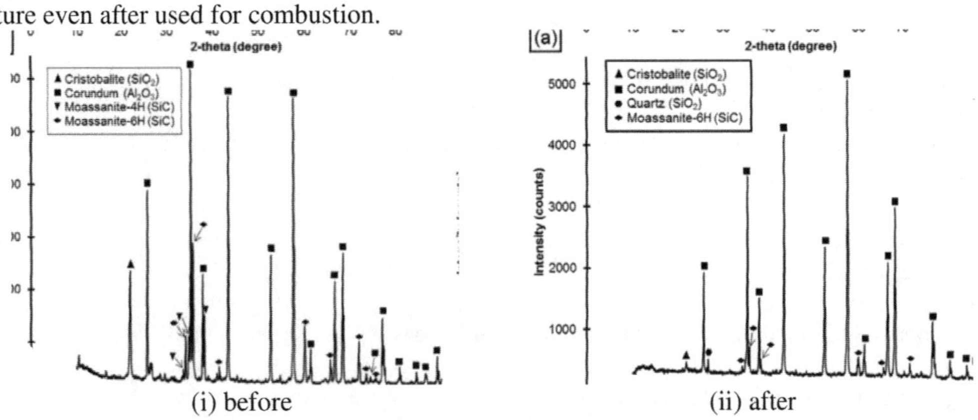

(i) before (ii) after

Figure 1. XRD patterns of porous Al2O3 substrate coated with SiC (before and after used for medium of combustion)[78]

5. Conclusion

Domestic porous medium burner has been intensively studied which most considered on the small scale burners and self-aspirating system. Based on the application of thermoelectrics and thermophotovoltaic for power generation in previous work, the application with porous medium burner can be realized. This type of porous burner which is used for domestic is very useful both for indoor and outdoor activities when necessary for electrical power output and heating purposes. The use of different fuels for porous medium combustion enhanced the combustion efficiency with less emission produced. Recent advances of domestic porous medium burner considered on many aspects including geometrical and different coating materials of porous medium to be used in combustion. The effect of coating on porous medium combustion has not been focused in previous work. This may reduce the cost of using purely catalytic materials as porous foam. The use or alumina as porous medium can be much improved with coating techniques.

References

[1] Tierney, C. and A.T. Harris, Materials design and selection issues in ultra-lean porous burners. Journal of the Australian Ceramic Society, 2009. **45**(2): p. 20-29.

[2] Wood, S. and A.T. Harris, *Porous burners for lean-burn applications.* Progress in Energy and Combustion Science, 2008. **34**(5): p. 667-684.

[3] Yu, B., et al., *Combustion characteristics and thermal efficiency for premixed porous-media types of burners.* Energy, 2013. **53**: p. 343-350.

[4] Gao, H.B., et al., *Methane/air premixed combustion in a two-layer porous burner with different foam materials.* Fuel, 2014. **115**(0): p. 154-161.

[5] Keramiotis, C. and M.A. Founti, *An experimental investigation of stability and operation of a biogas fueled porous burner.* Fuel, 2013. **103**(0): p. 278-284.

[6] Charoensuk, J. and A. Lapirattanakun, *On flame stability, temperature distribution and burnout of air-staged porous media combustor firing LPG with different porosity and excess air.* Applied Thermal Engineering, 2011. **31**(16): p. 3125-3141.

[7] Muthukumar, P., P. Anand, and P. Sachdeva, *Performance analysis of porous radiant burners used in LPG cooking stove.* International Journal of Energy and Environment, 2011. **2**(2): p. 367-374.

[8] Pantangi, V.K., et al., *Studies on porous radiant burners for LPG (liquefied petroleum gas) cooking applications.* Energy, 2011. **36**(10): p. 6074-6080.

[9] ABDULMUJEEBU, M., M.Z. Abdullah, and M. Zuber, *EXPERIMENT AND SIMULATION TO DEVELOP CLEAN POROUS MEDIUM SURFACE COMBUSTOR USING LPG.* Isi Bilimi ve Teknigi Dergisi/Journal of Thermal Science & Technology, 2013. **33**(1).

[10] Mujeebu, M., et al., *A mesoscale premixed LPG burner with surface combustion in porous ceramic foam.* Energy Sources, Part A: Recovery, Utilization, and Environmental Effects, 2011. **34**(1): p. 9-18.

[11] Cerri, I., G. Saracco, and V. Specchia, *Methane combustion over low-emission catalytic foam burners.* Catalysis today, 2000. **60**(1-2): p. 21-32.

[12]. Gao, H., et al. *Experimental Investigation of CH4 and C3H8 Combustion in a Packed Bed Burner.* in *ASME 2011 International Mechanical Engineering Congress and Exposition.* 2011: American Society of Mechanical Engineers.

[13]. Gao, H., et al., *Methane/air premixed combustion in a two-layer porous burner with different foam materials.* Fuel, 2014. **115**: p. 154-161.

[14]. Gao, H., et al., *Combustion of methane/air mixtures in a two-layer porous burner: A comparison of alumina foams, beads, and honeycombs.* Experimental Thermal and Fluid Science, 2014. **52**: p. 215-220.

[15]. Li, C. and P.F. Hsu, *Emissions and radiation efficiency for methane combustion within a porous ceramic burner.* Clean Air, 2006. **7**(3): p. 285-302.

[16]. Zhu, H., et al. *Catalytic partial oxidation of methane using RhSr- And Ni-substituted hexaaluminates*. 2007. Heidelberg.
[17]. Alavandi, S. and A. Agrawal, *Experimental study of combustion of hydrogen–syngas/methane fuel mixtures in a porous burner.* International journal of hydrogen energy, 2008. **33**(4): p. 1407-1415.
[18]. Dobrego, K.V., et al., *Overall chemical kinetics model for partial oxidation of methane in inert porous media.* Chemical Engineering Journal, 2008. **144**(1): p. 79-87.
[19]. Ma, P.Y., et al., *Experimental study on hydrogen production from methane rich combustion in built-in porous media Swiss-roll reactor.* Ranliao Huaxue Xuebao/Journal of Fuel Chemistry and Technology, 2010. **38**(5): p. 636-640.
[20]. Wang, Y., et al., *Homogeneous combustion of fuel ultra-lean methane–air mixtures: Experimental study and simplified reaction mechanism.* Energy & Fuels, 2011. **25**(8): p. 3437-3445.
[21]. Xu, J., et al., *Research on influencing factors of methane/air combustion in the ring porous medium burner*. 2012: Hefei. p. 1734-1738.
[22]. Toledo, M., E. Vergara, and A.V. Saveliev, *Syngas production in hybrid filtration combustion.* International Journal of Hydrogen Energy, 2011. **36**(6): p. 3907-3912.
[23]. Toledo, M., et al., *Hydrogen production in ultrarich combustion of hydrocarbon fuels in porous media.* International Journal of Hydrogen Energy, 2009. **34**(4): p. 1818-1827.
[24]. Xu, K., M. Liu, and P. Zhao, *Stability of lean combustion in mini-scale porous media combustor with heat recuperation.* Chemical Engineering and Processing: Process Intensification, 2011. **50**(7): p. 608-613s.
[25]. Xu, K., et al., *Porous combustor applying in thermophotovoltaic power system.* Kung Cheng Je Wu Li Hsueh Pao/Journal of Engineering Thermophysics, 2009. **30**(5): p. 887-889.
[26]. Shi, J.R., et al., *Experimental study on the low-velocity filtration combustion of lean premixtures.* Dalian Haishi Daxue Xuebao/Journal of Dalian Maritime University, 2008. **34**(2): p. 69-72.
[27]. Kakutkina, N.A., A.A. Korzhavin, and M. Mbarawa, *Filtration combustion of hydrogen-air, propane-air, and methane-air mixtures in inert porous media.* Combustion, Explosion and Shock Waves, 2006. **42**(4): p. 372-383.
[28]. Korzhavin, A.A., et al., *Selective diffusion during flame propagation and quenching in a porous medium.* Combustion, Explosion and Shock Waves, 2005. **41**(4): p. 405-413.
[29]. Vijaykant, S. and A.K. Agrawal. *Effect of porous media configuration on combustion of kerosene*. 2006. Reno, NV.
[30]. Periasamy, C. and S.R. Gollahalli, *Experimental investigation of kerosene spray flames in inert porous media near lean extinction.* Energy and Fuels, 2011. **25**(8): p. 3428-3436.
[31]. Periasamy, C. and S.R. Gollahalli, *Numerical modeling of evaporation enhancement of aviation-grade kerosene sprayin porous media combustors.* Journal of Porous Media, 2010. **13**(8): p. 707-723.
[32]. Wongwatcharaphon, K., P. Tongtem, and S. Jugjai, *Numerical and experimental study of late mixing porous burner.* Journal of the Energy Institute, 2013. **86**(1): p. 15-23.
[33]. Wongwatcharaphon, K., P. Tongtem, and S. Jugjai, *A Numerical and Experimental Investigation of Performance of a Nonsprayed Porous Burner.* Heat Transfer Engineering, 2014. **35**(10): p. 942-952.
[34]. Pastore, A. and E. Mastorakos, *Syngas production from liquid fuels in a non-catalytic porous burner.* Fuel, 2011. **90**(1): p. 64-76.
[35]. Pastore, A. and E. Mastorakos, *Rich n-heptane and diesel combustion in porous media.* Experimental Thermal and Fluid Science, 2010. **34**(3): p. 359-365.
[36]. Kaplan, M. and M.J. Hall, *The combustion of liquid fuels within a porous media radiant burner.* Experimental Thermal and Fluid Science, 1995. **11**(1): p. 13-20.

[37]. Kamal, M.M. and A.A. Mohamad, *Investigation of liquid fuel combustion in a cross-flow burner.* Proceedings of the Institution of Mechanical Engineers, Part A: Journal of Power and Energy, 2007. **221**(3): p. 371-385.

[38]. Trimis, D., et al., *A novel porous burner for liquid fuels based on porous media combustion and cool flame vaporization.* Combustion and Enviroment: XXIV Event of the Italian Section of the Combustion Institute, S. Margherite Ligure, Italy, 2001.

[39]. Krittacom, B., et al., *The pack-bed sphere liquid porous burner.* Proceedings of World Academy of Science, Engineering and Technology, 2011. **81**: p. 80-84.

[40]. Jugjai, S. and N. Polmart, *Down-flow Combustion of Liquid Fuels through Porous Media without Spray Atomization.* Asian J. Energy Environ, 2002. **3**(1-2): p. 79-109.

[41]. Jugjai, S. and C. Pongsai, *Liquid fuels-fired porous burner.* Combustion Science and Technology, 2007. **179**(9): p. 1823-1840.

[42]. Cinar, M., L.M. Castanier, and A.R. Kovscek, *Combustion kinetics of heavy oils in porous media.* Energy and Fuels, 2011. **25**(10): p. 4438-4451.

[43]. Mustafa, K., et al., *Comparative assessment of a porous burner using vegetable cooking oil–kerosene fuel blends for thermoelectric and thermophotovoltaic power generation.* Fuel, 2016. **180**: p. 137-147.

[44]. Mustafa, K., et al., *Experimental analysis of a porous burner operating on kerosene–vegetable cooking oil blends for thermophotovoltaic power generation.* Energy Conversion and Management, 2015. **96**: p. 544-560.

[45]. Mustafa, K.F., et al., *Experimental investigation of the performance of a liquid fuel-fired porous burner operating on kerosene-vegetable cooking oil (VCO) blends for micro-cogeneration of thermoelectric power.* Renewable Energy, 2015. **74**: p. 505-516.

[46]. Bakry, A., et al., *Low emission combustion of vegetable oils with the porous burner technology*, in *New and Renewable Technologies for Sustainable Development*. 2002, Springer. p. 541-554.

[47]. Brehmer, T., et al. *BIOFLAM Project: Application of liquid biofuels in new heating technologies for domestic appliances based on cool flame vaporization and porous medium combustion*. 2003.

[48]. Heeb, A., et al., *Development of a heating system with stageless output modulation for use with fuel oil EL and fatty acid methyl ester (FAME) mixtures within the setting of the BIOFLAM project*, in *Entwicklung eines Heizungssystems mit stufenloser Leistungsmodulation für die Anwendung mit Heizöl EL und FAME-Gemischen im Rahmen des Projektes BIOFLAM*. 2005. p. 653-658.

[49]. Bakry, A., et al., *Novel flame stabilization technique in porous inert media (PIM) combustion under high pressure and temperature.* Energy and Fuels, 2010. **24**(1): p. 274-287.

[50]. Barajas, P.E., R.N. Parthasarathy, and S.R. Gollahalli. *Combustion characteristics of biofuels in porous-media burners*. 2010. Montecatini.

[51]. de Araújo, W.C., et al., *BIOGAS COMBUSTION ON RECIPROCAL FLOW POROUS BURNER WITH ENERGY EXTRACTION.* 2013.

[52]. Gao, H., et al., *Experimental study of biogas combustion in a two-layer packed bed burner.* Energy & Fuels, 2011. **25**(7): p. 2887-2895.

[53]. Keramiotis, C. and M.A. Founti, *An experimental investigation of stability and operation of a biogas fueled porous burner.* Fuel, 2013. **103**: p. 278-284.

[54]. Trimis, D. and F. Durst, *Combustion in a porous medium-advances and applications.* Combustion Science and Technology, 1996. **121**(1-6): p. 153-168.

[55]. Zhou, X.Y. and J.C.F. Pereira, *Numerical study of combustion and pollutants formation in inert nonhomogeneous porous media.* Combustion Science and Technology, 1997. **130**(1-6): p. 335-364.

[56]. Bakry, A., et al., *CO and NOx emissions in porous inert media (PIM) burner system operated under elevated pressure and inlet temperature using a new flame stabilization technique.* Chemical Engineering Journal, 2010. **165**(2): p. 589-596.

[57]. Dobrego, K.V., et al., *Water purification of organic inclusions by the method of combustion within an inert porous media.* International Journal of Heat and Mass Transfer, 2010. **53**(11-12): p. 2484-2490.

[58]. Koznacheev, I.A., K.V. Dobrego, and E.S. Shmelev, *Water purification of organic inclusions in a reverse flow filtration combustion reactor.* International Journal of Heat and Mass Transfer, 2011. **54**(4): p. 932-937.

[59]. Cao, H.L., et al., *Diffusion combustion in a micro heat-recirculating combustor using porous media.* Ranshao Kexue Yu Jishu/Journal of Combustion Science and Technology, 2011. **17**(5): p. 394-400.

[60]. Cao, H.L., et al., *Investigation on combustion characteristics in the micro porous media combustor.* Kung Cheng Je Wu Li Hsueh Pao/Journal of Engineering Thermophysics, 2011. **32**(12): p. 2164-2167.

[61]. Jugjai, S. and S. Sanitjai, *Parametric studies of thermal efficiency in a proposed porous radiant recirculated burner (PRRB): a design concept for the future burner.* RERIC International Energy Journal, 1996. **18**: p. 97-112.

[62]. Pantangi, V.K., et al., *Performance analysis of domestic LPG cooking stoves with porous media.* International Energy Journal, 2007. **8**(2): p. 139-144.

[63]. Yoksenakul, W. and S. Jugjai, *Design and development of a SPMB (self-aspirating, porous medium burner) with a submerged flame.* Energy, 2011. **36**(5): p. 3092-3100.

[64]. Avdic, F., M. Adzic, and F. Durst, *Small scale porous medium combustion system for heat production in households.* Applied Energy, 2010. **87**(7): p. 2148-2155.

[65]. Abdul Mujeebu, M., et al., *Development of premixed burner based on stabilized combustion within discrete porous medium.* Journal of Porous Media, 2011. **14**(10): p. 909-917.

[66]. Mujeebu, M.A., M.Z. Abdullah, and A.A. Mohamad, *Development of energy efficient porous medium burners on surface and submerged combustion modes.* Energy, 2011. **36**(8): p. 5132-5139.

[67]. Laphirattanakul, P., A. Laphirattanakul, and J. Charoensuk, *Effect of self-entrainment and porous geometry on stability of premixed LPG porous burner.* Applied Thermal Engineering, 2016. **103**: p. 583-591.

[68]. Qiu, K. and A. Hayden, *Performance of low bandgap thermophotovoltaic cells in a small cogeneration system.* Solar energy, 2003. **74**(6): p. 489-495.

[69]. Champier, D., et al., *Thermoelectric power generation from biomass cook stoves.* Energy, 2010. **35**(2): p. 935-942.

[70]. Posthill, J., et al. *Portable power sources using combustion of butane and thermoelectrics.* in *Thermoelectrics, 2005. ICT 2005. 24th International Conference on*. 2005: IEEE.

[71]. Mueller, K.T., et al., *Super-adiabatic combustion in Al< sub> 2</sub> O< sub> 3</sub> and SiC coated porous media for thermoelectric power conversion.* Energy, 2013. **56**: p. 108-116.

[72]. Cedar, J.M. and A.H. Drummond, *Portable combustion device utilizing thermoelectrical generation.* 2012, Google Patents.

[73]. Hu, K. and C. Hu, *Gas stove with thermoelectric generator*. 2004, Google Patents.

[74]. Sark, W.v., *Feasibility of photovoltaic–thermoelectric hybrid modules.* Applied Energy, 2011. **88**(8): p. 2785-2790.

[75]. Yang, D. and H. Yin, *Energy conversion efficiency of a novel hybrid solar system for photovoltaic, thermoelectric, and heat utilization.* Energy Conversion, IEEE Transactions on, 2011. **26**(2): p. 662-670.

[76]. Singh, T., R. Marsh, and G. Min, *Development and investigation of a non-catalytic self-aspirating meso-scale premixed burner integrated thermoelectric power generator.* Energy Conversion and Management, 2016. **117**: p. 431-441.

[77]. Ismail, A.K., et al., *Application of porous medium burner with micro cogeneration system.* Energy, 2013. **50**(0): p. 131-142.

[78]. Ismail, AK., et al., *Effect of Ceramic Coating in Combustion and Cogeneration Performance of Al2O3 Porous Medium.* Journal of the Energy Institute, 2015.

[79]. Jamaludin, A.R., et al., *Fabrication of porcelain foam substrates coated with SiC, Ni, and Cr using the dip-coating technique.* Ceramics International, 2015. **41**(2, Part B): p. 2940-2947.

IOP Conf. Series: Materials Science and Engineering **370** (2018) 012005 doi:10.1088/1757-899X/370/1/012005

Life extension of Structural Repairs – A statistical approach towards efficiency improvement

Deepashri N V and Mohan Kalaiyappan
AXISCADES Engineering Technologies Ltd, Bangalore – 560024, India.

*Corresponding author: deepashrinv@gmail.com and kaymohan007@gmail.com

Abstract: The life extension program of aircraft is taken up whenever aircraft's intended life reaches close to its DSG (Design Service Goal). The Extended Service Goal (ESG) of an aircraft, in general, and structural repairs, in particular, is arrived at on the basis of F&DT (Fatigue & Damage Tolerance) analysis. Life extension program of aircraft consists of assessment of remaining life of all parts of the aircrafts including structural, mechanical, and electrical and avionics equipment and structural repairs. For life extension of stringer repair, as an example, it is required to re-assess the fatigue life of stringer in the presence of coupling under modified load spectrum. This is achieved by assessing the fatigue life of Web and Outer Flange (OF) part of stringers separately as per F&DT justification philosophy. Assessment of the fatigue life requires determination of stress concentration factor (Kt) for different combination of width, pitch, stringer thickness, coupling thickness and pad-up thickness of all stringer profiles available in different sections of fuselage. Determination of stress concentration factor for Web and Outer Flange of stringer profile covering entire ranges involves substantial number of Finite Element (FE) analysis. In order to optimise the number of FE runs, stress concentration factor is determined under worst repair factors combination (max. plate width; max. thickness; max. pitch; min. rivet dia.; and min. No. of rivets) resulting in conservative value. A parametric study of Web and Outer Flange data across stringer profiles were carried out and proven statistical techniques were used to find the optimal equation to predict stress concentration factor. This in turn reduced number of FE runs substantially for a given range of width, pitch, stringer thickness and so on. The use of optimal equation obtained through regression analysis is able to predict Kt within reasonable accuracy for a given range of inputs.

1. Introduction

Damage to aircraft structures is often caused by corrosion, erosion, incidents, accidents and mishaps. Sometimes modifications to aircraft structure require extensive structural rework. For example, the installation of winglets on aircraft not only replaces a wing tip with a winglet, but also requires extensive reinforcing of the wing structure to carry additional stresses [1].

Structural repair is defined as a design change to an aeronautical product intended to restore it to an airworthy condition and to ensure that the aircraft continues to comply with the design aspects of the airworthiness requirements used for the issuance of a Type certificate for that aircraft type after it has been damaged [2]. When a structural damage is detected, usually one of the following decisions is taken on the basis of severity of damage vis-a-vis structural repair manual:

- No-repair
- Cosmetic or sealing repair to correct minor damage.

- Structural repair due to reduction of strength below the design limits
- Un-economical for the repair leading to replacement

Structural repair plays a key role to ensure the structural integrity of aircraft throughout its intended operational life. When a group of lead aircraft approaches its DSG (Design Service Goal), life extension program of aircraft involving assessment of remaining life studies of all parts is taken up individually. Since structural repair also influences the structural integrity of aircraft, it is also mandatory to evaluate its life as per regulation. The Extended Service Goal (ESG) of an aircraft, in general, and structural repairs, in particular, is arrived at on the basis of F&DT (Fatigue & Damage Tolerance) analysis as well as tests under modified load spectrum.

The present paper provides an overview of methodology involved in the exercise of life extension of a typical structural repair under 3rd line item (from above) as an example to demonstrate how effectively the statistical approach helps in improving the efficiency substantially without compromising quality of results.

2. Structural Repair Manual

Structural Repair Manual (SRM) of aircraft provides general procedures, identification of structures, allowable damage limits and specific instructions for repair to those parts of the airframe that are most likely to be damaged [3].

SRM also gives the dimensional limit to the damage to aircraft structure and it provides the approved repair method or repair scheme and approved materials to be used to repair the damage occurred at different section as per ATA Chapters. Typical structural repairs that cover in SRM for fuselage are:

- External Structure – Skin
    - Skin Lightning Strike Repairs
    - External Skin Repairs (small, limited, unlimited, and conversion of temporary skin repairs)
    - External Skin Repair (Longitudinal joint, Circumferential joint)
    - Internal Skin repairs (small, limited, and conversion of temporary skin repairs)
    - Door Surround Skin Repair
    - Bush Skin Repair
    - Plug Repair
- Internal Structure
    - Stringer Repair
    - Frame Repair
    - Cross-beam Repair

Subsequent to catastrophic failure of aircraft happened in aviation history in the past due to in-appropriate repairs (available in open literature), following conclusions/amendments were made:

- In-service repairs can influence the damage tolerance of aircraft significantly
- Fatigue justifications are required for existing repairs of oldest aircraft
- Damage tolerance certification of repairs has become mandatory
- In-service repairs are subjected to inspection programmes as per those for the original un-repaired structure
- Regulations for repair are introduced
- Life extension of repair is also to be taken up as a part of aircraft life extension program

3. Extended Service Goal (ESG)

When an aircraft is approaching the Design Service Goal limitation, it is either withdrawn from service or the lease of life is provided based on detail analysis and tests undertaken during the Extended Service Goal program.

The Extended Service Goal will enable customers to earn additional revenue by flying their aircraft for longer, as well as benefiting from the increased residual value of their fleet. In addition, operators will have the opportunity to upgrade their fleet to keep in-service aircraft at the highest level of efficiency.

Life Extension activities include a general review of the fatigue and damage tolerance analysis prepared for type certification, interpretation of full scale fatigue test findings, tear down results and Service Bulletin (SB) review with respect to extended service goals. As a part of life extension of aircraft exercise program, following items are taken up individually:

- Airframe
- Landing Gear
- Aero Engines
- Rotables
    - Mechanical equipment
    - Electrical & avionics equipment
- Perishable Items
    - Rubber seals and hoses
- Structural Repairs

With regard to structural repairs are concerned, the impact of the ESG loads on existing repairs in the fuselage structure are re-assessed for extended service goal requirements. Generally Fatigue & Damage Tolerance (F&DT) justification is performed for ESG requirements.

ESG exercise of a typical stringer repair is considered in the present paper to demonstrate the usefulness of statistical approach towards efficiency improvement.

4. Methodology for Stringer Repair

Stringers, running in the longitudinal direction of the fuselage, are internal structures of the aircraft (Figure 1) and are fastened to the fuselage through a series of rivets.

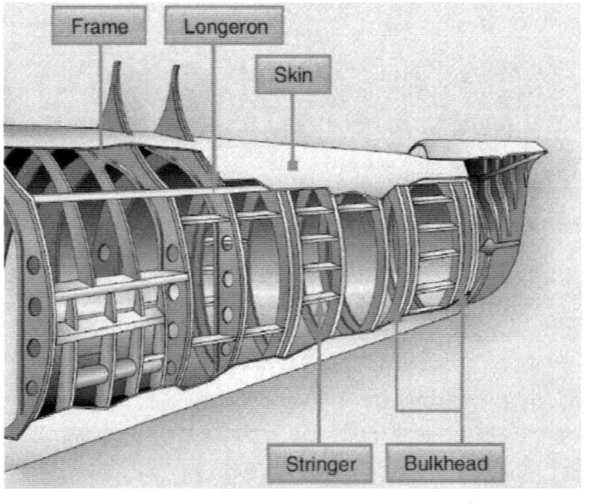

Figure 1. Stringer location on aircraft fuselage

As a part of life extension of stringer repair, it is required to re-assess the fatigue life of stringer in the presence of coupling under modified load spectrum (Figures 2 & 3)

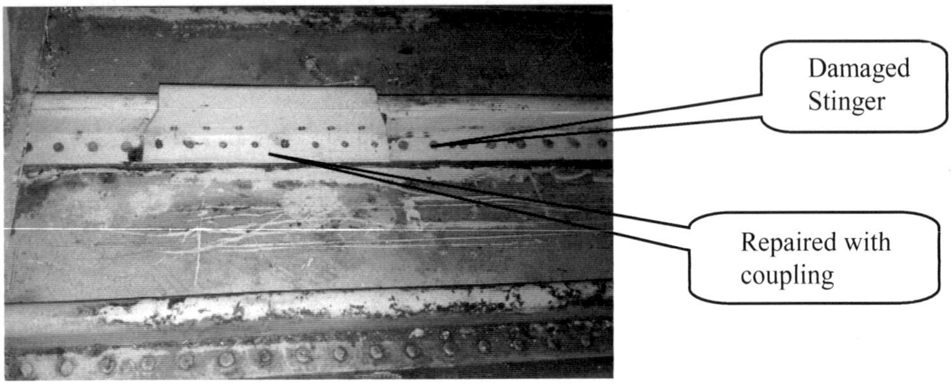

Figure 2. Repaired stringer with coupling

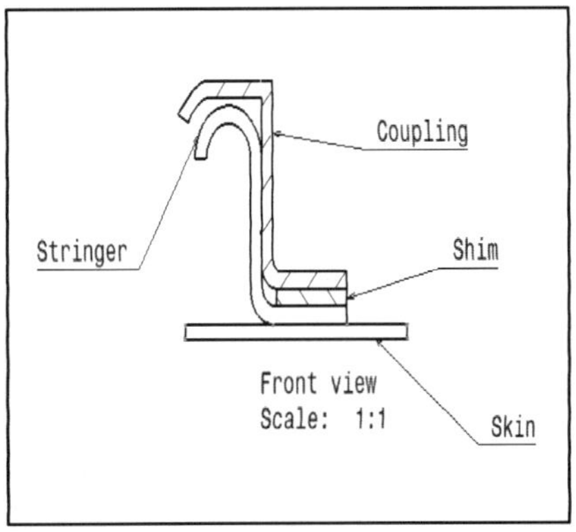

Figure 3. Side view of a stringer

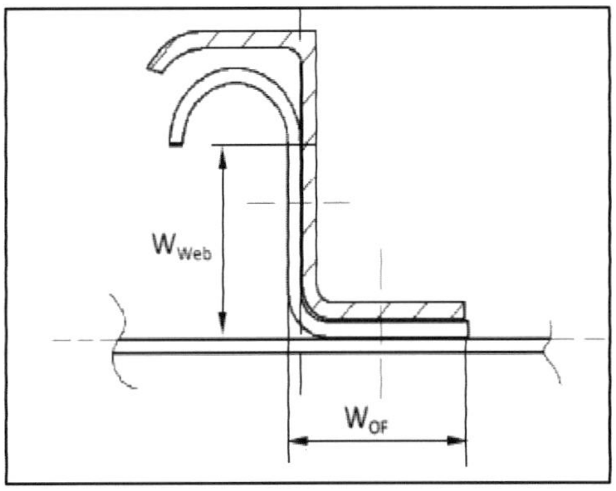

Figure 4. Web & OF parts of a stringer

Re-assessment of fatigue life is done for both Web and OF separately (Figure 4) as per F&DT justification philosophy. Determination of stress concentration factor (K_t) for a two plates system (for Web-stringer &coupling) or three plates system (for OF-skin, stringer &coupling) is one of the important parameters required for re-assessment of fatigue life under modified load spectrum.

As per the process for life extension program of stringer repair, it is required to determine K_t for different combination of width, pitch, stringer thickness, coupling thickness, rivet diameter and pad-up thickness of all stringer profiles available in different sections of fuselage. Moreover, the recalculated fatigue life should cover all combination of repair scenarios possible for a given repair at different location of the fuselage.

For the given case in the present study, fuselage of a particular passenger aircraft contains 60 plus different stringer profiles with number of variations in geometry parameters.

There will be a range of values existing in 2 different stringer materials of the web over 60 plus stringer profiles. Similarly, same order of variations are existing in OF corresponding to profiles.

The determination of stress concentration factor of 2 plates joint model (Web) or 3 plates joint model (OF) covering entire ranges involves substantial number of FE runs. In order to optimise the number of runs, stress concentration factor is determined under worst condition scenario based on parametric study of different parameters that affects stress concentration factor such as rivet diameter, width, pitch, stringer thickness, coupling thickness and in addition pad-up thickness in the case of OF. Typical worst scenario conditions that results in maximum stress concentration factor are maximum width, maximum pitch, maximum thickness combination of stringer & coupling (mean thickness), minimum rivets diameter and maximum pad-up thickness (in case OF only). In the process, the predicted life at most of the cases (locations) does not meet the target life due to use of conservative K_t value obtained under worst condition scenario.

In the present paper, efforts were made to make use of relevant statistical modelling and analysis for possibility of bringing in efficiency enhancement in terms of reducing number of FE analysis substantially. Also upon successful statistical modelling of stinger data across profiles for an aircraft, the same can be horizontally deployed for other aircrafts for a given set of pre-defined values for each geometric parameter.

5. Statistical approach towards efficiency improvement

5.1. Overview of the approach:

Aircraft structure getting damaged during its service is inevitable. A multimillion dollar aircraft's utilization after DSG is essential from cost optimization point of view and also to address gradual and rapid increase in volume of traffic. In such a situation, it is imperative that fatigue analysis is done for all internal and external components of aircraft and meet airworthiness requirement and qualify the aircraft for ESG. Fatigue life assessment was done using deterministic approaches. Instead of that, statistical approach in determining the stress concentration factor was used, which is more probabilistic in nature.

5.2. Statistical analysis:

Stringer data for aircrafts that were undergoing ESG captured over a period of time was taken up for a systematic study. This study included factors such as:

- Geometrical dimension of the stringer profile data
- Material used for coupling the damaged stringer
- Rivet diameter

The stingers have numerous profiles and various dimensional parameters like pitch, thickness, width that impacts stress concentration factor (K_t). The permutation and combination of these parameters ran into large number of data sets, hence stringers was chosen as the first part for this study as they are more in volume and present in every aircraft. Also upon successful statistical modelling of stinger data across profiles for an aircraft, the same can be deployed for other aircrafts for a given set of pre-defined dimensions for each parameter without the need for revalidation.

5.3. Objective of the statistical analysis of stinger data:

K_t is determined for stringer (with coupling) corresponding to each profile to determine the fatigue life of the aircraft structure. In order to optimise number of FE runs, K_t is determined under worst repair scenario which is a very conservative approach. In the process, the predicted life at most of the cases (locations) does not meet the target life due to conservative K_t value. Instead, based on parametric study and statistical analysis, optimum equation for K_t can be derived for a combination of geometry without compromising the level of accuracy of K_t and also reduce number of FE runs and thereby improve efficiency of the team involved in re-assessing of stringer repair's life through conventional approach.

Note: Regression analysis is a statistical technique that attempts to estimate the relationships among variables and relationship between a dependent and multiple independent variables (1Y and multiple

IOP Conf. Series: Materials Science and Engineering **370** (2018) 012005 doi:10.1088/1757-899X/370/1/012005

X's). It provides cause and effect relationship between X's (predictors) and Y (response) [4]. Understanding the relationship between X and Y can be further extended in having a meaningful use of the relation to predict the future

Tools used:
MS Excel, OEM specific tools for K_t generation for certain combination of geometrical parameters of web and outer flange and R tool was used for the statistical modelling, as it provides variety of analysis & is highly extensible.

5.4. Parametric study of stringer parts

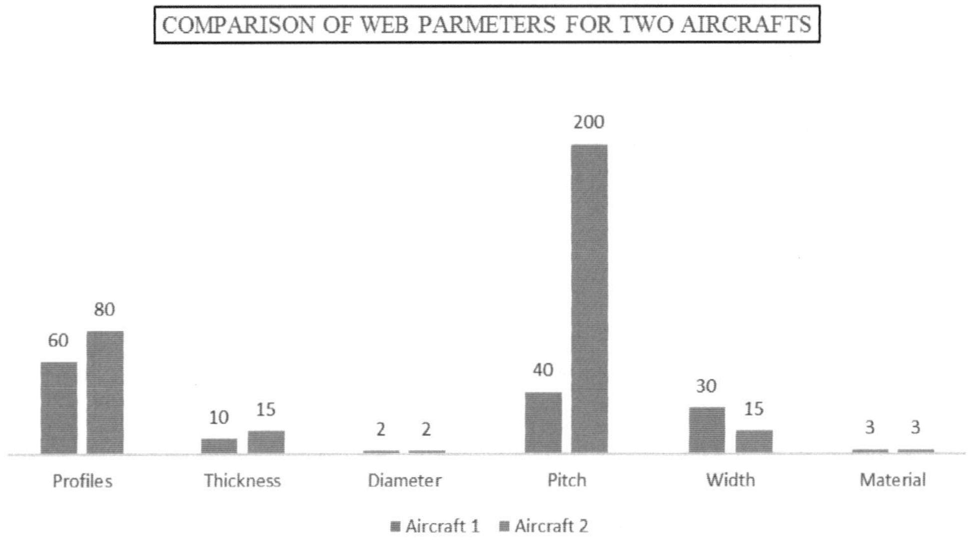

Figure 5. Comparison for WEB (indicative numbers)

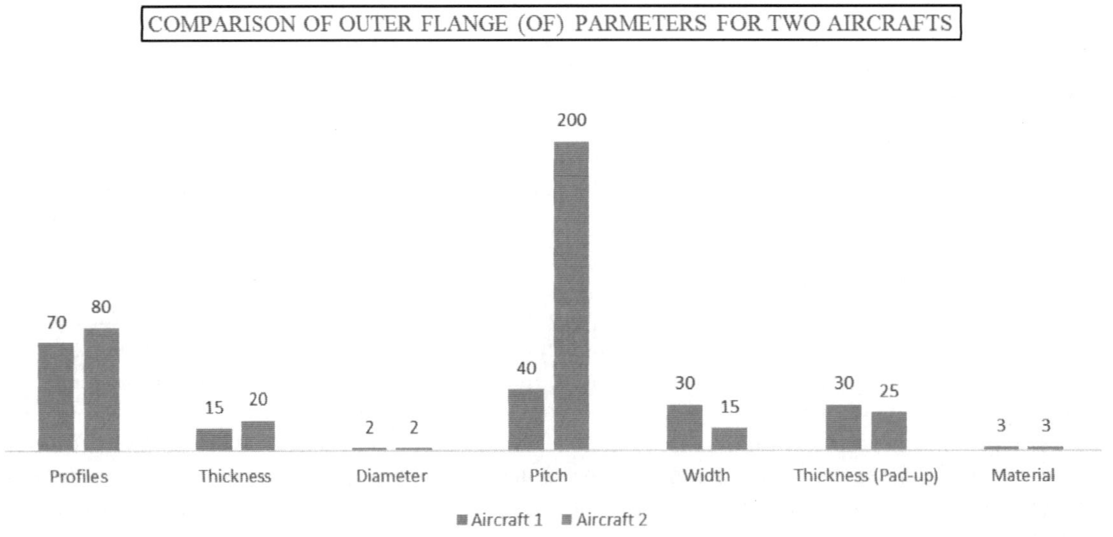

Figure 6. Comparison for OUTER FLANGE (indicative numbers)

5.5. Understanding key characteristics that impact Kt

In order to derive at an equation that predict the behaviour of K_t, it was important to understand which of the key characteristics that really impact from statistical angle.

All predictors were subjected to Hypothesis Test at every level of classification, i.e. does change in profiles, material, rivet pitch etc. impact K_t.

Via hypothesis test, it was evident that only geometrical dimensions impact K_t for a given fastener diameter, whereas material of stringer/coupling and stringer profile does not impact K_t

Prediction models were built using statistically qualified predictors. Available range of data was considered at minimum, average and maximum values for each parameters. K_t was determined for two combinations: Y = f(X1, X2, X3) for Web and Y = f(X1, X2, X3, X4) for OF, where Y=*Kt*

- X1 = Width
- X2 = Pitch
- X3 = Pad-up thickness (Skin)
- X4 = Mean Thickness (Average thickness between stringer (S) and coupling (C))

5.6. Regression output for WEB

From table 1, it is evident that range of width is split into two. This is done to ensure that prediction accuracy doesn't get compromised in order to accommodate the entire geometrical range for individual parameters.

Table 1. Outcome from regression for Web

Fastener Diameter in mm	Width in mm [X1]	Pitch in mm [X2]	Mean Thickness in mm [X3]	Coefficient of determination (R^2)
4	20, 30 & 40.9	20, 30 & 40	1.5 ,3.1 & 4.6	99.89%
4	41 & 50	20, 30 & 40	1.5 ,3.1 & 4.6	100%
4.8	20, 30 & 40.9	20, 30 & 40	1.5 ,3.1, & 5.1	99.99%
4.8	41 & 50	20, 30 & 40	1.5 ,3.1, & 5.1	99.99%

Table 2. Regression model output for rivet diameter 4 mm

Analysis of Variance					
Source	**DF**	**Sum of Squares**	**Mean Square**	**F Value**	**Pr > F**
Model	3	25.80220	8.60073	115195	<.0001
Error	14	0.00105	0.00007466		
Corrected Total	17	25.80325			

Root MSE	0.00864	**R-Square**	1.0000
Dependent Mean	7.40167	**Adj R-Sq**	1.0000

Variable	Parameter Estimate	Standard Error	t Value	Pr > |t|
Intercept	4.49633	0.02255	199.38	<.0001
Pitch	0.00567	0.00024944	22.72	<.0001
Mean Thickness	0.94508	0.00161	587.37	<.0001

IOP Conf. Series: Materials Science and Engineering **370** (2018) 012005 doi:10.1088/1757-899X/370/1/012005

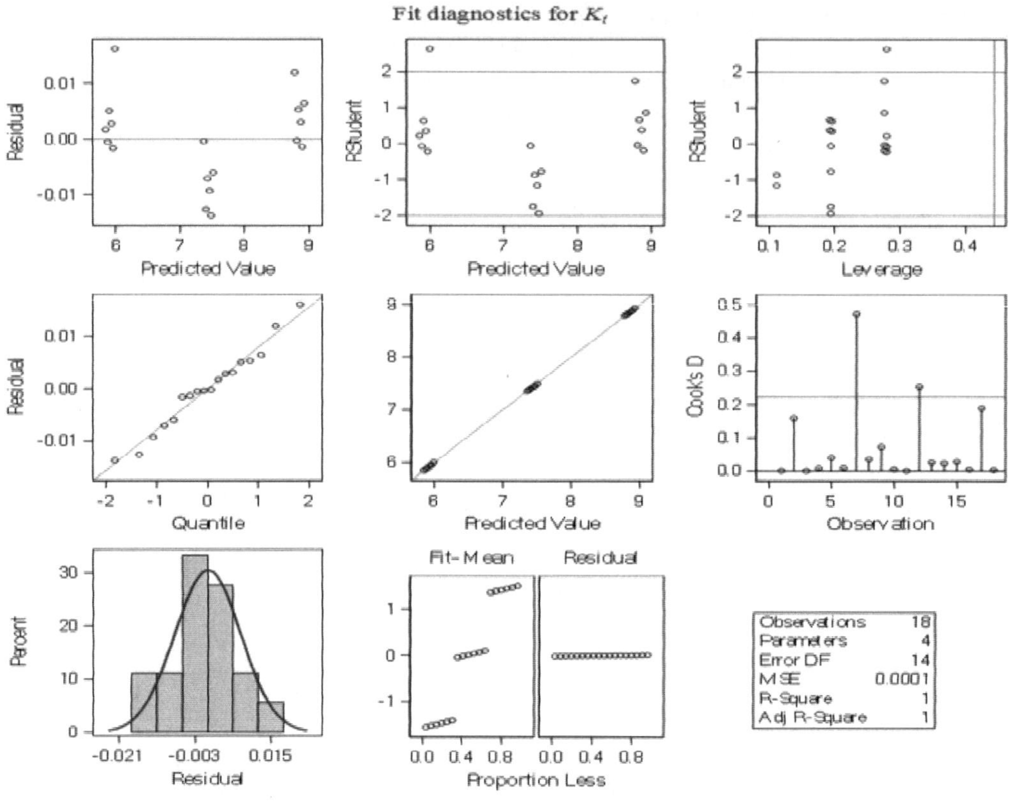

Figure 7. Model diagnostics

Inference:

- Result is achieved through multilinear regression with two-level-interaction between the predictors
- Co-efficient of determination R^2 was at close to 1, which implies best prediction.
- Overall P and individual P value of all predictors is < 0.05, which validates the hypothesis.
- It was possible to achieve < +/-1% error between observed and predicted for a given range of predictors, which is acceptable prediction from engineering accuracy point of view.
- Prediction accuracy % at less than +/-1 is critical considering that the fatigue life reduces exponentially with every decimal increase in prediction accuracy beyond +/-1%.
- Table 3 shows a typical comparison of predicted (using equation) and theoretical results (obtained through FE analysis) of Kt for different combination of width, pitch and mean thickness for the Web (4.0 mm rivet dia.) which are not a part of pre-defined data set.

Table 3. Comparison between predicted and theoretical Kt for validation

Width in mm [X1]	Pitch in mm [X2]	Mean Thickness in mm [X3]	K_t predicted	K_t through FE analysis	% Error
21	27	2.1	4.88	4.86	0.4%
21	27	4.1	6.09	6.06	0.4%
25	27	1.7	4.95	4.96	-0.2%
25	30	1.7	4.96	4.98	-0.3%
27	27	2.3	5.53	5.56	-0.5%
27	30	2.5	5.69	5.72	-0.5%

5.7. *Regression output for OUTER FLANGE*

Table 4. Outcome from regression for OF

Fastener Diameter in mm	Width in mm [X1]	Pitch in mm [X2]	Thickness (skin pad-up) in mm [X3]	Thickness - OF in mm [X4]	Coefficient of determination (R^2)
4	21,30,40.9	21,30 & 40	1.6,3,6	1.9,4.5	99.71%
4	40,50	21,30,40,50	6,12	1.9,3.5	99.46%
4.8	21, 30 & 40	21,30 & 40	2.2,6,12	1.9,4.5	99.62%

Inference:

- For the 3 plates system (OF), it was possible to achieve <+/- 2% error between observed and predicted for a given range of predictors.
- It is achieved through multilinear regression with two-level-interaction between the predictors
- Containing error % at less than +/-1 is critical, but it was not possible to achieve in present exercise as it is a 3 plates system with additional predictor influencing the K_t

Note: Similar to web, validation of observed and predicted K_t were carried out for outer flange for intermediate levels for each of the parameter and results were in line with pre-defined levels used for model building.

Table 5. Sample comparison of efforts resulting in saving due to statistical analysis approach

Part	Effort w/o Improvement [Hours]	Effort with Improvement [Hours]	Total Efforts [Hours]	No. of FE runs	No. of FE runs using regression analysis	% saving in efforts
Web	16	4	20	106 runs	45 runs	57%
OF	40	56	96	230 runs	108 runs	53%

For the given ESG program exercise for stringer repairs of an aircraft containing 60 plus profiles, 106 FE runs, based on the worst case scenario conditions were made which took 16 hours of efforts for re-calculation of fatigue life based on K_t results. Since FE analysis results in conservative K_t value due to consideration of worst case scenario conditions, some of locations did not meet the required target life. Therefore, additional efforts of 4 hours were spent in improving the results by re-calculating K_t in FE using actual geometry parameters at those locations. A total of 20 hours were spent to cover the entire profiles with all sort of geometry parameter combinations existing in the fuselage of the aircraft in the process of re-calculation of fatigue life. Percentage of effort saving is calculated based on the number of FE runs (45)required to carry out regression analysis to obtain the optimum equation compared to the number FE runs (106) made to obtain the conservative K_t value. The percentage effort saving stands at 57% for web. Efforts involved in the calculation of fatigue life is ignored as it involves excel based calculation sheet corresponding to all combinations of geometry parameters. Similar case of effort saving of 53% was witnessed for OF, which is very significant in terms of saving man hours & software usage).

6. Concluding Remarks

- Overview of methodology for life extension of stringer repair is outlined from aircraft industry practice point of view and F&DT justification is the basis for life extension of structural repairs

- Statistical tool is used to develop regression analysis for prediction of K_t based on limited number of FE runs required for determination of stress concentration factor
- The correlation between the predicted K_t and K_t obtained through FE analysis is found to be excellent with error well within 1% for Web for a given range of pre-defined data set
- The correlation between the predicted K_t and K_t obtained through FE run is found to be good with error well within 2% for OF for a given range of pre-defined data set
- Optimum equation for prediction of K_t for both Web & OF is achieved through multilinear regression with two-level-interaction between the predictor
- Substantial benefit, in terms of reduction in number of FE runs and thereby huge time saving can be realized while carrying out similar exercise for other a/c program provided they are within pre-defined data set
- In the present study, the effect of secondary bending moment in joint model is ignored as per F&DT justification philosophy.
- Error percent between predicted K_t and observed K_t of up to 2% is not considered to be exceptionally good outcome in case of outer flange since the impact of fatigue life is considerable with increase in error percentage, this will be considered for further refinement and analysis.

Acknowledgements
We would like to extend our thanks to subject matter experts who helped us in combining technical and statistical knowledge in an efficiency improvement exercise. Project team's support in getting relevant data for K_t generation was key to achieve desired results.

References

[1] Aircraft Metal Structural Repair, Aviation Maintenance Technician Handbook Volume 1, 2008
[2] AIRWORTHINESS ADVISORY CIRCULAR, AAC No. 2 of 2013 Dated 20th January, 2013, Government of India, Civil Aviation Department, Director General of Civil Aviation
[3] Structural repair manuals are customer specific, hence online references alone can be used.
[4] Understandable statistics concepts and methods by Brase

IOP Conf. Series: Materials Science and Engineering **370** (2018) 012006 doi:10.1088/1757-899X/370/1/012006

CFD Simulation of a Wing-In-Ground-Effect UAV

C T Lao[1] and E T T Wong[2]

[1] Interdisciplinary Division of Aeronautical and Aviation Engineering, The Hong Kong Polytechnic University, Hong Kong

[2] Department of Mechanical Engineering, The Hong Kong Polytechnic University, Hong Kong

[1] jack.lao@connect.polyu.hk
[2] mmttwong@connect.polyu.hk

Abstract. This paper reports a numerical analysis on a wing section used for a Wing-In-Ground-Effect (WIG) unmanned aerial vehicle (UAV). The wing geometry was created by SolidWorks and the incompressible Reynolds-averaged Navier-Stokes (RANS) equations were solved with the Spalart–Allmaras turbulence model using CFD software ANSYS FLUENT. In FLUENT, the Spalart-Allmaras model has been implemented to use wall functions when the mesh resolution is not sufficiently fine. This might make it the best choice for relatively crude simulations on coarse meshes where accurate turbulent flow computations are not critical. The results show that the lift coefficient and lift-drag ratio derived excellent performance enhancement by ground effect. However, the moment coefficient shows inconsistency when the wing is operating in very low altitude - this is owing to the difficulty on the stability control of WIG vehicle. A drag polar estimation based on the analysis also indicated that the Oswald (or span) efficiency of the wing was improved by ground effect.

1. Introduction

In the long history of aerodynamic research, it was proven that a lifting body operates proximately to a physical boundary (or ground) has greater aerodynamic efficiency than when it operates in freestream. 'Ground effect' is an enhanced aerodynamic performance phenomenon of a lifting body, which is evident while a lifting body operating in close proximity to the ground [1].

A Wing-In-Ground-Effect (WIG) vehicle needs some forward velocity to produce lift dynamically and the principal benefit of operating a wing in ground effect is to reduce its lift-dependent drag. The basic design principle is that the closer the wing operates to an external surface such as the ground, said to be in ground effect, the more efficient it becomes. As sea surface provides a more stable boundary in comparing to dry land, WIG vehicle is a suitable and eco-friendly solution on high speed marine transportation purpose. Some of researches have discussed the service limitations and capabilities of WIG vehicle and summarized that is a possible solution of niche transportation in multi-island countries such as Malaysia, the Philippines and Caribbean area [2][3].

In this paper, a CFD simulation was carried out to analyze the aerodynamic characteristics of a wing used in a WIG UAV. The analysis examines the performance of the wing in both freestream and ground effect air flow. The analysis results are significantly important for future WIG vehicle and WIG UAV developments in terms of flight dynamics and control.

2. Theory

In the early century, ground effect is also called as 'cushion effect' since it act as an air cushion which support the aircraft while landing. Theoretically, ground effect is caused by the following physical phenomena [4]:

1. Since the downward airflow movement (or downwash) is blocked by the boundary, more air molecules stay under the wing bottom surface and the pressure on these area increases. Due to this, the airflow spanwise vector would be strengthened and the wing effective span enlarges. Furthermore, the wing aspect ratio increases relatively with effective span, hence, lift force raises.

2. Due to the airflow blockage, wingtip vortex would not acuminate as the aircraft in freestream. When the vortex flows reach to the boundary, it would stop to distribute around the wingtip. The reduction of strength of wingtip vortex also reduces the induced drag and downwash.

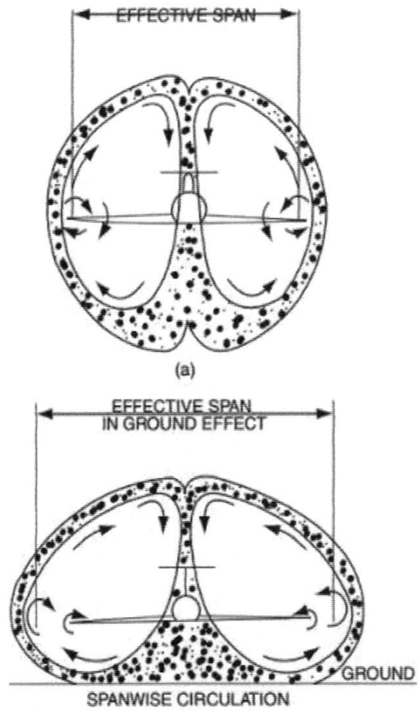

Figure 1. Flow illustration of wing with and without ground effect [4]

3. Research methodology

A wing prototype (Fig. 2) was constructed using SolidWorks and it was exported to ANSYS for creating a mesh and computational domain. To save the computer resources, only single side of the wing would be examined. The wing is a Lippisch style reversed delta wing which has a root chord of 300mm, a tip chord of 75mm and 300mm span. Airfoil shape for this wing is Clark-Y.

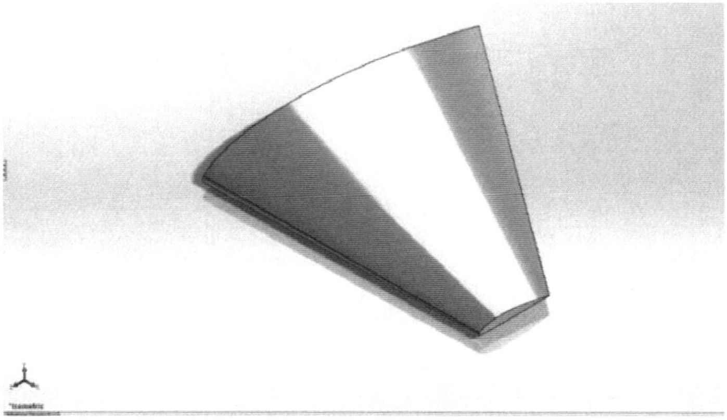

Figure 2. Wing geometry

The CFD simulation was carried out on ANSYS FLUENT, and the incompressible Reynolds-averaged Navier-Stokes (RANS) equations were solved with the Spalart–Allmaras (S-A) turbulence model at the Reynolds number of $2.05\text{x}10^5$ (based on the averaged chord length of wing chord)[5]. The Spalart-Allmaras model is a relatively simple one-equation model that does not require the calculation of a length scale related to the local shear layer thickness. The Spalart-Allmaras model was designed specifically for aerospace applications involving wall-bounded flows and has been shown to give good results for boundary layers subjected to adverse pressure gradients. The lift, drag and moment coefficients and lift-drag ratio (aerodynamic efficiency) were estimated in the freestream flow of 16 m/s for various angles of attack, viz. 0, 5, 10, 15, and 20 degrees.

For ground effect analysis, the operating altitude of the wing is defined as the ratio between height above ground and mean chord length (h/c). With lower h/c number, the wing is closer to the ground. The height is defined as the distance between the mean quarter chord and ground. In this analysis, the operating altitude of the wing in terms of height-chord ratio would be 0.5 (93.75mm above ground), 1.0 (187.5mm) and 1.5 (281.25mm). The ground effect can be negligible for h/c > 1 and the wing in this altitude can be considered as operating in freestream air.

4. Governing equations and turbulence model

Similar to the most cases of CFD, ANSYS FLUENT is governed by Reynolds-averaged Navier–Stokes (RANS) equations with various turbulence models which depend on the application.

The general form of continuity equation is written as:

$$\frac{\partial u_i}{\partial x_i} = 0 \tag{1}$$

And the momentum equation is written as:

$$\rho \bar{u}_j \frac{\partial \bar{u}_i}{\partial x_i} = \rho \bar{f}_i + \frac{\partial}{\partial x_j}\left[-\bar{p}\delta_{ij} + \mu\left(\frac{\partial \bar{u}_i}{\partial x_j} + \frac{\partial \bar{u}_j}{\partial x_i}\right) - \rho\overline{u'_i u'_j}\right] \tag{2}$$

In terms of turbulence model selection, various researches figured out that Spalart–Allmaras turbulence model would be the most suitable one for handling WIG effect problems due to result accuracy and integrity [6][7].

The one equational model is written as:

IOP Conf. Series: Materials Science and Engineering **370** (2018) 012006 doi:10.1088/1757-899X/370/1/012006

$$\frac{\partial\tilde{v}}{\partial t} + u_j\frac{\partial\tilde{v}}{\partial x_j} = C_{b1}[1 - f_{t2}]\tilde{S}\tilde{v} + \frac{1}{\sigma}\{\nabla[(v + \tilde{v})\nabla\tilde{v}] + C_{b2}|\nabla\tilde{v}|^2\} - \left[C_{w1}f_w - \frac{C_{b1}}{\kappa^2}f_{t2}\right]\left(\frac{\tilde{v}}{d}\right)^2 + f_{t1}\Delta U^2 \quad (3)$$

Where the terms involved are:

$$v_t = \tilde{v}f_{v1};\; f_{v1} = \frac{X^3}{X^3+{C^3}_{v1}};\; \mathrm{X} := \frac{\tilde{v}}{v};\; \tilde{S} \equiv \mathrm{S} + \frac{\tilde{v}}{\kappa^2 d^2}f_{v2},\; f_{v2} = 1 - \frac{X}{1+Xf_{v1}};\; f_w = g\left[\frac{1+{C^6}_{w3}}{g^6+{C^6}_{w3}}\right]^{1/6};$$

$$\mathrm{g} = \mathrm{r} + C_{w2}(r^6 - r);\; \mathrm{r} \equiv \frac{\tilde{v}}{\tilde{S}\kappa^2 d^2};\; f_{t1} = C_{t1}g_t\exp\left(-C_{t2}\frac{{w_t}^2}{\Delta U^2}\left[d^2 + {g_t}^2{d_t}^2\right]\right);\; f_{t2} = C_{t3}\exp(-C_{t4}X^2);$$

$$\mathrm{S} = \sqrt{2\Omega_{ij}\Omega_{ij}}$$

The rotation tensor is given by:

$$\Omega_{ij} = \frac{1}{2}\left(\partial u_i/\partial x_j - \partial u_j/\partial x_i\right) \quad (4)$$

The constants used in the analysis are:

$\sigma = \frac{2}{3}$; $C_{b1} = 0.1355$; $C_{b2} = 0.622$; $\kappa = 0.41$; $C_{w1} = \frac{C_{b1}}{\kappa^2} + \frac{1+C_{b2}}{\sigma^2}$; $C_{w2} = 0.3$; $C_{w3} = 2$; $C_{v1} = 7.1$; $C_{t1} = 1$; $C_{t2} = 2$; $C_{t3} = 1.1$; $C_{t4} = 2$

5. Mesh And Boundary Conditions

Once the CAD file was exported to ANSYS, a box of dimensions 0.5m × 0.5m × 1m is created to cover the wing which acts as a fluid chamber or a virtual wind tunnel. A subtract Boolean condition is applied where the box is target body and the wing is the tool body.

To simplify the procedure on CFD, except the minimum grid size, all meshing parameters would be the default values in ANSYS. All models have been meshed with tetrahedral, unstructured grids since the mesh can automatically be generated with default values, which reduces the complexity to create the computational domain. For all meshing files, the numbers of elements are at least 500,000 but not beyond the upper mesh limit (512,000 elements for ANSYS 17.1 Academic)

The boundary conditions were defined for various sections – these include the inlet, outlet, ground and sidewall. The inlet velocity would be 16 m/s, and the outlet gauge pressure would be 0 Pa. The turbulent viscosity ratio is 10 and the modified turbulent viscosity is 0.000146 m^2/s for all simulation. A moving wall conditions is applied to simulate the ground. It is a no-slip wall and slides along x direction with the same velocity of inlet flow. Finally, a symmetrical boundary condition was set on the sidewall to achieve a zero shear condition.

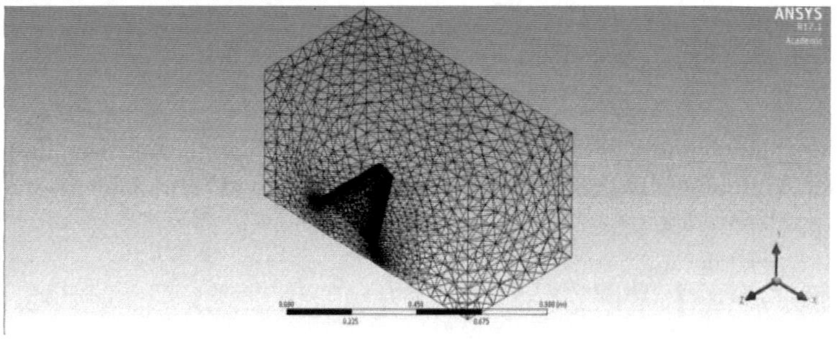

Figure 3. Geometry mesh

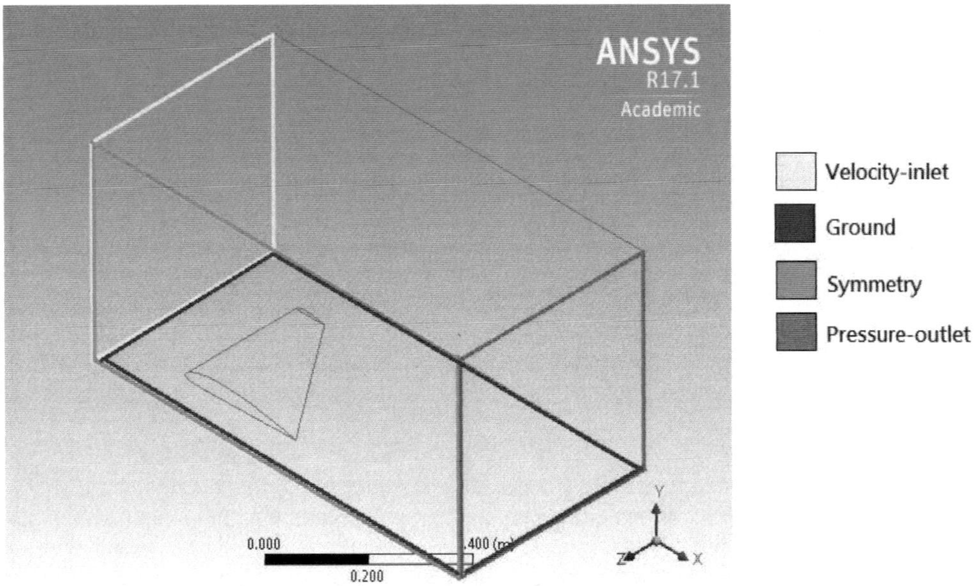

Figure 4. Boundary conditions

6. Result and discussion

Referring to Fig. 5, which showed that the enhancement by ground effect in terms of lift coefficient. The lift coefficients before stall when the wing flies at 0.5 height-chord ratio, is 1.85 times of it flies at 1.5 h/c. Also, the lift gradient is greater when the wing flies with assisting by ground effect. In terms of drag coefficient (fig. 6), the differences are not significant when the wing flies at different flight level unless in high angles of attack.

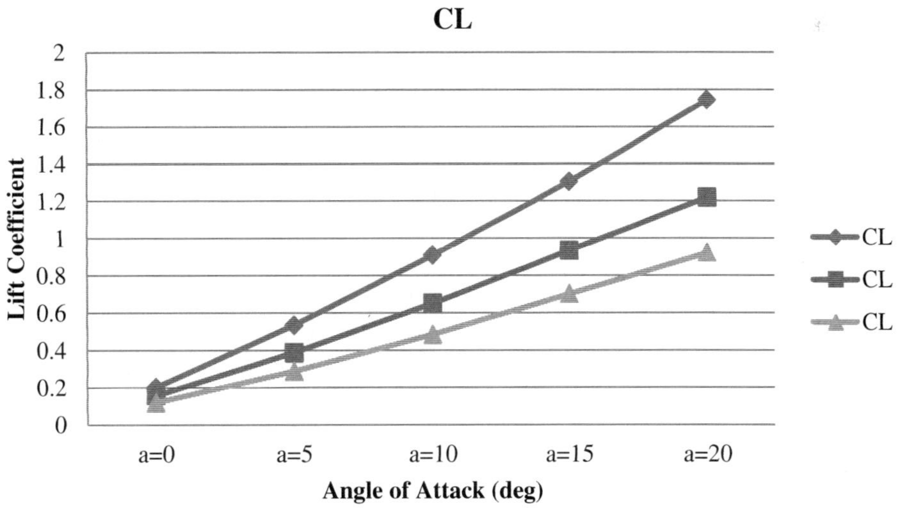

Figure 5. Lift coefficient

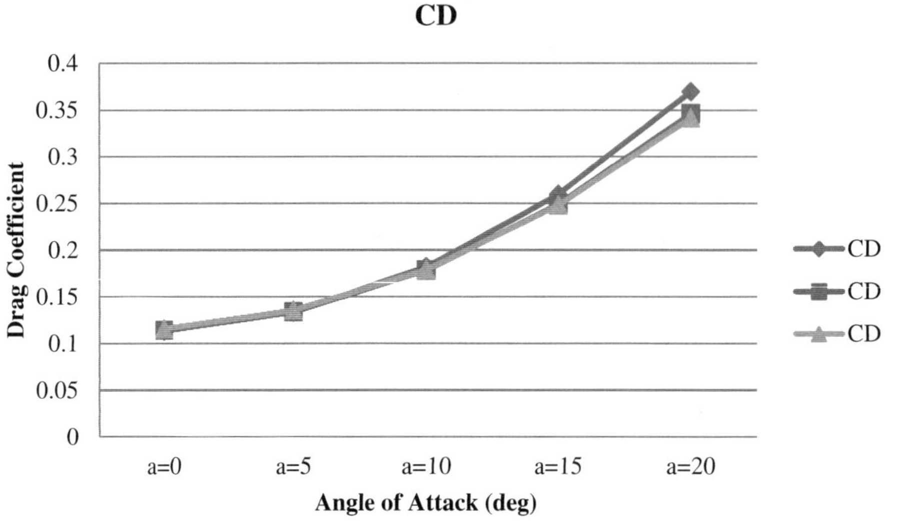

Figure 6. Drag coefficient

In terms of moment coefficient (Fig. 7), it is not consistent in different operation altitude as the wing flies in freestream air. It means the pitching moment will change significantly with very small operating altitude change. When the wing operates relatively closer to the ground, the moment coefficient is more sensitive with the wing angle of attack. This shows the difficulty on maintaining the longitudinal stability while operating a WIG vehicle.

Combining the results of lift and drag coefficients, it can be summarized that the lift-drag ratio (fig. 8) is greatly enhanced by ground effect.

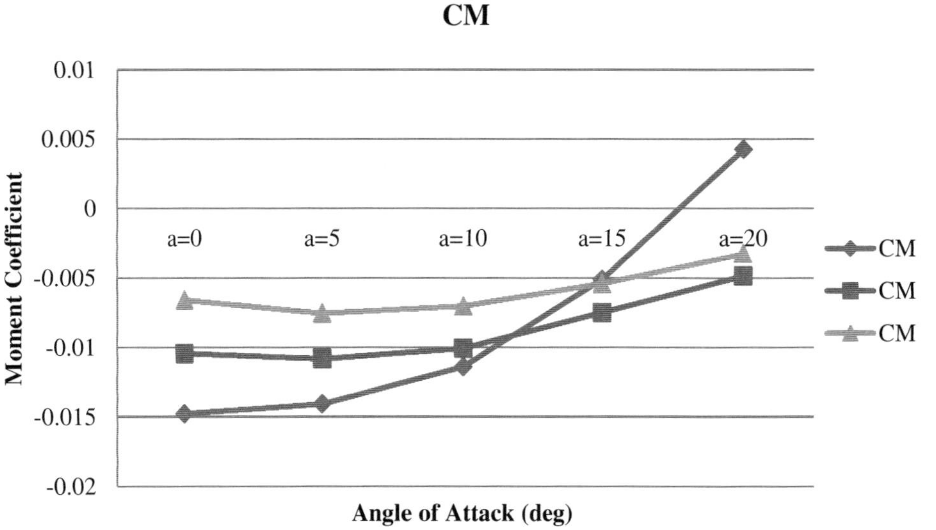

Figure 7. Moment coefficient

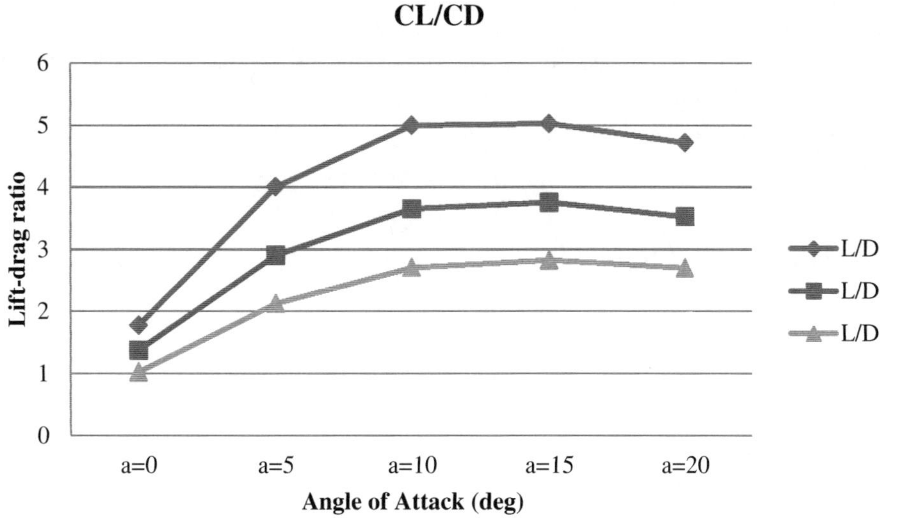

Figure 8. Lift-drag ratio

7. Mesh independence study

To validate the results generated, a mesh independence study was carried out for the wing angle of attack in 15 degree at various altitude. The numbers of element in this study are fairly close to 300,000 (ultra-coarse mesh), 400,000 (coarse) and 500,000 (fine) which are adjusted by changing minimum grid size. The result in lift-drag ratio are shown in the figure 9.

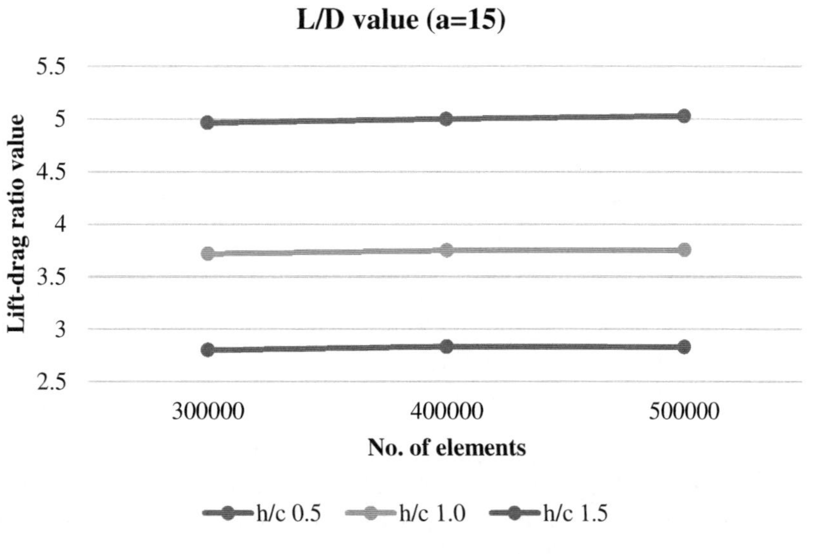

Figure 9. Variation of lift-drag ratio at variated mesh size

The study shows that the variation in terms of lift-drag ratio values have no significant changes at coarser mesh. The mean difference in result between ultra-coarse mesh and fine mesh is about 1% which is acceptable for the simulation. It was proven that the results obtained in this simulation are valid which have no dependency with number of mesh element.

8. Drag polar estimation

Drag is one of aerodynamic forces applied to the wing and it includes two major parts: parasite drag and induced drag. Drag polar is one of major aerodynamic considerations in terms of wing design which is the major factor of induced drag. The drag polar (K) of the wing is affected by the wing aspect ratio (AR) and the Oswald efficiency (e).

The drag coefficient (C_D) can be explained as:

$$C_D = C_{D,0} + KC_L{}^2, where\ K = \frac{1}{\pi eAR} \quad (5)$$

For the first derivative of C_D to $C_L{}^2$:

$$\frac{dC_D}{dC_L{}^2} = K \quad (6)$$

$$\frac{dC_L{}^2}{dC_D} = \pi eAR \quad (7)$$

For π and AR are constant for the wing, hence:

Table 1. Drag polar estimation.

	K (Drag polar)	e (Oswald efficiency)
h/c=0.5	0.085761	1.160459
h/c=1.0	0.158474	0.628004
h/c=1.5	0.312145	0.318834

In the CFD simulation, the wings of WIG vehicle were designed at low aspect ratio (3.2 for this design). A major drawback of this wing is that the induced drag coefficient plays the major role in total drag coefficient. The estimated results showed the drag polar in h/c=0.5 was only 27.4% of it in the freestream, which means the induced drag was reduced with aid of ground effect.

9. Concluding remarks

The increase of the numerically calculated lift coefficient in the ground mode was 85% when compared to the lift coefficient in a freestream. The estimated results also showed that the Oswald efficiency of the wing was improved by ground effect. These positive numerical results indicate that wider application of WIG UAV has a great potential. However, before building a real WIG UAV model, one should also study the problem of flight stability and control. Hence further works have been planned on the fabrication and flight testing of the model UAV.

References

[1] Cui E and Zhang X 2010 Ground Effect Aerodynamics *Encyclopedia of Aerospace Engineering* ed R Blockley and W Shyy (Hoboken: John Wiley & Sons, Ltd.) Cheapter 18 245-256

[2] Yang W and Czysz P A 2011 WIG craft serves niche transportation needs *World Review of Intermodal Transportation Research* **3** 395-406

[3] Taylor G K 2002 Are You Missing The Boat? The Ekranoplan in the 21ST Century - Its Possibilities and Limitations

[4] Yun L, Bliault A and Doo J 2010 WIG craft and ekranoplan *Ground Effect Craft Technology* (New York: Springer) 26-27

[5] ANSYS 2013 ANSYS Fluent Theory Guide

[6] Genua E 2009 A CFD Investigation into Ground Effect Aerodynamics

[7] Qu Q, Lu Z, Liu P and Agarwal R K 2014 Numerical study of aerodynamics of a Wing-in-Ground-Effect craft *Journal of Aircraft*

IOP Conf. Series: Materials Science and Engineering **370** (2018) 012007 doi:10.1088/1757-899X/370/1/012007

A review of phase separation issues in aviation gasoline fuel and motor gasoline fuels in aviation

K Thanikasalam[1], M Rahmat[1], A G Mohammad Fahmi[2], A M Zulkifli[1], N Noor Shawal[1] , K Ilanchelvi[3], M Ananth[3], R Elayarasan

[1]Faculty of Chemical and Energy Engineering, Universiti Teknologi Malaysia (UTM), 81310 Skudai, Johor, Malaysia
[2]Department of Aeronautical Engineering, Universiti Tun Hussein Onn Malaysia (UTHM), 86400 Batu Pahat, Johor, Malaysia
[3]Department of Chemical Engineering, Universiti Malaysia Sabah (UMS), 88400 Kota Kinabalu, Sabah, Malaysia

*Corresponding author: insp466@hotmail.com

Abstract. In an attempt to bring in sustainable energy resources into the current combustibles mix, recent European legislations make obligatory the addition of biogenic fuels into traditional fossil gasoline. The preferred biogenic fuel, for economic reasons, is predominantly ethanol. Even though likened to fossil gasoline constituents, ethanol has a dissimilar chemical formulation that may lead to a potentially hazardous physicochemical phenomenon, particularly in the presence of water. Owing to increased financially driven propensity to utilize motor vehicle gasoline as aviation gasoline fuel, this may result in potentially hazardous situations, specifically in running smaller or compact General Aviation aircraft. The potential risks posed by ethanol admixtures in aircraft are phase separation and carburettor icing. Gasoline mixed with ethanol is also prone to an increased vulnerability to vapor lock that happens when fuel turns into vapor in the fuel pumps due to high temperatures and lessened ambient pressure at high altitudes. This article provides a literature review on phase separation issues in aviation gasoline fuel and motor gasoline fuels in aviation.

1. Introduction

Esch et al. (2010) suggest that, currently, over 90% of all gasoline fuel sold in the European Union comprises a blend of ethanol. According to Esch et al. (2010), the usage of blended fuels increases the likelihood for a phenomenon called phase separation. Phase separation takes place when water gets into a tank that houses an ethanol-gasoline mixed fuel. Esch et al. (2010) claim that ethanol soaks up water and when gasoline becomes soaked, a film of water and ethanol, called phase separation, forms in the tank. The water-ethanol mixture has a heavier density compared with the gasoline phase, thus the mixture patches up at the base of the tank [2]. The mixture is not combustible and would result in an immediate starvation of the aviation engine if it were sucked in.

2. How Phase Separation Occurs

Boichenko and Lejda (2015) argue that the risk posed by a phase separation, although always reliant on the availability of water in the fuel tank, is a complex one. They suggest that as long as the water quantity does not surpass specific limits it is bearable. Numerous boundary circumstances investigated based on their significance in numerous tests influences this limit [1]. When an aircraft begins its service, the fuel's ceiling becomes frosty. The water carrying capacity depends on the fuel's temperature. Thus, flight experiment assists in the exploration of the cool-down effect. Boichenko and Lejda (2015) argue that, even if at the beginning of a flight there is a lack of adequate water content in the fuel tank to prompt a phase separation, an aircraft might gather additional water during the aerial navigation if it flies through chilly water vapour like fog or clouds. The aircraft may also cause the condensed water vapour to flow out of the ingested tank into the gasoline tank when descending from a freezing service height. Boichenko and Lejda (2015) also allege that when an aircraft is unused for a prolonged period the custom is to fill up its tanks to avoid corrosion. Since ethanol-admixed gasoline fuel is to some extent hygroscopic, this can result in a slow but progressive intake of environmental water.

Esch et al (2010) conducted an experiment to assess phase separation on a MORANE aircraft. The findings indicated that even though MORANE's tank is more thermally insulated likened to a wet wing system, the aircraft cools down considerably in a long flight. Owing to this, there is a real risk of phase separation if the gasoline encounters a considerable amount of water. This water, though not present when MORANE begins to fly, may flow into the fuel tank through the venting gap [2]. This threat exists in spite of the fact that minimal rates of water intake exists during rainy conditions, even when the aircraft passes through clouds or rains. Esch et al. (2010) argue that scientists should investigate further the quantity of potential humidity condensation in the descent stage from dry and cold heights to moist warm air in lower stratum. Esch et al. (2010) concluded that the risk of phase separation depends on the quantity of air ingested by the fuel tank's venting system, the particular crossover temperature gap bumped into during the last stage of descent, and the amount of moisture in the air.

Morgan (2014) suggests that an affinity to a buildup of water, even in powerful ethanol-admixed petrol (E-10, 10% of ethanol containing gasoline), is only identifiable once exposed fuel tank. When parked in a temperature-adjustable equilibrating hangar, the impact is significantly less. Morgan (2014) notes that experimentally established water bearing capabilities are larger compared to those indicated in existing literature, even though these pieces of literatures fail to provide principal clearance, and the concern by itself remains significant. Morgan (2014) also alleges that a phase separation in gasoline mixed with water may happen during the cool-down of the tank ingredients in the mid-flight. According to Morgan (2014), additional problems may emerge in the future if aircraft operators use automotive petrol without any direct pre-flight water content management. Given that currently there are no standardizing upper limits of water quantity for the ever-increasing ethanol-admixed gasoline, it would not be long before some competitive retailers are tempted to make the most of this boundary situation to a tip where motor vehicles with their almost stable fuel tank temperature would not endure any obvious drawback. A number of experiments using hydrous E-15 (E-15, 15% of ethanol containing gasoline) intended at saving the cost of making super-azeotropic ethanol before gasoline admixing process are already in this direction [4]. If such fuels with moderately high water content serve as aviation fuel, then the powerful cool-down in uncovered aircraft tanks might trigger the phase separation effect. Given that the absolute temperature reliance on the water tolerance rises considerably with increasing ethanol content, this phenomenon would progress considerably if aircraft operators exploit "non-disturbing water content" for commercial purposes.

Gramajo de Doz, Bonatti, and Sólimo (2004) assert that there is a limit to the quantity of water absorbed by ethanol-gasoline mixed fuel. According to Gramajo de Doz et al. (2004), this limit is reliant on the temperature and quantity of ethanol in the admixed fuel. If the water soaked up by ethanol-gasoline mixed fuel separates, a film of ethanol and water forms underneath the fuel tank. The film of ethanol and water blend is not an ideal fuel for the airplane, thus a fractional, if not absolute, power loss occurs. Gramajo de Doz et al. (2004) suggest that at 60°F, a one-gallon mixture of 90% gasoline and 10% ethanol absorb around 0.60 ounces of water before experiencing phase separation. If the temperature decreases to 32°F, 20% of the whole water contained in the fuel will phase separate.

Gramajo de Doz et al. (2004) note that with a 48-gallon fuel tank, around 6 ounces of water would amass and possibly flow through the pipe system. According to Gramajo de Doz et al. (2004), certification requirements only need 1 ounce per 20 gallons.

3. Effects of Phase Separation

Mukhopadhyay and Datta (2008) suggest that phase separation can cause harm to aviation petroleum infrastructure. Phase separation comprises a larger percentage of ethanol, some percentage of water, and a small percentage of gasoline. Mukhopadhyay and Datta (2008) allege that this mixture is very corrosive even compared with E-10 or pure water. There are some claims that phase separation might be even highly abrasive than 100% ethanol [5]. Some aviation petroleum tanks are incompatible to this ethanol-blended and corrosive liquid. Based on the content of the tank, phase separation soakedmay gradually depreciate the integrity of aviation tank walls and increase the threat of a leakage into the environment [5]. Other than the fuel tanks, other components of delivery and storage systems like pipes, dispensers, and submersible turbine pumps may also be in jeopardy when the level of phase separation rises to the position where pumping system can pick it up. Each of these concerns presents financial and environmental issues, as their substitute may be costly and harmful to the ecosystem if not detected [5]. Ideally, the aviation petroleum owner would want to identify any incidence of phase separation as quick as possible to allow them to speedily remediate and lessen the exposure to potential corrosion.

Other than damaging the aviation petroleum infrastructure, phase separation also has the potential to damage aircraft applications. Nakama, Kusaka, and Daisho, (2008) suggest that when using ethanol-gasoline mixed fuel like E10 in aircraft applications, possible problems exist. The more the concentration of ethanol, the higher the chances that the aircraft will experience issues that comprise;

- harm to fused gasoline tanks and rubber gaskets. Ethanol-gasoline mixed fuel is not as compatible as pure gasoline to these apparatuses [6].
- Phase separation also causes corrosion issues with metal gasoline tanks, electric fuel siphons, and other fuel system apparatuses. Ethanol, contained in the fuel, readily soaks up water.

Nakama et al. (2008) claim that ethanol may even soak up a noteworthy quantity of water from the environment during humid conditions. According to Nakama et al. (2008), when too much water is soaked up, phase separation occurs resulting in ethanol and water falling to the bottom of the gasoline tank. The combined water and ethanol can be caustic to metal gasoline tanks and fuel system apparatus, particularly if allowed to stay for a longer period.

Like Nakama et al. (2008), Strauss and Gonzalez (1989) claim that phase separation may damage aircraft's components. When separation occurs in the gasoline tank, there are higher chances that they will go through into the pumping system and enter the piping system, dispensers, and ultimately into an aircraft engine. Strauss and Gonzalez (1989) allege that phase separation causes many aircraft engines to stall. The affected aircraft will need maintenance leading to financial costs and losses.

4. Solutions to Address Phase Separation

Yüksel (2004) claims that if correctly installed with fuel tank sump drains, an aircraft should be able to drain off phase separation leaving only gasoline. However, Yüksel (2004) suggests that in such instances the remaining fuel will have somewhat reduced octane levels, less by 2-3 points based on the antiknock index (AKI) ranking technique. According to Yüksel (2004), the Rotax 912-ULS engine, two-stroke 618 engines, and the turbocharged 914 engines need 91-octane fuel using the AKI ranking procedure. The 912UL engine and the two-stroke 582, 503, and 447 engines operate on 87-octane.

Totten, Westbrook, and Shah (2003) note that if an aircraft's fuel tanks do not have sump drains installations and aircraft operator finds water at the gascolator, the operator should contemplate depleting a considerable quantity of fuel from every tank via the gascolator. Thereafter, the technician should assess the drained fuel for any signs of water contamination. The technician should carry on depleting the fuel up until the water is completely removed [9]. The gascolator ought to be the lowest tip in the fuel system. Thus, an aircraft ought to be level to ensure any water in the fuel tanks positions at the fuel tank pickups. The tip in the tank where the fuel lines sketch ought to be the lowest point if gasoline tanks lack sump drains installation [9]. This is where water will amass supposing the aircraft stays leveled.

Olah, Goeppert, and Prakash (2011) suggest that since ethanol has a tendency to absorb water, aircraft operators should ensure that they utilize the freshest E10 fuel whenever possible, check their fuel tanks for any water prior adding fresh gasoline, and do not stock up E10 fuel in containers several weeks unless in a dry place or climate. In addition, Olah et al. (2011) indicate that aircraft owners should only purchase fuel when there is an intention to use it and have fuel tank sump drains or gascolator installed in their planes. These gadgets will act as water collectors and filters. Olah et al. (2011) also point out that when any water is found on the gascolator, there are higher chances that phase separation has already taken place. This implies that there is likelihood for a considerable quantity of ethanol and water mix in the aircraft's fuel system. The aircraft's engine cannot operate using the phase separation, so it ought not to be permitted to find its way into the engine [7]. In such circumstances, the technicians should investigate the fuel system for any water before trying to take off.

5. Conclusion

According to Esch et al. (2010), in their Safety Implication Of Biofuels In Aviation 2010 report, there is an actual threat of phase separation if the gasoline contains a significant amount of water. This water, even if not already present at the time of start, may enter the fuel tank by its venting opening. This danger exists despite the fact that low rates of water ingestions are observed in rain conditions, even if the aircraft traverses clouds or strong rain. The amount of potential humidity condensation during the descent phase from cold and dry heights to humid warm air in lower strata should be scrutinized more thoroughly still. It will depend on the amount of air ingested by the tank venting system, the specific crossover temperature gap encountered during the last phase of descent, and the humidity of the air Esch et al. (2010). A phase separation in a water charged gasoline may occur during the cool-down of the tank content in mid-flight. Here another potential problem may arise, especially in future years, if automotive gasoline should be used uncritically and without at least an immediate pre-flight water content control: Since at present there is no normative upper limit of water content for the increasingly ethanol-admixed gasolines, only the absence of haze is required, some very competitive vendors may be tempted to deliberately exploit this boundary condition to a point where vehicles with their almost constabt gasoline temperature will not experience any obvious disadvantage Esch et al. (2010).

Acknowledgements

The authors express their gratitude to the Universiti Teknologi Malaysia (UTM), Skudai, Johor, Malaysia for their financial support.

References

[1] Boichenko, S. and Lejda, K., 2015. Modification of aviation gasoline with aliphatic alcohols additives: an analytical review of prospects. *International Journal of Sustainable Aviation*, 1 (4), 324.

[2] Esch, T., Funke, H. and Roosen, P., 2010. *SIoBiA-Safety Implication Of Biofuels In Aviation.*

[3] Gramajo de Doz, M., Bonatti, C. and Sólimo, H., 2004. Water Tolerance and Ethanol Concentration in Ethanol-Gasoline Fuels at Three Temperatures. *Energy & Fuels*, 18 (2), 334-337.

[4] Morgan, N., 2014. Aviation biofuels: Getting ready for take-off. *Biofuels, Bioproducts and Biorefining*, 8 (6), 739-742.

[5] Mukhopadhyay, R. and Datta, S., 2008. *Engineering Chemistry*. New Delhi: New Age International Pvt. Ltd., Publishers.

[6] Nakama, K., Kusaka, J. and Daisho, Y., 2008. Effect of Ethanol on Knock in Spark Ignition Gasoline Engines. *SAE International Journal of Engines*, 1 (1), 1366-1380.
[7] Olah, G., Goeppert, A. and Prakash, G., 2011. *Beyond Oil and Gas: The Methanol Economy*. Weinheim: Wiley-VCH.
[8] Strauss, K. and Gonzalez, C., 1989. *Future fuels for general aviation*. Philadelphia, Pa: American Society for Testing and Materials.
[9] Totten, G., Westbrook, S. and Shah, R., 2003. *Fuels and lubricants handbook*. West Conshohocken, Pa: ASTM International.
[10] Yüksel, F., 2004. The use of ethanol-gasoline blend as a fuel in an SI engine. *Fuel and Energy Abstracts*, 45 (5), 352.

A review of vapour lock issues during motor gasoline or automotive gasoline usage in piston engine aircraft

K Thanikasalam[1*], M Rahmat[1], A G Mohammad Fahmi[2], A M Zulkifli[1], N Noor Shawal[1], K Ilanchelvi[3], M Ananth[3], R Elayarasan[1]

[1]Faculty of Chemical and Energy Engineering, Universiti Teknologi Malaysia (UTM), 81310 Skudai, Johor, Malaysia
[2]Department of Aeronautical Engineering, Universiti Tun Hussein Onn Malaysia (UTHM), 86400 Batu Pahat, Johor, Malaysia
[3]Department of Chemical Engineering, Universiti Malaysia Sabah (UMS), 88400 Kota Kinabalu, Sabah, Malaysia

*Corresponding author: insp466@hotmail.com

Abstract. Since there is a developing practice of utilizing automotive fuels as flight fuel, there are higher chances of dangerous scenarios, particularly in the operation of piston aircraft engines. The use of motor vehicle gas (MOGAS) or aviation gas (AVGAS) in the operation of aviation piston engine increases the risk of vapour locking. A statistical examination of European aviation industry indicates that around 20,000 aircraft are affected either specifically or conceivably by the different negative impacts of gasoline blended with ethanol. Particularly, for most contemporary carburettor engines, there are risks associated with ethanol-admixed fuels that have potential to upset engine operation. The danger of vapour locking, which is the generation of gas bubbles inside the fuel system causing an impairment of fuel movement in the engine, is well documented particularly by studies on aircraft using MOGAS. Contrasted with AVGAS, MOGAS is inclined to demonstrate this phenomenon. Vapour lock is perhaps the leading serious problem that ought to be addressed if MOGAS is to be used as a substitute for AVGAS. Vapour lock problem is critical because it causes malfunctions to aircraft engines. Thus, an understanding of vapour handling ability of small aircraft is essential to establish safe operating confines at existing fuel temperature and pressures.

1. Introduction - an overview of piston-engine aircraft fuel system

It is necessary that prior to the analysis of vapour lock issue, a brief outline of the piston-engine aircraft fuel system be presented. Wiley (2009) says that the fuel system structure varies from one engine to another and from one aircraft to another. The differences are intended to enable these systems to suit particular designs. According to Wiley (2009), the fundamental part of aircraft fuel systems is fuel tanks, filters, fuel pumps, and fuel lines connecting these components together. The fuel tank serves as the source of the fuel supply. It is made of self-sealing material and is it not exposed to the atmosphere [20]. The filter prevents foreign materials that may be there in the gasoline from entering the engine. On the other hand, the fuel pump draws gasoline from the tank and distributes it to the carburettor. Ferrara and Wares (1988) points out that vapour lock is perhaps the leading serious problem that should be addressed if MOGAS is to be used as a substitute for AVGAS. The fuel's pressure and temperature influences vapour lock. Temperature and pressure affect gasoline's vapour forming qualities, the capability of the system to hold vapour, in addition to the operating state of the engine. According to Ferrara and Wares (1988), vapour bubbles may crop up at any part of the fuel system. However, the most crucial segments are typically the fuel pump. At the fuel pump, a bulk of heat transfer to the gasoline occurs. Here, pump suction lessens the pressure that increases vapour development. The precise vapour's volume under fuel- system temperatures is around 160 times compared to the volume of liquid [5]. Based on this, it is apparent that a pump with a maximum volume flow pace will not be able to administer a sufficient mass flow rate of gasoline to the carburetor for maximum power if a bulk of the fuel has turned into vapour.

2. Vapour Lock in Aircraft Fuel System

According to Cheremisinoff (2001), vapour lock in aviation fuel systems presents in form of the malfunction of the fuel flow into the aircraft carburettors owing to the formation of fuel bubbles in the fuel system leading to engine malfunctioning. Cheremisinoff (2001) suggest that this phenomenon occurs in altitudes with low atmospheric pressures facilitating the formation of vapour bubbles in the fuel system. The occurrence of vapour does not necessarily imply that there is no fuel flowing to the carburettor. It rather lessens the fuel pressure to the extent that the engine ceases to operate as required.

Hsia (1993) maintains that the issue of vapour lock is not recent. The automobile technicians/engineers have experienced the problem for decades. However, the setting or climate conditions that aircraft operate in worsen the phenomenon in aeroplanes compared to automobiles. Hsia (1993) says that the issue was, nonetheless, unrecognised in pioneering days of aircraft designs. Possibly, it was during a voyage over the "hump" in Burma by a United States army pilot that the problem was manifested for the first time [7]. The physical and climatic conditions of this route have created many complimentary factors for the issues to build up and become a huge threat to the users of that route.

According to Hsia (1993), gasoline can vaporise when it heated by the engine or because of a lower boiling point at elevated altitudes. In areas where higher volatility gasoline is utilised to enhance engine performance, the use of these fuels in summer may result in vapour lock. Hsia (1993) claims that vapour lock is more common to piston engine because they incorporate a low-pressure fuel pump powered by the engine. This pump is positioned at the engine cubicle and feeds a carburettor. When vapour lock is drawn into the fuel pump, it disrupts the fuel pressure enabling the float chamber inside the carburettor to drain impairing the flow of fuel into the engine.

Ferrara and Wares (1988) points out that vapour lock is perhaps the leading serious problem that should be addressed if MOGAS is to be used as a substitute for AVGAS. The fuel's pressure and temperature influences vapour lock. Temperature and pressure affect gasoline's vapour forming qualities, the capability of the system to hold vapour, in addition to the operating state of the engine. According to Ferrara and Wares (1988), vapour bubbles may crop up at any part of the fuel system. However, the most crucial segments are typically the fuel pump. At the fuel pump, a bulk of heat transfer to the gasoline occurs. Here, pump suction lessens the pressure that increases vapour development. The precise vapour's volume under fuel-system temperatures is around 160 times compared to the volume of liquid [5]. Based on this, it is apparent that a pump with a maximum volume flow pace will not be able to administer a sufficient mass flow rate of gasoline to the carburettor for maximum power if a bulk of the fuel has turned into vapour.

Morrison, Sung, and Patterson (1984) allege that the current upsurge in the numbers of small aircraft will certainly result in a universal use of these types of aircraft. This trend also implies that various operations under extreme weather and working conditions will be made, increasing the affinity on the part of several users to use MOGAS as an alternative for AVGAS owing to its extensive availability. Both these aspects enhance the issue of fuel vapour lock. Thus, Morrison et al. (1984) suggest that it is ideal for aircraft engineers to tests small aircraft fuel systems to understand their vapour handling capacity under different conditions. Besides, Morrison et al. (1984) put forward that the issue of vapour lock will be predominant during hot summers specifically at elevated locations.

Vapour lock should not be tolerated because it causes malfunctions to aircraft engines resulting in an emergency landing. In this regard, an understanding of vapour handling ability of small aircraft is essential to establish the tolerable safe operating confines at existing fuel temperature and pressures.

Patterson (1980) suggests that the operating limits are subject to the vapour pressure of the particular gasoline used. In history, in addition to the current, the vapour pressure of any AVGAS has been reserved consistently at seven psi. However, of late a trend has emerged leading to the use of MOGAS to power piston-engine aircraft [12]. The use of MOGAS implies that the fuel vapour pressure will no longer be consistent at seven psi. Thus, it increases the chances of development of vapour lock.

Like Patterson (1980), Patterson, Morrison, Remondino, and Slopsema (1980) allege that when temperatures increase during summer, incidences of vapour lock swell, particularly for pilots flying at high elevations. This phenomenon transpires when gasoline, usually in liquid form, changes to vapour following an increased temperature, reduced pressure, elevated Reid Vapor Pressure (RVP) of the gasoline, or a blend of any of these aspects. Patterson et al. (1980) point out that the RVP represents a physical quality of petroleum associated with its volatility. Fuel manufacturers based on air quality set of laws adjust this quality. It is expressed using pressure value (in PSI) based on where the gasoline starts to switch from liquid form to vapour form at 100° Fahrenheit. Patterson et al. (1980) contend that an RVP of 14.7 implies that the fuel will vaporise at sea level pressure at 100° Fahrenheit.

The less the RVP value, the less possible that gasoline will transform to its vapour form, except if the temperature surpasses 100° or the fuel pressure reduces as it does when an aircraft ascends. Given that the temperature inside the bonnet of an engine is more likely to be rather higher, the confined temperatures surrounding the fuel connector, gascolator, fuel pump, carburettor, and injector are crucial other than the ambient setting outside the aircraft. In this regard, aircraft manufacturers should protect fuel from hot gears like exhaust system and cylinder heads. They should also provide routed cooling air tubes and blast baffles to ensure that the fuel remains in its liquid state. Through this approach, the baffles in the engine are maintained to ensure blast tubes exist at their planned targets.

Wallington, Kaiser, and Farrell (2006) claim that based on Bernoulli's theory "the pressure of liquids/fluid reduces with the square of its velocity". Any limitations on the fuel lines, gasoline pumps, or injectors will speed up the fuel velocity and reduce its pressure. Wallington et al. (2006) argue that it is because of this reason that engines installed with gasoline pump and injection systems, like piston engines, are more vulnerable to vapour lock. However, the issue can affect even the naturally aspirated engine when conditions get very hot or when the RVP becomes very high affecting the conditions of fuel systems. Besides, a pinch on a fuel line amid the carburettor and the gascolator can also trigger localised vapour lock.

Wallington et al. (2006) say that the V/L (vapour to liquid) ratio represents a parameter that illustrates the proportion of vapour to a liquid of fuel at a particular temperature. V/L is helpful when forecasting vapour-locking predisposition. According to Wallington et al. (2006), scientists may evaluate V/L ratio in a laboratory using a temperature-V/L tool or with the help of the ASTM D439-Xl.2 procedure utilising distillation and vapour pressure information. Wallington et al. (2006) suggest that the "Reid vapour pressure" is a rationally appropriate forecaster of the vapour-locking propensity in aircraft. The "Reid vapour" evaluation is undertaken at a V/L ratio of four. This ratio is near the V/L tolerance of many light aircraft. According to Wallington et al. (2006), V/L curves plotted using the ASTM distillation information, when fuel temperature increases past the temperature matching to the V/L tolerance of a particular aircraft vapour lock transpires.

Byrnes, Cavage, and Ferrara (1987) assert that the V/L tolerance of many vehicles falls between 15 and 25. Quite a lot of experiments show that with a number of aircraft, a V/L ratios of less than one (17,22) may affect the working of a fuel pump in addition to fuel metering, with elevated ratios triggering stalling [2]. When V/L ratios are lower, they also affect the performance of an engine because the discharge of dissolved air in the gasoline becomes an essential aspect. Byrnes et al. (1987) point out that the quantity of air, which may dissolve in fuel, relies on the pressure of fuel vapour. The summation of the fractional pressures of fuel and dissolved air, at equilibrium, ought to match atmospheric pressure. Byrnes et al. (1987) allege that air is discharged with altitude and temperature swells consequently.

Byrnes et al. (1987) claim that numerous experiments have been conducted to assess the impact of altitude (falling pressure) on the temperature V/L correlation. The findings revealed that an aircraft with specified V/L tolerance vapour locks at 10,000 feet under a gasoline temperature of 16°F lesser compared to at sea level. The findings also indicated the pace at which atmospheric temperature decreases with altitude. According to Byrnes et al. (1987), 10,000 feet increase in altitude lessens the ambient temperature to around 35°F. Thus, if the fuel supply/tank temperature could attain equilibrium at the same rate as the ambient temperature the two impacts would be likely to cancel, and the propensity of vapour lock would not amplify with altitude. Gasoline temperatures lag the ambient temperatures with a predisposition to worsen vapour lock qualities characteristics when ascending.

Strauss and Gonzalez (1989) maintain that when deliberating on the use of MOGAS in an aeroplane, it is significant to be acquainted with the V/L tolerance of the particular aircraft. According to Strauss and Gonzalez (1989), this is the ratio point above which an engine malfunctions and develops an intolerable reduction in power. The way to establish this is to evaluate temperature together with pressure at the point vapour lock is anticipated in the fuel system. If the V/L against temperature statistics is obtainable for the specific fuel, the restraining V/L ratio may be calculated. Strauss and Gonzalez (1989) claim that after measuring the temperature increase beyond ambient for several operating states, the restraining fuel volatility, which can be applied without coming across vapour lock may be predicted. Operating conditions influencing vapour lock propensity are preliminary fuel temperature, altitude, and pace of climb, in addition to ambient temperature. Strauss and Gonzalez (1989) imply that if the V/L ratio tolerance calculated for a particular aircraft is much low, fuel system adjustment is needed. The modifications could comprise enlarging the fuel pump volume by the installation of a larger fuel pump or the lessening of heat transmission to the vital fuel-system parts. According to Strauss and Gonzalez (1989), one possible adjustment to think about would be the removal of any engine-driven diaphragm pumps. Installing centrifugal-fuel pumps inside the tank lessens the temperature ascend of the fuel as pressurisation of the fuel system continues.

Turner (1969) undertook a study fuel management and fuel systems. The study illustrates that a "bottleneck point" frequently crops up at the inlet of the fuel pump. Many carburettors can tolerate V/L ratios of up to 45, but excessively rich air fuel mixture under this state will lead to unnecessary consumption of fuel. Turner (1969) emphasises that the low-pressure fuel injection system used in many aircraft is moderately intolerant to vapour and is most probable to be the bottleneck point. The fuel tends to get warm at the motor driven fuel pump because of the pump's substantial surface area and the low fuel speed through it. The pressure also decreases at the inlet check valve, thus increasing the build-up of vapour [17]. The fuel that is unutilized by the engine bypasses the pump and enters the inlet, additionally increasing the fuel's temperature. To address the issue of vapour locking, automotive pumps are designed to bear V/L proportions of between 15 and 25 and installed in an upstream position to limit heat pick up [17]. Contemporary aeroplane fuel pumps are not designed to tolerate V/L proportions much over four since the V/L curve for aviation gas begins to flatten at this level. If fuel temperatures surpass this level, huge V/L proportions are promptly created. Turner (1969) also says that the V/L curves associated with MOGAS do not level off in the same way as those of avgas. In this regard, expanding pump capacity is more successful. Turner (1969) points out that other solutions employed in automotive fuel systems are directing of gasoline lines far from heat sources and installing vapour separators.

ASTM (2009) cautions that it ought to be remembered that small contrasts in the ASTM 10%, 20% and 50% distillation levels may have a considerable impact on the V/L versus temperature curve. ASTM (2009) indicates that the steepness of this curve is very significant. If the curve is almost level, the contrasts in vital fuel system temperature between the V/L ratio that causes the engine to stop and V/L ratio that triggers leaning is minimal. This implies that there is even minimal forewarning time to distinguish the signs of vapour lock. According to ASTM (2009), experiments/tests to assess the effect of the use of MOGAS on small aircraft fuel systems are desirable. ASTM (2009) proposes that simulated fuel systems may be created in a laboratory setting prior subjecting them to diverse pressure, vacuum, and temperature circumstances. Other aircraft may be evolved to tolerate this issue by adding appropriate intake pumps. ASTM (2009) proposes that more prototype systems should be developed and assessed. Solutions seem probable, but fuel system alteration is needed, particularly in sensitive aircraft like the low-wing models.

3. Practical Methods of Improvement

Totten, Westbrook, and Shah (2003) contemplate that currently, a universal solution to address the issue of vapour lock in piston engine aircraft is yet to be identified. As the aviation altitude keeps increasing, a point will eventually be attained when gasoline will begin to boil. Totten et al. (2003) note that current mitigation measures only focus on increasing the maximum altitude of operation. Use of auxiliary fuel pump, a submerged booster pump, and pressurising the fuel pump are some of the standard methods, highlighted in the existing literature, to address the issue of vapour lock.

Concerning the use of an auxiliary pump, Moir and Seabridge (2008) suggest that an addition of an extra fuel pump into the fuel system enhances the operational upper limit of the system. According to the preliminary statistics, an increase of up to 10,000 feet on failure attitude is possible with the use of an auxiliary pump [9]. This pump also acts as an emergency pump whenever an engine-driven pump fails. The integration of an auxiliary pump does not affect the actual course of air and vapour discharge. Instead, it improves the system's efficiency enabling the failure attitude to be close to the theoretical one [9]. In a conventional fuel system, if a primary pump relies on the engine to operate it is positioned close to the carburettor. Thus, in such engines, it is ideal to place the auxiliary pump close to the gasoline tank. It is appropriate to position the pumps far away from one another.

Regarding the use of submerged booster pump, Vacuum (1979) says that this pump serves the same purpose as the auxiliary pump. This centrifugal pump is designed to be installed inside the gasoline tank, implying that it works under submerged conditions. Its impeller has a propensity to drive the fuel vapour and air far off from the pump and consequently increasing the upper limit altitude of the system. Given that this type of pump has been established to be more efficient compared to auxiliary pump, regarding amplifying the operational altitude, it is preferable. Besides, it can also serve as an emergency pump in incidences where the engine-driven pump fails [18].

Concerning the pressurising the fuel tank, Rohatgi (1995) claims that because the vapour generation from the gasoline results from a decrease in pressure in the surrounding air, the fuel tank should be constructed in a way that it is sealed from the atmosphere. Through this, the altitude will not influence the pressure contained in the tank. Similarly, vapour generation will not intensify in intensity following an increase in altitude as the aircraft ascends. According to Rohatgi (1995), this approach is hypothetically possible, but practically it is very challenging. The difficulty stems from the sound construction needed to endure the tank's internal pressure with consideration of the ambient atmosphere. Reinforced construction imparts additional weight to the aircraft, whereas designers aim to lessen the weight of aircraft to enhance their performance. This indicates that pressurising of the tank is only achievable within certain limits based on ambient atmospheric pressure. Rohatgi (1995) suggests that this approach is useful. However, it is restricted by the structural strength of the fuel tank. It is usually used jointly with automatic valves installed on the vent lines. The valves close so long as the reservoir's pressure does not surpass certain level referenced with the ambient atmospheric pressure. The valves open automatically if the tank's pressure exceeds this limit.

4. Conclusion

Mogas normally has a much higher vapour pressure, which varies seasonally. With a high Reid Vapour Pressure (RVP) fuel the risk of vapour lock during take-off and climb increases, particularly if the aircraft had been parked in high ambient temperatures and does not have a gravity-fed fuel system [14]. The higher volatilities and vapour pressures of motor gasoline (MOGAS) may overtax the vapour handling capabilities of some airframe and engine fuel systems leading to vapour lock and carburettor icing. The higher volatilities also increase the risk of fire hazard. While the tendency of avgas to vapor lock increases with volatility, fuel overheating is the main cause of vapor lock. Local temperatures in the fuel system are determined by how hard the engine is working and how well the fuel system is isolated from the heat of the engine. Fuel residence time in the hot sections of the system, mechanical vibration, and other factors also play a significant role in vapor lock behavior. The altitude at which the engine is operating has two opposing influences: ambient temperatures are lower at higher altitudes, which should improve fuel system cooling; but ambient pressures are also lower, making vaporization easier [6]. The design of an aircraft fuel system must take all the above factors into account to ensure that liquid fuel, with little or no free vapor, is delivered to the engine's fuel metering system.

Acknowledgements
The authors express their gratitude to the Universiti Teknologi Malaysia (UTM), Skudai, Johor, Malaysia for their financial support.

References

[1] ASTM, 2009. Annual book of ASTM standards. West Conshohocken: ASTM International.
[2] Byrnes, H., Cavage, W. and Ferrara, A., 1987. Autogas in General Aviation Aircraft. Ft. Belvoir: Defense Technical Information Center.
[3] Cheremisinoff, N., 2001. Polymer characterization. Westwood, N.J: Noyes Publications.
[4] Esch, T., Funke, H. and Roosen, P., 2010. SIoBiA-Safety Implication Of Biofuels In Aviation.
[5] Ferrara, A. and Wares, R., 1988. The performance of alternate fuels in general aviation aircraft. Atlantic City International Airport, N.J.: FAA Technical Center.
[6] Greg Hemighaus, T. B. (2007). Aviation Fuels Technical Review. San Ramon, CA: Chevron Products Company.
[7] Hsia, C., 1993. Vapour lock in aircraft fuel system. *Science Paper*, 45 (2), 21-54.
[8] Martyr, A. and Plint, M., 2012. Engine testing. Oxford: Butterworth-Heinemann.
[9] Moir, I. and Seabridge, A., 2008. Aircraft systems. Chichester, West Sussex, England: Wiley.
[10] Morrison, K., Sung, N. and Patterson, D., 1984. Some Characteristics of Automotive Gasolines and their Performance in a Light Aircraft Engine. Ft. Belvoir: Defense Technical Information Center.
[11] Nakama, K., Kusaka, J. and Daisho, Y., 2008. Effect of Ethanol on Knock in Spark Ignition Gasoline Engines. *SAE International Journal of Engines*, 1 (1), 1366-1380.
[12] Patterson, D., 1980. Light aircraft engines, the potential and problems for use of automotive fuels. Atlantic City Airport, N.J.: U.S. Federal Aviation Administration Technical Center.
[13] Rohatgi, U., 1995. Sizing of an Aircraft Fuel Pump. *Journal of Fluids Engineering*, 117 (2), 298.
[14] Safety Regulation Division Irish Aviation Authority. (23 October 2010). Use Of Motor Gasoline (Mogas) In Microlight Aircraft. Dublin 2: Irish Aviation Authority.
[15] Strauss, K. and Gonzalez, C., 1989. Future fuels for general aviation. Philadelphia, Pa: American Society for Testing and Materials.
[16] Totten, G., Westbrook, S. and Shah, R., 2003. Fuels and lubricants handbook. West Conshohocken, Pa: ASTM International.
[17] Turner, H., 1969. Fuel Management in Aircraft. Aircraft Engineering and Aerospace Technology, 41 (1), 27-28.
[18] Vacuum, E., 1979. New vapour booster pump fluid. *Vacuum*, 29 (6-7), 259.
[19] Wallington, T., Kaiser, E. and Farrell, J., 2006. Automotive Fuels and Internal Combustion Engines: A Chemical Perspective. *ChemInform*, 37 (25).
[20] Wiley, J., 2009. Aircraft fuel systems. Chichester: Wiley.
[21] Yüksel, F., 2004. The use of ethanol-gasoline blend as a fuel in an SI engine. *Fuel and Energy Abstracts*, 45 (5), 352.

Ethanol content concerns in motor gasoline (mogas) in aviation in comparison to aviation gasoline (avgas)

K Thanikasalam[1*], M Rahmat[1], A G Mohammad Fahmi[2], A M Zulkifli[1], N Noor Shawal[1], K Ilanchelvi[3], M Ananth[3], R Elayarasan[1]

[1]Faculty of Chemical and Energy Engineering, Universiti Teknologi Malaysia (UTM), 81310 Skudai, Johor, Malaysia
[2]Department of Aeronautical Engineering, Universiti Tun Hussein Onn Malaysia (UTHM), 86400 Batu Pahat, Johor, Malaysia
[3]Department of Chemical Engineering, Universiti Malaysia Sabah (UMS), 88400 Kota Kinabalu, Sabah, Malaysia

*Corresponding author: insp466@hotmail.com

Abstract. Mogas has been an alternative to leaded fuel since 1964 when Experimental Aircraft Association (EAA) began testing on it. However, in order for mogas to be used in aircraft engines and air frame modification, approval via the Supplemental Type Certificate (STC) authorization from Federal Aviation Administration (FAA) is mandatory. Cessna on 01.06.2010 evaluated alternative fuels with ethanol based fuels approved by FAA STCs for use in some single engine airplanes. However, Cessna's tests discovered that ethanol based gasoline cannot be viewed as an option to 100LL avgas. The test likewise proposed that operational safety might be in jeopardy if usage of these fuels containing ethanol is continued. Cessna outlined a few problems in MOGAS; MOGAS needs fuel flow increase of 40% compared to AVGAS, MOGAS fuel is incompatible with some fuel system components, possible hazardous influence of electric fuel pumps by adding internal wear causing unexpected spark generation, MOGAS is incompatible with some fuel gauging systems and cause be able to incorrect fuel amount signs on the indicator, dissolve large amounts of water at conditions down to -77°F, impeding detection and removal of water from the fuel system, possible blockage of fuel filters and fuel flow and possible heavy losses from evaporation. This paper reviews concerns when using MOGAS in aircraft.

1. Introduction

Aircraft engines designed for gasoline only face compatibility issues with chemically different constituents in the fuel. In these engines, the whole combustion system is gasoline oriented. Majority of the equipment was manufactured many years ago and the materials used were not oriented to diverse fuels. Although some smaller parts of the fuel system such as fittings, horses, seals and fuel pumps can be replaced by others made from newer materials, some original parts will not be compatible with these replacements [5]. The design for new parts must bear in mind the closely fitted tanks, carburetors, custom integrated fuel pumps or aluminium cylinder head parts that work under very hot liquid or evaporated fuel.

Ethanol admixtures significantly alter the physico-chemical properties of fuels. Consequently, various aspects of both the fuel and the ethanol have to be scrutinized separately to determine the safe operation of the aircraft. Modern flex-fuel automotive engines incorporate complex adaptive fuel conditioning, and therefore, can operate on a wide range of combustibles [9]. If this adaptive fuel conditioning happens in aircraft engines, they should be able to operate on the low octane gasoline portion.

According to EASA (2010), although modern aircraft engines such as those used in ultra-light aircrafts are certified to use automotive gasoline and are compatible with respective ethanol admixtures, they do not have the flex-fuel conditioning. If the fuel does not exhibit phase separation, the engine will operate as it should because all the other challenges related to avgas properties are taken care of. However, if the phase separation occurs, the modern engines will not be in a position to cope.

Fuels admixed with ethanol have a larger enthalpy of evaporation, which produces larger temperature drop of the ingested air in the carburetor. This leads to faster ice deposits, and requires a stronger pre-heating of air. For aircrafts that are prone to carburetor icing, using ethanol admixed fuel increases the threat [5].

If ethanol blended fuels are unintentionally mixed, this may lead to elevated vapor pressures. With these high vapor pressures, there is reduced margin of acceptable heat-up of the fuel system parts relative to the conventional non-admixed MOGAS before an engine stifling vapor lock may occur [12]. For engines known to experience vapor lock with former MOGAS qualities, vapor pressure above this level will be detrimental. These and more concerns on ethanol admixed fuels are discussed below.

2. Water induced phase separation in gasoline-ethanol Mixture

One concern that has been reported regarding addition of ethanol to MOGAS is the tendency for phase separation. This concern is not as prevalent as when the MOGAS is used for automobiles; however, when used in aircraft, fuel is affected by rapid (more than automobiles) changes in altitude [12]. When the aircraft climbs, the fuel was rapidly cools as the temperature aloft falls. This rapid change in temperature triggers what is known as phase separation, which is where the oxygenated ethanol separates from the hydrocarbon based gasoline. The concern is that, "when the alcohol separates from the gasoline, it may carry water that has been held in solution and that cannot be handled by the sediment bowl" [12]. When this phase separation occurs, it could lead to the water freezing in the fuel lines, causing fuel starvation, or a slug of water could pass into the engine at just a point in time when engine power is needed, causing the engine to stop running. Again if the ethanol is separated from the water, the gasoline would no longer have any anti-knock, or anti-detonation, protection. This can lead to severe engine knocking, resulting in the potential for a catastrophic engine failure in flight.

Although a spontaneous disintegration of a formerly homogeneous phase into two distinct ones will usually occur in the form of cloudiness of the blended fuel, the condition should be avoided by all means because it is not assured how long the physically meta-stable suspension will remain in existence [5]. In addition, this suspension does not exhibit the same physical properties as the homogeneous combined phase as the microscopically distinct phases may behave differently at surfaces of small bores and in filters and thus create new problems.

EAA (2009) explains that as the level of ethanol admixture to the base gasoline increases, the emerging phase of alcohol and water phase occupies a significant space in the tank if a phase separation has taken place and the heavier hydrous phase is settling. With an E-5 gasoline this would yield approximately 5 per cent of volumetric occupancy. Therefore, if there is a low positioned tank outlet, it would fall into this range, delivering predominantly the alcohol phase to the engine which will choke on it.

EASA (2010) and EAA (2009) state a biggest challenge as being no simple method of extracting the dissolved water from the fuel. Even a small filter that may well separate even emulgated water will not remove it. As a result the companies that manufacture and trade with ethanol-admixed gasoline

IOP Conf. Series: Materials Science and Engineering **370** (2018) 012009 doi:10.1088/1757-899X/370/1/012009

fuels know of the necessity to keep any kind of water (liquid and gaseous) from entering the fuel storage vessels. They commonly use venting systems with water traps to achieve this. It is for this reason that the ethanol or ethanol admixed fuel cannot be transported by ship or by pipelines [5].

3. Vapor Lock Concerns

The vapor lock involves the creation of vapor bubbles in the fuel system resulting to stagnation of fuel flow to the engine. In aviation, this is a well-known phenomenon is well known especially to those operating their aircraft on MOGAS. Motor gas has a comparatively high vapor pressure, as compared to AVGAS, and is therefore more likely to exhibit vapor lock. Blending gasoline with ethanol adds a new dimension to this problem as the mixture will display a non-linear vapor pressure increment effect in a case of mixing [3].

Pure ethanol has a temperature-dependent, but clearly defined vapor pressure, as opposed to gasoline mixtures that can, as multi-component mixtures, only sensibly be described by boiling curves [2]. When the engine is running, thermal transients of varying time scales are observed, though. Evaporation of the fuel in the engine, which substantially influences the ignitability of the fuel air mixture, takes a very short time, governed by the total enthalpy of evaporation. On the other hand, the fuel supply to the carburetor and to the fuel injectors is a slow process that leads the liquid through various potentially heated aggregates with extended times of exposition. In this case, the vapor pressure of the light fuel components plays a major role with respect to the danger of spontaneous and fuel flow choking vapor bubble creation.

Generally, overall volatility of the gasoline is altered when ethanol is added to it. This could leads to vapor lock issues when trying to start an aircraft engine. Aviation gasoline has a very tight range for volatility (i.e. vapor pressure) if the vapor pressure is outside the standard band of 5.5psi to 7.1psi [1], the fuel might not be volatile enough to start an engine in extremely cold or extremely hot conditions. When using MOGAS that is ethanol blended, there is a concern on the corrosive nature of ethanol. The FAA explains that aircraft fuel system components are not made to withstand the corrosive nature of the added ethanol. Also, since some aircraft have been flying for many years, the natural rubber components that are designed for robust, ethanol free, hydrocarbon-based fuels are unable to withstand the abrasive action the ethanol imparts on them. These problems lead to reduced fuel system performance and eventually failure.

Helder (2000) notes that although it is not a direct concern of ethanol blended fuels, there is potential influence of vibration induced cavitation and boiling which also leads to a vapor lock in the fuel system. This is aggravated by the possible increase in vapor pressure in case of a casual unlucky mixture. The physical effects of cavitation and boiling are usually not treated in combination, although addressing almost the same physical phenomenon of spontaneous creation of a vapor phase out of the liquid one. The rising temperature of the liquid also raises the tendency of cavitation, induced by dynamic pressure drops. This effect is experienced by both fast moving parts in pumps or larger vibrations of fuel lines induced by mechanical coupling to the engine [11]. However, according to EASA (2010), the theoretical and experimental considerations mostly focus on pure and mostly clean substances to form a basis of scientific reproducibility. There has been limited information available on pressure drop induced evaporation of gasoline-like non-equilibrium mixtures under elevated temperature conditions. However, it can be noted that the heat input into cavitating increases, which raises the fact that the fuel lines and with them the fuel may experience additional heating.

4. Decreased energy Content

A biggest concern of ethanol blended MOGAS is that when ethanol is used, it decreases the overall energy content in each gallon of the fuel. The SIAB points out that "Methanol has approximately 55 percent of the energy content of gasoline, and ethanol has approximately 73 percent of the energy content of automobile gasoline." The FAA states that the "greater the amount of alcohol in the automobile gasoline, the greater the reduction in the aircraft's range" [6]. For an automobile a decrease in range is generally, only an annoyance of more frequent stops at gas stations. For an aircraft a decrease in range brings about the difference between making it to the destination and crashing somewhere away from an airport. The biggest problem is that pilots cannot compensate for this shortened range, because they may have no idea what type of fuel they may have on board [9]. Again, unlike automobiles where people routinely fill up the gas tank at a fuel stop, it is routinely necessary for aircraft to carry only the required amount of fuel for the distance needed, and a safety reserve. Often aircraft may not carry full tanks of fuel due to the inability to carry the extra weight of the fuel, when fully loaded with cargo and passengers.

The lower heat content of the ethanol blended fuel reduces the engine power, which results to inability of the aircraft to climb as the performance charts for the aircraft indicates. If the aircraft is not able to climb properly, it will not be able to clear obstacles at the end of the runway, or climb over mountains or adverse weather. All of these concerns are individually extremely hazardous to safe aircraft operation and likely fatal for flight crews and passengers [2].

5. Compatibility with Fuel system and Engine materials

There are different materials that are exposed to the fuel of a combustion engine system. The first in the system is the fuel tank and its built-in fitments such as breathers, tank caps, level indicators, and sealers. The next components are fuel hoses, filters and fuel pumps. On the end, are cylinder heads and combustion chamber valves, both of which the individual materials and material combinations are subject to fuel interaction [5]. These materials and their combination have been in the past chosen with the best possible resistance and material properties suitable for pure for gasoline. They therefore, fully comply with AVGAS. Earlier brands of MOGAS gasoline with ethanol contents limited usually to less than one percent for aviation purposes, do not deviate substantially from this behavior [5]. Although larger quantities of ethanol admixtures to the gasoline are allowed already for a while, it has not been possible to reach the upper limit due to economic reasons. Having the compulsory biogenic admixtures will change this and it may well be assumed that at least a larger share of commercial gasoline will contain ethanol up to the legally defined value.

From George (2005), it is evident the metallic components have a concern of increased corrosion induced by a slightly higher electrical conductivity of ethanol-admixed gasoline in combination with slightly increased water content. Another challenge is the formation of aluminium ethanoate. When liquid ethanol comes into contact with uncovered aluminium under higher temperatures, such a specific corrosion is likely to take place. According to EASA (2010), this situation is catalyzed by the fact that the corrosion can already be induced by just one faulty fueling and will not stop even if the engine is operated on standard non-alcoholic fuel thereafter. The pits corroded by the ethanol fuel will continue to dig into the affected surface and finally destroy it.

EASA (2010) also discusses other parts of the fuel system made of non-metal components. Such elements include tubes, filters, reservoirs, fittings and tanks, among others. These components are made of different materials and hence are affected the fuel constituents in different ways. The usage and hence the required properties of the respective component dictates the choice of the available materials. Adhesives are commonly found in the composite expendable items where different material types must be joined. The fuel and oil filters are the mostly affected as their usual properties such as mechanical stability, permeability to a liquid, and retaining of undesired contents, are obtained by different functional constituents.

The greatest issue is the ethanol admixture compatibility with elastomers. Majority of the elastic horses used in aviation are produced from aviation certified Nitrile Rubbers (NBR) of the various kinds. For pure gasoline or with only small quantities of admixed ethanol (less than E-5) they are considered suitable with respect to their material longevity [8]. If they interact with higher ethanol contents they become brittle and may swell, depending on the individual type of elastomer [8]. A solution to this general problem would be the alternative deployment of fluorinated rubbers. The cost of fluorinated rubbers is about ten times that of NBR based materials and hence there has been no commercial interest in using them. Consequently they were not put to work in certified General Aviation constructive parts.

Hard plastics, also called thermoplastic materials a similar behavior as elastomers. It is possible to distinguish the types on the basis of the molecular structure. The available materials may be classified into semi-crystalline and amorphous types. With the amorphous kinds of plastics, there is a general hydrolysis threat issue, if a larger amount of ethanol is present in the gasoline and the construction part made from the material is operated at very high temperatures, as frequently encountered in the vicinity of the engine [8]. As a result, only the semi-crystalline materials should be considered for fuel system component manufacturing.

6. Certification Concerns

To facilitate the use of MOGAS, aircraft has to be approved on an individual basis to be able to use the fuel. The approval involves a design change process, which a regulatory method is controlled by the FAA to allow aircrafts to have their original manufacturers' designs altered to accommodate some design changes. In this case the design change has to accommodate the use of MOGAS in an aircraft certified for aviation gasoline [6]. For some aircraft models the design approval calls for little more than the addition of a line in the pilot handbook allowing and a label on the fuel filler cap calling out the approval of the MOGAS. For other models of aircraft the requirements to satisfy the FAA of safe flight on MOGAS requires more extensive changes. In general, however, the FAA only approves these changes for aircraft known to be able to run acceptably on MOGAS. Throughout the years since the MOGAS projects started for aircraft, groups like EAA and Petersen Aviation, Inc. began securing various STC's for multiple aircraft/engine combinations and reselling them to aircraft owners. Over the years many aircraft were modified to MOGAS, and the aircraft performed acceptably. So for one class of aircraft there did seem to be an unleaded fuel solution.

7. Insurance related concerns

If the aviators choose to continue to fly on ethanol containing MOGAS, there remains risk beyond those of safety of flight. All the insurance companies require that an insured aircraft should be airworthy and also comply with the set regulations if the aircraft has to be insured. If the aircraft does not operate in compliance with these regulations, or if the aircraft is not maintained and operated within the accepted methods, the insurance companies will often not compensate in case of an accident [12]. The owner of an aircraft or the pilot can suddenly find it difficult to secure insurance if there is an accident and the aircraft has ethanol in the fuel tanks. Some of the aircraft owners might opt to accept the risk of the loss of their aircraft. However, they do not realize that this leads to losing any protection against damages to other people or property. There are standard ethanol test kits, which are a standard tool in the tool kit for aircraft insurance adjusters. This is one of the first things tested for by the insurance company for any accident claim.

Acknowledgements

The authors express their gratitude to the Universiti Teknologi Malaysia (UTM), Skudai, Johor, Malaysia for their financial support.

References

[1] ASTM International. (2007). Standard Specification for. In A. *International, Annual Book of ASTM Standards*. West Conshohocken, PA: ASTM International.

[2] B. Engelen. Guidelines for blending and handling motor gasoline containing up to 10% v/v ethanol. *Concawe Report*, 3, 2008.

[3] Christopher Earls Brennen. Cavitation and Bubble Dynamics. Oxford University Press, 1995.

[4] E. Brandes, D.-H. Frobese, and M. Mitu. Safety characteristics of ethanol / automotive petrol mixtures. *OIL GAS European Magazine*, 4:199–202, 2006.

[5] EASA. (2010). EASA Safety implication of biofuels in Aviation; *EASA Report No. EASA.2008.*

[6] FAA - Small Airplane Directorate. (1998, Jue 4). Autogas Use. Kansas City, MO: Federal Aviation Administration.

[7] Kenneth Kar. Measurement of vapor pressures and enthalpies of vaporization of gasolin. *SAE special publication,* 2187:155–167, 2008.
[8] S. Bauer, J. Krüger-Venus, M. Kulikowski, and O. Büchner. Alternative fuels test program plastics, elastomers and glue connections. *Mineral Oil Based and Alternative Fuels*, 2009.
[9] Clark, A. Q. (2007, October 5-11). Aviation Gasoline: History and Future. *IASH 2007, the 10th International Conference on Stability, Handling and Use of Liquid Fuels*. Tucson, Arizona.
[10] George, C. E. (2005). Effects of ethanol as a fuel additive on general aviation aircraft fuel system electrochemical corrosion.
[11] Helder, D. (2000, April 28). Ethanol Based Fuels for Aviation. Brookings, SD, United States.
[12] EAA. (2009). FAA to STC Holders: No Ethanol in Auto Fuel. Retrieved February 11, 2011, from EAA Autofuel: http://www.eaa.org/autofuel/autogas/test_kit.asp

Piston Aviation Fuel Initiative (PAFI) – A Review

K Thanikasalam[1], M Rahmat[1], A G Mohammad Fahmi[2], A M Zulkifli[1], N Noor Shawal[1], K Ilanchelvi[3], M Ananth[3], R Elayarasan

[1]Faculty of Chemical and Energy Engineering, Universiti Teknologi Malaysia (UTM), 81310 Skudai, Johor, Malaysia
[2]Department of Aeronautical Engineering, Universiti Tun Hussein Onn Malaysia (UTHM), 86400 Batu Pahat, Johor, Malaysia
[3]Department of Chemical Engineering, Universiti Malaysia Sabah (UMS), 88400 Kota Kinabalu, Sabah, Malaysia

*Corresponding author: insp466@hotmail.com

Abstract. Aviation gasoline (Avgas) has remained unchanged for seventy years and the existing fleet of piston aircraft was designed to be compatible with its chemical and physical properties to achieve superior levels of safety. Tetra-ethyl lead (TEL) is an octane-enhancing metal additive used in aviation gasoline to prevent knocking. Studies have shown that lead causes brain damage in children reducing their IQ and cardiovascular difficulties and kidney failure in adults. Friends of the Earth (FOE) petitioned the Environmental Protection Agency (EPA) in 2006 to make a finding that lead emissions from general aviation (GA) aircraft cause to public health endangerment or carry out studies and issue a report on its findings. PAFI was set up by Federal Aviation Administration (FAA) to find most suitable unleaded replacements for Avgas to recognize best unleaded fuel that have the capacity to in fact satisfy the requirements of the present aircraft fleet while additionally considering the creation, dispersion, cost, availability, environmental impacts. This study will technically review PAFI and broaden the limited knowledge on piston aviation fuels in Malaysia by giving a comprehensive analysis and possible gap in reciprocation aviation engine market in Malaysia.

1. Introduction and Overview of Piston Aviation Industry

Aviation gasoline (AVGAS) standard represents a fundamental part of the piston engine-aircraft safety setup. Around 230,000 aircraft internationally depends on 100 low-lead (100LL) AVGAS in their operation [9]. 100LL (100 Low Lead) represents the only remaining aviation fuel in the UK that has the additive tetraethyl lead (TEL). The AVGAS utilized currently has its beginning in the advancement of high power aircraft engines needed to facilitate reliable and inexpensive commercial and military flight [1]. Milner (2006) says that aviators have been using TEL as an AVGAS chemical addition for decades to generate very high-octane levels needed to thwart engine knock, specifically in high power aircraft engines. Notably, operation with insufficient fuel octane may lead to engine failure and consequently aircraft catastrophes.

In their article, Atwood and Canizales (2004) note that environmental conservation organizations against use of lead-containing AVGAS have appealed and taken legal actions calling for environmental protection agencies to establish regulatory measures to abolish or lessen lead emissions from aircraft. Similar regulatory actions are under deliberation internationally. These appeals and legal actions raise issues regarding the continual accessibility and use of AVGAS blended with lead. Similarly, international indecision and worry exist among piston aircraft equipment producers, AVGAS manufacturers, AVGAS sellers, and aircraft owners. The fear concerns: (1) future effectiveness and value of contemporary aircraft, (2) accessibility and price of AVGAS to uphold feasible business operations, (3) validation of future aviation product- development and (4) rationale of new aircraft acquisitions.

With the existing number of piston engine aircraft internationally being 200 times larger compared with the annual new aircraft manufacturing, the turnover rate of the current aircraft in use is minimal. As evidenced by Pei and Hou (2016), the low turnover rate exposes the owners of piston engine aircraft to a depreciation of their airplanes should an unleaded substitute AVGAS be unsuited with the current fleet. This susceptibility, coupled with the sluggishness of new aircraft sales in addition to a general weakening economic status in the aviation industry, has generated a sense of necessity on the growth and use of an unleaded AVGAS, which meets the operation demands of the existing fleet.

The Federal Aviation Administration (FAA) (2012) together with the Environmental Protection Agency's (EPA) share concerns regarding lead emissions by small aircraft. Proprietors and users of over 167,000 piston engine powered aircraft in use in the US depend on AVGAS to run their aircraft. Chen, Liang, Liu, Ding, and Li (2017) note that Avgas is the only residual lead-blended aviation fuel. AVGAS blended with lead prevent engine knock and blowing up preventing sudden engine failure in small aircraft. According to Chen et al. (2017), lead is a poisonous substance, which can be ingested, absorbed, or inhaled into the bloodstream. In an attempt to end the use of lead blended AVGAS, FAA, EPA, and other industry players are collaborating to eliminate the contaminant from avgas.

In reaction to the swiftly increasing issues articulated by the general aviation stakeholders, Federal Aviation Administration (2012) says that the Piston Aviation Fuel Initiative (PAFI) was chartered to build up a path forward for the classification, assessment, and use of the most potential unleaded substitute for 100LL low-lead aviation gasoline. The responsibility of PAFI is to assess a candidate for unleaded alternate fuels and classify the alternatives best competent to technically meet the requirements of the current aircraft fleet at the same time also taking into account the production, supply, cost, accessibility, and health and environmental impacts of these fuels. Increasing economic and ecological demands call for a shift to unleaded fuel. Regrettably, the contemporary government directives and policies together with gasoline and aviation market do not uphold a regular and cost-effective feasible fleet-wide shift to new fuels, thus the necessity for a joint airline industry and mutual government program called PAFI.

2. Background to the emergence of PAFI

AVGAS has remained mostly unchanged for more than 70 years, and the in use fleet of piston engine aircraft was premeditated to be attuned to its physical and chemical properties to attain better levels of safety, consistency, stability, and performance [5]. The Federal Aviation Administration (FAA) and European Aviation Safety Agency's (EASA) certification measures and supporting strategies have not concentrated on certifying contemporary fleet of aircraft to the latest fuel or assessing the aspects and performance of gasoline themselves. Instead, Lovestead and Bruno (2009) observe that they have focused on guaranteeing the airworthiness of engines running on known fuels in compliance with long-established stipulations.

Currently, the only pathway for granting a new fuel for use in in-use engines is permitting an original equipment manufacturer (OEM) to revise its type certificate (TC) or for a third force/party to acquire a supplemental-type-certificate (STC) from the EASA or the FAA. They attain this through a procedure aimed at ensuring flight safety of an in use aeronautical engine when running on a particular gasoline to which it was tested [5]. This approval procedure calls for a separate stipulation that every engine and aircraft conforms to every airworthiness principle when running on a new fuel. Industry players have acknowledged this process as being very expensive and incompetent to be successful in shifting the whole in use aircraft to any new fuel, chiefly because an active manufacturer no longer maintains a number of these aircraft. It has also been noted that contemporary evaluation pathways that inspect the airworthiness of the aviation products are not anticipated to appraise the chemistry and attributes of the fuel [8]. Whereas there are substitutes accessible for "approved model list supplemental type certificates" (AML-STC), which may cover a variety of engine types and aircraft models, such an authorization course may be multifaceted. Thus, it would not possibly facilitate an orderly fleet-wide change over needed to uphold the economic feasibility of the existing piston engine aircraft fleet [5]. Other available paths for authorization, like the amended TC or the disbursement of manufacturer service instructions approving the usage of a new fuel by an assortment of models, present same hurdles and impediments to a regular and inclusive changeover. They also do less to tackle the issue of piston engine aircraft no longer serviced by an existing manufacturer.

Aviation fuel commercial expansion and operation in the past seven decades have depended on industry organizations consisting of a diverse grouping of industry stakeholders. These stakeholders have an understanding and practical knowledge in power plant technology, fuel system design engineering, chemical engineering, combustion engineering, toxicology, emissions, and fuel manufacturing and supply [5]. Prove these industry players necessitate that a new fuel ought to both to perform safely by all in use piston engine aircraft, and have to be manufactured and disseminated across contemporary infrastructure safely and competently [5]. Therefore, it is acknowledged that considerable further information other than those stipulated by FAA and EASA's airworthiness approvals are needed to bring fuel into efficient production and dispersed as a product in the market.

Fuels move flawlessly across the globe due to a broad-based approval and understanding of the merchandise, their characteristics and behaviours, and cohesion between manufacturing, supply, and testing methods [5]. Such support is essential to guarantee a prevalent, consistent, and financially viable production, supply and use of any aviation fuel needed. This international approval is a product of open consensus-reliant procedures, which allow peer review and much standardization between both the merchandise and their particular testing process and provisions.

Contemporary EASA and FAA certification measures like STCs, revised TCs, and service letters, represent closed review processes between fuel developers and the FAA/ EASA offices, and OEM [5]. These measures also depend on information subject to intellectual property. Whereas this may work for EASA or FAA airworthiness authorization ensuing in the capability to burn a specific fuel in a particular engine or aircraft, it does less to address the barricades to the broad acceptance needed for fleet-wide realization by the petroleum, aviation, chemical, and insurance industries. The FAA and EASA certification processes also fail to tackle the issues raised by environmental and health advocacy factions and watchdogs who have a responsibility in monitoring the toxicology and emissions of latest unleaded fuel [5]. Further peer review, test, data gathering, and the creation of industry consensus regulations are all indispensable footsteps beyond EASA and FAA airworthiness authorization to allow fuel to enter the market.

Esler (2015) says that PAFI was conceived and instituted to address the challenges impeding entry into the aviation fuel market. The organization establishes a procedure that would assess all of the properties and conditions required for extensive production, supply, and use of any new unleaded aviation fuel, and speedily create data needed to support EASA and FAA authorization of a bulk of the in use fleet of piston engine aircraft to run on that fuel. Esler (2015) points out that PAFI was also established to undertake a number of the testing required for fuel production and supply approval and fleet acceptance using usual test amenities, measures, and industry consent directives resulting in an extensive marketplace adoption and acceptance. According to Esler (2015), the PAFI procedure is needed to adopt best practices from fuel developers and advance them beyond EASA and FAA approvals in limited requests to full fleet endorsement and broad based adoption in the market.

The PAFI process comprises a two stage-testing course [4]. Phase 1 assesses candidate fuels for any possible hindrance affecting issues in the manufacturing, supply, and operation sectors prior considerable investments are made in obtaining FAA or EASA design authorization. Among these footsteps is an assessment of the physiochemistry of the fuel in addition to suitability for its planned purpose [4]. Owing to the significantly different chemical structure of a variety of candidate unleaded fuels together with their expected departure from the physiochemistry and attributes of conventional aviation fuels, tests needed to institute a fuels' suitability for use under all practically envisioned settings could differ from one gasoline to another. The physiochemistry and performance attributes of these fuels will mostly influence the required assessment and appraisal process needed, which increases in complexity and scope with rising departure from the characteristics and properties of the contemporary gasoline for which the in use fleet was premeditated and certificated. The PAFI course is anticipated to evaluate the composition of a variety of fuels and create realistic and peer-reviewed assessment procedures for determining essential fit-for-purpose information [7]. Since the PAFI practice has an extensive buy-in transversely the aviation and petroleum and industries and is supervised by a sovereign, mutual government/industry group of technical specialists with no profit cause or stake in the end product, their findings are deemed objective, helping to guarantee broad approval of successful candidate gasoline.

Other critical assessments undertaken during Phase 1 comprise evaluating the toxicology and emissions attributes and ensuing impacts, assessing whether gasoline can be manufactured and supplied broadly and inexpensively. They also involve establishing if the particular fuel will perform satisfactorily across its anticipated full compositional array in the in use engines and aircraft and effectively guaranteeing that it will be assessed under unfavourable fuel constitution and operating environment [7]. The PAFI course also anticipates examining the business case for any candidate fuels based on the projected manufacturing, accessibility, and supply models with the intentions to determine if the fuel would be readily producible and obtainable at a convenient cost.

After proving the technical, ecological, and business case qualities of anticipated unleaded substitutes to 100LL in Phase 1, gasoline/fuel established to be promising are endorsed to enter Phase 2 [7]. These types of gasoline are all set to be appraised now at the aircraft and engine level with the intention of adopting them across as many in use fleet as possible. The PAFI program aims to achieve this by providing financial support for phase 2 of the engine and aircraft testing program. The data and findings from this government-funded test plan will not only enhance approval of the gasoline but also create data, which can be applied to endorse the fleet-wide approval of engines and aircraft. This stage is significant to tackling the deployment of gasoline in the market in a methodical and all-inclusive manner [7]. FAA participation in this stage of the procedure is vital not only to guarantee that the whole fleet is tackled but also to boost the reliability of assessment methodology and data needed for aviation industry customers to acknowledge and implement the fuel across the board.

Gillette (2017) notes that PAFI procedure is not planned to be a barricade to entry for candidate fuels but instead are intended to facilitate the most capable gasoline to undertake the needed sovereign peer review and information collection. The above is required to attain broad-based regulatory, industry, and customer approval resulting in the manufacturing, supply, and sale of the fuel in the whole aviation market. Gillette (2017) says that history proves that the FAA and producer airworthiness authorization of gasoline only does not guarantee recognition by users and industry players. Some STCs and producer service directives have been publicized endorsing gasoline, which has never been triumphant in attaining broad manufacturing, supply, and user community approval. The causes for this are multifaceted and diverse. However, the PAFI process is intended to conquer them.

In coming up with the PAFI, the FAA, the EPA, and the aviation industry are applying the knowledge and lessons acquired from the past determinations to endorse new piston aviation gasoline and help in conquering the hurdles to new fuels shifting from being theoretical concepts to extensive production, supply, and sale in the market. The information is beneficial for all stakeholders concerned with the safety and impacts of AVGAS. AviationPros (2017) says that it is everyone's aim that all piston engine aircraft moves effectively and inexpensively to a possible and secure unleaded future. The PAFI process offers a sound procedure to guarantee that this objective is realized with the least amount of disturbance to the aviation industry and with the unlimited possibility of marketplace achievement.

3. PAFI Organization

PAFI is organized as an industry-FAA alliance. The FAA funds and provides administrative help for PAFI directors and funds other advisors as required [2]. This administrative support comprises the creation and sustenance of PAFI's website. The membership of PAFI encompassed stakeholders from the aviation community like FAA, aviation associations, and industry traders. The members are anticipated to facilitate with the in-kind support needed to perform the responsibilities required for PAFI to accomplish its roles as described in their mission [2]. Members allot resources to sustain exclusive PAFI responsibilities like the creation of job aids and to facilitate industry tasks associated with the development and authorization of unleaded AVGAS.

4. PAFI Fuel Development Stages

The PAFI's roles, obligations, resources, financial support, and schedule necessities are passed through three separate stages that are pre-arranged to aid the integrations with FAA's fuel assessment program together with the AVGAS development procedure [2]. The development stages are:

4.1 Preparatory Stage

This step precedes the commencement of the FAA fuel assessment program other evaluation processes meant for candidate fuels. Job aids are created in this phase by PAFI to sustain the succeeding stages. Job aids comprise technological, logistical, fiscal and other AVGAS-allied industry information that is necessary for the PAFI to carry out testing in collaboration with the FAA fuel testing plan. Besides, job aids provide reference data for prospective fuel manufacturers, potential financiers, and government organizations that play a future responsibility in commercializing the unleaded AVGAS. At this stage, FAA is required to set up a national aviation fuel-certification office.

4.2 Project Stage

During this juncture, FAA issues a solicitation for potential unleaded AVGAS manufacturers to present their fuel products for testing. The FAA then picks a few numbers of the most promising candidates to be tested at the FAA Technical Centre. The information obtained through this testing program will aid the simultaneous ASTM specification progress and FAA authorization activity in this phase. As suitable, PAFI participants may also support and endorse both government and private financing prospects to aid this initiative.

4.3 Deployment Stage

This phase begins after the conclusion of fuel assessment, specification improvement, and FAA authorization activities. PAFI facilitates professional support to help the production, supply, and deployment of a fleet-wide performance of the new unleaded fuel.

5. FAA Integration

In the preparatory stage, the PAFI steering group helps in the creation of job aids, which FAA uses in to facilitate screening and assessment of candidate fuels. The FAA uses these job aids to generate "Request for Proposals" (RFPs) used in soliciting new fuels to be tested at the FAA Tech Centre [6]. The data generated may be utilized by the applicants to facilitate fuel approval. The FAA will then institute a FAA review board, which will use these job aids to appraise promising candidate gasoline for admittance at the FAA Test Program. The board will need the technical proficiency essential to assess fuel qualities and structure data to establish the viability of the candidate fuel [6]. Furthermore, the FAA will create a national certification office. During the Project phase, the fuel testing programs are undertaken at the FAA Tech Centre.

6. Recent Program Updates

6.1 FAA issues appeal replacements of lead blended Avgas

On June 10 of the year 2013, FAA publicised an appeal for candidate fuel manufacturers to forward their unleaded fuels to be assessed as a substitute for 100LL [6]. The announcement marked an important landmark in a government/industry joint endeavour to establish an unleaded substitution gasoline for the aviation sector. The appeal for candidate fuels triggered a multi-year research and development program, which will assist in choosing the most appropriate unleaded fuels with the least effect on the general aviation fleet.

6.2 General Aviation Caucus Update

This briefing was held on November 21, 2013. During the meeting, EPA, FAA, and other industry stakeholders met with Senator Johanns and Senator Begich, who were the co- leaders of the "Senate General Aviation Caucus Briefing" [6]. The meeting informed the attendees on the history of AVGAS and an update on advancements made in the shift to unleaded fuels. This briefing incorporated a talk of the "FAA Modernization and Reform Act" of the year 2012 and FAA's research and development program (Federal Aviation Administration 2017). The meeting also included a talk on the "Unleaded Avgas Transition Aviation Rulemaking Committee" (UAT ARC), and undertakings made since acceptance of that team's commendations and report.

6.3 PAFI Update at ASTM Conference

On December 9, 2013, PAFI presenters attended the ASTM conference in Tampa, Florida [6]. In the meeting, they presented two presentations. The first talk was done jointly by industry stakeholders and the FAA to discuss PAFI, their achievements to date, the synopsis of phase 1 and phase 2's evaluation programs, and the fleet-wide certification difficulty. The FAA presented the second talk. The presentation provided an update on Screening Information Request (SIR) [6]. The discussion provided background, directives, and targets for the SIR. Many people attended the conference, and many concerns were raised and argued.

6.4 PAFI Steering Group Meeting

The meeting occurred on January 14-15, 2014. Monthly PAFI Steering Group gatherings were launched in the year 2013 and continued up to the year 2014 [6]. The team was tasked with completing the UAT ARC commendations. In the meeting, the participants talked on $6M grants from the Congress intended to support the PAFI research program at the FAA Technical Centre.

6.5 FAA accepts Unleaded Avgas Fuel submission

The FAA's appeal for fuel manufacturers to forward unleaded avgas to substitute 100LL ended on July 1st, 2014 [6]. On this day, the FAA received nine fuel submissions from five fuel manufacturers. Following the presentation, the FAA was now ready to evaluate the feasibility of the candidate fuels concerning their effect on the in use fleet, the manufacturing and supply infrastructure, their influence on the environment, toxicology, and their effect on aircraft operations. Some fuels selected were further evaluated in phase 1 of the test program. During this period, FAA estimated 100 gallons of every fuel candidate using lab and rig test program [6]. The Phase 1 evaluation program lasted for almost one year. Through this, FAA assessed the fuels for continual partaking in Phase 2 of the evaluation.

Two to three of the fuels tested in the first phase were chosen to participate in the 2nd phase [6]. After that, the suppliers were requested to submit 10,000 gallons of their candidate fuel. The second phase lasted for roughly two years and generated data needed to acquire an ASTM production certification for the fuels and to authorize many of the in use fleet to run on these fuels.

6.6 FAA Picks Four Unleaded Fuels for Testing

On September 8th, 2014, the FAA picked four unleaded fuels for assessment. The FAA picked four fuels: one from Shell, one from TOTAL, and two from Swift Fuels [6]. After that, the FAA started working with the manufacturers to identify the formulations to be forwarded to the Phase 1 of the test program. The Phase 1 process took approximately one year. Through this period, FAA evaluated the fuels to establish if they were appropriate to proceed to the second step of the test program.

6.7 PAFI Phase 1 Test Program Status

This test was undertaken on November 20, 2015. Earlier, on August of 2014, PAFI appraised 17 proposals from six firms proposing substitutes for unleaded fuel for the aviation fleet [6]. After that, PAFI provided its recommendation before requesting four formulations from 3 firms on September of the year 2014. An inclusive Phase 1 assessment program comprising lab, engine, and rig tests started in March 2015. A number of these tests are by now complete, and findings and reports have been forwarded to FAA for appraisals.

6.8 Unleaded Avgas Progress Update

The latest unleaded avgas progress update was on December 20, 2016 [6]. PAFI Phase 1 lab, rig, and materials suitability assessment were concluded in December of the year 2015. Concerns that emerged in phase 1 assessment program are being further tested in Phase 2. Phase 2 comprises of additional testing processes intended to establish the differences and better measure their effects on the wide-ranging aviation fleet, gasoline production, and supply infrastructure. Mitigation stratagems and operation plans are being created to reduce impacts and enhance smooth shift to unleaded petrol.

Engine assessment and aircraft flight test development is in progress at the FAA's Technical Centre [6]. Similarly, the FAA is in quest of a new authority to perform engine and aircraft authorization for the PAFI program. Currently, PAFI is creating action plans to tackle many deployment concerns like regulatory and legislative needs, manufacturing ability, supply system and airport operation, engine and aircraft adaptation, and training and communications. Besides, the FAA and PAFI continue to support other unleaded gasoline endorsements by working jointly with other fuel manufacturers in quest of unleaded avgas authorizations through conventional procedures. The FAA remains determined on delivering substitutes to leaded avgas via conventional processes and through the PAFI program [6]. The PAFI program also continues to be on the agenda with an anticipated attainment of all assessments before mid-2018 and publication of every final test findings by the end of the year 2018.

7. Conclusion

In conclusion, it should be noted that around 230,000 aircraft internationally depends on 100 low-lead (100LL) AVGAS in their operation. There are appeals for the establishment regulatory measures to abolish or lessen lead emissions from aircraft. The FAA and the EPA share worry regarding lead emissions by small planes. In reaction to the speedily increasing issues articulated by the general aviation stakeholders, the PAFI was established to draft a path for the classification, assessment, and use of the most potential unleaded substitute for 100LL low-lead aviation gasoline. The PAFI process comprises a two stage testing course. Phase 1 assesses candidate fuels for any possible hindrance affecting issues in the manufacturing, supply, and operation sectors prior considerable investments are made in obtaining FAA or EASA design authorization. Phase 2 comprises of additional testing processes intended to establish the differences and better measure their effects on the wide-ranging aviation fleet, gasoline production, and supply infrastructure. The establishment of PAFI is beneficial for all stakeholders concerned with the safety and impacts of AVGAS. It is everyone's aim that all piston engine aircraft moves effectively and inexpensively to a possible and secure unleaded future. The process offers a sound procedure to guarantee that this objective is realized with the least amount of disturbance to the aviation industry and with the unlimited possibility of marketplace achievement.

Acknowledgements

The authors express their gratitude to the Universiti Teknologi Malaysia (UTM), Skudai, Johor, Malaysia for their financial support.

References

[1] Atwood, D. and Canizales, J., 2004. *Full-scale engine knock tests of 30 unleaded, high-octane blends*. Washington, D.C.: Office of Aviation Research, Federal Aviation Administration.

[2] AviationPros, 2017. *Unleaded Avgas Put to the Test by FAA, Industry*. [online] Available from: http://www.aviationpros.com/article/12220281/unleaded-avgas-put-to-the-test-by-faa-industry [Accessed 26 Jul. 2017].

[3] Chen, L., Liang, Z., Liu, H., Ding, S. and Li, Y., 2017. Sensitivity analysis of fuel types and operational parameters on the particulate matter emissions from an aviation piston engine burning heavy fuels. *Fuel*, 202, 520-528.

[4] Esler, D., 2015. *Getting the Lead Out: The Future of Avgas*. [online] Available from: http://aviationweek.com/bca/getting-lead-out-future-avgas [Accessed 26 Jul. 2017].

[5] Federal Aviation Administration, 2012. *FAA UAT ARC Final Report Part I Body Unleaded AVGAS Findings & Recommendations*. [online] Available from: https://www.faa.gov/regulations_policies/rulemaking/committees/documents/media/Avgas.ARC.RR.2.17.12.pdf [Accessed 26 Jul. 2017].

[6] Federal Aviation Administration, 2017. *Press Release – FAA Announces Finalists Working to Get the Lead out of General Aviation Fuel*. [online] Available from: https://www.faa.gov/news/press_releases/news_story.cfm?newsId=20154 [Accessed 26 Jul. 2017].

[7] Gillette, W., 2017. *Development of Unleaded Aviation Gasoline*. [online] Available from: http://file:///C:/Users/Userr/Downloads/Gillette-ANERS2017%20(1).pdf [Accessed 26 Jul. 2017].

[8] Lovestead, T. and Bruno, T., 2009. Application of the Advanced Distillation Curve Method to the Aviation Fuel Avgas 100LL. *Energy & Fuels*, 23 (4), 2176-2183.

[9] Millner, P., 2006. *AVGAS 2020: The future of fuel*. [online] Available from: http://file:///C:/Users/Userr/AppData/Local/Temp/Temp1_order553771.zip/AVGAS_2020.pdf [Accessed 26 Jul. 2017].

[10] Pei, X. and Hou, L., 2016. Secondary flow and oxidation coking deposition of aviation fuel. *Fuel*, 167, 68-74.

IOP Conf. Series: Materials Science and Engineering **370** (2018) 012011 doi:10.1088/1757-899X/370/1/012011

Simulation Study of the Effect of Anti-Icing on the Nacelle Lip-skin Material

M H Y Syed[1], M A Ismail[1*], Q Azam[1], P Rajendran[2] and N M Mazlan[2]

[1] School of Mechanical Engineering, Universiti Sains Malaysia, Penang, Malaysia
[2] School of Aerospace Engineering, Universiti Sains Malaysia, Penang, Malaysia

Corresponding email: *azmi_meche@usm.my

Abstract. One of the key concerns in aircraft flight is the accumulation of ice on the wing leading edge and nacelle lip-skin. It is well known that ice accretion may dangerous to the aircraft since it severely degrades the aerodynamic performance of aircraft. As a consequence, the piccolo tube anti-icing system is installed by the aircraft manufacturer inside wing leading edge and nacelle lip-skin in order to prevent ice accumulation. However, the hotspots from piccolo tube anti-icing system potentially defect or destroy material of nacelle lip-skin. Therefore, in present work, the numerical investigation with one-way fluid-structure interaction (FSI) has been carried out to study the effect of Reynolds number based on effective-impingement-surface of piccolo tube anti-icing system on the maximum thermal stress and strain of nacelle lip-skin for several aluminum series. The simulation results reveal that the maximum strain increases with the Reynolds number. For the stress analysis, the maximum stress increases to the peak value, then it gradually decreases with Reynolds number. Among the materials have been studied, aluminium 7 series shows the highest maximum stress and maximum strain for all Reynolds number and hotspot temperature, however, aluminium 1 series produces the lowest maximum stress and maximum strain.

1. Introduction

Aircraft icing is the accretion of supercooled liquid which forms on the aircraft during flight. Accretion of ice onto vital parts of the aircraft components can bring severe effect to the flight performance and aircraft conditions [1]. This accumulation of ice on the inner part surface of the lip-skin can be hazardous to flying aircraft. The ice will break and hit the compressor. As a result, the compressor was broken and aircraft might happen. Therefore, it is very important for any aircraft to be installed with anti-icing system. Besides, ice accumulation on nacelle lip-skin can possess a major problem on any flying aircraft as it greatly alters the aerodynamic properties of the aircraft which in turn can lead to higher fuel consumption due to increased drag force generated by the non-aerodynamic shape present on the lip skin [2].

There are many types of aircraft anti-icing system which is now commercially used [3]. One of a good example and most used anti-icing system is the Piccolo Tube Anti Icing (PTAI) [4]. Piccolo tube is a series of in line and staggered holes, is placed inside the wing leading edge and near to its surface. Besides PTAI system, some researchers studied alternative anti-icing system. In 2015, Ismail et al. [12] studied temperature distribution of nacelle lip-skin by using swirl anti-icing (SAI) system. According to their result, SAI system provides much better temperature distribution than PTAI. Besides, SAI system has several advantages including fewer components, simple plumbing, is light and inexpensive compared to PTAI [13-15]. In addition, BAL may possibly be installed on the inner skin of the nacelle lip, SAI system has the potential to be used together with BAL since it does not produce hotspots with the same level of temperature as the hotspots of PTAI system along the inner skin.

In PTAI system, hot air is bled from the engine compressor at high pressure and temperature and is conducted forward to the fixed leading edge [5]. The high pressure and temperature of air from the tube flow out through the holes/nozzles at a high-velocity jet, then it impinges on the inner surface of the wing and nacelle leading edge. This heat transfer by conduction from the inner part surface to the outer part surface then results in maintaining the temperature of the outer part surface leading edge in order to avoid ice formation on the surface [6]. Most of the anti-icing study focuses on the thermal performance only [4], meanwhile the damage study of nacelle lip-skin only focus on the external object like a bird strike or debris strikes during flight [7]. Therefore, the presented study of the effect of hotspot temperature on the thermal strain and thermal stress on the nacelle lip-skin.

2. Methodology

2.1 Geometry

First of all, the geometry was drawn using ANSYS workbench. The geometry used in this study is as shown in Figure 1. Note the diameter, d in this study is 2mm.

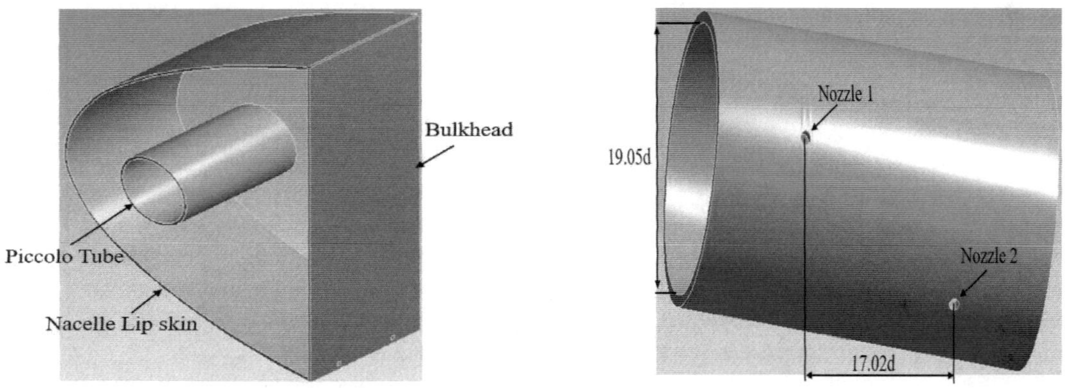

Figure 1. The Overall geometry of nacelle lip skin and Details of piccolo tube showing nozzle 1 and nozzle 2.

2.2 Meshing and Boundary Conditions

Using ANSYS workbench, the mesh was created which includes naming the surface and boundary conditions for the faces. 3 million hexahedron elements were used in the present study.

Figure 2. Meshing of nacelle lip-skin

For boundary condition as shown in Figures 3 and 4, pressure far field with 0.1 Mach number was used on the outer surface of the nacelle lip skin. Other boundary condition includes the inlet for hot gas which is called hot gas in and a pressure outlet for the hot gas which is called exhausted hot gas.

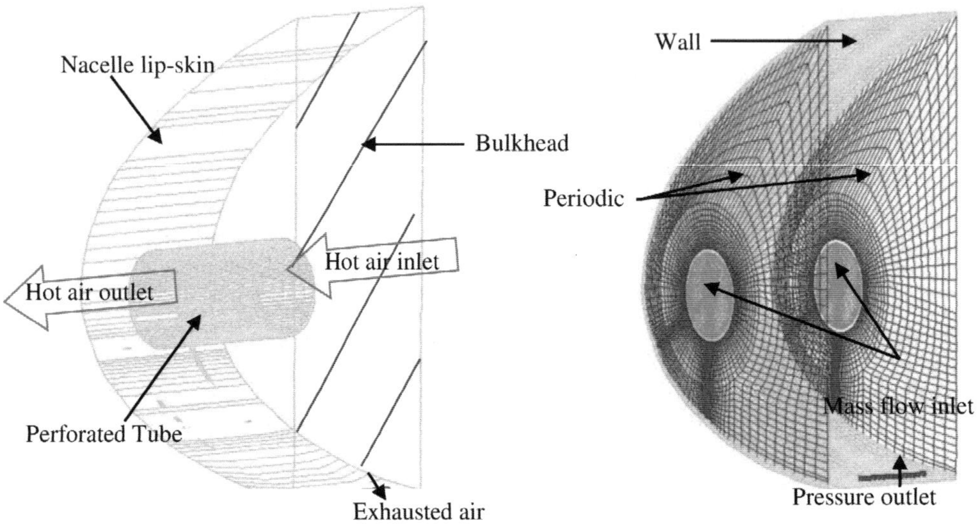

Figure 3. Boundary condition for PTAI inside D-chamber

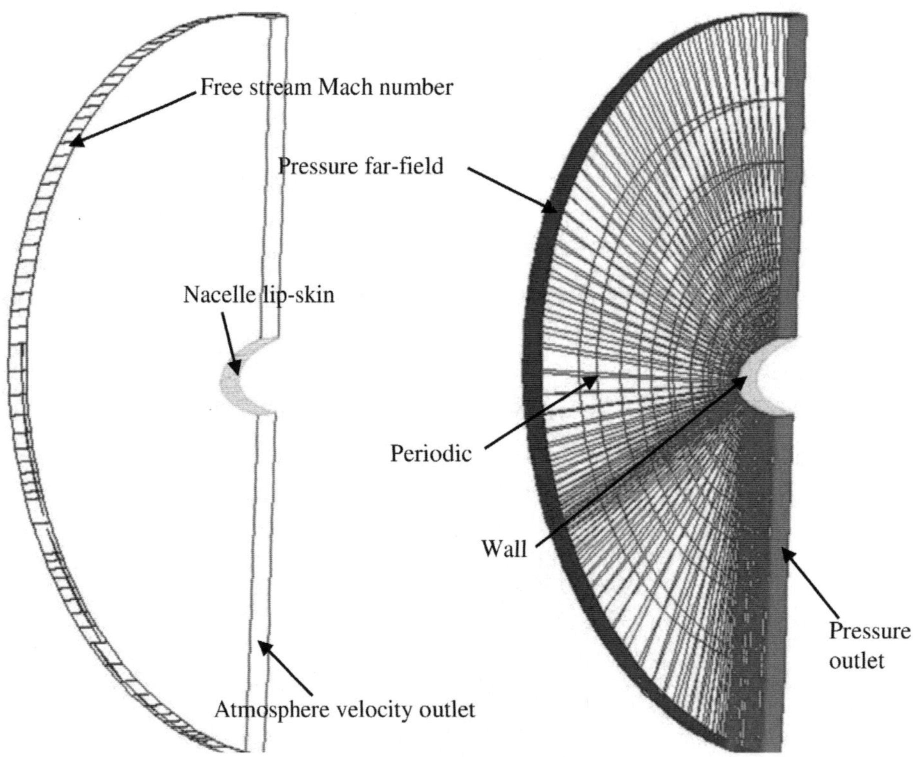

Figure 4. Boundary condition for PTAI inside D-chamber

2.3 ANSYS FLUENT Simulation

The simulation uses a turbulent model which is the shear-stress transport (SST) k-ω model and also the energy equation. SST k-ω model is an effective and accurate formulation of the k-ω model in the far field. The SST model incorporates a damped cross diffusion derivative term in the ω equation. Hence SST k-ω is able to provide a better accuracy of simulation result [8].

The boundary values are then inserted, operating condition in this simulation is 68050 Pa [9-11]. As for hot gas in and hot gas out the boundary. Next, the value of the parameter to be studied, the Reynolds number based on effective-impingement-surface (Re_G) of the hot air varies from 60 to 380. The ambient temperature and hot gas temperature of the study are 3°C and 272°C respectively. Re_G in present work is calculated by the following equation:

$$Re_G = \left(\frac{4Gd}{\pi\mu}\right) \tag{1}$$

Where

G	=	Hot air mass flow per unit area of effective-impingement-Surface
d	=	Diameter of Hole
μ	=	Air dynamic viscosity

2.4 Static Structural

In static structural, nacelle lip-skin was set as a solid domain. Since present work using FSI, the CFD results especially pressure and temperature were automatically exported to the Static structural software. After that, only one mechanical boundary conditions applied. Imported temperature data from CFD was applied on the nacelle lip-skin. For present work, one-way FSI is enough to predict stress and strain of nacelle lip-skin.

For this project, the material used and their composition are as follows:
AL 1100-O: 99% Al, 0.12 Cu, Aluminium 1st series
AL 2024-T4: 93.5% Al, Cu 4.4, Mn 0.6, Mg 1.5, Aluminium 2nd series
AL 7075-T6: 90% Al, Zn 5.6, Mg 2.5, Cu 1.6, Cr 0.23, Aluminium 7th series

3. Stress And Strain Analysis for Different Temperature on the Impingement Area

The temperature contour obtained from CFD simulation is shown in Figure 5 and the thermal stress and thermal strain contours on the nacelle lip-skin are shown in Figure 6. In Figure 5, the hotspot temperature occurs on the lower part of nacelle lip-skin, and the upper part of nacelle lip-skin has a lower temperature than the lower part of nacelle lip-skin. Figure 6 reveals that the highest strain and high stress occur on the hotspot area.

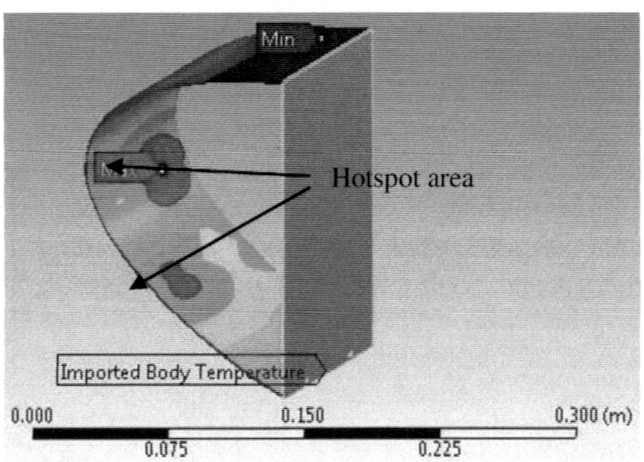

Figure 5. Temperature contour of nacelle lip-skin at ReG = 0.01362944

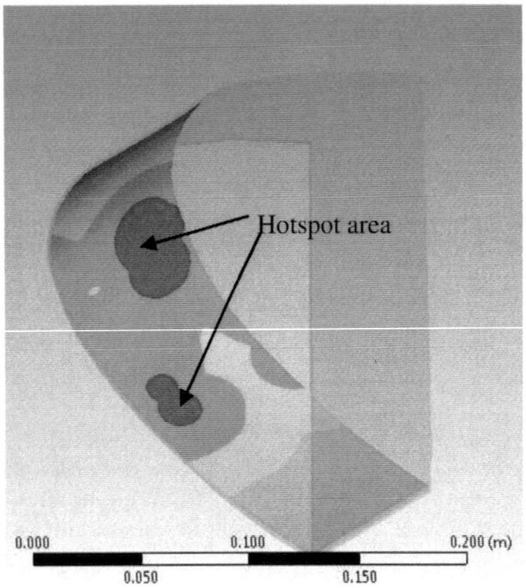

(i)

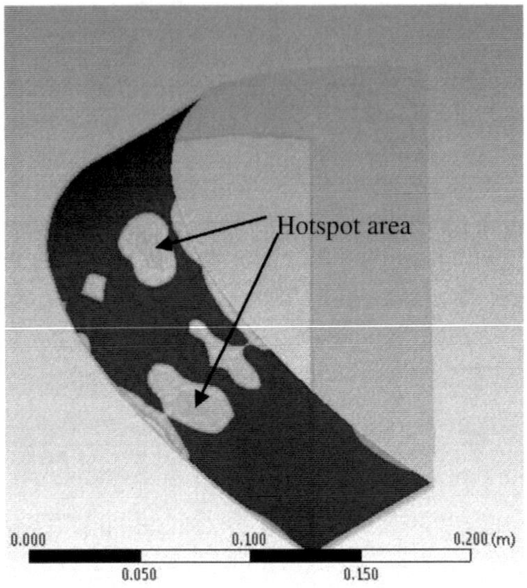

(ii)

Figure 6. (i) Thermal strain contour and (ii) Thermal stress contour on nacelle lip-skin at Re_G = 0.013629447

Figure 7 shows the relationship between maximum stress and Re_G. As shown in Figure 7, the maximum stress increases tremendously with Re_G until it reaches peak value for all materials. Afterwards, the maximum stress decreases with Re_G. The figure also shows that the Aluminium 7 series has the highest maximum stress along Re_G, followed by Aluminium 2 series and Aluminium 1 series.

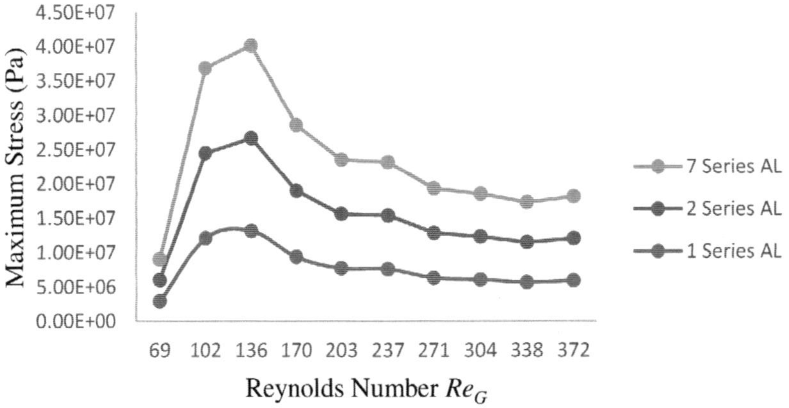

Figure 7. Graph of maximum stress against Reynolds number Re_G for all Aluminum series

The effect of Re_G on the maximum strain is shown in Figure 8. According to the figure, the maximum strain increases along Re_G for all materials studied. Similar to Figure 7, Aluminium 7 series shows the highest maximum strain and the lowest maximum strain is produced by Aluminium 1 series.

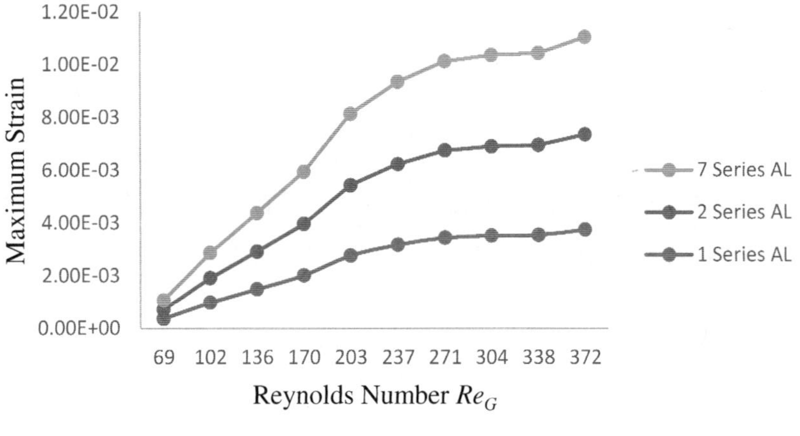

Figure 8. Graph of maximum strain against Reynold Number ReG for all Aluminum series

4. Conclusion

This present work studies the effect of hotspot temperature on the maximum thermal stress and thermal strain of nacelle lip-skin. The simulation has been done by the one-way fluid-structure interaction of ANSYS software. CFD part has been conducted by FLUENT and Mechanical ANSYS is used to predict maximum stress and strain. According to the results, maximum stress increases with Re_G until peak value, then it decreases with Re_G. Meanwhile, maximum strain continuously increases with Re_G. The simulation results show that Aluminium 1 series has the lowest maximum stress and maximum strain and the highest maximum stress and strain is produced by Aluminium 7 series. The lowest thermal stress and strain in present work are 40Mpa and 0.011 respectively.

Acknowledgement

This publication was supported by Universiti Sains Malaysia Grant No. 304/PAERO/ 60315002.

References

[1] J. W. Moe, J. J. Wunsch, and M. S. Sperling 2009 *Method and apparatus for noise abatement and ice protection of an aircraft engine nacelle inlet lip* ed: Google Patents.
[2] Q. Azam, M. A. Ismail, N. M. Mazlan, and M. Bashir 2016 *International Review of Mechanical Engineering (IREME)* **10** 390
[3] R. B. Rutherford 2001 *De-ice and anti-ice system and method for aircraft surfaces* ed: Google Patents.
[4] M. Ismail and M. Abdullah 2015 *Indian Journal of Science and Technology* **8**
[5] T. C. Currie, P. M. Struk, J.-C. Tsao, D. Fuleki, and D. C. Knezevici 2012 *4th AIAA Atmospheric and Space Environments Conference* 6.2012
[6] S. K. Thomas, R. P. Cassoni, and C. D. MacArthur 1996 **33** 841
[7] J. Thorpe, "100 Years of fatalities and destroyed civil aircraft due to bird strikes 2012 *International Bird Strike Committee, Stavanger* 1
[8] S. H. Park and J. H. Kwon 2004 *AIAA journal* **42** 1348
[9] M. Pascotto, A. Armellini, L. Casarsa, P. Giannatasio, and C. Mucignat 2012, *Effects of rotation and channel orientation on the flow field inside a trailing edge internal cooling channel* ASME Paper No. GT2012-68050
[10] E. D. Holstein *PEH: Reservoir Simulation*
[11] B. Srinivasan, A. Dhamarla, C. Jayamurugan, and A. B. Rajan 2014 *Numerical studies on effect of channel orientation in a rotating smooth wedge-shaped cooling channel* in ASME Turbo Expo 2014: Turbine

Emissions of piston engine aircraft using aviation gasoline (avgas) and motor gasoline (mogas) as fuel – a review

K Thanikasalam[1], M Rahmat[1], A G Mohammad Fahmi[2], A M Zulkifli[1], N Noor Shawal[1], K Ilanchelvi[3], M Ananth[3], R Elayarasan[1]

[1]Faculty of Chemical and Energy Engineering, Universiti Teknologi Malaysia (UTM), 81310 Skudai, Johor, Malaysia
[2]Department of Aeronautical Engineering, Universiti Tun Hussein Onn Malaysia (UTHM), 86400 Batu Pahat, Johor, Malaysia
[3]Department of Chemical Engineering, Universiti Malaysia Sabah (UMS), 88400 Kota Kinabalu, Sabah, Malaysia

E-mail: insp466@hotmail.com

Abstract. There are two categories of aircraft engines, namely, piston and gas turbine engines. Piston engine extracts energy from a combustion compartment through a piston and crank apparatus that engages the propellers, which in turn, provides an aircraft the needed momentum. On the other hand, gas turbine engine heats a compressed air in the combustion compartment resulting in propulsion that drives an aircraft. Piston engine aircrafts might appear small but together thousands of piston engine aircraft, which encompasses a bulk of the general aviation fleet, present a considerable health threat. That is because these aircraft, which depend on avgas and mogas to run, comprise major remaining sources of lead emissions. People exposed to even small levels of lead, particularly children, have tendencies to suffer from cognitive and neurological harm. Dissimilar from commercial airliners that do not utilize leaded fuels, piston engine aircraft account for nearly half of the lead discharge in skies. But, what is the extent of the impact caused by these airborne emissions on the country's economy and public health? To answer this query, a thorough literature review on emissions of piston engine aircraft ought to be undertaken. This article conducts a literature review on emissions of piston engine aircraft using avgas as fuel and mogas as fuel.

1. Introduction

Piston engines are the improvements of steam engines that first emerged during the 17th century. Czarnigowski et al. 2010 provide a brief background on piston engines. Czarnigowski et al. 2010 say that steam engines utilize the heat generated to create steam pressure that is then used to produce a rotating motion. Steam engines were in time applied to drive nearly all mechanical apparatus including the first railway trains and aircraft. In contemporary aircraft piston engine, gas propulsion substitutes steam in producing pressure to drive the engine. According to Czarnigowski et al. 2010,one or more cylinders characterise the piston engine. The cylinders contain a comfortably fitting piston that is capable of moving freely forward and rear inside the cylinder. Every piston attaches to a crankshaft through a connecting rod.

Czarnigowski et al. 2010 write that when fuel enters and ignites inside the cylinder, the scorching gases generated inflate with incredible force and drives the piston frontward in the cylinder that in turns move the piston rod forward causing the crankshaft to revolve. The full revolution of the crankshaft thrust the piston in the rear direction into the cylinder before the cycle commences again. Czarnigowski et al. 2010 claim that it is the crankshaft, which transfers the reciprocating and linear movement of the piston into the rotary motion that pushes the propeller. Czarnigowski et al. 2010 claim that piston engines were employed in powering all aircraft up until jet engines were made during the early 20th century, motivated by the need for an aircraft that could fly at superior speeds and elevated altitudes. Currently, piston engine aircraft are still popular for both private and commercial use owing to their reduced cost of entry and reliable performance.

Federal Aviation Administration (2005) provides a brief summary of significant issues concerning aviation emissions. Aircraft generate similar types of emissions like an automobile. According to Federal Aviation Administration (2005), piston aircraft jet engines, like automobiles engines, emit carbon dioxide, water vapour, nitrogen oxides, and carbon monoxide, and hydrocarbon emissions. However, unlike automobile engines, piston aircraft jet engines generate lead emissions.

Federal Aviation Administration (2005) data suggests that aircraft engine emissions comprise around 70% carbon dioxide, roughly 29% water vapour, and less than 1% of carbon monoxide and nitrogen oxide emissions. Aircraft emissions, based on whether they happen near the earth surface or at higher altitudes, are mainly deemed local air quality contaminants or greenhouse gas emissions, respectively. Water released by aircraft engines at altitude can trigger greenhouse effect. Around 10% of aircraft emissions of every form, apart from hydrocarbons and carbon dioxide, are generated during ground level manoeuvres and during takeoff and landing processes. According to Federal Aviation Administration (2005), a considerable amount of aircraft emissions, nearly 90% happen at higher altitudes. Concerning hydrocarbons and carbon monoxide, the split is around 30% emissions at ground level and 70% emissions at higher altitudes. Federal Aviation Administration (2005) argues that aviation emissions reveal the intensity of general aviation activity. The expansion of air travel in the last numerous decades has been very speedy. The need for travel services both cargo and passenger hauling is increasing considerably. Federal Aviation Administration (2005) argues that this also implies that aviation emissions have increased over time.

2. Aircraft Piston Engine Exhaust Emissions Svmposium

Lewis Research Center (1976) focuses on piston engine emissions. The study notes that the published literature has little information on piston engine aircraft emissions. Thus, they suggest that more researches are required to fill the knowledge gap in piston engine aircraft emissions. This symposium report notes that piston aircraft jet engines, like automobiles engines, emit carbon dioxide, water vapour, nitrogen oxides, and carbon monoxide, and hydrocarbon emissions. Nevertheless, contrasting automobile engines, piston aircraft jet engines generate lead emissions. The findings by Lewis Research Center (1976) indicate that piston engine aircraft emissions standards fail to meet the standards set by Environmental Protection Agency (EPA) lead, hydrocarbon, and carbon monoxide.

This symposium report also claims that people exposed to even small levels of lead, particularly children, have tendencies to suffer from cognitive and neurological harm [9]. Dissimilar from commercial airliners that do not utilize leaded fuels, piston engine aircraft account for nearly half of the lead discharge in skies. To understand the extent of the impact caused by these airborne emissions on a global economy and public health, Lewis Research Center (1976) suggests that more research should be undertaken on piston engine emissions.

3. Aircraft engine exhaust emissions and other airport-related contributions to ambient air pollution

Masiol and Harrison (2014) also claim that engines for both general and civil aviation are normally categorized as piston or gas turbine engines. Aviation piston engines operate on AVGAS as fuel. Masiol and Harrison (2014) asserts that piston engines are usually installed in small-sized aircraft normally associated to private use, flying associations, flight instruction, farming, and tourism. Piston engines operate using the same fundamental principles used by spark ignition engines for automobiles but normally need higher performance. According to Masiol and Harrison (2014), the major variation between gas turbine engines and piston engines is that combustion is incessant in gas turbine engines and alternating in piston engines.

The accessibility of dependable data on fuel consumption is necessary to make authoritative approximations of aviation emissions at international and national scales. Masiol and Harrison (2014) claim that a number of approximations of aviation fuel utilization presented in existing literature generally refer only to gas turbine engines since piston-powered engines are assumed to account for nearly 2% of propeller fuel burn. Masiol and Harrison (2014) approximated that up to 169 x 10^{12} g fuel was used internationally in the years 2000. Of this, nearly 152 x 10^{12} g was used in civil flights. In addition, Masiol and Harrison (2014) indicated that AERO2k international aviation emissions inventories recorded that 176 x 10^{12} g of kerosene were utilized in the year 2002. After methodically reviewing existing literature, Masiol and Harrison (2014) concluded that the existing information on the effect of emissions of piston engine aircraft is insufficient and the impacts of future expansion in the quantity of air traffic are very problematic to predict. Masiol and Harrison (2014) concluded that quantification of the effect of emissions of piston engine aircraft on domestic air quality is very challenging owing to the intricacy of air emissions and the existence of considerable levels of contamination from other sources.

4. Modelling and measurement of aircraft engine emissions inside the airport area

Zaporozhets and Synylo (2015) explore the role aviation emissions contribute towards greenhouse gas issues. This study notes that a number of emissions from piston engine aircraft occur at higher altitudes. These emissions generate greenhouse gases that likely contribute to climate change. Zaporozhets and Synylo (2015) also suggest that under specific conditions, piston engine exhaust may generate contrails. Internationally, scientists are exploring the possible effects of these contrails to understand if they have a considerable effect on the greenhouse effect. Zaporozhets and Synylo (2015) argue that concerns about greenhouse gas emissions have increased worldwide. While there are numerous greenhouse gases, this study suggests that carbon dioxide and nitrogen oxide emissions are usually most pertinent from an aviation viewpoint.

Zaporozhets and Synylo (2015) also illustrate greenhouse gas emissions based on countries. The analysis indicates that land transportation emissions contribute 27% of greenhouse gas inventory, while aviation emissions contribute around 2.7%. In addition, the analysis illustrates that global greenhouse gas emissions have increased in the last decade and are anticipated to increase further in the coming years. Zaporozhets and Synylo (2015) point out that there many ongoing initiatives aimed at reducing emissions both at the airport and in the national aviation system. First, there are numerous voluntary ventures in progress at airports to lessen emissions from ground support apparatus and other airport automobiles. For instance, FAA is developing a pilot program in collaboration with DOE and EPA, aimed at demonstrating air quality enhancement with substitute fuel ground support apparatus. Zaporozhets and Synylo (2015) also reveal that many airports are independently taking action to lessen emissions from buses, tractors, trucks, taxis, and other on-road automobiles, which operate around the airports. Concerning piston engine modifications, Zaporozhets and Synylo (2015) suggest that there exist multifaceted emission interrelationships. This makes it hard to alter its design as a mitigation approach because it introduces a tradeoff among particular contaminants, emissions, and noise.

5. Aircraft piston engine emissions summary report - Switzerland

Federal Office of Civil Aviation (2007) suggests that aircraft piston engines use aviation AVGAS as fuel. It claims that aircraft piston engines are commonly referred as "small propeller aircraft". On an international scale, piston engine aircraft use a comparatively small share of fuel. Federal Office of Civil Aviation (2007) notes that in Switzerland the yearly fuel use by aircraft piston engines is fewer compared to fuel used in gardening activities. Federal Office of Civil Aviation (2007) also points out that emissions generated by aircraft have often attracted little attention. Globally, experts have expressed little efforts to come up with emission standardization for such engines. Federal Office of Civil Aviation (2007) also notes that information regarding emissions from aircraft piston engines is practically missing. This shortage has occasionally caused challenges, especially when producing a comprehensive aircraft emissions catalogue or when providing reliable data on an ecological impact assessment. Federal Office of Civil Aviation (2007) undertook an exploratory experiment to fill this gap of knowledge.

Federal Office of Civil Aviation (2007) findings indicate the contribution of piston engine aircraft discharges on Switzerland's civil aviation emission. The findings disclose that piston engine aircraft contribute less than 1% of the total civil aviation carbon dioxide emissions. Federal Office of Civil Aviation (2007) also reveals that these engines contribute 1% share on nitrogen oxides, 10% share of hydrocarbons, 40% share of Carbon monoxide, and 100% share of the lead. In the findings, Federal Office of Civil Aviation (2007) claims that carbon monoxide and hydrocarbon emissions are significant during landing and take-off phase. In spite of a fuel burn, which is more than a few magnitudes lesser, carbon monoxide and hydrocarbon emissions might reach the amount produced by large aircraft. Federal Office of Civil Aviation (2007) argues that this phenomenon has been caused by aircraft piston engine's technology that has been at a standstill ever since the 1960s.

6. Exhaust Emissions Characteristics for a General Aviation Light-Aircraft Avco Lycoming IO-360-B1BD Piston Engine

Becker (1979) conducted emission tests on a general aviation piston engine to determine emission characteristics, establish the impact of leaning-out of gasoline metering system on exhaust discharges, authenticate the suitability of test processes, testing procedures, and instrumentation. The tests also sought to identify a drop in operating confines and safety boundaries ensuing from fuel adjustments appraised for enhanced piston engine exhaust emissions attributes [1]. The tests were done at "National Aviation Facility Experimental Centre," USA.

The findings by Becker (1979) reveal that engine IO-360-BlBD failed to meet the 1979/1980 standards set by Environmental Protection Agency (EPA) on hydrocarbon and carbon monoxide. The engine only met the EPA standards on nitrogen oxides. Becker (1979) also suggests that the engine gasoline metering system could be attuned on the test plunk to lessen the current carbon monoxide exhaust emission levels, although not to levels needed by EPA standards, under the most strict LTO cycle necessities. It was also noted that the engine could be attuned on the test plunk to lessen the unutilized hydrocarbon exhaust emission levels, however not to levels needed by EPA [1]. After testing engine IO-360-BlBD, Becker (1979) concluded that simple fuel management modification like varying of fuel schedule, do not offer the sole capacity to safely lessen piston engine exhaust emissions. The test data also revealed that fuel management alterations ought to be integrated with engine cooling variations prior safe and optimal low-emission aircraft-engine blend is attained. Becker (1979) also noted that spark settings, excluding the 25° BTC setting, fail to generate considerable beneficial enhancements in exhaust emissions. In addition, the EPA carbon monoxide limit of 0.0420 lb/cycle/rated BHP seemed unachievable when hot-day take-off and ascending necessities are affected by aircraft's heavy gross weight plus the requirement to pay vigilant consideration to CHT confines.

7. Exhaust emissions from in-use general aviation aircraft

Yacovitch (2016) also claims that piston engine aircraft emissions data either are missing or have not been autonomously confirmed. Yacovitch et al. (2016) sought to fill this gap of knowledge by providing the results from an emissions test. For this test, the researchers assessed emissions and calculated emission indices for numerous in-use aircraft. Yacovitch et al. (2016) document 47 complete engine tests comprising 10 engines from a category of a top 20 general piston engines. The researchers noted that numerous factors may influence emissions. Yacovitch et al. (2016) explored a number of sources for high inconsistencies in emission indices. The research team noted that pilot's way of thinking coupled with his or her operation experience of an aircraft influences combustion factors. Yacovitch et al. (2016) also explored a number of technical information linked with piston engines aircraft emissions like trends in particle mass and unpredictability, the impact of fuel additives, and thermal generation of nitrogen oxides and its relations to lean combustion.

The findings by Yacovitch et al. (2016) suggest that piston and gas turbine engines have very dissimilar emissions, with respect to their magnitude and power trends. The disparities may be understood in regards to the higher combustion performance in contemporary gas turbine engines opposed to piston engines. Thus, piston engines release more hydrocarbons and carbon monoxide emissions and a smaller amount nitrogen oxide compared with gas turbine engines. Yacovitch et al. (2016) also reveal warped distributions of emission indices recorded from an experiment assessing emissions of hydrocarbons, nitrogen oxides, and impure matter from a piston engine. The findings indicated that the most general emission index is not equivalent to the approximate emission index. Notably, Yacovitch et al. (2016) claim that carbon monoxide emissions' distributions were not skewed.

8. Advance Notice of Proposed Rulemaking on Lead Emissions from Piston-Engine Aircraft Using Leaded Aviation Gasoline - Regulatory Announcement

Environmental Protection Agency (2010) represents an advance notice to illustrate information at present available and those being collected, which will be applied by the proprietor to issue an ensuing proposal concerning whether, in the proprietor's decision, piston aircraft lead emissions cause or add to air pollution that might reasonably be expected to jeopardize public health or well-being. In this notice, Environmental Protection Agency (2010) describes and requests observations on the existing data for appraising lead emissions, atmospheric concentrations, and possible exposure to lead through continuous use of leaded aviation gasoline in piston engine aircraft. The agency also describes and requests commentaries on further data being collected, which will update any future decision.

The findings by Environmental Protection Agency (2010) indicate that lead emissions from piston engine aircraft consist of roughly half of the countrywide catalogue of lead released into the air. Environmental Protection Agency (2010) suggests that there are around 20,000 airport services in the US that use leaded aviation gases. The agency asserts that positive contribute findings on lead emissions prompt EPA's responsibility to create emission standards. In deliberating emission standards, the agency asserts that EPA should deliberate controlling piston engines reliant on AVGAS. Environmental Protection Agency (2010) proposes that EPA in collaboration with FAA should consider evaluating the technical viability of a probable phase-down or abolition of leaded AVGAS. One alternative to consider, for instance, is an emissions standard, which will order all newly-designed aviation piston engines to run successfully on unleaded AVGAS. However, Environmental Protection Agency (2010) recognizes that a huge challenge exists in addressing lead emissions from the in-use fleet. Renovating in-use aircraft piston engines to run on unleaded AVGAs will require a considerable logistical and technical challenge.

9. Non-road Large Spark-Ignition Engines: Exhaust and Evaporative Emission Standards

Office of Transportation and Air Quality (2016) focuses on evaporative and exhaust emission standards. The agency argues that numerical emission sets for hydrocarbons ought to be met with respect to the following forms of hydrocarbon discharges for engines operating on (1) non-methane hydrocarbon fuels, (2) total hydrocarbon equivalent and (3) total hydrocarbons for natural gas, alcohol, and other fuels respectively. Similarly, the agency provides "Voluntary Blue Sky" standards required for spark-ignition (SI) aviation engines. Office of Transportation and Air Quality (2016) suggests that engines that have displacement below 1,000 CC with maximum power below 30 kW should be categorized and certified as small SI engines. According to Office of Transportation and Air Quality (2016), emission standards rely on evaluation over a steady-state duty-cycle. The agency asserts that the Tier-1 hydrocarbon plus nitrogen oxide emission standard used during the evaluation is 5.40 grams per kW-hour.

In addition, Office of Transportation and Air Quality (2016) express piston engine useful life in terms of years and hours, which are the least useful life necessities. Concerning severe-duty engines, the least useful life is 1500 hours or seven years of service. Office of Transportation and Air Quality (2016) suggests that an extended useful life in hours is needed if: (1) the engine is intended to run longer than the least useful life reliant on the suggested rebuild interval; or (2) the fundamental engine warranty is longer compared with the least useful life. An extended warranty period of 3, 500 hours or five years of service is applicable for expensive warranted parts. Office of Transportation and Air Quality (2016) also claims that optional engine certification is permitted based on the following formula: "HC+NOx) × CO0.784 ≤ 8.57."

10. Emissions of air pollutants from civil aviation in the Netherlands

Hulskotte (2017) investigates air population emissions in one airport in Netherlands. The investigation uses CLEO model, a methodology derived from an internationally established approach used by EPA. The model approximates emissions from takeoff and landing cycles using aboard auxiliary power units. Emissions occurring at high altitudes were not considered part of the tests. The findings by Hulskotte (2017) illustrate that nitrogen oxide emissions attributed to civil aviation have increased by nearly 140% between the years 1990 to 2015. Hulskotte (2017) also notes that between 1990 and 1995, hydrocarbon emissions have lessened by 14%. 380 tons of hydrocarbon emission occurred in the year 2015. Around 25% of these emissions are attributed to fuel handling. Carbon monoxide emissions also reduced by 4% over the same period. In the year 2015, 3.5 tonnes of carbon monoxide were recorded. These findings indicate that current aircraft engines release less hydrocarbon and carbon dioxides but have barely improved with respect to nitrogen oxide emissions.

In addition, the findings by Hulskotte (2017) indicate that piston engine aircraft operating on AVGAS contribute a small portion of nitrogen oxide emissions, a share of less than 1%. However, the investigations show that piston engine aircraft contribute more considerably to carbon monoxide and hydrocarbon emissions, a share of more than 5%. Furthermore, the use of AVGAS by piston engine aircraft proved to be a key source of aviation lead emissions. The investigations by Hulskotte (2017) revealed that up to 0.8 tonnes of lead emissions were recorded in the years 2015. Piston engine aircraft emissions elements were established using measurement information from US EPA in collaboration with Swiss Federal Office for Civil Aviation [6].

11. New Methodology for Modeling Annual Aircraft Emissions at Airports

Woodmansey and Patterson (1994) presented a methodology for assessing yearly aircraft emissions in airports. This methodology can be used in testing annual piston engine aircraft emissions. Woodmansey and Patterson (1994) assert that accurate approximations of total- aircraft emissions are vital components of any environmental impact appraisal done for any proposed airport expansion. To establish the volume of emissions created by piston engine aircraft, in particular, airport, using the model presented by Woodmansey and Patterson (1994), it is essential to know the emission characteristics of piston engines. However, they point out that the published literature has little information on piston engine aircraft emissions. Thus, they suggest that new methodologies are required to assist in approximating annual emissions from piston engine aircraft at airports. Woodmansey and Patterson (1994) proposed a linear regression equation linking a number of emissions with aircraft weight. Using this approach, they claim that total annual emissions for piston engine powered aircraft like carbon monoxide, nitrogen oxides, carbon dioxide, and lead emissions can be tabulated. The regression model presented by Woodmansey and Patterson (1994) is comparatively simple, quick, and low-priced way of calculating a yearly emission catalogue for an airport.

12. Reduction emission level of harmful components exhaust gases by means of control of parameters influencing on spraying process of biofuel components for aircraft engines

Jankowski (2011) undertook an exploratory study to test fuel additives, that lessen the magnitude of atomised fuel drops, by altering particular that affect the atomisation procedure. These parameters encompass surface tension, thickness, viscosity index, and density. Jankowski (2011) noted that the magnitude of bio-fuel drops was much larger likened with those of hydrocarbon fuels. By altering the chemical and physical structures of bio-fuels, magnitudes of drops of atomised fuel streams become lesser. Jankowski (2011) claim that these magnitudes play a key role in the level of carbon monoxide and hydrocarbon emissions, in addition to nitrogen oxides emissions. Jankowski (2011) argues that the exploration studies on emissions of lethal elements of fuel are comparatively superior in the field of piston engines, particularly automobile applications.

However, Jankowski (2011) argues that the dynamic expansion of both civic and general aviation presents more pressure on the problem of poisonous emissions by aircraft engines. According to Jankowski (2011), the extent of toxic emissions aircraft engines could be a thousand times more compared to the level of piston engine aircraft emissions. Thus, the problem of how bio-fuel additives may influence the procedure of fuel atomisation and facilitate control over the atomisation in obtaining the least likely drops and subsequently less nitrogen oxide emissions is still a new issue. Jankowski (2011) notes that the reduced nitrogen oxide emission, with respect to the use of bio-fuels, is of greatest implication since new data indicates that its emission levels are on the rise.

13. Piston engine emissions: results of tests and development of test methods

Klueg, Salmon, Becker, and Imbrogno (1978) explain the procedures created by the Federal Aviation Administration in the latest years for testing, gauging, and computing exhaust emissions from piston engines aircraft. This study outlines instrumentation recommended for monitoring emissions piston engine aircraft. The instrumentation comprises apparatus for assessing amounts of carbon monoxide, carbon dioxide, nitric oxides, and hydrocarbons.

Klueg et al (1978) suggest that to identify mass emissions during aircraft landing and takeoff a calculation procedure that utilizes the calculated exhaust concentrations and calculated fuel flow and airflows is adopted. Klueg et al (1978) warn that caution ought to be taken when choosing a computation approach from the published procedures in existing literature to guarantee that the method is valid over the whole operating range of piston engine aircraft and that any suppositions made are precise. The findings by Klueg et al (1978) noted that any effort to enforce engine exhaust emission restrictions without giving due deliberation to engine's CC, head temperature confines and nacelle cooling may lead to severe troubles. This implies that piston engine modifications present multifaceted emission interrelationships, which make it hard to alter its design as a mitigation approach because it introduces a tradeoff among particular contaminants, emissions, and noise.

14. Conclusion

Based on the above literature review, it should be noted that there is an increasing concern regarding the emissions of piston engine aircraft. Piston engine aircraft might appear small but together thousands of these machines present a considerable health threat. People exposed to even small levels of lead, particularly children, have tendencies to suffer from cognitive and neurological harm. Dissimilar from commercial airliners that do not utilize leaded fuels, piston engine aircraft account for nearly half of the lead discharge in the skies. The literature review also indicates that there is a knowledge gap on emissions of piston engine aircrafts. Thus more researchers should be undertaken on this topic.

Acknowledgement

The authors express their gratitude to the Universiti Teknologi Malaysia (UTM), Skudai, Johor, Malaysia for their financial support.

References

[1] Becker, E., 1979. *Exhaust Emissions Characteristics for a General Aviation Light-Aircraft Avco Lycoming IO-360-B1BD Piston Engine.*[online] Available from: http://www.dtic.mil/get-tr-doc/pdf?AD=ADA066589 [Accessed 20 Jul. 2017].

[2] Czarnigowski, J., Jakliński, P. and Wendeker, M., 2010. Fuelling of aircraft radial piston

engines by ES95 and 100LL gasoline. *Fuel*, 89 (11), 3568-3578.

[3] Environmental Protection Agency, 2010. *Advance Notice of Proposed Rulemaking on Lead Emissions from Piston-Engine Aircraft Using Leaded Aviation Gasoline - Regulatory Announcement.* [online]. Available from: http://realneo.us/content/advance-notice-proposed-rulemaking-lead-emissions-piston-engine-aircraft-using-leaded-aviati [Accessed 20 Jul. 2017].

[4] Federal Aviation Administration, 2005. *Aviation & Emissions A Primer.* [online] Available from:https://www.faa.gov/regulations_policies/policy_guidance/envir_policy/media/AEPRIMER.pdf [Accessed 20 Jul. 2017].

[5] Federal Office of Civil Aviation, 2007. *Aircraft piston engine emissions summary report.* [online] Available from: http://www.hjelmco.com/upl/files/2425.pdf [Accessed 20 Jul. 2017].

[6] Hulskotte, I., 2017. *Emissions of air pollutants from civil aviation in the Netherlands.* Princeton: Princeton University.

[7] Jankowski, A., 2011. Reduction emission level of harmful components exhaust gases by means of control of parameters influencing on spraying process of biofuel components for aircraft engines. *Journal of KONES*, 18 (3), 129-134.

[8] Klueg, E., Salmon, R., Becker, E. and Imbrogno, S., 1978. *Piston engine emissions: results of tests and development of test methods.* New Jersey: Federal Aviation Administration.

[9] Lewis Research Center, 1976. *Aircraft piston engine exhaust emissions svmposium.* [online] Available from: https://ntrs.nasa.gov/archive/nasa/casi.ntrs.nasa.gov/19770010138.pdf [Accessed 20 Jul. 2017].

[10] Masiol, M. and Harrison, R., 2014. Aircraft engine exhaust emissions and other airport-related contributions to ambient air pollution: A review. *Atmospheric Environment*, 95 (7), 409-455.

[11] Ng, J. and Gan, S., 2012. Development of emissions predictor equations for a light-duty diesel engine using biodiesel fuel properties. *Fuel*, 95, 544-552.

[12] Office of Transportation and Air Quality, 2016. *Nonroad Large Spark-Ignition Engines: Exhaust and Evaporative Emission Standards.* [online] Available from: cta.ornl.gov/data/tedb35/Spreadsheets/Table12_24.xls [Accessed 20 Jul. 2017].

[13] Rae, J. and Heron, S., 1962. History of the Aircraft Piston Engine. A Brief outline. *Technology and Culture*, 3 (2), 209.

[14] Woodmansey, B. and Patterson, J., 1994. New Methodology for Modeling Annual- Aircraft Emissions at Airports. *Journal of Transportation Engineering*, 120 (3), 339-357.

[15] Yacovitch, T., 2016. *Exhaust emissions from in-use general aviation aircraft.* Washington, D.C.: The National Academies Press.

[16] Zaporozhets, O. and Synylo, K., 2015. Modelling and measurement of aircraft engine emissions inside the airport area. *Proceedings of National Aviation University*, 63 (2).

IOP Conf. Series: Materials Science and Engineering **370** (2018) 012013 doi:10.1088/1757-899X/370/1/012013

Validation of Harris Detector and Eigen Features Detector

K Y Kok and P Rajendran

School of Aerospace Engineering, Universiti Sains Malaysia, Engineering Campus, Nibong Tebal, Pulau Pinang, Malaysia.

Email: aeparvathy@usm.my

Abstract. Harris detector is one of the most common features detection for applications such as object recognition, stereo matching and target tracking. In this paper, a similar Harris detector algorithm is written using MATLAB and the performance is compared with MATLAB built in Harris detector for validation. This is to ensure that rewritten version of Harris detector can be used for Unmanned Aerial Vehicle (UAV) application research purpose yet can be further improvised. Another corner detector close to Harris detector, which is Eigen features detector is rewritten and compared as well using same procedures with same purpose. The simulation results have shown that rewritten version for both Harris and Eigen features detectors have the same performance with MATLAB built in detectors with not more than 0.4% coordination deviation, less than 4% & 5% response deviation respectively, and maximum 3% computational cost error.

1. Introduction

Features detection is a process to detect significant curves, edges or corners from an image. These features have necessary information for motion tracking and pattern recognition through feature extraction and matching with features from other frame [1, 2]. Hence, features detection is very useful and crucial in the development of UAV application as it can make UAV to be more robust and intelligent.

Nonetheless, precision and speed are the main requirements for implementation of features detection in real applications. The computational cost of detection needs to be low and features must be detected consistently in each frame to maintain tracking or recognition in various viewpoints and orientations [3].

Among features detector, the Harris detector is one of the most widely used method due to its simplicity, reliability and invariance of lighting, transformation and rotation [4]. Generally, this detector considers corner points as features by determining whether there is large intensity change in all directions for each pixel. Shi and Tomasi [5] proposed Eigen features detector and it has the same concept as Harris detector in some ways. The advantage of this method is that it is more stable, but the computational cost is higher than Harris detector.

In this paper, the algorithm of Harris detector and Eigen features detector are made using MATLAB, and both algorithms' performance will be compared with its MATLAB version built-in function detectors for validation. This is crucial so that both written algorithms can be used, analyzed and further improvised for UAV application research if they are validated. The performance comparison is analyzed in the respect of features' coordinates, response value and overall computational cost.

2. Methodology

In this section, the introduction of Harris detector and Eigen features detector is elucidated. In addition, the principle of both detectors is briefly explained.

2.1. *Harris detector*

Harris detector is proposed by Harris and Stephens [6], which is basically to determine the corner points of an image [7] and to have improved performance than the Moravec's corner detector [8]. There are three aspects in terms of improvement. Firstly, Morevec's corner detector only consider eight discrete directions in obtaining corner points while Harris detector minimize the dependency on the direction [9]. Next, Harris detector is utilizing the Gaussian function which avoids interference that occurs in Moravec's corner detector. Lastly, Harris detector decreases the probability of false detection to edge by having various direction grey measurement [9].

In general for the Harris detector, the gradient of image in x and y direction, I_x and I_y will be estimated for each pixel. Then, the correlation matrix of each pixel is formed [10, 11] as shown in equation (1).

$$M = \begin{bmatrix} I_x^2 & I_x I_y \\ I_x I_y & I_y^2 \end{bmatrix} \tag{1}$$

The elements in the matrix are smoothed with Gaussian function in order to increase the robustness of the detection. The response of each pixel can be calculated using equation (2) [12],

$$H = \lambda_1 \lambda_2 - k * (\lambda_1 + \lambda_2)^2 \tag{2}$$

whereby λ_1 and λ_2 are the eigenvalues of M and k is the sensitivity factor which is usually within [0.04, 0.06]. The pixel is considered as corner, only when both eigenvalues λ_1 and λ_2 are large as shown in figure 1 [13].

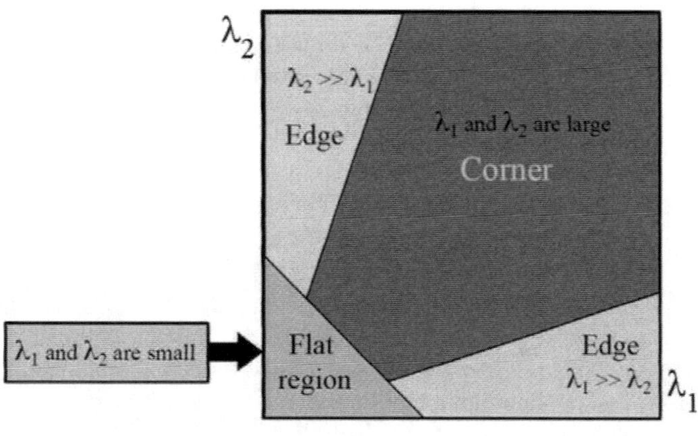

Figure 1. Classification of image point based on eigenvalues of the autocorrelation matrix.

Still, the computation cost of calculating eigenvalues is high, thus, an approximation using equation (3) having the determinant and the trace of M is developed to replace equation (2).

$$H = \det(M) - k * trace(M)^2 \tag{3}$$

If the response is negative, this means the pixel is an edge, while if the response is small in value; the pixel is in flat region. Only when the response is positive and large in value, the pixel might be considered as corner point [14]. This is confirmed by comparing the value of response with threshold value as shown in equation (4) and those with higher value than threshold are corner points.

$$Corner = true, only\ if\ H > threshold \tag{4}$$

2.2. *Eigen features detector*

This method has the same steps as Harris detector in the beginning by forming correlation matrix with intensity gradient for each pixel. The components of the matrix will be smoothed using Gaussian function as well. Nevertheless, the response of each pixel is estimated using equation (5) as shown below [5]. Only when eigenvalues are greater than threshold value, the pixel will be considered as corner point, which is illustrated in light yellow region in Figure 2.

$$R = \min(\lambda_1, \lambda_2) \tag{5}$$

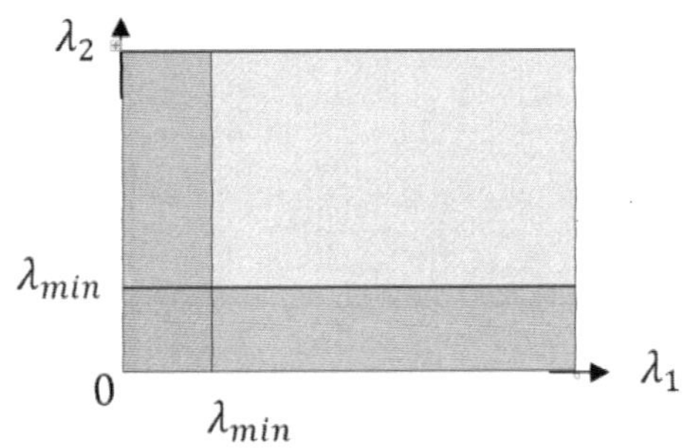

Figure 2. Classification of image point based on eigenvalues of the autocorrelation matrix.

3. Results ad Discussions

In this section, a simulation comparison test is done on MATLAB built in detectors and rewritten version detectors using 6 various images. In terms of Harris detectors, the sensitivity factor, k is set as 0.04 in both versions of the detector. The threshold of both Harris and Eigen features detectors is set as 1% of maximum response value in the image. Figures 3-6 display two out of the 6 case studies images used to assess both Harris and Eigen features detectors.

Figure 3. MATLAB built in (left) and the rewritten version of Harris detector

Figure 4. MATLAB built in (left) and the rewritten version of Eigen features detector

Figure 5. MATLAB built in (left) and the rewritten version of Harris detector

Figure 6. MATLAB built in (left) and the rewritten version of Eigen features detector

The comparison includes the difference of x and y coordinates, the response value, and the computational cost for each corner points between the MATLAB built in detectors and the rewritten version detectors. The comparison results are shown in table 1 and 2. Based on table 1 and 2, the error difference comparison between x and y coordinates in both Harris and Eigen features detector is less than 0.4% for all case studies. In addition, both rewritten detectors are able to find the same feature points as in MATLAB built in detectors.

Table 1. Comparison of x and y coordinates, response and computational cost of Harris detector

Image	1	2	3	4	5	6
X coordinate error (%)	0.30	0.10	0.32	0.24	0.19	0.16
Y coordinate error (%)	0.24	0.08	0.39	0.15	0.18	0.10
Response error (%)	4.3	5.1	3.7	4.3	4.6	5.1
Computational cost error (%)	0.7	1.5	1.6	1.5	0.2	1.6

Table 2. Comparison of x and y coordinates, response and computational cost of Eigen detector

Image	1	2	3	4	5	6
X coordinate error (%)	0.28	0.13	0.28	0.25	0.33	0.19
Y coordinate error (%)	0.22	0.07	0.32	0.22	0.20	0.23
Response error (%)	3.7	3.1	3.6	3.7	4.1	3.8
Computational cost error (%)	1.8	1.1	2.5	1.6	2.3	2.9

Furthermore, the maximum difference of response value is around 5% in Harris detector comparison, while in Eigen features detector comparison is around 4%. Although these two values are at least 10 times higher than coordinates error, it is still small in value and acceptable. Besides, it can be observed that small deviation in coordinates can lead to large deviation in corner point response value. Also, the difference of computational cost is less than 3% in both Harris and Eigen features detectors case studies, which is satisfactory.

4. Conclusion

In conclusion, the difference in detectors of MATLAB built in and the rewritten version in terms of coordinates are less than 0.4%. In terms of response value are around 5% for Harris detector and 4% for Eigen features detector. While in terms of computational costs are less than 3% in both detectors. Thus after analysing the corner points' coordinates, response value, and computation cost, the rewritten version of detectors is validated and can be considered to have similar performance as MATLAB's built in detectors. Both rewritten detectors can be used in research, either as comparison with other detectors, or for modification and improvement in detectors.

Acknowledgement

This publication was supported by Universiti Sains Malaysia Grant No. 304/PAERO/6315002.

References

[1] Verkeenko M 2014 Development of an algorithm for fast corner points detection *Journal of Computer and Systems Sciences International* **53** 392-401

[2] Sinzinger E D 2008 A model-based approach to junction detection using radial energy *Pattern Recognition* **41** 494-505

IOP Conf. Series: Materials Science and Engineering **370** (2018) 012013 doi:10.1088/1757-899X/370/1/012013

[3] Vasco V, Glover A and Bartolozzi C 2016 Fast event-based Harris corner detection exploiting the advantages of event-driven cameras. In: *Intelligent Robots and Systems (IROS), 2016 IEEE/RSJ International Conference on*: IEEE) pp 4144-9

[4] Parks D and Gravel J-P 2004 Corner detection *International Journal of Computer Vision* 1-17

[5] Shi J 1994 Good features to track. In: *Computer Vision and Pattern Recognition, 1994. Proceedings CVPR'94., 1994 IEEE Computer Society Conference on*: IEEE) pp 593-600

[6] Harris C and Stephens M 1988 A combined corner and edge detector. In: *Alvey vision conference*: Manchester, UK) p 10.5244

[7] Karthik O, Varun D and Ramasangu H 2016 Localized Harris-FAST interest point detector. In: *India Conference (INDICON), 2016 IEEE Annual*: IEEE) pp 1-6

[8] Li J and Allinson N M 2008 A comprehensive review of current local features for computer vision *Neurocomputing* **71** 1771-87

[9] Wu B, Zhou S and Ji K 2016 A novel method of corner detector for SAR images based on Bilateral Filter. In: *Geoscience and Remote Sensing Symposium (IGARSS), 2016 IEEE International*: IEEE) pp 2734-7

[10] Přibyl B, Chalmers A, Zemčík P, Hooberman L and Čadík M 2016 Evaluation of feature point detection in high dynamic range imagery *Journal of Visual Communication and Image Representation* **38** 141-60

[11] Elfakiri Y, Khaissidi G, Mrabti M, Chenouni D and El Yacoubi M 2016 Word spotting for handwritten Arabic documents using Harris detector. In: *Information Technology for Organizations Development (IT4OD), 2016 International Conference on*: IEEE) pp 1-4

[12] Orguner U and Gustafsson F 2007 Statistical characteristics of harris corner detector. In: *Statistical Signal Processing, 2007. SSP'07. IEEE/SP 14th Workshop on*: IEEE) pp 571-5

[13] Hassaballah M, Abdelmgeid A A and Alshazly H A 2016 *Image Feature Detectors and Descriptors*: Springer) pp 11-45

[14] Hauler F, Furtado H, Jurisic M, Polanec S, Spick C, Laprie A, Nestle U, Sabatini U and Birkfellner W 2016 Automatic quantification of multi-modal rigid registration accuracy using feature detectors *Physics in medicine and biology* **61** 5198

Effect of engine location on flutter speed and frequency of a tapered viscoelastic wing

Y S Matter[1], T T Darabseh[1] and A-H I Mourad[1]

[1] Mechanical Engineering Department, United Arab Emirates University, P.O. Box 15551, Al Ain, UAE

E-mail: 201570005@uaeu.ac.ae

Abstract. The flutter of a tapered viscoelastic wing carrying an engine and subjected to a follower thrust force is investigated. The wing is considered as a cantilever tapered Euler-Bernoulli beam, made of a linear viscoelastic material where Kelvin-Voigt model is assumed to represent the viscoelastic behaviour of the material. In addition, quasi-steady and unsteady aerodynamic forces models are introduced along with the follower thrust force. The mass and inertia of the engine are modelled in order to achieve more realistic behaviour of the engine upon flutter characteristics of the system. Moreover, the governing equations of motion are derived through the Extended Hamilton's Principle. The generalized function theory is used to more accurately consider the contribution of the mass and its follower force in the governing equations. The resulting partial differential equations are solved by Galerkin method along with the classical flutter investigation approach. Parametric studies highlighting the sensitivity of the chord-wise engine location, the span-wise engine location, and the vertical engine location on the flutter speed and flutter frequency are reported. It is found that the location of the engine in the three directions play an important role in the dynamic stability of the wing.

1. Introduction

Aeroelasticity is the term used to denote the field of study concerned with the interaction among aerodynamics, elasticity, inertia forces, and the phenomena that can result. Aeroelastic phenomena are usually classified as being either static (as the divergence) or dynamic (as the flutter). Divergence is a phenomenon that occurs when the moments resulting from aerodynamic forces overcome the elastic restoring forces due to structural stiffness. The flutter is defined as a dynamic lack of stability that occurs in a flexible structure subjected to aerodynamic loads at certain speed and frequency - called the flutter speed and the flutter frequency - which cause the structure to undergo divergent oscillations.

Bending-torsion aeroelastic instabilities have been investigated by many authors. References [1] and [2] studied the flutter phenomenon of a uniform wing by analysing a set of partial differential equations governing the motion of the wing. The use of quasi-steady aerodynamic theory for aeroelastic analysis of the lifting surfaces is a good approximation in the field of aeroelasticity as found in [3-6], and others. In Reference [7], a systematic approach based on Galerkin method was developed to investigate the flutter speed and frequency for a wing subjected to quasi-steady aerodynamic forces. The quasi-steady aerodynamic model can be used for low frequencies with acceptable results as investigated in [8]. However, Reference [9] observed that the quasi-steady

IOP Conf. Series: Materials Science and Engineering **370** (2018) 012014 doi:10.1088/1757-899X/370/1/012014

aerodynamic theory give inaccurate flutter results for a lifting surface under incompressible flow compared to the unsteady aerodynamic model.

Viscoelastic materials are typically used for enhancing damping and reduce structural vibration. Reference [10] analysed the dynamic response of Kelvin-Voigt viscoelastic simply supported beam. It was shown that the flutter speed associated with the viscoelastic wing might be greater or lower than that associated with the elastic wing. In addition, it was pointed out that the use of viscoelastic materials has either stabilization or destabilization contributions on the response of the structure, as in [11-15].

The engine's thrust, which acts as a non-conservative follower force on a mass attached to the wing, may affect the behaviour of the wing vibration. Many authors have studied bending-torsional flutter of elastic systems exposed to such non-conservative follower forces. For elastic systems, two books have been presented discussing comprehensively the dynamic stability of conservative and non-conservative elastic systems [16, 17]. By applying a method of expansion of fractional power of parameters, Reference [18] used a cantilever viscoelastic bar subjected to a tip follower force to show that, for realistic damping behaviour, there is a part of quasi-stability region should be added to the instability region. References [19] and [20] examined the effects of a transverse follower force on stability regions of HALE wing; however, they did not account for the inertia properties of the engine and considered only one location along the wing. The dynamic stability of wings carrying external stores and subjected to a lateral follower force was examined by [21]. The study observed that the engine mass, thrust, and location are of great influence on the dynamic stability of the aircraft wing. The dynamic instability of a cantilever composite wing with an attached mass subjected to follower force representing the thrust of the engine was studied in [22]. Therein, it was reported that the ply angle, engine location, the magnitude of engine mass and thrust have significant effects on the aeroelastic stability of the composite wings.

The objective of this work is to examine the dynamic response and determine regions of stability and flutter speeds and frequencies of a tapered viscoelastic wing carrying an engine and subjected to a follower thrust force modelled as a cantilever tapered Euler-Bernoulli beam. Furthermore, a non-dimensional parametric study is conducted to investigate the effect of the engine location on the flutter speed and frequency. The governing equations of motion are developed using Hamilton's principle with the quasi-steady and unsteady models for the aerodynamic forces. The equations of motion are solved using Galerkin method and classical flutter investigation procedure.

2. Formulation

2.1. Wing Model

The cantilever viscoelastic tapered wing carrying a concentrated engine mass and subjected to follower thrust force and aerodynamic loading is shown in Figure 1(a) with x_e denotes the engine location along the span. Furthermore, the deformed typical section of the wing is modelled in Figure 1(b) with y_a is the chord-wise distance between the wing centre of gravity (Cg_w) and the elastic axis and y_0 is the chord-wise distance from the wing's leading edge to the elastic axis. The chord-wise distance between the engine centre of gravity (Cg_e) and the elastic axis of the wing is denoted by y_e and the vertical distance from the engine centre of gravity to the wing chord line is denoted by z_e. The points *AE* and *AC* refer to the wing elastic axis and the aerodynamic centre of the wing, respectively.

The coordinate axes x, y, z are fixed on the wing root in which the x-axis lies exactly on the elastic axis and directed along the length of the wing in the span-wise direction. The orthogonal axes x\`, $y$\`, z\` are attached to represent the wing after deformation. The external loads cause the wing to be deformed such that the elastic axis of the wing is moved by an amount of h in the z-direction. Moreover, the wing rotates about its elastic axis by an angle of θ. Therefore, the system has two degrees of freedom.

It is important to note that the engine thrust (denoted here by P) is exerted exactly on the engine Cg and directed along the chord-wise direction of the wing. This gives the ability to recognize the thrust of the engine as a transverse follower force.

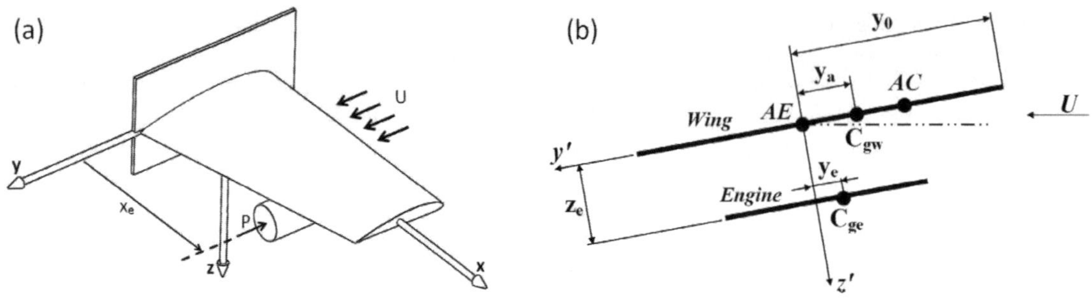

Figure 1. (a) The wing configuration and (b) its typical section model.

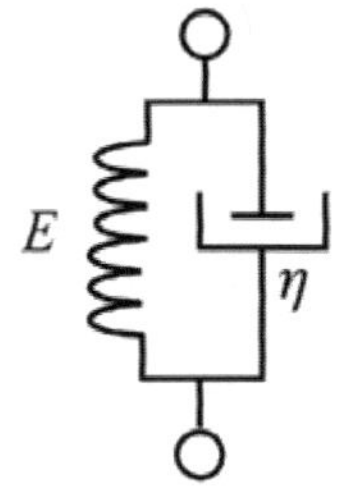

Figure 2. Kelvin-Voigt model.

2.2. Equations of Motion

The wing is made of a linear viscoelastic material, where the stress is linearly proportional to the strain history. Various models were constructed to represent the viscoelastic behaviour of the material. It was predicted that the spring–dashpot models are useful to conceive how viscoelastic behaviour can arise. Kelvin-Voigt model (as shown in Figure 2) is used to describe the behaviour of the viscoelastic material. The Kelvin–Voigt model consists of a spring and dashpot connected in parallel. Because the two elements are subjected to the same strain, the total stress is the sum of stress in each element, so

$$\frac{\sigma_n}{\varepsilon} = \{E + \eta_E \partial_t\} = E^* \tag{1}$$

Similarly

$$\frac{\sigma_s}{\gamma} = \{G + \eta_G \partial_t\} = G^* \tag{2}$$

where $\partial_t = \frac{\partial}{\partial t}$, σ_n, and σ_s are the normal and shear stresses, respectively; E and G are Young's and the shear moduli, respectively; η_E, and η_G are the coefficients of viscous damping forces in bending and torsion, respectively; and ε and γ are the normal and the shear strains, respectively.

The equations of motion are derived via the variational Hamilton's principle that can be expressed as

$$\int_{t1}^{t2} [\delta U - \delta T_w - \delta T_e - \delta W_A - \delta W_F] = 0 \tag{3}$$

where U here is the potential energy (not the air speed), T_w is the kinetic energy of the wing, T_e is the kinetic energy of the engine, W_A is the virtual work of the distributed aerodynamic loads, W_F is the virtual work of the concentrated engine thrust, and δ is the variational operator.

The first variation of the potential energy of the wing is given by

$$\delta U = \frac{1}{2}\int_0^l [2G^*J\theta'\delta\theta' + 2E^*Ih''\delta h'' + 2P(x_e - x)H(x_e - x)\theta\delta h'' + 2P(x_e - x)H(x_e - x)h''\delta\theta]dx \tag{4}$$

where E^*I and G^*J are the wing bending and torsional rigidities, respectively. $H(x_e - x)$ is the Heaviside function, which is used in order to account for the location of the engine thrust force.

The first variation of the wing kinetic energy is given by

$$\delta T_w = \int_0^l \frac{1}{2}[2m\dot{h}\delta\dot{h} + 2my_a\dot{\theta}\delta\dot{h} + 2my_a\dot{h}\delta\dot{\theta} + 2I_{EA}\dot{\theta}\delta\dot{\theta}]dx \tag{5}$$

where m is the wing mass per unit span, I_{EA} is the wing moment of inertia about the elastic axis.

The first variation of the engine kinetic energy is given by

$$\delta T_e = \int_0^l [M_e(z_e^2\dot{h}'\delta\dot{h}' + z_e^2\dot{\theta}\delta\dot{\theta} + \dot{h}\delta\dot{h} + y_e\dot{h}\delta\dot{\theta} + y_e\dot{\theta}\delta\dot{h} + y_e^2\dot{\theta}\delta\dot{\theta}) + I_{Me}\dot{\theta}\delta\dot{\theta}]\delta_D(x_e - x)dx \tag{6}$$

where M_e is the engine mass, I_{Me} is the engine moment of inertia, and $\delta_D(x_e - x)$ is the Dirac-Delta function, which is used in order to precisely account for the location of the engine mass along the wing span.

The variation of the virtual work of the distributed aerodynamic loads is given by

$$\delta W_A = \int_0^l (L\delta h + M\delta\theta)\,dx \tag{7}$$

where L and M are the aerodynamic lift force and twisting moment per unit span, respectively.

The variation of the virtual work of the concentrated engine thrust is given by

$$\delta W_F = \int_0^l \{P\delta_D(x_e - x)\theta\delta h + [(Py_e\theta - Pz_e)\delta_D(x_e - x)]\delta\theta\}\,dx \tag{8}$$

Using the previous expressions for the variation of the potential and kinetic energies and the variation of the virtual work as well as using Kelvin-Voigt model to represent the viscoelastic behaviour of the material, the equations of motion are obtained as

$$m\ddot{h} + my_a\ddot{\theta} + (EIh'')'' + (\eta_E I\dot{h}'')'' + P(x_e - x)H(x_e - x)\theta'' - 2PH(x_e - x)\theta' + [M_e\ddot{h} + M_e y_e\ddot{\theta} - M_e z_e^2\ddot{h}'' - P\theta]\delta_D(x_e - x) = -L \tag{9}$$

$$I_{EA}\ddot{\theta} + my_a\ddot{h} - (GJ\theta')' - (\eta_G J\dot{\theta}')' + P(x_e - x)H(x_e - x)h'' + \left[M_e y_e\ddot{h} + (I_{Me} + M_e(z_e^2 + y_e^2))\ddot{\theta} + Pz_e - Py_e\theta\right]\delta_D(x_e - x) = M \tag{10}$$

2.3. Aerodynamic Models

In this study, the quasi-steady and unsteady models for subsonic 2-dimensional flow are considered to represent the aerodynamic forces about the elastic axis. The lift and moment equations based on the quasi-steady model are [3]

$$L_{QS} = \frac{\rho_\infty U^2}{2}c\frac{dC_L}{d\theta}\left[\theta + \frac{\dot{h}}{U} + \frac{c}{U}\left(\frac{3}{4} - \frac{y_0}{c}\right)\dot{\theta}\right] \tag{11.a}$$

$$M_{QS}=\frac{\rho_\infty U^2}{2}c^2\left\{-\frac{c\pi}{8U}\dot{\theta}+\left(\frac{y_0}{c}-\frac{1}{4}\right)\frac{dC_L}{d\theta}\left[\theta+\frac{\dot{h}}{U}+\frac{c}{U}\left(\frac{3}{4}-\frac{y_0}{c}\right)\dot{\theta}\right]\right\} \tag{11.b}$$

Here, c is the chord length and ρ_∞ and U are the density and speed of air, respectively. The term $dC_L/d\theta$ is considered to be constant, with a value of 2π obtained theoretically for incompressible flow.

In the unsteady aerodynamic model, the lift and moment equations are [23, 24]

$$L=\frac{\pi\rho_\infty c^2}{4}\left[\ddot{h}+U\dot{\theta}-\left(y_0-\frac{c}{2}\right)\ddot{\theta}\right]+\frac{\rho_\infty U^2}{2}c\frac{dC_L}{d\theta}C(k)\left[\theta+\frac{\dot{h}}{U}+\frac{c}{U}\left(\frac{3}{4}-\frac{y_0}{c}\right)\dot{\theta}\right] \tag{12.a}$$

$$M=\frac{\pi\rho_\infty c^2}{4}\left[\left(y_0-\frac{c}{2}\right)\ddot{h}-U\left(\frac{3c}{4}-y_0\right)\dot{\theta}-c^2\left[\frac{9}{32}+\frac{y_0}{c}\left(\frac{y_0}{c}-1\right)\right]\ddot{\theta}\right]$$
$$+\frac{\rho_\infty U^2}{2}c^2\frac{dC_L}{d\theta}C(k)\left(\frac{y_0}{c}-\frac{1}{4}\right)\left[\theta+\frac{\dot{h}}{U}+\frac{c}{U}\left(\frac{3}{4}-\frac{y_0}{c}\right)\dot{\theta}\right] \tag{12.b}$$

where $C(k)$ is the Theodorsen's function, which can be approximated by [23]

$$C(k)=1-\frac{0.165}{1-\frac{0.045}{k}i}-\frac{0.335}{1-\frac{0.30}{k}i},\qquad k\le 0.5, \tag{13.a}$$

$$C(k)=1-\frac{0.165}{1-\frac{0.041}{k}i}-\frac{0.335}{1-\frac{0.32}{k}i},\qquad k>0.5 \tag{13.b}$$

where i is the imaginary unit and k is the reduced frequency, given by

$$k=\frac{\omega c}{U} \tag{14}$$

The final governing equations of motion, for the tapered viscoelastic cantilever wing subjected to bending and torsion loading governed by the aerodynamic strip theory, can be obtained for both aerodynamic models.

For the Quasi-Steady model, substitute equation (11.a) in (9) and equation (11.b) in (10). For the Unsteady model, substitute equation (12.a) in (9) and equation (12.b) in (10).

3. Solution procedure

In this study, a beam model that tapers in one plane, namely the xy-plane, is considered. Therefore, the general equations for the cord length $c(x)$, the wing mass per unit length $m(x)$, the elastic axis location $y_0(x)$, the offset between the elastic axis and the wing centre of gravity $y_a(x)$, the moment of inertia $I(x)$, the polar moment $J(x)$, the bending rigidity $EI(x)$, the torsional rigidity $GJ(x)$, and the mass moment of inertia per unit length $I_{EA}(x)$ of the wing are given by

$$\beta(x)=\beta_r\left(1-c_t\frac{x}{l}\right) \tag{15}$$

where β represents the parameter, the subscript r refers to the value of that parameter at the wing root and c_t represents the taper ratio.

Throughout the numerical analysis, the following non-dimensional parameters are introduced

$$\omega_\theta = \sqrt{\frac{GJ_r}{I_{EA_r} l^2}} \qquad \omega_h = \sqrt{\frac{EI_r}{m_r l^4}} \qquad r_a = \sqrt{\frac{I_{EA_r}}{m_r c_r^2}}$$

$$\sigma = \frac{\omega_h}{\omega_\theta} \qquad \lambda = \frac{EI_r}{GJ_r} \qquad \omega^* = \frac{\omega}{\omega_\theta}$$

$$\xi = \frac{x}{l} \qquad \mu = \frac{m_r}{\rho c_r^2} \qquad AR = \frac{l}{c_r}$$

$$x_e^* = \frac{x_e}{l} \qquad y_e^* = \frac{y_e}{c_r} \qquad z_e^* = \frac{z_e}{l}$$

$$M_e^* = \frac{M_e}{m_r l} \qquad P^* = \frac{Pl^2}{\sqrt{EI_r GJ_r}} \qquad V = \frac{U}{c_r \omega_\theta}$$

$$\eta_E^* = \frac{\eta_E I_r}{\sqrt{EI_r\, m_r l^4}} \qquad \eta_G^* = \frac{\eta_G J_r}{\sqrt{GJ_r I_{EA_r} l^2}}$$

Due to the complexity of the partial differential equations of motion, a closed form solution cannot be found. However, the solution can be investigated via Galerkin approximate solution technique. At the flutter boundary, the equations of h and θ are

$$h(\xi,t) = \bar{h} f(\xi) e^{i\omega t} \tag{16}$$

$$\theta(\xi,t) = \bar{\theta}\phi(\xi) e^{i\omega t} \tag{17}$$

where $\bar{h}$ and $\bar{\theta}$ are the amplitudes of motion, and are dimensionally the same as h and θ respectively. ω represents the frequency of harmonic oscillations occurring at the flutter boundary. $f(\xi)$ and $\phi(\xi)$ (given below) represent the uncoupled bending and torsion mode shapes of a cantilever beam, respectively.

$$f(\xi) = [\cosh(1.875\xi) - \cos(1.875\xi)] - 0.734[\sinh(1.875\xi) - \sin(1.875\xi)] \tag{18}$$

$$\phi(\xi) = \sin\left(\frac{\pi\xi}{2}\right) \tag{19}$$

Substituting equations (16) and (17) into equations (9) and (10), respectively and multiplying equation (9) by f and equation (10) by ϕ and integrating over the wing span, a set of two algebraic equations are obtained; which can be written in matrix form as

$$\begin{bmatrix} (A_1) & (B_1) \\ (A_2) & (B_2) \end{bmatrix} \begin{bmatrix} \bar{h} \\ \bar{\theta} \end{bmatrix} = 0 \tag{20}$$

The coefficients A_i and B_i depend on which aerodynamic model is used. Therefore, there are two sets of the coefficients A_i and B_i as given below. The determinant of the coefficient matrix has to vanish for nontrivial solution. With the determinant being complex in general, both its real and imaginary parts must vanish. This leads to two equations with two unknowns U_{cr} and ω, which are the flutter speed and flutter frequency, respectively.

- For the quasi-steady aerodynamic model:

$$A_1 = a_{11} + i\omega a_{12} - \omega^2 a_{13} - \omega^2 a_{14} + \omega^2 a_{15} + i\omega U_{cr} a_{16}$$

$$B_1 = -\omega^2 b_{11} + b_{12} - b_{13} - \omega^2 b_{14} - b_{15} + U_{cr}^2 b_{16} + i\omega U_{cr} b_{17}$$

$$A_2 = \omega^2 a_{21} - a_{22} + \omega^2 a_{23} + i\omega U_{cr} a_{24}$$

$$B_2 = b_{21} + i\omega b_{22} + \omega^2 b_{23} + \omega^2 b_{24} + b_{25} + U_{cr}^2 b_{26} + i\omega U_{cr} b_{27}$$

- For the unsteady aerodynamic model:

$$A_1 = a_{11} + i\omega a_{12} - \omega^2 a_{13} - \omega^2 a_{14} + \omega^2 a_{15} + i\omega U_{cr} C(k) a_{16} - \omega^2 a_{17}$$

$$B_1 = -\omega^2 b_{11} + b_{12} - b_{13} - \omega^2 b_{14} - b_{15} + U_{cr}^2 C(k) b_{16} + i\omega U_{cr} C(k) b_{17} + i\omega U_{cr} b_{18} + \omega^2 b_{19}$$

$$A_2 = \omega^2 a_{21} - a_{22} + \omega^2 a_{23} + i\omega U_{cr} C(k) a_{24} - \omega^2 a_{25}$$

$$B_2 = b_{21} + i\omega b_{22} + \omega^2 b_{23} + \omega^2 b_{24} + b_{25} + U_{cr}^2 C(k) b_{26} + i\omega U_{cr} C(k) b'_{27} - i\omega U_{cr} b_{28} + \omega^2 b_{29}$$

where the coefficients a_{ij} and b_{ij} are given in the appendix.

Galerkin method can be effectively applied to aeroelastic analysis because of its versatility. The implementation of Galerkin-method-based aeroelastic analysis is developed entirely within a numerical code. The numerical values of the dimensions and material mechanical properties of the model used in this study are listed in Table 1 as found in [1].

Table 1. Properties of Goland's wing at root section.

Parameter	Value
Wing span	20 ft.
Chord	6 ft.
Bending rigidity	23.6 x 10^6 lbf.ft^2
Torsional rigidity	2.39 x 10^6 lbf.ft^2
Mass of the wing per unit length	0.746 slugs/ft.
Mass moment of inertia	1.943 slugs.ft^2/ft.
Elastic axis position	33% of the chord
Inertial axis position	43% of the chord
Air density	0.00237 slugs/ft^3

4. Results and discussion

4.1. Validation

To verify the accuracy of the model, the results for a clean elastic straight wing (in the absence of the engine, taper ratio, and viscoelastic damping of the material) are validated against those in Reference [9] for both aerodynamic models, and good agreement is observed as seen in Table 2. The slight difference is due to the fact that they used 'p' and 'p-k' methods in their study whereas the determinant method is utilized in this work.

Table 2. Validation of flutter speed and frequency for a clean straight wing.

Aerodynamic Model		Velocity (ft/sec)	Frequency (rad/sec)
Ref. [9]	Quasi-Steady	110	93
	Unsteady	449	70
Present work	Quasi-Steady	110.36	94
	Unsteady	464.9	68.6
Exact [1]	Unsteady	451	70.7

4.2. Effect of span-wise engine location

The influence of the span-wise engine location (x_e) on the non-dimensional flutter speed and frequency for a straight elastic wing ($c_t = \eta_E = \eta_G = 0$) is shown in Figure 3. The engine's Cg is located exactly on the elastic axis of the wing ($y_e = 0$) with $z_e = 0$, $P = 1$, and $M_e = 1$. Figure 3 reveals that, for the unsteady aerodynamic model, moving the engine from the wing root to almost 40% of the wing span ($x_e = 0.4$) decreases the flutter speed. As the engine slides further towards the wing tip, the flutter speed dramatically increases. For the quasi-steady model, the flutter speed slightly decreases as the engine moves from the wing root to around 25% of the span ($x_e = 0.25$), further sliding towards the wing tip increases the flutter speed. These results show agreement with those in [22]. The flutter frequency decreases by moving the engine towards the wing tip for the two aerodynamic models. It is also observed that the quasi-steady aerodynamic model provide more conservative results than the unsteady model although the two models give almost the same behaviour.

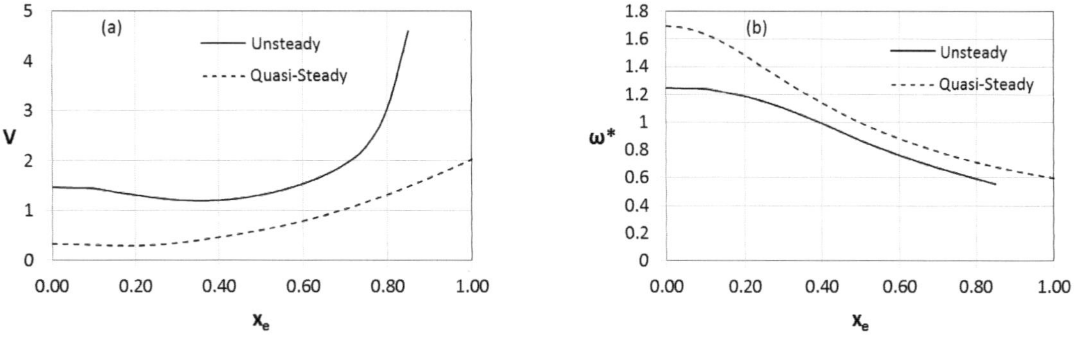

Figure 3. Effect of span-wise engine location on (a) non-dimensional flutter speed and (b) non-dimensional flutter frequency for straight elastic wing ($P = 1$, $M_e = 1$).

4.3. Effect of chord-wise engine location

Figure 4 demonstrates the effect of the chord-wise location of the engine (y_e) on the non-dimensional flutter speed and frequency for a straight elastic wing ($c_t = \eta_E = \eta_G = 0$). The engine is located at quarter-span ($x_e = 0.25$) with $z_e = 0$, $P = 1$, and $M_e = 1$. The plot shows that, for the unsteady aerodynamic model, the flutter speed slightly increases as the engine slides from the wing leading edge up to 10% of the chord before the wing elastic axis ($y_e = -0.1$). Moving the engine further towards the wing trailing edge decreases the flutter speed. In addition, the quasi-steady aerodynamic model provides that the flutter speed decreases as the engine moves from the leading edge to the trailing edge. For both of the aerodynamic models, the flutter frequency increases as the engine slides from the leading edge to the trailing edge. Indeed, similar behaviour was obtained in [21] and [22], where in the former it was pointed out that moving the engine from trailing edge to the leading edge in chord-wise direction makes the wing more stable.

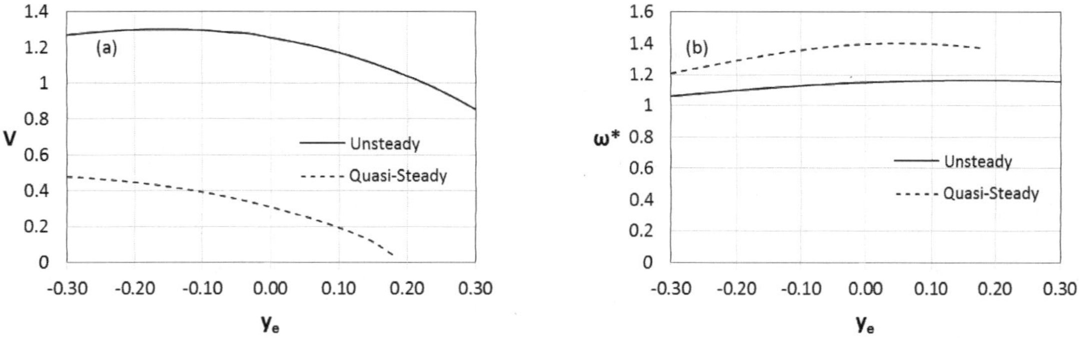

Figure 4. Effect of chord-wise engine location on (a) non-dimensional flutter speed and (b) non-dimensional flutter frequency for straight elastic wing ($P = 1$, $M_e = 1$).

4.4. *Effect of vertical engine location*

The effect of the vertical location of the engine (z_e) on the non-dimensional flutter speed and frequency for a straight elastic wing ($c_t = \eta_E = \eta_G = 0$) is illustrated in Figure 5. The engine is located at quarter-span ($x_e = 0.25$) with $y_e = 0$, $P = 1$, and $M_e = 1$. The quasi-steady aerodynamic model provides that the wing becomes more stable as the engine goes further below the wing. However, different behaviour is observed when the unsteady aerodynamic model is considered, where the wing becomes less stable as the engine goes further below the wing.

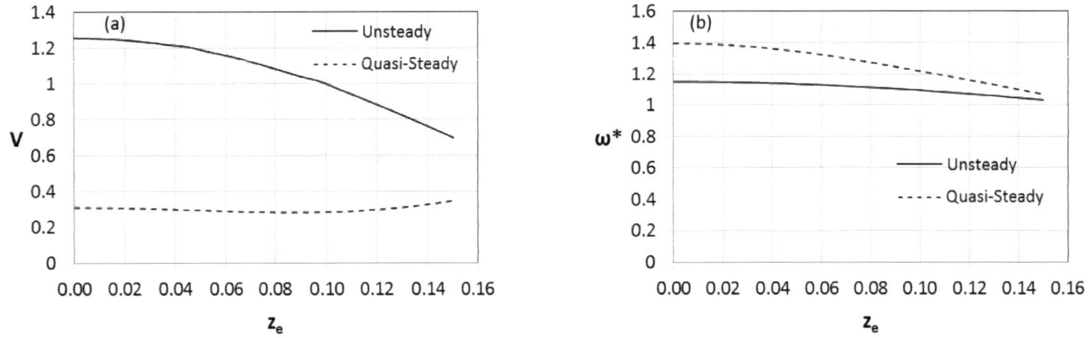

Figure 5. Effect of vertical engine location on (a) non-dimensional flutter speed and (b) non-dimensional flutter frequency for straight elastic wing ($P = 1$, $M_e = 1$).

5. Conclusions

Flutter is one of the significant phenomena of Aeroelasticity. This dangerous aeroelastic phenomenon can occur to any flexible structure subjected to aerodynamic forces such as aircraft wings, bridges, buildings, etc. It is important to analyse the flutter in order to predict the speed and frequency at which it occurs so that structural damages and failures can be avoided. By incorporating the effects of the engine location in the three directions, this work is a growing out of the aeroelastic research pertains the quasi-steady and unsteady aerodynamic formulations.

The purpose of this work is modelling and analysing a tapered viscoelastic wing carrying an engine and subjected to a follower thrust force as well as to aerodynamic forces from an aeroelastic viewpoint. The effects of the engine location in the three directions on the wing flutter characteristics have been studied.

The unsteady aerodynamic model provides more reliable results than the quasi-steady model which offers more conservative predictions. In most cases, the two models yield the same behaviour or at least the same trend. In some other cases, however, they give diverse results.

Parametric studies show that the engine location along the span, the chord-wise engine location, and the vertical location of the engine have significant effects on the flutter speed and frequency of the aircraft wing. Therefore, all these parameters must be taken into consideration to get accurate stability investigation results.

According to this work, the wing will become more stable if the engine is located away from the fuselage towards the wing tip, the engine centre of gravity is moved towards the wing leading edge and/or the engine is placed right below the wing (i.e. the vertical distance between the engine centre of gravity and the wing chord-line is minimal).

References

[1] Goland M 1945 The flutter of a uniform cantilever wing *J. Applied Mechanics* **12** pp A197-A208
[2] Goland M and Luke Y L 1948 The flutter of a uniform wing with tip weights *J. Applied Mechanics* **15** pp 13-20
[3] Fung Y C 1955 *An Introduction to the Theory of Aeroelasticity* 2nd ed. (New York: Dover Publications)

[4] Dowell E H 1967 Nonlinear oscillations of a fluttering plate II *AIAA J.* **5** pp 1856-62
[5] Dowell E H and Voss H M 1965 Theoretical and experimental panel flutter studies in the Mach number range 1.0 to 5.0 *AIAA J.* **3** pp 2292-304
[6] Meirovitch L 1975 *Elements of Vibration Analysis* (New York: McGraw-Hill)
[7] Moosavi M R, Oskouei A N and Khelil A 2005 Flutter of subsonic wing *Thin-walled structures* **43** pp 617-27
[8] Acum W E A 1959 *The Comparison of Theory and Experiment of Oscillating Wings* vol 2 AGARD manual on aeroelasticity
[9] Haddadpour H and Firouz-Abadi R D 2006 Evaluation of quasi-steady aerodynamic modeling for flutter prediction of aircraft wings in incompressible flow *Thin-walled structures* **44** pp 931-6
[10] Jia-ju S and Ke-hwa J 1981 Dynamic response of viscoelastic beam *Applied Mathematics and Mechanics* **2** pp 255-64
[11] Hilton H H 1957 Pitching instability of rigid lifting surfaces on viscoelastic supports in subsonic or supersonic potential flow *Proc. 3rd Midwestern Conf. on Solid Mechanics (Michigan)* pp 1-19
[12] Hilton H H 1960 The divergence of supersonic, linear viscoelastic lifting surfaces, including chordwise bending *J. Aerospace Sciences 27* pp 926-34
[13] Hilton H H 1991 Viscoelastic and structural damping analysis *Proc. on Damping'91 (Ohio)* Air Force Technical Report WL-TR-91 **3078** pp 1-15
[14] Ungar E E 1971 Chapter 14 in *Noise and Vibration Control* ed L L Beranek (New York: McGraw-Hill)
[15] Yi S Ahmad M F and Hilton H H 1996 Dynamic responses of plates with viscoelastic free layer damping treatment *J. vibration and acoustics* **118** pp 362-7
[16] Bolotin V V 1962 *The Dynamic Stability of Elastic Systems* vol 1 (California: Aerospace Corp El Segundo)
[17] Bolotin V V 1963 *Nonconservative Problems of the Theory of Elastic Stability* (Oxford: Pergamon Press)
[18] Bolotin V V and Zhinzher N I 1969 Effects of damping on stability of elastic systems subjected to nonconservative forces *Int. J. Solids and Structures* **5** pp 965-89
[19] Hodges D H 2001 Lateral-torsional flutter of a deep cantilever loaded by a lateral follower force at the tip *J. Sound and Vibration* **247** pp 175-83
[20] Hodges D H, Patil M J and Chae S 2002 Effect of thrust on bending-torsion flutter of wings *J. Aircraft* **39** pp 371-6
[21] Fazelzadeh S A, Mazidi A and Kalantari H 2009 Bending-torsional flutter of wings with an attached mass subjected to a follower force *J. Sound and Vibration* **323** pp 148-62
[22] Amoozgar M R, Irani S and Vio G A 2013 Aeroelastic instability of a composite wing with a powered-engine *J. Fluids and Structures* **36** pp 70-82
[23] Theodorsen T 1935 *General Theory of Aerodynamic Instability and the Mechanism of Flutter* NACA TR 496 (Washington: National Advisory Committee for Aeronautics)
[24] Hodges D H and Pierce G A 2011 *Introduction to Structural Dynamics and Aeroelasticity* 2nd ed. (New York: Cambridge University Press)

International Conference on Aerospace and Mechanical Engineering (AeroMech17) IOP Publishing
IOP Conf. Series: Materials Science and Engineering **370** (2018) 012015 doi:10.1088/1757-899X/370/1/012015

A Critical Review on Tubercles Design for Propellers

A Seeni, P Rajendran and H A Kutty

School of Aerospace Engineering, Universiti Sains Malaysia, Engineering Campus, Nibong Tebal, Pulau Pinang, Malaysia.

Email: aeparvathy@usm.my

Abstract. The design of propellers in aeronautics is still earning high interest. This has led to the application of biomimetic for improving aerodynamic performance. One such design is the usage of tubercles in either leading or trailing edge of airfoils. So the usage of tubercles has so far been in the design of wing or wind-turbines. Recent studies have aimed to study tubercles for the design of a propeller. The studies have so far gained reputations in improving the aerodynamic performance on a few but not all of the airfoil shapes. Similar results have been identified for propeller designs with improvement in thrust at low torque. We have so far led research works on developing new propeller designs with the aim to improve the thrust and torque performance of propeller. A review of recent studies has been performed to study the state-of-the-art. The results of this work will be discussed in this paper.

1. Introduction

Significant research has been witnessed in aircraft propeller design in recent years. Much of the research's focus can be attributed to straight-edged propellers [1]. Modern propeller research has focused on providing modifications to existing designs [2]. This has been achieved by development of new designs and modifications to either leading or trailing edge of propeller surface which are made up of airfoil cross-sections. One of these modifications amongst others is tubercles.

Tubercles are sinusoidal wavy formations on the leading edge (LE) that takes the form throughout the surface from leading to trailing edge (TE). The implementation of tubercles on modern airfoil surfaces originated from the works of Fish and Battle in 1995 [3]. Tubercles are essentially designs based on biomimetic. The pectoral flippers of humpback whales are found to provide improved aerodynamics with reduced drag while diving and catching prey (Figure 1). The tubercle design has so far been adapted to many but not limited to design of aircraft, marine propeller as well as wind turbine blades.

In this paper, a review of recent studies performed in the topic of tubercle propeller designs will be made. Although many studies have reported on this particular development in aircraft components such as wing, new studies on its usage on propeller design have emerged. Current research calls for an improvement in propeller design with high thrust, low torque characteristics. With this in mind, a detailed survey has been undertaken in order to review the state-of-the-art. The discussion on the same is divided into the following sections in this paper.

IOP Conf. Series: Materials Science and Engineering **370** (2018) 012015 doi:10.1088/1757-899X/370/1/012015

In section 2, recent research made in the field of designing propellers with tubercles will be discussed. In addition, a comprehensive survey of various studies related to tubercles that are not only limited to propellers but also on other airfoil applications will be provided. In section 3, a discussion of results from the various studies will be provided. In section 4, new improved designs will be suggested with a goal to provide future improvements. Section 5 provides conclusive remarks.

Figure. 1. Illustration of humpback whale's pectoral flipper with tubercles on leading edge [4].

2. Methodology

Most of the studies on tubercles so far have been on non-thrust generating designs of fixed aircraft wing and wind turbine and very few studies exist that have been implemented directly on wither aircraft or marine propellers. Studies on non-thrust generating platform try to improve lift and stall angle whereas propellers with tubercles have been studied to improve thrust generation. Recent research in this topic will be discussed subsequently.

2.1. Propellers designs

The study by Ibrahim and New [5] is on marine propellers designed with tubercles. Tubercles are implemented on LE of the airfoil. The wavelength, amplitude, airfoil, propeller diameter and blade number are fixed for a single design. The analysis is performed numerically in CFD for different advance ratios. The amount of performance in thrust and torque and aerodynamic efficiency improvements are predicted and compared with a baseline propeller. The baseline propeller is typically a propeller with straight LE of similar dimensions. The results of the study are provided in Table 1.

Table 1. Description of study by Ibrahim and New [5].

Propeller design	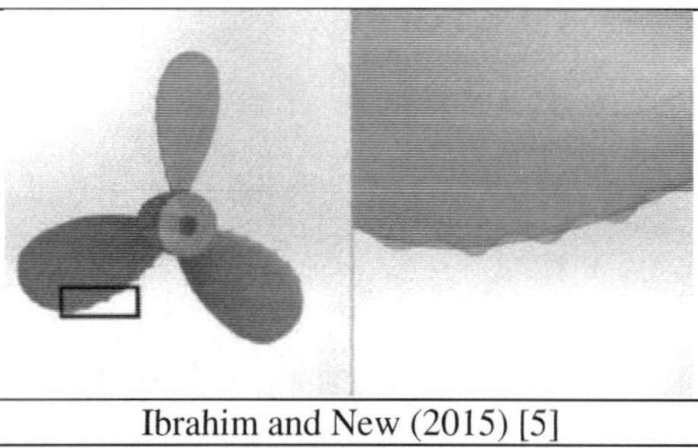
Author	Ibrahim and New (2015) [5]
Study type	Numerical
Enhancement	1.5% ↑ in thrust co-efficient (C_T)

The study by Moore and Ning [6] focuses on studying performance improvements of tubercles in small scale propellers. The design comprises of Clark-Y airfoil cross-sections. LE tubercular structure is in the shape of oblate spheroid. The sizes of the oblate spheroid are varied as well as the spacing between them and tested for 9 different cases in total. The widths are assumed to be 5, 6.25 and 8 percent of propeller radius. The tubercle intervals are assumed to be 7.14, 8.33, and 10 percent of the effective propeller radius. The height of the spheroid tubercle is assumed to be 2/3 of widths. All the 9 cases were tested for similar Reynolds number (RN) of $5x10^4$. The results of the study are summarized in Table 2.

Table 2. Description of study by Moore and Ning [6].

Propeller design	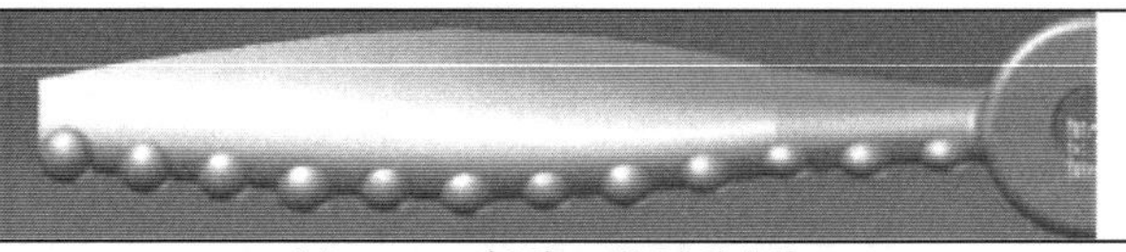
Author	Moore and Ning (2016) [6]
Study type	Experimental
Enhancement	C_T lower than baseline; Efficiency lower than baseline

2.2. *Other studies*

Other studies done so far were on wing and wind turbine models. The studies have tried to examine the aerodynamic performance improvement such as lift and drag and efficiency over unmodified straight-edged models. These studies will be discussed subsequently.

In 2001, Watts and Fish [7] have performed a numerical study on aircraft wing with tubercle LE modifications. Both aerodynamic performance and efficiencies are compared to a baseline model. The model with tubercles is assumed to be similar to that of the humpback whale flipper's cross-section resembling a NACA 634-021 airfoil. A free stream velocity (V_{inf}) of 1 m/s is assumed for the analysis. The results from the study are provided in Table 3.

Table 3. Description of study by Watts and Fish [7].

Propeller design	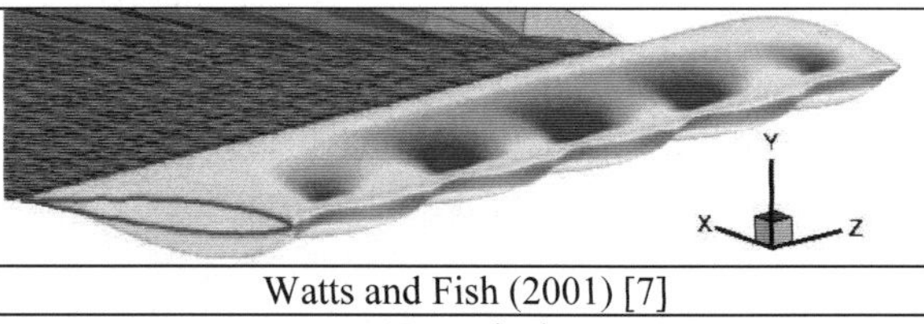
Author	Watts and Fish (2001) [7]
Study type	Numerical
Enhancement	4.8% ↑ in lift, a 10.9% ↓ in induced drag, 17.6% ↑ in lift to drag ratio

The study by Pedro and Kobayashi [8] is on studying tubercles for designing of wings of small Unmanned Aerial Vehicles (UAVs). The study consists of both numerical and experimental tests of two tubercled models, also known as flippers, namely scalloped and smooth. The model uses tapered wing tips. The tests were performed for a RN of $5x10^5$. The description and results of the study are summarized in Table 4.

Hansen et al. [9] has performed an experimental study on LE tubercles of a full and finite wing. A NACA 0021 airfoil section is assumed. The lift coefficients (C_L) and drag coefficients (C_D) are found for wings with tubercle wavelength variations of λ = 7.5, 15, 30 and 60. Also amplitude variation of 2 and 4 is considered for a fixed wavelength. Both C_L and C_D variations are studied and compared with unmodified airfoil. Also both 2D and 3D airfoil is considered with fixed RN at $1.2x10^5$. The results of the study are summarized in Table 5.

Table 4. Description of study by Pedro and Kobayashi [8].

Propeller design	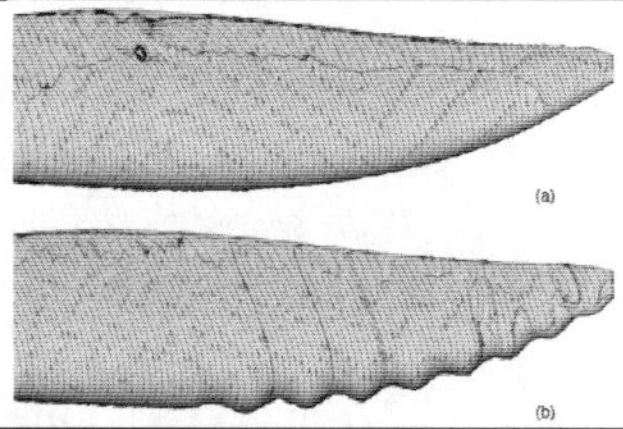
Author	Pedro and Kobayashi (2008) [8]
Study type	Numerical and Experimental
Enhancement	Scalloped flipper exhibited better performance than smooth model

Table 5. Description of study by Hansen et al. [9].

Propeller design	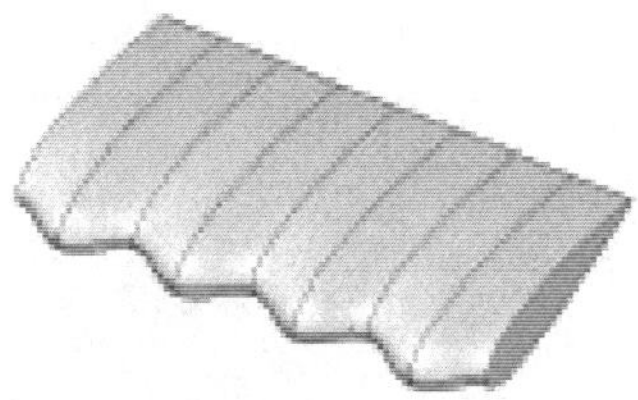
Author	Hansen et al. (2010) [9]
Study type	Experimental
Enhancement	$\uparrow$ in performance of tubercles for RN $\geq 5x10^5$; $\downarrow$ in performance for RN < $3x10^5$

The study by Lohry et al. [10] is on studying the aerodynamic performance of a 3D wing model with tubercled LE. NACA 0020 airfoil section is used in this study. The RN assumed for flow condition is from $6.25x10^4$ to $5x10^5$. The details of this study are provided in Table 6. The study by Xingwei et al. [11] is on studying tubercles for gliding and flapping flight of Micro Air Vehicles. A CFD approach is followed for studying tubercled airfoils with NACA 0014 cross-sections. A low RN of $1x10^4$ is assumed. The results of the study are also summarized in Table 7.

Table 6. Description of study by Lohry et al. [10].

Propeller design	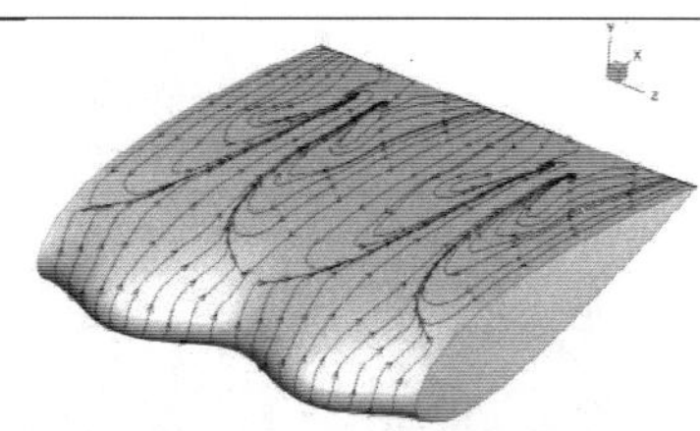
Author	Lohry et al. (2012) [10]
Study type	Numerical
Enhancement	4.8% $\uparrow$ in lift, a 10.9% $\downarrow$ in induced drag, 17.6% $\uparrow$ in lift to drag ratio

The study by Zhang et al. [12] is on the aerodynamics of LE protuberances with models operating at a V_{inf} of 7.5 m/s and RN of $5x10^4$. The models are of NACA 63_4-021 airfoil cross-sections. The aerodynamic loads, stall characteristics and stream-wise vortices were studied experimentally with fixed amplitude and wavelength. The details of this study are provided in Table 8. The study by Ibrahim et al. [13] is on studying both tubercles and slotted blades for wind turbine blades. Tubercles are implemented at the TE in this study. The test model consists of a blade with NACA 4412 sections and diameters of 60 cm. Different Vinf from 4 to 8 m/s are assumed. The results from the study are provided in Table 9.

Table 7. Description of study by Xingwei et al. [11].

Propeller design	
Author	Xingwei et al. (2013) [11]
Study type	Numerical
Enhancement	Gliding flight: Tubercles provide negligible effect compared to baseline at zero Angle of attack (AoA). At higher AoA, tubercles improve flight performance.

Table 8. Description of study by Zhang et al. [12].

Propeller design	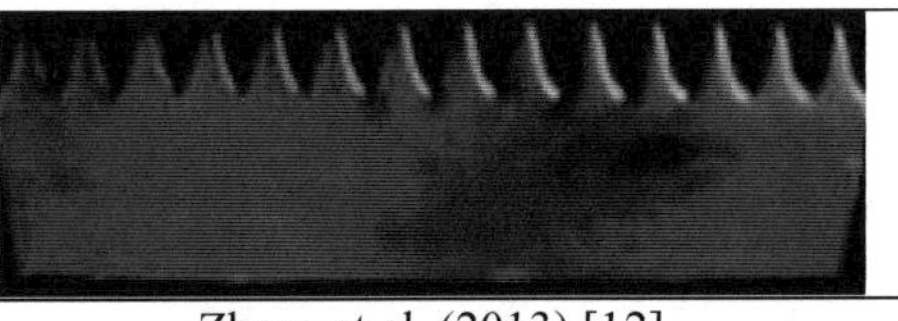
Author	Zhang et al. (2013) [12]
Study type	Numerical
Enhancement	25% ↑ in C_L. 20% ↑ in CD. 39.2 % ↑ in L/D. Aerodynamics performance decreased i.e. lessened lift in near stall region due to complicated flow modification compared to baseline.

Table 9. Description of study by Ibrahim et al. [13].

Propeller design	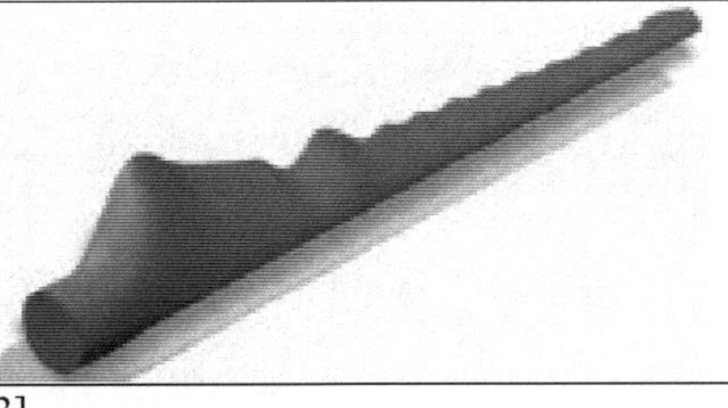
Author	Ibrahim et al. (2015) [13]
Study type	Experimental
Enhancement	Improved steady behavior for V_{inf} changes compared to baseline; Stall occurs at higher wind speed than baseline

The study by Rostamzadeh et al. [14] is on predicting performance using CFD on the effects of RN on two different tubercled models and straight-edged model. Both models are identical with one a scaled up version of the other and having similar NACA 0021 cross-sections. The models are compared to a straight-edged model of similar sizes. The study was performed for flow conditions of $1.2x10^5$ and $1.5x10^6$ RN. The results are provided in Table 10. The study by Serson et al. [15] on tubercles is on an infinite wing for a very low RN of $1x10^3$. The analyses was performed for 9 designs with different wavelengths and amplitude and compared to a baseline design. The results are provided in Table 11.

Table 10. Description of study by Rostamzadeh et al. [14].

Propeller design	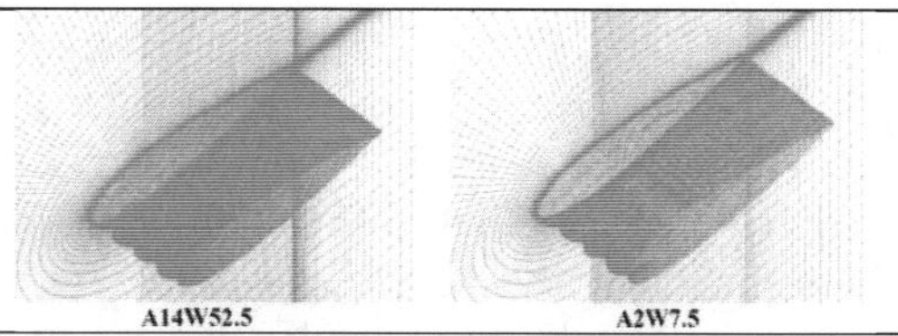
Author	Rostamzadeh et al. (2017) [14]
Study type	Numerical
Enhancement	Inferior performance of tubercled model with lower C_L post-stall, as compared to straight-edged models. In turbulent regime, the baseline experiences gradual stall and produces more lift than the tubercled model. Baseline model yields higher C_L at post-stall angles.

Table 11. Description of study by Serson et al. [15].

Propeller design	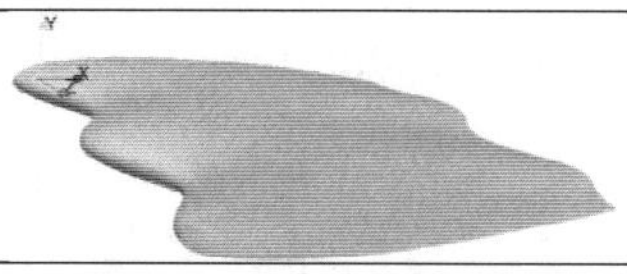
Author	Serson et al. (2017) [15]
Study type	Numerical
Enhancement	Weakening of suction peak causes reduction in AoA. Lower effective AoA caused by LE waviness leads to reduction in lift.

3. Discussion

The above discussed studies are categorically summarised in tabular form and compared in Table 12. The discussion of the results is two-fold. The aerodynamic improvement values for the designs considered are provided in terms of augmentation in thrust or power and torque. Studies that have discussed design of aircraft wing or turbine blades, the maximum lift co-efficient (C_{Lmax}) and C_D achieved in simulation or experiments is provided. The discussion on the same is provided subsequently:-

- Watts and Fish [7] and Zhang et al. [12] have performed tests on tubercles with NACA 63_4-021 airfoil sections. The results of both the studies indicate positive maximum increment of C_L of 4.8% and 25% respectively compared to baseline. Similarly, the C_D has shown a 10.9% and 20% decrement. A value of RN was not provided but tests measured at a V_{inf} of 1 m/s. The numerical results of Zhang et al. [12] were obtained for RN of $5x10^4$.
- The study by Miklosevic *et al.* [16] and Lohry et al. [10] is on tubercled models with NACA 0020 sections. Miklosevic *et al.* [16] has studied two different variants – scalloped and smooth whereas Lohry et al. [10] has performed analysis on point design. A C_{Lmax} of 0.93 and 0.88 is achieved for scalloped and smooth flipper model cases respectively by Miklosevic *et al.* [16]. In the case of Lohry et al. [10], the C_{Lmax} is approximately 0.9. 13% decrease in C_L is found at AoA 12^0 for smooth case. Both models have shown to develop higher C_D at 34% and 69%. Therefore lower aerodynamic performances have been observed in both. Irrespective of RN in both these studies, the above results has been found.
- The studies by Hansen *et al.* [9] and Rostamzadeh *et al.* [14] are on models with NACA 0021 airfoil cross-section. Both studies have been conducted for a RN assumption of $1.2x10^5$, while Rostamzadeh et al. [14] has considered high additional RN values of $1.2x10^5$ for a different model variant. Both C_L and C_D have shown a large decrement compared to baseline in the study.
- Xingwei et al. [11] have conducted tests on NACA 0014 airfoil cross-sectional models for flapping and gliding of MAVs. A higher C_{Lmax} was reached for tubercled model as compared to baseline. The RN assumed in $1x10^5$. The study by Serson et al. [15] also has similarity in these

observations, however for a lower order RN of $1\text{x}10^4$. A marginally higher C_{Lmax} is observed for AoA (α) of 21°.

- The study by Pedro and Kobayashi [8] on flipper models with hypothetical airfoil cross-section have yielded higher C_L for scalloped model as compared to smooth, although the stall AoA is delayed to a higher value which confirms observations made by Miklosevic *et al.* [16].
- The study by Ibrahim et al. [5] is on the usage of tubercles on TE. A pressure co-efficient (C_P) of 0.39 is achieved marginally higher over straight edged.
- Only two studies have studied the direct implementation of tubercles on propellers. Ibrahim et al. [5] studied tubercled LE propellers for marine applications on marine propellers. Clark-Y sections are considered. A 1.5% increase in C_T and 6% increase in Torque levels were observed. Moore and Ning [6] have focused on studying small scale propellers identified for UAVs. A C_T of 0.75 at advance ratio, *J*=0.1 and ~0% increase of C_T for all Advanced ratio (*J*) is observed. Both studies seem to agree that with increase in J the C_T gradually decreases. This is not so the case with propeller efficiency. As per Ibrahim et al. [5], the propeller efficiency increased with increasing *J* and the efficiency exceeds that of the baseline at *J*=0.85. However, the results seem to show negative feasibility of tubercles for UAV.

Table 12. Summary of results from studies.

Study	RN	Variant	Improvement				
			C_T / C_T	$C_{L\,max}$	$C_{D\,max}$	ΔC_L	ΔC_D
Watts and Fish [7]	V_{inf}=1m/s	Baseline	NA	-	-	NA	NA
		Tubercle	NA	-	-	4.8% ↑	10.9% ↓
Miklosevic *et al.* [16]	$5.05\text{x}10^5$ - $5.2\text{x}10^5$	Baseline	NA	-	-	NA	NA
		Scallop flipper	NA	0.93	-	-	34% ↑
		Smooth flipper	NA	0.88	-	13% ↓ at $\alpha=12^0$	69% ↑
Pedro and Kobayashi [8]	$5\text{x}10^5$	Scallop	NA	<1 at $\alpha=18^0$	0.25 at $\alpha=21^0$	-	-
		Smooth	NA	~0.9 at $\alpha=12^0$	~0.225 at $\alpha=18^0$	-	-
Hansen *et al.* [9]	$1.2\text{x}10^5$	Baseline	NA	>1 at $\alpha=12^0$	-	NA	NA
		Tubercle	NA	~0.95 at $\alpha=14^0$	-	↓ C_L	↓C_D
Lohry et al. [10]	$6.25\text{x}10^4$ - $5\text{x}10^5$	Tubercle	NA	~0.9 at $\alpha=15^0$	~0.24 at $\alpha=20^0$	-`	-
Zhang et al. [12]	$5\text{x}10^4$	Baseline	NA	-	-	NA	NA
		Tubercle	NA	-	-	25% ↑	20% ↑
Xingwei et al. [11]	$1\text{x}10^4$	Straight	NA	~0.125 at $\alpha=15^0$	<0.08 at $\alpha=15^0$	NA	NA
		Tubercle	NA	>0.2 at $\alpha=15^0$	<0.1 at $\alpha=15^0$	-	-
Ibrahim et al. [5]	$2.9\text{x}10^4$ - $5.3\text{x}10^4$	Tubercle	1.5% ↑ C_T 6% ↑ Torque	NA	NA	NA	NA
Moore and Ning [6]	$5\text{x}10^4$	Tubercle	↑ C_T for all *J*	NA	NA	NA	NA
Rostamzadeh et al. [14]	$15\text{x}10^5$	Baseline	NA	0.588	0.010	NA	NA
	$1.2\text{x}10^5$	A2W7.5	NA	0.165	0.020		
	$15\text{x}10^5$	A14W52.5	NA	0.588	0.010	0	0
Serson et al. [15]	$1\text{x}10^3$	Baseline	NA	>0.8 at $\alpha=21^0$	>0.4 at $\alpha=21^0$	NA	NA
		Tubercle	NA	>0.8 at $\alpha=21^0$ (>baseline)	>0.4 at $\alpha=21^0$ (>baseline)		
Ibrahim et al. [16]	Varied	Straight	Mean C_P 0.36	NA	NA	NA	NA
		Tubercle	Mean C_P 0.39	NA	NA	NA	NA

c=chord, ΔC_L=Lift co-efficient difference, ΔC_D=Drag co-efficient difference, NA=Not Applicable

From the above discussions, the following remarks can be made. It can be concluded that the usage of tubercles,

1. Is suitable for delaying stall and therefore higher AoA can be reached for $C_{L\,max}$.
2. Is for maintaining laminar flow during gust conditions. Therefore better performance during turbulence than straight-edged.
3. There is no concrete evidence of tubercles providing significant improvement in design for propeller design. For NACA 0020 and NACA 0021 sectional models, the lift and drag performance are found to be inferior compared to baseline straight-edged models. Only for NACA 63_4-021 airfoil sectional designs, a relatively higher lift and lower drag values have been achieved. This does not preclude the fact that for a propeller, the results showed only a marginal change in C_T over baseline in all test cases.
4. The design of propellers is driven by a large number of factors. More detailed studies are required in order to provide concrete evidence on the usage of tubercles for propeller design.

The usefulness of tubercles for propeller design is not yet well-established. Other design modifications such as slots on airfoil surface could be used as high lift, low drag devices. Slots are devices that employ modification of flow control. Slotted designs of propellers have not yet been studied extensively. New innovative designs in slots could be useful for improving existing designs of propeller.

4. Conclusion

A review of studies based on tubercled LE has been conducted. The most significant results that were obtained so far have been discussed. The results primarily present a not-so-highly optimistic case for tubercled propellers. Future work on propeller will be directed based on these findings and conclusion.

Acknowledgement

This publication was supported by Universiti Sains Malaysia Grant No. 304/PAERO/6315002.

References

[1] Kutty H A and Rajendran P 2017 Review on Numerical and Experimental Work of Conventional and Unconventional Propeller Blade Design International Review of Aerospace Engineering Volume 10 No 2.

[2] Kutty H A and Rajendran P 2017 3D CFD Simulation and Experimental Validation of Small APC Slow Flyer Propeller Blade Aerospace Volume 4 Issue 10.

[3] Fish, Frank; Battle J 1995 Hydrodynamic design of the humpback flipper J. Morphol. 225 51–60.

[4] Favier J, Pinelli A and Piomelli U 2012 Control of the separated flow around an airfoil using a wavy leading edge inspired by humpback whale flippers Comptes Rendus - Mec. 340 107–14.

[5] Ibrahim I H and New T H 2015 A numerical study on the effects of leading-edge modifications upon propeller flow characteristics Ninth International Symposium on Turbulence and Shear Flow Phenomena pp 1–6.

[6] Moore K R and Ning A 2016 Aerodynamic Performance Characterization of Leading Edge Protrusions on Small Propellers 54th AIAA Aerosp. Sci. Meet.

[7] Watts P and Fish F E 2001 The influence of passive, leading edge tubercles on wing performance Proc. Twelfth Intl. Symp. Unmanned Untethered Submers. Technol., Durham New Hampsh. 2–9.

[8] Carreira Pedro H and Kobayashi M 2008 Numerical Study of Stall Delay on Humpback Whale Flippers 46th AIAA Aerosp. Sci. Meet. Exhib. 7–10.

[9] Hansen K, Kelso R and Dally B 2010 An investigation of three-dimensional effects on the performance of tubercles at low reynolds numbers Australas. Fluid Mech. 1–4.
[10] Lohry M W, Clifton D and Martinelli L 2012 Characterization and Design of Tubercle Leading-Edge Wings 7th Int. Coneference Comput. Fluid Dyn. 1–11.
[11] Xingwei Z, Chaoying Z, Tao Z and Wenying J 2013 Numerical study on effect of leading-edge tubercles Aircr. Eng. Aerosp. Technol. 85 247–57.
[12] M.M., Zhang; G.F.., Wang, J.Z. X 2013 Effect of Humpback Whale Like Leading Edge Protuberances On the Low-Reynolds Number Airfoil Aerodynamics Fluid-Structure-Sound Interact. Control 107–13.
[13] Ibrahim M, Alsultan A, Shen S and Amano R S 2015 Advances in Horizontal Axis Wind Turbine Blade Designs : Introduction of Slots and Tubercle J. Energy Resour. Technol. 137 1–6
[14] Rostamzadeh N, Kelso R M and Dally B 2017 A numerical investigation into the effects of Reynolds number on the flow mechanism induced by a tubercled leading edge Theor. Comput. Fluid Dyn. 31 1–32.
[15] Serson D, Meneghini J R and Sherwin S J 2017 Direct numerical simulations of the flow around wings with spanwise waviness at a very low Reynolds number Comput. Fluids 146 117–24.
[16] Miklosovic D S, Murray M M, Howle L E and Fish F E 2004 Leading-edge tubercles delay stall on humpback whale (Megaptera novaeangliae) flippers Phys. Fluids 16.

IOP Conf. Series: Materials Science and Engineering **370** (2018) 012016 doi:10.1088/1757-899X/370/1/012016

Microvane in controlling noise in open cavity flow

M Z M Aliashak, M. R. Saad[1], A.T. Latif, A. Che Idris and M. R. A. Rahman

Department of Mechanical Engineering, Faculty of Engineering, Universiti Pertahanan Nasional Malaysia, Kem Perdana Sungai Besi, Kuala Lumpur 57000, Malaysia

[1]rashdan@upnm.edu.my

Abstract. The cavity can be commonly found in aircraft and they are used to store weapons, cargo and also landing gear. Despite the importance of cavity, they can cause vibrations and noise that can disturb the aircraft performance. In order to overcome these problems, the microvane had been studied in controlling noise on open cavity flow. The investigation was carried out at a flow speed of 64.97 meters per second with Reynolds number of 4.3×10^6 per meter on cavity with length over depth ratio of 5. The study tested two configurations of microvane which are counter-rotating and co-rotating orientation. The results show that counter-rotating microvane performed better than co-rotating microvane to suppress cavity noise. The use of counter-rotating microvane managed to reduce the sound pressure level throughout the cavity by 3 % with the highest reduction of 24 decibels. However, the co-rotating microvane cause overall increase of sound pressure level by 0.3 % with the highest reduction of 7 decibels.

1. Introduction

Nowadays, modern higher performance aircraft serve as a carriage for cargo components that are placed inside the cavities embedded in the fuselage. For modern tactical fighter aircraft, the weapons such as missiles, air-to-ground bombs and guns are stored inside the bay. Due to the importance of cavity, it has been a subject of interest in the past and also recent years.

There are three types of cavity flow which are open, transitional and closed [1]. The authors also described the boundaries at which each type occur at supersonic flow. For open cavity flow, it usually occurs at cavities with maximum length over depth *L/D* ratio less than 10. Meanwhile, closed cavity flow occurs at cavity starting from *L/D* ratio of 13 and transitional cavity flow can occur between closed and open flow cavity at *L/D* ratio from 10 to 13.

Even though the presence of cavities are common in an aircraft such as weapon bays [2-4] and landing gear [5], flow over cavity can have detrimental effects on the aircraft performance. For example, flow inside the open cavity is highly unstable and cause noise and vibrations that can lead to fatigue [6-9]. Rowley [8] and Izawa [9] stated that optical distortion and separation problems can occur inside weapon bay in a high-speed flow. Meanwhile, the flow in the closed cavity can increase drag and more susceptible to pitching up moments due to the adverse pressure gradient [1].

In order to overcome the disadvantages of cavity flow, there are two types of flow control method that can be used which are active and passive flow control. According to Jansen [10], active flow control involves additional of external energy added into the system especially into flow separation region to overcome the problems. Among the active flow control devices that have been studied are jets [11, 12-14], pulsed energy deposition [15, 16] and oscillating fences [17]. Nonetheless, this technique is expensive and complicated in design and will not be discussed further in this paper.

Passive flow control is a method that does not require additional energy to control the cavity flow separation. The passive flow control is usually rigid geometric structure was placed at the incoming airflow and then interrupts the airflow without adding additional energy to the system [18]. In addition, this flow control technique is inexpensive and simple in design. The examples of passive flow control devices included static fences [17], spoiler [8, 14] and transverse rod [14, 19]. The purposes of passive control devices are to deflect or thicken the shear layer over the cavity to deduce its interaction with the cavity trailing edge [5, 18, 20]. Furthermore, Rowley et al [19] stated that as the amount of impingement of the shear layer, the amplitude of the resonant acoustic feedback mechanism could be reduced.

Even though there are a lot of research regarding passive control devices, they are only efficient in certain flow conditions and can cause an increase in drag. In order to overcome this issue, the use of micro vortex generator (MVG) had been proposed. MVG is a device that has height is only a fraction of a boundary layer height, δ [21]. Research conducted by Lin [22] recorded that vane shaped MVG or microvane is the best in preventing flow separation where the flow separation can be delayed up to 30 %. Furthermore, microvane also eliminate shock induced separation by producing vortices that have higher energy and much further apart that facilitate the energising of the boundary layer [23].

There are two orientations of microvane that can be commonly found which are co-rotating and counter-rotating [22, 24]. Co-rotating microvane consist of VG's arranged parallel to each other with a respective angle such as 30°, 45° or 60°.Meanwhile, the arrangement of counter-rotating microvane needs two couples of VGs. The first VGs and third VGs have the same angle while the second VGs and fourth VGs have the same angle but in opposite way [24].

The microvane had been used in many applications such as aerofoil, fuselage and also to control flow in aircraft engine inlet system [21]. Despite the wide use of microvane in different aircraft component, there are still not many research regarding the use of microvane to control cavity flow. Therefore, this research will focus on the capabilities of microvane to control noise on closed cavity flow.

2. Methodology

The methodology of this research can be divided into several parts. The following are the details of the steps taken in conducting this research.

2.1. Design of the cavity model

SOLIDWORKS 2015 was used to model the cavity for the flow simulation. The model was designed with a curve ramp and slot where the VG will be placed. The cavity model had an *L/D* ratio of 5 with the length set 25 cm and depth of 5 cm. There are 10 points that had been marked on the cavity floor that were used for pressure measurement. Details on the dimensions of the model is shown in Figure 1.

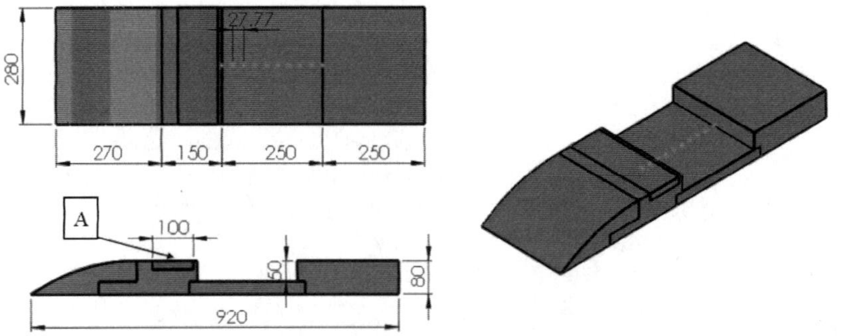

Figure 1. Dimensions of the cavity model.

IOP Conf. Series: Materials Science and Engineering **370** (2018) 012016 doi:10.1088/1757-899X/370/1/012016

2.2. Details of computation

The computational domain used was taken from [25] is shown in Figure2 and all of the solid surfaces were imposed with the no-slip condition. For all of the cases, the flow speed used was 64.97 m/s and the Reynolds number based on the cavity length was 4.3×10^6/m. The ambient pressure and temperature were set to 101.325 kpa and 20 °C. The model for computation was SOLIDWORKS turbulence model that was based on modified k-ε turbulence model with damping functions proposed by Lam & Bremhorst [26] where the intensity used was 0.1 % with a length of 0.0001 m. The global meshing was done using automatic settings with the highest resolution, 8 and level 4 refinement. Furthermore, the local initial mesh was also applied on all of the model top surface and inside the cavity with same resolution and level of refinement. Figure 3 and Figure 4 shows the details of the global mesh and local initial mesh where each level of meshing was labelled with letter A for coarse, B for medium and C for fine. The meshing for the simulation was set to adaptive where the size of the cell will automatically refine on the critical areas of the flow during calculation.

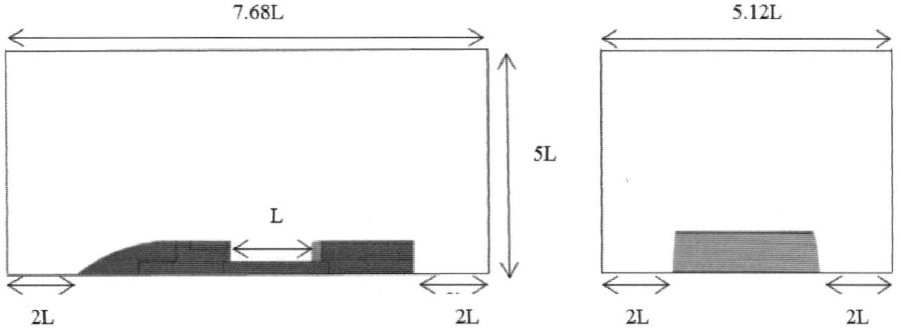

Figure 2. Size of the computational domain.

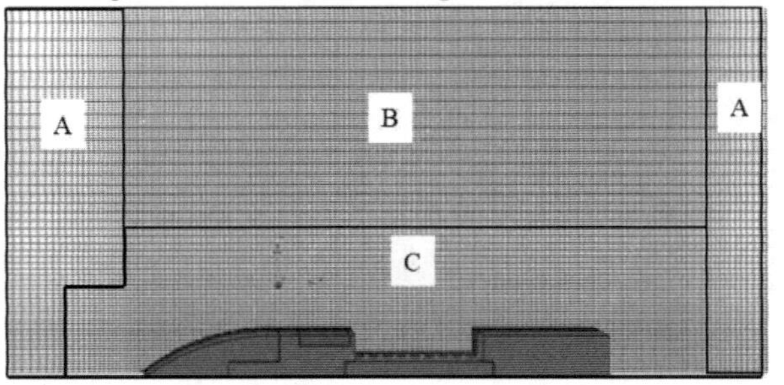

Figure 3. Simulation global mesh.

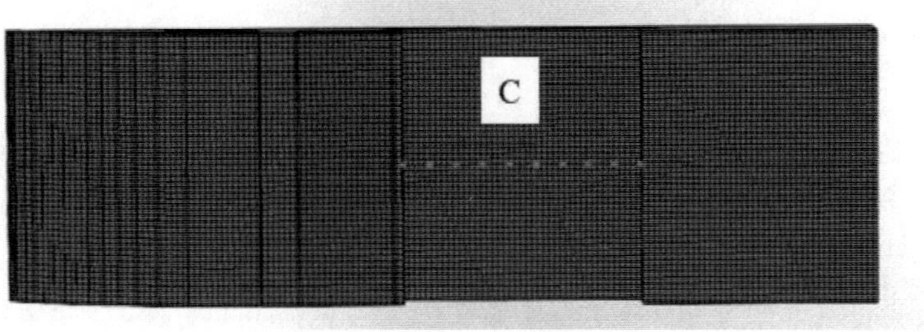

Figure 4. Simulation local initial mesh.

2.3. Microvanes parameters

The simulations were done in both counter-rotating and co-rotating microvane. Before the microvanes were tested, simulation of the clean cavity was done to determine the boundary layer thickness, δ that was used to design the microvane. Figure 5 shows the design of microvanes and the details of the microvane dimensions are shown in Table 1.

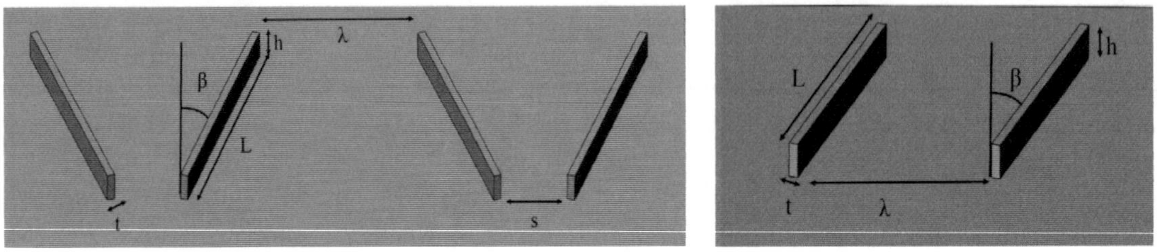

(a) (b)

Figure 5. Orientation of microvanes (a) counter-rotating (b) co-rotating.

Table 1. Dimensions of microvanes.

	h	β	t (mm)	d (mm)	L	λ	s
Counter-rotating	0.4δ	18°	0.8	20	9h	6h	2.5h
Co-rotating	0.4 δ	18°	0.8	20	9h	6h	-

In Table 1, the following letters h, β, t and L represents the height, angle, thickness and the microvane length. In addition, λ, s and d represent the distance between microvane, the spacing between counter-rotating microvane and also the location of microvane away from the leading edge. Details of the microvane position are shown with letter A in Figure 1.

2.4. *Measurement of sound pressure level*

The sound pressure level (SPL) had been chosen to quantify the ability of the microvane to suppress noise inside an open cavity. The value was computed along the cavity floor for all cases. 10 static pressure measurement along the cavity flow were taken by using the point goals function in SOLIDWORKS. Then, the pressure measurement was used to calculate the SPL using Formula 1 [27].

$$SPL\ (dB) = 20\ \log_{10}(\frac{P}{p_{ref}}) \quad (1)$$

The value of p is the pressure measured at a pressure point in Pascal and p_{ref} is the international standard for minimum audible sound 2×10^{-5} Pa.

3. Results and discussion

3.1. *Validation of simulation*

Validation of the flow simulation was conducted by replicating the analysis conducted by Lawson & Barakos [25] using SOLIDWORKS 2015 flow simulation software. The paper was chosen due to the similarity in the type of cavity used, the parameter used to quantify the cavity noise suppression. The cavity model was positioned at zero yaw and pitch angle with the Mach number of the flow set to 0.85. The cavity had a length of 20 inches and depth of 4 inches with the *L/D* ratio of 5. The simulation was conducted without using any flow control devices and the pressure was measured at ten points along the floor of the cavity.The data obtained from the validation were collected and compared with the experimental data. The pressure readings obtained were used to calculate the SPL. Figure6 shows the comparison between the SPL on cavity floor from the simulation and experiment Lawson & Barakos [25].

Figure 6 shows that the magnitude of the SPL was different for four points along the cavity between x/L = 20 % to x/L = 60 %. This might be caused by the turbulence model used by SOLIDWORKS were different compared to the one used by Lawson & Barakos [25].Even though, the simulation shows some discrepancies, the overall margin of error between the simulation and experiment were within 3% which is in line with the research conducted by the authors [25].

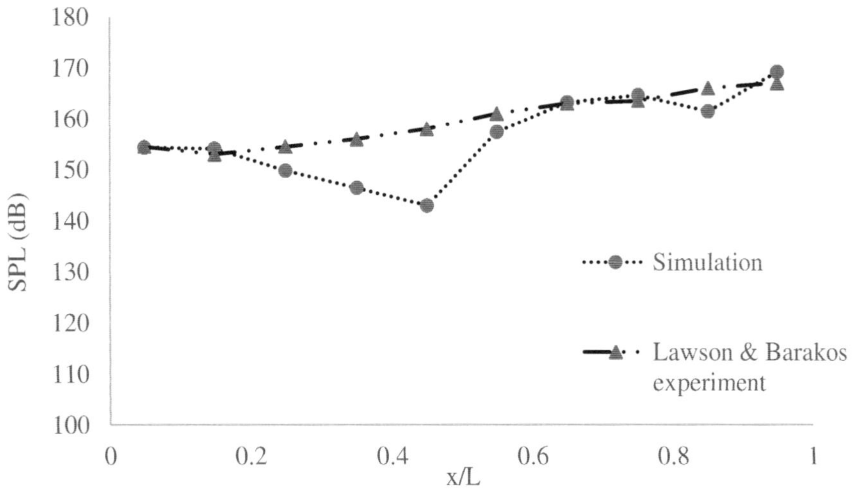

Figure 6. Comparison of SPL distribution across cavity floor.

Furthermore, the validation was also done qualitatively on the model of the cavity baseline. The streamline plot for the simulation was compared the description of the open cavity from Lada & Kontis [11]. Research from the authors was used as a comparison due to the same type of cavity and also similar flow speed. The comparison between the cavity flows is shown in Figure 7.

From Figure 7, the simulation shows similarities with Lada & Kontis [11] where there is huge recirculation of flow inside the cavity. Even though the experiment conducted by the authors have slightly higher flow speed at 100 m/s and lower *L/D* ratio of 2 both experiments shows that there is huge recirculation of flow inside the cavity. The simulation also shows that there was a high-pressure region at the trailing edge of the cavity as described in by the authors [11]. This indicates that the simulation conducted manage to simulate the flow inside an open cavity accurately. From the quantitative and qualitative analysis, it was concluded that the simulation conducted in SOLIDWORKS flow simulation was valid

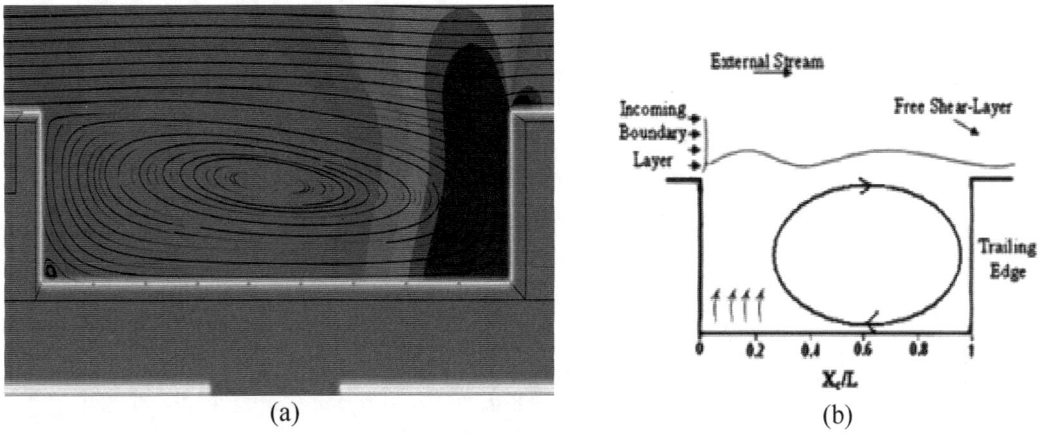

Figure 7. Comparison between open cavity flow (a) simulation (b) Lada & Kontis [11].

IOP Conf. Series: Materials Science and Engineering **370** (2018) 012016 doi:10.1088/1757-899X/370/1/012016

3.2. Baseline cavity without microvane

Simulation with the baseline was conducted to determine the boundary layer thickness. It was found that the region where the microvane was located had a boundary layer ranging from 5 mm to 7.5 mm. Meanwhile, the SPL distribution of clean cavity is shown in Figure 8. From the figure, it can be seen that the distribution of the SPL gradually increasing up to x/L = 0.44 where the SPL started to drop until it reaches the lowest point at x/L = 0.67 with 125 dB. After that, the reading of the SPL increased rapidly until x/L = 1 where the highest reading recorded with 154 dB.

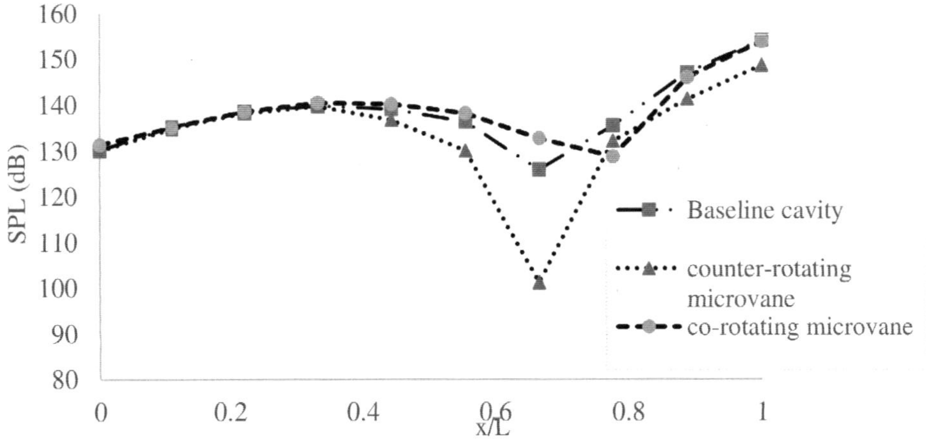

Figure 8. SPL distribution across cavity floor for all cases tested.

3.3. Effect of counter-rotating microvane on cavity flow

Figure 8 also shows the comparison between the cavity SPL with and without the counter-rotating microvane. All of the points inside the cavity with counter-rotating microvane shows a drop of SPL. A significant drop of SPL can be seen from x/L = 0.55 until x/L = 1 where the highest drop can be seen at x/L = 0.67 where the SPL was reduced by 24 dB. On the overall, the SPL value drops by 3.7 % compared to the clean cavity. The reduction of SPL occurs due to the counter-rotating vortices generated by the counter-rotating microvane that can be seen in Figure 9. Figure 10 suggested that the vortices generated by the counter-rotating microvane reduce the impingement on the trailing wall of the cavity which was marked with the letter A. The counter-rotating microvane also caused another recirculation inside the cavity that was shown in Figure 11 and was marked with the letter B. This increase of circulation reduced the pressure throughout the cavity which in turn causes reduction in the SPL.

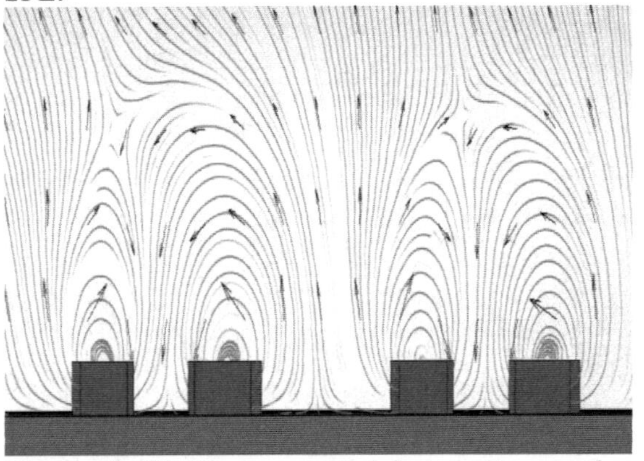

Figure 9. Vortices generated by counter-rotating microvane.

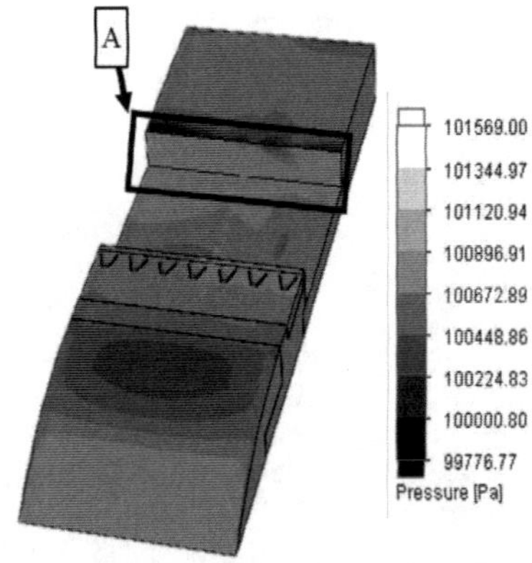

Figure 10. Pressure surface plot for counter-rotating microvane.

IOP Conf. Series: Materials Science and Engineering **370** (2018) 012016 doi:10.1088/1757-899X/370/1/012016

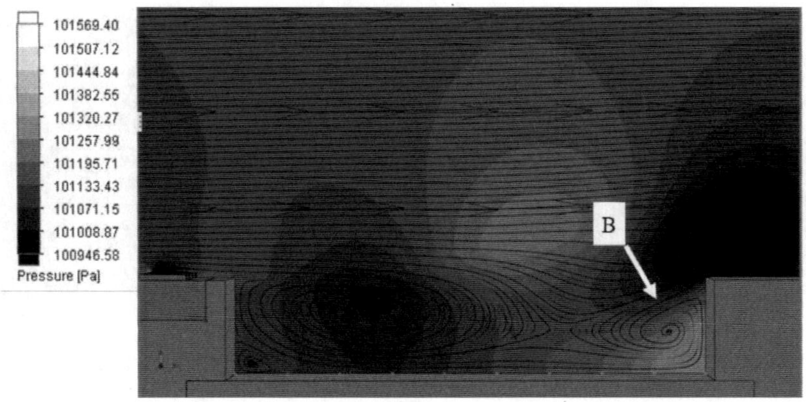

Figure 11. Pressure contour plot for counter-rotating microvane.

3.4. *Effect of co-rotating microvane on cavity flow*

The co-rotating microvane does not significantly improve the SPL of the cavity. From Figure 8, the region where reduction of SPL can be seen from x/L = 0.78 until x/L = 1 with the highest decrease of SPL occurred at x/L = 0.78 where it was reduced by 7 dB. However, the SPL at x/L = 0 and at x/L = 0.33 until x/L = 0.67 were higher than clean cavity. On overall, the use of co-rotating microvanes increases the cavity SPL by 0.3 %. Figure 12 shows that co-rotating microvane also generated vortices similar to counter-rotating microvane but in co-rotating direction. However, Figure 13 shows that the impingement of the shear layer on the cavity trailing wall was much intense than counter-rotating microvane which was marked with the letter A. Figure 14 also shows that the flow condition inside the cavity does not differ from the baseline. Therefore, it was assumed that the vortices generated by co-rotating microvane were not effective in altering the flow and controlling the noise inside the cavity.

3.5. *Comparison between counter-rotating and co-rotating microvane on cavity flow*

The counter-rotating performed better than co-rotating microvane in reducing the SPL. The SPL recorded for the cavity with counter-rotating microvane were lower on every point except the eight point which located on x/L = 0.78. In total, the counter-rotating microvane performed better by 3.8 % in reducing the cavity SPL. This shows that the vortices generated by counter-rotating microvane are more effective at energising shear layer for a cavity with *L/D* ratio of 5.

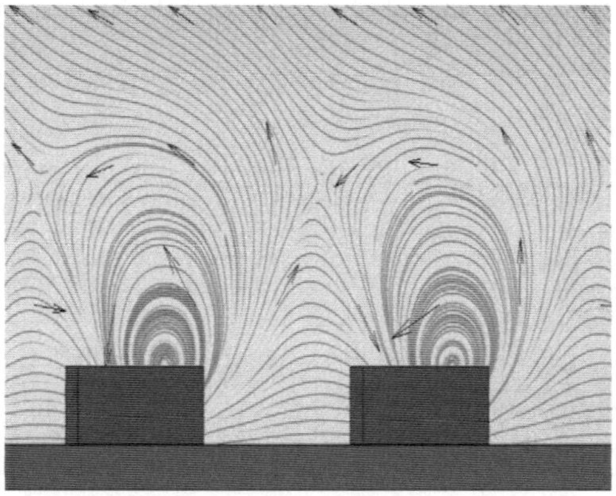

Figure 12. Vortices generated by co-rotating microvane.

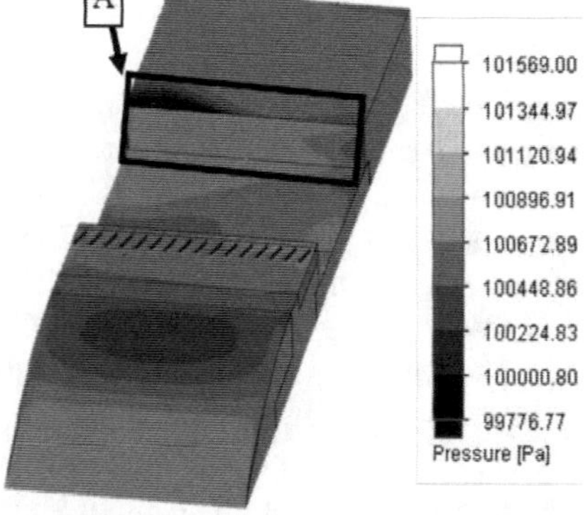

Figure 13. Pressure surface plot for co-rotating microvane.

Figure 14. Pressure contour plot for co-rotating microvane.

4. Conclusions

In this study, the effects of microvane to control cavity flow with *L/D* ratio of 5 at Reynolds number of 4.3×10^6 /m had been conducted. The results show that counter-rotating microvane performed better than co-rotating microvane in controlling cavity flow. Counter-rotating microvane generated vortices that have the capability to reduce the SPL throughout the cavity. For co-rotating microvane, the SPL increase until x/L = 0.67 and started to decrease until the trailing edge of the cavity. The results also indicate that vortices generated by co-rotating microvane do not have a significant impact on altering the flow inside the cavity. This shows that the counter-rotating microvane have a high potential to be used as an alternative to other passive control devices to suppress noise in an open cavity.

References

[1] Maureen B, Tracy, and Plentovich EB. Cavity unsteady-pressure measurements at subsonic and transonic speeds, NASA TP-3669,1997.

[2] Shaw, L., Clark, R., and Talmadge, D., F-111 Generic Weapons Bay Acoustic Environment, *Journal of Aircraft*, **Vol. 25**, No. 2, 1988, pp. 147–153.

[3] Cattafesta, III L, Shukla D, Garg S, Ross J. Development of an adaptive weapons-bay suppression system. *In5th AIAA/CEAS Aeroacoustics Conference and Exhibit* 1999 (p. 1901).

[4] Bower WW, Kibens V, Cary AW, Alvi FS, Raman G, Annaswamy A, Malmuth NM. High-frequency excitation active flow control for high-speed weapon release (HIFEX). AIAA paper. 2004 Jun;2513:2004.

[5] Mitchell, Douglas. Control of High-Speed Cavity Flow Using Plasma Actuators. Diss. The Ohio State University, 2006.

[6] Tracy, Maureen B.; and Stallings, Robert L., Jr. Coupling of Acoustic Environment in Rectangular Cavity With Store Vibration Characteristics During Simulated Separation in Supersonic Flow. NASA TP-2986, 1990

[7] Shaw, L. L.; and Shimovetz, R. M. Weapons Bay Acoustic Environment. Impact of Acoustic Loads on Aircraft Structures, AGARD-CP-549, Sept. 1999

[8] C. W. Rowley. Modelling, simulation and control of cavity flow oscillations. PhD thesis, Institute of Technology Pasadena, California, 2002.

[9] S. Izawa. Active and passive control of flow past a cavity. Wind Tunnel and Experimental Fluid Dynamics Research, Tohoku University, Japan, 2011.

[10] D. P. Jansen. Passive flow separation control on an airfoil flap model the effect of cylinders and vortex generators. Master thesis, Deflt University of Technology, Netherland, 2012

[11] Lada C, Kontis K. Microjet Flow Control Effectiveness of Cavity Configurations at Low Speeds. *Journal of Aircraft*. 2014 Sep 24.

[12] Vakili AD, Gauthier C. Control of cavity flow by upstream mass-injection. *Journal of Aircraft*. 1994 Jan 2;31(1):169-74.
[13] Bueno P, Unalmis O, Clemens N, Dolling D. The effects of upstream mass injection on a Mach 2 Cavity Flow. *In40th AIAA Aerospace Sciences Meeting & Exhibit 2002* (p. 663).
[14] Ukeiley LS, Ponton MK, Seiner JM, Jansen B. Suppression of Pressure loads in resonating cavities through blowing. AIAA Paper. 2003 Jan;181:2003.
[15] Lazar ES, Elliott G, Glumac N. Control of the shear layer above a supersonic cavity using energy deposition. Diss. University of Illinois at Urbana-Champaign, 2008.
[16] Adelgren RG, Elliott GS, Crawford JB, Carter CD, Donbar JM, Grosjean DF. Axisymmetric jet shear-layer excitation by laser energy and electric arc discharges. *AIAA journal*. 2005 Apr 1;43(4):776-91.
[17] Sarno RL, Franke ME. Suppression of flow-induced pressure oscillations in cavities. *Journal of Aircraft*. 1994 Jan 2;31(1):90-6.
[18] Danilov PV, Quackenbush TR. Flow Driven Oscillating Vortex Generators for Control of Cavity Resonance. *In49th AIAA Aerospace Sciences Meeting including the New Horizons Forum and Aerospace Exposition* 2011 (p. 1219).
[19] C. W. Rowley, V. Juttijudata and D. R. Willicas. Cavity for control simulation and experiments. AIAA Paper 2005-0292, 2005.
[20] Smith B, Welterlen T, Maines B, Shaw L, Stanek M, Grove J. Weapons bay acoustic suppression from rod spoilers. AIAA Paper. 2002 Jan 14;662:2002.
[21] Lin JC. Review of research on low-profile vortex generators to control boundary-layer separation. *Progress in Aerospace Sciences*. 2002 Jul 31;38(4):389-420.
[22] Lin J. Control of turbulent boundary-layer separation using micro-vortex generators. *In30th Fluid Dynamics Conference* 1999 (p. 3404).
[23] Holden HA, Babinsky H. Vortex generators near shock/boundary layer interactions. AIAA paper. 2004 Jan 8;1242:2004.
[24] Jansen DP. Passive Flow Separation Control on an Airfoil-Flap Model. Deft University. 2012 Aug.
[25] Lawson SJ, Barakos GN. Assessment of passive flow control for transonic cavity flow using detached-eddy simulation. *Journal of Aircraft*. 2009 May 1;46(3):1009.
[26] Lam, C K G and Bremhorst, K A Modified Form of Model for Predicting Wall Turbulence. *ASME Journal of Fluids Engineering*, 1981, **Vol. 103**, pp. 456-460
[27] Ruth F Weiner and R Matthews*Environmental Engineering fourth edition* Butterworth Heineman 2003 p 428

Acknowledgements
The authors would like to thank the Ministry of Education Malaysia for the financial support under the RAGS (RAGS/1/2014/TK09/UPNM/1), RACE (RACE/F3/PK6/UPNM/13) and UPNM GPJP (UPNM/2015/GPJP/2/TK/09) research grant schemes. Also an appreciation to the technical staffs of the Faculty of Engineering, UniversitiPertahanan Nasional Malaysia for providing the assistances to the equipment and computing facilities.

IOP Conf. Series: Materials Science and Engineering **370** (2018) 012017 doi:10.1088/1757-899X/370/1/012017

Identifying Inconel 718 Fasteners Failure Using Structured Problem Solving Method

K N M Tohit[1], N Zulkifli[2], S Sulaiman[3] and B T H T Baharudin[4]
Project Quality Manager, Aker Solutions Malaysia Sdn. Bhd.[1]
Department of Mechanical and Manufacturing Engineering, Universiti Putra Malaysia 43400 Serdang, Selangor[2,3,4]

*Corresponding author: khairul.nizam@akersolutions.com, norzima@upm.edu.my

Abstract: This paper is about the development of problem solving method to identify failures in inconel 718 mainly applied in the subsea environment. A3 analysis with support from fishbone diagram and 5 whys analysis techniques were introduced to identify the actual cause of failures. Upon identification of the actual cause of failure, improvement actions were taken to eliminate the actual cause. In this research, the identification of the actual cause of failure related with the Inconel 718 fasteners that were used for Methanol Injection Tree Valve (MITV) was performed. Lab analysis on fresh and failure fasteners was performed by comparing the mechanical properties, chemical composition and microstructure analysis. The data was then compared to understand the characteristics of the failure. Future state and target in A3 stage required a mitigation action to eliminate the cause of failure. This was performed by conducting new material analysis i.e. L7M fasteners for future reference. Standard template for subsea's problem solving method was established upon completion the analysis for future use.

1. Introduction

Many organizations nowadays consider quality as the first and foremost priority in their products and services within their management systems [1]. Fasteners used for critical applications in subsea industry provide a challenge in terms of material selection, quality control, traceability and documentation for manufacture and delivery. Criticality of subsea fasteners are defined as both pressure containing and primary load bearing members that may affect the functionality of valves, equipment and instruments for safe and reliability [2]. As reported on Wall Street Journal in July 2016, the bolt issue could affect more than 2,400 platforms and oil rigs in the Gulf of Mexico, as well as of 23 off the coast of California, and one active rig on the outer continental shelf in Alaska.

Hydrogen induced stress cracking (HISC) or hydrogen embrittlement (HE) was identified as the major failure mechanism experienced during the last decades that was resulting from the hydrogen charging conditions of the cathodic protection (CP) system. The main bulk of fasteners used subsea is of low alloyed steel grade type which relies on electrical continuity to the cathodic protection system to avoid sea water corrosion such as ASTM 320 Grade L7, L7M, L43 bolts with ASTM 194 Grade 4, 7M, 7 nuts, that will have a possibility prone to HISC/HE if the fastener hardness and strength level exceed the specified and established limits [1].

A leak was detected on the Production Wing Block (PWB) of Subsea Tree (XMT #2) which has been installed on P201 Well for three years since 2014. Further investigation reveals that a stud mounting of the Methanol Injection Tree Valve (MITV) has broken off. A portion of the broken stud from XMT#2 was retrieved from subsea and was sent out for lab analysis. Figure 1 shows the actual condition on what were happened in subsea during this incident. Fluid was observed came out from the MITV due to one out four stud bolts were broken off.

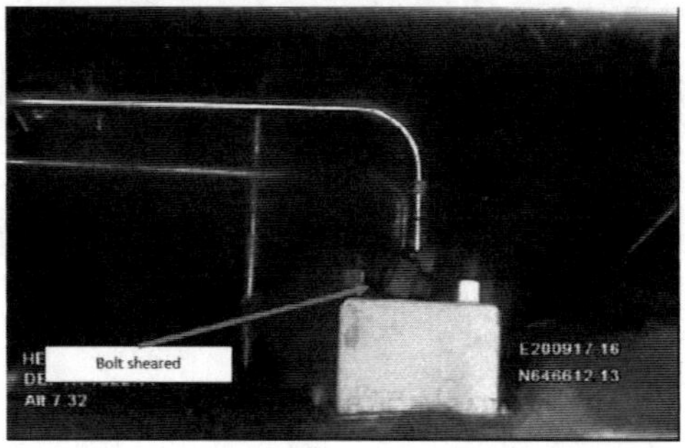

Figure 1. Fluid observed coming out from this point

2. Methodology

2.1 Problem Solving and A3 Technique

T.C Lin et. al [3] mentioned that problem solving competence is largely determined whether the members can effectively utilize knowledge resources located within the team. Based on the trans active memory concept and following traditional wisdom, that knowing the location of knowledge allow better problem-solving competency. There was a quality approach refers as Total Quality Management (TQM) introduced to the organization aims to guarantee organization's success. Powel [4] defines this approach as a systematic framework which enhances operational activities of an organization to fulfil customers' expectations. There are eight steps in the A3 tool that needs to be completed prior to identify the actual cause of the issues or problems. The steps are designed to ensure systematic approached is applied to understand the situation correctly. Deming [5] was highlighted that the top management has the responsibility to assign the employees a clear work standard. This will provide us with the correct improvement actions i.e. short and long-term solutions. M.S. D Caro et. al [6] mentioned on his research that the successfulness of the problem solving methods were based on (a) the demands of the task being performed, (b) the characteristics of the individual performing the task and (c) the constraints of the skill execution environment. Anderson [7] mentioned the importance of Deming theory on the organizational system creation that encourages on learning and cooperation in enhancing the management practices implementation. Below are the eight steps that will be applied in A3 tool approach described in Figure 2.

Please take note that, the A3 implementation technique is only a problem solving tool that requires full support from respective functional area namely Engineering, Manufacturing and Quality to work hand on hand to ensure actual cause of failure is identified. The main objective of performing this activity to the team is to identify the underlying root cause and not only focusing on the primary root cause.

Below description describe each column expectation in order to complete the analysis.

1. Observed Problem
   Describe the problem so that it is understandable for the reader
   The problem should be based upon a (performance) gap between the goals or customer needs
2. Current State
   What do we know about the problem to be resolved
   What characteristics the system that the problem is a part of (the process, the product, the organisation etc.)

Which facts and data can we gather to understand the problem better, and what do these facts tell us about the problem

Facts or outcome of analysis shall be visually illustrated

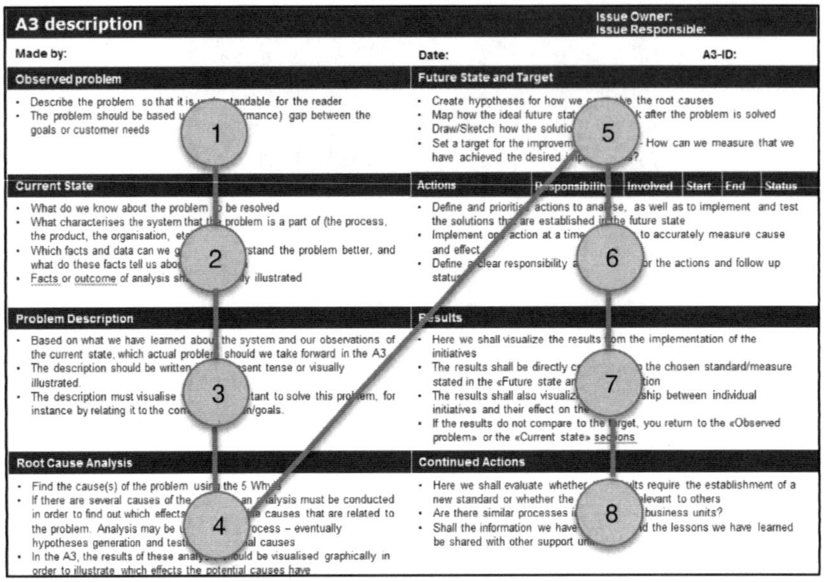

Figure 2. A3 tool process flow descriptions

3. Problem Description

   Based on what we have learned about the system and our observations of the current state, which actual problem should we take forward in the A3

   The description should be written in the present tense or visually illustrated

   The description must visualise why it is important to solve this problem, for instance by relating it to the company's vision/goals

4. Root Cause Analysis

   Find the cause(s) of the problem using 5 Why's

   If there are several causes of the problem, an analysis must be conducted in order to find out which effects of the possible causes that are related to the problem. Analysis may be used for this process – eventually hypotheses generation and testing of potential causes

   In the A3, the results of these analyses should be visualized graphically in order to illustrate which effects the potential causes have

5. Future State and Target

   Create hypotheses for how can we solve the root cause

   Map how the ideal future state should look after the problem is solved

   Draw/sketch how the solution should be

   Set a target for the improvement initiative – How can we measure that we have achieved the desired improvement?

6. Actions and responsibility

   Define and prioritise actions to analyse, as well as to implement and test that solutions that are established in the future state

   Implement one action at a time to be able to accurately measure cause and effect

   Define a clear responsibility and deadline for the actions and follow up status

7. Results

   Visualize the results from the implementation of the results

The results shall be directly comparable to the chosen standard/measure stated in the future state and target section

The results shall also visualize the relationship between individual initiatives and their effect on the result

If the results do not compare to the target, return to the Observed Problem or the Current State sections

8. Continued Actions

To evaluate whether the results require the establishment of a new standard or whether A3 may be relevant to others

Are there similar processes in our or other business units?

Shall the information gathered and the lessons learned be shared with other support units.

2.2 Root Cause Analysis

For Root Cause Analysis column, two (2) existing and famous methods will be used to identify the possible causes for a problem especially when a team's thinking tend to fall into ruts. These two methods are

1. The Cause and effect diagram/Fishbone diagram
2. 5 Whys analysis technique

The fishbone diagram is identifies many possible causes for an effect or problem. It will be used to structure a brainstorming session and immediately sorts ideas into useful categories. Below principles will be followed through during the brainstorming session to ensure team members are fully contributed to the process:

1. Agree on a problem statement (effect)
2. Brainstorm the major categories of causes of the problem. This will be performed by referring to generic headings:
    - Methods
    - Machines (equipment)
    - People (manpower)
    - Materials
    - Measurement
    - Environment
3. List down the categories of causes as branches from the main arrow.
4. Brainstorm all the possible causes of the problem. 5 Whys Analysis techniques will be applied during this stage. As each idea is given, the facilitator/team leader needs to list down it as a branch from the appropriate category. Causes can be written in several places if they relate to several categories.
5. Again ask "why does this happen?" about each cause. Write sub–causes branching off the causes. Continue to ask "Why?" and generate deeper levels of causes. Layers of branches indicate causal relationships.
6. When the group runs out of ideas, focus attention to places on the chart where ideas are few.

This fishbone diagram was drawn by a manufacturing team to try to understand the source of periodic iron contamination. The team used the five generic headings to prompt ideas. Figure 3 illustrates the fishbone diagram example.

The root cause analysis will be supported by another quality tool called the 5 Whys Analysis technique. The 5 Whys technique is true to this tradition, and it is most effective when the answers come from people who have hands-on experience of the process being examined. It is remarkably simple: when a problem occurs, you uncover its nature and source by asking "why" no fewer than five times.

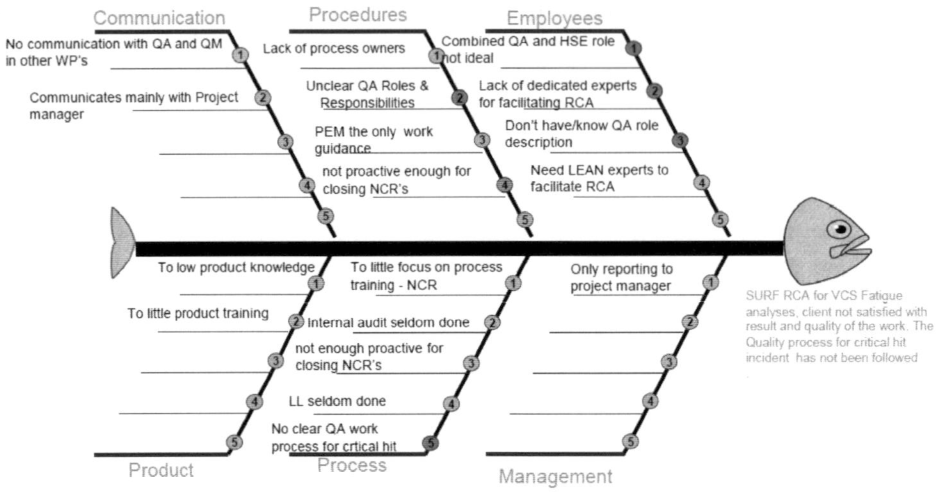

Figure 3. Cause and effect diagram example

Here it is in action:
Problem: Customer is refusing to pay for the leaflets that were printed out for them.

i. Why? The delivery was late, so leaflets couldn't be used
ii. Why? The job took longer than planned
iii. Why? Printer's ink shortage
iv. Why? Sudden additional quantity ordered
v. Why? Not enough printer's ink in stock

As a countermeasure from the above analysis, team has to identify and appoint a supplier who can deliver the printer's ink at very short notice so that will help the team to reduce inventory, waste and response to customer demand in a timely manner. Figure 4 shows the 5 Whys analysis technique templates that will be used to complete this research.

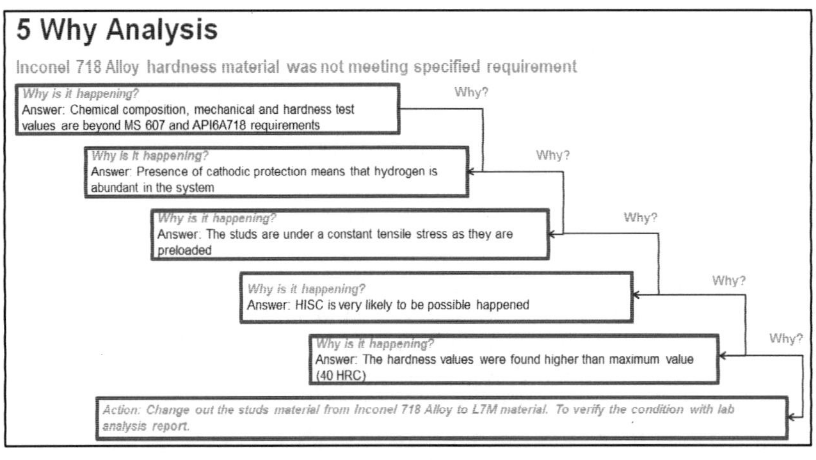

Figure 4. 5 Whys analysis technique template

The A3 tool is team work effort and requires full support and commitment from each team member. Therefore, team work is very crucial to ensure the fruitful of the brainstorming and analysis conducted. It is very important to have hands-on team members that were experiencing the issue. There will be a lot of questions and answers session during the brainstorming stage until there is no answer to explain on why the issue happened. Time frame to complete the analysis will be different for each issue depending on the criticality and severity of the issue. Follow-up meeting is always requiring after the initial start-up. This is to ensure updated and recent findings especially on the validation of the improvement actions are captured and reviewed again. Therefore, this approach can be considered as a continuous improvement effort.

3. Results and Discussion

3.1 Current State

A portion of the broken stud from XMT#2 was retrieved from subsea and was sent to material laboratory for further material analysis where the analysis was conducted previously following a similar failure that occurred on XMT #1 (10127822; SN: 190131-001). As there were some studs of the same heat no. in current stock, one of the stud was sent to material laboratory so that it can be compared against. Due to the similarity in many aspect of the material analysed, it can be concluded that the two failed studs and the fresh stud from stock are indeed from the same batch of material.

The fact that the fresh stud has never been torqued exposed to seawater and cathodic protection reflects the condition of the stud material in the as supplied condition. Work hardening or exposure to sea water and cathodic protection has not contributed to the increase in hardness of the failed studs. Reference made to mill certificate supplied by the supplier in the Manufacturing Record Book (MRB), confirms that the material has been treated to ASME-SB-637-93a which will result in a high tensile strength material. Due to the condition of the material, the material is considered highly susceptible to Hydrogen Induced Stress Cracking (HISC).

HISC is very likely to be the possible cause of failure due to the following reasons:

1. Strength and hardness of material that is significantly more than the requirement of MS-607 as well as API Standard 6A718
2. The studs are under a constant tensile stress as due to the preload
3. Presence of cathodic protection means that hydrogen is abundant
4. Intergranular brittle failure was observed as the crack initiation and propagation
5. Crack path is sharp and singular profile without any significant branching indicates the unlikeliness of Stress Corrosion Cracking (SCC)
6. No corrosion product was observed on the fracture surface.

In this section, the implementation of another quality tools namely Fishbone Diagram/ Cause and Effect Diagram and 5Whys Analysis template are used to identify the actual cause of the failure. The decision has been made to send out the broken stud bolt that was failed subsea operation to the material lab for further analysis. Another fresh stud bolt from the current stock with the same heat no was sent out together to be compared with. Scope of Works of this analysis is defined in the next section of the analysis. Figure 5 shows the Cause and Effect Diagram established to analyze the potential cause of the failure.

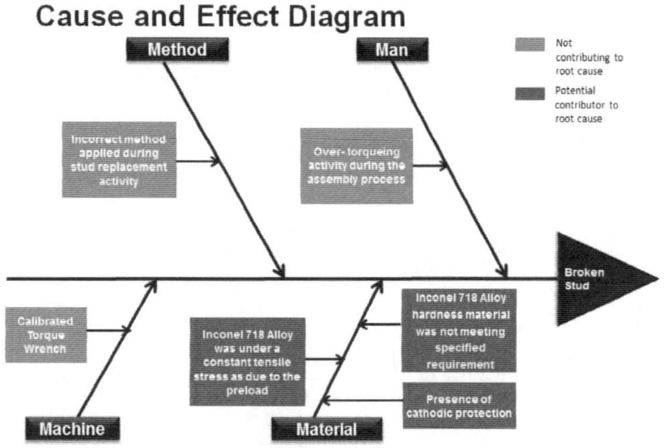

Figure 5. Cause and effect diagram performed for the failure analysis

There are four criteria's of the potential causes of the failures that required being focus on. There are

1. Method
2. Man
3. Machine
4. Material

Those criteria's also required the definition whether the identified potential cause will/maybe contributed to the actual failure or not. Therefore, the definition of below categories is defined clearly into the template.

i. Not contributing to root cause (label in orange)
ii. Potential contributor to root cause (label in red)

Incorrect method applied during the stud replacement activity was identified as not contributing factor to root cause in the method category. This is due to proper monitoring/witnessing and record was performed earlier during the execution of the activity. Figure 6 shows on the actual record – Serial/Batch no of new stud was recorded into section 3.2 CITV1 Stud Replacement while performing this activity earlier.

3.2. CITV1 Stud Replacement

Step	Description	Date/Sign/Remarks
1	Record Serial/Batch no. of new stud (P/N: 10060680) and nut (P/N: BB17-MX0407-13) to be installed onto CITV1. Stud Serial /Batch No : 1110350001 : 1110350001 : 1110350001 : 1110350001 Nut Serial/Batch No : 1110350002 : 1110350002 : 1110350002 : 1110350002	1/12/11
2	Apply grease on stud prior to installation. Use ALCO-EP-73 PLUS or equivalent.	1/12/11

Figure 6. Serial/batch no recorded during the studs replacement

Over-torqueing activity during the assembly process was identified as not a contributing factor to the root cause in the Man category. The decision was made after review the actual torque application value record performed earlier. Figure 7 shows the actual torque value application on the replacement of four studs to 305ft-lbf.

3	Replace stud on CITV1. Tighten 4 stud on MITV to 305ft-lbf. CAUTION: If possible, do not disconnect the chemical injection line. Hydraulic control line to be left connected. NOTE: CITV1 stud shall be removed one at a time on diagonal pattern to maintain gasket position on CITV-FIV Block interface. NOTE: Existing Gasket to be left in place.	1/12/11

Figure 7. Torque value application during the studs replacement

Torque Wrench that was used during the torqueing application was ensured to be calibrated prior to use. This has eliminated the potential of having incorrect machine/equipment usage into Machine criteria. The analysis was continued by focusing into material criteria. There are three potential causes identified into the Material criteria as below:

i. Inconel 718 Alloy hardness material was not meeting the specified requirement
ii. Inconel 718 Alloy was under a constant tensile stress as due to the preload
iii. Presence of cathodic protection

Further analysis is required for Item i and ii above by using the 5 Whys analysis technique to understand the actual cause of the failure or underlying cause. Please take note that, by completing the cause and effect diagram analysis only the primary cause of failure is identified. Therefore, the analysis is continued further by implementing the 5 Whys analysis technique. The objective of performing further evaluation is to dig out the details actual cause of failure that has contributed to the issue. By identifying the actual cause then only the correct countermeasure i.e. containment and corrective actions can be suggested and implemented to ensure no re-occurrence for future. Presence of cathodic protection was set aside as the Subsea Tree was designed to be applied with the cathodic protection criterion. Thus, necessary actions are required to ensure those requirement are fulfilled for the Subsea Tree deployment. 5 Whys analysis was performed to identify the potential cause related with the Inconel 718 Alloy hardness material issue that not meeting to the specified requirement. This is shown in Figure 8. This action was performed upon received the lab test analysis report that focus on below scope of works.

i. Visual/Macro Examination
ii. Cross-sectional Metallographic Examination
iii. Chemical Composition Analysis
iv. Hardness Measurement

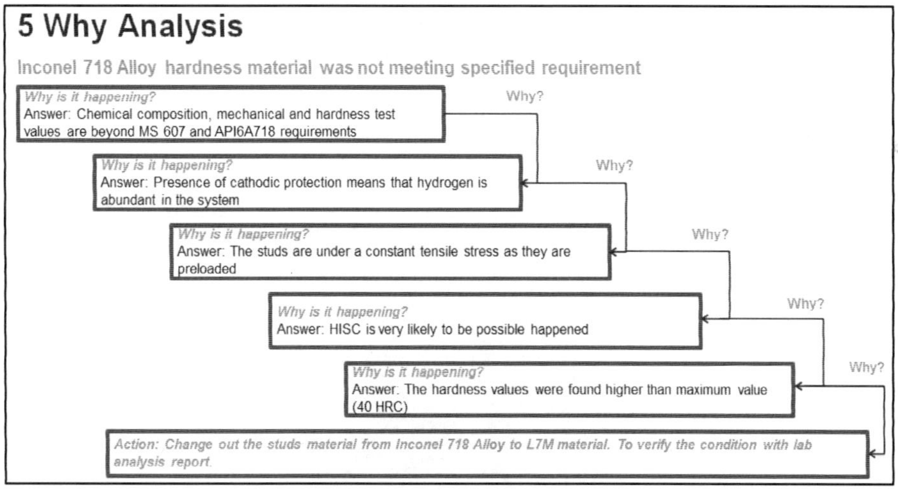

Figure 8. 5 why analysis on the Inconel 718 alloy hardness material was not meeting specified requirement

One broken stud bolt with 17mm Outer Diameter (OD) and 35mm length (L)) of MITV connection was analyzed to understand the characteristics and behavior of the material consists of mechanical and chemical composition analysis. The objective of the analysis was to establish the possible cause (s) of the failure of stud bolt:

1. The stud bolt had been broken in brittle mode with intergranular profile in association with torsional loading in addition to the normal tensional loading

2. The crack initiated from circumferential areas of the thread root where higher stress had exerted
3. Intergranular fracture was likely attributed to hydrogen induced stress cracking (HISC) in the exposure of cathodic protection
4. The chemical compositions of the broken stud bolt met the requirements of MS-607 specification of material Inconel 718 as well as API standard 6A 718-2009
5. Microstructure of stud base material reveal fine grain austenite matrix with large carbides/nitrides particles.

3.2 Visual/Macro Examination

The broken stud bolt was undergone the visual inspection upon received at the material laboratory. Figure 9 showed the visual appearance of the broken stud bolt as at received condition. The remaining length of the stud was measured and estimated around 30mm.

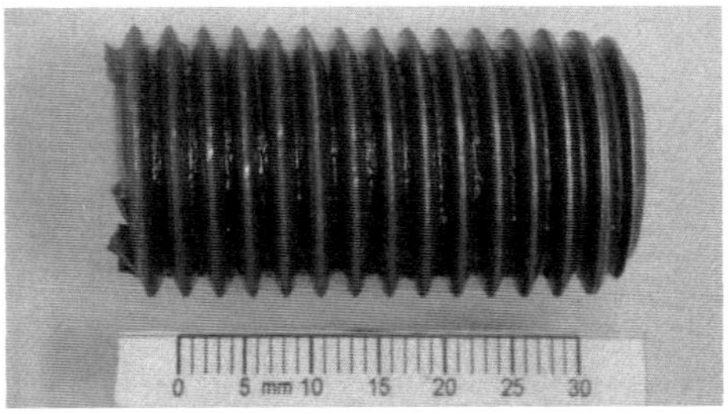

Figure 9. As-received condition of broken stud bolt

The result from the visual inspection activity was recorded. Figure 10 showed the visual appearance of the broken stud bolt. It was observed that the stud bolt had broken at the thread root, which was the high stress region. Fracture surface revealed flat and little plastic deformation with rough surface appearance, indicative of brittle fracture. It also revealed multiple initiation planes distributed around the bolt circumference. There was a location, which displayed fibrous and shear lip features suggesting final consequential fracture area. This fracture appearance indicated that the bolt had experienced torsional loading in addition to the normal tensional loading.

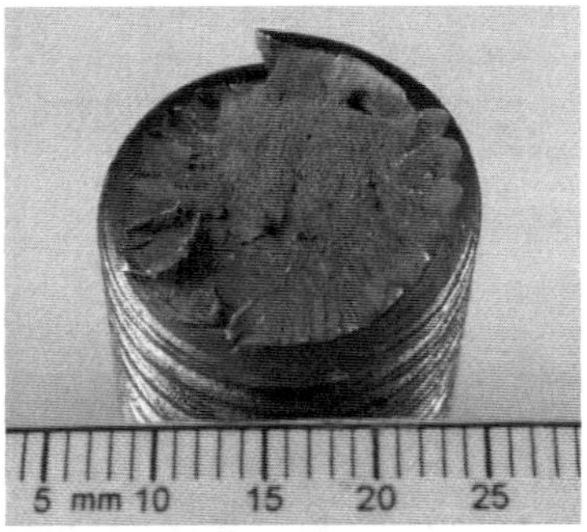

Figure 10. Fracture surface of broken stud bolt

3.3 Cross-sectional Metallographic Examination

Cross sectional metallographic examination was conducted across the fracture surface of the broken stud bolt. Metallographic sections were prepared and etched according to ASTM E3-11 and ASTM E407-07. The Kalling's No.2 reagent was used for etching the Inconel alloy 718 of bolt material. Figure 11 shows a cross section macrograph which revealed nearly straight fracture profile with little macro plastic deformation observed, evident of brittle fracture path [8].

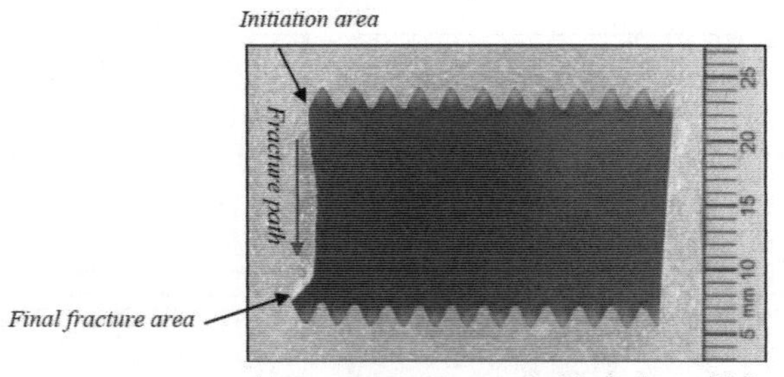

(a) Cross section macrograph of the broken stud bolt

Figure 11. Cross section macrograph of the broken stud bolt

The cross-sectional activity was continued to inspect the rolled thread profile of broken stud. Figure 12 displayed the rolled thread profile of broken stud bolt.

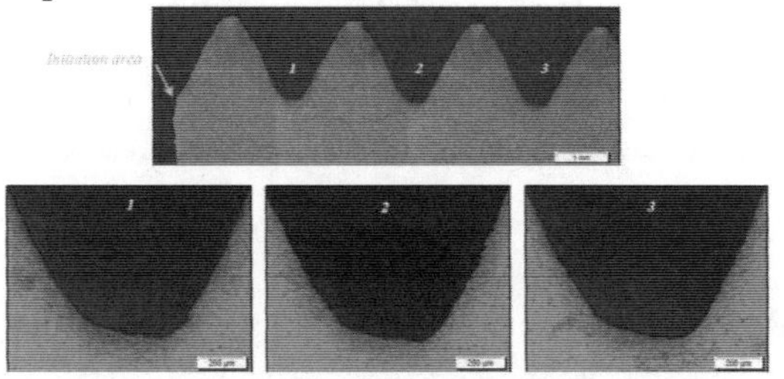

Figure 12. Rolled-thread profile of broken stud bolt

Incipient crack was identified upon performed a cross section micrograph at initiation area. Figure 13 was shown a cross section micrograph at initiation area.

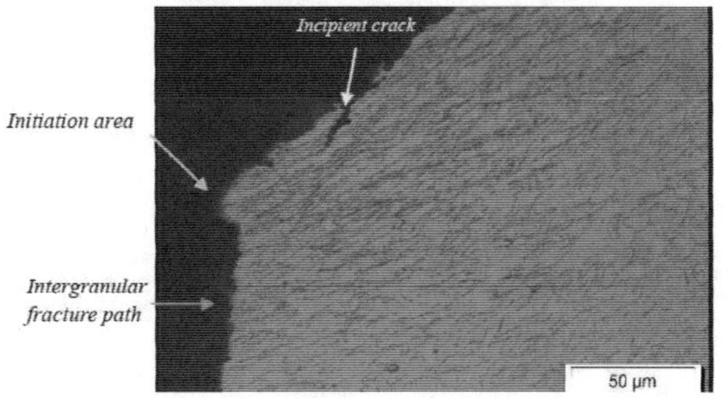

Figure 13. Grain deformation at the thread root of initiation area intergranular fracture path

A cross section of SEM micrographs was performed along the threads next to broken area. This is shown in Figure 14 and it was revealed that the grain deformation along the threads resulting from work hardening processing i.e. thread rolling operation.

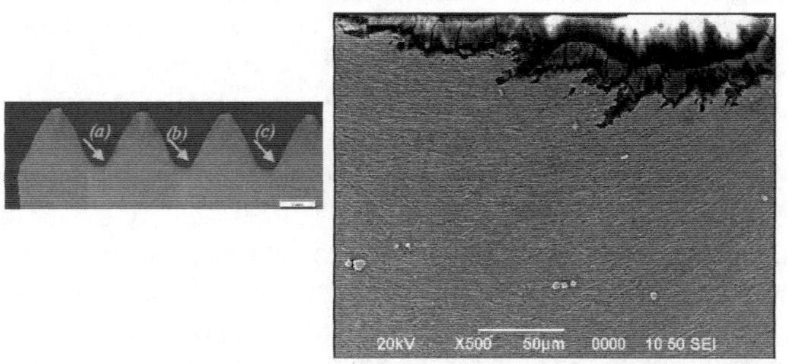

Figure 14. Cross section SEM micrographs revealed grain deformation at the thread roots.

3.4 Chemical Composition Analysis

Chemical analysis was conducted on the base material of broken stud bolt sample using an Optical Emission Spectroscopy (OES) Analyzer. The results were tabulated in Table 1 as below. It was shown that the chemical compositions of the broken stud bolt met the requirements of MS-607 specification of material Inconel 718 as well as API standard 6A718-2009.

Table 1. Broken bolt chemical composition analysis results

Element	Weight %	MS-607 Spec	API Standard API7186A-2009
Carbon, C	0.02	0.045 max	0.045 max
Silicon, Si	0.1	0.35 max	0.35 max
Manganese, Mn	0.06	0.35 max	0.35 max
Phosporus, P	0.01	0.01 max	0.01 max
Sulphur, S	<0.001	0.01 max	0.01 max
Chromium, Cr	17.63	17-21	17-21
Molybdenum, M	3	2.8 - 3.3	2.8 - 3.3
Nickel, Ni	54.03	50 -55	50 -55

3.5 Hardness Measurement

Rockwell hardness measurement (HRC) was performed on the base material of broken stud bolt sample by using a Zwick/Rowell ZHR machine. The results were indicated in figure 15 as below:

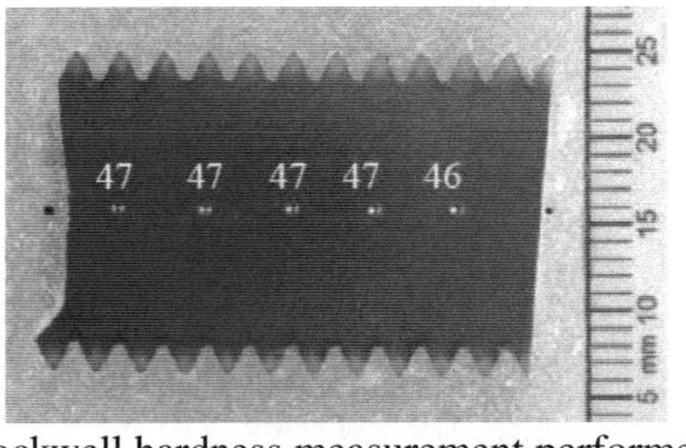

Figure 15. Rockwell hardness measurement performed on the base material of broken stud bolt

The hardness values are slightly 18% higher than maximum value (40 HRC) specified in MS-607 specification of material Inconel 718 as well as API standard 6A718-2009. Visual appearance of flat fracture surface with little plastic deformation and intergranular surface morphology under SEM examination proved that the stud bolt had broken in brittle mode. Fracture surface showed multiple initiation planes distributed along the circumferential areas of the thread root with the final fibrous shear lip area. This observation suggested that the stud bolt had experienced certain torsional loading in addition to the normal tensional loading. Cross section examination revealed straight and intergranular fracture path which was consistent with brittle fracture mode. It was plausible that intergranular fracture path was contributed by either grain boundary precipitation or hydrogen induced stress cracking (HISC). Magnified microstructure of the base material of the broken stud bolt clearly revealed austenite grain boundary with twin structure suggesting that intergranular fracture due to grain boundary precipitation was of less possibility.

It was reported that the broken stud bolt was installed in Subsea Tree#2 (XMT#2) which was exposed to cathodic protection (CP). CP is one of the sources rendering to hydrogen charging to the respective component in the offshore equipment and to subsequent HISC. From the analysis, it was found that the broken stud bolt material was high strength and high hardness material. It is well known that the higher the strength and the hardness, the higher the susceptibility to HISC. Besides, cross section examination displayed the grain deformation along the threads indicating that the threads had been undergone rolled threading process. This might possibly cause inherent residual stress at the threaded areas.

4. Conclusion

It was concluded that the objective of this research was successful by achieving below objectives.

1. To develop a problem solving method for identifying the actual cause of failure of the Alloy 718 studs on the MITV connection. The development of A3 problem solving technique with implementation of the fishbone diagram and 5 Whys analysis technique was successfully identified the actual cause of failure was due to high hardness material value (more than 35 HRC) than the specification.
2. To validate the effectiveness of using the A3 problem solving method to be used in subsea's industry for future issue. The A3 template with fishbone diagram and 5Whys analysis technique was used successfully.
3. Validation was done through the establishment of the containment and corrective actions as below.
   a. Containment actions
      i. Concession to de-rated the Subsea Tree#2 (XMT #2) to 5,000psi and change out the studs material from Alloy 718 to L7M
      ii. Clamping down MITV valve to allow additional 4 unit of stud bolts to be added during the refurbishment stage
   b. Corrective actions
      i. Change out stud bolt material to from Alloy 718 to L7M during the refurbishment stage
      ii. To consider the use of L7M stud bolt for future design Subsea Tree for the same application. This has been added up into Global Subsea Lessons Learned database

References

[1] Mallur, S.B., Hiregrouder, N.L., & Sequeira, A.H. (2011). The Relationship between TQM Practices and Business Excellence in Small and Medium Sized Manufacturing Enterprises of North Karnataka Region. SDM 1MD *Journal of Management*, 2(1): 18-31.

[2] Kirkemo, F., Fasteners – Strength and Quality Requirements, PVP2014-28719.

[3] T.C Lin, C.M Chen, J.S.C Hsu and T.W Fu (2015). The Impact of Team Knowledge on Problem Solving Competence in Information Systems Development Team, Volume 33, Issue 8, *International Journal of Project Management*

[4] Deming, W.E. (1986). Out of the Crisis. Cambridge: Massachusetts Institute of Technology Press.

[5] Powell, T.C. (1995). Total quality management as competitive advantage: A review and empirical study. *Strategic Management Journal*, 16(1): 15.

[6] M.S.D Caro, M Wieth and S.L Beilock (2007). Methodologies for Examining Problem Solving Success and Failure, volume 42, Issue 1.

[7] Anderson, M., & Soha, A.S. (1999). A study of relationship between quality management practices and performance in small businesses. *International Journal of Quality and Reliability Management* , 16 (9): 859-877.

[8] 10000559172-PDC-000 – Aker Solutions Subsea Fasteners Global Procedure, Version 6, 2014

IOP Conf. Series: Materials Science and Engineering **370** (2018) 012018 doi:10.1088/1757-899X/370/1/012018

Comparative study of MYSat attitude stability effect on power generation and lifetime

Norilmi Amilia Ismail[1], Ahmad Shaqeer Mohamed Thaheer[1,*] and Mohd. Izmir Yamin[2]

[1]School of Aerospace Engineering, Universiti Sains Malaysia, Engineering Campus, Nibong Tebal, Pulau Pinang, Malaysia.
2 Independence-X Aerospace Sdn. Bhd., Innovation Incubation Centre, Technology Park Malaysia, Kuala Lumpur, Malaysia.

[*]ashaqeer@gmail.com

Abstract. Universiti Sains Malaysia Space System Lab (USSL) is currently developing a 1U cubesat named MYSat. The satellite mission is to measure electron-density in the Ionosphere E-Layer. Power generation from a solar panel is limited due to a small area of the satellite. Apart from that, the satellite is expecting to continuously spinning and tumbling throughout the mission lifetime as the satellite will be launched without an attitude control system. This paper compares the effect on power generation and the lifetime of MYSat of two conditions; first is with attitude controll where satellite pointing to nadir and later is uncontrol attitude of the satellite. The analysis has been conducted using Analytical Graphics, Inc. (AGI) Systems Tool Kit (STK) software. This study assumed the satellite used a hexagonal solar cell with a theoretical efficiency of 29% identical to an Ultra Triple-Junction (UTJ) solar cell. The simulation is done in one year duration on different attitude configuration. The worst-case condition, where the Earth is positioned at apogee, has been chosen for the comparative study and the lifetime of the satellite is also simulated and compared.

1. Introductions

Space projects becoming a trend in recent years with the development of small-satellites. These small-satellites concepts has decreased the development and mission cost [1]. Cubesat deployer [2], [3] such as P-POD and J-SSOD have enabled further reductions in the launch costs acting as a standardized deployment system. The typical small-satellite system usually uses commercially off-the-shelf (COTS) components and mainly operates for scientific research purposes and to prove new technology [4]. Nevertheless, constraint in weight and size challenges small-satellites developer in all aspects as it limits power generation, computational performance, and accurate attitude measurements and control [5].

Despite these limitations, small-satellites can be equipped with ambitious scientific payloads and technology for in-orbit demonstrations that lead Universiti Sains Malaysia (USM) to build an in-house small-satellite namely MYSat or Malaysian Youth Satellite. This small-satellite is planned to be deployed using J-SSOD (KIBO) module in the International Space Station (ISS) with the mission of measuring electron density in the Ionosphere E-layer. The 1U satellite will have dimensions of 10 cm x 10 cm x 10 cm and a maximum weight of 1.3 kg as standardized by the deployer requirements [3].

Nevertheless, the need of having attitude control in MYSat is abandoned due to power and thermal issues. Thermal issues are expected to occur when one side of the solar panel is facing the sun at all time. This will induce thermal heating to the MYSat [6]. Power generation is clearly affected by the withdrawing of the attitude control system here as the satellite will continuously spinning which in turn solves the thermal issues. The attitude control usually applied on a satellite that requires pointing accuracy [7], [8] such as FITSAT-1 [9], COMPASS-1 [10], and OUFTI-1 [11]. With the mission requirement does not require pointing accuracy, the needs of having control are rendered moot. Typical spin rate for cubesat is 1 deg/s [12] with 10 deg/s being the worst-case condition [1].

Power generated from the solar cell is a function of their efficiency, area, cell density, and temperature, Orbital parameters and satellite architecture (e.g. deployed panel configurations) and orientation all affect the anticipated power generation levels for solar panel systems [13]. Nevertheless, most of the solar energy is transformed into heat or rejected off the surface of the PV-cell. The efficiency of the PV-cell refers to percentages of the solar energy that is successfully transformed to electrical energy [14]. It is determined by the doping characteristics of the semiconductors.

The paper discusses the methods used in the analysis in Section 2 then continues with the results and discussion in Section 3.

2. Methodology

MYSat is expected to be deployed using J-SSOD (KIBO module) in ISS, thus the specifications is laid out in the JEM handbook documented by JAXA engineers have to be followed [3]. MYSat has a 10 cm x 10 cm x 11.35 cm in dimension with a maximum mass of 1.3 kg. Figure 1 depicts the MYSat structure and configuration. Each four sides of MYSat is covered with solar panels. The top and bottom surface are equipped with Electron Density and Temperature Probe (TeNeP) science unit and antenna deployment mechanism, respectively. Figure 1 show assembly and exploded view of MYSat.

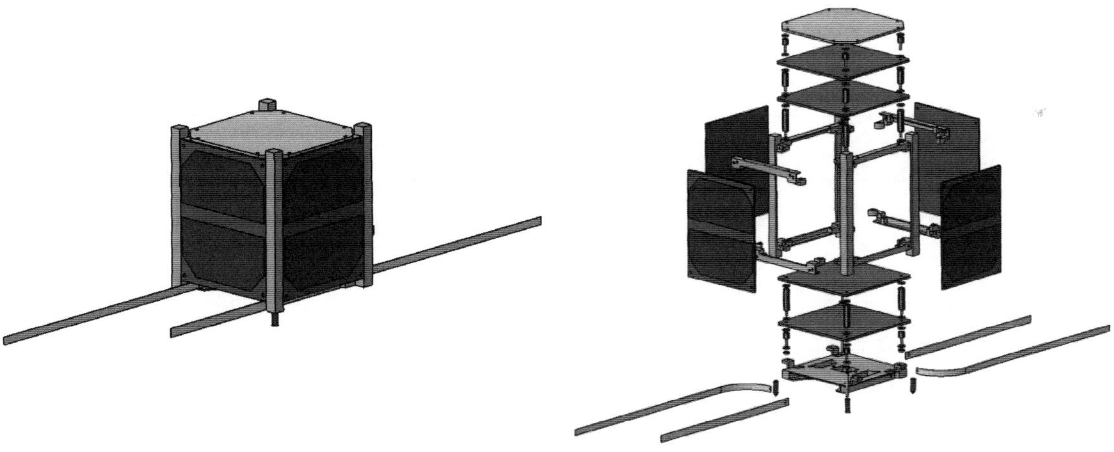

(a) (b)

Figure 1. (a) MYSat assembly CAD design. (b) MYSat exploded view.

In order to perform the analysis, the CAD model has been exported in *.dae format to define the solar panel area. Then, it was save in *.anc format to set the efficiency of the solar panel. The orbit is propagated using the Simplified General Perturbations Model 4 and analysis conducted using Analytical Graphics (AGI) System Tool Kit (STK) [15], [16]. It considers Earth magnetic field (International Geomagnetic Reference Field 11), third body gravity, and Earth atmosphere (NRLMSISE-2000). Atmospheric density model NRLMSISE-2000 is chosen due to its latest model in modeling atmospheric density. The model have statistical accuracies of about 15 % and that there has been no significant improvement in density models over the last two decades [17].

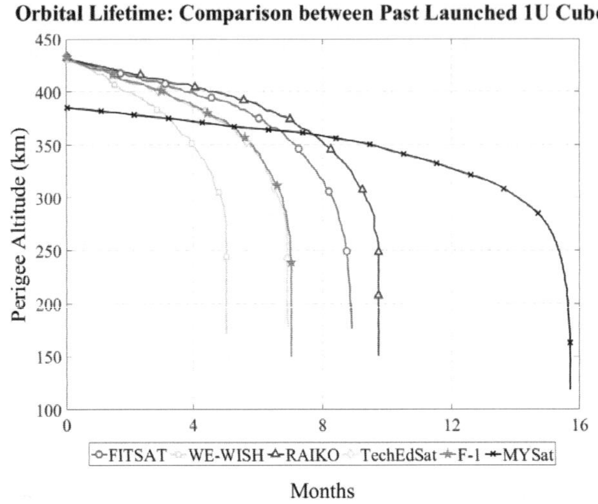

Figure 2. Comparisons of MYSat STK lifetime results and previously launched 1U cubesat

Previously launched Cubesat only manage to orbit for 6 to 12 months based on Figure 2, the analysis predicts that MYSat will be on-orbit for more than a year [18]. Choi Ha-Yeon et. al [19] gives the similar trend on their research. It is an utmost importance to remember that the result is an estimation. Simulated satellite lifetime has an approximately 10% error due to difficulty in accurately predict environmental effects such as solar and geomagnetic activities, atmospheric density and others [20]–[22]. The satellite lifetime can be increased by increasing its dry mass, reducing its cross-sectional area, or increased deployed altitude [23].

Next, the attitude is set Nadir oriented or spinning in all direction. The case study settings are presented in the Table 1 and are set in the STK's object properties window.

Table 1. Orbit settings

	Orbit
Propagator	SGP4
Step Size	60 s
	Attitude: Standard
Type	Nadir alignment with ECI velocity constraint Spinning
Spinning	**Body** Type: Cartesian X: 0.57735, Y: 0.57735, Z: 0.57735 **Inertial** Type: PR Angles Pitch: 0 deg, Roll: 0 deg Spin Rate: 1,3,5,7,10 deg/s Spin Offset: 0 deg Epoch: 1 Jan 2018 00:00:00.000 UTCG
	Model
	*.dae and *.anc file

The settings setup for solar panel analysis is shown in Table 2. Lifetime analysis is done using STK with the setting the parameters shows as in Table 3.

Table 2. Solar Panel analysis setup

Parameters	Descriptions
Obscuring Objects	Facility (optional) Bound Radius: 0.1 m
Visualization	Solar Panel Groups: Plus-Y, Plus-X, Minus-X, Minus-Y (user defined)
Data	Start: 1 Jan 2018 00:00:00.000 UTCG Stop: 1 Jan 2019 00:00:00.000 UTCG Time Step: 60 s
Data Reporting	Type: Power Generate: Report/Graph

Table 3. Lifetime analysis setup

Parameters	Description
Atmospheric Drag, C_d	2.2
Solar Radiation Pressure, C_r	1.5
Drag Area	0.01m^2
Area Exposed to Sun	0.01m^2
Mass	1 kg
Atmospheric Density	NRLMSISE-2000
Solar Flux File	SolFlx_CSSI.dat

The summary of the analysis is represented in the flowchart shown in Figure 3.

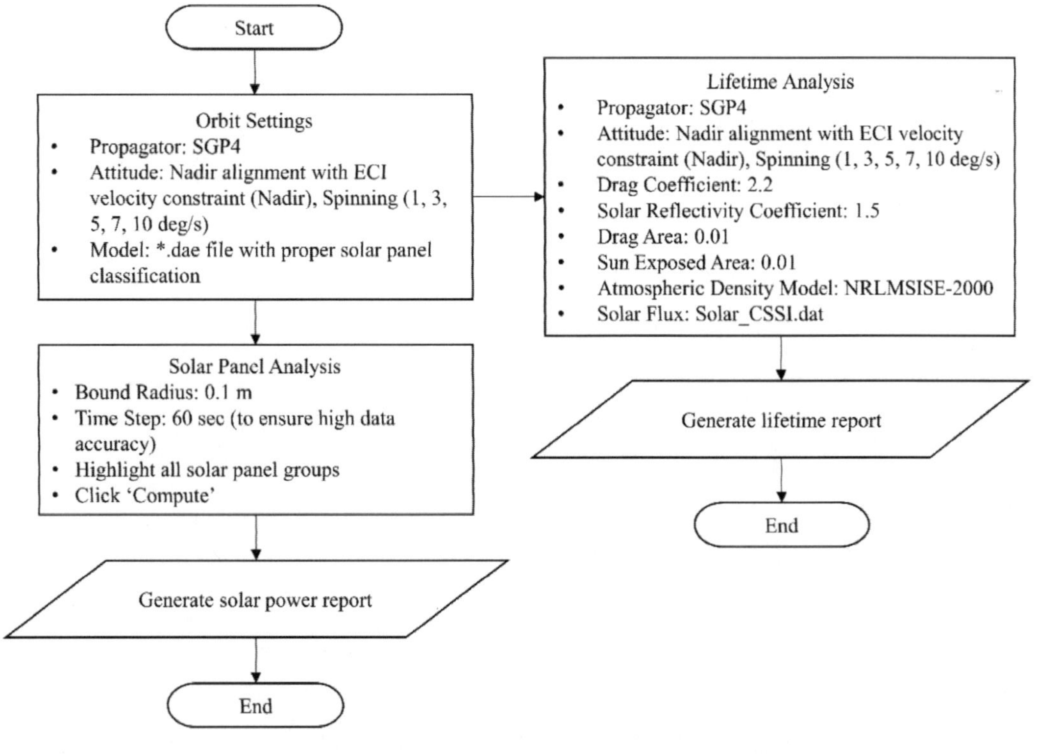

Figure 3. Attitude and lifetime analysis flowchart

3. Results and Discussion

3.1. Power Generation

Table 4 shows different shape of the solar panel and cell area resulted in the different amount of power generation. Hexagonal cell shape produces most solar power as it has bigger surface area. The solar cell area is based on the allowable size of the solar panel.

Table 4. Power generation based on different solar cell shapes

Cell Shape	Maximum (W)	Mean (W)
Hexagon	3.46	1.65
Square	2.81	1.34
Triangular	2.82	1.35

The solar cell efficiency plays an important role in selecting the best solar panel in the market [24]. This effect is shown in Table 5 where, higher efficiency will produce higher solar power output. This includes a safety margin of 20% from the theoretical efficiency to simulate actual solar cell performance. Note that, theoretical efficiency assumed the incident light is perpendicular to the solar panel while in actuality, the orientation changes.

Table 5. Comparison of power output on different solar cell material and efficiency

	GaAs (UTJ)	GaAs (SJ)	GaAs (MJ)	Thin Sheet Amorphous Si	Indium Phosphide	Crystalline Si
Efficiency	30.32%	22.00%	26.08%	11.20%	18.24%	20.24%
Max. Power	3.46	2.54	3.01	1.27	2.10	2.34
Avg. Power	1.65	1.21	1.43	0.60	1.00	1.11

The total solar power generated at any instances are from the solar power from each panel. Figure 4 shows the solar output from each solar panel with MYSat spins at 1 deg/s in all direction. Throughout one orbital period, there will a high and low output or no output at all from the panel depending location of MYSat with respect to the sun.

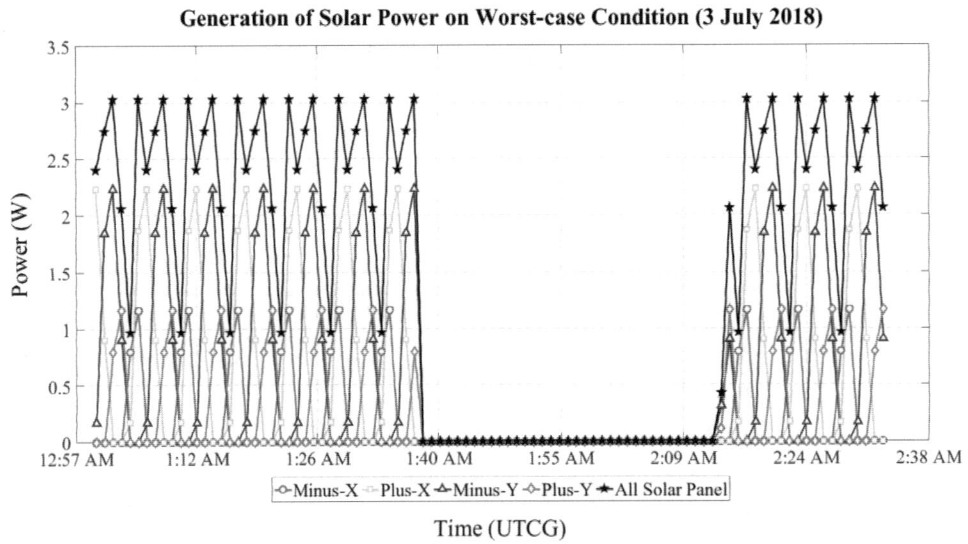

Figure 4. Individual solar panel output

Based on Figure 4, solar panel on -Y and +X produces the highest solar power while +Y and -X panel produces the lowest. The sum between all four panels produces the 'All Solar Panel' results. It can be seen to that, -Y and +X panel manage to produce about 2.3 W of power while +Y and -X produces 1.3 W. This shows, if only one solar panel is used, there will be power issues.

In addition, from 1:40 AM to 2:13 AM, the power graph drops to zero. This indicates that there is no power generation due to the eclipse. On this time, the battery will not be charged and battery is essential as it provides power to sustain the components used. Thus, power management needs to be properly set to avoid power loss [25].

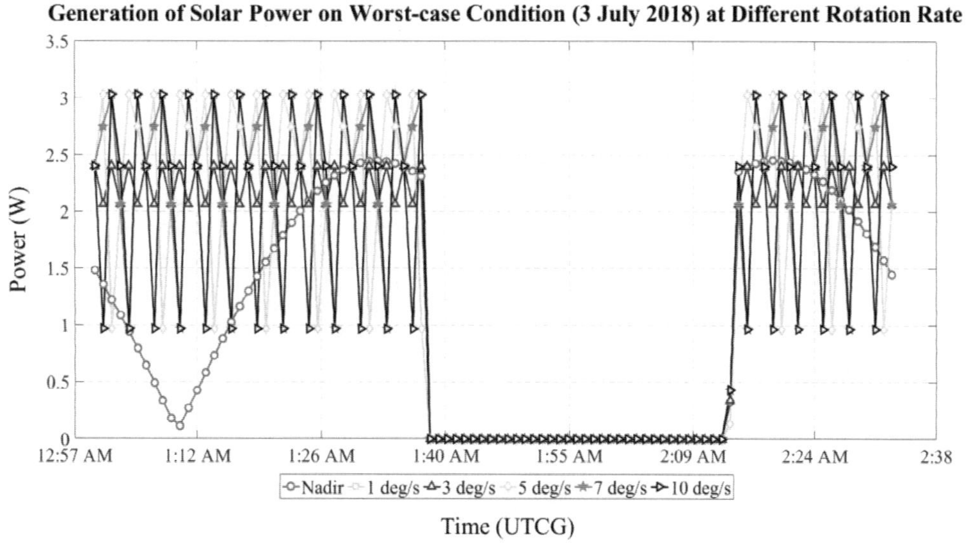

Figure 5. Comparison of generated solar power on different attitude spin rates

Figure 5 depicts the comparison of MYSat spin rate on total power generation. The power generated is oscillating in a short period of time due to rapid changes of the solar panel orientation while the Nadir pointed MYSat has a longer period. This is because of each solar panel able to spend a longer time in the sun. Thus, creating a constant power generation. However, the drawback is the power generated is limited by the efficiency of the solar cell creating a lower power output [26].

With referencing to Figure 5, different spin rate produces a same power output pattern but with lead-lag circumstances. 1 deg/s and 10 deg/s spin rate results produce the same output and probably due to STK's computation while 5 deg/s spin rate leads 7 deg/s spin rate results as it spun slower and have more time receiving sun rays. Interestingly, the spin rate of 3 deg/s produces a power generation oscillation with small differences compared to other spin rates. Overall, the power generated is approximated at 2 W to 2.7 W. In reality, MYSat cannot maintain its attitude spin rate without attitude control and external disturbance might induce the attitude spin rates [27].

3.2. *Lifetime Results*

As shown in
Table **6**, the spin rate does not affect the orbital decay date. This is because lifetime computation in STK only considers, drag coefficient, solar reflectivity coefficient, drag area, sun exposed area, atmospheric density model, and solar activity cycle [15]. In a circumstance where the attitude is unknown such as this research holds, a mean cross-sectional is assumed that produces a uniform attitude behavior which relative to the velocity direction [28]. The more detailed analysis shown in Figure 6 where the changes of perigee altitude are identical regardless of attitude behavior.

Table 6. Evaluation of decay date and duration on attitude spin rates

Spin Rate (deg/s)	Decay Date	Duration (years)
0.00 (Nadir)	21/4/2019	1.3
1.00	21/4/2019	1.3
3.00	21/4/2019	1.3
5.00	21/4/2019	1.3
7.00	21/4/2019	1.3
10.00	21/4/2019	1.3

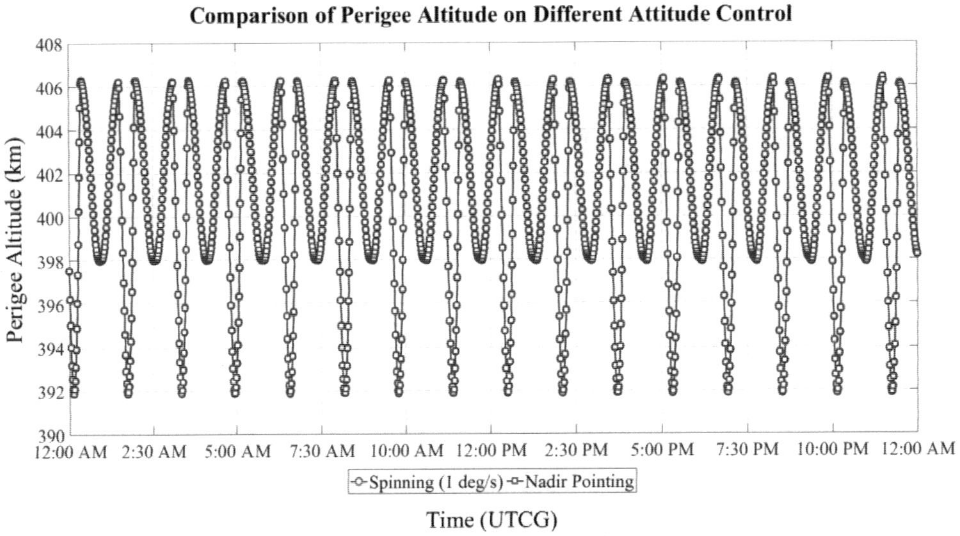

Figure 6. Assessment on perigee altitude on attitude control behavior

4. Conclusion

The effect on power generation and the lifetime of MYSat of two attitude behaviors i.e. Nadir oriented and continuously spinning condition for one year is done using Analytical Graphics, Inc. (AGI) Systems Tool Kit (STK) software. The study assumed the satellite used a hexagonal solar cell with a theoretical efficiency of 29% identical to an Ultra Triple-Junction (UTJ) solar cell. It can be seen that, power generated occurs in oscillation with Nadir oriented has a longer period compare to other spin rates. This is due to rapid changes of the solar panel orientation. Different spin rate produces a same power output pattern but with lead-lag circumstances. 1 deg/s and 10 deg/s spin rate results produce the same output while 5 deg/s spin rate leads 7 deg/s spin rate results as it spun slower and have more time receiving sun rays. The spin rate of 3 deg/s produces a small oscillation compared to other spin rates. The power generated is approximated at 2 W to 2.7 W. As for lifetime analysis, the spin rate does not affect the orbital decay date because of lifetime computation only considers, drag coefficient, solar reflectivity coefficient, drag area, sun exposed area, atmospheric density model, and solar activity cycle.

References

[1] T. Ito et al., "Reliable and Robust Implementation of Attitude Determination and Control Subsystem and Initial Flight Operation Results of 50-kg Class Interplanetary Spacecraft Procyon," in International Symposium on Space Flight Dynamics (ISSFD) 2015, 2015.

[2] J. Puig-Suari, C. Turner, and W. Ahlgren, "Development of the standard CubeSat deployer and a CubeSat class PicoSatellite," in 2001 IEEE Aerospace Conference Proceedings (Cat. No.01TH8542), 2001, vol. 1, p. 1/347-1/353.
[3] JAXA, "Small Satellite Deployment Interface Control Document," Chofu, Japan, JX-ESPC-101133-B, 2015.
[4] K. Woellert, P. Ehrenfreund, A. J. Ricco, and H. Hertzfeld, "Cubesats: Cost-effective science and technology platforms for emerging and developing nations," Adv. Sp. Res., vol. 47, no. 4, pp. 663–684, Feb. 2011.
[5] A. Addaim, A. Kherras, and E. Bachir, "Design of Low-cost Telecommunications CubeSat-class Spacecraft," in Aerospace Technologies Advancements, InTech, 2010.
[6] D. Q. Dinh, "Thermal Modelling of Nanosat," San José State University, 2012.
[7] J. Cutler, M. Bennett, A. Klesh, H. Bahcivan, and R. Doe, "The Radio Aurora Explorer - A Bistatic Radar Mission to Measure Space Weather Phenomenon," in 24th Annual AIAA/USU Conference on Small Satellites, 2010.
[8] S. Rawashdeh and J. Lumpp, "Nano-Satellite Passive Attitude Stabilization Systems Design by Orbital Environment Modeling and Simulation," in AIAA Infotech@Aerospace 2010, 2010.
[9] "Development and operations of nano-satellite FITSAT-1 (NIWAKA)," Acta Astronaut., vol. 107, pp. 112–129, Feb. 2015.
[10] A. Scholz et al., "COMPASS-1 Phase A Study," Aachen, Germany, 2003.
[11] S. Galli, "Mission Design for the CubeSat OUFTI-1," Université de Liège, Liège, Belgium, 2007.
[12] D. M. Erb, S. A. Rawashdeh, and J. E. Lumpp Jr., "Evaluation of Solar Array Peak Power Tracking Technologies for CubeSats," in 25th Annual AIAA/USU Conference on Small Satellites, 2011.
[13] I. T. Osman, "Design and Implementation of EPS (Electrical Power System) of a Cubesat," University of Khartoum, 2012.
[14] K. O. Skyttemyr, "Design and Implementation of the Electrical Power System of the CubeSTAR Satellite," University of Oslo, 2013.
[15] Analytical Graphics Inc., "Systems Tool Kit Software," 2017. [Online]. Available: https://www.agi.com/products/stk/. [Accessed: 06-Feb-2017].
[16] D. Vallado, P. Crawford, R. Hujsak, and T. S. Kelso, "Revisiting Spacetrack Report #3," in AIAA/AAS Astrodynamics Specialist Conference and Exhibit, 2006.
[17] O. Montenbruck and E. Gill, Satellite Orbits: Models, Methods and Applications, 1st ed. Berlin, Germany: Springer-Verlag, 2000.
[18] A. S. Mohamed Thaheer, N. A. Ismail, S. H. Md. Yusoff, M. I. Yamin, and N. A. Razak, "Mission Analysis and Design of MYSat - A 1U Cubesat for Electron-density Measurement," in 45th Student Conference - International Astronautical Congress (IAC), 2017.
[19] H.-Y. Choi, H.-D. Kim, and J.-D. Seong, "Analysis of Orbital Lifetime Prediction Parameters in Preparation for Post-Mission Disposal," J. Astron. Sp. Sci., vol. 32, no. 4, pp. 367–377, Dec. 2015.
[20] A.-A. L. C. Cojuangco, "Orbital Lifetime Analysis of Pico- and Nano-Satellites," University of Florida, Florida, USA, 2007.
[21] V. Baukelaers, "From Mission Analysis to Space Flight Simulation of the OUFTI-1 Nanosatellite," University of Liège, 2009.
[22] L. Qiao, C. Rizos, and A. G. Dempster, "Analysis and Comparison of CubeSat Lifetime," Surv. Geospatial Eng. (SAGE), Univ. New South Wales, p. 12, 2012.
[23] M. A. Afful, "Orbital Lifetime Predictions of Low Earth Orbit Satellites and the Effect of a DeOrbitSail," University of Stellenbosch, South Africa, Stellenbosch, South Africa, 2013.
[24] I. Vertat and A. Vobornik, "Efficient and Reliable Solar Panels for Small CubeSat Picosatellites," Int. J. Photoenergy, vol. 2014, pp. 1–8, 2014.

IOP Conf. Series: Materials Science and Engineering **370** (2018) 012018 doi:10.1088/1757-899X/370/1/012018

[25] C. Clark and S. Kirk, "Off-the-Shelf, Deployable Solar Panels for CubeSats." Clyde Space Ltd., 2012.
[26] I. Vertat and A. Vobornik, "Efficient and Reliable Solar Panels for Small CubeSat Picosatellites," Int. J. Photoenergy, vol. 2014, pp. 1–8, 2014.
[27] R. teja Nallapu and J. Thangavelautham, "Precise Pointing of Cubesat Telescopes: Comparison Between Heat and Light Induced Attitude Control Methods," 2017.
[28] D. Oltrogge and K. Leveque, "An Evaluation of CubeSat Orbital Decay," in Proceedings of the AIAA/USU Conference on Small Satellites, 2011, vol. SSC11-VII-, p. 12.

IOP Conf. Series: Materials Science and Engineering **370** (2018) 012019 doi:10.1088/1757-899X/370/1/012019

Rate of Perceived Exertion of Female Aviation Workers Pertaining to Selected Aviation Workspace and Recommended Fitness Regime to Reduce Exertion

Mohd Harridon

Universiti Kuala Lumpur Malaysian Institute of Aviation Technology
Lot 2891, Jalan Jenderam Hulu, Jenderam Hulu, 43800 Dengkil, Selangor.

*Corresponding author: mdharridon@unikl.edu.my

Abstract. In aviation maintenance, workers usually work in tight spaces to maintain or repair aircraft. These constricted spaces provide uncomfortable feeling to these workers and in return the workmanship of the workers declined in terms of quality and thus affect the safety of the aircraft. However, some workers are fit enough to go through these conditions without hassle or without much pain. This paper focused upon female aviation workers and collected data on their rate of perceived exertion pertaining to two selected aviation workspaces. The data and analyses showed that those who are fit felt less exertion while those who are not fit felt more exertion. This paper then recommended several physical exercises to increase their fitness and in turn reduce the exertion felt.

1. Introduction

The aviation world involves around safety where safety is the uttermost important. In the field of aircraft maintenance, aviation workers indulge in hours of works fixing and maintaining fleets of aircrafts. Those who are fit and are not subjected to rigorous situation would produce optimum results maintaining the aircraft. Work outputs which are mediocre or bad would compromise the safety of the aircrafts and plausibly lead to accidents or incidents.

Aircraft Maintenance Textbook authored by Federal Aviation Administration (FAA) stated that 80% of aircraft accidents were due to human factors [1]. The handbook goes further by stating that "not conducive" working environment is a contribution to stress which also one of the attributes of human factors [2]. While "not conducive" working environment such as tight spaces in an aircraft is hard to eliminate, the stress value associated with it could be reduced if one is fit. A fit person with certain muscle mass (for example at the legs) and optimum body weight could withstand kneeling, for say around 30 seconds, to fix an aircraft component at a tight space.

In this paper we had chosen two working conditions which were likely causes of stress for maintenance workers. Our respondents or test subjects are female students of aviation maintenance studying at Universiti Kuala Lumpur Malaysian Institute of Aviation Technology (UniKL MIAT). These students, in their course of the aviation maintenance programme, had gone through numerous practicals (tight and open spaces) in fixing or maintaining aircraft. This paper recorded their RPE (Rate of Perceived Exertion) which indicates the level of their stress, actuate analysis upon these RPE, and recommended appropriate physical exercise to decrease these RPE levels.

2. Literature Review of Rate of Perceived Exertion (RPE)

The United State Health Department stated that RPE is a tool that could measure the intensity of a physical activity [3]. Perceived Exertion is a feeling that one get whether it is easy or hard, depending upon one's fitness and the gravity of the physical activity. The feeling encompasses heart rate, perspiration, fatigue of muscle, and breathing rate.

RPE is based upon one's perception, thus is it subjective. But the US Health Department reiterated that it still provides a good measure of one's capability. RPE was invented by Gunnar Borg of University of Stockholm. Borg's intention was to quantify one's strain or aches and this is done using a scale [4]. This scale provides a tool for comparison, regardless of gender, age, race, and others.

Borg acknowledged that there's a drawback in making comparison between individuals but this drawback is countered by categorizing the scale with categories such as "Nothing at all", "Very very weak", "Very weak", "Weak", "Moderate", "Somewhat strong", "Strong", "Very Strong", and "Very very strong". There exist in the variation of the scale in terms of the terminologies used in the categorization but the basis is the same. The RPE scale that's used in this paper is based upon the Cleveland Clinic in the United States with a ranging from 0 till 10 (0 being the most easiest and 10 being the most hardest)

Borg further explained that the scale could also be used to rate one's difficulties in breathing. This is quite interesting as maintenance workers, in the author's opinion based upon heuristic, who are subjected to intense working condition such as tight space reported slight changes in their breathing rate. According to a report by Kane from the United States Air Force, there is a need to measure an employees stress level in lieu with their working conditions and investigation upon their perceptions, through RPE, is one way to gauge stress level [5]. Kane recommended that employees to be "conditioned" (via physical exercise) to allow them to cope or decrease their stress level and thus increase safety [5].

3. Methodology

The methodology of this project is shown in Figure 1. 50 respondents which are students from UniKL MIAT were chosen. The chosen subjects were all females and chosen randomly. The scope of this project is confined only for females as they are a minority in the first line aviation maintenance workforce. According to Clark, only 2.3% of the certified aircraft mechanic workforce in the United States are women [6]. The author is interested on how the female workers fare in their roles, in terms of fitness and workplace exertion (strain), thus the existence of this project.

Two scenarios of aircraft maintenance activity were selected where these two activities are performed at tight spaces. Its a norm that avionics rack in an aircraft is located at the cargo bay of the aircraft and workers need to transverse tight spaces in order to actuate maintenance upon the rack. Workers are then required to kneel or sit in order to tinker upon the rack. Thus the two scenarios used in this project are : "Kneeling for 8 Seconds" (Scenario 1) and "Crawling a Distance of 2 Meters" (Scenario 2).

The crawling of 2 meters is based upon the exact length that one has to transverse in the Boeing 727 in order to gain access to the avionics rack [7]. For other mid size aircrafts the crawling length is roughly similar. As for kneeling for 8 seconds it was based upon real life maintenance works upon avionics rack of the Boeing 727 where rough measurements of the duration of the workers kneeling was taken by the author. To gain realism, the two scenarios of this project were actuated at the avionics rack located at the cargo bay of the Boeing 727 that's currently parked at UniKL MIAT.

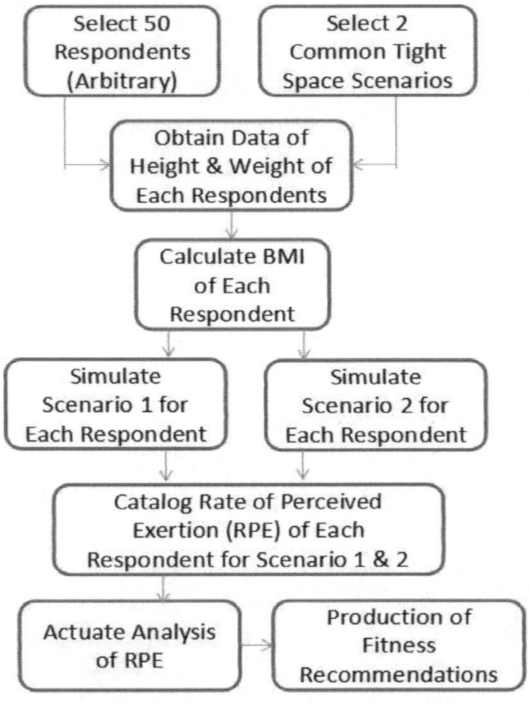

Figure 1. The Methodology of the Project

Each respondent was measured in terms of height (meter) and weight (kg). A measuring tape was used to measure the height and a digital scale was used to measure the weight. Using equation (1), the Body Mass Index (BMI) of each respondent was calculated.

$$BMI = Weight\ (kg) / [Height\ (m)*Height\ (m)] \quad (1)$$

Then, Scenario 1 was simulated for each respondent. After the end of Scenario 1, each respondent was asked to rate the exertion that she had encountered using the RPE scale shown in Figure 2 (this RPE scale was taken from the notes of Eric Kim) [8]. These were then repeated for Scenario 2.

Figure 2. The RPE Scale that Was Used

The rated exertions of each respondent for each scenario were then recorded and cataloged in an excel spreadsheet. Analysis and calculation were then actuated from these data. The analysis and calculation are:

1. Average RPE for Situation 1
2. Average RPE for Situation 2
3. 2D Graphical Trend of RPE Situation 1
4. 2D Graphical Trend of RPE Situation 2
5. The Frequency of RPE Values for Situation 1
6. The Frequency of RPE Values for Situation 2
7. Distribution of Respondents in Terms of BMI Values.

The final step was to give recommendations in terms of physical exercise in order to decrease the RPE values of respondents. The recommendations were based upon the analysis of the RPE and the recommended appropriate exercises were based upon literature review or best practices in fitness regime.

Figure 3. Tight Spaces where Aviation Workers need to Kneel and Crawl in Order to Maintain Aircraft Components

4. Results

Figure 4 shows the RPE values among 50 respondents pertaining to Situation 1. The RPE values among respondents are mixed with high, middle, and low values. Meanwhile Figure 5 shows the RPE values among 50 respondents pertaining to Situation 2. The RPE values among respondents are also mixed with high, middle, and low values. Figure 6 shows the average value of RPE pertaining to Situation 1. Figure 7 shows the average value of RPE pertaining to Situation 2. Figure 8 shows the frequency of RPE values pertaining to Situation 1. Figure 9 shows the frequency of RPE values pertaining to Situation 2. Figure 10 shows the distribution of respondents in terms of BMI values.

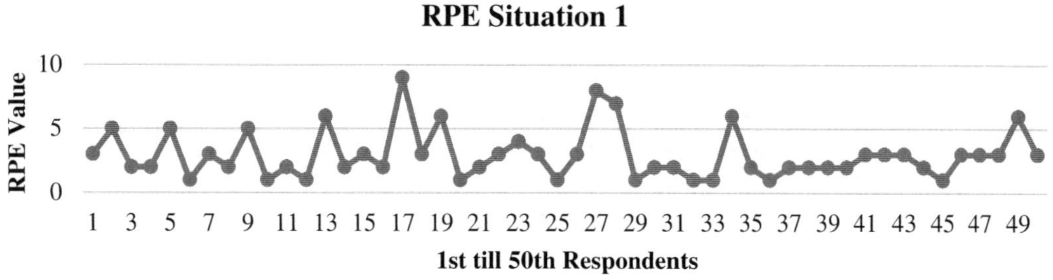

Figure 4. RPE for Situation 1

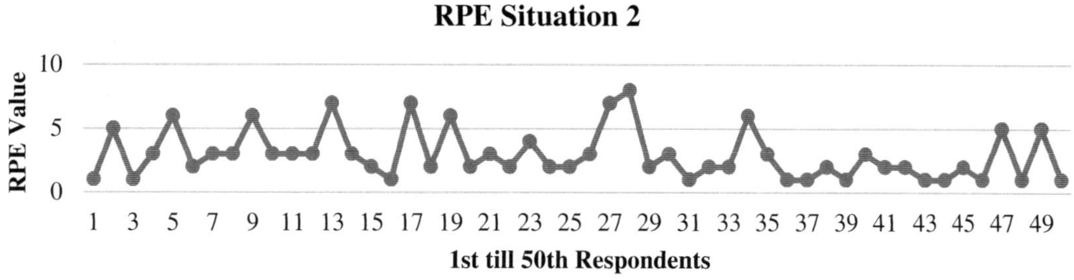

Figure 5. RPE for Situation 2

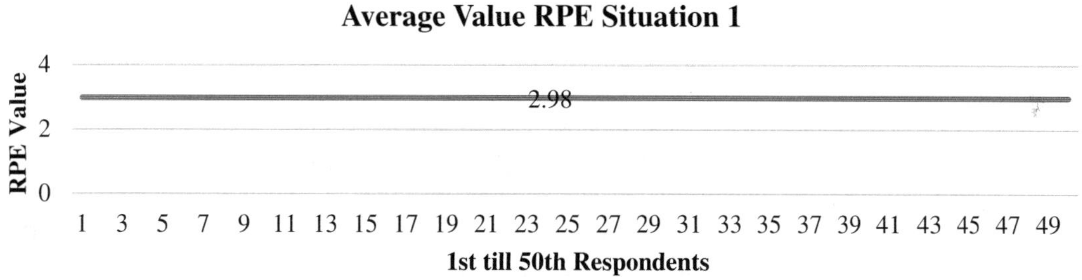

Figure 6. Average Value of RPE pertaining to Situation 1

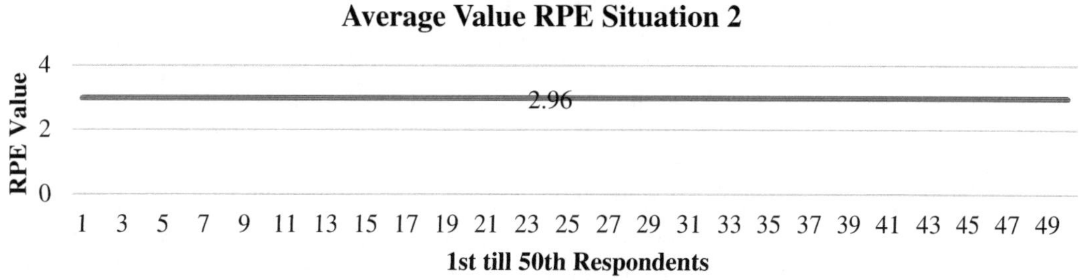

Figure 7. Average Value of RPE pertaining to Situation 2

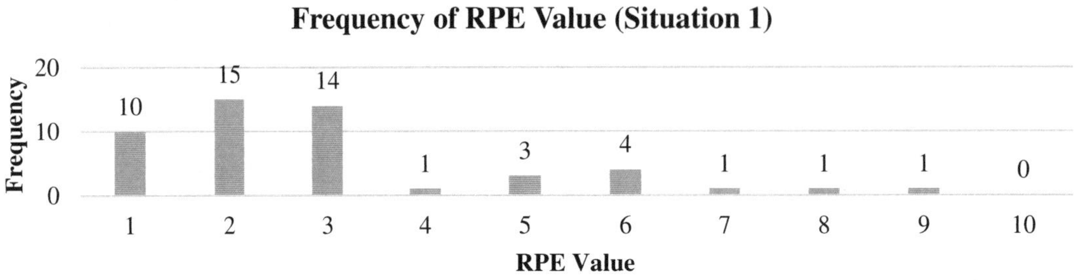

Figure 8. Frequency of RPE values pertaining to Situation 1

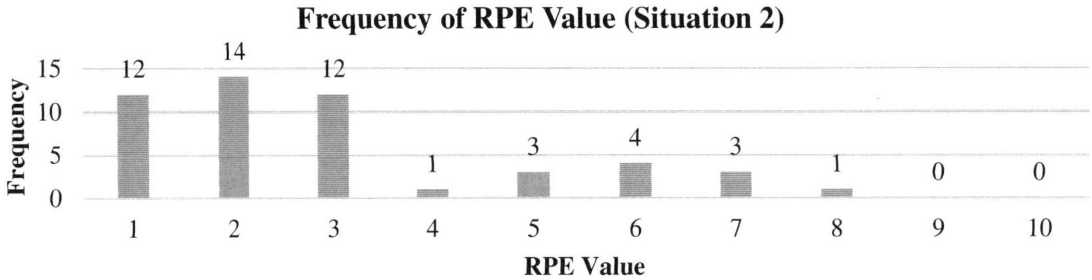

Figure 9. Frequency of RPE values pertaining to Situation 2

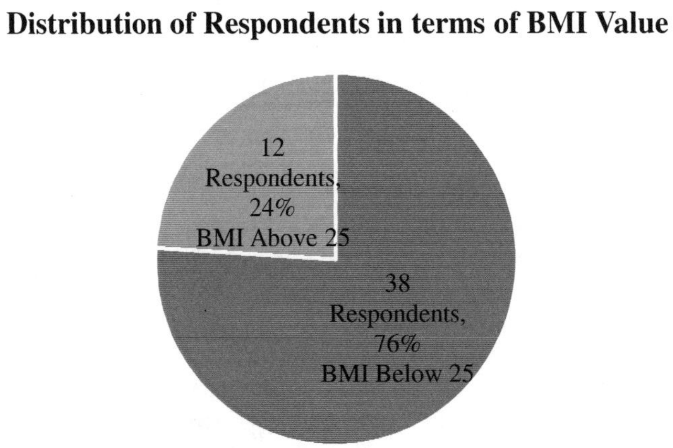

Figure 10. Distribution of Respondents in terms of BMI Values

5. Analysis of Results and Discussion

Pertaining to Figure 4, we can observe that the respondents gave mixed reactions with the highest RPE being 9 and the lowest RPE being 1. Situation 1, which is kneeling for 8 seconds, was deemed Really Really Hard by 1 respondent, as shown in Figure 8. This 1 respondent was respondent number 17. Peering the data sheet, the BMI for this respondent was 29.4758. According to the BMI Scale, a value of 25 above (for female) is considered overweight and obese. It is not surprising that this 1 respondent has a BMI value of above 25 and thus this "overweight" contributed to her feeling or perceiving that Situation 1 was Really Really Hard.

Situation 1 was deemed Really Easy by 10 respondents, as shown in Figure 8. Peering the data sheet, the BMI for these respondents were below the value of 25. According to the BMI Scale, a value below 25 (for female) is considered healthy weight. This healthy weight may have contributed to them feeling or perceiving that Situation 1 was Really Easy.

Looking at the RPE Scale from Figure 2, the "hard" category starts from RPE value of 4 (sort of hard) till 10 (maximal). We are interested to observe how many respondent perceive Situation 1 as "sort of hard" till "maximal". Peering at Figure 8, 11 respondents had indicated that Situation 1 is hard (ranging from sort of hard till maximal). That's 22% out of the total respondents. We were interested to know whether these 11 respondents have BMI above 25 or not. Peering at the excel sheet, all of the 11 respondents indicated above have BMI above 25.

Looking at the RPE Scale from Figure 2, the "easy" category starts from RPE value of 0 (rest) till 3 (moderate). We are interested to observe how many respondent perceive Situation 1 as "rest" till "moderate". Peering at Figure 8, 39 respondents had indicated that Situation 1 is easy (ranging from rest till moderate). That's 78% out of the total respondents. We were interested to know whether these 39 respondents have BMI below 25 or not. Peering at the excel sheet, 38 respondents indicated above have BMI below 25 while only 1 has a BMI above 25 (the BMI value of this one respondent is 25.711).

Pertaining to Figure 5, we can observe that the respondents gave mixed reactions with the highest RPE being 8 and the lowest RPE being 1. Situation 2, which is crawling a distance of 2 meters, was deemed Really Hard by 1 respondent, as shown in Figure 9. This 1 respondent was respondent number 28. Peering the data sheet, the BMI for this respondent was 26.3702. According to the BMI Scale, a value of 25 above (for female) is considered overweight and obese. It is not surprising that this 1 respondent have a BMI value of above 25 and thus this "overweight" contributed to her feeling or perceiving that Situation 2 was Really Hard.

Situation 2 was deemed Really Easy by 12 respondents, as shown in Figure 9. Peering the data sheet, the BMI for these 12 respondents were between 19 till 24.5603. According to the BMI Scale, a value between 19 and 24.5603 (for female) is considered healthy weight. This healthy weight may have contributed to them feeling or perceiving that Situation 2 was Really Easy.

Again looking at the RPE Scale from Figure 2, the "hard" category starts from RPE value of 4 (sort of hard) till 10 (maximal). We are interested to observe how many respondent perceive Situation 2 as "sort of hard" till "maximal". Peering at Figure 9, 12 respondents had indicated that Situation 2 is hard (ranging from sort of hard till maximal). That's 24% out of the total respondents. We were interested to know whether these 12 respondents have BMI above 25 or not. Peering at the excel sheet, all 12 respondents have BMI above 25.

Looking at the RPE Scale from Figure 2, the "easy" category starts from RPE value of 0 (rest) till 3 (moderate). We are interested to observe how many respondent perceive Situation 2 as "rest" till "moderate". Peering at Figure 9, 38 respondents had indicated that Situation 2 is easy (ranging from rest till moderate). That's 75% out of the total respondents. We were interested to know whether these 38 respondents have BMI below 25 or not. Peering at the excel sheet, all of the 38 respondents indicated above have BMI below 25.

Situation 2 seems to be just slightly harder to the respondents in comparison to Situation 1 based upon the fact that 24% of the respondents indicated Situation 2 as hard as compared to 22% of the respondents which indicated Situation 1 as hard. On average Situation 1 has an RPE of 2.98 which falls in the category of easy. This indicates Situation 1 is still a favorable working condition but it is advisable for the respondents that perceived it as hard to go through recommended physical exercises to reduce the strains they encountered during Situation 1.

On average Situation 2 has an RPE of 2.96 which falls in the category of easy. This indicates Situation 2 is also still a favorable working condition but its advisable for the respondents that perceived it as hard to go through recommended physical exercises to reduce the strains they encountered during Situation 2.

24% of the respondents have unhealthy BMI which is slightly alarming as maintenance crews have to work in tight spaces most of the time. Hence actuating the recommended physical exercises would reduce their BMI to a range which is lower than 25.

A majority of respondents deemed Situation 1 and 2 as easy (78% and 75% respectively) which is good in terms of workers output as a worker which is in a environment which is not strenuous would produce optimum results.

6. Recommended Physical Exercises

Situation 1 which is kneeling for 8 seconds and Situation 2 which is crawling a distance of 2 meters require an individual to press the knee caps onto the floor for a stipulated duration of time (for kneeling) and move the knees at a distance (crawling). This requires the individual to have strong knees and there are several exercises which focuses upon the knees. Based upon the data collected and analysis of the results, its recommended, though not compulsory, that those with BMI above 25 to lose weight first before actuating the knees exercise. This is to prevent strain upon the knees as those with excessive weight would induce high amount of forces upon the knees.

One such way to decrease the BMI is through the implementation of High Intensity Interval Training (HIIT). HIIT program requires the individual to go through short intense exercise with short rest in between. And this cycle is repeated several time where the total duration is usually less than 30 minutes. Below is a typical HIIT program based upon the basis from Stoppani [9].

1. Sprint with 100% high intensity for 15 seconds
2. Rest or walk for 60 seconds
3. Repeat the cycle above another 10 times
4. Last Sprint with 100% high intensity for 15 seconds
5. Total Time of the above program is 14 minutes

The above program should be implemented for 2 till 3 weeks depending upon the outcome of the program. If weight loss is not imminent, extend the duration till 4 weeks or more. The design of the program is not rigid with extension allowable.

Individuals with BMI below 25 or those that already done HIIT and had achieved BMI less than 25 should then proceed to the knees exercise. Below are basic knees exercises which strengthen the muscle around the knee joints which in turn decreases stress upon the knees (this prevents injuries to the knees as well) [10].

To actuate Partial Squats, one has to stand about 12 inches away from the front chair with one's feet about hip-width apart and toes forward (look at Figure 11). Bending at the hips, one has to slowly lower one's self halfway down to the chair. One has to keep one abs tight and ensure the knees are behind the toes. Then lift oneself up again. One has to actuate this 10 till 12 repetitions and 3 sets till 4 sets with around 2 till 4 minutes rest in between set. It is advisable for one to actuate this exercise 2 or 3 times a week.

Another knee exercise that could be actuated is the Step-Ups as shown in Figure 12. Using an Aerobic Step Bench or a staircase, one has to step up onto the step with one's right foot. One's left foot is then tap on the top of the step and then lower it down on the floor. As one step up, the knee should be directly over the ankle. The process is then repeated for the left foot. One has to actuate this 10 till 12 repetitions and 3 sets till 4 sets with around 2 till 4 minutes rest in between set. It is advisable for one to actuate this exercise 2 or 3 times a week.

Figure 11. Partial Squats

Figure 12. Step-Ups

Another knee exercise which is good for strengthening the knees is the Calf Raises. This is shown in Figure 13. Using a chair or wall for balance, one stand with one's feet about hip width apart and toes straight ahead. Then one slowly lift one's heels off the floor,rising up onto one's toes. Hold this position and then lower back to the floor. One has to actuate this 10 till 12 repetitions and 3 sets till 4 sets with around 2 till 4 minutes rest in between set. It is advisable for one to actuate this exercise 2 or 3 times a week.

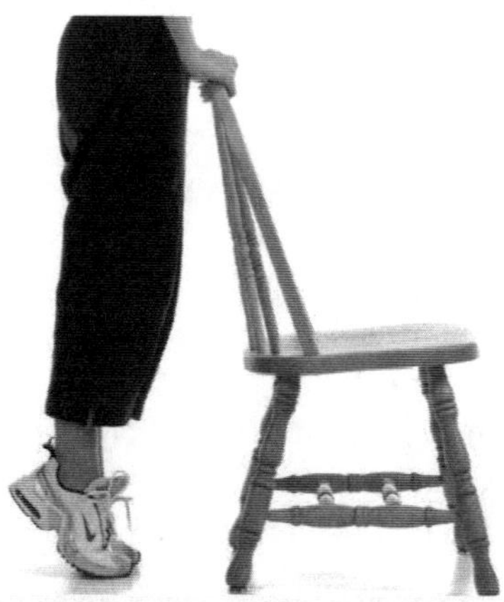

Figure 13. Calf Raises

7. Conclusion

It is imperative that workers in the aviation field are fit as their fitness would influence the outcome of their work. As shown in the data collected, workers (respondents) that have healthy BMI have less exertion imparted on them making them "comfortable" during the actuation of their jobs. This perhaps would make them produce optimum results in their work, thus improving safety.

Analysis of the data also showed only a marginal percentage (22% and 24%) of the workforce (respondents) had a hard time actuating the simulated maintenance task. And data shown that these percentage had a hard time due to their unhealthy BMI.

The author had recommended several concentrated exercises to be actuated by all workers in order for them to retain fitness or to increase their fitness (at the same time lower their BMI to acceptable level). These prescribed exercises are based upon the data collected which indicates concentration should be upon the surrounding muscle of the knees. Overall, the situation is not alarming as a majority of the sampled aviation workers (78% and 75% respondents) had an easy time actuating the simulated maintenance task.

The author opinionated that there is room for improvement in terms of workers fitness and we do hope the recommended exercises could increase their fitness and thus produce optimum work. The author currently in the midst of collecting data of the fitness level of these workers after they had gone through the prescribed exercises.

8. Acknowledgment

The author would like to acknowledge the contribution of the respondents (aviation students of UniKL MIAT) in participating in this project.

References

[1] Federal Aviation Administration (2011). Aviation Maintenance Technician Handbook - General. Chapter 14, Addendum Human Factors, page 28 of Chapter 14.

[2] Federal Aviation Administration (2011). Aviation Maintenance Technician Handbook - General. Chapter 14, Addendum Human Factors, page 23 of Chapter 14.

[3] US Health Department (2015). Perceived Exertion (Borg Rating of Perceived Exertion Scale). Centers for Disease Control & Prevention. Atlanta, Georgia.

[4] Borg, G. (1982). Psychophysical Bases of Perceived Exertion. Medicine and Science in Sports and Exercise. Volume 14, No. 5, page 377. Stockholm Sweden.

[5] Kane, W. (1988). Stress and Aircraft Maintenance Performance in a Combat Environment. Logistics and Human Factor Division. Wright Patterson Air Force Base, Ohio. 45433-6503. August 1988.

[6] Clark, P. (2015). Overcoming Gender Barriers in Aircraft Maintenance : Women's Perceptions in the United States. Collegiate Aviation Review. Fall 2015, Volume 33, Number 2. University Aviation Association.

[7] Continental Airlines (1980). Electrical Avionic Systems Boeing 727. Book 1. Maintenance Training Department. Houston.

[8] Kim, E. (2017). Math of Fitness. Wellness Resources. Andrews University, Berrien Springs, MI.

[9] Stoppani, J (2016). The Ultimate 8 Week HIIT for Fat Burning Program. BodyBuilding.com, Boise Idaho, July 24th 2016.

[10] Stanten, M. (2015). The Best and Worst Exercises for Bad Knees. Prevention Magazine. July 30th 2015, Rodale Incorporated, Harlan IA.

Evaluation of reaction time performance and subjective drowsiness during whole-body vibration exposure

A Azizan*[1], Z Zali[1], and H Padil[1]

[1]Universiti Kuala Lumpur-Malaysia Institute of Aviation Technology. Lot 2891 Jalan Jenderam Hulu, 43800 Dengkil, Selangor, Malaysia

*Corresponding author: mohdamzar@unikl.edu.my

Abstract. Despite the automotive industry's interest in how vibration affects the level of human comfort, there is little focus on the effect of vibration on drowsiness level. Thus, this study involves eighteen healthy male participants to study the effect of exposure to vibration on the drowsiness level. Prior to the experiment, the total transmitted vibration measured at interfaces between the seat pan and seat back to the human body for each participant was modified to become 0.2 ms^{-2} r.m.s and 0.4 ms^{-2} r.m.s. During the experiment, the participants were seated and exposed to 20-minutes of Gaussian random vibration with frequency band 1-15 Hz at two level of amplitude (low vibration amplitude and medium vibration amplitude) on separate days. The level of drowsiness was measured using a PVT test prior and after exposure to the vibration while participants rated their subjective drowsiness by using the Karolinska Sleepiness Scale (KSS). The significant increase in the number of lapse and reaction time because of the exposure to vibration in both conditions provide strong evidence of drowsiness. In this regard, the medium vibration amplitude shows a more prominent effect. All participants have shown a steady increase of drowsiness level in KSS. Meanwhile, there are no significant differences found between low vibration amplitude and medium vibration amplitude in the KSS. These findings suggest that human alertness level is greatly affected by the exposure to vibration and these effects are more pronounced at higher vibration amplitude. Both findings indicate that the presence of vibration promotes drowsiness, especially at higher vibration amplitude.

1. Introduction

Over the last decades, researchers have focused on early prediction and assessment on the drowsiness level of seated occupants' drowsiness in a range of transportation vehicles. There are many factors leading to drowsiness, including influences from substances like alcohol as well as sleep deprivation [1, 2]. In this regard, numerous national and international safety standards have been drafted which help describe the health risks of perceived vibration to the human body. These standards establish guidelines and thresholds, such as weighting filters, that can use to approximate the comfort of the human ride, however, how drowsiness is caused by the vibration of a vehicle is yet to investigated and understood. In this regard, drowsiness is a multifactorial aspect that influences not only the notions of quality perception (comfort) and judgment (safety) but also health as long-term exposure to vibration could lead to muscular-skeletal injuries [2]. Past studies had predicted and developed an understanding of how the human body subjectively responds to vibrations [3]. However, there is still limited quantitative study which investigated the influence of vibration on seated occupants' drowsiness level.

Studies had shown that drowsiness and prolonged sleep deprivation could lead to declining response speed (1/RT), increase in reaction time (RT) and the increase in the number of lapses (RT > 500 ms) [4, 5, 6], but the attempt to establish the relations between psychomotor deficit such as drowsiness through measuring the reaction time performances, has not yet been able to fully investigate the drowsiness caused by exposure to vehicle vibration. The experiment is based on the ISO2631-1, 'Evaluation of human exposure to whole-body vibration' weightings and test procedures [7] guidelines which stipulate that regardless of the frequency content, the fixed frequency weighted r.m.s value of vibration transmitted to the seated human occupants will show a constant subjective comfort rating, in order to conduct an objective and subjective measurement of human participants' drowsiness level based on human perceived vibration.

2. Methodology

2.1. Human participant

The study involved eighteen healthy, randomly selected male university students, with the mean age of 23.0 ± 1.3 years. The average height for the participants was 168.2 ± 4.0 cm, and their average weight was 69.3 ± 9.88 kg, with the average BMI of 22.6 (SD=2.54) kg/m^2. Prior to the experiment, the Epworth Sleepiness Scale (ESS) was used to detect any abnormalities in sleepiness among the participants [8]. A score of > 10 indicates excessive sleepiness, and participants with this score were exempted from the experiment [9]. The participants were instructed to have enough sleep a day prior, and based on the results, the participants' total ESS score ranged from 0 to 24. The scores below 7 are perceived as normal; 8 to 10 are considered as moderate while participants with scores of 11 to 15 are defined as elevated risks and finally, those with scores exceeding 16 illustrate severe excessive daytime sleepiness (EDS) [10], showing a high risk for them to fall asleep in a range of monotonous situations [11].

2.2. Objective Measure-Psychomotor Vigilance Test (PVT)

The PC-based Psychomotor Vigilance Test (PVT-192: Ambulatory Monitoring Inc., Ardsley, New York) was used to measure reaction time [12]. The test comprises of a 10–minutes visual reaction time task evaluating the sustained attention in two conditions; before vibration exposure and after vibration exposure. Participants were asked to click a mouse as fast as they can to respond to the appearance of a visual stimulus. The stimulus was in the form of a diode which emits a red light that displays time in milliseconds. A visual stimulus would appear in 2-10 seconds variable intervals of during each 10-minutes session. A software program was used to obtain three PVT performance metrics during each PVT condition; these metrics are mean reaction time, the number of lapses and median reaction time. A valid response would be > 100 ms and response less than 100 ms is considered as a false signal [5]. The participants were exposed to a Gaussian random vibration, with 1-15 Hz frequency bandwidth, for 20-minutes. During vibration condition, the participants were asked to sit comfortably while their feet as firmly fixed at the footrest, their back on the backrest and hands on their lap. They were also were asked to restrict any physical movement. The footrest was isolated from the vibration and is not connected to the vibration table and. The participants also endured the similar experimental procedures during the no-vibration condition. Here, after the first PVT test was completed, the participants were asked to sit for 20-minutes and the second PVT test will be carried out straight away after 20-minutes of sitting.

2.3. Subjective measure-Karolinska Sleepiness Scale (KSS)

The participants' subjective sleepiness was rated using Karolinska Sleepiness Scale (KSS). The scale was used prior to vibration exposure, every 5-minutes of vibration and after exposure to vibration. The test leader would initiate the rating by the uttering KSS. The scale had been used in many past studies. consists of the following indicators: 1 = extremely alert, 2 = very alert, 3 = alert, 4 = rather alert, 5 = neither alert or sleepy, 6 = some sign of sleepiness, 7 = sleepy, but no effort to stay awake, 8 = sleepy, some effort to stay awake, 9 = very sleepy, great effort to stay awake [13]. The participants and the test leader were not allowed to speak to each other, unless in the event of an emergency.

2.4. Experiment Procedures

An experimental setup was developed for the purpose of this study. The study had used an actual vehicle seat with adjustable headrest. The seat was affixed on a cast aluminum table, while the table was affixed on four air mountings. Air mountings were used to eliminate the table's rigid mass. The inclination angle of the seatback was set at 15° to the vertical direction. Meanwhile, the servo-controlled hydraulic actuator (5 kN) that was placed vertically at the corner of the table was used to determine the table's excitation input force prior to drowsiness measurement. The measurement of total transmitted vibration conducted on each participant has was based on ISO 2631-1 (1997) [7]. By using accelerometer pad, the measurement was conducted to modify the hydraulic input force required for every participant (Fig. 1). Each participant was instructed maintain a regular sleep pattern a week prior to the experiment and to avoid any caffeine intake. The experiment started at 9.00 AM, and prior to the experiment, participants were assessed by using Epworth Sleepiness Scale (ESS). Those with a score exceeding 10 will be exempted from the experiment [8]. A practice session was also given to all participants to minimize the learning effect. Participants were required to complete two randomly organized vibration conditions;

(a) Low amplitude vibration	0.2 ms^{-2} r.m.s.
(b) Medium amplitude vibration	0.4 ms^{-2} r.m.s.

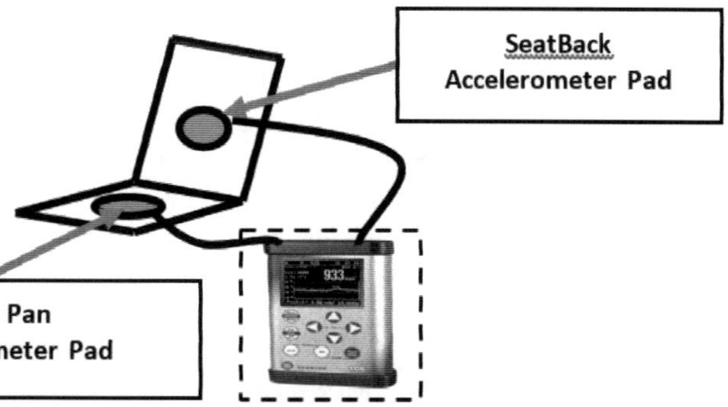

Figure 1. The vibration transmitted to the human body was measured by using two accelerometer pads that are connected to the seat back and seat pan

3. Results

3.1. Objective Measure

Table 1 presents the PVT results which measured the effects of vibration amplitude (ms^{-2} r.m.s) on the participants' drowsiness level. Three PVT metrics were assessed, which are (1) minor lapse, (2) mean RT and (3) median RT. The comparison between the PVT metrics prior and after exposure to vibration indicates the how a 20-minutes exposure to vibration could have significant ($p < 0.05$) influences on participant's RT in both conditions (low vibration amplitude - 0.2 ms^{-2} r.m.s and medium vibration amplitude - 0.4 ms^{-2} r.m.s). Furthermore, significant increases in the RT and number of lapses were observed in all eighteen participants.

The data in Table 1 shows that after a 20-minute exposure to vibration, all of the participants showed increase in the average number of PVT lapses (mean ± SEM) in both conditions participants from (1.67 ± 0.36 to 3.67 ± 0.58, $p = 0.0018$ in low vibration amplitude and from 1.71 ± 0.36 to 4.86 ± 0.46, $p = 0.0002$ in medium vibration amplitude which show a decrease in the level of alertness. When these two results were compared, there is a greater mean difference in medium vibration amplitude (0.4 ms-2 r.m.s) compared to low vibration amplitude (0.2 ms^{-2} r.m.s). Such apparent changes demonstrate that the effect of vibration amplitude show that the more pronounced lapse of attention in medium vibration amplitude (0.4 ms^{-2} r.m.s), and the result is significant ($p = 0.0024$).

The mean RT for eighteen participants in both vibration conditions show statistical significance, and there is an increase in the reaction time participants. Due to exposure to vibration, the participants mean RT for the participants had increased from (283.3 ± 6.24 to 320.2 ± 10.68, $p < 0.0001$ in low vibration amplitude and from 281.1 ± 6.56 to 360.6 ± 11.50, $p < 0.0001$ in medium vibration amplitude). In comparison, the mean difference of mean RT in medium vibration amplitude (0.4 ms^{-2}) is larger than low vibration amplitude (0.2 ms^{-2}). This shows that it is statistically significant ($p = 0.0027$) that exposure to medium vibration amplitude (0.4 ms^{-2}) will greatly influence the level human drowsiness level, as shown in the participants' mean RT. Table 1 also indicates the increase of PVT metrics based on the median RT as a result of the exposure to low and medium vibration amplitude. The participants' median RT had increased from 268.5 ± 7.21 to 301.3 ± 9.68, $p = 0.0020$ in low vibration amplitude and from 268.3 ± 7.89 to 329.1 ± 10.13, $p < 0.0001$ in medium vibration amplitude. The bigger difference in the mean for medium vibration amplitude (0.4 ms^{-2}) shows that higher vibration amplitude has a more apparent effect of drowsiness on a human. Meanwhile, the effect sizes for all PVT metrics were measured and shown in Table 2. In this light, the magnitude of the reaction time changes between prior and after the exposure to vibration in both conditions is indicated by the increase of effect sizes along with the increase in the magnitude of the difference between the participants and the decline of effect size as the variability of the differences increases. According to the definition above,

PVT metrics was arranged based on effect sizes and susceptibility to vibration. During the low vibration amplitude condition, the median RT indicated high effect size (ES = 0.838) of the difference between prior and after 20-minutes of exposure to vibration. This shows higher susceptibility to vibration followed by mean RT (ES = 0.796) and minor Lapse (ES = 0.632). The moderate statistical and the clinical difference between two variables are evident when the effect sizes exceed 0.5. Meanwhile, during the medium vibration amplitude condition, there was a high magnitude of the differences prior and after vibration exposure with the mean RT (ES = 0.848), followed by minor lapse (ES = 0.821) and median RT (ES = 0.784). Consequently, both results were compared and the result shows the significant difference between low vibration amplitude conditions and medium vibration amplitude conditions, and the effect is more significant in medium vibration amplitude which shows higher effect sizes. Further analysis indicates that the mean RT is a sensitive PVT metrics in both vibration amplitude conditions.

Table 1. Comparison of PVT metrics between low vibration amplitude and medium vibration amplitude is shown. A significant increase of reaction time and minor lapse (RT > 500 ms) can be observed following exposure to vibration. However, the effect is more pronounced in medium amplitude vibration.

	Low Vibration Amplitude 0.2 ms^{-2} r.m.s.	
	Before exposure	**After exposure**
Minor lapse (n)	1.67 ± 0.36	3.67 ± 0.58
Mean RT (ms)	283.1 ± 6.24	320.2 ± 10.68
Median RT (ms)	268.5 ± 7.21	301.3 ± 9.68

	Medium Vibration Amplitude 0.4 ms^{-2} r.m.s.	
	Before exposure	**After exposure**
Minor lapse (n)	1.71 ± 0.36	4.86 ± 0.46
Mean RT (ms)	281.1 ± 6.56	360.6 ± 11.50
Median RT (ms)	268.3 ± 7.89	329.1 ± 10.13

Table 2. The effect size of PVT metrics in both vibration conditions is shown. Higher effect size will indicate larger statistical and the clinical difference between two variables

Low Vibration Amplitude 0.2 ms^{-2} r.m.s.		
Rank	**PVT metrics**	**Effect size**
1	Median RT	0.838
2	Mean RT	0.796
3	Minor Lapse	0.632
Medium Vibration Amplitude 0.4 ms^{-2} r.m.s.		
Rank	**PVT metrics**	**Effect size**
1	Mean RT	0.848
2	Minor Lapse	0.821
3	Median RT	0.784

3.2. Subjective Measure
Fig. 2. presents the plot of Karolinska Sleepiness Scale (KSS) score for low and medium vibration amplitude against time. Based on the repeated measures-ANOVA test ($p < 0.0001$) of all eighteen participants in both conditions, it can be observed that there are significant increases in KSS score after every subsequent 5-minutes of exposure to vibration compared to prior to exposure. Fig. 2 illustrate a definite decrease in the level of as reflected by the continuous increase in subjective sleepiness score during exposure to vibration in both vibration amplitude conditions (0.2 ms^{-2} r.m.s. and 0.4 ms^{-2} r.m.s.). The figure indicates that the average KSS score (mean ± SEM) prior to vibration exposure for low and medium vibration amplitude was 2.56 ± 0.16 and 2.72 ± 0.19.

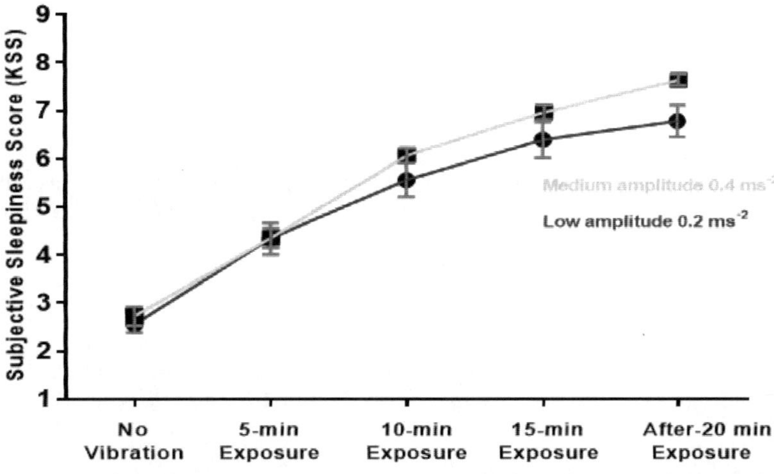

Figure 2. The mean of subjective sleepiness scores (KSS) for all the participants in two vibration conditions (low vibration amplitude and medium vibration amplitude)

However, there is significant statistical significance difference ($p > 0.05$) between low and medium vibration amplitude conditions prior to vibration exposure. Meanwhile, being exposed to vibration for 10-minutes, the average KSS score for all participants had consistently increased to 5.56 ± 0.35 (mean ± SEM) in low vibration amplitude and 6.06 ± 0.15 (mean ± SEM) in medium vibration amplitude. In this light, there is only a slight difference between values obtained for low and medium vibration amplitude. ($p > 0.05$). From the figure, it can be observed that there is a more pronounced drowsiness after the participants were exposed to the vibration for 20 minutes and the KSS values were 6.78 ±

0.33 (mean ± SEM) in low vibration amplitude and 7.61 ± 0.12 (mean ± SEM) in medium vibration amplitude. This clearly shows the depleting level of awareness in all participants after they were exposed to vibration for 20 minutes. One-way repeated measures-ANOVA was carried out to investigate the statistical significance, and the results show highly significant intra-individual and interindividual differences for all participants ($p < 0.0001$) for each vibration condition. Meanwhile, there are no significant differences between low vibration amplitude and medium vibration amplitude.

4. Discussion

This study hypothesized that when the participants are drowsy, they will not be able to provide a good judgment on their drowsiness level. The study has shown significant effects of vibration on human reaction time. Past studies have shown the impact of low vibration amplitude on the drowsiness of seated humans. [14]. These studies have shown a significant relationship between drowsiness level and vibration, as well as the significant increase in the reaction time after being exposed to vibration for 20 minutes [15, 16]. This study's finding is parallel to Azizan [14] which postulated that the increase in vibration amplitude could more significantly impair the level of human alertness. The present study had compared low vibration amplitude (0.2 ms^{-2} r.m.s) with medium vibration amplitude (0.4 ms^{-2} r.m.s) and indicated that the degree of drowsiness, as measured by PVT, was more significant when the vibration amplitude was double. In this regard, the increase in the transmitted vibration to the human body will make the drowsiness level more apparent. This finding shows that the link between exposure to vibration and the level of drowsiness. On the other hand, the study did not find any significant difference in the subjective measurement (KSS) in both vibration conditions. In this regard, this finding is consistent with the existing literature [17]. This study's finding provides a significant implication in the development a drowsiness contour which offers a comprehensive guideline on whole-body vibration area. On the hand, before the association between vibration exposure, and drowsiness level could be clearly understood, there is a need for more research on this topic^{-}

5. Conclusion

This present study is aimed to identify the impact of vibration on the reaction time and drowsiness level of seated occupants by conducting a comprehensive, objective measurement test (PVT) and subjective evaluation score (KSS). The finding supports the hypothesis that human drowsiness level could be considerably influenced by the presence of low-frequency vibration (between 1 -15 Hz). The results show that compared to low vibration amplitude (0.2 ms^{-2} r.m.s), medium vibration amplitude (0.4 ms^{-2} r.m.s) create a higher level of perceived drowsiness among the participants. On the other hand, the KSS result shows that there is no significant difference subjective drowsiness level found in both conditions of vibration while the PVT test shows a substantial increase in the reaction time and the number of lapses in both conditions, and medium vibration amplitude shows a higher degree of error. Thus, it is concluded that the increasing amount of vibration amplitude level would lead to the slower reaction time by the participants, indicating an elevated level of drowsiness.

References

[1] Arnedt, G. J. Wilde, P. W. Munt, and W. MacLean,"How do prolonged wakefulness and alcohol compare in the decrements they produce on a simulated driving task?, *Accident; Analysis and Prevention*, vol. 33, no. 3, pp: 337–44, 2001.

[2] P. Philip, P. Sagaspe, N. Moore, J. Taillard, A. Charles, C. Guilleminault, and B. Bioulac. Fatigue, sleep restriction and driving performance., *Accident; Analysis and Prevention*, vol. 37, no. 3, pp: 473–478 , 2005.

[3] M. Amzar, M. Fard. Effects of vehicle seat dynamics on ride comfort assessment, in *International Congress on Noise and Vibration*, 2013.

[4] C. Anderson, A. W. J. Wales and J. A. Horne. PVT lapses and the eyes PVTlapses differ according to eyes open, closed, or looking away, *Sleep*, 33 (2), pp: 197-204, 2010.

[5] M. Basner and D. F. Dinges. Maximizing sensitivity of the psychomotor vigilance test (PVT) to sleep loss., *Sleep*, vol. 34, no. 5, pp: 581–591, 2011.

[6] M. L. Jackson, R. J. Croft, G. A. Kennedy, K. Owens, and M. E. Howard. Cognitive components of simulated driving performance: Sleep loss effects and predictors., *Accident; Analysis and Prevention*, vol. 50, pp: 438–444 , 2013.

[7] International Standard. ISO 2631-1 Mechanical vibration and shock - Evaluation of human exposure to whole-body vibration, 1997.

[8] M. W. John. A new method for measure daytime sleepiness the Epworth Sleepiness Scale. American Sleep Disorder Association and Sleep Research Society, vol. 14(6), pp: 540–545, 1991.

[9] N. L. Shattuck and P. Matsangas. Psychomotor vigilance performance predicted by Epworth Sleepiness Scale scores in an operational setting with the United States Navy., *Journal of sleep research*, vol. 24, pp: 174-180, 2015.

[10] M. Karimi, J. Hedner, C. Lombardi, W. T. Mcnicholas, T. Penzel, R. L. Riha, D. Rodenstein, and L. Grote. Driving habits and risk factors for traffic accidents among sleep apnea patients-a European multi-centre cohort study., *Journal of sleep research*, vol. 23, no. 6, pp: 689–699 , 2014.

[11] M. E. Howard, A. V. Desai, R. R. Grunstein, C. Hukins, J. G. Armstrong, D. Joffe, P. Swann, D. A. Campbell, and R. J. Pierce. Sleepiness, sleep-disordered breathing, and accident risk factors in commercial vehicle drivers., *American Journal of Respiratory and Critical Care Medicine*, vol. 170, no. 9, pp: 1014–1021, 2004.

[12] M. Y. Khitrov, S. Laxminarayan, D. Thorsley, S. Ramakrishnan, S. Rajaraman, N. J. Wesensten, and J. Reifman. PC-PVT: a platform for psychomotor vigilance task testing, analysis, and prediction., *Behavior Research Methods*, vol. 46, no. 1, pp: 140–147, 2014.

[13] M. Gillberg, G. Kecklund, and T. Akerstedt. Relations between performance and subjective ratings of sleepiness during a night awake., *Sleep*, vol. 17, no. 3, pp: 236–241, 1994.

[14] M. Amzar, M. Fard. The influence of vibrations on vehicle occupant fatigue, *Internoise Conference*, vol. 62, no. ISO 2631, pp: 1–15, 2014.

[15] Y. Satou, H. Ando, M. Nakiri, K. Nagatomi, Y. Yamaguchi, M. Hoshino, Y. Tsuji, J. Muramoto, M. Mori, K. Hara, and T. Ishitake. Effects of short-term exposure to whole-body vibration on wakefulness level., *Industrial Health*, vol. 45, pp: 217–223, 2007.

[16] Y. Satou, T. Ishitake, H. Ando, K. Nagatomi, M. Hoshiko, Y. Tsuji, H. Tamaki, A. Shigemoto, M. Kusano, M. Mori, and K. Hara. Effect of short-term exposure to whole body vibration in humans: relationship between wakefulness level and vibration frequencies., *The Kurume Medical Journal*, vol. 56, no. 1–2, pp: 17–23, 2009.

[17] J. A Horne and S. D. Baulk. Awareness of sleepiness when driving., *Psychophysiology*, vol. 41, no. 1, pp: 161–165, 2004.

IOP Conf. Series: Materials Science and Engineering **370** (2018) 012021 doi:10.1088/1757-899X/370/1/012021

Effects of High and Low Temperature on the Tensile Strength of Glass Fiber Reinforced Polymer Composites

S. Kumarasamy[1], M. Shukur Zainol Abidin[1, 2], M. N. Abu Bakar[1], M.S. Nazida[1] Z. Mustafa[3] and A. Anjang[1, 2*]

[1] School of Aerospace Engineering, Universiti Sains Malaysia, 14300 Nibong Tebal, Penang, Malaysia.
[2] Cluster for Polymer Composites, Science and Engineering Research Centre, Universiti Sains Malaysia, 14300 Nibong Tebal, Penang, Malaysia.
[3] Faculty of Manufacturing Engineering, Universiti Teknikal Malaysia Melaka, Durian Tunggal, 76100 Melaka.

*Corresponding author: aeaslina@usm.my

Abstract. In this paper, the tensile performance of glass fiber reinforced polymer (GFRP) composites at high and low temperature was experimentally evaluated. GFRP laminates were manufactured using the wet hand lay-up assisted by vacuum bag, which has resulted in average fibre volume fraction of 0.45. Using simultaneous heating/cooling and loading, glass fiber epoxy and polyester laminates were evaluated for their mechanical performance in static tensile loading. In the elevated temperature environment test, the tension mechanical properties; stress and modulus were reduced with increasing temperature from 25°C to 80°C. Results of low temperature environment from room temperature to a minimum temperature of -20°C, indicated that there is no considerable effect on the tensile strength, however a slight decrease of tensile modulus were observed on the GFRP laminates. The results obtained from the research highlight the structural survivability on tensile properties at low and high temperature of the GFRP laminates.

1. Introduction

Fiber reinforced polymer (FRP) composites has been widely accepted in many engineering applications that requires high performance structures. FRP composites holds a huge advantage over traditional or conventional material in term of its mechanical and chemical properties. It has a high fatigue resistance due to its excellent tension handling and can withstand high load tension. FRP composites have a non-elastic structure which can absorb vibration and dissipate it. This property is suitable for bridge structure and aircraft as they are constantly subjected to vibration from the wind. One of the biggest advantage of using FRP composites is their high strength to weight ratio [1]. This will enable manufacturer to produce high strength materials while reducing its overall weight. With the manufacturing cost of FRP composites are decreasing, many manufacturer and engineers are opting FRP composite as a replacement for traditional materials [2]. Because of this FRPs are being considered in the aerospace, marine, construction, automotive, oil and gas industries as well as in wind turbine application.

There are two type of FRP composite that are widely used in the industries which are carbon fiber reinforced polymer (CFRP) and glass fiber reinforced polymer (GFRP). Both of this FRP have their pros and cons but for this research, we will mainly focus on GFRP composites since this materials is low cost and readily available locally compared to CFRP while still having excellent mechanical properties. GFRP has also been extensively used in many applications from the household industry to a high performance structure in the aerospace industry [3]. Despite many advantages offered by the GFRP, one of the critical limitations on the use of GFRP is in high temperature applications. The increasing use of GFRP at low temperature has also draw interest for researcher in studying the behavior and effect of the FRP specimens.

The purpose of this paper is to investigate the effect of hot and cold environment under tensile loading of GFRP epoxy and polyester laminates. The strength of FRP is strongly dependent on temperature, with tensile strength and modulus rapidly decreasing once the temperature exceeds the glass transition temperature [4, 5]. The tensile behavior of FRP composites at high temperature and fire was studied by many researchers using several experimental approach and several fiber/matrix combinations [6-13]. Similar to high temperature, as the composite is exposed to low temperature environment, the strength and modulus might changed. Few studies has been performed to evaluate the mechanical properties of FRP composites at low temperature environment and the results are varies depending on the loading, experimental setting and type of specimens tested [14-17].

This paper evaluates both hot and cold environment on the GFRP specimens made using wet lay-up assisted by vacuum bagging method using an available testing machine fitted with a hot/cold box chamber. The results will provide important insight on the effect of both hot and cold environment to GFRP laminates.

2. Materials and experimental techniques

2.1. GFRP laminates

The GFRP laminates were manufactured from E-glass plain woven fabric (800 g/m^2) and epoxy resin (Bisphenol-a) or polyester resin. The resin did not contain flame retardant fillers or additives. GFRP laminates were manufactured using the wet hand lay-up assisted by vacuum bag. The laminates were then left to cure at room temperature and went through post-cure process in the oven at 80°C for 1 hour. The average fibre volume fraction of the laminates was 0.45 and was determined through burn-off test method following ASTM D2584.

2.2. High and low temperature environment test

The static tensile properties were obtained using a 50 kN Testometric M500-30 range computer controlled universal materials testing machine. To simulate a hot/cold temperature environment, a close chamber box was attached to the Testometric machine. The hot/cold box is capable to simulate a hot condition at maximum temperature of 80°C and cold condition at a minimum temperature of -20°C. The size of the specimen was 140 mm x 20 mm with a gauge length of 50 mm. To reduce the slippage problem, tabs from sand paper were introduced on the GFRP specimens. The tensile test was conducted under position control with a crosshead speed of 2 mm/min. The GFRP laminates specimen was tested at room temperature and at both hot/cold environment under simultaneous tensile loading. A total number of 3 specimens were tested at each condition for both GFRP laminates.

3. Results and discussion

The experimental results obtained from the high and low temperature tests for the GFRP polyester and epoxy laminates are presented next.

3.1. High temperature tensile behaviour

From the tensile test results, the stress strain curves were drawn. Figure 1a and 2a show these curve for the GFRP laminates at the four different temperatures considered. The laminate exhibits a linear elastic behaviour at room temperature until breakage while at elevated temperatures; the curves show a non-linear region before breakage. The slope of the stress strain curve decrease significantly as the temperature increases.

The tension strength and elastic modulus as a function of temperature has been plotted to evaluate the effect of elevated temperature for both types of GFRP laminates tested. Figure 1b and 2b show the relationship of the tensile strength and modulus of the surrounding temperature for the two types of GFRP laminates respectively. The curves clearly show that the tensile strength and modulus decreases as the temperature of the surrounding increases. The loss in tensile strength is mainly due to the resin softening of the GFRP laminates. The stiffness loss is attributed to the straightening of woven fibers as the matrix softens.

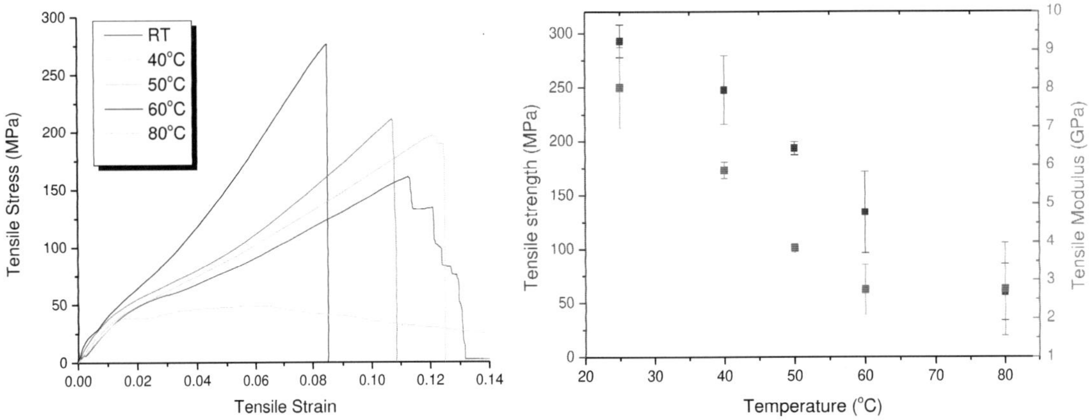

(a) (b)

Figure 1. GFRP epoxy laminate specimens at elevated temperatures (a) Tensile stress vs strain curves (b) Tensile strength and modulus

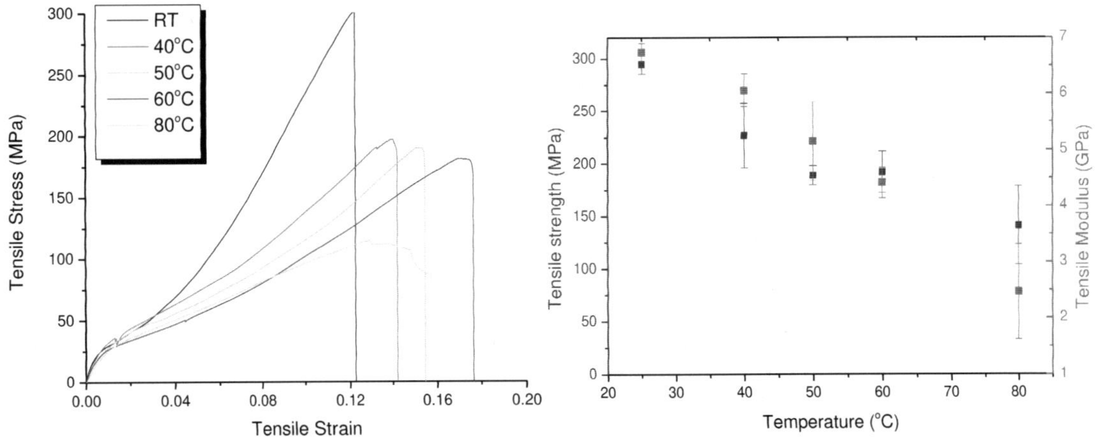

(a) (b)

Figure 2. GFRP polyester laminate specimens at elevated temperatures (a) Tensile stress vs strain curves (b) Tensile strength and modulus

3.2. Low temperature tensile behaviour

From the tensile test results, the stress strain curves are plotted. Fig. 3a and 4a show these curve for the GFRP laminate at the four different temperatures considered. The GFRP laminates exhibits a linear elastic behavior until breakage. The slope of the stress strain curve increases slightly as the temperature decreases. Both types of GFRP specimens show brittle behaviour as the strain to failure decrease at lower temperature.

The tension strength and elastic modulus as a function of temperature has been plotted to evaluate the effect of low temperature environment for both types of GFRP laminates tested. Figure 3b and 4b show the relationship of the tensile strength and modulus of the surrounding temperature for the two types of GFRP laminates respectively. The curves show that there is no considerable effect on the tensile strength. However, a slight increase of tensile modulus from room temperature to -5°C was observed and slight decrease from -5°C to -20°C was observed on the GFRP laminates.

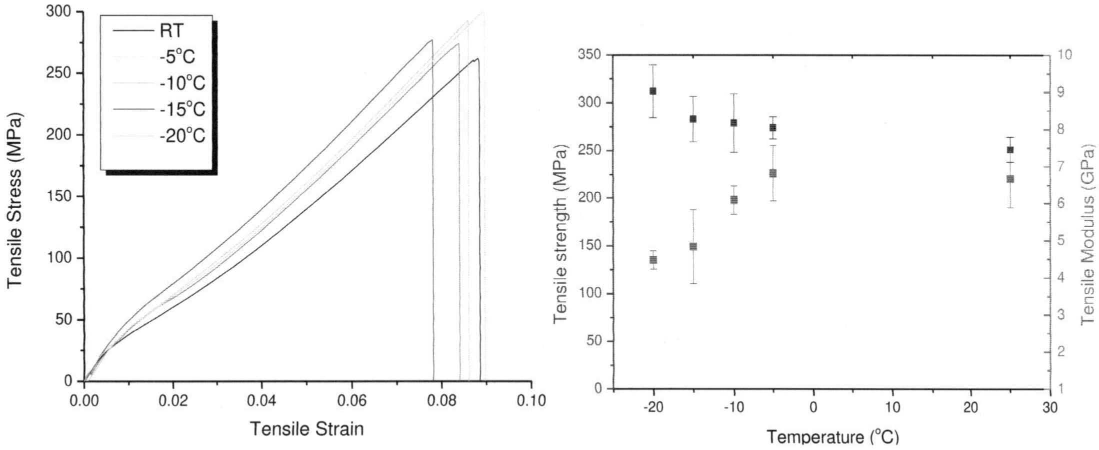

(a) (b)

Figure 3. GFRP epoxy laminate specimens at low temperature surrounding (a) Tensile stress vs strain curves (b) Tensile strength and modulus

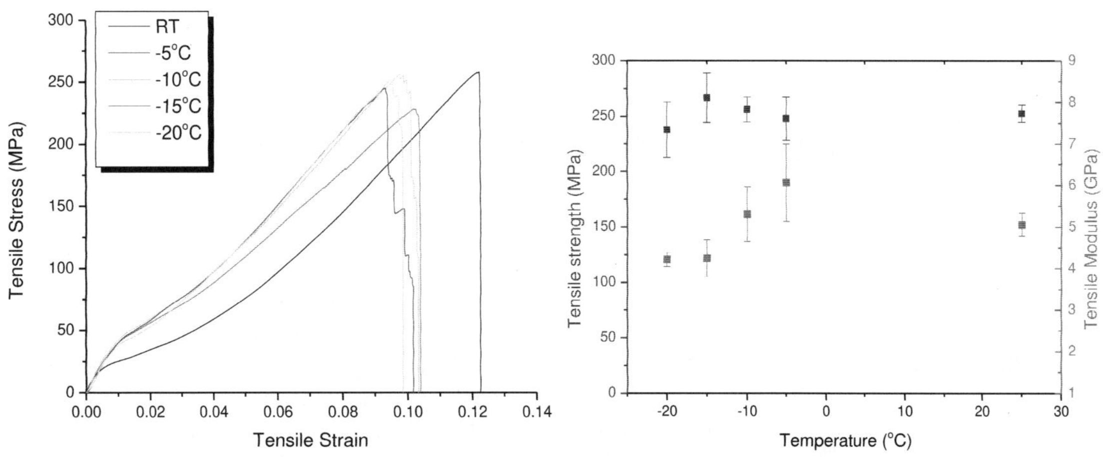

(a) (b)

Figure 4. GFRP polyester laminate specimens at low temperature surrounding (a) Tensile stress vs strain curves (b) Tensile strength and modulus

4. Conclusions

The high and low temperature properties of GFRP laminates during tensile loading have been evaluated in this paper. Both tensile strength and modulus of the GFRP composites decrease as the temperature of the surrounding increases. The loss in tensile strength is mainly due to the resin softening of the composites tested. The stiffness loss is attributed to the straightening of woven fibers as the matrix softens. Results of low temperature environment indicated that there is no considerable effect on the tensile strength tested between room temperature to a minimum of -20°C. However, a slight increase of tensile modulus from room temperature to -5°C was observed and slight decrease from -5°C to -20°C was observed on the GFRP laminates. GFRP laminates also have shown a brittle behavior as the surrounding temperature decreases. The results obtained from the research highlight the structural survivability on tensile properties at low and high temperature environment of the GFRP laminates.

Acknowledgements
The project work was supported by the Short Term Grant of Universiti Sains Malaysia (Grant No. 304/PAERO/6315057) and Bridging Grant of Universiti Sains Malaysia (Grant No. 304/PAERO/6316087). The technical assistance of Mr. Hasfizan Hashim and Mr. Mohd Shahar Che Had for the specimen's fabrication and testing is gratefully acknowledged.

References
[1] Bagherpour S 2012 Fibre Reinforced Polyester Composites. In: *Polyester,* ed D H E-D S (Ed.) (https://www.intechopen.com/books/polyester/fibre-reinforced-polyester-composites: InTech) pp 135-66
[2] Bakis C E, Bank L C, Brown V L, Cosenza E, Davalos J F, Lesko J J, Machida A, Rizkalla S H and Triantafillou T C 2002 Fiber-Reinforced Polymer Composites for Construction 2014;State-of-the-Art Review *Journal of Composites for Construction* **6** 73-87
[3] T P S, Satheeshkumar S and J N 2014 *Glass fiber-reinforced polymer composites - A review* vol 33
[4] Mouritz A P and Gibson A G 2006 *Fire Properties of Polymer Composite Materials*: Springer
[5] Mouritz A P, Feih S, Kandare E, Mathys Z, Gibson A G, Des Jardin P E, Case S W and Lattimer B Y 2009 Review of fire structural modelling of polymer composites *Composites Part A: Applied Science and Manufacturing* **40** 1800-14
[6] Gibson A G, Torres M E O, Browne T N A, Feih S and Mouritz A P 2010 High temperature and fire behaviour of continuous glass fibre/polypropylene laminates *Composites Part A: Applied Science and Manufacturing* **41** 1219-31
[7] Lattimer B Y, Ouellette J, Trelles J 2011 Thermal response of composite materials to elevated temperatures *In: Modeling of Naval Composite Structures in Fire* 1-49
[8] Gardiner C P, Mathys Z and Mouritz A P 2002 Tensile and Compressive Properties of FRP Composites with Localised Fire Damage *Applied Composite Materials* **9** 353-67
[9] Feih S, Mouritz A P, Mathys Z and Gibson A G 2007 Tensile Strength Modeling of Glass Fiber—Polymer Composites in Fire *Journal of Composite Materials* **41** 2387-410
[10] Feih S and Mouritz A P 2012 Tensile properties of carbon fibres and carbon fibre–polymer composites in fire *Composites Part A: Applied Science and Manufacturing* **43** 765-72
[11] Feih S, Mathys Z, Gibson A G and Mouritz A P 2007 Modelling the tension and compression strengths of polymer laminates in fire *Composites Science and Technology* **67** 551-64
[12] Anjang A, Chevali V S, Kandare E, Mouritz A P and Feih S 2014 Tension modelling and testing of sandwich composites in fire *Composite Structures* **113** 437-45
[13] Anjang A, Mouritz A P and Feih S 2017 Influence of fibre orientation on the tensile performance of sandwich composites in fire *Composites Part A: Applied Science and Manufacturing* **100** 342-51
[14] Pan Z, Sun B and Gu B 2016 Thermo-mechanical numerical modeling on impact compressive damage of 3-D braided composite materials under room and low temperatures *Aerospace Science and Technology* **54** 23-40
[15] Ma H-l, Jia Z, Lau K-t, Leng J and Hui D 2016 Impact properties of glass fiber/epoxy composites at cryogenic environment *Composites Part B: Engineering* **92** 210-7

[16] Islam M S, Melendez-Soto E, Castellanos A G and Prabhakar P 2015 Investigation of woven composites as potential cryogenic tank materials *Cryogenics* **72, Part 1** 82-9
[17] Kim M-G, Kang S-G, Kim C-G and Kong C-W 2007 Tensile response of graphite/epoxy composites at low temperatures *Composite Structures* **79** 84-9

Optimization of Heat Transfer on Thermal Barrier Coated Gas Turbine Blade

Abdul Aabid[1]**, and S. A. Khan**[2]
Department of Mechanical Engineering, Kulliyyah of Engineering, International Islamic University Malaysia

Corresponding author: [1]aabidhussain.ae@gmail.com, [2]sakhan06@gmail.com

Abstract- In the field of Aerospace Propulsion technology, material required to resist the maximum temperature. In this paper, using thermal barrier coatings (TBCs) method in gas turbine blade is used to protect hot section component from high-temperature effect to extend the service life and reduce the maintenance costs. The TBCs which include three layers of coating corresponding initial coat is super alloy-INCONEL 718 with 1 mm thickness, bond coat is Nano-structured ceramic-metallic composite-NiCoCrAIY with 0.15 mm thickness and top coat is ceramic composite-$La_2Ce_2O_7$ with 0.09 mm thickness on the nickel alloy turbine blade which in turn increases the strength, efficiency and life span of the blades. Modeling a gas turbine blade using CATIA software and determining the amount of heat transfer on thermal barrier coated blade using ANSYS software has been performed. Thermal stresses and effects of different TBCs blade base alloys are considered using CATIA and ANSYS.

1. Introduction

The blades are responsible for extracting energy from the elevated temperature, high pressure gas produced by the combustor. To survive in this difficult environment, turbine blades often use exotic materials like super alloys and many different methods of cooling, such as internal air channels, boundary layer cooling, and thermal barrier coatings. Thermal barrier coatings (TBCs) as a kind of temperature-resistance material have been widely applied in super elevated temperature components in aircraft engines due to its low thermal conductivity. Prediction of the stress distribution on a 3D turbine blade coated with thermal barrier coatings (TBCs) plays a key role in analyzing the failure of TBCs [1]. Due to their low thermal conductivity, a temperature drop can be up to 200 °C through thermal isolation in TBCs as well as an inner cooling system [2, 3]. Thermally grown oxide on the stress distribution and evolution under cyclic thermal loading, a three-dimensional finite element model of a turbine blade with thermal barrier coatings is developed, in which the coating deposition process and elevated temperature behavior are considered [4].

The aging and degradation mechanisms of a present standard TBC system (AM1 single crystal super alloy / NiPtAl bond coat / 8YPSZ EB-PVD top coat), advanced thermal barrier systems operating at higher temperatures and for very long durations (commercial aircraft applications for example) needs further developments [5]. Effects of Compositional Changes on the Performance of a TBCs System; As part of continuing effort to develop thermal barrier systems for advanced aircraft gas turbine engine components, systems consisting of Ni-base bond coatings containing about 16Cr, 6A1, and from 0.15 to 1.08Y (all in wt.%) and zirconium oxide layers containing from 4.0 to 24.4Y_2O_3 (in w.%) were evaluated. The data indicate that the best thermal barrier systems consisted of combinations involving the Ni-16.4Cr-5.1Al-0.15Y and Ni-17.0Cr- 5.4Al-0.35Y bond coatings and the 6.2Y_2O_3 and 7.9Y_2O_3 (all in wt. %) stabilized zirconium oxide layers. In furnace, cyclic tests these thermal barrier systems withstood 1500 1-hour cycles between 990° - 280° C and over 100 1-hour cycles between 1095° - 280° C. In a natural gas-oxygen torch rig these systems on solid specimens withstood approximately 590 to 790 1-hour cycles between 1200° - 100° C. On air-cooled blades, the NiCrA1-O. l5Y/ZrO-6.2Y_20_3 (in wt. %) at 1580° C surface temperature and the NiCrAl-0.35Y/ZrO-7.9Y_2O_3 (in wt. %) at 1550° C surface temperature withstood over 1300 and 1500 l-hour cycles without failure, respectively [6]. Non-destructive evaluation of as-sprayed APS TBCs was carried out by pulsed thermography. The results showed that pulsed thermography can be employed to evaluate the quality [7].

The restrictions of phase compatibility with the thermally grown aluminum oxide and the TBC coating itself place severe constraints on the possible luminescence sensor materials and suggest that the best chromophores are likely to be those that are incorporated into the crystal structure of the thermal barrier coating. For both YSZ and the rare-earth pyrochlore coatings, rare-earth dopants can serve as chromophores and several concepts for visualization and quantitative luminescence sensing are demonstrated based on the optical properties of these coating materials and the luminescence properties of the rare-earth dopants. For temperature sensing, the choice of rare-earth dopants is much more restrictive as will be described elsewhere [8]. TBCs have evolved from the laboratory to negligible risk turbine section application and then on to an integral part of engine design. These coatings applied to advanced air-cooled, super alloy components will be the materials system of choice in advanced engines for the foreseeable future. Even as non-super alloy components gradually came into service, these new materials will still require the protection of TBCs [6]. The distribution patterns of existing elements both in the matrix of base alloy and in the successive coatings of used and refurbished turbine blades were determined by scanning proton microscopy. The results of these analyses revealed the cluster distribution of rather enormous amounts of Si in the refurbished coating layer. The cluster distribution of Si, which could reduce the lifetime of the component, is more likely due to the preparation of blades by sandblasting prior to the deposition of the final protective layer [9].

The problem is to study steady state conjugate heat transfer through different layers of TBC and to find the best TBC combination of different layers.

2. Approach

Using ANSYS parametric design language tool the different layers of TBC over a blade base super alloy is modeled and boundary conditions are defined and analyzed. The results obtained from the computation compared with the combination of different layers of TBC and best combination for the

TBC is determined. The following method used for simulation as shown in below figure 1.

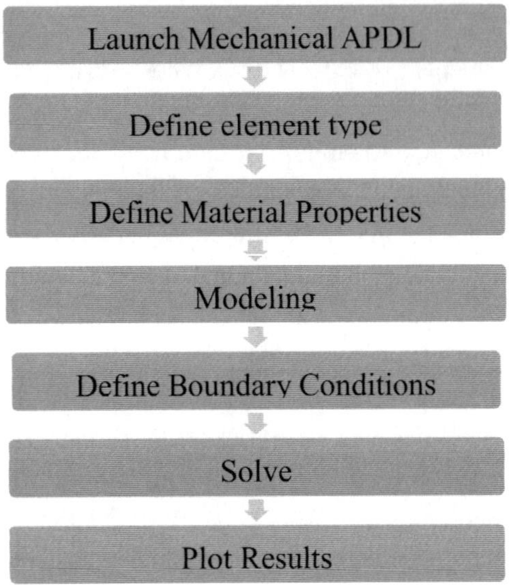

Figure 1. Method for Simulation

The different layers of TBC for this case are listed below. The thickness of the respective layer along with their thermal conductivities is mentioned. The film co-efficient and the temperature of the turbine inlet gas is 2028 w/m2k and 1373 K.

Table 1. Coating layers' parameters

Coating layers	Thickness [mm]	Thermal conductivity (W/m K)
Convective Element [Hot Gas]	0.01	-
Top Coat [$La_2Ce_2O_7$]	0.15	0.6
TGO	0.01	4.0
Bond Coat [NiCoCrAlY]	0.09	17.0
Super Alloy [Inconel 718]	1	25.0

3. Gas Turbine Blade Material

Gas turbine blade materials have always been a major concern and which are highly capable to withstand elevated temperature service and hence the engine efficiency; for the materials with elevated temperature to weight ratio helps in weight reduction [10–13].

3.1. Inconel

Inconel is a family of austenitic nickel-chromium-based super alloys. Inconel alloys are oxidation and corrosion resistant materials well suited for service in extreme environments subjected to pressure and heat.

3.2. Thermal Barrier Coating

Progress in Gas Turbine Engine (GTE) manufacturing is continuously linked with a rise of operating temperature and stresses of engine gas path elements, especially the turbine parts. To provide simulation of loading conditions for the hot engine parts with a TBC under service conditions, the test procedure shall ensure the possibility of cyclic surface heating of the object under testing (simulating its heating in hot gas flow) up to temperatures of 11500 C and more at heating rates of 150-2000 C/s and subsequent cooling [14].

3.3. *Structure of TBCs*

Generally, TBC is a two-layer's system which incorporates about 250 μm thickness layer of ceramic top coating applied to the outer surface of the substrate and about 150 μm thickness underlying of metallic bond coating. The metallic bond coating performs two functions:

- To provide oxidation resistance and
- To adhere the ceramic to the super alloy substrate physically and chemically.

The oxide that is commonly used is Zirconia oxide (ZrO_2) and Yttrium oxide (Y_2O_3). The metallic bond coat is an oxidation/hot corrosion resistant layer. The bond coat empirically represented as MCrAlY alloy.

Where;

M - Metals like Ni, Co or Fe.
Y - Reactive metals like Yttrium.
Cr-Al - Base metal.

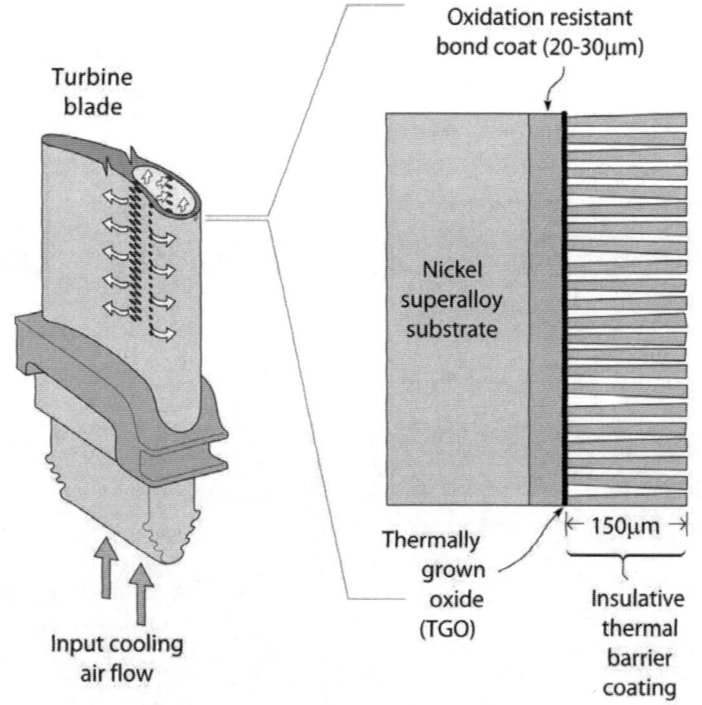

Figure 2. Structure of thermal barrier coating

Bond coat: The bond coat is typically a metallic layer made of a Nano-structured ceramic-metallic composite-NiCoCrAlY, the layer to the metal substrate and is responsible for generating the second

IOP Conf. Series: Materials Science and Engineering **370** (2018) 012022 doi:10.1088/1757-899X/370/1/012022

coating layer of thermally grown ceramic oxide, which occurs when the coating is subjected to an elevated temperature.

When Nano-particles of aluminum oxides and nitrides are distributed throughout the bond coat or along its surface, the formation of thermally grown oxides is catalyzed. This ceramic layer is responsible for forming a uniform, thermally protective barrier by acting as an oxygen diffuser which prevents the substrate from becoming thermally oxidized.

Top coat: The last layer of the coating is the ceramic top coat, which is made of top coat is ceramic composite-$La_2Ce_2O_7$. The top coat protects the substrate by keeping the other coating layers at a lower temperature than the surface.

3.4. *Fabrication Method*

Method used to find the presence of TBC is Coloration method

- Masking
- Degreasing
- Vapour blasting
- Heating the blade at 650 degrees Celsius for 1 Hr.

3.5. *Heat Transferred Blade*

Heating the blade at 650 degrees Celsius for 1 Hr

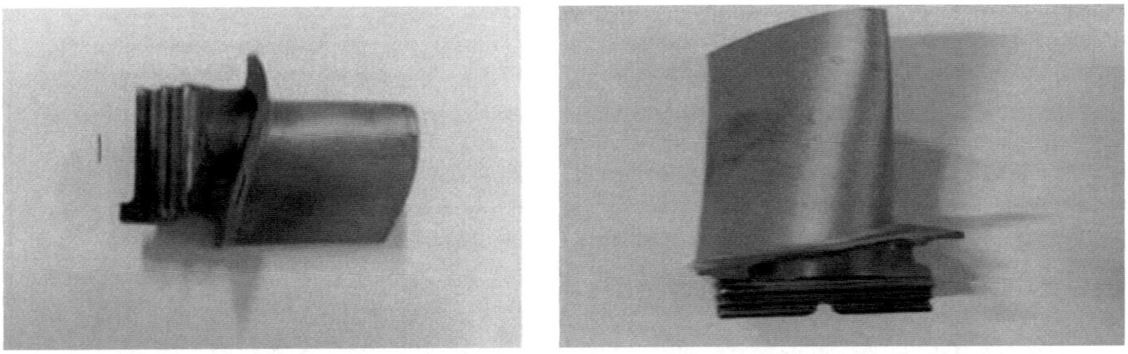

Figure 3. Turbine blades before heat treatment

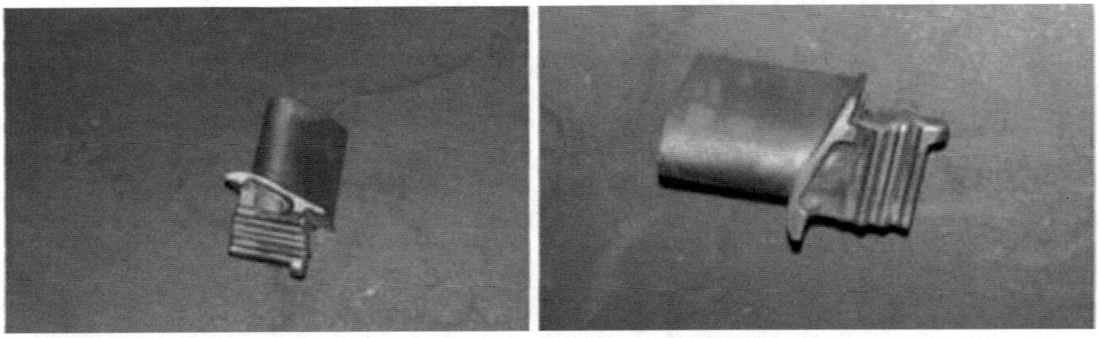

Figure 4. Turbine blades after heat treatment

In figure 3 and figure 4 are Illustrates the two-different heated blade. The difference between the two blades are colour slightly changed after one hour of heating and the property of the material is also changed. The material temperature resisting capacity is observed and it is quite good compared to the normal blade. The selection of material for three layers of coating which chosen and it is affected.

3.6. Air Plasma Method (APS)

The APS process can obtain high deposition rates and the capability to spray varied geometries. Plasma spraying also allows composition control and can produce coatings with high mechanical strength and durability. In the plasma spraying process, a high intensity arc is operated between a stick-type cathode and a nozzle-shaped water-cooled anode. Gas, introduced along the cathode, is heated by the arc to plasma temperatures and then exits the anode nozzle as a plasma jet. Powder injected into the plasma jet is accelerated and heated to a molten state. As the molten particles in the jet strike the selected substrate, they form splats that build up, particle by particle, into a co-mating. Among the many spraying parameters, the standoff distance, powder size distribution, power level, and arc gas selection strongly influence the microstructure of coatings. The particle velocity at substrate impact is governed by the plasma velocity and density, which are dependent on the type and volume of the gases being used and the energy being input to the gases [15]. The figure 5 shows fully coated turbine blade.

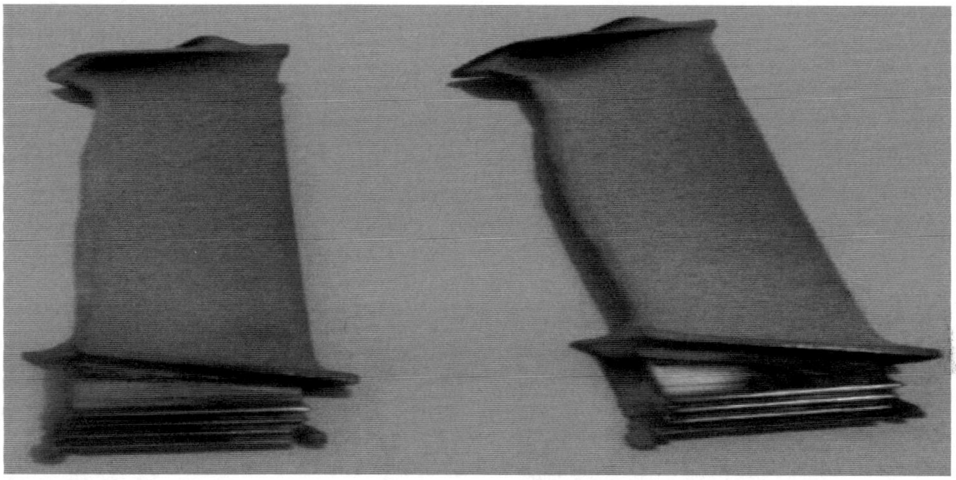

Figure 5. Coated turbine blades

4. Analytical Study of Heat Transferred

4.1. Hot Air Blade Surface (Convective Heat Transfer)

$$Q = h_C A dT \tag{1}$$

Q = Heat transferred per unit time (W)
A = Heat transfer area of the surface
h_c= Convective heat transfer coefficient of the process, W/m^2K or $W/m^2\,^\circ C$
dT =Temperature difference between the surface and the bulk fluid (^0K / °C)

$$h_c = 10.45 - V + 10 \times V^{0.5} \quad (2)$$

V= the relative speed of the object through the air.

4.2. TBC to Blade Base Alloy (Conduction Heat Transfer)

$$HT = \frac{Q}{t} = \frac{(KA(T_{hot} - T_{cold}))}{d} \quad (3)$$

$$dT = T_{hot} - T_{cold} \quad (4)$$

K= Thermal conductivity
A= Surface area of the blade
d= TBC Thickness

4.3. Faurier's Law

$$q = -KA\frac{dT}{dx} \quad (5)$$

q = Heat flow
K = Thermal conductivity
A = Cross sectional area

$\frac{dT}{dx}$ = temperature gradient in the direction of flow.

5. Results and Discussion

5.1. Outlines for Temperature Distribution

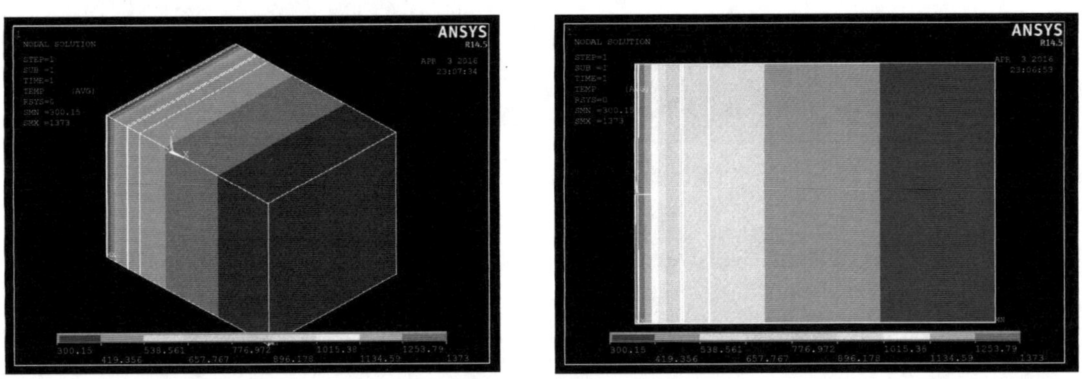

Figure 6. Temperature Distribution

Figure 6 illustrate the outline of temperature distribution on the sample of coated blade, as per the experimental work, applied to the three-different material coat for validation purpose in simulation and observed with the different temperature variation. In simulation, our aim is to find out the temperature effect on coated material blade and figure 5 is only modeled as per the coated material thickness not with material blade dimension.

5.2. Heat Transfer Analysis by Changing Coating Material

Case I: Top Coat [$La_2Ce_2O_7$] and Super Alloy [Inconel 718]

When applied a $La_2Ce_2O_7$ and Inconel 718 on the top coat the temperature variation on the blade which resist up to 1400k which is good for normal blade.

Case II: Top Coat [3YSZ] and Super Alloy [Chromium Steel]

The heat transfer on the coated material blade with 3YSZ and Chromium Steel, it is seen that variation in the temperature is from minimum temperature 500k to the maximum temperature up to 1400k.

Case III: Top Coat [3YSZ] and Super Alloy [Inconel 625]

Super alloy the to Inconel 625 and it is observed that the change from the super alloy the to Inconel 625 from Chromium Steel the temperature changes are appreciable, however, the range of the temperature is similar as discussed earlier but variation of middle temperature is different.

Case IV: Top Coat [$La_2Zr_2O_7$] and Super Alloy [Inconel 718]

It is coated with $La_2Zr_2O_7$ and Inconel 718 and it has been observed that from the previous case as well as the present case the temperature changes is almost in the same range.

5.3. Comparison of all Cases

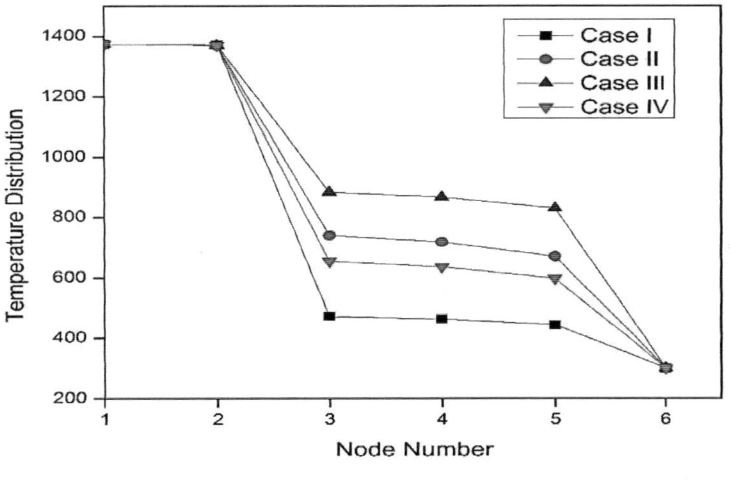

Figure 7. Nodal temperature distribution for all cases

Figure 7 illustrates the comparison of all four cases which has different material coating for top coating and super alloy. From the results, it seen that the coating of the gas turbine blade can resist the maximum temperature up to 1400 k and it is good for the case of thermal barrier coating using air plasma method on gas turbine blade. It is also seen that the TBC at node five is 444.02 K. Experimental work of this case still in progress and results which observed from the test in thermal conductivity test which is the TBC of temperature 450K and in the simulation case it is 444.02 K.

6. Conclusions

Base on the above discussions we can draw the following conclusions:

- Heat Transfer Analysis/simulation was performed by using ANSYS Mechanical APDL software by changing the coated material in four different cases.

- Out of these four cases, the finest result of coated gas turbine blade could be done by using air plasma method with three different layer initial coat-Inconel 718, bottom coat-NiCoCrAIY and top coat-$La_2Ce_2O_7$.
- When thermal conductivity test was performed the temperature was 450K. From the analytical results, it was concluded that the result of case one has performed well because of the TBC temperature at node number five is very less i.e. 444.02 K compared to other cases. Hence this combination of TBC on gas turbine blade is best suited for higher operating temperature.

References

[1] W. Zhu, J. W. Wang, L. Yang, Y. C. Zhou, Y. G. Wei, and R. T. Wu 2017 *Surf. Coatings Technol.* **315** 443.
[2] M. Bialas 2008 *Surf. Coatings Technol.* **202 (24)** 6002–6010.
[3] A. Moridi, M. Azadi, and G. H. Farrahi 2014 *Surf. Coatings Technol.* **243** 91.
[4] W. Zhu, M. Cai, L. Yang, J. W. Guo, Y. C. Zhou, and C. Lu 2015 *Surf. Coatings Technol.* **276** 160.
[5] M. P. Bacos, J. M. Dorvaux, R. M. O. Lavigne, C. R. M. Poulain, and M. H. Vidal-Setif 2011 *J. Aerosp. Lab* **3** 1.
[6] R. A. Miller 1997 *J. Therm. Spray Technol.* **6 (1)** 35.
[7] S. Zhao, C. Zhang, N. Wu, and H. Wang 2011 *Prog. Nat. Sci. Mater. Int.* **21 (4)** 301.
[8] M. M. Gentleman and D. R. Clarke *Surf. Coatings Technol* 2004 vol. **188–189 (1-3)** SPEC.ISS 93.
[9] O. Kakuee, V. Fathollahi, P. Oliaiy, D. Agha-Aligol, and M. Lamehi-Rachti 2015 *Bull. Mater. Sci.* **38 (2)** 511.
[10] J. Rosler, M. Baker, and M. Volgmann 2001 *Acta Mater.* **49 (18)** 3659.
[11] J. Aktaa, K. Sfar, and D. Munz 2005 *Acta Mater.* **53 (16)** 4399.
[12] M. T. Hernandez, A. M. Karlsson, and M. Bartsch 2009 *Surf. Coatings Technol.* **203 (23)** 3549.
[13] S. R. Choi, J. W. Hutchinson, and A. G. Evans 1999 *Mech. Mater.* **31 (7)** 431.
[14] J. H. Liu, Y. B. Liu, X. He, and L. Liu 2016 *Case Stud. Therm. Eng.* **8** 250.
[15] N. P. Padture, M. Gell, E. H. Jordan 2002 *Science* **296 (5566)** 280.

IOP Conf. Series: Materials Science and Engineering **370** (2018) 012023 doi:10.1088/1757-899X/370/1/012023

A Critical Review on Slotted Design for Propellers

A Seeni, P Rajendran and H A Kutty

School of Aerospace Engineering, Universiti Sains Malaysia, Engineering Campus, Nibong Tebal, Pulau Pinang, Malaysia.

Email: aeparvathy@usm.my

Abstract. The usage of slots has gained renewed interest in aerospace particularly on propeller design. Most of the works have focused on improving the aerodynamic performance and efficiency. Modern research on propeller design aims to design propellers with high thrust performance under low torque conditions without any weight penalty. This paper aims to review recent studies made in slotted designs of aerospace structures as well as other applications such as wind turbines. A review on the usage of slots is performed in order to understand the state-of-the-art in current technology. A review of the various studies has been made and general recommendations are provided in order to perform future research in propeller design.

1. Introduction

The use of propellers as a thrust generation device has not diminished ever since the first flight of Wright Brothers. Propellers are axial rotating devices that impart forward motion for an aircraft. Initial designs modifications have so far been in increased thrust performance [1-2]. From 1920s to 1950s several inventions were patented in propeller design. The slotted propeller was a new development during this period such as those of Fairey (1938) [3] and Barnett (1939) [4].

Slots are devices that enable flow control modification around airfoil surface. This slotted concept is based on the principle that when an internal flow is introduced within the blade, the velocity of air is increased at the slot exit, disrupts the streamline flow and creates a flow separation reducing fluid velocity. The reduction in velocity increases pressure underneath the blade, thus generating more lift.

Slotted designs can be either active or passive. Active slots employ modification of flow control through blowing or suction or jets. Stored fluids are pumped and blown either tangential or vertical to the surface. Passive slots do not employ any additional active devices but rather use free stream air velocity. An illustration of a passive slot is given in Figure 1.

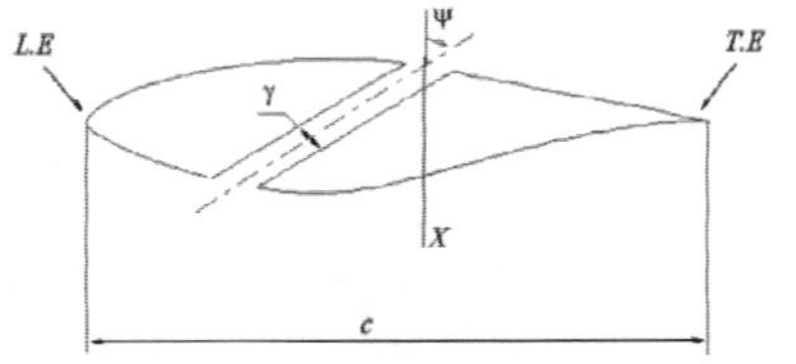

Figure 1. Illustration of a slotted airfoil cross-section [5].

Although patented inventions on slots are available decades earlier, very few studies exist in literature on slotted propeller design. Current research calls for an improvement in propeller design with high thrust, low torque characteristics. With this in mind, a detailed survey has been undertaken in order to understand the state-of-the-art.

One of the earliest studies that examined the positive effect of slots on propellers was of Braslow in 1981[6]. The study examined the feasibility of slots as circulation control (CC) devices through the development of "Coanda effect". A propeller with 2.13m diameter fitted with active CC is assumed. The airfoil shapes studied are elliptical and ellipses with modified trailing edge (TE) shapes. The TE of airfoils is blunted in order to improve lift. The Reynolds number (RN) assumed for the test case is $5x10^5$.

The study demonstrated an increase in aerodynamic performance with the presence of slots. However the aerodynamic efficiency suffered. The study also suggested that blowing requires more engine power at cruise speed and altitude than baseline, Blowing causes reduction in propeller diameter needed for cruise; Blowing can provide additional thrust for climb and acceleration at lower speeds; Blowing also increases noise levels due to jet velocity at low altitudes; The study concluded that there is greater than 7% reduction in fuel consumption with CC.

Since then only a few more studies have been witnessed. Most of the studies are on improving the aerodynamic performance of aircraft wing and wind turbines. The cross-sections of a wing, wind-turbine and propeller are similar and therefore recent studies that have so far used slotted designs on either will be discussed in this paper. The discussion will be made henceforth. In Section 2, recent research made in the field of designing propellers with straight-edge will be discussed.

In addition, a comprehensive survey of various studies related to slots that are not only limited to propellers but also on other airfoil applications will be provided. In Section 3, a discussion of results from the various studies will be provided. New improved designs will be suggested with a goal to provide future direction. Section 4 provides conclusive remarks.

2. Methodology

Most of the studies on slot so far have been on non-thrust generating designs of fixed aircraft wing and wind turbine and very few studies exist that have been implemented directly on either aircraft or marine propellers. Studies on non-thrust generating platform try to improve lift and stall angle. Propellers with slots have been studied to improve thrust generation. Recent research in this topic will be discussed subsequently.

2.1. Passive slot

The study by Kweder et al. [7] is on performing a design feasibility study on using passive slots for small scale propellers. The propellers consisted of Clark-Y airfoil sections. The diameter of tested propeller is 0.609 mm which is small-scale and applicable for UAVs. The design is based on the principle of using the forward velocity of the aircraft to pressurize the internal CC plenum. The system is shown to provide 5.12% increase in aerodynamic efficiency over all range of flight envelope cases. The results of the study are provided in Table 1.

2.2. Other studies

2.2.1. Active slots. Huang [8] studied the suction and blowing flow control techniques on a NACA 0012 airfoil. A RN of $5x10^5$ is assumed in this study. The results are briefly described in Table 2. Shojaefard et al. [9] presents a numerical investigation concerning flow control by suction and injection. The investigation consists of a subsonic airfoil with four suction and injection slots on the suction side of the airfoil. Five different AoA where 0°, 5°, 10°, 13.3° and 20° at a Mach number of 0.15 were studied. The effects of suction and injection on aerodynamic performance for each AoA are studied. The results of the study are provided in Table 3.

Table 1. Description of study by Kweder et al. (2014) [7].

Propeller design	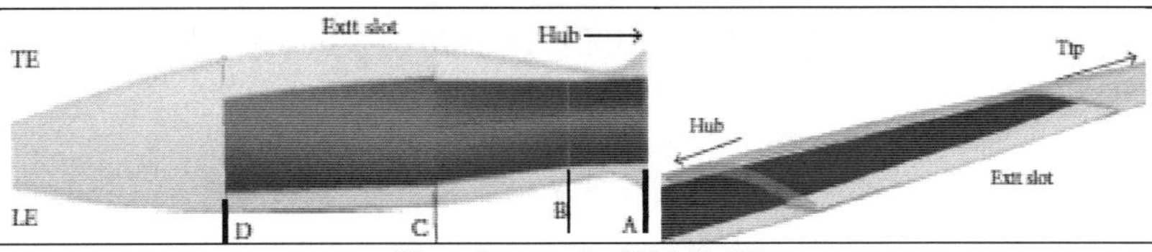
Author	Kweder et al. (2014) [7]
Study type	Experimental
Enhancement	5.12% ↑ in aerodynamic efficiency on all flight envelope cases; Output thrust force marginally higher than baseline for all advance ratios (*J*); Power and torque co-efficient are ↑ than baseline design; As *J* increases, net negative thrust output is generated leading to power taken from free stream producing "wind milling" effect.

Table 2. Description of study by Huang [8].

Propeller design	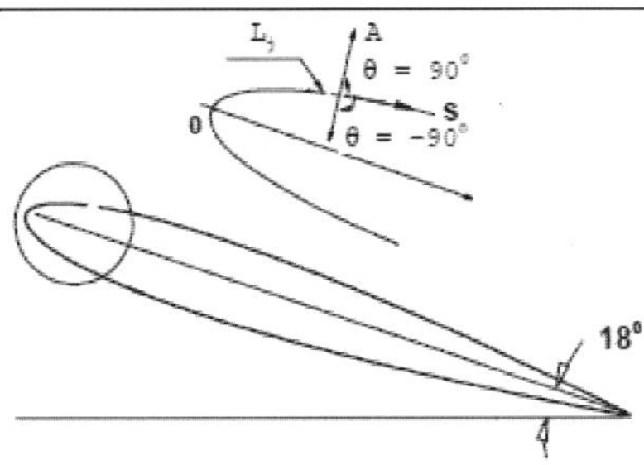
Author	Huang (2004) [8]
Study type	Numerical
Enhancement	When jet location and angles of attack (AoA) were combined, perpendicular suction at the leading edge (LE) increased lift coefficient (C_L) better than other suction situations; Tangential blowing at downstream locations was found to lead to the maximum increase in the C_L value.

Table 3. Description of study by Shojaefard et al. [9].

Propeller design	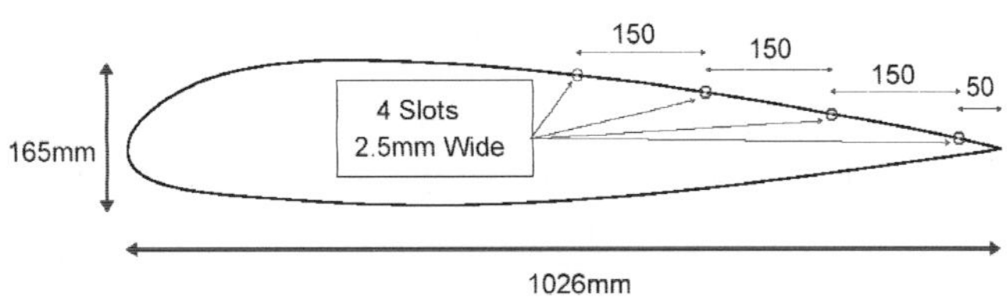
Author	Shojaefard et al. (2005) [9]
Study type	Numerical
Enhancement	Suction increases the C_L by 10% over baseline. The injection mechanism causes reduction in friction drag.

The study by Goodarzi et al. [10] is on the impact of slots on aerodynamic performance. The model is a 2D NACA 0015 airfoil sections and has assumed to have a slot with a width of 2.5% chord (*c*). Three slot positions, 10% *c*, 30% *c* and 50% *c* are considered. The analysis was performed for AoA between 12° and 17°. The flow conditions are assumed to have a RN of 4.55×10^5. The results of the study are provided in Table 4.

Table 4. Description of study by Goodarzi et al. [10].

Propeller design	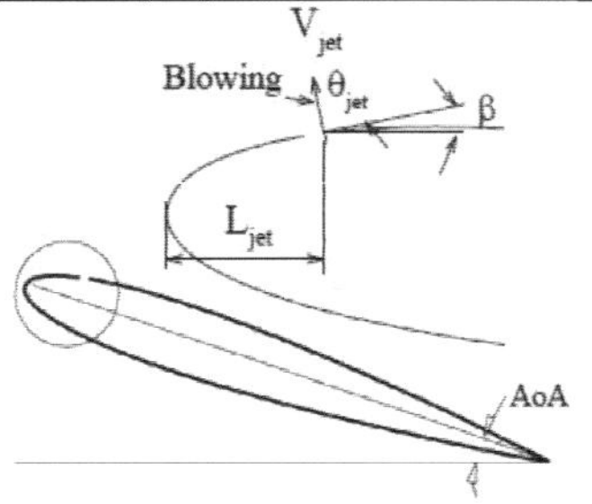
Author	Goodarzi et al. (2012) [10]
Study type	Numerical
Enhancement	Blowing increases the amount of lift and reduces drag. At higher AoA, the separation is delayed and improves the performance of the airfoil.

Yousefi *et al.* [11] performed a numerical study on slotted design with tangential and perpendicular blowing at the TE of a NACA 0012 airfoil. Perpendicular suction at the LE was applied on the airfoil upper surface. The jet widths were varied from 1.5% to 4% of the *c* and the jet velocity (Jet_A) varied between 0.3 and 0.5 of the free stream velocity. The performance improvements are provided in terms of lift-to-drag (L/D) ratio instead of C_L and drag coefficient (C_D) increase or decrease. The results of the study are provided in Table 5.

Table 5. Description of study by Yousefi et al. [11].

Propeller design	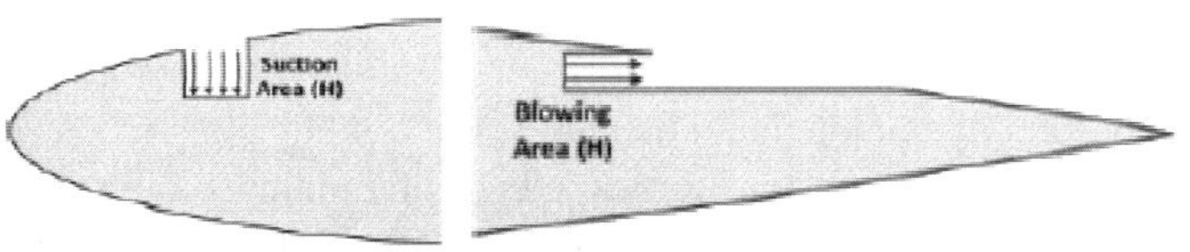
Author	Yousefi et al. [11]
Study type	Numerical
Enhancement	As the blowing jet width ↑, the L/D ratio ↑ continuously in tangential blowing and ↓ quasi-linearly in perpendicular blowing; The L/D ratio improves when the suction jet width ↑ and reaches its maximum value at 2.5% *c*.

Eggert and Rumsey [12] have studied the impact of slots as an effective means of flow control on NACA 0018 airfoil. CFD method is used in this study. The flow control device consists of a blowing slot located on the upper surface of an NACA 0018 airfoil near the LE. The analyses were performed for RN of 2.5×10^5. Numerical analyses were performed using RANS and experimental study was performed in a wind-tunnel. The results of the study are provided in Table 6.

2.2.2. Passive slots. A study was conducted by Kang and Park [13] for a wind turbine with S809 airfoil sections. The RN assumed in this study is at 2×10^6. Five different AoA conditions, 0°, 1.02°, 5.13°, 9.22°, 14.24° and 20.15° were studied. The results of this study are described in Table 7. Xie et al. [14] has studied a flow control mechanism called split blade that is identical to slot. In this study, the effect of slots was studied on S809 airfoil that is investigated both numerically and experimentally.

The results show that at lower AoA from 0° to 10° the flow is unchanged. The C_L is increased and C_D decreased from AoA 15° to 20°. The results of this study are described in Table 8.

Table 6. Description of study by Eggert and Rumsey [12].

Propeller design	
Author	Eggert and Rumsey (2017) [12]
Study type	Numerical and Experimental
Enhancement	Blowing with high momentum coefficient (C_μ) will ↑ the lift of the airfoil and delay flow separation. Slot with low C_μ will ↓ the lift and induce separation even at low AoA.

Table 7. Description of study by Kang and Park [13].

Propeller design	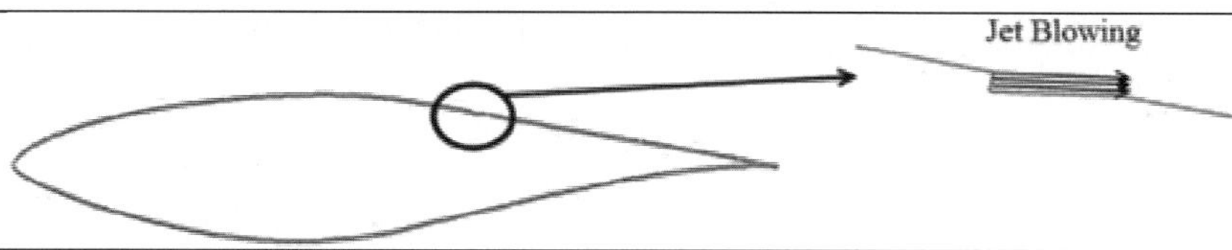
Author	Kang and Park (2013) [13]
Study type	Numerical
Enhancement	At 0.6*c* slot location, maximum lift coefficient (C_{Lmax}) is observed at AoA 14.24°. At 0.024*c*, a stall delay of 14.24° and 20.15° AoA is found.

Table 8. Description of study by Xie et al. [14].

Propeller design	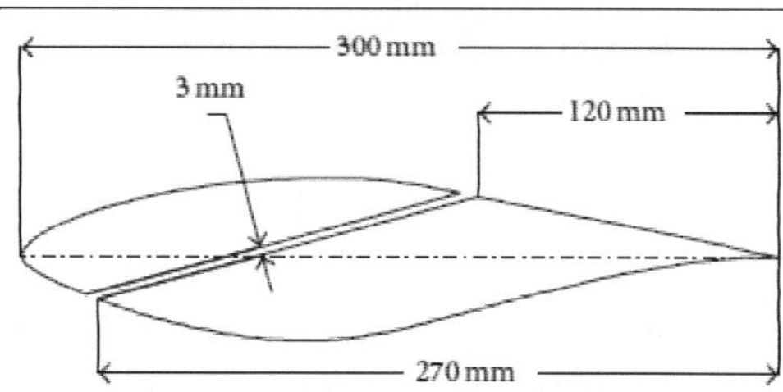
Author	Xie et al. (2013) [14]
Study type	Numerical and Experimental
Enhancement	4.9% ↑ in C_L at AoA 20^0, 1.6% ↑ at AoA 15^0

Belamadi et al. [5] has studied the effect of slots as a passive flow control mechanism on wind turbines. The blade is assumed to be made of S809 airfoil sections. CFD analysis is performed for different configurations by varying slot location, slot width and slot slope. The analysis was performed on 2D models for varying AoA. The RN for all analyses is $1x10^6$. The results of this study are described in Table 9. In addition to tubercles design, slotted designs are studied by Ibrahim et al. [15]. The test model consists of a blade with NACA 4412 sections and a diameter of 60 cm. Different free stream velocities from 4 to 8 m/s are assumed. The results from the study are provided in Table 10.

Table 9. Description of study by Belamadi et al. [5].

Propeller design	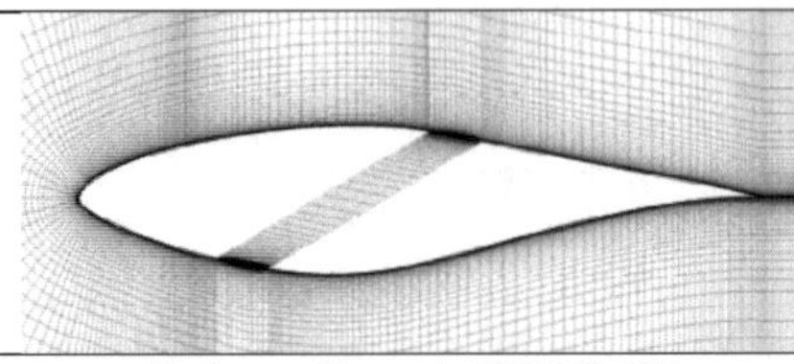
Author	Belamadi et al. (2016) [5]
Study type	Numerical
Enhancement	Aerodynamic improvement can be found only on specific range of AoA from 10° to 20°. At lower AoA there is significant drag penalty.

Table 10. Description of study by Ibrahim et al. [15].

Propeller design	
Author	Ibrahim et al. (2015) [15]
Study type	Experimental
Enhancement	Slotted blade was able to generate up to 60% more power at a lower speed compared to the straight blade

3. Discussion

The various studies discussed have been summarized and the results are compared in Table 11. The discussion of the results is two-fold. The aerodynamic improvement values for the propeller designs considered are provided in terms of augmentation in thrust or power and torque. The discussions are provided subsequently:-

- The study by Kweder et al. [7] confirmed that both aerodynamic performance and efficiency can be increased using slots. This is an improvement on the old design by Braslow [6]. The study by Kweder et al. [7] is different as it is on passive control on small scale propellers of UAVs with Clark-Y sections. Here a significant improvement in thrust co-efficient (C_T) values of 50% is achieved only with a small 17% ↑ in power co-efficient (C_P) for J>0.25 typical of normal operating conditions of UAVs. A 5.12% ↑ in aerodynamic efficiency is derived through this passive control.
- Huang [8] and Yousefi et al. [11] have studied slots with active jets on NACA 0012 airfoil models. The studies were conducted for a RN of 5×10^5. The results indicate C_L ↑ and C_D ↓ over baseline for all test cases of section, perpendicular and tangential blowing. A high C_L ↑ of 75% and 45% ↓ in C_D was found by Yousefi et al. [11] for suction case. The study by Goodarzi et al. [10] is on NACA 0015 employing active blowing and assuming a RN of 4.55×10^5. Here an 80% ↑ and 45% ↓ in C_L and C_D values respectively has been achieved.
- Kang and Park [13], Xie et al. [14] and Belamadi et al. [5] have investigated passive flow control on S809 airfoils with different *c* and RN. Still, all studies show a positive improvement in aerodynamic performance over unmodified baseline airfoil case. Xie et al. [14] has achieved a ↑ C_{Lmax} of 4.9% at AoA 19^0. Belamadi et al. [5] has achieved a relatively low of 3.75% ↑ C_{Lmax} for AoA 16^0. A 6.12% ↓ in C_D at AoA 19^0 is also found.
- The study by Shojaefard et al. [9] on Aerospatiale A-airfoil with 1.026 m *c* indicates a 10% ↑ ΔC_L for suction case and 23% ↑ for injection.

- In the case of wind turbines, the results from the study by Ibrahim et al. [15] with slotted blades provided a 60% ↑ in power compared to baseline. Slotted blades are shown to provide a high performance.
- A different type of study was by Eggert and Rumsey [12] in which different blowing conditions at different coefficients were studied on a NACA 0018 test model using CFD. The results showed for C_μ =5%, the C_L is much higher at 1.75 compared to 0% C_μ.
- Only two studies have so far been undertaken for employing slotted designs on propellers. Braslow [6] showed that a 7% ↓ in fuel consumption during cruise could be achieved. Kweder et al. [7] showed a 5.12% ↑ in aerodynamic efficiency.

Table 11. Summary of results from studies.

Study name	Type	RN	Variant	Improvement					
				C_L/ $C_L$base	C_D/ $C_D$base	C_Lmax	C_Dmax	ΔC_L	ΔC_D
Braslow [6]	Active	5x10^5	Blowing	Propeller: >7% reduction in fuel consumption during cruise					
Kweder et al. [7]	Passive	Varied	Not Applicable	8-10% ↑ in torque at J<=0.2; 50% ↑ bin C_T and 17% ↑ in C_P for J>0.25 and J<0.55. 8-10% more power required for all *J*					
Huang [8]	Active	5x10^5	Suction	1.243	0.805	-	-	-	-
			Perpendicular blowing	1.023	0.902	-	-	-	-
			Tangential blowing	1.048	0.925	-	-	-	-
Shojaefard et al. [9]	Active	M=0.15	Suction	-	-	5 at AoA 10^0	>0.4 at AoA 20^0	10% ↑ at AoA 5^0	
			Injection	-	-	>4 at AoA 13.3^0	0.4 at AoA 20^0	23% ↑ at AoA 5^0	
Goodarzi et al. [10]	Active	4.55x10^5	Blowing	-	-	-	-	80% ↑ at AoA 17^0	45% ↓ at AoA 17^0
Yousefi et al. [11]	Active	5x10^5	Suction	1.295	0.708	<1.8 at AoA 14^0	≤0.18 at AoA (0-18^0)	75% ↑ at AoA 18^0	56% ↓ at AoA 18^0
			Perpendicular blowing	1.021	0.897	<1.2 at AoA 14^0	<0.4 at AoA (0-18^0)	8.5% ↓ at Jet_A=0.3 14.5% ↓ at Jet_A =0.5	8% ↑ at Jet_A =0.3 5% ↑ at Jet_A =0.5
			Tangential blowing	1.070	0.926	<1.2 at AoA 14^0	<0.2 at AoA (0-18^0)	7% ↑	7% ↓
Eggert and Rumsey [12]	Active	2.5x10^5	Blowing	-	-	>1.75 at AoA 18^0 >1.2 at AoA 15^0	-	-	-
Shojaefard et al. [9]	Passive	2x10^6	Not Applicable	-	-	~1.4 at AoA 14^0	~0.24 at AoA 20^0	-	-
Xie et al. [14]	Passive	1x10^5	Not Applicable	-	-	0.688 at AoA 20^0	~0.075 at AoA 20^0	4.9% ↑ at AoA 20^0, 1.6% ↑ at AoA 15^0	-
Belamadi et al. [5]	Passive	1.8x10^6	Not Applicable	-	-	~1.12 at AoA 5^0	-	3.75% ↑ at AoA 16^0	6.12% ↓ at AoA 19^0
Ibrahim et al. [15]	Passive	Varied	Not Applicable	Wind turbine: 60% more power generated by slotted compared to straight-edged turbine blade.					

From the above, the following remarks can be made.

- Slotted designs provide a high C_L regardless of airfoil shape and RN. Using blowing, a maximum ΔC_L over unmodified baseline of up to 80% is realistically achievable.
- Using suction, a maximum ΔC_L over unmodified baseline of up to 75% is realistically achievable.
- Active slotted designs have so far provided better ΔC_L increment to baseline. Therefore, active designs provide better performance enhancement over passive designs.
- Improvements in performance and efficiency for propeller design can be realistically achievable even by using passive designs. The thrust levels could be improved even at low torque inputs. The improved efficiency may reduce the total fuel consumption and take-off weight.

4. Conclusion

A review of studies based on slotted designs has been concluded. The most significant results that were obtained so far have been discussed. It can be found that slots provide significant aerodynamic improvements. Only very few studies exist that has studied slots directly for propeller blade design. New slotted designs will be studied extensively in the future.

Acknowledgement

This publication was supported by Universiti Sains Malaysia Grant No. 304/PAERO/6315002.

References

[1] Kutty H A and Rajendran P 2017 Review on Numerical and Experimental Work of Conventional and Unconventional Propeller Blade Design International Review of Aerospace Engineering Volume 10 No 2.
[2] Kutty H A and Rajendran P 2017 3D CFD Simulation and Experimental Validation of Small APC Slow Flyer Propeller Blade Aerospace Volume 4 Issue 10.
[3] Fairey C R 1938 BLADE FOR AIRSCREWS AND THE LIKE 2–7
[4] Barnett T B 1939 Propeller 11–2
[5] Belamadi R, Djemili A, Ilinca A and Mdouki R 2016 Aerodynamic performance analysis of slotted airfoils for application to wind turbine blades J. Wind Eng. Ind. Aerodyn. 151 79–99
[6] Braslow A 1981 AERODYNAMIC EVALUATION OF CIRCULATION CONTROL PROPELLERS
[7] Kweder, J; Zeune, C; Geiger, J; Lower, A; Smith J 2014 Experimental Evaluation of an Internally Pressurized Passively Circulation Control Propeller J. Aerodyn. 2014 1–21
[8] Huang L 2004 Optimization of Blowing and Suction Control on Naca 0012 Airfoil Using Genetic Algorithm with Diversity Control (University of Kentucky)
[9] Shojaefard, M; Noorpoor, A; Avanesians, A; Ghaffarpour M 2005 Numerical Investigation of Flow Control by Suction and Injection on a Subsonic Airfoil Am. J. Appl. Sci. 2 1474–80
[10] Goodarzi M, Rahimi M and Fereidouni R 2012 Investigation of Active Flow Control over NACA0015 Airfoil Via Blowing Int. J. Aerosp. Sci. 1 57–63
[11] Yousefi K, Saleh R and Zahedi P 2014 Numerical study of blowing and suction slot geometry optimization on NACA 0012 airfoil J. Mech. Sci. Technol. 28 1297–310
[12] Eggert C A and Rumsey C L 2017 (NASA Prelim)CFD Study of NACA 0018 Airfoil with Flow Control 2017–219602
[13] Kang T and Park W-G 2013 Numerical investigation of active control for an S809 wind turbine airfoil Int. J. Precis. Eng. Manuf. 14 1037–41
[14] Xie Y, Chen J, Qu H, Xie G, Zhang D and Moshfeghi M 2013 Numerical and Experimental Investigation on the Flow Separation Control of S809 Airfoil with Slot Math. Probl. Eng. 2013
[15] Ibrahim M, Alsultan A, Shen S and Amano R S 2015 Advances in Horizontal Axis Wind Turbine Blade Designs : Introduction of Slots and Tubercle J. Energy Resour. Technol. 137 1–6

IOP Conf. Series: Materials Science and Engineering **370** (2018) 012024 doi:10.1088/1757-899X/370/1/012024

An Assessment on Temperature Profile of Jet-A/Biodiesel Mixture in a Simple Combustion Chamber with Plain Orifice Atomiser

Ng, W. X.[1], Mazlan, N. M.[2], Ismail M. A.[4], Rajendran, P.[3]

[1,2,3] School of Aerospace Engineering, Engineering Campus, Universiti Sains Malaysia, 14300 Nibong Tebal, Pulau Pinang, MALAYSIA
[4] School of Mechanical Engineering, Engineering Campus, Universiti Sains Malaysia, 14300 Nibong Tebal, Pulau Pinang, MALAYSIA

Corresponding email: nmusfirah@usm.my

Abstract. The preliminary study to evaluate influence of biodiesel/kerosene mixtures on combustion temperature profile is explored. A simple cylindrical combustion chamber configuration with plain orifice atomiser is used for the evaluation. The evaluation is performed under stoichiometric air to fuel ratio. Six samples of fuels are used: 100BD (pure biodiesel), 100KE (pure Jet-A), 20KE80BD (20% Jet-A/80% Biodiesel), 40KE60BD (40% Jet-A/60% Biodiesel), 60KE40BD (60% Jet-A/40% Biodiesel), and 80KE20BD (80% Jet-A/20% Biodiesel). Results showed that the oxygen content, viscosity, and lower heating value are key parameters in affecting the temperature profile inside the chamber. Biodiesel is known to have higher energy content, higher viscosity and lower heating value compared to kerosene. Mixing biodiesel with kerosene improves viscosity and caloric value but reduces oxygen content of the fuel. High oxygen content of the biodiesel resulted to the highest flame temperature. However the flame temperature reduce as the percentage of biodiesel in the fuel mixture reduces.

1. Introduction

The evaluation of fuel properties is essential in determining the suitability of the fuel for gas turbine engines. Fuel properties are highlighted as important requirements in order to use new fuels in gas turbine engines [1]. Fuel volatility will affect combustion efficiency, ignition, exhaust temperature distribution, and safety. The increase in carbon/hydrogen (C/H) ratio is found to affect the tendency of carbon to deposit in the combustion chamber. Vapour pressure, viscosity and density are found to influence the evaporation and atomisation of the liquid fuel [2]. Fuel heating value is known to affect the amount of fuel consumed by the gas turbine and the thrust produced from the gas turbine engine [3-5]. In another study, heating value affects the gradient of flame temperature, which consequently influences the increase in nitrogen oxide (NOx) [5]. The heating value of the fuel is related to the molecular weight of the fuels. The heaviest fuels normally have the highest heating value, and thus the fuel consumption is reduced. Aside from heating value, heat capacity also contributes to the production of engine thrust [4]. The same study also observed an increment increase in fuel consumption as the density of the fuel increases.

Due to the fuel shortage crisis and the escalation of environmental pollution, a massive study on the effectiveness of using biofuel as a substitute for conventional kerosene has been performed on aircraft application [3-6] as well as other gas turbine engine that generates power [7]. However, the challenge in using biofuel in the gas turbine engine is steep. The combustion chamber has to be redesigned to obtain higher combustion efficiency and lower NOx [8]. In aviation perspective, for the near and short-term application, biofuel has to blend with the aviation turbine for up to 50% [9] to ensure the standard requirement so that the aviation fuel is achieved. Hydrotreated renewable jet fuel and bio-synthetic paraffinic kerosene types of fuel are suitable to be used as drop-in fuel in the aircraft engine. The capability of these fuels in generating thrust and low NOx generated from the aircraft engine is promising [3, 5, 9-11]. However, incomplete combustion, which consequently produces high carbon monoxide (CO), is observed.

Instead of powering aircraft, empowering gas turbine engine with biofuel for generating power has been imposed and reported vastly. Mustafa et al. [7] elucidated the effects of fuel-air equivalence ratios on combustion driven thermoelectric (TE) and thermophotovoltaic (TPV) power systems using several blends of vegetable-cooking oil. The use of biofuel increases CO, however, the generation of CO for TE system is found to be lower than TPV system, while the electrical efficiencies of TE and TPV system were improved at rich mixtures. The performance and emission characteristics of biofuel and biofuel blends in a 30kW gas turbine engine have been studied by Habib et al. [12]. Biofuels used in this study were soy methyl ester, canola methyl ester, recycled rapeseed methyl ester, hog-fat biofuel, and their 50% (volume) blends in Jet A. Decreases on CO and NOx pollutant have been observed with greater reduction, and they were found if blended fuel was used. Higher thermal efficiencies of pure biofuels are found owing to the presence of oxygen molecules in the biofuel. Meanwhile, Li Zhou et al. [13] performed an analysis on the influence of ethanol and methanol on the combustion characteristics and engine performance system. The study pointed out that viscosity and caloric value are key factors affecting the atomisation and combustion characteristics of the biofuels. Due to low fuel viscosity, low initial ejection temperature without preheating is necessary to obtain a better atomisation. The study found out that the mass flow rate of the biofuel needs to be increased in order to obtain the same magnitude of thrust as conventional fuel. Most of the studies above have concerned on CO generation, which associates with incomplete combustion due to poor fuel spray formation process.

Understanding the spray formation process is an important step to increase combustion efficiency and reduce pollutant emissions [14]. Fuel atomisation can be affected by many factors such as droplet velocity, spray cone angle, and spray tip penetration [15], which was influenced by the property of the fuel [15, 16] an ambient condition of the combustion chamber [6, 15, 17] as well as nozzle design [17]. Fuel with low viscosity, such as bioethanol, improves fuel evaporation and the breakup process [15]. Lujaji et al. [16] found that liquid density, heating value, and viscosity are the major contributors to spray droplet size when different liquid flow rates are set to maintain the same energy input. The higher viscosity of atomised liquid tends to create a larger droplet to minimise the growth rate of instabilities. A higher energy density causes lesser amounts of fuel exiting the nozzle and better atomisation.

A complete combustion can also be achieved by having fuel with high oxygen content such as biodiesel and vegetable oil [18]. Biodiesel contains 11% oxygen by weight [18]. The oxygen provides higher heat release during the premixed phase combustion, offers rapid breakage of hydrocarbon, and contributes to a hotter combustion process for complete combustion [19, 20]. Mustafa et al. [7] observed that fuel with high oxygen content provides advantageous post flame oxidation and flame speed during air-fuel interactions. This fuel promotes better oxidation during the combustion process and produces low CO.

Most of the works mentioned above pointed out the importance of fuel properties on NOx and CO generation. However, works prior to transient behaviour of the biofuel inside the combustion chamber are lacking and remained persistently challenging. In this paper, Jet-A, biodiesel, and biodiesel/kerosene mixtures are chosen to be tested in a simple cylindrical shape of a combustion chamber with plain orifice atomiser. Temperature profile, velocity contours and oxygen concentrations of the selected fuels are discussed and a comparison between the fuels is presented.

2. Methodology

2.1. Model Description

The study uses a simple combustion chamber with plain orifice atomiser. The length of the chamber is 310mm, air inlet and outer diameter of 70mm and 60mm respectively, and fuel inlet diameter of 10mm. The geometric model of the chamber is shown in Figure 1(a) while Figure 1(b) shows 62530 cells of non-structured mesh was performed. Table 1 shows dimension for each zone in the combustion chamber.

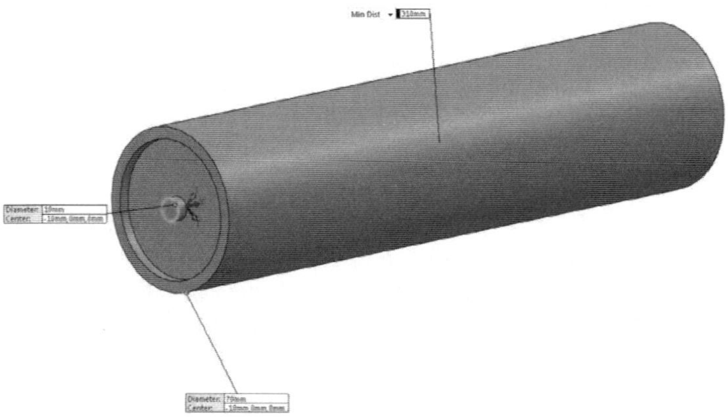

(a) Geometric model

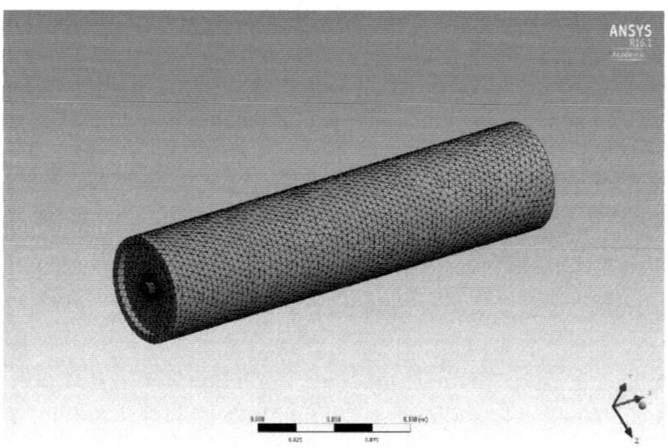

(b) Mesh

Figure 1. Geometric model (a) and mesh (b) of the combustion chamber

Table 1. Main dimension of combustor chamber

Zone	Flow Area (m^2)	Reactor Length (m)
Flame front (FF)	0.015	
Primary (PZ)	0.015	0.019
Secondary (SZ)	0.015	0.056
Dilution (DZ)	0.015	0.31

Biodiesel has been recognised as one of the potential substitutes for conventional fuels in a combustion system. Biodiesel consists of a mixture of methyl esters derived from transesterification process of either vegetable oils [21], recycled waste, or animal fats [22]. Thermal stability issues and the tendency of biodiesel to freeze at normal cruise temperature make it inappropriate to be used in aircraft engine [23], but it has been successfully used as blends with diesel in compression-ignition engines [24].

Six samples of fuels are used: 100BD (pure biodiesel), 100KE (pure Jet-A), 20KE80BD (20% Jet-A/80% Biodiesel), 40KE60BD (40% Jet-A/60% Biodiesel), 60KE40BD (60% Jet-A/40% Biodiesel), and 80KE20BD (80% Jet-A/20% Biodiesel). The chemical properties of the fuel mixtures such as density, kinematic viscosity, flash point and heating value of the fuel mixtures were measured experimentally. However due to limitation of the equipment available in the lab, specific heat capacity, surface tension, latent heat of vaporisation, vaporisation temperature, and boiling point were predicted using Equation 1.

$$X = aA + bB \tag{1}$$

where $\boldsymbol{X}$ is fuel properties of the mixtures, $\boldsymbol{a}$ and $\boldsymbol{b}$ is the percentage of kerosene and biodiesel respectively (%), while $\boldsymbol{A}$ and $\boldsymbol{B}$ are the properties of kerosene and biodiesel obtained from literature respectively. Density of the fuel mixture was determined according to ASTM D1298 using a hydrometer and a water bath. The density was measured from 30 °C to 80 °C. Kinematic viscosity of the fuels is obtained by dividing dynamic viscosity with density (Equation 2).

$$v = \frac{\mu}{\rho}, \tag{2}$$

where v is kinematic viscosity (mm^2/s), μ is dynamic viscosity (kg/mms), and ρ is density of fuel (kg/mm^3). Dynamic viscosity of the fuels is measured according to ASTM D445 using Brookfield DV-III Ultra's viscometer. The dynamic viscosity is measure for temperature ranging temperature ranging from 30 °C to 80 °C.

An elemental analyser (Perkin Elmer 2400 CHN Element) was used to analyse the chemical components contain in 100BD. Results obtained from the analysis were used to predict the chemical components of the fuel mixtures by using Equation (3).

$$C = (D \times a) + (E \times b) \tag{3}$$

where C is the number of carbon/hydrogen/oxygen in fuel mixtures, a is percentage of kerosene in the mixtures (%), b is percentage of biodiesel in the mixtures (%), D is number of carbon/hydrogen in kerosene, and E is number of carbon/hydrogen/oxygen in biodiesel. Table 2 summarises the properties of the fuel mixtures used in this study.

In evaluating the fuel mixtures, user defined function (UDF) was used due to unavailability of the properties of the fuel mixture in the CFD library. Properties of the fuel as listed in Table 2 will be used as input into the UDF.

2.2. Simulation Method

Temperature profile of all tested fuels in the designed combustion chamber is assessed at stoichiometric, lean, and rich air to fuel ratio. The simulation was performed by assuming the combustion to be non-premixed while the flow was taken as steady and incompressible. Temperature and pressure of the combustor was set to be room temperature and atmospheric pressure respectively.

Table 2. Summary of Fuel Properties for Different Fuel Samples Measured via Experimental setup or Prediction

Properties	100KE	20KE/ 80BD	40KE/ 60BD	60KE/ 40BD	80KE/ 20BD	100BD	Measurement method
Density (kg/m³)	795	850	840	830	810	860	ASTM D1298
Specific Heat Capacity (J/kgK)	2090	1966.8	1948.32	1954.48	1960.64	1936*	Prediction
Dynamic Viscosity (kg/ms)	0.001	0.004	0.0025	0.002	0.0015	0.0055	ASTM D445
Surface Tension (N/m)	0.0263	0.0253	0.0251	0.0252	0.0252	0.025*	Prediction
Latent Heat of Vaporization (MJ/kg)	226	237.2	234.4	231.6	228.8	240*	Prediction
Vaporization Temp (K)	341	574.6	516.2	457.8	399.4	633*	Prediction
Boiling Point (K)	477	579.4	553.8	528.8	502.6	605*	Prediction
Caloric value (MJ/kg)	30.6	31.5	32.0	34.4	35.4	42.6	ASTM D4809
Molecular Formula	$C_{12}H_{23}$	$C_{10}H_{33}O_2$	$C_{11}H_{30}O$	$C_{11}H_{28}O$	$C_{12}H_{25}$	$C_{10}H_{35}O_2$	Prediction

*Reference Value [25]

Boundary condition was set according to Eq. (3) to Eq. (5). The boundary condition was defined by assuming constant power generated from the combustion. Fuel mass flow rate was calculated using Eq. (3) by taking into account the relationship between the power generated through combustion, and the caloric value of the fuel samples. The caloric value of the tested fuels used for mass flow rate calculation was obtained from the test experiment conducted earlier. The velocity of the fuel was calculated using Eq. (4) by assuming that the system is incompressible. The oxidizer (air) inlet rate was calculated using Eq. (5) by multiplying the fuel rate with the respected air to fuel ratio. The turbulent density and turbulent viscosity ratio were both set as 5%. Details of the parameters used in the evaluation are shown in Table 3.

$$\dot{m}_f = \frac{P}{C_v} \tag{3}$$

$$\mathrm{V}_f = \frac{\dot{\mathrm{m}}_f}{\rho \times \mathrm{A}} \tag{4}$$

$$V_o = \mathrm{V}_f \times \mathrm{AFR} \tag{5}$$

where $\dot{m}_f$ is fuel mass flow rate (kg/s), V_f is fuel velocity rate (m/s), V_o is oxidiser velocity rate (m/s), P is power (0.6MJ/s), A is fuel inlet area (0.0000785 m^2), C_v is caloric value (MJ/kg), ρ is fuel's density (kg/m^3) and AFR is air to fuel ratio.

Table 3. Inlet Boundary Condition for Fuel and Oxidizer at Stoichiometric, Lean and Rich AFR

	100KE	20KE80BD	40KE60BD	60KE40BD	80KE20BD	100BD
Fuel Mass Flow Rate (kg/s)	0.014	0.019	0.0187	0.017	0.0169	0.0196
Fuel Velocity Rate (m/s)	0.223	0.287	0.285	0.268	0.267	0.290
Air-Fuel Ratio	15.6	13.92	14.34	14.76	15.18	13.5
Air Velocity Rate (m/s)	3.479	3.976	4.080	3.955	4.047	3.917

3. Results and Discussion

3.1. *Temperature distribution of the fuel mixtures along the chamber*

Temperature profile of the designed combustion chamber is predicted for all the fuel mixtures. The effects of the fuel mixture were analysed for three different cases: stoichiometric, lean and rich air to fuel ratio. The analysis was performed under non-premixed combustion, incompressible flow, under room temperature, and at atmospheric conditions. Desired power output generated through combustion is assumed constant to identify the influence of fuel properties during combustion. An equivalence ratio of one (1) was used for stoichiometric, the equivalence ratio of 0.7 was used for lean and the equivalence ratio of 1.5 was used for rich.

The distribution of temperature for all the tested fuels at stoichiometric, lean and rich fuel-air equivalence ratio test cases is illustrated in Figure 2, Figure 3, and Figure 4 respectively. From the figures, it can be seen that the combustion took place at the upstream of the combustion section. The temperature reduces and getting constant downstream. Figure 5 shows comparison of flame temperature for all fuels at different test cases. It is noticed that for all type of fuels, maximum temperature is observed at stoichiometric ratio where the both fuel and air has ideal ratio in order to burn completely without having excessive air. Temperature of the flame is lower at rich while at lean the lowest flame temperature is observed. At stoichiometric, 100BD is observed to have the highest flame temperature compared to other fuels. The presence of oxygen in biodiesel was found to be a factor to the highest flame temperature. Oxygen content in 100BD promoted to the higher heat release during the combustion, and thus it resulted to the higher gas temperature in the combustion chamber. The result was consistent with Palash et al. [26] and Mustafa et al. [7], who observed the similar behaviour for the oxygenated type of fuel. Additionally, the flame temperature reduced as the percentage of kerosene in the mixture increased due to the reduction of oxygen molecules in the mixture.

In addition, the highest flame temperature of 100BD is due to the amount of fuel injected into the combustion chamber. As in this study, all fuels are assumed to generate same power output as 100KE. Since LHV of 100BD is lower than 100KE, more fuel has to be injected thus produce higher combustion temperature.

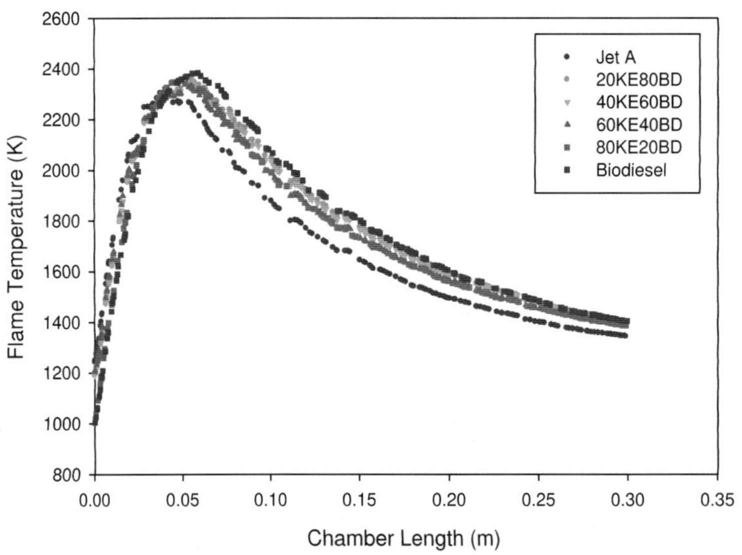

Figure 2. Temperature contours of different fuel mixtures at stoichiometric air to fuel ratio

3.2. Comparison of flame temperature at different combustion zones

Figure 6 to Figure 8 show the comparison of flame temperature at different combustion zones as defined in Table 2. For all test cases, the highest flame temperature is observed to be located at the secondary zone of the chamber. It is worth to mention here that the effect of cooling air is not counted in this study. Therefore highest temperature in this region is expected. In the real combustion chamber, an amount of air will be extracted from compressor bleed to be entered into secondary and dilution zone. This air is used to ensure the turbine section can tolerate high temperature that exit from combustion chamber.

Meanwhile, at PZ, 100KE is found to have the highest flame temperature compared to 100BD. This is because of the lowest boiling temperature of the fuel which eventually provides faster vaporisation process will be able to mix with air homogenously for combustion. Similarly due to highest boiling temperature, 100BD requires longer time to fully evaporate thus only be able to fully combust in SZ. As a result, higher temperature is observed at DZ which consequently requires more cooling air is required to be extracted from compressor bleed.

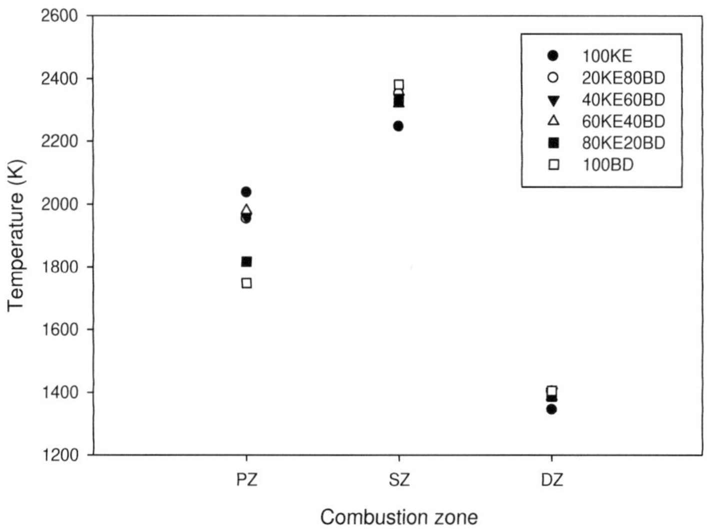

Figure 3. Temperature of the tested fuels at Primary, Secondary and Dilution zones for Stoichiometric

4. Conclusion

Five blends of BD – JP4 blended fuels: 20KE80BD, 40KE60BD, 60KE40BD, and 80KE20BD were tested in a simple cylindrical shape of the combustion chamber with plain orifice atomiser at stoichiometric air to fuel ratio. The combustion characteristics were discussed in terms of temperature aspect. The effects of fuel properties on combustion characteristics of the tested fuels were observed. Combusting 100BD produces the highest flame temperature compared to other fuel mixtures. The temperature reduces as the percentage of biodiesel in the mixtures reduces. High oxygen content in biodiesel promotes the oxidation process during combustion, and thus it increases the flame temperature. Additionally due to lower LHV of 100BD, more fuel has to be injected in order to maintain the similar power output as 100KE. As a result the highest flame temperature is observed. 100BD also has highest boiling temperature and more volatile compare to the other fuels. Consequently, 100BD requires more time to evaporate completely and finally burnt at SZ. As a result 100BD provides highest temperature at DZ, therefore requires more cooling air to be extracted from compressor bleed to reduce the temperature.

Acknowledgement
This research and publication was supported by Universiti Sains Malaysia Grant No. 304/PAERO/ 60315002.

References

[1] Sharp J 1951 *Aircraft Engineering and Aerospace Technology* **23** 2
[2] Mazlan N, Savill M and Kipouros T 2011 *Computational Evaluation and Exploration of Combustion Performance for Liquid Jet Fuels Derived from Biomass presented* In: 3rd CEAS Air & Space Conference
[3] Mazlan N M, Savill M and Kipouros T 2017 *Proceedings of the Institution of Mechanical Engineers, Part G: Journal of Aerospace Engineering* **231** 859
[4] Mazlan N M, Savill M, Kipouros T and Li Y-G 2012 *A numerical study into the effects of bio-synthetic paraffinic kerosine blends with jet-A fuel for civil aircraft engine*. In: ASME Turbo Expo 2012: Turbine Technical Conference and Exposition: American Society of Mechanical Engineers 165
[5] Mazlan N M, Savill M and Kipouros T 2015 *Aircraft Engineering and Aerospace Technology: An International Journal* **87** 437
[6] Azami M H and Savill M 2016 *Fuel* **180** 514
[7] Mustafa K F, Abdullah S, Abdullah M Z and Sopian K 2016 *Fuel* **180** 137
[8] Yamasaki Y, Okada Y, Iijima K and Kaneko S 2009 *Operation of micro gas turbine system employing two stage combustion of biomass gas.* In: ASME Turbo Expo 2009: Power for Land, Sea, and Air: American Society of Mechanical Engineers) 459
[9] Rahmes T F, Kinder J D, Henry T M, Crenfeldt G, LeDuc G F, Zombanakis G P, Abe Y, Lambert D M, Lewis C and Juenger J A 2009 *Sustainable bio-derived synthetic paraffinic kerosene (Bio-SPK) jet fuel flights and engine tests program results* Report No. AIAA 7002
[10] Timko M T, Yu Z, Onasch T B, Wong H W, Miake-Lye R C, Beyersdorf A J, Anderson B E, Thornhill K L, Winstead E L, Corporan E, DeWitt M J, Klingshirn C D, Wey C, Tacina K, Liscinsky D S, Howard R and Bhargava A 2010 *Energy & Fuels* **24** 5883
[11] Beyersdorf A J, Timko M T, Ziemba L D, Bulzan D, Corporan E, Herndon S C, Howard R, Miake-Lye R, Thornhill K L, Winstead E, Wey C, Yu Z and Anderson B E 2014 *Atmos. Chem. Phys.* **14** 11
[12] Habib Z, Parthasarathy R and Gollahalli S 2010 *Energy* **87** 1701
[13] Zhou L, Liu Z-w and Wang Z-x 2015 *Applied Thermal Engineering* **91** 399
[14] Yu Y, Li G, Wang Y and Ding J 2016 *Applied Mathematical Modelling* **40** 268
[15] Park S H, Suh H K and Lee C S 2009 *Energy & Fuels* **23** 4092
[16] Lujaji F C, Boateng A A, Schaffer M, Mtui P L and Mkilaha I S N 2016 *Thermal and Fluid Science* **71** 146
[17] Suh H K and Lee C S 2016 *Renewable and Sustainable Energy Reviews* **58** 1601
[18] Xue J, Grift T E and Hansen A C 2011 *Renewable and Sustainable Energy Reviews* **15** 1098
[19] Mueller C J, Boehman A L and Martin G C 2009 *SAE Int. J. Fuels Lubr* **2** 789-816
[20] Ye P and Boehman A L 2010 *Energy & Fuels* **24** 4215
[21] Chong C T and Hochgreb S 2014 *Fuel* **115** 551
[22] Mofijur M, Masjuki H, Kalam M, Hazrat M, Liaquat A, Shahabuddin M and Varman M 2012 *Renewable and Sustainable Energy Reviews* **16** 5007
[23] Daggett D, Hadaller O, Hendricks R and Walther R 2006 *Alternative fuels and their potential impact on aviation National Aeronautics and Space Administration*, NASA/TM 214365
[24] Arai M, Saito T and Furuhata T 2008 *Journal of propulsion and power* **24** 603
[25] Sundarapandian S and Devaradjane G 2007 *Journal of Engineering, computing and Architecture* **1** 1
[26] Palash S M, Kalam M A, Masjuki H H, Masum B M, Rizwanul Fattah I M and Mofijur M 2013 *Renewable and Sustainable Energy Reviews* **23** 473

IOP Conf. Series: Materials Science and Engineering **370** (2018) 012025 doi:10.1088/1757-899X/370/1/012025

Improving Coolant Effectiveness through Drill Design Optimization in Gundrilling

K.S. Woon[1], G.L. Tnay[2], M. Rahman[1]

[1]Department of Mechanical Engineering, National University of Singapore, Singapore
[2]Machining Technology, Singapore Institute of Manufacturing Technology, Singapore

*Corresponding author: mpewks@nus.edu.sg

Abstract. Effective coolant application is essential to prevent thermo-mechanical failures of gun drills. This paper presents a novel study that enhances coolant effectiveness in evacuating chips from the cutting zone using a computational fluid dynamic (CFD) method. Drag coefficients and transport behaviour over a wide range of Reynold numbers were first established through a series of vertical drop tests. With these, a CFD model was then developed and calibrated with a set of horizontal drilling tests. Using this CFD model, critical drill geometries that lead to poor chip evacuation including the nose grind contour, coolant hole configuration and shoulder dub-off angle in commercial gun drills are identified. From this study, a new design that consists a 20° inner edge, 15° outer edge, 0° shoulder dub-off and kidney-shaped coolant channel is proposed and experimentally proven to be more superior than all other commercial designs.

1. Introduction

Gundrilling is an invaluable deep hole drilling technology to produce aerospace components made of exotic materials like Inconel and titanium alloys. Successful implementation of gundrilling relies on effective coolant application, especially in evacuating chips from the cutting zone. With continuous production of chips from the process, failure to evacuate chips effectively leads to chip packing, recutting of chips, build-up of intense heat and increase in cutting force. Ultimately, the drills will fail [1] and the parts will be damaged - resulted in costly material scrap and down time. Such issue calls for a total chip control system [2] where effective chip evacuation begins with adequate chip breaking.

Simple techniques like peck drilling [3] have long been a popular choice while coupling vibration to gundrilling [4] is also recently attempted. Nevertheless, the abruptness nature of pecking has adverse impacts on tool life especially when difficult materials are involved. Vibration-assisted drilling requires rigid carbide drills to transfer vibration energy effectively, which makes it impossible for depths of drilling beyond 1 meter.

To avoid auxiliary processes, high-pressure delivery of coolant through drill shafts is still a standard practice. Be it oil-based or water-based, an ideal coolant evacuates chips from the drill path while lubricates bearing surfaces and cools cutting edges. Essentially, these abilities are derived from optimum concentration of the base medium special additives, in view of the Rehbinder effect, surface energy and Marangoni effect [5]; as well as the flow behaviour and rheological properties governed by drill designs. Initial attempt to understand the latter began with the study of flow energy loss with a closed pressurized hydraulic circuit [6]. Later, a new concept known as 'hydraulic diameter' was introduced to predict friction loss in drill shanks. Further losses at the bottom hole clearance due to flow interaction and hydraulic resistance [7] were detected from pressure measurement. An analytical method was then proposed to improve drill design for better chip evacuation and tool life [8].

Nevertheless, previous studies based on pressure measurement have several drawbacks. Firstly, unavailability of miniature gauges limited most investigations to big drills. Secondly, construction of pressure conduits and mounts artificially alters the coolant flow and chip transport behaviour. Thirdly, measuring pressure discretely on the drill describes scalar quantities of the coolant flow but not its rheological characteristics in vector. These, perhaps, explain the limited success in tackling more extreme deep-hole applications of this era [9] with existing knowledge in gundrilling.

In this paper, a novel CFD model that reveals the actual behaviour of coolant flow and chip transport is presented. The governing geometries of a gun drill including the nose grind contour, coolant hole configuration and shoulder dub-off angle were subjected to fluid flow, mass transfer and heat exchange analyses. All findings were validated with drilling experiments.

2. Methodology

Three distinct stages are involved in developing the model: (i) Determine drag coefficient of gundrilling chips with vertical drop test; (ii) Set up CFD model and simulate the drop tests for comparison; (iii) Calibrate chip motion with drilling experiments.

2.1. *Empirical drag coefficient*

Chip transport behaviour is governed by a dimensionless parameter known as drag coefficient C_D, which measures resistance of chips within a particular coolant. To quantify C_d of gundrilling chips comprehensively in water-miscible coolants, a series of drop tests was conducted with Fuchs Ecocool 701 (5%- 85%) using a 0.1m×0.1m×1.0m transparent tube. The corresponding dynamic viscosity μ_f was measured with the Anton Paar MCR301 rheometer.

Coolants were filled and rested in the tube until no air bubble was visible, before the drop of a gundrilling chip was made. Travel distance and time duration for the chips to reach zero acceleration were recorded with the Photron Fastcam SA5 to compute terminal velocity U_t. This occurs when the weight of the chip F_g equals the drag force F_d. Hence, the net force F_g-F_d acting on the chip is zero. With V_{ter}, the particle Reynolds number Re_p and drag coefficient C_D are determined as follows:

$$Re_p = \frac{d_{sph} u_t \rho_f}{\mu_f} \tag{1}$$

$$C_D = \frac{4}{3}\frac{g d_{sph}}{u_t^2}\frac{(\rho_c - \rho_f)}{\rho_f} \tag{2}$$

where $d_{sph} = 2\left(\frac{3V_c}{4\pi}\right)^{\frac{1}{3}}$ is the equivalent spherical diameter of the chip [10], V_c is chip volume; A_c is chip surface area; u_t is chip terminal velocity; g is gravity acceleration; ρ_c is chip density; ρ_f is coolant density; and μ_f is coolant dynamic viscosity.

Each condition was repeated for three times with different chip samples. The resultant Re_p and C_D for water-miscible coolants 5%-85% in concentration are 0.06-36.1 and 28.9-2150.

2.2. *CFD model development*

Next, the CFD model was developed based on finite volume method and shear stress transport turbulence model in order to handle vortexes in compressible water-miscible coolant at high pressure [11]. Gundrilling chips were reproduced in the flow domain as static solids; and together with the physical properties of the coolants were input into ANSYS Workbench 14.0. Complex shape of the chips was meshed with 0.01mm tetrahedral elements, contained within the flow domain with hexahedral meshing. Five prismatic layers were included to capture the boundary flow. Refer to Fig. 1 for the model setup.

To simulate the drop test, the boundary condition of the flow outlet was set to 1atm. boundary condition of flow outlet inflow velocity was varied for water-miscible coolants while the outlet. The drag force F_d and net force F_g-F_d can be derived from the CFD model. When F_g-F_d for a particular coolant variant was reduced to zero, the corresponding V_{ter} was determined and thus, the Re_p and C_D can be computed respectively. Fig. 2 shows the simulated and experimental results.

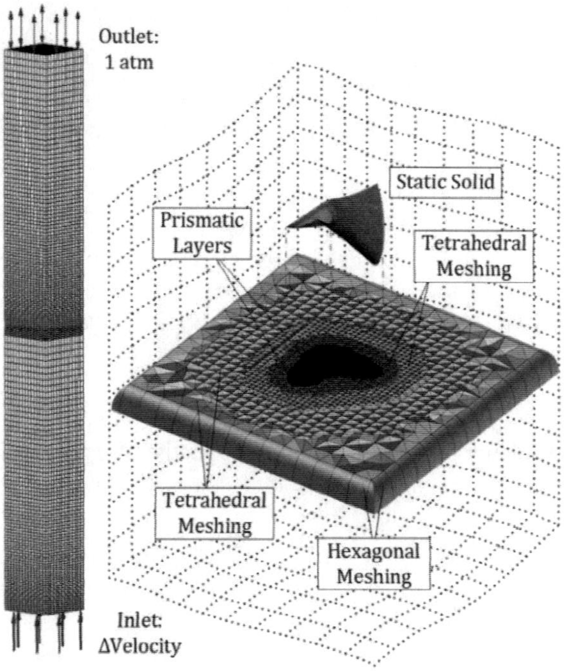

Figure 1. Setup of the base CFD model for deep hole gundrilling.

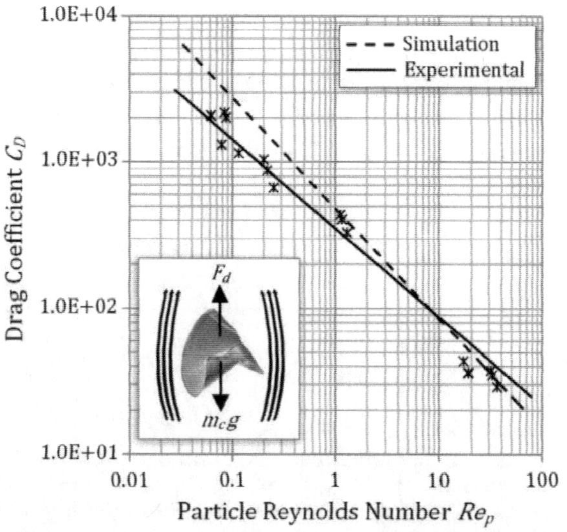

Figure 2. Drag coefficients of gundrilling chips for water-miscible coolants 5%-85% concentration.

2.3. Chip motion calibration

As chips are flushed from the drill point, the general motion of chip flow across water-miscible coolant is governed by the conservation law of momentum [12]. In differential form, the translational and angular motions are expressed as:

$$\frac{d\boldsymbol{x}}{dt} = \boldsymbol{u}_i; \; m_c\frac{d\boldsymbol{u}_i}{dt} = \sum \boldsymbol{F}_i$$

$$\frac{d\boldsymbol{\theta}}{dt} = \boldsymbol{\omega}_i; \; \boldsymbol{I}_c\frac{d\boldsymbol{\omega}_i}{dt} = \sum \boldsymbol{T}_i \tag{4}$$

where $\boldsymbol{x}$ is the position vector, $\boldsymbol{u}_i$ is the initial chip translational velocity, m_c is the chip mass, F is the total force acting on the chip, $\boldsymbol{\theta}$ is the angle of rotation, $\boldsymbol{\omega}_i$ is the initial chip angular velocity, $\boldsymbol{I}_c$ is moment of inertia and $\boldsymbol{T}$ is the total torque acting on the chip.

Transporting from a blind hole, chips collide and rebound on the hole and V-channel of the gundrill along the course of evacuation to the chip collection box. For such impacts with minimum elastic deformation [13], the normal u_{nr} and tangential u_{tr} velocity components of the chip are:

$$u_{nr} = e_n u_{ni}; \; u_{tr} = e_t u_{ti} \tag{5}$$

where e_n and e_t are the coefficients of restitution in the respective directions. The corresponding forces acting on the chip are essentially impulses from the impact:

$$F_n = m_c(u_{ni} - u_{nr}); \; F_t = m_c(u_{ni} - u_{nr}) \tag{6}$$

while the resultant angular velocity is

$$\omega_r = \frac{F_t d_{sph}}{2m_c k^2} = \frac{5(u_{ni} - u_{nr})}{d_{sph}} \tag{7}$$

where $k = \sqrt{\frac{2}{5}\frac{d_{sph}}{2}}$ is the radius of gyration of the chip.

For the CFD model to predict chip motion accurately, a series of calibration drilling test was performed with commercial gun drills on the DMU 80 P duoBLOCK. Ø8mm Inconel 718 bars were mounted in hardened acrylic tubes of high transparency. Fuchs Ecocool 701 with 12%wt oil was applied at 20-40bar while drilling speeds and feeds were set at 700-1000rpm and 5-10mm/min respectively. Chip motion was recorded at 6000fps with the Photron Fastcam SA5. Then, by deriving the location and orientation discrepancies between the experimental and numerical results, governing parameters such as coefficients of restitution, numerical time step are refined accordingly. Such calibration was repeated until the error is reduced below 20%. A sample of the final outcome is shown in Fig. 3.

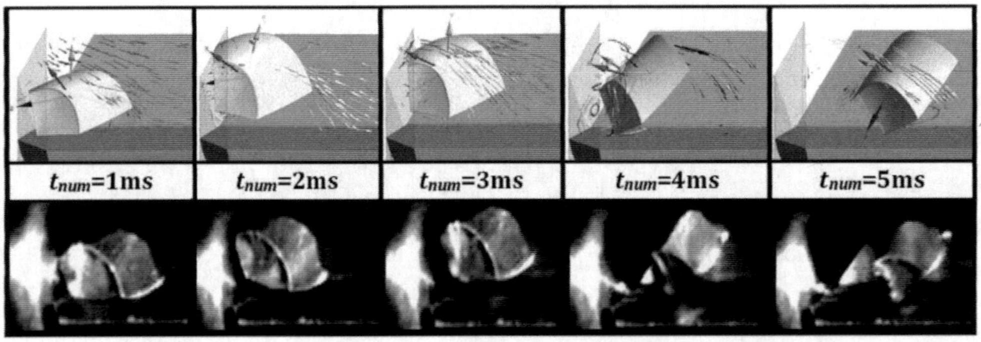

Figure 3. Calibration of chip motion with horizontal drilling tests.

3. Methodology

Gun drill design governs chip transportation efficiency through the influence of the coolant's rheological behaviour and characteristics, alongside properties of the coolants. The gun drill geometries involved are nose grind contour, coolant hole configuration and shoulder dub-off angle. Using the CFD model derived from Sec. 2.3, evaluation of the governing geometries was conducted for Ø8mm drilling of Inconel 718 according to the scope in Fig. 4. In all cases, the drilling speed and feed were 800rpm and 8mm/min respectively, with water-miscible coolant 12% operating at 40bar.

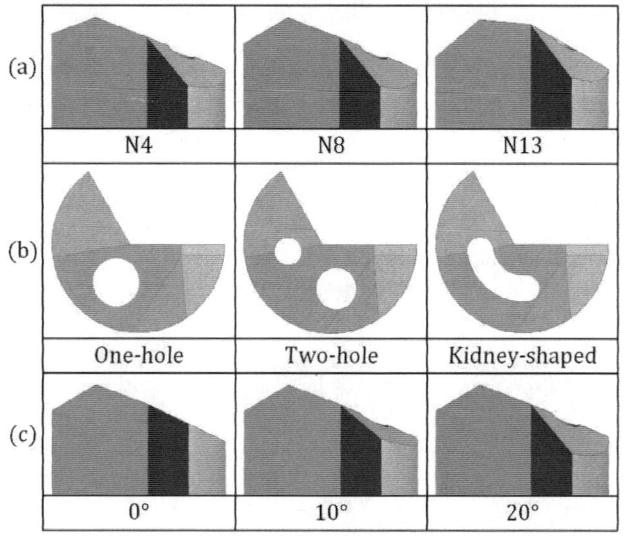

Figure 4. Scope of evaluation (a) nose grind contour; (b) coolant hole configuration; and (c) shoulder dub-off angle.

3.1. *Effects of nose grind contour*

The nose grind contour defines the most prominent geometry of the gun drill that determines the mechanics and performance of the process. It consists of the outer cutting edge, inner cutting edge and drill apex. Different contour designs are classified according to their variations in the outer edge angle ϕ_o, inner edge angle ϕ_i and drill apex offset A_p, as a combination. The most commonly found designs are N4 (ϕ_o =15°, ϕ_i =20°, A_p = D/4); N8 (ϕ_o =30°, ϕ_i =20°, A_p = D/4); and N13 (ϕ_o =40°, ϕ_i =5°, A_p = D/4). Different nosegrind contours are intended for different materials but the guidelines have never been clear. In this study, we evaluate the coolant flow characteristics of N4, N8 and N13 collectively with the two-hole coolant configuration and shoulder dub-off angle of 20°.

From the CFD results in Fig. 5a, it is noticeable that delivery of coolant to the drill point deteriorates significantly from N4 to N8 and N13. With N4 and N8, large amount of pressurized coolant is able to reach the vicinity of the drill point and penetrate the cutting edges on the rake faces. This posts a stark contrast to N13 where most of the coolant that rebounds from the bottom hole, escapes directly into V-channel without being able to access the cutting edges where chips are formed. The difference between the conditions of N4 and N8 is least significant.

3.2. *Effects of coolant hole configuration*

In gundrilling, coolant that is supplied under high pressure travels along the internal conduits of drill shafts and drill tips before ejected through coolant hole(s) on the face of a gun drill to reach the drill point. Configuration of the coolant holes thus plays an important role in through-shaft coolant delivery of the process chiefly, in regulating the coolant volume and pressure. Although three types of configuration namely one-hole, two-hole and kidney-shaped are commercially available, the two-hole gun drills are most frequently popular. In order to compare the coolant performance of these configurations, the N8 nose grind contour and shoulder dub-off angle of 20° were applied.

Fig. 5b depicts the results of the CFD analysis. In general, the performance involved in delivering high-pressure coolant to the drill point improves from 1-hole to 2-hole and with kidney-shaped being the best. Having the single largest orifice with an effective area A_{orf} of 4.81mm^2, the kidney-shaped configuration is capable of 'flooding' the entire rake faces at high pressure. For the 2-hole configuration with a top and bottom A_{orf} of 1.24mm^2 and 2.83mm^2 respectively, reasonable amount of coolant can be found at the drill point although pressure of the combined flow is reduced. As a comparison, one-hole configuration with a A_{orf} of 3.78mm^2 produces a strong stream of coolant but the flow is deflected back into the V-channel, resulted in poor access of coolant to the cutting edges.

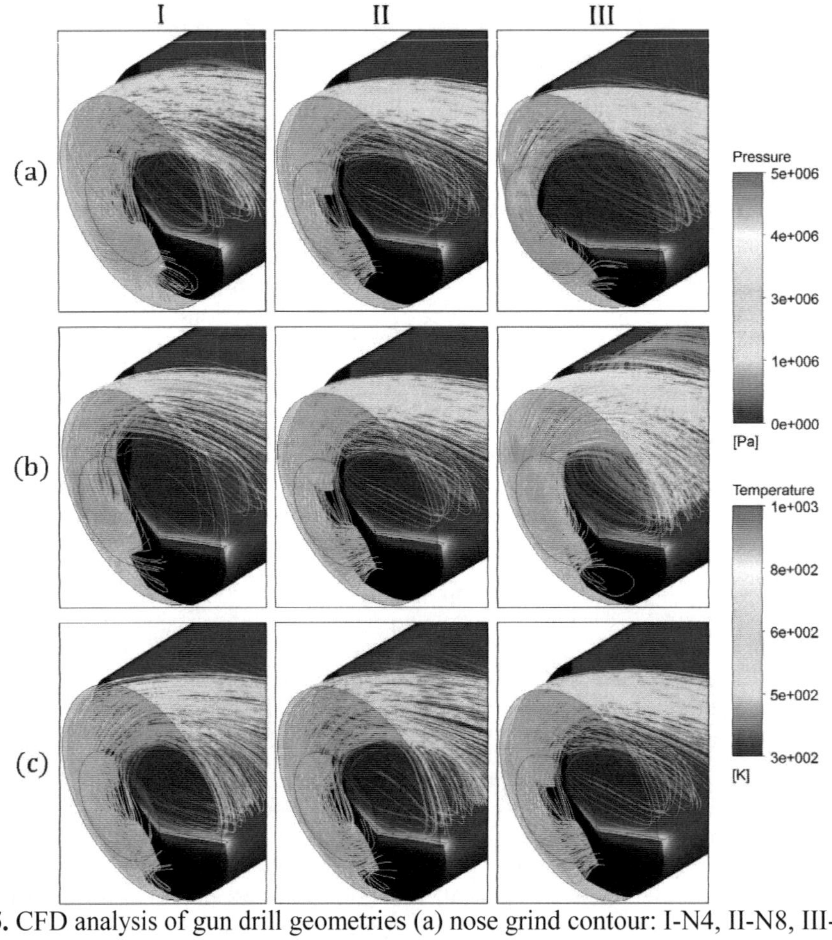

Figure 5. CFD analysis of gun drill geometries (a) nose grind contour: I-N4, II-N8, III-N13; (b) coolant hole configuration: I-One hole, II-Two hole, III-Kidney shaped; and (c) shoulder dub-off angle: I-0°, II-10°, III-20°.

3.3. Effects of shoulder dub-off angle

The shoulder dub-off of gun drills has two major roles in the process. Firstly, it is one of the relief geometries of the drill to prevent burnishing on the flank surfaces. Secondly, it diverts rebounding coolant from the bottom hole to the cutting edges. The latter contributes to the effectiveness of coolant application, to a great extent that the overall tool life is strongly governed. According to our experience, most of the commercial gun drills are shipped with a shoulder dub-off angle φ_d of 20°, disregard of the applications involved. The effects of shoulder dub-off was studied with the CFD model for φ_d ranging from 0° to 10° and 20°. The two-hole configuration and N8 nose grind contour were used.

As evidently found from the CFD analysis in Fig. 5c, the ability to deliver coolant to the drill point increases significantly by reducing φ_d from 20° to 10° and 0°. With small φ_d at 0°, delivery of the coolant is capable to penetrate deep into the inner and outer rake faces where heat is generated from chip formation and chip flow. Increasing φ_d to 10° and 20° leads to the drop in both hydraulic pressure and volume of the coolant flow and resulted in deterioration of cooling and chip evacuation.

4. Discussions

To improve the effectiveness of coolant application in gundrilling requires good knowledge of the coolant's rheological properties and characteristics, in the context of commercially available solutions. First and foremost, one should understand the active role of nose grind contour is cutting. When materials are removed from the hole with the advancement of the drill, the bottom of the hole generated bears the same geometry of the drill's nose grind contour, in a directly inverse manner. The bottom hole deflects high-pressure coolant onto the rake faces to evacuate chips and remove heats from the cutting edges. With the combination of a large inner edge angle (ϕ_i =20°) and a small outer edge angle (ϕ_o =15°) of the N4 nose grind contour, the large ϕ_i allows higher volume of coolant flow while the small ϕ_o deflects the coolant to the cutting zone. Therefore, chip evacuation and heat removal in N4 is superior.

Secondly, in order to achieve better results, N4 nose grind contour has to be coupled with the correct coolant hole configuration. Of the designs available in the market, the reason to use the kidney-shaped design is obvious. Having the largest, single orifice that spans across the drill face, high volume of coolant with good conservation of pressure is deflected uniformly over the rake face is an assurance for high effectiveness. Although the effective orifice area of the kidney-shaped and two-hole configurations is comparable, the difference in their performance is overwhelming due to the lost pressure and volume of the coolant supplied from the upper hole of the two-hole design.

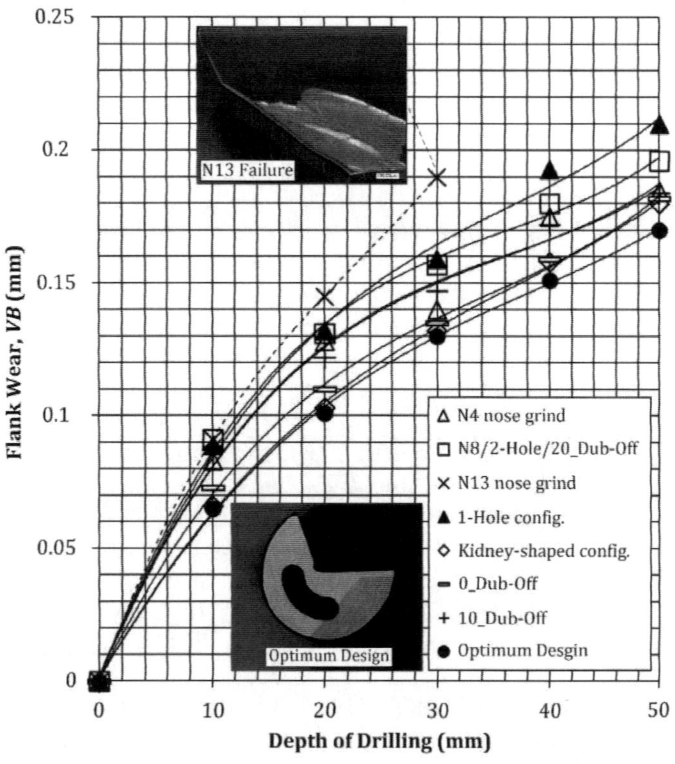

Figure 6. Tool life validation for various combinations of drill geometries.

Thirdly, with N4 nose grind contour and kidney-shaped configuration, the shoulder dub-off angle φ_d of the optimum drill design is set at 0°. Using the smallest possible φ_d, the narrow passage enables the build-up of sufficient hydraulic resistance for the deflected coolant to maintain its high pressure, according to the Bernoulli's principle. This is especially critical for deep penetration of coolant delivery into the cutting zone to remove excessive heat from the rake faces. On the contrary, the wrongfully use of a large φ_d, 20° for instance in most commercial gun drills impairs the coolant performance with the loss of hydraulic pressure due to the sudden expansion of coolant flow that is deflected from the bottom hole.

With deeper understanding of the effects of commercial gun drill geometries on rheological properties and characteristics of high-pressure coolant, an optimum drill design for effective chip evacuation and cooling is proposed. Essentially, combining the N4 nose grind contour, kidney-shaped coolant hole configuration and shoulder dub-off angle φ_d of 0° as a whole, one should achieve better results with the salient merits of each geometry. To validate this hypothesis and the CFD analyses in Sec. 3, a series of gundrilling tests on Inconel 718 were carried out on the DMU 80 P duoBLOCK with the same operating conditions of the CFD simulations. All designs were subjected to the drilling of 50mm in depth. Each drilling cycle was 10mm. By the end of every cycle, the flank wear *VB* was recorded. The results as summarized in Fig. 6 show apparent agreement with the CFD analyses.

5. Conclusions

In this paper, a CFD model that is capable to simulate rheological characteristics of coolant, flow transport behaviour of chips and heat exchange on rake face is proposed and experimentally substantiated. Based on the analyses of this model, the following conclusions are drawn:

- Tool life is governed by nose grind contour, coolant hole configuration and shoulder dub-off angle in gundrilling.
- Commercial drill designs with large outer angle and small inner angle, single coolant hole configuration and large shoulder dub-off angle adversely affects coolant delivery and deteriorates chip evacuation and cooling of cutting edges.
- An optimum design that combines small inner angle and large outer angle, kidney-shaped coolant hole configuration and small shoulder dub-off angle produces the best tool life.

References

[1] Woon KS, Chaudhari A, Rahman M, Wan S, Senthil Kumar A (2014) The Effects of Tool Edge Radius on Drill Deflection and Hole Misalignment in Deep Hole Gundrilling of Inconel-718. *CIRP Annals–Manufacturing Technologies* 63: 125-128.

[2] Jawahir IS, van Luttervelt CA (1993) Recent Developments in Chip Control Research and Applications. *CIRP Annals–Manufacturing Technologies* 42: 659-693.

[3] Stephenson DA, Agapiou JS (2006) Metal Cutting Theory and Practice 2nd Ed. CRC Press, Boca Raton.

[4] Heisel U, Wallaschek J, Eisseler R, Potthast C (2008) Ultrasonic Deep Hole Drilling in Electrolytic Copper ECu 57. *CIRP Annals–Manufacturing Technologies* 57: 53-56.

[5] Brinksmeier E, Meyer D, Huesmann-Cordes AG, Herrmann C (2015) Metalworking Fluids – Mechanisms and Performance. *CIRP Annals–Manufacturing Technologies* 64: 605-628.

[6] Astakhov VP, Subramanya PS, Osman MOM (1995) An Investigation of the Cutting Fluid in Self-Piloting Drills. *International Journal of Machine Tools and Manufacture* 35: 547-563.

[7] Latinovic V, Osman MOM (1986) Friction Losses in Coolant Flow through Kidney Shaped Gundrill Shank. *International Journal of Production Research* 24: 1319-1329.

[8] Astakhov VP, Galitsky VV, Osman MOM (1995) A Novel Approach to the Design of Self-Piloting Drills with External Chip Removal Part 2. *Journal of Engineering for Industry* 177: 464-474.

IOP Conf. Series: Materials Science and Engineering **370** (2018) 012025 doi:10.1088/1757-899X/370/1/012025

[9] Woon KS, Chaudhari A, Rahman M, Wan S, Senthil Kumar A, (2014) The Effects of Tool Edge Radius on Drill Deflection and Hole Misalignment in Deep Hole Gundrilling of Inconel-718. *CIRP Annals–Manufacturing Technologies* 63: 125-128.
[10] Haider A, Levenspiel O (1989) Drag Coefficient and Terminal Velocity of Spherical and Non-Spherical Particles. *Powder Technology* 58: 63-70.
[11] Massey BS, Ward-Smith J (2001) Mechanics of Fluids, Nelson Thornes, UK.
[12] Mandø M, Rosendahl L (2010) On the Motion of Non-Spherical Particles at High Reynolds Number. *Powder Technology* 202: 1-13.
[13] Kharaz AH, Gorham DA, Salman AD (2001) An Experimental Study of the Elastic Rebound of Spheres. *Powder Technology* 120: 281-291.

IOP Conf. Series: Materials Science and Engineering **370** (2018) 012026 doi:10.1088/1757-899X/370/1/012026

Load Lifting and the Risk of Work-Related Musculoskeletal Disorders among Cabin Crews

Nurhayati Mohd Nur[1*], Mohd Amirul Syazwan Mohamed Salleh[1], Mulia Minhat, Nurul Zuhairah Mahmud Zuhudi[1]

[1] Aerospace Department, Universiti Kuala Lumpur, Malaysian Institute of Aviation Technology, 43800 Dengkil, Selangor, Malaysia

*Corresponding author: nurhayatimn@unikl.edu.my

Abstract. Lifting and loading passengers' hand-carry luggage into the overhead compartment in the flight cabin is one of the tasks performed by the cabin crews. The hand-carry luggage weight varies and can be up to 10 kg. The task involved load lifting over the head, awkward posture and repetitive that could lead to the risk of work-related musculoskeletal disorders (WMSDs). Hence, further investigation of those factors stimulates this study. An industrial survey was conducted and WMSDs symptom data were collected using Nordic Musculoskeletal Questionnaire (NMQ). Rapid Upper Limb (RULA) assessment was performed to analyse the posture during the load lifting task. The results of the industrial survey indicated that 88.3% of the workers experienced WMSD symptoms in various regions of the body. The highest symptoms reported at lower back (80.0%), followed by the neck (68.0%) and shoulder (59.0%). The result is supported with the result of RULA with final score 7, which indicated the load lifting task performed involved awkward posture, needs changes and further improvement should be immediately implemented. In conclusion, the cabin crews exposed to the risks of WMSDs and the results of this study can be used as references by the aviation industry to mitigate WMSDs issues among the cabin crews specifically and contribute to the improvement of the society well-being generally.

1. Introduction

WMSDs involved pain and injuries which are commonly experienced in the neck, cervical spine, muscle tendons and joints in the body [1]. There are many factors that contribute to the risk of contracting WMSDs. The demands at work and the capacity of the worker to perform work-related activities may play a role in the development of this pain. WMSDs are known to be associated with work-related physical risk factors such as force, posture and repetition [2]. The trend of WMSDs increased every year [3] and its adverse effect on the workers' health and performance over time is significant [4]. Even though the number of the cases reduced about 3% in 2015 but the occurrence rate is still high [5]. The occurrence of WMSDs is costly due to high compensation cost and lost wages [6].

WMSDs provide a significant threat to the employees in many industries including aviation industry. It is important to reduce the work-related physical risk factors in order to reduce the risk of WMSDs. It is the major challenge to design tasks to prevent or reduce work-musculoskeletal disorders (WMSD). In performing manual handling work, workers are exposed to the work-related physical risk factors whereby the tasks involve repetitive work, high force, extreme or awkward posture, short cycles, high movements and lack of recovery [2]. In a real work, these factors are less considered that could lead to the increasing number of WMSDs cases per year.

Lifting and loading hand-carry luggage into the overhead compartment in the flight cabin is one of the common manual handling tasks performed by the cabin crews. The hand-carry luggage weight varies and can be up to 10 kg, depending on the airline's policy. As stated in the passenger boarding policy of airlines inside their website, the hand-carry luggage allowed on premium airlines is up to 10 kg for each passenger on board which usually limited to one laptop bag and one cabin bag which both total up the approved weight by the airlines. For the low-cost airlines, technically they are applying maximum 7 kg hand-carry luggage policy to all their passengers. However, most of the passengers carried a hand-carry luggage more than approved weight into the cabin which means that the cabin crews may handle the hand luggage which is more than 7 kg or less. The lifting tasks might involve the upper limb body parts and the previous study found that the risk of WMSDs in arm, shoulder and neck region are prevalent even though with light exertion [7].

The load lifting tasks performed involved work-related physical risk factors which are known as force, repetition and posture. These factors either single factor or combined could lead to the risk of WMSDs. The load lifting task is not totally the responsibility of the cabin crews and sometimes the passengers performed the task themselves. Is the passenger also vulnerable to the WMSDs risk? The answer depends on the exposure to the work-related physical risk factors involved as discussed earlier. In this study, the scope is limited to the cabin crews since based on the preliminary assessment, most of the cabin crews performed the task and might be exposed to the risk of contracting WMSDs. In addition, there is a lack of studies conducted to evaluate the load lifting task which involved hand-carry luggage and the prevalence of WMSDs among the cabin crews. Therefore, it is timely to investigate this issue. The result of this study could provide knowledge and awareness to the airline companies specifically and aviation industry generally to mitigate this issue from becoming catastrophic.

2. Methodology

2.1 Industrial Survey

The Nordic Musculoskeletal Questionnaire (NMQ) is used to collect data on the risk of WMSDs. It consists of structured, forced, multiple-choice questions and can be used as a self-administered questionnaire as an on-paper interview method. The subjects involved in this study are 59 respondents from the cabin crew profession which are 30 stewards and 29 stewardesses.

2.2 Experimental Tasks,

The tasks involved the lifting and placement of different weight of hand-carry luggage into the aircraft overhead compartment. The luggage weights used are 5kg and 7kg. The maximum luggage weight allowed by the low-cost airlines is used in this experimental study. There are 10 subjects involved (5 males, 5 females). The subjects are required to lift and place the luggage every 2 minutes for 20 minutes, as depicted in Figure 1. The subjects' height and weight are between 160 -180 cm and 40 – 70 kg respectively. The subjects fulfill the characteristics required as cabin crews in terms of physical, height weight and normal body mass index (BMI).

Figure 1. Load lifting task

The posture of the subject during load lifting tasks is evaluated using Rapid Upper Limb Assessment (RULA) method. After the task completion, the subjects will rate their discomfort level using Borg's Scale [8] as shown in Table 1.

Table 1. The Borg's Scale

		The Borg General Scale
0	---	nothing at all
0.5	---	extremely weak (just noticeable)
1	---	very weak
2	---	weak
3	---	moderate
4	---	somewhat strong
5	---	strong
6	---	
7	---	very strong
8	---	
9	---	
10	---	extremely strong (almost maximal)

3. Results

3.1 Industrial Survey

Complete responses were obtained from 59 cabin crews from two main Airlines in Malaysia. The demographic data of the respondents are summarized in Table 2. It can be seen that 50.8% of the respondents are males whereas the remaining 49.2% are females. Based on age distribution, it can be seen that majority of the respondents fall within an age range of 18-33 years old, giving a value of 91.5% and the remaining 8.5% fall within an age range of more than 33 years old.

Table 2. Demographic Data of the Respondents

Gender	Male	30 (49.2%)
	Female	29 (50.8%)
Age	18 – 22	2 (3.4%)
	23 – 27	40 (67.8%)
	28 – 33	12 (20.3%)
	.> 33	5 (8.5%)
Flying hours in a week	1 – 20	14 (23.7%)
	21 – 40	36 (61.0%)
	41 – 60	9 (15.3%)
Number of placing loads per flight	1 – 10	9 (15.3%)
	11 – 20	14 (23.7%)
	21 – 30	15 (25.4%)
	31 – 40	17 (28.8%)
	41 – 50	4 (6.8%)

The demographic data show that the highest percentage of respondents have to work with flying hours between 21-40 hours, giving a value of 61.0%. The number of respondents with 1-2 flying hours and 41-60 hours constitute 23.7 and 15.3%, respectively. It can be observed that 61.0% of the respondents have to do repetitive load lifting task (placing the luggage in the overhead compartment) more than 20 times per flight while the remaining 39.0 %, the frequency of repetitive load lifting per flight is 20 times and less.

3.1.1 WMSDs for the last 12 months

The prevalence of WMSD symptoms among cabin crews over the past 12 months is summarized in Table 3. It can be observed that the occurrence of WMSD symptoms varies from 6.8 – 68.0%, depending on the region of the body. The results indicate that 88.3% of the cabin crews experienced WMSD symptoms in various regions of the body. The prevalence of musculoskeletal symptoms is highest at the lower back, followed by the neck, shoulder, upper back, knee, ankle, hand wrists/elbow and least of all, the thigh/hip.

Table 3. Prevalence of WMSD symptoms over the past 12 months.

Body Region	Frequency (N)	Percentage (%)
Neck	40	68.0
Shoulder	35	59.0
Elbow	8	13.6
Hand/Wrists	18	30.5
Back (U)	30	51.0
Back (L)	47	80.0
Thigh/Hip	4	6.8
Knee	22	37.0
Ankle	19	32.0

3.1.2 WMSDs for the last 7 days

The prevalence of WMSD symptoms among cabin crews over the past 7 days is summarized in Table 4, and it can be seen that the percentage of occurrence ranges from 3.3 – 29.0%, depending on the body region.

Table 4. Prevalence of WMSDs symptoms over the past 7 days

Body Region	Frequency (N)	Percentage (%)
Neck	17	29.0
Shoulder	12	20.3
Elbow	5	8.5
Hand/Wrists	17	28.9
Back (U)	17	28.8
Back (L)	19	32.2
Thigh/Hip	2	3.3
Knee	8	13.5
Ankle	8	13.5

It is found that 54.2% of the cabin crews experienced WMSDs symptoms in various regions of the body. The prevalence of WMSD symptoms is highest at the lower back, followed by the neck, hand/wrist, upper back, shoulder, knee, ankle, elbow and least of all, the thigh/hip.

3.1.3 Experimental Task

The experimental tasks involved 10 subjects which are volunteered among university students, which consist of 5 males and 5 females subjects. Each subject performed load (luggage) lifting tasks of 2 different weight of hand-carry luggage which are 5 kg and 7 kg. They repetitively lift the luggage for every 2 minutes for 20 minutes duration. The posture of the subjects during repetitive task was assessed using Rapid Upper Limb Assessment method and the result is presented in Table 5 and 6. In Table 5, the muscles involved are arm and wrist, which contributed to Part A score, 8.

Table 5. RULA results for Part A Score

Body Parts	Scores
Upper Arm	**4**
Lower Arm	**3**
Wrist twist	**2**
Wrist	**4**
Posture A	**6**
Muscle Use	**0**
Force/Load	**2**
Wrist&Arm (Posture A + Muscle use + Force/load)	**8**

In Table 6, the muscles involved are the neck, trunk and legs, which give results for Part B score, 8.

Table 6. RULA results for Part B score

Body Parts	Scores
Neck	**3**
Trunk	**4**
Legs	**1**
Posture B	**6**
Muscle use	**0**
Force/load	**2**
Neck, Trunk, Leg score (Posture B + muscle use + force/load)	**8**

The RULA scores for part B and Part A is then further analysed and assessed as presented in Table 7, It can be found in Table 7 that the RULA final score produced is 7.

Table 7. RULA Final Score Assessment

Wrist and	Neck, Trunk and Leg Score						
Arm Score	1	2	3	4	5	6	7+
1	1	2	3	3	4	5	5
2	2	2	3	4	4	5	5
3	3	3	3	4	4	5	6
4	3	3	3	4	5	6	6
5	4	4	4	5	6	7	7
6	4	4	5	6	6	7	7
7	5	5	6	6	7	7	7
8+	5	5	6	7	7	7	7

The participants were required to rate their perceived discomfort level after the task completion and the result of the discomfort level is presented in Table 8. It can be seen that the participants felt more discomfort while performing the task with a heavier load and the most discomfort area is found at the neck, with discomfort level is rated as somewhat strong.

Table 8. Perceived discomfort level

Luggage weight	Discomfort Level (mean)								
	Neck	Shoulder	Upper Back	Elbow	Wrist	Lower Back	Hips/ Tight	Knee	Ankle/ Fit
5kg	3.3	2.0	2.0	2.0	2.3	2.0	1.7	2.0	1.7
7kg	4.0	3.0	2.0	2.2	3.3	2.3	2.5	2.0	2.5

4. Discussion

The aim of this study is to investigate the risk of WMSDs among cabin crews. Two approaches had been performed in this study which involved an industrial survey and experimental task. An industrial survey is conducted to investigate the prevalence of WMSDs among cabin crews while the experimental task is conducted to assess the work posture and the risk of WMSDs during lifting and loading the hand-carry luggage into the overhead compartment in the flight cabin.

The results of the industrial survey showed that the cabin crews experienced WMSD symptoms, whereby approximately 88.3% of the cabin crews experienced pain and discomfort in various regions of the body parts over the past 12 months. The most affected parts are lower back, followed by the neck and the shoulder. The results also revealed that the prevalence of WMSD symptoms ranges from 3.3 –32.2% for different body areas over the past 7 days. It can be observed from the results that the prevalence of WMSDs is highest at the lower back followed by the neck and the hand/wrists. The cabin crews felt more pain at the lower back and neck due to the heavy load lifting tasks since these muscles are the most affected muscles. More pain experienced on the particular muscles during the tasks indicated that the muscles are the most affected muscles [9].

Cabin crews performed load lifting tasks repetitively, involved force (load) and also awkward posture. Based on the industrial survey data, it is found that 61.0% of the cabin crews performed repetitive load lifting tasks more than 20 times per flight. The repetitive load lifting task done contributed to the prevalence of WMSDs risk among the cabin crews as showed by the high percentage of complaints (88.3%). Several studies have reported that repetitive task is one of the work-related physical risk factors that could lead to the risk of WMSDs [10–13] which supported the result of this study. In addition, the combination of force (load) with repetitive task tends to give more risk of WMSDs. The cabin crews lifting and loading the heavy hand-carry luggage (up to 10 kg) repetitively and the tasks involved a combination of force and repetition factors that could lead to the higher risk of WMSDs. The previous study also discovered that as the level of force and repetition increased, the level of pain and discomfort also increased, which indicated the risk of WMSDs [14].

The posture of the cabin crews while lifting and placing the load is further analysed with rapid upper limb assessment (RULA) method. The assessments involved two parts (Part A and Part B). The score of both parts is then calculated and compared with the guideline provided for RULA assessment. The final RULA score obtained for both forces (load) used is at level 7. The results indicated that the load lifting task performed in an awkward posture, the task needs changes and further improvement should be implemented [15] in order to reduce it. The result is aligned with the previous studies result which found that awkward postures lead to the risk of WMSDs [16]. On top of that, it is found that the mean perceived discomfort level while placing the heavier load (7 kg) is found 25.0% higher compared to the lighter load (5kg). The result indicated the higher risk of WMSDs when lifting heavier hand carry luggage with the awkward posture.

The overall results showed that load lifting tasks performed involved work-related physical risk factors (force, repetition and posture) that contributed to the risk of WMSDs among the cabin crews.

Therefore, it is important to re-attend the existing method of performing the load lifting tasks to reduce the effects of those work-related physical risk factors and the risk of contracting WMSDs.

5. Conclusion

Based on the results obtained from this study, the cabin crews performed repetitive load (hand-carry luggage) lifting task with the weight varies from 5-10 kg are found to expose to the risk of work-related musculoskeletal disorders. The results of the industrial survey indicated that 88.3% of the workers experienced WMSD symptoms in various regions of the body parts. The highest symptoms for the last 12 months was reported at lower back (80.0%), followed by the neck (68.0%) and shoulder (59.0%). Further investigation on the work posture (RULA analysis) showed that the load lifting task performed involved awkward posture and the task needs changes on the implementation. The overall results showed that the work-related physical risk factors (force, repetition and posture) involved in the load lifting tasks lead to the risk of WMSDs among cabin crews. Therefore, it is important to re-attend the load lifting tasks to reduce the effects of those factors in order to mitigate the WMSDs risks. The outcome of this study could be used as references by the aviation industry to mitigate WMSDs issues among the cabin crews specifically and contribute to the improvement of the society well-being generally.

References

[1] NIOSH 1997 Musculoskeletal Disorders and Workplace Factors. *NIOSH Publ No 97B141*
[2] Nordander C, Hansson GA, Ohlsson K, Arvidsson I, Balogh I, Strömberg U, Rittner R and Skerfving S 2016 Exposure-response relationships for work-related neck and shoulder musculoskeletal disorders - Analyses of pooled uniform data sets. *Appl Ergon* **55** 70–84.
[3] SOCSO 2014 SOCSO Annual Report 2014
[4] Walt R, Winnie I and James M. 2013 Musculoskeletal Disorders, Five Common Management Approaches. *Pers Saf* 35–42.
[5] SOCSO 2015 SOCSO Annual Report 2015.
[6] Joseph, B., Naveen, B., Suguna, A., and Surekha A 2016 Prevalence, Pattern and Factors Associated with Work-related Musculoskeletal Disorders (WRMD) among Housekeeping Workers in a Private Tertial Care Hospital in Bangalore. *J Health Manag* **18** 545–54.
[7] Bosch T, Mathiassen SE, Visser B, de Looze MP and van Dieën JH 2011 The effect of work pace on workload, motor variability and fatigue during simulated light assembly work. *Ergonomics* **54** 154–68
[8] Borg G 1998 Borg's Perceived Exertion and Pain Scales Human Kinetics Publisher
[9] Akmal, K., Norzaimi, M. C. A., and Hairizal MO 2014 Analysis on the Work- Related Musculoskeletal Disorders (WMSD) Based on Ergonomic Study in Case of Industry Study. *Int J Eng Res* **5013**190–5.
[10] Bosch T, de Looze MP, Kingma I, Visser B and van Dieën JH 2008 Electromyographical manifestations of muscle fatigue during different levels of simulated light manual assembly work. *J Electromyogr Kinesiol* **19** 246-56
[11] Gooyers CE and Stevenson JM 2011 The impact of an increase in work rate on task demands for a simulated industrial hand tool assembly task. *Int J Ind Ergon* **42** 80–9.
[12] Nordander C, Ohlsson K, Akesson I, Arvidsson I, Balogh I, Hansson GA, Strömberg U, Rittner R and Skerfving S 2009 Risk of musculoskeletal disorders among females and males in repetitive/constrained work. *Ergonomics* **52** 1226–39.
[13] van Rijin RM, Huisstede BM, Koes BW and Burdof A 2009 Association between work-related factors and specific disorders at the elbow : A systematic literature review. *Rheumatology* **48** 528–36.
[14] O'Sullivan, L. and Clancy P 2007. Guideline threshold limit values (TLVs) for discomfort in repetitive assembly work. *Hum Factors Ergon Manuf* **17** 423–34.
[15] Middlesworth M 2015 Rapid upper limb assessment (RULA) A Step-by-step guide. *Ergon*

IOP Conf. Series: Materials Science and Engineering **370** (2018) 012026 doi:10.1088/1757-899X/370/1/012026

Plus http://ergo-plus.com/rula-assessment-tool-guide/ (accessed April 15, 2017).
[16] Carey EJ and Gallwey TJ 2002 Effects of wrist posture, pace and exertion on discomfort. *Int J Ind Ergon* **29** 85–94

IOP Conf. Series: Materials Science and Engineering **370** (2018) 012027 doi:10.1088/1757-899X/370/1/012027

Study of Theoretical and Numerical Fluid Characteristics of Plain Wing with Winglets

Mohamed B.W. Nabhan

University of Bahrain, College of Engineering, Mechanical Engineering Department, PO Box 32038, Isa Town, Kingdom of Bahrain, Tel.+973-17876215, Fax.+97317681234

Email: mnabhan@uob.edu.bh

Abstract. Aerodynamic characteristics of plain wing designed for Light Sport Aircraft has been studied. The fluid characteristics include induced drag and lift to drag ratio. Then, winglets are added to reduce the induced drag and increase the lift to drag ratio which are affected by the wing tip vortices. The theoretical and numerical approaches are used to verify the results. A rectangular untwisted 9.528 m wing spans with an Airfoil NACA 4412 was used for the basic design. Winglets are added with a tip airfoil of NACA 0012, side angle of 65° and new projected area of 10.328 m^2. Lift and drag coefficients are used as means to measure the improvement of the aerodynamic characteristics. The wing tip vortices increase the induced drag and spoil the lift over the wing's surface. The winglets design main objectives are to decrease the induced drag, decrease the fuel consumption, and increase the flight safety, especially in take-off condition. The wing with winglets model was simulated first using 3-D Fluent ANSYS version 14 at 50 m/s velocity and (0°, 5°, and 10°) angles of attack with laminar flow and standard atmospheric conditions at 15°C, and 101 kPa and all other flow parameters as well. The second verification method was to simulate the 3-D model using the 3-D Foil Multi-Surfaces code again with the same flow parameters. Finally, the last verification method was to solve the problem theoretically using the theoretical governing equations. The theoretical solutions were used as a base line for all other results. The total drag reduction observed from the calculation is about 2% to 14.5% during the takeoff regime, where the induced drag contributes about 60% of total drag of the wings. The lift to drag ratio improved also in our designed model wing with winglets by a maximum of 18.6% from the plain wing design.

1. Introduction

Winglets belong to the class of wingtip devices aimed to reduce induced drag. Selection of the wingtip device depends on the specific situation and the airplane model. In the case of winglets, the reduction of the induced drag is accomplished by acting like a small sail whose lift component generates a traction force, draining energy from the tip vortices (Bento Mattos, Antonini Macedo, & Durval Silva Filho, 2003) [1].Winglets are a type of wing tip devices, which are used to enhance aerodynamic efficiency of wings.

In general a winglet produces a side force, which has a forward component because of the side wash produced by the wing tip circulation. This forward component of the lift acts as a thrust force reducing the aircraft induced drag, which is about 40% of the total drag in cruise and 80-90% of the total drag in the second climb segment (Kroo, 2005) [2]. A research tried to optimize winglet for minimum drag, while defining some constraints to avoid a huge weight penalty (Hantrais-Gervois et al., 2009) [3]. These constraints usually limit the wing root bending moment. However, using root bending moment is not a good indicator for the wings total weight, because it does not account for the effect of chord wise changes in the platform on the bending weight.

Some researchers used low fidelity semi-empirical methods (Ning & Kroo, 2010) [4] or high fidelity finite element analysis (Jansen, Perez, & Ra Martins, 2010) [5] for wing weight estimation. In this research, a medium level of fidelity aerodynamic solver together with a medium level of fidelity weight estimation tool is used for winglet analysis. Using a medium level of fidelity allows reducing the computational time comparing to higher level of fidelity tools while keeping the level of accuracy high enough.

Drag is caused by the induced drag that is created by the downwash of airflow onto the upper surface of the wing at the wingtips thus producing vortices that trail in the aircraft's wake. The most important wingtip devise's functions is reduce the induced drag and consequently, the trailing vortex strength. By minimizing the induced drag, and thus the wingtip vortices produced by an aircraft's wing, the energy required to create the tip vortices can be conserved and the total drag on the wing reduced (Yahaya & Sheriff, 2012) [6].

A research by NASA Langley research center developed the concept of winglet technology. According to Gimmestad, winglet could be described as the small wing like vertical structures which extends from the wingtip, aiming at reduction in induced drag when compared to other wing tip devices or extensions. He also claimed in his research that the winglet showed 20% reduction in induced drag when compared to tip extension and also improved lift-to-drag ratio. Unfortunately, his theory could not reduce the overall drag of aircraft despite reducing the induced drag. The increase in viscous drag during cruise conditions outruns the reduction in induced drag (Gimmestad, 1981) [7].

CFD has been used by various researchers for studying winglet performance, as it provides a reliable solution for modeling viscous effects as well. For example Azlin et al. (Azlin, Taib, Kasolang, & Muhammad, 2011) [8] examined the performance of winglets at low subsonic flow. They studied two types of design, elliptical and semi-circular winglets with different angles, attached on a wing at various angles of attack. Takenaka et al. also used CFD in order to perform a multidisciplinary design exploration technique regarding winglets for a commercial aircraft (Takenaka, Hatanaka, Yamazaki, & Nakahashi, 2008) [9].

The aim of this work is to study the aerodynamic characteristics in low subsonic speeds (incompressible) laminar/ turbulent flow over airfoil NACA 4412 and straight rectangular wing which are designated to an Australian Light Sport Aircraft developed as touring/training aircraft and manufactured by Jabiru Aircraft Company. The renowned aircraft company (Jabiru) manufactures a light training aircraft without winglets installed on the wings.

2. Wing Theory

2.1. Plain wing

The induced drag on a plain wing at an angle of attack α due to lift is shown in the figure below.

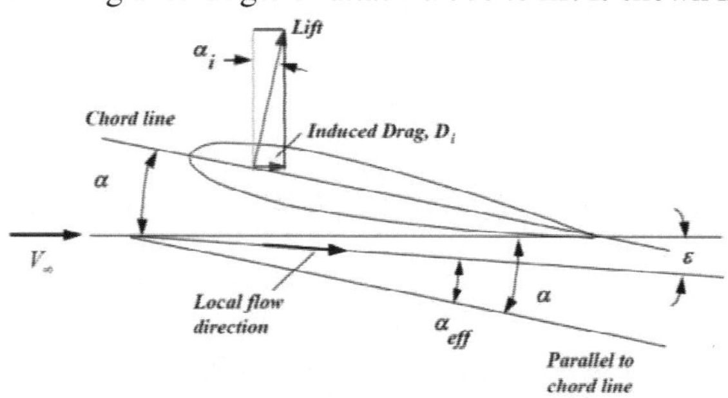

Figure 1. Induced drag definition

The lift, and drag coefficients for an airfoil are given by the following relations (J.D. Anderson, 2003) [10];

$$C_L = L / qS \quad (1)$$

$$C_D = D / qS \quad (2)$$

The reduction in effective angle of attack due to downwash decreases lift at any given α_{eff} and delays stall to higher values of α.

Fig.1 represents the induced drag as a component of the three-dimensional lift in the drag direction;

$$D_i = L sin\varepsilon \text{ or } C_{Di} = C_L sin\varepsilon \quad (3)$$

It can be shown that the induced angle of attack everywhere along the span of wings with elliptical lift distributions is given by (J. D. Anderson, 2003);

$$\varepsilon = \frac{C_L}{\pi AR} rad = \frac{57.3 C_L}{\pi AR} deg \quad \text{and} \quad AR = \frac{S^2}{A} \quad (4)$$

The summation of parasite and induced drag results in the total drag coefficient;

$$C_D = C_L + C_{Di} \quad (5)$$

The induced drag coefficient C_{Di} is proportional to square of the lift coefficient C_L, and inversely to the aspect ratio AR and the shape of the wing e. The relation between the induced drag coefficient and the lift coefficient is:

For ε small, sin $\varepsilon \approx \varepsilon = \alpha_i$ (in radians) and;

$$C_{Di} = C_L \varepsilon = C_L \frac{C_L}{\pi eAR} = \frac{C_L^2}{\pi eAR} \quad (6)$$

The geometry factor, e for all wing types elliptical, rectangular, swept back and trapezoidal is given by;

$$e = \frac{1}{1+\delta} \quad (7)$$

where δ is the induced drag factor.

Therefore, the lift coefficient of the wing is equal to the lift slope of the wing multiplied by the difference of the angles between the wing's angle of attack and the wing's zero lift angle of attack as follow:

$$C_L = a\,(\alpha - \alpha_{L=0}) \quad (8)$$

where a is the lift slope of the wing.

The induced angle of attack for the airfoil NACA 4412 can be calculated from;

$$\alpha_i = \frac{C_L}{\pi AR} \quad (9)$$

As the effective angle of the wing is affected by the induced angle, then the effective angle will decrease by the calculated induced angle above to obtain;

$$\alpha_{eff} = \alpha - \alpha_i \quad (10)$$

Finally, the lift coefficient of the airfoil is determined as follows;

$$C_L = a_0\,(\alpha_{eff} - \alpha_{L=0}) \quad (11)$$

where a_o is the slope for a finite wing.

2.2. *Wing with Winglets Theoretical Work*

Fig.2 shows the various forces acting on the wing tip with winglet. Figure (a) shows a view in direction of flight. Figure (b) is the plan view, and Figure (c) shows the forces acting on winglets. Outboard of the tip, the flow is nearly circular as air from beneath the wing flows outward along the span, around the tip, and inward on the upper surface. The velocities induced by the wing are shown. To these the free-stream velocity is vectorially added. The magnitudes of the induced velocities generally increase toward the tip. At a given span wise location, the induced velocities are highest close to the surface of the wing, just outside the boundary layer (Mccormick, Barens W., 1995) [11].

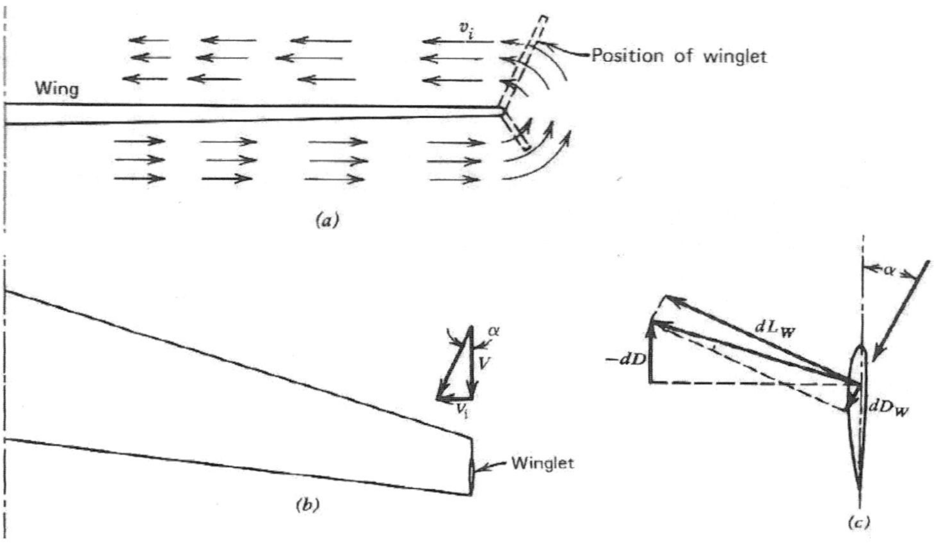

Figure 2. Generation of negative drag by winglets (a) Direction of flight (b) Plan view (c) Forces acting on winglet (Source: Mccormick, Barens W, 1995)

The severe limitations inherent in the assumptions leading to equation (13) below must be recognized (McCormick, Barens W, 1995) [17]. For a given value of wing's wetted area to the whole wing's planform area (S_w/ S), it would appear that increasing the reference area (A_w) would always result in a smaller constant of proportionality, K.

$$C_D = C_d + (C_{Di} + \Delta C_{Di}) \quad (12)$$

As C_{Dwo} is the drag coefficient of the wing at zero lift, hence an increment in induced drag coefficient value ΔC_{Di} becomes;

$$\Delta C_{D_i} = -\frac{S_W}{S}\left[2\pi\left(\frac{A_W}{A_W+2}\right)^2 K^2 C_L^2 - C_{D_{Wo}}\right] \quad (13)$$

The approximate analysis by (McCormick, Barens W.) indicates that:

1. The reduction in C_D increases linearly with C_L^2.
2. At low C_L values, C_D will be increased by the addition of a winglet.
3. High winglet aspect ratio is desirable.

3. Numerical Simulation

The base line of our study is the theoretical solution. The governing aerodynamics equations are used to analyze the original wing parameters, then the modified design with winglets. We have selected 0° and 5° as the common angles of attack for the steady level flight or cruise flight. While 10° angle of attack is the take-off angle of attack. The numerical analysis was performed using two types of software: 3-D Foil-Multi surface code and Fluent ANSYS 14 code. Firstly, to analyze the original wing (Plain wing), and then to analyze the wing with winglet, the iteration process is used to obtain converged solution.

3.1 Computational Simulation

The NACA airfoils are airfoil shapes for aircraft wings developed by the National Advisory Committee for Aeronautics (NACA). The shape of the NACA airfoils is described using a series of digits following the word "NACA". The parameters in the numerical code can be entered into equations to precisely generate the cross-section of the airfoil and calculate its specifications. The NACA four-digit wing sections define the profile by first digit describing maximum camber as percentage of the chord, second digit describing the distance of maximum camber from the airfoil leading edge in tens of percent's of the chord and last two digits describing maximum thickness of the airfoil as percent of the chord. The airfoil used in Jabiru aircraft, which we are studying in this work is NACA 4412, and the wing is rectangular with no aerodynamic twist and of constant chord and without winglets. The main objectives of the plain wing with winglets design is to:

1. Decrease induced drag resulting in a total drag decrease and increases the lift to drag ratio.
2. Reduce the strength of the tip vortex.
3. Decrease fuel consumption and increase the flight range.

3.2 Foil Multi-Surfaces Code (3-D)

It is simple version software which uses the thin airfoil theory model. Simplicity of this code allowed us to design the wing with winglet model. The suggested design needed around ten iterations to obtain the optimum design of the winglets. Then we proceeded to test and verify it numerically, and theoretically. The parameters changed in the design iteration using the 3-D Multi-Surface code are:

1. The span wise slope angle of the winglets to obtain an optimum angle at $\beta = 65^{o}$.
2. The length of the winglets to be found optimum at 0.4 m.
3. Fixed chord of the airfoils from tip to the root of the winglets.
4. NACA 4412 airfoil for the winglets root, and NACA 0012 for the winglets tip.

After obtaining the optimum design from the 3-D Foil Multi-Surfaces code, Solid Works is used to draw the wing with winglet final model ready to export to ANSYS Fluent 14 code. Fig.3 (a) shows top view of the design, while Fig.3 (b) presents the 3-D wing with winglet design.

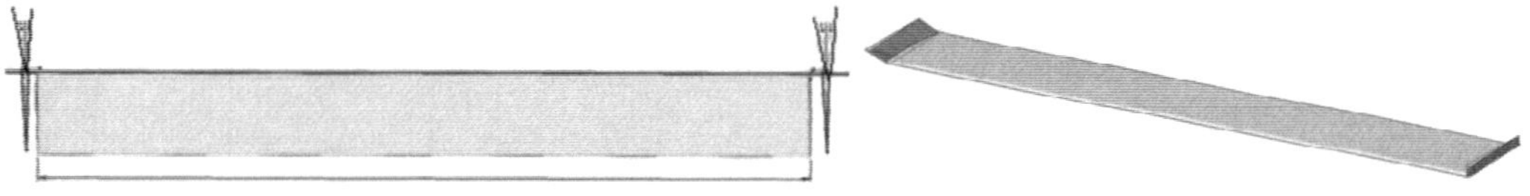

(a) (b)

Figure 3. (a) Top view and (b) Isometric view for wing with winglets by Solid Works

4. Theoretical Results

The lift and drag coefficients results are calculated at 0°, 5°, and 10° angles of attack. The total drag is a combination of Parasite drag and induced drag. All the boundary conditions are measured at sea level, ambient temperature and Reynolds Number Re = 3.52 E^{6} for free stream velocity of 50 m/s. Reynolds numbers are used in airfoil design to determine "Scale Effect" when computing flow characteristics. Aerodynamics Reynolds number, Re, $Re = Vc/v$, where V is the flow velocity, c is the chord length, and v is the kinematic viscosity of air surroundings the airfoil, which is 1.460×10^{-5} m^2/s for the air at sea level. By using 3-D Foil Multi-Surfaces and Solid Works codes, the final drawings for the two models plain wing and wing with winglet were obtained. For a constant airfoil NACA 4412, the plain wing is 9.528 m span and, and 1 m chord. While the wing with winglets as seen from top (reference area view) is 10.286 m span, 1m chord, with the side angle of the winglets is $\beta = 65^{\circ}$, with a constant airfoil NACA 4412, and the tip of the winglets airfoil is NACA 0012, and the length of the winglet is 0.879 m for each side. Fig.4 shows the top view of both models. Both wings are merged in one view to simplify the dimensions of the original and designed models. In the front view, the plain wing with 9.528 m span is seen. The winglets with the optimum side angle â = 65° as determined from 3-D Foil Multi-Surface code was added, but with a different winglets tip airfoil. Both wings have a constant maximum camber through all the geometry of 0.12 m. Fig.5 shows the front view of the combined models.

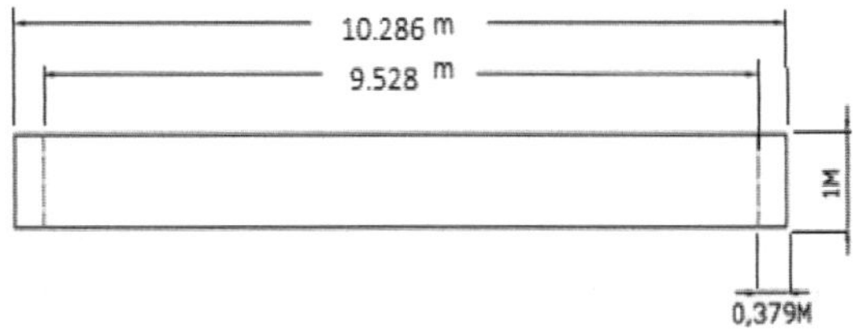

Figure 4. Top view for the plain wing and the wing with winglets

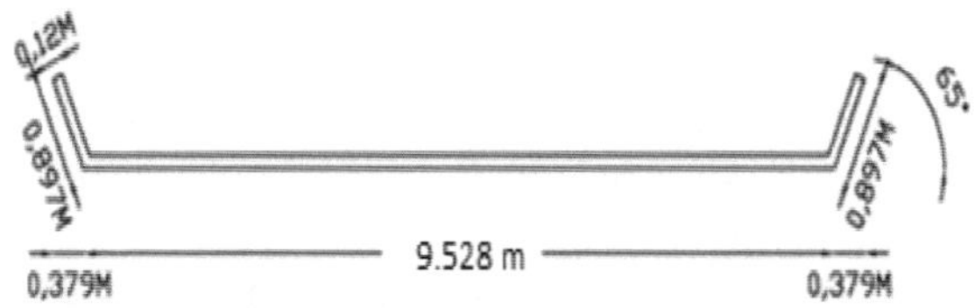

Figure 5. Front view for the plain wing and the wing with winglets

5. Discussion of Results

The comparison of the results between plain wing and the designed wing with winglets has shown that the induced drag has decreased with the plain wing with winglets in comparison to the plain wing model for all numerical methods as depicted in Fig.6 and Fig.7 at the various studied angles of attack. It is also noticed that the induced drag increases with angles of contact for all solutions. Using the Fluent code 14 has shown closer induced drag values to the mathematical (theoretical) solution for both plain and with winglets wing.

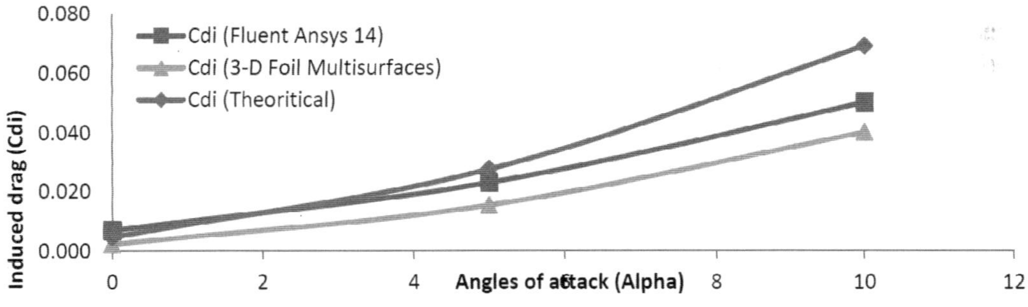

Figure 6. Induced drag (C_{Di}) for plain wing at different angles of attack (α)

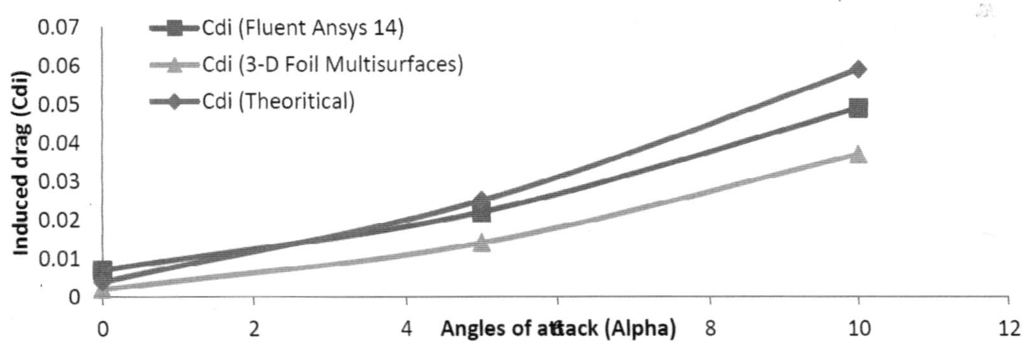

Figure 7. Induced drag (C_{Di}) for wing with winglets at different angles of attack (α)

The percentage reduction in induced drag between plain wing and wing with winglets results determined for all verification methods at the different angles of attack are; 14.9% at $\alpha = 0^{\circ}$, 9.1% at $\alpha = 5^{\circ}$, and 14.5% at $\alpha = 10^{\circ}$ using the theoretical model, and 0% at $\alpha = 0^{\circ}$, 20% at $\alpha = 5^{\circ}$, and 2% at $\alpha = 10^{\circ}$ using Fluent ANSYS 14 simulation code. Moreover, the percentage reduction simulated using the 3-D Multi-Surfaces code is 13% at $\alpha = 0^{\circ}$, 9% at $\alpha = 5^{\circ}$, and 7.5% at $\alpha = 10^{\circ}$. The designed wing with winglets improved the aerodynamic characteristics of the wing by decreasing the induced drag values at different angles of attack. The second main objective of our study is to show that the winglets design has a significant effect in improving the lift to drag ratio. This ratio has increased especially at high angles of attack due to the reduction of the amount of the induced drag, which allow the total drag coefficient to decrease and the lift coefficient to increase at the same time. Different numerical results obtained from each working method for plain wing are shown in Fig.8 and for the designed wing with winglets in Fig.9.

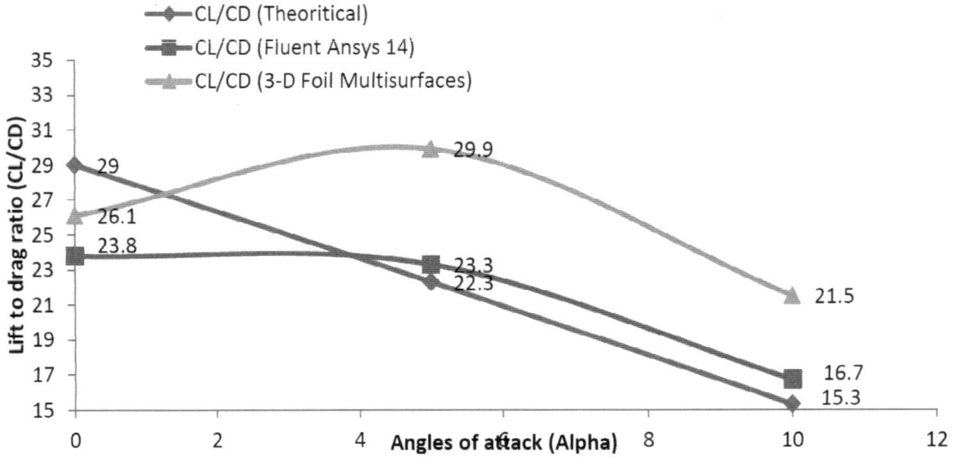

Figure 8. Lift to drag Ratio (C_L/C_D) comparison for plain wing at different angles of attack

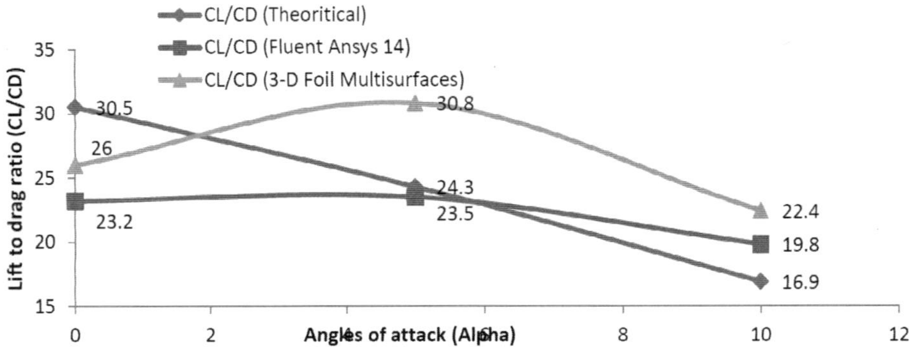

Figure 9. Lift to Drag Ratio (C_L/C_D) Comparison for Wing with Winglets at Different Angles of Attack

After processing the solution of the designed wing with winglets in Fluent ANSYS 14, it is shown that the flow characteristics of the wing will be improved after applying the winglets with the optimum winglet's geometry. The vortices are shifted to the edge of the winglet as seen in Fig.10. This shift moves all the vortices to the winglets instead of the wing tips. Because of this effect, the stability and maneuverability increase at high angles of attack. The Induced drag decreases due to that as well as the lift to drag ratio increases. The tip vortices in this case will not spoil the lift on the wing tips and the resistance will decrease.

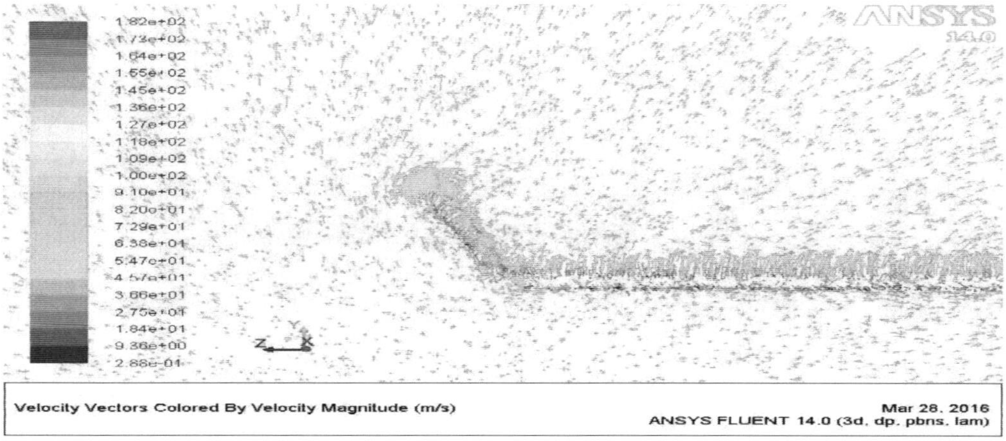

Figure 10. Velocity vectors for wing with winglets at á=10°

The total pressure of the wing is a major factor in all aerodynamics studies. The airfoil shape of the wing's cross section divides the pressure into two separate zones; the lower surface of the wing is a high pressure zone, and the upper surface of the wing is a low pressure zone. The difference in the pressure between these zones creates the lift. In the chord's direction (the 2-D direction) it is noticed that the major effect of the winglets is smoothing the pressure gradient on both airfoil surfaces. As seen in Fig.11, the lower surface of the plain wing airfoil has significant fluctuations increased at the leading and trailing edges as depicted from the ANSYS 14 Code simulation. In Fig.12, for wing with winglets both airfoil surfaces show stabilization in the pressures. This stabilization will allow getting the maximum benefit of the pressure difference to generate the lift. This phenomenon is more noticeable at the upper airfoil surface.

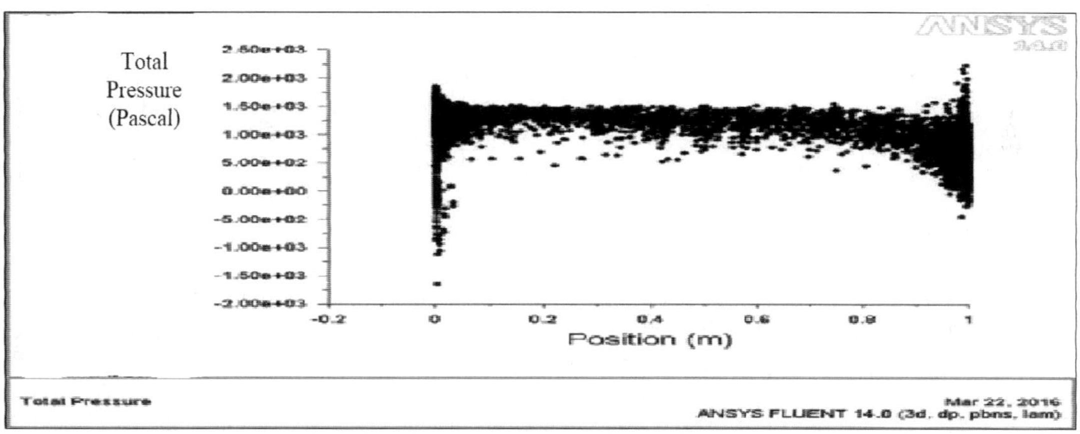

Figure 11. Plain wing chord's direction total pressure

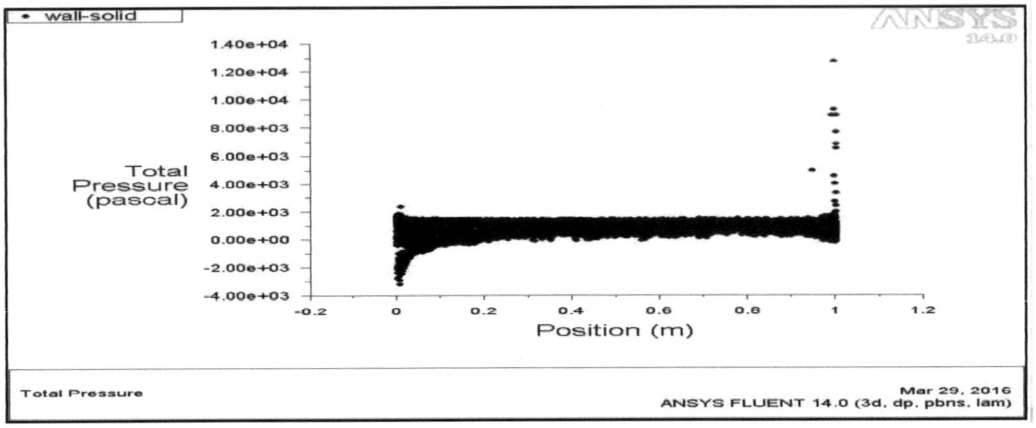

Figure 12. Wing with winglets chord's direction total pressure

6. Conclusions

Theoretical and numerical work has been performed to examine the efficiency of winglets mounted on an airfoil in subsonic flows. The winglets that are considered here are placed upward at the tip of the wing in order to improve the wing efficiency; decreasing the induced drag and increasing the lift to drag ratio. The main results achieved in this study can be described in the following points:

1. The theoretical work shows that decrement in the values of induced drag achieved for our designed wing with winglets is 14.9 % at 0o angle of attack, 9.1 % at 5° angle of attack, and 14.5 % at 10° angle of attack. The values of the lift to drag ratio increased by 5.1 % at 0° angle of attack, 9 % at 5° angle of attack, and 10.5 % at 10° angle of attack.
2. The numerical work was performed using two different codes. First, Fluent ANSYS 14 is used to verify the results by the most realistic values in simulating the plain wing and wing with winglets. The values of induced drag for the designed wing with winglets decreased 20 % at 5o angle of attack and decreased by only 2 % at 10° angle of attack, but remained constant at 0° angle of attack. The lift to drag ratio increased at 5° angle of attack by 0.9 %, and increased by 18.6 % at 10° angle of attack.

The second numerical verification method used is 3-D Foil Multi-Surfaces Code. The values of the induced drag decreased 13 % at 0° angle of attack, 9 % at 5° angle of attack and 7.5 % at 10° angle of attack. The lift to drag ratio increased by 3 % at 5° angle of attack and 4.2 % at 10° angle of attack.

3. The stability and maneuverability improved in our designed wing with winglets. This is noticeable at high angles of attack at low speeds. This presents the condition in take-off, where the induced drag contributes by nearly 60 % of the total drag ratio (Jones, R., 1990) [12]. This benefit emphasizes the safety required for this type of aircraft under study.
4. As a result of decreasing the induced drag and increasing the lift to drag ratio, this will add an increment in the range, which will add a benefit to the ALS aircraft which is mainly owned by its users because it is an economical aircraft.
5. There are some differences between the theoretical and numerical methods used in the verification work due to several reasons, namely, the governing equations used by each numerical method is different due to the designer's code concept and also the selected boundary conditions.

References

[1] Bento Mattos, Antonini Macedo, and Durval Silva Filho (2003). Consideration about winglet design, 21st Applied Aerodynamics Conference, 23-26 June 2003, Florida, U.S.A.

[2] I. Kroo, (2005). Nonplanar wing concepts for increased aircraft efficiency. VKI lecture series on innovative configurations and advanced concepts for future civil aircraft, June 6-10, 2005, pp.1-29.

[3] Hantrains-Gervois et al., (2009). Effect of gurney flaps and winglets of the performance of HAWT, chapter 6.

[4] Ning A., Kroo L, (2010). Multi-disciplinary considerations in the design of wings and wing tip devices, Journal of aircraft, vol.47, pp.534-543.

[5] Jansen, Perez, and Ra Martins, (2010). A python-based object-oriented frame work for non-linear constrained optimization, Journal of structural and multi-disciplinary optimization, vol.45, issue 1, pp.101-118.

[6] Nurulain, Y., Jamaluddin, S. (2012). Flow behavior around winglets. *Journal Mechanical,* 34, 95-100.

[7] Gimmestad, (1981). Technology reshaping aviation partners winglets, Aviation partners, internet website.

[8] Azlin M.A., Maltaib C.F., Kasolang S., and Muhammed F.H., (2010). CFD analysis of winglets at low subsonic flow, proceedings of the world congress on engineering, WCE, vol.1, pp.87-91.

[9] Takenaka K., Hatanaka, K., Yamazaki W., and Nakahashi K., (2008). Multidisciplinary design exploration for a winglet, Journal of aircraft, vol 45, 5, pp.1601-1611.

[10] Anderson, J. D. (2003). Fundamentals of Aerodynamics (4th Ed.). New York: McGraw-Hill.

[11] McCormick, Barens W. (1995). Aerodynamics Aeronautics and Flight Mechanics (2nd Ed.). John Wiley & SONS, INC.

[12] Jones, R. (1990). Wing theory. Princeton, NJ: Princeton University Press.

Hybrid Method for Power Control Simulation of a Single Fluid Plasma Thruster

S. Jaisankar[1] and T. S. Sheshadri[2]

Department of Aerospace Engineering, Indian Institute of Science, Bangalore, India

Corresponding e-mail: [1]jshankar.s@gmail.com, [1]jshankar@aero.iisc.ernet.in, [2]tss@aero.iisc.ernet.in

Abstract. Propulsive plasma flow through a cylindrical-conical diverging thruster is simulated by a power controlled hybrid method to obtain the basic flow, thermodynamic and electromagnetic variables. Simulation is based on a single fluid model with electromagnetics being described by the equations of potential Poisson, Maxwell and the Ohm's law while the compressible fluid dynamics by the Navier Stokes in cylindrical form. The proposed method solved the electromagnetics and fluid dynamics separately, both to segregate the two prominent scales for an efficient computation and for the delivery of voltage controlled rated power. The magnetic transport is solved for steady state while fluid dynamics is allowed to evolve in time along with an electromagnetic source using schemes based on generalized finite difference discretization. The multistep methodology with power control is employed for simulating fully ionized propulsive flow of argon plasma through the thruster. Numerical solution shows convergence of every part of the solver including grid stability causing the multistep hybrid method to converge for a rated power delivery. Simulation results are reasonably in agreement with the reported physics of plasma flow in the thruster thus indicating the potential utility of this hybrid computational framework, especially when single fluid approximation of plasma is relevant.

1. Introduction

Magneto-plasma dynamic (MPD) thruster is an electric propulsion device having applications ranging from near-Earth orbit raising maneuvers to long duration interplanetary high impulse space missions [1]. Such a thruster in its simple form consists of a cylindrical cathode surrounded by a concentric anode and the arc between the electrodes ionizes a gaseous propellant. The interaction of the current with the self-induced magnetic field accelerates the plasma to produce thrust. Numerous experimental investigations leading to plasma thruster development have been reported and so also on the simulation of single or two fluid plasma using cartesian or cylindrical co-ordinate based models. A single fluid simulation with a detailed model including viscosity, Hall effect, electrostatic forces etc., along with rated power control is seldom, which is being attempted here.

Here, a hybrid solver is proposed in which the electromagnetic equations in steady form and fluid equation evolving in time are solved separately, so as to have a reasonable combination of robustness and speed of convergence. The hyperbolic treatment in the magnetic transport is employed following

Heiermann et al.[2]. The overall method uses generalized finite differences [3] which is particularly useful in simplifying the computation when cross derivatives are involved with arbitrary shaped mesh. The separation of fluid solver simplified the coding efforts as up-gradation for plasma propulsion is done just by the addition of source terms.

2. Governing Model

The basic electrodynamics equations include (i) the Poisson equation (1) for potential ($\emptyset$) written as

$$\nabla^2\emptyset = -\frac{q}{\epsilon_0} \tag{1}$$

with q and ϵ_0 being the space charge distribution and free space permittivity, respectively (ii) the electric field distribution given by

$$\vec{E} = -\Delta\emptyset \tag{2}$$

and the magnetic transport equation (5) which could be derived by combining Maxwell's equations (3)

$$\Delta.\vec{B} = 0;\ \ \Delta \times \vec{B} = \mu_0\vec{J};\ \ \Delta \times \vec{E} = -\frac{\partial \vec{B}}{\partial t};\ \ \Delta \times \vec{E} = -\frac{q}{\epsilon_0} \tag{3}$$

along with (iv) the generalized Ohm's law (4) which relates current $(\vec{J})$to the quantities of plasma velocity $(\vec{v})$, electric field $(\vec{E})$and magnetic field $(\vec{B})$ as

$$\vec{J} = \sigma[\vec{E} + (\vec{v} \times \vec{B})] - \frac{H_p}{B}(\vec{J} \times \vec{B}) \tag{4}$$

Following (3) and (4), divergence free magnetic flux is expressed as

$$\frac{\partial \vec{B}}{\partial t} = rot(\vec{v} \times \vec{B}) - rot\left(\frac{rot\,\vec{B}}{\mu_0\sigma} + \frac{H_p}{\sigma B} rot\vec{B} \times \vec{B}\right) \tag{5}$$

with the Hall parameter H_p for a degree of ionization D and propellant molecular weight Wbeing defined as

$$H_p = 10^{-11}\frac{WT^{3/2}}{D\rho}B \tag{6}$$

The fluid model being employed here is that of two dimensional cylindrical Navier-Stokes (7).

$$\frac{\partial U}{\partial t} + \frac{\partial A}{\partial z} + \frac{\partial B}{\partial r} + \frac{C}{r} + D = 0 \tag{7}$$

where

$$U = \begin{pmatrix} \rho \\ \rho u \\ \rho v \\ \rho E \end{pmatrix}\ \ D = \begin{pmatrix} 0 \\ (\vec{J} \times \vec{B} + q\vec{E})_z \\ (\vec{J} \times \vec{B} + q\vec{E})_r \\ \vec{J}\cdot\vec{E} \end{pmatrix}\ \ A = \begin{pmatrix} \rho u \\ p + \rho u^2 - \tau_{zz} \\ \rho uv - \tau_{rz} \\ pu + \rho uE + q_z - u\tau_{zz} - v\tau_{rz} \end{pmatrix} \tag{8}$$

$$B = \begin{pmatrix} \rho v \\ \rho uv - \tau_{rz} \\ \rho vv + p - \tau_{rr} \\ pv + pvE + q_r - u\tau_{rz} - v\tau_{rr} \end{pmatrix}\ \ C = \begin{pmatrix} \rho v \\ \rho uv - \tau_{rz} \\ pvv - \tau_{rr} \\ pv + \rho vE + q_r - u\tau_{rz} - v\tau_{rr} \end{pmatrix} \tag{9}$$

The molecular stress tensor components are

$$\tau_{zz} = \frac{2\mu}{3}\left(2\frac{\partial u}{\partial z} - \frac{\partial v}{\partial r} - \frac{v}{r}\right);\ \tau_{rr} = \frac{2\mu}{3}\left(2\frac{\partial v}{\partial r} - \frac{\partial u}{\partial z} - \frac{v}{r}\right);\ \tau_{rz} = \mu\left(\frac{\partial u}{\partial r} + \frac{\partial v}{\partial z}\right) \tag{10}$$

with the diffusive heat-flux vector described as

$$q_z = -k\frac{\partial T}{\partial z} \text{ and } q_r = -k\frac{\partial T}{\partial r} \tag{11}$$

The plasma viscosity μ and thermal conductivity k are computed using logarithm law with temperature while electrical conductivity σ is calculated using Spitzer-Harm formulation [4]. Charge density q is computed using the divergence of electric field (3).

2.1 Numerical Solution

Solution of the governing model consists of obtaining the voltage potential across the electrodes using the Poisson equation, from the divergence of which the electric field distribution could be obtained. Electric field is then used in magnetic transport equation and general Ohm's law (4) which yields a combined elliptic-hyperbolic equation. Each of these elliptic potential Poisson (1) and steady magnetic transport (5) equations can be cast in a general form as

$$\gamma_1\frac{\partial^2\psi}{\partial r^2} + \gamma_2\frac{\partial\psi}{\partial r} + \gamma_3\frac{\partial\psi}{\partial z}\gamma_4\frac{\partial^2\psi}{\partial z^2} + \gamma_5\psi + \gamma_6 = 0 \tag{12}$$

Finite difference discretization of (12) is used to solve for ψ. Central differences for second order derivatives and sign of co-efficients(of γ_2,γ_3) based backward or forward differences for the first derivatives are applied. ψ either as a potential or magnetic field is obtained iteratively using successive relaxation till their residual reached just below 10^{-3}. Current, magnetic and electric fields thus obtained define the source term of the fluid solver. As the source could cause stiffness in the flow solution, very low CFL number is used for time marching in addition to source terms being incrementally added with iterations to avoid blow up of the code. The fluid solver adopted is a mesh-free method [5] in which the flux derivatives are discretized as (13) and (14).

$$\left(\frac{\partial\Psi}{\partial z}\right)_p = \frac{\sum(\Delta r_m)^2\sum\Delta z_m\Delta\Psi_m - \sum\Delta z_m\Delta r_m\sum\Delta r_m(\sum\Delta\Psi_m)}{\sum(\Delta z_m)^2\sum(\Delta r_m)^2 - (\sum\Delta z_m\Delta r_m)^2} \tag{13}$$

$$\left(\frac{\partial\Psi}{\partial z}\right)_p = \frac{\sum(\Delta z_m)^2\sum\Delta r_m\Delta\Psi_m - \sum\Delta z_m\Delta r_m\sum\Delta z_m(\sum\Delta\Psi_m)}{\sum(\Delta z_m)^2\sum(\Delta r_m)^2 - (\sum\Delta z_m\Delta r_m)^2} \tag{14}$$

where $\Delta z_m = z_m - z_p$; $\Delta r_m - r_p$ and $\Delta\Psi_m = (\Psi_m - \Psi_p)$

Flux derivative at a point $'p'$ is obtained using the neighborhood points, say $= 1..8$. Ψ_m is the mid-point state between points $'p'$ and $'i'$and defined as per the physics of the term. For convective terms, Ψ_m represents the flux and being hyperbolic is defined here as $\Psi_m = \left[\frac{1}{2}\left(\Psi_p + \Psi_i - |\alpha|\left(U_p - U_i\right)\right), \forall_i\right]$, following the diffusion regulated flux [6, 7]. α is the numerical viscosity co-efficient and U_i, U_p are conserved variables in (8). For diffusive terms (thermal and viscous) Ψ (which could be velocity or temperature) $\Psi_m = \left[\frac{1}{2}(\Psi_p + \Psi_i), \forall_i\right]$ is the mean of two states in the stencil. A schematic of the steps involved in computation is given in fig. 1.

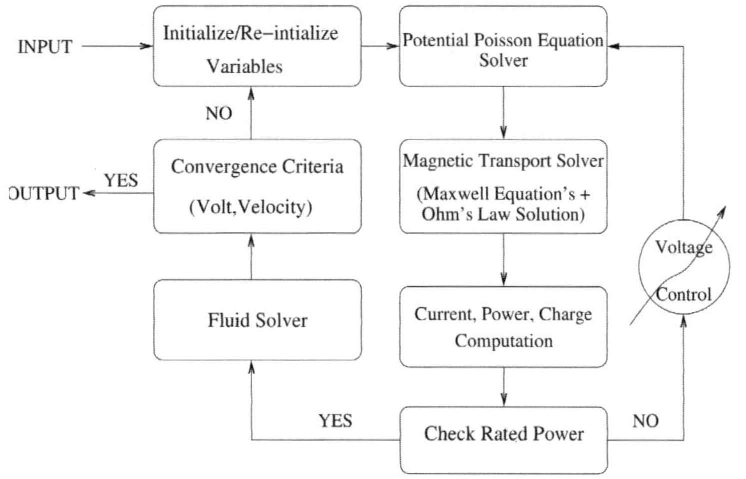

Figure 1. Flow Chart of Steps in the Hybrid Computational Method

The thruster is evaluated for thrust generation, specific impulse and efficiency. Thrust is evaluated by integrating as $F = \int_e 2\pi r dr \dot{m} V_e$. Specific impulse ($I_{sp}$) and efficiency ($\eta$) are calculated as

$$I_{sp} = \frac{1}{g}\frac{\int_e 2\pi r \dot{m} V_e dr}{\int_e 2\pi r \dot{m} dr}; \qquad \eta = \frac{\int_e \left(p_e + \frac{1}{2}\rho_e {V_e}^2\right) V_e \, dA_e}{(V_a - V_c)\int_a J_a dA_a} \tag{15}$$

with subscripts 'e', 'c' and 'a' denoting exit, cathode and anode, respectively. V_e is the exit velocity, $V_a$and V_c are voltage at the electrodes.

2.2 Power Control

In this simulation with constant voltage supply, magnetic field is obtained by solving each of Poisson equation and magnetic transport flux equation separately. Current is then computed using the Ohm's law to obtain the integrated power over the volume of the thruster. The relevant source terms are included in the fluid solver and the whole computational sequence is repeated till convergence as depicted in fig. 1. When a constant magnetic field through constant total discharge current is not applied, there is a possibility of nonphysically more current getting drawn into the system as current and electric field are functions of fluid variables which are computed later in the multistep computational method used here. This can cause a runaway in the simulation and hence power supply is to be controlled. Power drawn by thruster can be controlled by any of the three ways (i) direct control of external energy in fluid solver (ii) by regulating the current through conductivity or temperature or (iii) by voltage control. Direct energy control in conservation law made a disproportionate source or widely varying scales between energy and momentum term which imposed strong restrictions on the marching time of the fluid solver. Power control by regulating current drawn at every discretized volume by modifying electrical conductivity and hence the magnetic field is found extremely slow in convergence and required multiple control points. This is due to space and time varying current being highly non-linear function of magnetic field and electrical conductivity. But here, power control by regulating the voltage across electrode is attempted using (16)

$$\Delta\phi_{new} = w'\Delta\phi' + w\Delta_\phi \quad \text{and} \quad \Delta\phi' = \left(\frac{Rated\ Power}{Drawn\ Power}\right)\Delta\phi \tag{16}$$

where w and w' are the weights of $\Delta\phi$ and $\Delta\phi_{new}$, respectively. Control strategy (16) is simple as electric field in energy term is linear to potential difference which is a constant quantity. Control of supply voltage is depicted in Fig. 1. Voltage control of power is found to be repetitively converging and faster than the other methods. Constant power thruster in the entire of literature is achieved by a constant total discharge current, which is quite different from rated power operation attempted here.

3. Simulation

Simulation of plasma flow through the classical cylindrical-conical thruster geometry [8, 9] shown in Fig. 2 is studied. A hybrid method for the simulation of fully-ionized argon plasma thruster with a single fluid model is used. The scheme features constant controlled power input and is general for any cylindrical-conical geometries and physical input parameters. Each of the solvers for Poisson, magnetic transport and cylindrical compressible flow are allowed to converge for a fall in numerical residue up to 10^{-3} or lesser. The computational domain is divided into (axial radial) nodes and the solution sequence comprises of potential Poisson computation, followed by magnetic transport equations to obtain current and magnetic field, and hence source terms for the fluid solver. Fluid equations with added source are then solved to reach steady state. The whole procedure is iterated till the residual due to potential distribution and exhaust velocity reached a very low value (say 10^{-5}). The solving of steady magnetic transport equation requires an initial value, more so it should be a careful guess as the electromagnetic solver is called several times in the power control loop. After some initial experiments, $B_\theta(z,r,0) = -\mu_0\sigma E_r\Delta Z$ is taken, where $\Delta Z = Z - Z_l$, with Z_l being length of cathode. In every subsequent call of electromagnetic solver only E_r is to be changed.

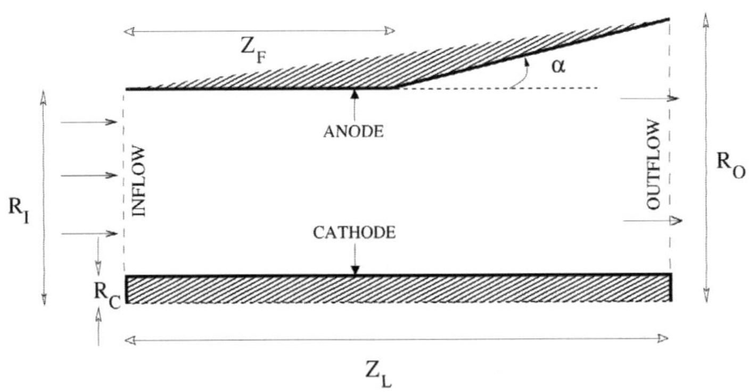

Figure 2. Geometry of Cylindrical Conical Diverging Plasma Thruster: $R_I = 7mm, R_O = 12.2mm, and\ R_c = 2mm$ – radius of inley, outlet and cathode, (ii)

3.1 Boundary Conditions

Flow and solid electrodic boundaries are marked in fig. 2. Propellant mass flow rate and density are specified at inlet while the static pressure is evaluated by the first-order extrapolation at entrance due to subsonic flow. Velocity at inlet follows a developed annular flow [10] while temperature is set by ideal gas law. At the exit of the thruster, the boundary condition is governed by whether the outflow is supersonic or subsonic. If supersonic, all the independent variables are evaluated by Neumann condition with zero gradient. The flow in the boundary layer is possibly subsonic, however, this region has little influence on the entire flow characteristics and that too when the objective is steady-state solution.

On the electrode boundaries, no slip condition for velocity and zero normal gradient for pressure are imposed. The electrode surface is treated as a adiabatic thermal boundary with temperature being set to that of contacting plasma. The discharge voltage is imposed across the electrode surfaces and controlled iteratively so that the calculated total power amounts to the rated value. At the inlet, center axis, and exit boundary, the electric field perpendicular to the boundary is set zero. Upstream and downstream boundaries for all electromagnetic variables are set for no current entering or leaving the thruster along with the fluid flow, which translates to $E_z - H_p E_r = -H_p v_z B_\theta$

3.2 Results and Discussion

Simulation results for argon flow in MPD thruster presented in figs. 3-10 exhibit the following on relevant physical and computation related parameters.

Electromagnetics: Electric field in fig. 3b is established normal to the equipotential lines of fig. 3a. The magnetic field as seen in fig. (4)a is negative maxima at the entrance which gradually reached minimum magnitude at the thruster exit. Cathodic current as seen in fig. 4b is concentrated close to the entrance and exit, similar to the trend reported [11].

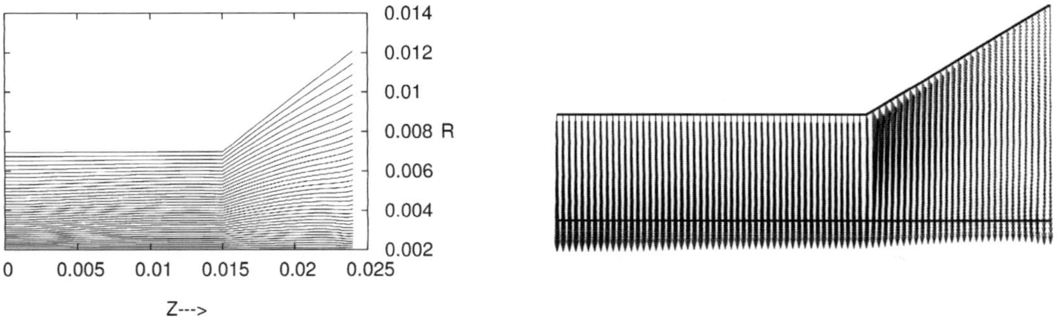

(a) (b)

Figure 3. (a) Constant voltage lines and (b) Electric field

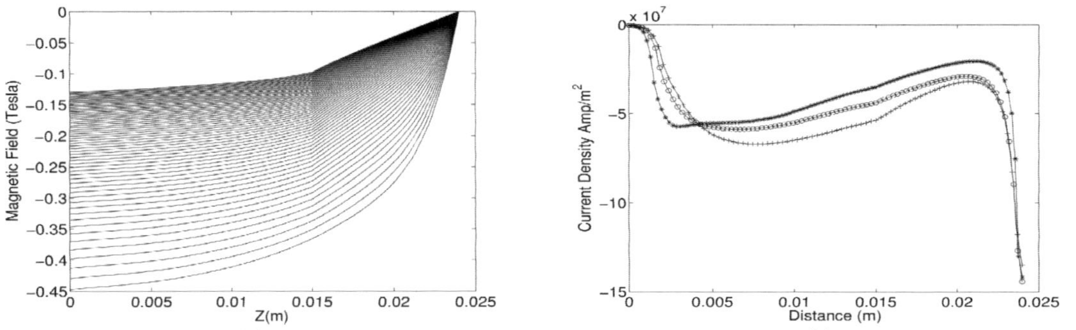

(a) (b)

Figure 4. (a) Magnetic field intensity and (b) Cathodic radial current for mesh sizes 80 x 60 (+), 100 x 75 (o) and 120 x 90 (*)

Hydrodynamics: All the flow related variables experience a discontinuity close to the entrance which is the zone of concentrated current and magnetic field (fig. 4), hence that of maximum added electromagnetic energy (fig. 6). The maximum acceleration seen through velocity vector in fig. 5b and pressure maxima in fig. 6 are consistent with the understanding on maximum energy being added close to the cathodic entrance. This is also reflected in high gradient in density closer to the cathode entrance. Fig. 7 show the argon plasma being accelerated upto 10 km/s, as expected out of a high specific impulse operation in a MPD device.

IOP Conf. Series: Materials Science and Engineering **370** (2018) 012028 doi:10.1088/1757-899X/370/1/012028

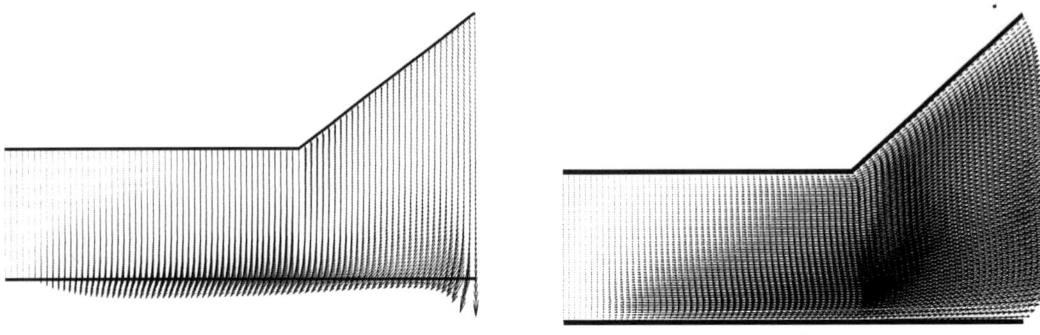

(a) (b)

Figure 5. Vector plot of (a) Current and (b) Velocity

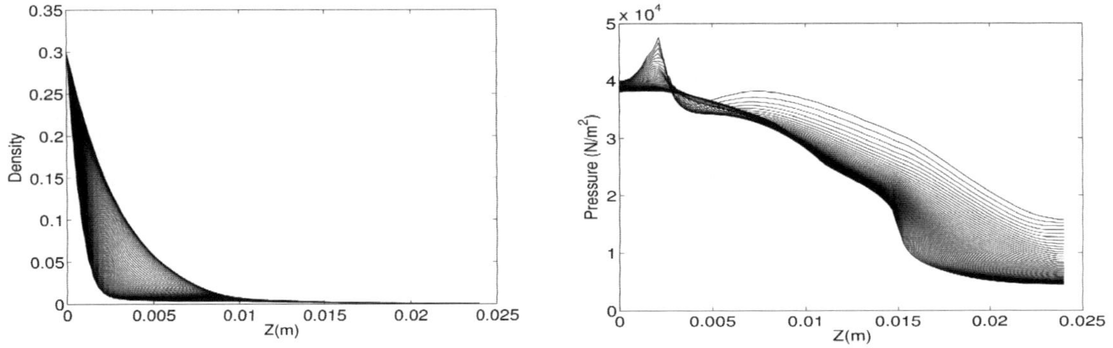

(a) (b)

Figure 6. (a) Density and (b) Pressure variation along the length of the thruster

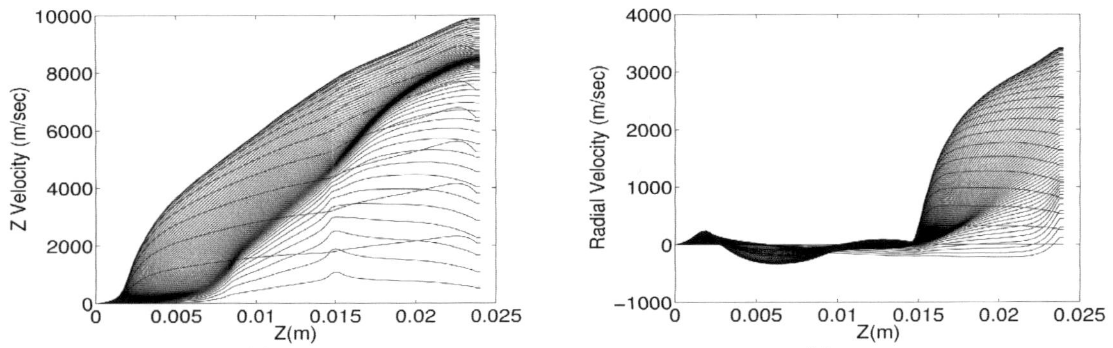

(a) (b)

Figure 7. Velocity variation along the length of the thruster

Computational: Current continuity observed in fig. 8 in 4 - 5 iterations of the full loop solver indicate the solver stability and efficiency. Thrust and efficiency in fig. 9 are quite comparable with the reported performance of such a MPD device [8, 12]. Figs. 8 to 10 show how the solution and performance variables are converging to their respective steady values in about 10 iterations, along with stability on varying mesh size.

In general, the physical understanding as indicated by the above simulation are reasonably consistent with the published literature on theory and experimental findings [11-13]. In the proposed hybrid multistep computation, with each part of the solver being allowed to converge up to a residue of 10^{-3}, it takes about 15-20 iterations for the combined loop residue to reach the desire level of

10^{-5}. Thus, the results show consistency with the known physics of the problem, the convergence of each part of the solver and grid stability of solution.

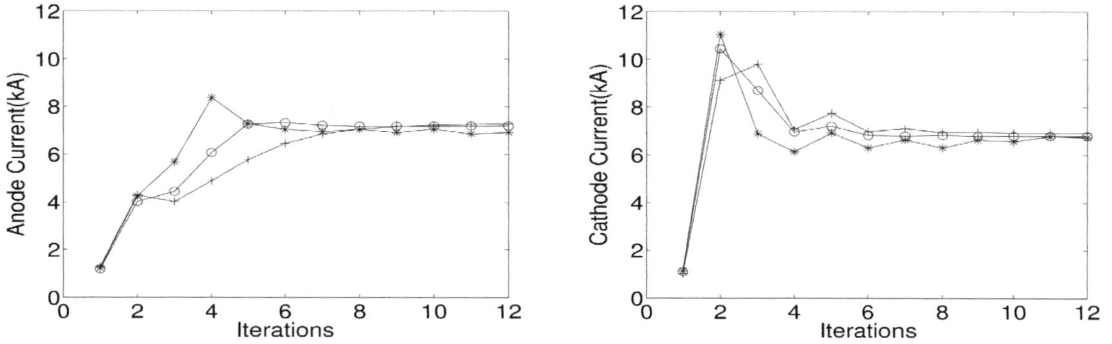

(a) (b)

Figure 8. Convergence of Current for mesh sizes $80 \times 60\,(+), 100 \times 757\,(o)$ and $120 \times 90\,(*)$

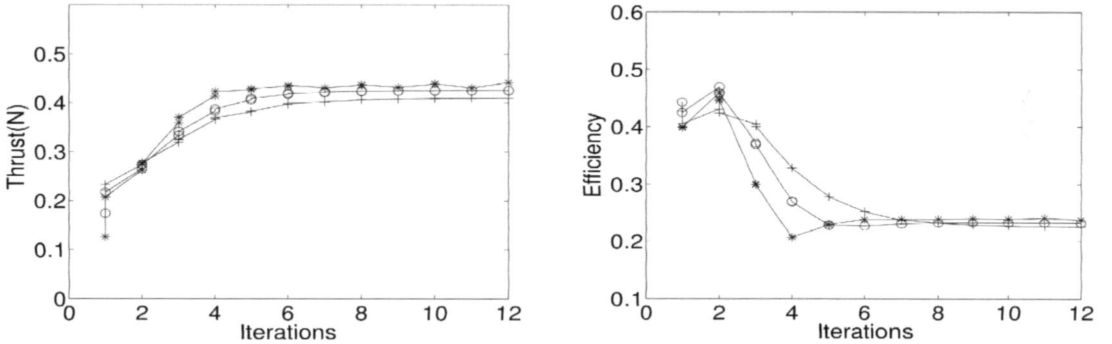

(a) (b)

Figure 9. Convergence of Thrust (L) and Efficiency (R) for mesh sizes $80 \times 60\,(+), 100 \times 757\,(o)$ and $120 \times 90\,(*)$

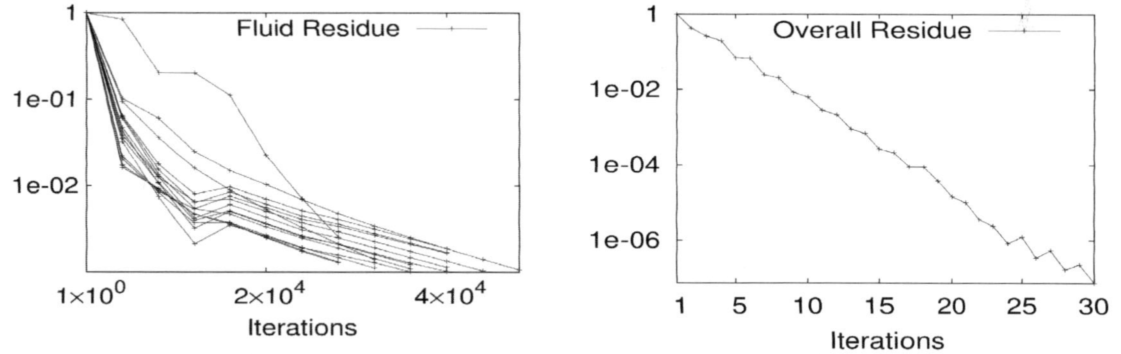

(a) (b)

Figure 10. Convergence of fluid (L) and overall solver (R)

4. Summary

A hybrid multistep method for the simulation of plasma thruster with a single fluid model is presented. The procedure is developed to solve for electromagnetic aerothermodynamics of internal flows in MPD thrusters driven by an external power source at a constant voltage. The scheme is general for any cylindrical-conical geometries and physical input parameters. The computational method involved solving for potential Poisson equation as a starter, followed by magnetic transport with Ohm's law and the cylindrical Navier Stokes, the above sequence to be repeated along with voltage control till convergence. Results of simulation for fully ionized argon plasma indicate residue of each computational module and that of the full solver converging to desired levels. Variables of electric field distribution, magnetic field intensity, current distribution, current at and near the electrodes and their effects on the fluid density, velocity and pressure are all observed to be in consistent with the known and reported physics of the problem. Performance variables of thrust, specific impulse and efficiency are quite comparable to those reported. In short, the method exhibited physical consistency, convergence as well as grid stability for the problem studied. The procedure accounts for only single fluid, its computational requirements are relatively low and hence, can be potentially explored as a reduced model for constant voltage driven plasma thruster simulation.

References

[1] LaPointe M R 1991 *AIAA Paper* **2341**

[2] Heiermann J and Auweter-Kurtz M 2004 *International Journal of Numerical Methods for Heat and Fluid Flow* **14** 559572

[3] Jaisankar S, Shivashankar K and R Rao S V 2007 *AIAA Paper* **3946**

[4] Spitzer Jr L and Ha¨rm R 1953 *Physical Review* **89** 977

[5] Jaisankar S and Sheshadri T S 2010 *AIAA Paper* **7022**

[6] Jaisankar S and Rao S V R 2007 *Journal of Computational Physics* **221** 577 599

[7] Jaisankar S and Sheshadri T S 2013 *Journal of Computational Physics* **233** 83 99

[8] Yoshikawa T, Kagaya Y, Yokoi Y and Tahara H *17th International Electric Propulsion Conference, Tokyo*, 1984 IEPC-84-58

[9] Jaisankar S, Sushma T V and Sheshadri T S 2012 *AIAA Paper* **0191**

[10] Schlichting H 1968 *Boundary-layer theory*, Sixth Edition (McGraw-Hill, New York)

[11] Wolff M, Kelly A and Jahn R *17th International Electric Propulsion Conference, Tokyo*, 1984 IEPC-84-32

[12] Jahn R G 1968 *Physics of Electric Propulsion* (McGraw-Hill, New York)

[13] Kubota K, Funaki I and Okuno Y 2009 *IEEE Transactions on Plasma Science* **37** 23902398

IOP Conf. Series: Materials Science and Engineering **370** (2018) 012029 doi:10.1088/1757-899X/370/1/012029

Multiobjective Simultaneous Topology, Shape and Sizing Optimization of Trusses Using Evolutionary Optimizers

Teerapol Techasen, Kittinan Wansasueb, Natee Panagant, Nantiwat Pholdee, Sujin Bureerat*

Sustainable and Infrastructure Research and Development Center, Department of Mechanical Engineering, Faculty of Engineering, Khon Kaen University, Khon Kaen, 40002, Thailand.

*Corresponding author. sujbur@kku.ac.th

Abstract. This paper presents design of two-dimensional (2D) trusses to achieve their simultaneously optimal topology, shape and size. The optimization problems are posed to search for structural topology, shape and sizing such that multiobjective functions consisting of mass and compliance are minimized while stresses and displacements are assigned as design constraints. The design approach is based on a ground structure approach meaning that a structure having all possible truss element connection is initiated. Design variables determine how to remove or maintain those elements and at the same time nodal positions are varied. Truss optimization problems are assigned whereas a number of multiobjective evolutionary algorithms (MOEAs) are implemented to solve the problems. Based on the hypervolume indicator, it is found that Hybridization of Real-Code Population-Based Incremental Learning and Differential Evolution (RPBIL-DE) is the best performer. This study gives the baseline results for implementing MOEAs for simultaneous topology, shape and sizing optimization of trusses.

1. Introduction

Presently, design optimization of engineering structures [1-4] is essential since lower structural weight as well as higher strength structure is expected. For truss design [5, 6], design engineers need to find the best design that is realizable and practical. Design of such structures is carried out in such a way that their topology, shape and sizes are optimized [7-13]. Design objectives usually include structural weight and cost. Nevertheless, it has been realized that the consequence of having a structure with low weight is reduction of structural strength. Apart from that, structural reliability is one important issue that should be taken into consideration. Therefore, there should be another design objective used to cope with this concern. Structural compliance can be used to somewhat measure structural strength and reliability. Thus, to have effective truss design, the optimization problem should have at least two objectives i.e .mass (or cost) and compliance.

Since there are two objective functions, the optimizer that can deal with such a problem efficiently is multiobjective evolutionary algorithms (MOEAs) [14]. Using such optimizers is advantageous since they are simple to understand, code, and use. The methods are derivative-free leading them to be capable of solving many types of optimization problems. Most importantly, they can explore the set of optimal solutions for multiobjective optimization which is usually called a Pareto optimal set within one optimization run. Over the past few decades, there have been numerous MOEAs proposed to solve a wide variety of engineering and other practical optimization problems. Some recent MOEAs are Multi-Objective Evolutionary Algorithm Based on Decomposition (MOEA/D) [15], Multiobjective Harmony Search (MOHS) [16], Unrestricted Population Size Evolutionary Multiobjective Optimisation Algorithm (UPS-EMOA) [17], Hybridsation of Real-Code Population-Based Incremental Learning and Differential Evolution (RPBIL-DE) [18], Differential Evolution for Multiobjective Optimization (DEMO) [19,20] and Multiobjective Particle Swarm Optimization (MPSO) [21]. The use of those MOEAs for tackling a design problem of trusses with simultaneous topological, shape and sizing design variables has not been widely investigated. As a consequence, their comparative performance for this type of structural design should be studied and reported.

This paper, therefore, presents the comparative studies of using a number of MOEAs for solving simultaneous topology shape and sizing design of two dimension trusses. The structure used as test problems are 15-bar truss, 39-bar truss and 45-bar truss . The problems are minimization of structural mass and compliance subject to stress and displacement constraints. MOEAs employed to explore Pareto fronts are MOEA/D, MOHS, UPS-EMOA, RPBIL-DE, DEMO and MPSO. The results obtained from the various optimizers are compared and discussed.

2. Optimization Problem Formulation

2.1 Test problems

In this paper, the two-objective design problem for trusses with design variables of simultaneous topology, shape and sizing can be shown as follows (1):

$$\textbf{Minimize: } f_1(\mathrm{x}), f_2(\mathrm{x}) \quad (1)$$
$$\textbf{subject to } \quad |\sigma_{max}| \leq \sigma_{al}$$
$$u_{max} \leq u_{al}$$

where $f_1(\mathrm{x})$ is structural mass of truss, $f_2(\mathrm{x})$ is structural compliance, σ_{max} is the maximum stress on truss elements, $\sigma_{al}$is the allowable stress relying on material used for buliding a truss, u_{max} is maximum nodal displacement of the structure due to applied forces and u_{al} is allowable nodal displacement. The compliance value is a measure of structural global stiffness which can be computed as a dot product of structural displacement and external force vectors.

Three truss optimization problems are selected**,** the first test problem is a simultaneous topology shape and sizing (TSS) optimization problem adapted from the 15-bar truss 2D structure problem from Ahrari, Atai and Deb in 2014 [8] and Rahami, Kaveh and Gholipour in 2008 [22]. Toplogical design variables are used to either remove or maintain those element while sizing variables are used to specify element sizes. Shape variables on the other hand will determine the nodal positions of some selected structural nodes**.** It should be noted that, during an optimization process, structural nodes of a groud structure in Figure 1 can be merged together.

The second test problem is the two-tier 39-bar truss problem, is TSS optimization by Ahrari, Atai and Deb in 2014 [8] and Deb and Gulati in 2001 [23] where the ground structure is shown in Figure **2.** Toplogical, shape and sizing design variables are similar to the 15-bar truss case but this case has a symmetric ground structure which all possible ground members are illustrated.

45-bar truss is the third test problem shown by Ahrari, Atai and Deb in 2014 [8] and Deb and Gulati in 2001 of [23]. The 45-bar truss ground structure is depicted in Figure 3. It should be noted that all truss dimensions are given in inch unless otherwise specified. All nodes of the structure are connected with the all-to-all scheme or all pair-wise interconnection.

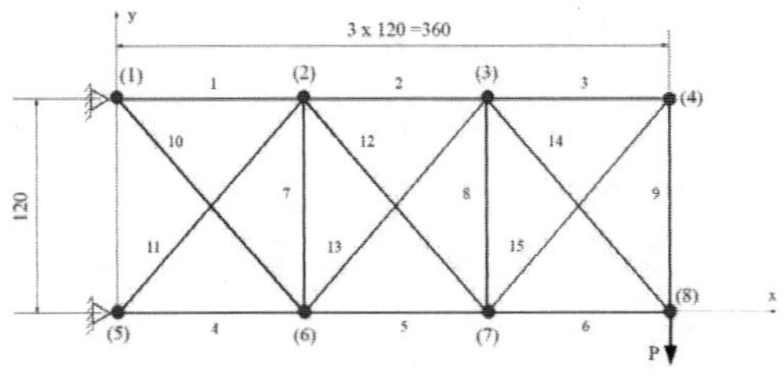

Figure 1. 15-bar truss ground structure.

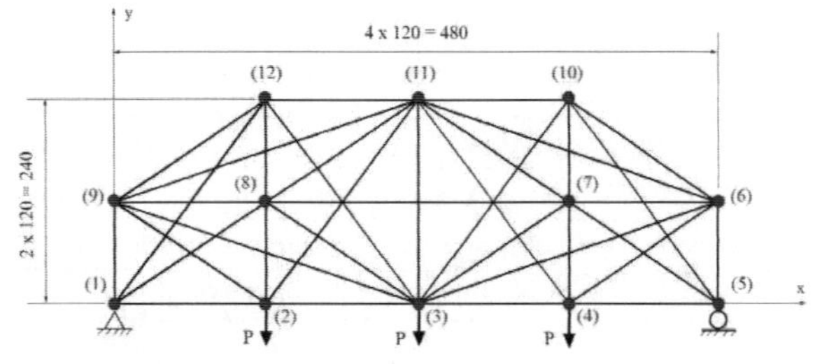

Figure 2. 39-bar truss ground structure.

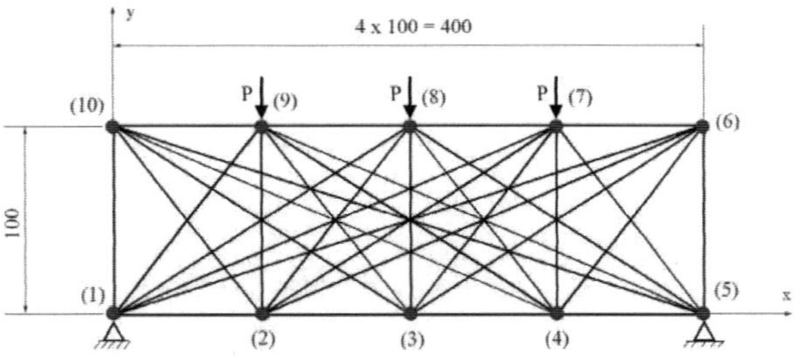

Figure 3. 45-bar truss ground structure.

2.2 Design variables and boundary conditions

Table 1. Simulation data of two-dimensions for 15-bar truss problem.

Design Variables	Shape(8)	$x_2{=}x_6$; $x_3{=}x_7$; y_2; y_3; y_4; y_6; y_7; y_8		
	Topology, Size(15)	A_i, i=1,2,…,15		
Constraints	Stress	$(\sigma_c)_i \leq$ 25 ksi; $(\sigma_t)_i \leq$ 25 ksi, i=1,2,…,15		
	Displacement	None		
Search Range	Shape Variables	100 in.$\leq x_2$, y_2, $y_3 \leq$140 in.; 220 in.$\leq x_3 \leq$260 in.; 50 in.$\leq y_4 \leq$90 in. ; 20 in.$\leq y_8 \leq$60 in.; -20 in.$\leq y_6$, $y_7 \leq$20 in.		
	Size Variables	A_i is set of **S**, i=1,2,…,15 **S** = {0.111, 0.141, 0.174, 0.22, 0.27, 0.287, 0.347, 0.44, 0.539, 0.954, 1.081 , 1.174, 1.333, 1.488, 1.764, 2.142, 2.697, 2.8, 3.131, 3.565, 3.813, 4.805 , 5.952, 6.572, 7.192, 8.525, 9.3, 10.85, 13.33, 14.29, 17.17, 19.18 } (in.^{2})		
Loading		Nodes	F_x (kips)	F_y (kips)
	Case I	8	-	-10
Mechanical Properties		Modulus of elasticity: E=10,000 ksi Density of the material is 0.1 lb/in.^{3}		

Table 1 shows the data required for optimization of the 15-bar truss. There are 23 design variables for all three variable types. Table 2 gives the details of design variables, structural loads and material properties for optimization of the two-tier 39-bar truss where the ground structure is shown in Figure 2. In the table x_i and y_i mean nodal positions of node i in x and y directions respectively. It should be noted that the structure is treated to be symmetric horizontally. There are 28 design variables for all three variable types. For the third case, details of design variables, structural loads and properties of material are presented in Table 3. There are 45 design variables for all two variable types. In this case considered toplogical and sizing design variables.

Table 2. Simulation data of two-dimensions for 39-bar truss problem.

Design Variables	Shape(7)	Δy_{11}; Δx_6=-Δx_9; Δy_6=Δy_9; Δx_7=-Δx_8; Δy_7=Δy_8; Δx_{10}=-Δx_{12}; Δy_{10}=Δy_{12}		
	Topology, Size(21)	Ai, i=1,2,…,21 (Geometry is symmetric in a-a axis)		
Constraints	Stress	$(\sigma_c)_i \leq$ 20 ksi; $(\sigma_t)_i \leq$ 20 ksi, i=1,2,…,39		
	Displacement	$u_j \leq$ 2 in., j=1,2,…,3*(n_{node})		
Search Range	Shape Variables	Horizontal coordinates may many very within ± 120 in. of their initial value.		
		Vertical coordinates may many very within ± 120 in. of their initial value.		
	Size Variables	Ai is set of [0.05, 2.25] (in.^{2}) , i=1,2,…,39		
Loading		Nodes	F_x (kips)	F_y (kips)
	Case I	2	-	-20
		3	-	-20
		4	-	-20
Mechanical Properties		Modulus of elasticity: E=10,000 ksi		
		Density of the material is 0.1 lb/in.^{3}		

Table 3. Simulation data of two-dimensions for 45-bar truss problem.

Design Variables	Topology, Size(15)	A_i, i=1,2,…,45		
Constraints	Stress	$(\sigma_c)_i \leq$ 25 ksi; $(\sigma_t)_i \leq$ 25 ksi, i=1,2,…,45		
	Displacement	$u_j \leq$ 2 in., j=1,2,…,3*(n_{node})		
Search Range	Size Variables	A_i is set of [0.09,1] (in.), i=1,2,…,45		
Loading		Nodes	F_x (kips)	F_y (kips)
	Case I	7	-	-10
		8	-	-10
		9	-	-10
Mechanical Properties		Modulus of elasticity: E=10,000 ksi		
		Density of the material is 0.1 lb/in.^{3}		

2.3 Optimization Parameter Settings

The optimization parameter settings of MOEAs used to solve the test problems are assigned as:

- MOEA/D using number of neighboring weight vectors are 6 which is real codes, mutation probabilities and crossover being 0.1 and 1.0 respectively [15].
- UPS-EMOA using crossover probability, scaling factor for Differential Evolution operator, probability of choosing elements from an offspring in crossover, minimum population size and number of solution to be used as parents are set as 0.7, 0.8, 0.5, 10, and 25 respectively [16].
- MOHS using harmony memory considering rate being 0.5, min pitch adjustment rate being 0.2, max pitch adjustment rate being 2, min bandwidth rate being 0.45 and max bandwidth rate being 0.9 [17].
- RPBIL-DE using N_T= 40 which real codes where each probability tray produces three design solutions. Crossover probability is set as 0.7, scaling factor is set as 0.8 and probability is set as 0.5 which is choosing elements from an offspring in crossover for Differential Evolution operators [18].
- DEMO using crossover probability, scaling factor for Differential Evolution (DE) operator, probability of choosing elements from an offspring in crossover are set as 0.7, 0.8 and 0.5 respectively [19,20].

- MPSO using the starting inertia weight is set as 0.5, ending inertia weight is set as 0.01, cognitive learning factor is set as 0.5 and social learning factor being 0.5. The best solution is randomly selected from Pareto archive [21].

Each optimizer is employed to solve each optimization problem 30 runs .The population size is set to be 100 while the total number of iterations is set to be 250 for all optimizers .

3. Results and discussions.

For multiobjective evolutionary algorithms (MOEAs), the non-dominated archive size is set as 100 which are equal to the population size. The hypervolume indicator will be used to measure the performance of MOEAs .The reference points used for computing the hypervolume for all test problems are given in Table **4.**

Table 4. Reference points (Wmax) for hypervolume calculation.

	15-bar truss	39-bar truss	45-bar truss
f_1(x) (lb)	3899.100	1960.900	906.254
f_2(x) (kip.inch)	51.256	76.909	25.993

The results of optimizing the 15-bar truss obtained from the various optimizers are compared in Table 5. According to the mean of the hypervolume from 30 runs, for the 15-bar truss, the best method is RPBIL-DE while the second best method is UPS-EMOA. The worst is MOEA/D. The best runs of all optimizers are found and the corresponding Pareto fronts are plotted as shown in Figure 4. From Table 5 the front that gives the highest hypervolume is from RPBIL-DE. This implies that RPBIL-DE gives a Pareto front with better front extension and front advancement more than other optimizers. Figure 5 shows some selected optimum solutions of the 15-bar truss problem obtained from the best Pareto front of RPBIL-DE. It can be seen that various structures can be obtained from running the optimizer once.

The same conclusion can be said in cases of the 39-bar truss problem as given in Table 6. The best runs of all optimizers are found and the corresponding Pareto fronts are plotted as shown in Figure 6.

RPBIL-DE is the best optimizer based on the hypervolume value. For the 39-bar truss, the best method is RPBIL-DE while the second best is UPS-EMOA. The worst is MPSO. Figure 7 shows some selected optimum solutions of the 39-bar truss problem obtained from the best run of RPBIL-DE.

Likewise, in case of 45-bar truss problem, similar conclusion can be said as given in Table 7. The best method still is RPBIL-DE while the second best still is UPS-EMOA and the worst is MPSO. The best runs of all optimizers are found and the corresponding Pareto fronts are plotted as shown in Figure 8. Figure 9 shows some selected optimum solutions of the 45-bar truss problem obtained from the best run of RPBIL-DE.

Table 5. Hypervolume of 15-bar truss.

Optimizers	Mean	(±)Std	Minimum	Maximum
MOEA/D	128317.305	4201.075	119059.589	137192.826
MOHS	160588.718	2379.642	156592.515	165291.290
UPS-EMOA	170529.799	4514.161	159594.869	178442.317
RPBIL-DE	178742.758	1166.760	176111.685	180326.015
DEMO	168372.467	3828.054	157347.706	180176.778
MPSO	130163.338	5136.202	119310.029	140318.421

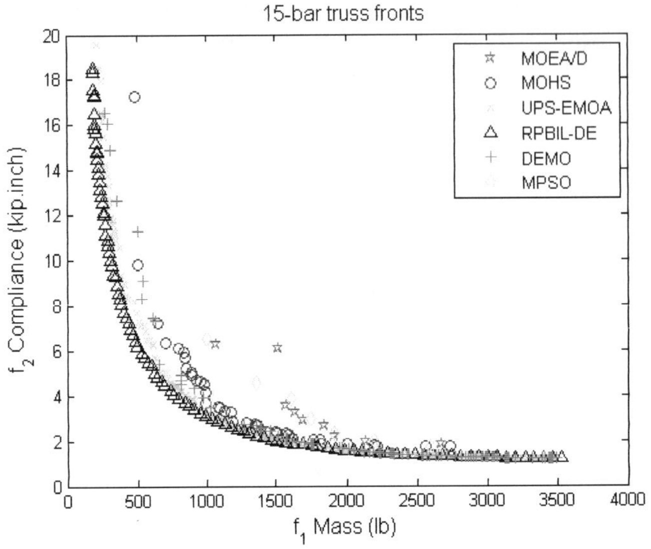

Figure 4. 15-bar truss Pareto fronts from MOEAs.

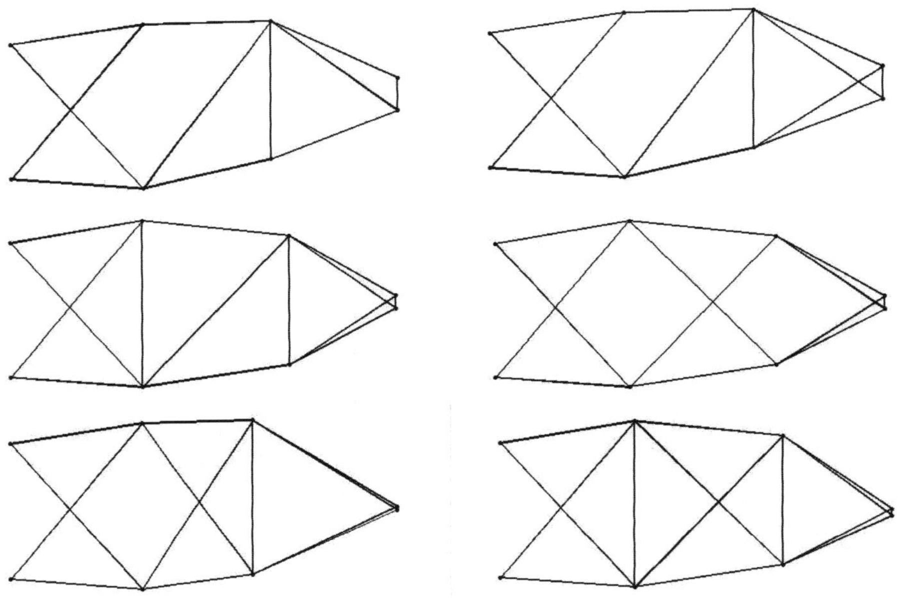

Figure 5. Some design solutions of the 15-bar truss problem from RPBIL-DE.

Table 6. Hypervolume of 39-bar truss.

Optimizers	Mean	(±)Std	Minimum	Maximum
MOEA/D	76300.680	1703.256	72775.722	80166.593
MOHS	81153.007	941.083	79469.077	82961.035
UPS-EMOA	95826.720	758.176	93729.321	97111.048
RPBIL-DE	97079.814	303.243	96265.062	97593.684
DEMO	93629.169	1987.434	88554.128	99218.717
MPSO	75345.682	1809.990	71945.227	78828.173

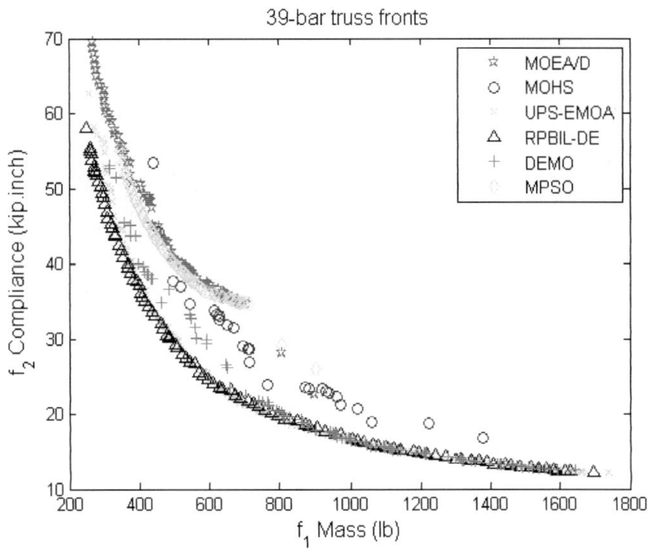

Figure 6. 39-bar truss Pareto fronts from MOEAs.

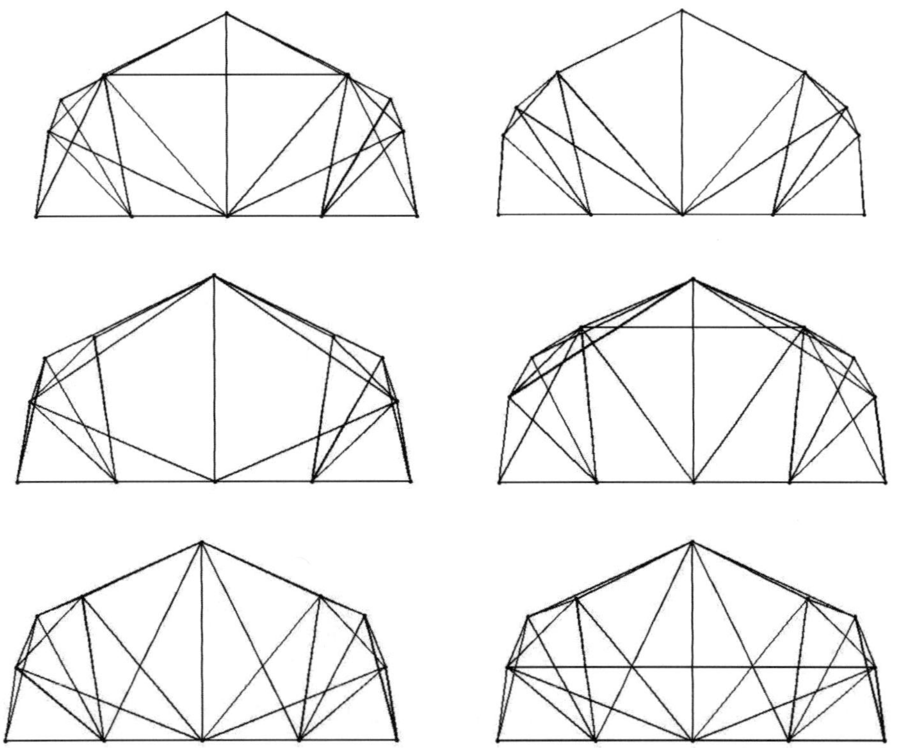

Figure 7. Some design solutions of the 39-bar truss problem from RPBIL-DE.

Table 7. Hypervolume of 45-bar truss.

Optimizers	Mean	(±)Std	Minimum	Maximum
MOEA/D	12243.775	241.263	11694.991	12701.859
MOHS	11622.509	146.357	11268.308	11883.705
UPS-EMOA	14322.093	236.615	13492.802	14626.433
RPBIL-DE	14461.992	58.705	14271.463	14563.805
DEMO	13531.746	560.496	12069.836	14973.128
MPSO	12170.609	391.988	11552.289	13045.755

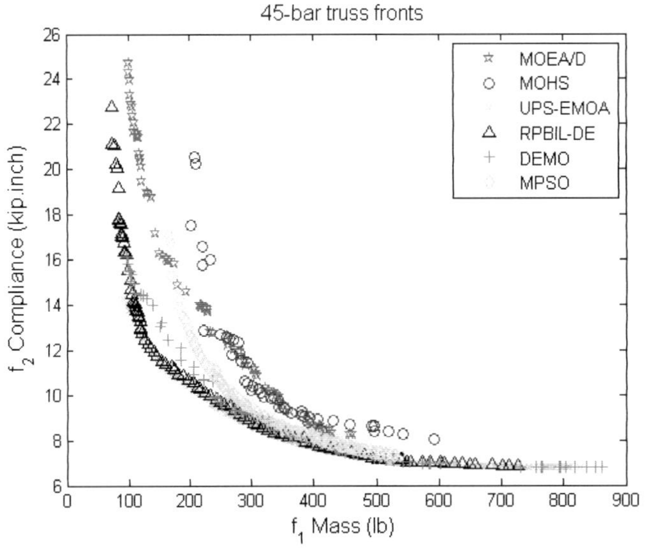

Figure 8. 45-bar truss Pareto fronts from MOEAs.

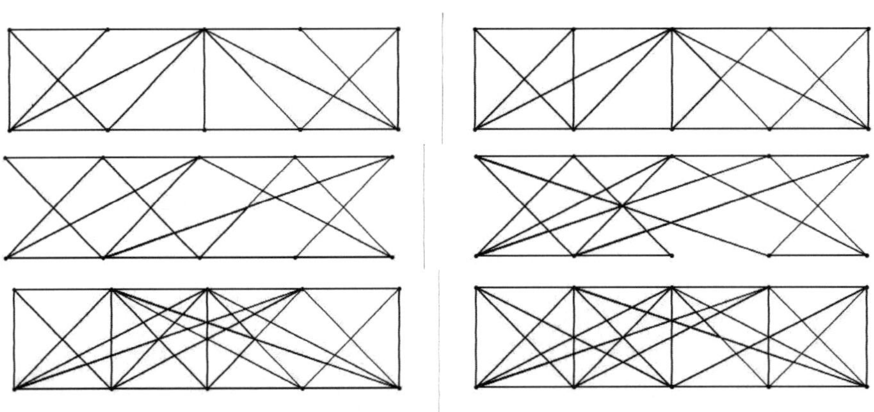

Figure 9. Some design solutions of the 45-bar truss problem from RPBIL-DE.

According to the mean value and standard deviation (Std) of the hypervolumes from 30 runs for all design problems, the best method is RPBIL-DE as shown in Table 8. The best parameter settings for this study are crossover probability as 0.7, a scaling factor as 0.8 and *CR* as 0.5, which are the same as those used in [18].

Table 8. The best MOEA based on the hypervolume indicator.

Trusses	Optimizers	Mean	(±)Std	Minimum	Maximum
15-bar	RPBIL-DE	178742.758	1166.760	176111.685	180326.015
39-bar	RPBIL-DE	97079.814	303.243	96265.062	97593.684
45-bar	RPBIL-DE	14461.992	58.705	14271.463	14563.805

4. Conclusions

A new design problem for truss TSS optimization is proposed. The optimization problem can optimize both structural mass (or cost) and reliability (compliance) .The results in this study are set as the initial performance baseline for future development of multiobjective optimizers to be used with these design problems. Based on the hypervolume indicator and the best Pareto front obtained, it can be said that RPBIL-DE is the best performer in terms of Pareto front advancement and extension. Using the hypervolume alone, however, may not be sufficient for measuring the performance of MOEAs, as a result, other indicators should be implemented.

Acknowledgement

This work was supported by the Graduate Engineering Camp Fund, Faculty of Engineering, Khon Kaen University, Thailand and the Thailand Research Fund (BRG5880014), Thailand.

References

[1] Bureerat S 2013 *Optimization for Mechanical Engineering* (Khon Kaen: Khon Kaen University Press)

[2] Srisomporn S and Bureerat S 2008 Geometrical design of plate-fin heat sinks using hybridization of MOEA and RSM *IEEE Transaction on Components and Packaging Technologies* **31** 351-9

[3] Kunakote T, Sombudha S and Bureerat S 2009 *Symp. Conf. on ME-NETT23 (Chiang Mai)* vol 23 (Khon Kaen: ME-NETT) p 373

[4] Zitzler E and Thiele L 1999 Multiobjective Evolutionary Algorithms :A Comparative Case Study and the Strength Pareto Approach *IEEE Tran.on Evolutionary Computation* **3** 257-71

[5] Bureerat S 2013 *Truss Finite Elements Analysis and Design* (Khon Kaen: Khon Kaen University Press)

[6] Chadchotsak P 2005 Analysis of Roof Truss using Timoshenko Beam Model *KKU Engineering Journal* **32** 829-40

[7] Noilublao C and Bureerat S 2009 Simultaneous Topology Shape and Sizing Optimisation of Skeletal Structures Using Multiobjective Evolutionary Algorithms *Evolutionary Computation* ISBN 978-953-307-008-7 pp 487–580

[8] Ahrari A, Atai A A and Deb K 2014 Simultaneous Topology Shape and Size Optimization of Truss Structures by Fully Stressed Design Based on Evolution Strategy *Engineering Optimization* 1-22

[9] Zhou M, Pagaldipti N, Thomas H L and Shyy Y K 2004 An integrated approach to topology, sizing, and shape optimization **26** 308-17

[10] Noilublao N and Bureerat S 2011 Simultaneous topology, shape and sizing optimisation of a three-dimensional slender truss tower using multiobjective evolutionary algorithms *Computers and Structures* **89** 2531-8

[11] Greiner D and Hajela P 2012 Truss topology optimization for mass and reliability considerations—co-evolutionary multiobjective formulations *Struct Multidisc Optim* **45** 589-613

[12] Pholdee N and Bureerat S 2012 Performance enhancement of multiobjective evolutionary optimisers for truss design using an approximate gradient *Computers and Structures* **106**

115-124

[13] Noiluplao N and Bureerat S 2013 Simultaneous Topology, Shape and Sizing Optimisation of Plane Trusses with Adaptive Ground Finite Element Using MOEAs *Hindawi Publishing Corporation Mathematical Problems in Engineering* 1-9

[14] Zhou A, Qu B Y, Li H Zhao S Z, Suganthan P N and Zhang Q 2011 Multiobjective evolutionary algorithms: A survey of the state of the art *Swarm and Evolutionary Computation* **1** 32-49

[15] Ma X, Liu F, Qi Y, Li L, Jiao L, Liu M and Wu J 2014 MOEA/D with Baldwinian learning inspired by the regularity property of continuous multiobjective problem *Neurocomputing* **145** 336-52

[16] Mohamed F, Santawy E and Ahmed A N 2012 A Multi-Objective Chaotic Harmony Search Technique for Structural Optimization *Computing Science* **1** 9-12

[17] Aittokoski T and Miettinen K 2008 *Int. Conf. on Engineering Optimization (Rio de Janeiro)* (Rio de Janeiro: EngOpt)

[18] Pholdee N and Bureerat S 2013 Hybridisation of real-code population-based incremental learning and differential evolution for multiobjective design of trusses *Information Sciences* **223** 136-52

[19] Das S and Suganthan P N 2011 Differential evolution :a survey of the state-of-the-art *IEEE Transactions on Evolutionary Computation* **15** 4-31

[20] Mlakar M, Petelin D, Tušar T and Filipi B 2015 GP-DEMO :Differential Evolution for Multiobjective Optimization based on Gaussian Processmodels *Operational Research* **243** 347-61

[21] Sierra M R and Coello C A C 2006 Multi-Objective Particle Swarm Optimizers: A Survey of the State-of-the-Art *Computational Intelligence Research* **2** 287-308

[22] Rahami H, Kaveh A and Gholipour Y 2008 Sizing, geometry and topology optimization of trusses via force method and genetic algorithm *Engineering Structures* **30** 2360–9

[23] Deb K and Gulati S 2001 Design of truss-structures for minimum weight using genetic Algorithms *Finite Elements in Analysis and Design* **37** 447-65

IOP Conf. Series: Materials Science and Engineering **370** (2018) 012030 doi:10.1088/1757-899X/370/1/012030

Preliminary Study on GF/Carbon/Epoxy Composite Permeability in Designing Close Compartment Processing

A.M Ya'acob[1], D.A Razali[2], U.A Anwar[3], A.H Radhi[4], A.A Ishak[5], M.Minhat[6], Mohd Aris K.D[7], M.K.Johari[8] and T.Casey[9]

[1,2,3,4,5,6,7,8] Malaysian Institute of Aviation Technology, Universiti Kuala Lumpur, Selangor, Malaysia.
[9]TenAsia Corporation Sdn Bhd, Lot 10701, Jalan Permata 1, Arab Malaysian Industrial Park, 71800 Nilai, Negeri Sembilan, Malaysia

Corresponding author: abdulmalek@unikl.edu.my

Abstract. This project involves discovering how the permeability effect inside a close compartment in processing. After the appropriate pressure range was found, the close compartment was designed by studying the relationship between pressure output and the flow rate. A variety of pressure ranges have been used in this test to determine the effective pressure range that can be applied to the manufacturing process. Based on the results, the suitable pressure ranges were found between 55 psi to 75 psi. These pressures have been chosen based on the area covered on the product surfaces and time taken to penetrate the proposed area. The relationship between pressure and flow rate have been found to be directly proportional until 75 psi only. In conclusion, 70 psi for the proposed design of close compartment mould is suitable to be used to fulfill the required area of 120 mm square within 90 seconds.

1. Introduction

This was an industrial-related project as an initial study for pull-winding process which would be employed to manufacture fiberglass or carbon, composite poles or tubing. On-site observation showed high inconsistency quality of composite part, rejection rate of composite products and excessive mixed resin waste. One of the suspected reasons being was the processing environmental condition such as high humidity. Hence, a close compartment processing concept was designed to solve this industry problem – an automated resin infusion-fiber drying system (RIFiDS) was proposed. One of the goals of this study was to create a new technology that can be integrated to the current pull-winding process line. Nevertheless, the most significant outcome of this preliminary study is that such proposed close compartment system has never been utilized commercially in the current pull-winding process and because of that, the proposed system is a novel technology.

The causes to these industry problems are schematically illustrated in Fig.1 below:

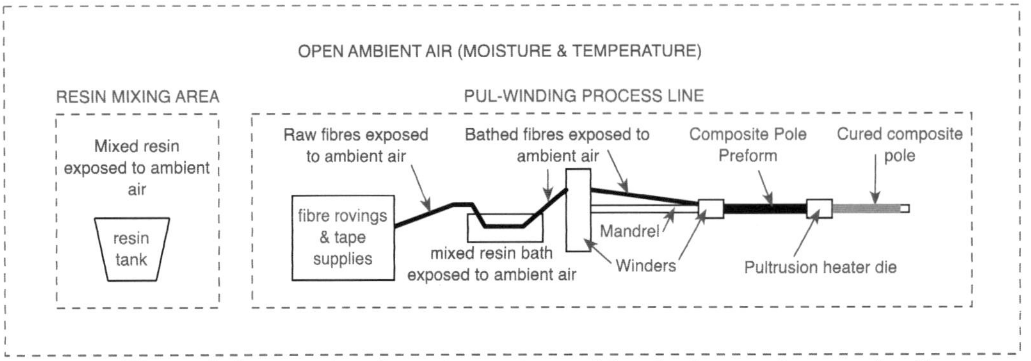

Figure 1. Suspected moisture ingression problem in the actual processing line

The resin flow in a laminate is one of the major contributors to void formation, transportation, and growth (Kardos, Duduković, & Dave, 1986). During processing, because of pressure application and decreased viscosity of resin, excess resin between fibers in tows and adjacent plies move, causing voids to swipe out of the laminate (Shan, Hongqiang, Zhengping, & Hao, 2014). The resin flow mainly occurs due to applied pressure gradients and viscosity changes. The resin flow rate is controlled by the resin and surrounding pressure of the laminate, resin viscosity, specific permeability, porosity of the fiber bed and laminate dimensions (Sánchez, Artero-Guerrero, Varas, & López-Puente, 2016). Resin flow in laminate is quasi-static and transient in nature (Sánchez, Artero-Guerrero, Varas, & López-Puente, 2016) and (Schaefer, Werner, & Daniel, 2014). The initial viscosity of resin in liquid stage is influenced by size of molecular extension (α) and temperature (T) hence as resin approaches the gelling point, viscosity rises rapidly and resin flow ceases (Van Velthem, et al., 2015).

There are two major mathematical models for resin flow in laminate. The first one is based on the theories of soil mechanics, considering the fiber bed of the laminate as a porous solid. The other follows lubrication theory (Van Velthem, et al., 2015). Models are based on soil science, which assumes a pressure gradient across the vertical and horizontal direction of the laminate. While (Grigoriev, Krasnovskii, & Kazakov, 2014) assumed a linear pressure gradient, (Sánchez, Artero-Guerrero, Varas, & López-Puente, 2016) and (Ghnatios, Chinesta, & Binetruy, 2013) considered nonlinear pressure gradients. Resin flow in a laminated composite is laminar and low Reynolds number flow (Sánchez, Artero-Guerrero, Varas, & López-Puente, 2016). The porous medium combined with a low Reynolds number ($R_e \leq 10$) flow can be characterized by Darcy's Law (Partha, Vimal, & Mishra, 2016). As a porous medium, flow inside the laminate is described by Darcy's law and requires the knowledge of fiber network permeability, viscosity, and pressure gradient of the resin during cure. According to Darcy's law, resin velocity (V) inside laminate is as per equation below:

$$V = \frac{k}{u}\frac{\Delta P}{L} \tag{1}$$

where μ is the viscosity, ΔP is pressure gradient of the resin, k is the fiber bed permeability and L is the length of fiber bed. Although Darcy's law is suitable for one dimensional flow, this equation can be applied to three-dimensional flow as well (Sánchez, Artero-Guerrero, Varas, & López-Puente, 2016). (Grigoriev, Krasnovskii, & Kazakov, 2014) and (Sánchez, Artero-Guerrero, Varas, & López-Puente, 2016) use Darcy's law to model resin flow in laminates during cure. The fiber bed permeability (k) is analogues to the conductivity of the porous media. The more permeable the fiber bed, the less resistance resin faces to flow. According to soil science, permeability is the specific discharge per unit hydraulic gradient (Sánchez, Artero-Guerrero, Varas, & López-Puente, 2016). Permeability of the fiber bed may change with time due to an applied external load that changes the structures and texture of the porous bed. As the fiber volume fraction changes during compaction, therefore, fiber bed permeability is a function of fiber volume fraction. (Sánchez, Artero-Guerrero, Varas, & López-

Puente, 2016) described the permeability of laminate fiber bed using Carman-Kozeny equation as follows:

$$k = \frac{r_f}{4K}\frac{(1-V_f)^3}{V_f{}^2} \quad (2)$$

Where r_f and K are the fiber radius and Kozeny constant, respectively. Anisotropy of the fiber bed permeability is determined by the Kozeny constant in three principle directions. The principle directions of the laminate are shown in Fig. 2. Since fiber bed is assumed to be transversely isotropic, permeability in fiber (1,x) and transverse (2,y) directions are different. (Ghnatios, Chinesta, & Binetruy, 2013) found axial and transverse Kozeny constant to be K_1=0.7 and K_2=17.9 for different fiber volume fractions. (Rouhi, Wysocki, & Larsson, 2015) reported a Kozeny constant for transverse permeability of K_2=11 for $0.5< V_f <0.75$ and observed substantial decrease in K_2 at higher fiber volume fractions from their permeability experiment. They obtained axial permeability K_1=0.35 and 0.68 for silicone oil and water respectively, indicating the effect of fluid properties on fiber bed permeability. Since laminate contains layers of varying permeability, layers having the highest permeability govern the resin flow (Schaefer, Werner, & Daniel, 2014). Combining the void ratio and effective stress relationship with Carman-Kozeny equation provides a relationship of permeability with the effective stress. Fig. 3 showing fiber bed permeability decreases with the increase in the effective loading.

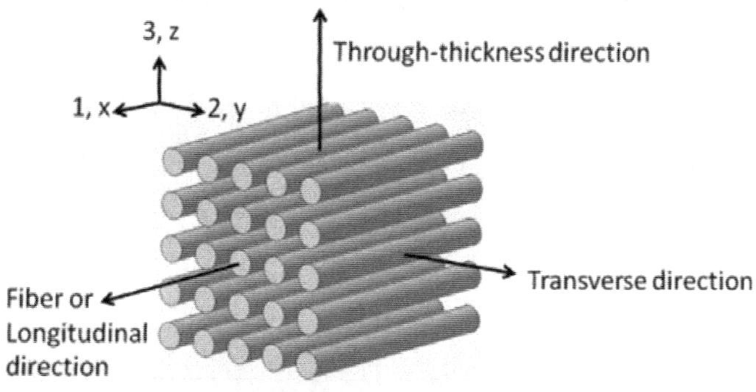

Figure 2. Unidirectional fiber bundle with direction convention. (Ghnatios, Chinesta, & Binetruy, 2013)

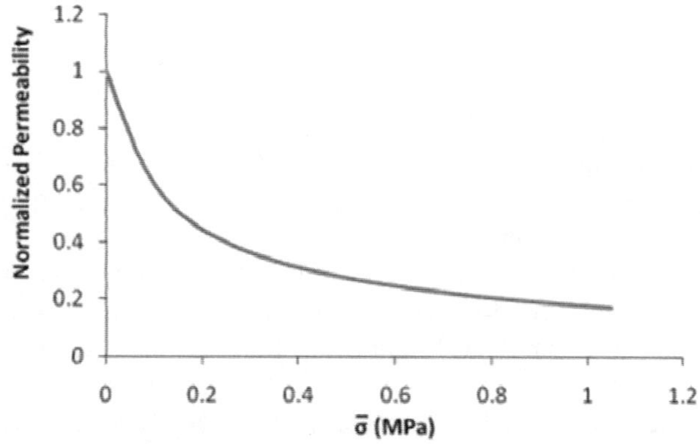

Figure 3. Permeability and effective stress relationship. (Sánchez, Artero-Guerrero, Varas, & López-Puente, 2016)

Many factors can affect the liquid drop size such as liquid properties, nozzle capacity, spray pressure and spray angle. Nozzles should be placed as close as possible to the beginning of the transfer point as possible (Prakash & Jhawar, 2014). The force of the moving preform in this study helps the resin in penetrating the preform as it moves through the transfer point (De La Haza, Andrey, & Samokrutov, 2013). Fig.4 showing the flat fan type is the best type for resin penetration and sharp uniform pattern to the preform. The liquid is shaped into a fan-shaped sheet of fluid. This can be comprised of droplets or a more or less coherent sheet of water like a waterfall. Flat fans can have a spray angle of between 15 and 145 degrees depending on the nozzle design. Another study also showed the flat fan pattern that creates a flat sheet of spray, with medium velocity and large drop size. It is excellent for covering an area when there is a relative motion between the nozzle and target such as pull-winding (Tseng, Raudensky, & Lee, 2016). This preliminary study utilizes the nozzle to break the liquid into droplets and form the spray pattern (Lujaji, Boateng, Schaffer, Mtui, & Mkilaha, 2016) Another study shows that the standard flat-fan nozzle normally operates between 30 psi and 60 psi – with an ideal range of 30 to 40 psi.

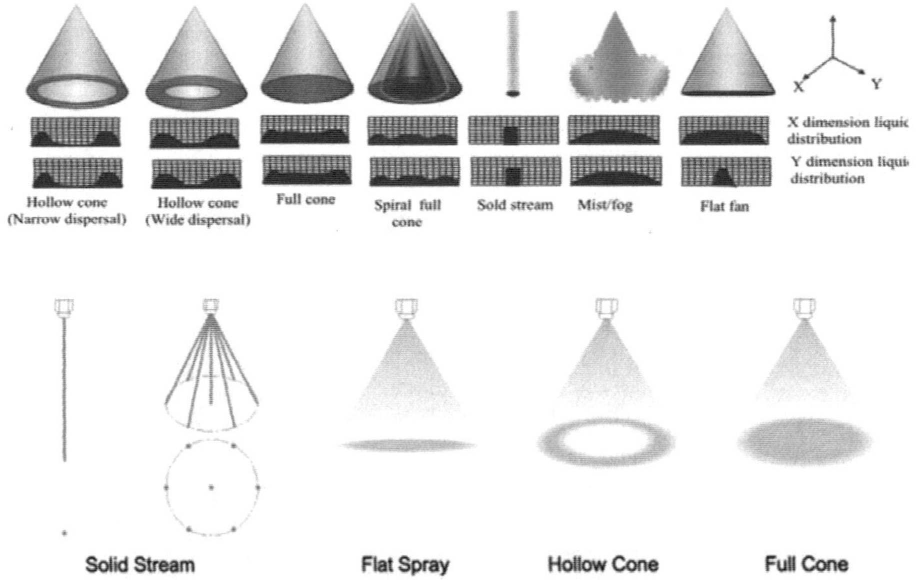

Figure 4. Spray pattern (BETE Fog Nozzle, Inc., 2017)

2. Methodology

Basically, the work requires the resin to be automatically metered and mixed so that consistent batch of resin quality would be obtained and then, the mixed resin was infused or injected or compressed directly to the composite preforms without excessively exposing the preform to the ambient air. This will eliminate moisture absorption problem that was causing inconsistent quality of fabricated composite part. Furthermore, the mixed resin will be infused as per required only; hence, it eliminates the excessive mixed resin waste. The raw material of glass fiber carbon fiber has been provided by TenAsia Corp. Sdn Bhd. The stacking sequence of fibers and epoxy were also based on recommendation from TenAsia Corp. Sdn Bhd. Both, the raw material identification and stacking sequence was considered as confidential by TenAsia Corp. Sdn Bhd. The sample dimension in this study was 12 inch by 12 inch sample. Initially, the fiber was laid flat on the glass, and then a metal plat mold was put on the fiber surface to avoid the preform from shifting. The spray process takes place in the mock-up close compartment. The spray gun was used to spray and the nozzle position was placed 3 inch high from the fiber surface. The container was design to prevent splashing of excessive resin.

The pressure used was 50, 55, 60, 65, 70, 75 and 80 psi respectively. Each of different pressure testing was conducted for 2 minutes. Metal plate was used to position the nozzle at 3 inches away from the preform. Hence, in this study a fan spray pattern was introduced by adjusting the nozzle. This spray pattern was been recorded every 30 second for 2 minutes. Fig. 5 showing steps in samples preparation.

(a) (b) (c)

Figure 5. (a) Preform preparation, (b) Spray pattern adjustment and (c) Permeability behaviour of the preform sample.

The close compartment was designed initially by using Perspex material (temporary mold) then later will be fabricated using metal. The mold has been designed in order to avoid preform exposure to the environment. The close compartment was also designed to reduce waste and emission for a cleaner and healthier environment. In this study an injection spray resin and compression resin concept was also introduce in the close compartment. The mold was designed to ensure resin penetration to the sample. Hence, this design was chosen to avoid environmental exposure to the fabrication. The internal design of the compartment was design in a circular form to avoid resin residual and also for ease of mold maintenance by the operator. In order to avoid the human error during the assembly of the mold, the guide pin was designed as a reference. In this design, a drain outlet was located at the bottom of the mold in order to remove the excessive resin during the resin spray stages. The excessive resin will flow down to the trap chamber to be reused back to the system. Expected drawback of this close compartment design is the possibilities of having resin clog at the drainage outlet and also at the nozzles area will be expected in long run. Continuously maintenance must be done to this close compartment to avoid the clog. The nozzle was designed at less than 0.4 inch as such to produce the flat fan pattern spraying.

3. Result and Discussion

Preliminary findings showing for every 120 second at 50 psi the resin spread will cover 110 cm^2 area, 55 psi will cover 137.5 cm^2 area, 60 psi will cover 146.4 cm^2, 65 psi will cover 162.5 cm^2 area, 70 psi will cover 184.6 cm^2 area, 75 psi will cover 129.8 cm^2 area and at 80 psi the resin spread will cover 103 cm^2 area. The best range of pressure was between 60 to 70 psi. In this study, 75 and 80 psi are not used as the resin will change into mist thus preventing it from being absorbed by the preform. Findings are shown as per Fig. 6 below.

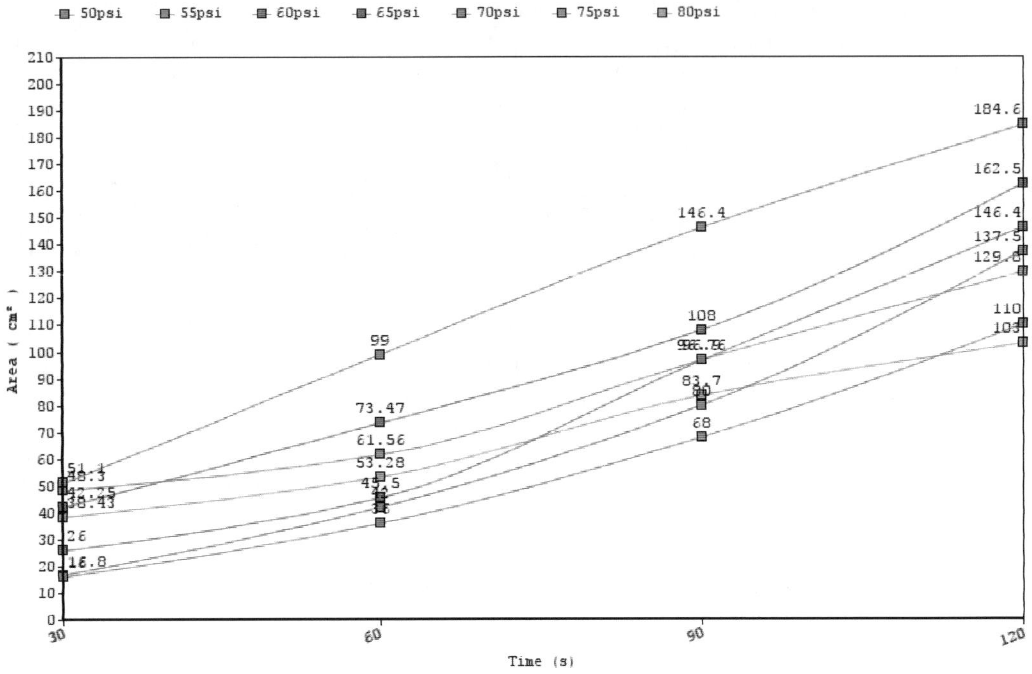

Figure 6. Relationship between pressure and area

Fig.7 show preliminary findings specifically for the first 30 second of the composite fabrication. At 50 psi the resin coverage was 16 cm^2, 55 psi was 16.2 cm^2, 60 psi was 26 cm^2, 65 psi was 42.25 cm^2, 70 psi was 51.1 cm^2, 75 psi was 48.3 cm^2 and at 80 psi the resin coverage was 38.43 cm^2. The pressure ranges suggested based on preliminary findings was 60 to 70 psi. The drop however happens between 75 to 80 psi because resin becomes mist.

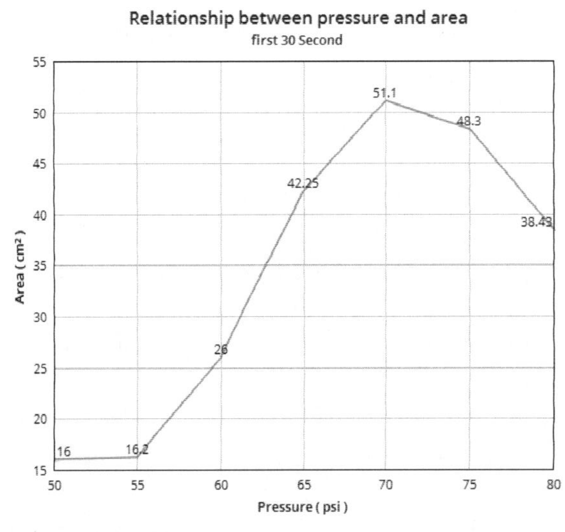

Figure 7. Pressure output observation at every 30 seconds

Overall finding was shown in Fig.8. Based on the preliminary results for duration of 120 seconds, the resin coverage at 50 psi was 118 cm^2, increasing the pressure at 55 psi resulted in resin coverage of 137.5 cm^2, at 60 psi was 146.4 cm^2, 65 psi was 162.5 cm^2, at 70 psi was 184.6 cm^2, at 75 psi was 121.8 cm^2 and at 80 psi was 103 cm^2. The suggested pressure range for 120 seconds was 60 to 70 psi. However, at both suggested pressure ranges the coverage area was already exceeding the preform product requirement.

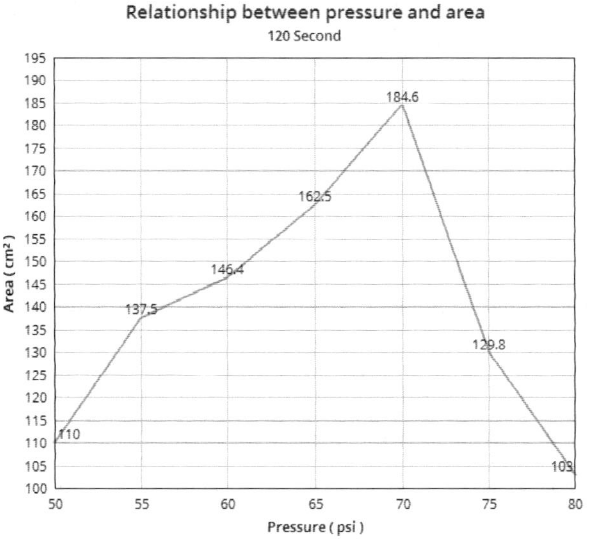

Figure 8. Preliminary findings for 120 seconds processing time

Fig.9 showing the close compartment concept, circular shape was designed for ease of cleaning and maintenance after used. This product has different type of diameter and the mold was designed separately into three parts for easy manageable change the mold size for specific product diameter. Circular shape was also designed to prevent cured resin retained after the processing completed in Fig.10. The size of circular shape was designed with small gap to ensure pressurize system will force-push in the resin into the fabric efficiently compared with large gap.

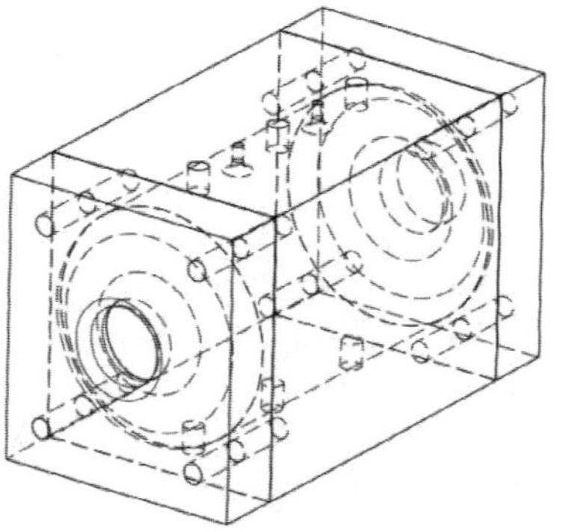

Figure 9. Close compartment with injected spray resin system for pull-winding process.

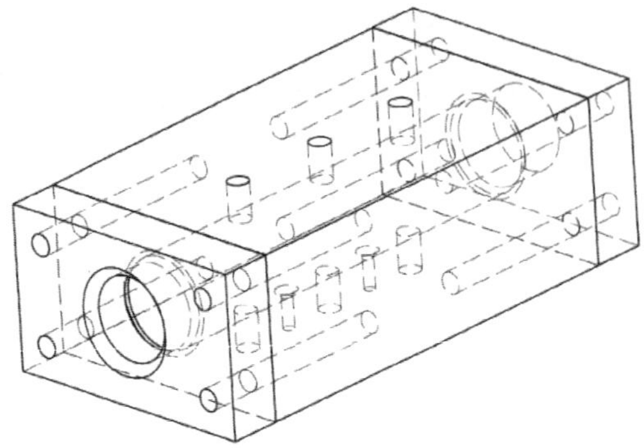

Figure 10. Close compartment with pressurize resin system for pull-winding process.

4. Conclusion

In conclusion, the appropriate pressure range in this preliminary study was identified between 55 psi to 75 psi. The design of the close compartment mold was to avoid exposure of the preform to the environment. Future investigation will involve the integration of close compartment e.g. Perspex mold, with fully operational pull-winding production line. In addition, the next stage of investigation will involve findings and observation comparison between experimental data to the actual condition of the manufacturing pull-winding system.

Acknowledgements

Appreciation and utmost gratitude to Mr Casey Teh of TenAsia Cooperation Sdn Bhd for giving consultation and material support for this project. This project has been supported financially by Ministry of Higher Education under the Demand–Driven Innovation Project Public-Private Research Network (PPRN) grant no. KPT.600-6/2/18 Jld25(1).

References

[1] BETE Fog Nozzle, Inc. (2017). BETE. Retrieved September 1, 2017, from http://www.bete.com/spraypatterns/fan_nf4015.html

[2] De La Haza, A., Andrey, A., & Samokrutov, P. (2013). Assessment Of Concrete Structures Using The Mira and Eyeconultrasonic Shear Wave Devices and The SAFT-C Image Reconstruction Technique. *Construction and Building Materials*, 1276-1291.

[3] Ghnatios, C., Chinesta, F., & Binetruy, C. (2013). *International Journal of Material Forming*. 3D Modeling of Squeeze Flows Occurring in Composite Laminates, 78-83.

[4] Grigoriev, S., Krasnovskii, A., & Kazakov, I. (2014). The Impact of Pre-Heating on Pressure Behavior in Tapered Cylindrical Die in Pultrusion of Large-Sized Composite Rods. *Materials Science and Technology*, 120-127.

[5] Kardos, J., Duduković, M., & Dave, R. (1986). Void Growth and Resin Transport During Processing of Thermosetting Matrix Composites. *Epoxy Resins and Composites IV*, 101-123.

[6] Lujaji, F., Boateng, A., Schaffer, M., Mtui, P., & Mkilaha, I. (2016). Spray Atomization Of Bio-Oil/Ethanol Blends With Externally Mixed Nozzles. *Experimental Thermal and Fluid Science*, 146-153.

[7] Partha, K., Vimal, K., & Mishra, I. (2016). Experimental and Numerical Investigation Of Fluid Flow Hydrodynamics In Porous Media: Characterization Of Pre-Darcy, Darcy and Non-Darcy Flow Regimes. *Powder Technology*, 278-291.

[8] Prakash, A., & Jhawar, A. (2014). Bubble Column With Internals: Effects On Hydrodynamics and Local Heat Transfer. *Chemical Engineering Research and Design*, 25-33.

[9] Rouhi, M., Wysocki, M., & Larsson, R. (2015). Experimental Assessment Of Dual-Scale Resin Flow-Deformation In Composites Processing. *Composites Part A: Applied Science and Manufacturing*, 215-223.

[10] Sánchez, J., Artero-Guerrero, J., Varas, D., & López-Puente, J. (2016). Experimental Analysis Of Ice Sphere Impacts On Unidirectional Carbon/Epoxy Laminates. *International Journal of Impact Engineering*, 1-10.

[11] Schaefer, J., Werner, B., & Daniel, I. (2014). Strain-Rate-Dependent Failure of a Toughened Matrix Composite. *Experimental Mechanics*, 1111-1120.

[12] Shan, L., Hongqiang, Y., Zhengping, F., & Hao, W. (2014). Effect of Graphene Nanosheets On Morphology, Thermal Stability and Flame Retardancy of Epoxy Resin. *Composites Science and Technology*, 40-47.

[13] Thomas, A., Pavel, S., & Suresh, G. (2013). Resin Film Impregnation in Fabric Prepregs With Dual Length Scale Permeability. *Composites Part A: Applied Science and Manufacturing*, 118-128.

[14] Tseng, A., Raudensky, M., & Lee, T.-W. (2016). Liquid Sprays for Heat Transfer Enhancements. *Heat Transfer Engineering*, 1401-1417.

[15] Van Velthem, P., Ballout, W., Daoust, D., Sclavons, M., Cordenier, F., Henry, E., . . . Bailly, C. (2015). Influence of Thermoplastic Diffusion On Morphology Gradient and On Delamination Toughness Of RTM-Manufactured Composites. *Composites Part A: Applied Science and Manufacturing*, 175-183.

Flatwise Compressive Properties of Kenaf/Polypropylene Honeycomb Core

Nabihah Sallih[1,2,*], Peter Lescher[1], Debes Bhattacharyya[1]

[1]Centre for Advanced Composite Materials, Department of Mechanical Engineering, The University of Auckland, Private Bag 92019, Auckland, New Zealand
[2]Department of Mechanical Engineering, Universiti Teknologi Petronas, 32610, Perak, Malaysia

Corresponding author [*]: nabihah.sallih@utp.edu.my

Abstract. Light weight and sustainable sandwich panels are increasingly sought for an extensive range of applications, from structures in airliners to wall linings in buildings. A study on flatwise compressive properties of honeycomb cores manufactured from kenaf/polypropylene sheets has been conducted. The influences of directionalities, the thickness-to-length ratio of cell walls and the depth to cell size ratio on the stiffness and strength of the cores have been examined. Cell wall thickness, core depth, directionality, and the interaction between core depth and cell wall thickness influenced both absolute and specific compressive strengths of the cores. Directionality and cell wall thickness as well as interaction between cell wall thickness and core depth influenced the absolute compressive modulus of the cores. However, only directionality and interaction between cell wall thickness and core depth affected specific compressive modulus.

1. Introduction

Sandwich construction is an attractive structural design concept, where panels having high stiffness- and strength-to-weight ratios can be achieved by appropriate choice of geometry even with moderate-quality constituent materials. The panels are formed by adhering strong and stiff face sheets to a thick, relatively weak yet light core material in proportions such that optimum stiffness and strength can be achieved with minimum structural weight. When a sandwich panel is subjected to transverse loads, the face sheets carry most of the bending moments and in-plane extensional stresses, while the core carries transverse shear and compressive stresses [1].

Figure 1 illustrates the compressive performances of various core materials used in sandwich structures [2-11]. Honeycomb cores and axially-oriented balsa wood exhibit significantly higher compressive strengths in comparison to various polymeric foams over a wide range of core densities. Because of these, honeycomb core structures are of interest in this study.

Currently, the majority of honeycomb sandwich structures are made of materials and polymers that are difficult to recycle and reuse. Due to increasing environmental concerns on the impacts of materials used, from production to end of life, a considerable research has been conducted on the development of new sandwich structures which utilise naturally sourced and recyclable materials. For core materials, traditional metallic structures are being replaced by composite- and polymer-based foams, corrugations and honeycombs due to cost, weight and manufacturing considerations. Some examples of natural-fibre based cores include sisal/polypropylene (PP) [12], sawdust/PP [13], flax/PP [14], jute/vinylester [15] and plywood [16].

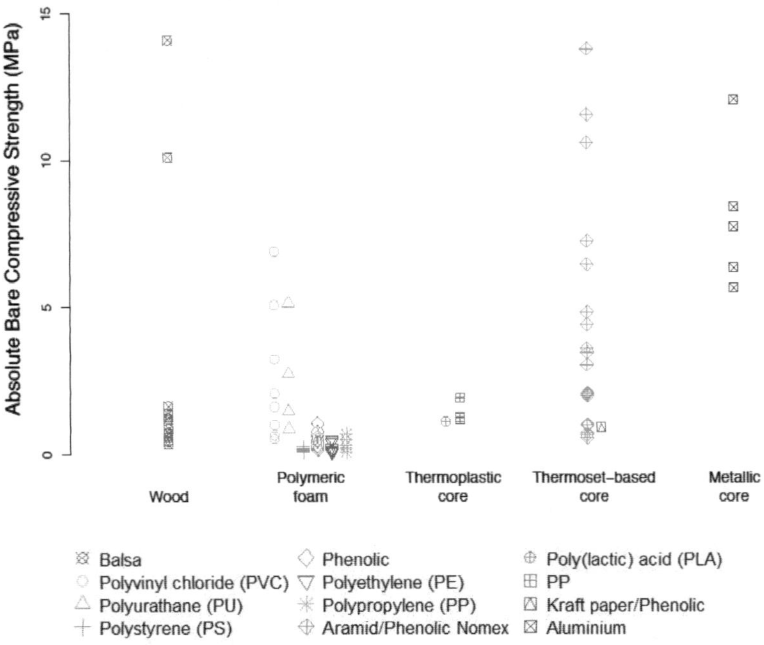

Figure 1. Comparison of compressive strengths of various wood, foams, and metallic and composite honeycomb cores published in the literature.

To date, there is no extensive study on how this a class of kenaf/PP hexagonal-cell honeycomb core materials would behave under compressive loadings. This paper investigates the influences of directionalities, cell wall thickness and core depth on the stiffness and strength of the structures under flatwise bare compression.

2. Experimental Details

2.1. Materials

Kenaf yarns were purchased from Jute-Bangladesh Ltd., Bangladesh. PP homopolymer HP422M (8.5 g/10 min) was supplied by Lyondell Basell. Licocene PP MA 6452, a maleic-anhydride modified PP (acid number: 43 mg KOH/g) supplied by Clariant, was chosen as a compatibiliser for PP.

2.2. Production of Honeycomb Core

Kenaf/HP422M PP/MAPP (40/57/3 wt%) composite sheets of two different thicknesses (0.6 mm and 1.2 mm) were manufactured using a twin-screw extruder. The mechanical properties of the composite sheets are listed in Table 1. The sheets were then thermoformed into half-hexagonal corrugations. The corrugations were joined together using an ultrasonic welder to form honeycomb cores.

2.3. Testing

The flatwise bare compressive properties of honeycomb core materials were determined using ASTM C365 with the cross-head speed set to 0.5 mm/min. The specimens consisted of seven and six cells in directions parallel and transverse to ribbon, respectively. The width of the specimens (in ribbon direction) was set to 101 mm. The lengths of the specimens (in direction transverse to ribbons) vary depending on the cell wall thickness of the specimens; the lengths of specimens with cell wall thicknesses of 0.6 mm and 1.2 mm were 96–98 mm and 106 mm, respectively.

IOP Conf. Series: Materials Science and Engineering **370** (2018) 012031 doi:10.1088/1757-899X/370/1/012031

Table 1. Mechanical properties of Kenaf/HP422M PP/MAPP (40/57/3 wt%) composite sheets of two different thicknesses (0.6 mm and 1.2 mm). MD and TMD refer to pulling axis being parallel and transverse to machine direction, respectively. B-TMD and B-MD refer to bending axis (cylindrical axis around which the sample curves) being parallel and transverse to machine direction, respectively.

Properties	Directions	0.6 mm	1.2 mm	Units
Tensile strength	MD	46.4 ± 0.87	46.3 ± 0.87	MPa
	TMD	24.4 ± 2.76	27.4 ± 0.36	
Tensile modulus	MD	5.64 ± 0.33	6.00 ± 0.40	GPa
	TMD	2.76 ± 0.24	2.63 ± 0.06	
Flexural strength	B-TMD	87.5 ± 1.64	88.0 ± 1.91	MPa
	B-MD	47.7 ± 0.93	51.6 ± 1.96	
Flexural modulus	B-TMD	5.30 ± 0.12	5.26 ± 0.12	GPa
	B-MD	2.67 ± 0.09	2.58 ± 0.04	
Out-of-plane shear strength		41.9 ± 0.66	40.5 ± 0.85	MPa

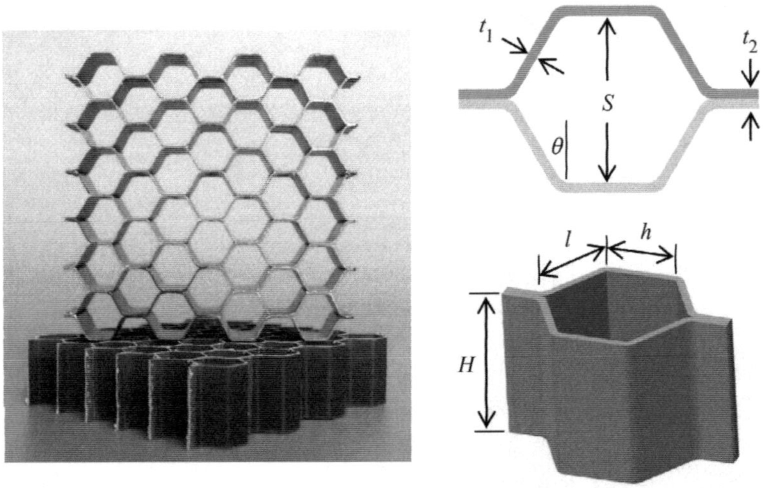

Figure 2. Samples (7 by 6 cells) for bare flatwise compression test, and nomenclature for a typical unit cell.

2.3.1. Experimental Design.

The purpose of the tests was to determine the sensitivity of the core materials' stiffness and strength under compressive loadings to the microscopic and macroscopic directionalities, the thickness-to-length ratios of cell walls and the core height-to-cell size ratios. Three factors were identified as likely to impact compressive properties. A full factorial Taguchi L_8 design, with eight experimental runs in which each factor was alternated between a "high" and "low" level, was used to determine if a change within those levels had an effect on mechanical properties.

Factor 1 - Directionality (A)

Low (A_1): F-MD, High (A_2): F-TMD

Kenaf/PP/MAPP composite sheets are anisotropic in nature due to the fibre and PP alignments in the extrusion direction [17]. During thermoforming, the resistance to forming deformation is, therefore, expected to be much higher when the bending axis is transverse to the extrusion direction (here refers to F-TMD) as opposed to being parallel (F-MD). Minor differences in conformability are expected in corrugated sheets. Furthermore, the strength and stiffness of the honeycomb are dependent on the loading directions and structural configurations (i.e. macroscopic directionality) with respect to the fibre and PP alignments (i.e. microscopic directionality). The ribbons in the hexagonal cores are approximately twice the thickness of the free walls, and thus the former are expected to be stiffer than the latter especially when flexed in the direction perpendicular to bending direction [18-20]. At the moment, with the use of anisotropic composites as parent materials, the interaction effects of the fibre and PP alignments (i.e. mechanical properties of parent materials) with respect to the forming axes (i.e. formability of corrugated sheets) and loading directions, when coupled with principal cell orientations (i.e. ribbons or free walls) are not precisely known.

Factor 2 - Cell wall thickness (B)
Low (B_1): 0.6 mm, High (B_2): 1.2 mm
The lower limit of cell wall thickness has been set to 0.6 mm because the minimum thickness of composite sheets that can be successfully produced using 1 mm thick sheet die is 0.6 mm. The upper limit of 1.2 mm has been chosen to represent a reasonably high cell wall thickness which can be thermoformed easily and to avoid in producing cores with higher densities.

Factor 3 - Core height (C)
Low (C_1): 20 mm, High (C_2): 30 mm
Although core height does not influence compressive properties, the shear strength of honeycomb cores reduces with an increase in core height. The lower limit of 20 mm has been set to provide enough depth for separating face sheets in later applications so as to significantly increase the stiffness and strength of sandwich panels in bending. The upper limit of 30 mm has been chosen so as to avoid significant reduction in the shear strength of the cores. Note that the thickness of 20 mm and 30 mm cores ended up being 18.55 mm and 28.27 mm, respectively, after surface milling process which was meant to ensure that faces of the cores were parallel and flat.

A Taguchi full factorial experimental design, as shown in Table 2, was used to investigate the effects of these factors on the flatwise bare compressive and shear properties.

Table 2. Two-level Taguchi full factorial experimental layout. (Example: Trial one consists of a low directionality, cell wall thickness and core depth).

Trial	Direction (A)		Cell wall (B)		Core depth (C)		AB		AC		BC	
	LO	HI	LO	HI	LO	HI	LO	HI	LO	HI	LO	HI
1	✓		✓		✓			✓		✓		✓
2	✓		✓			✓		✓	✓		✓	
3	✓			✓	✓		✓			✓	✓	
4	✓			✓		✓	✓		✓			✓
5		✓	✓		✓		✓		✓			✓
6		✓	✓			✓	✓			✓	✓	
7		✓		✓	✓			✓	✓		✓	
8		✓		✓		✓		✓		✓		✓

3. Results and Discussion

The typical compressive stress-strain plots for bare honeycomb core specimens are shown in Figure 3. It should be noted that the stress is based on the projected area of the honeycomb. In comparison to any typical metallic or thermoset-based honeycomb cores, kenaf/PP cores demonstrated a significant drop in compressive stress after the peak stress has been attained. This shows that kenaf/PP cores lack of energy absorption capability. This is possibly related to the lack of reinforcing effect contributed by neighbouring cells as they are folded during crushing stage simply because the cell size in the honeycombs used is larger than the typical size used in the literature (3.2–9.5 mm).

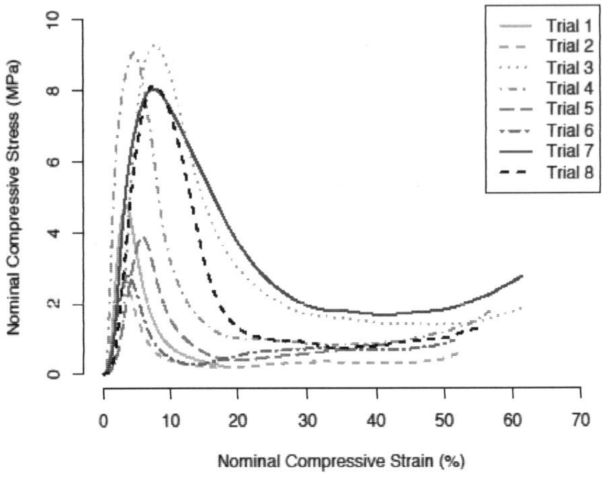

Figure 3. Typical compressive stress-strain plots for bare honeycomb cores of various combinations of directionality, cell wall thickness and core depth.

Figure 4 illustrates the sequence of microstructural change during the compressive deformation of honeycomb cores. The deformation process varies depending on the cell wall thickness of honeycomb cores. The initial response was stiff and nearly linear. The compressive modulus was calculated based on stress-strain response at this region. As the strain increases, the free walls of honeycomb cores were observed to buckle.

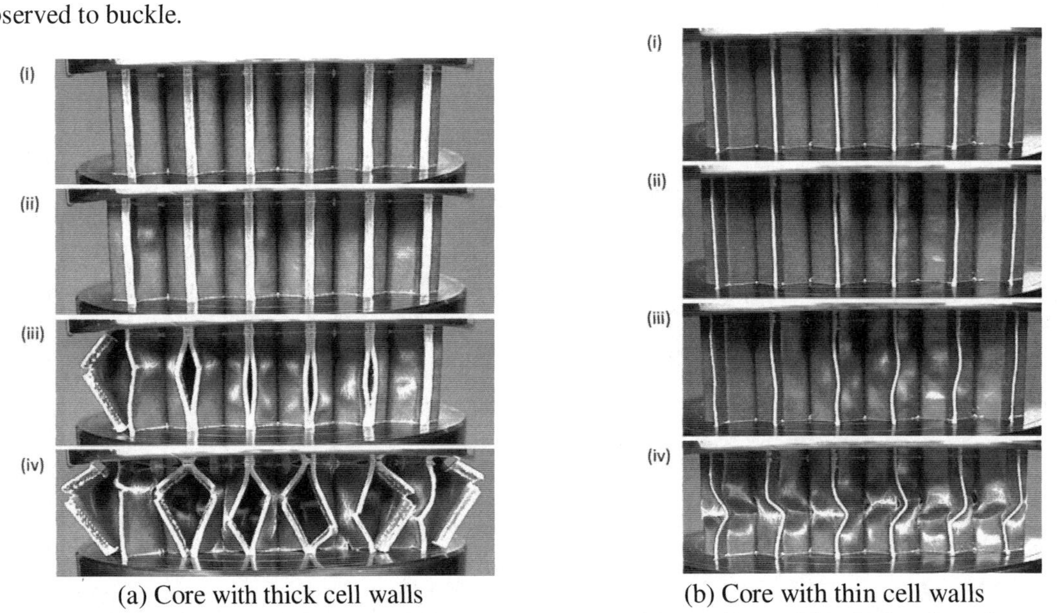

(a) Core with thick cell walls (b) Core with thin cell walls

Figure 4. Sequence of photographs of honeycomb cores with (a) thin and (b) thick cell walls at various stages of compression: (i) initial condition. (ii) onset of failure by buckling at the free walls. (iii) for (a): buckling at the ribbons and for (b): compressive failure at the ribbons, which is initiated by debonding at the cell wall interfaces. (iv) further material yielding and fracture. The plane is viewed from the direction transverse to ribbon direction.

Further increase in strain resulted in two different responses depending on the cell wall thickness — the ribbons in cores with thin (0.6 mm) cell walls started to buckle while the ribbons in cores with thick (1.2 mm) cell walls demonstrated compressive failure, which was initiated by debonding at cell wall interfaces of walls at the ends of the specimens. The maximum stress achieved represents the compressive strength of the honeycomb cores and the onset of crushing. As compression continued, the stress dropped due to material yielding and consequently fracture as the cells folded.

3.1. Compressive Strength

The effects of each factor on absolute compressive strengths of bare honeycomb cores are shown in Figure 5(a). The grand mean of absolute strength is 6.1 MPa. Since honeycomb cores are mainly applied in lightweight structures, their compressive strength relative to their weight (refers to here as specific compressive strength) is also important. The grand mean of specific strength is 0.0458 MPa/kg.m^{-3}.

ANOVA shows that the cell wall thickness (B), core depth (C), directionality (A) as well as interaction between core depth and wall thickness (BC) are significant at the 95% confidence level for both absolute and specific compressive strengths. Corrugated sheets formed in F-MD are preferable because kenaf fibres and PP molecular alignments are mainly oriented in the same direction as the compressive loading in comparison to those formed using F-TMD. Under uniaxial compression, cell wall's resistance towards buckling deformation is dependent on the cell width-to-wall thickness (l/t or h/t) and the cell width-to-core depth (l/H or h/H), as described by Timoshenko and Gere [21] — the resistance of the cell walls towards buckling deformation increases with the decrease in the cell width-to-wall thickness ratio and with the increase in the cell width-to-core depth ratio. This explains why cores with thicker cell walls (for the same width of cell walls) and shallower cores resulted in higher compressive strength in comparison to their counterparts.

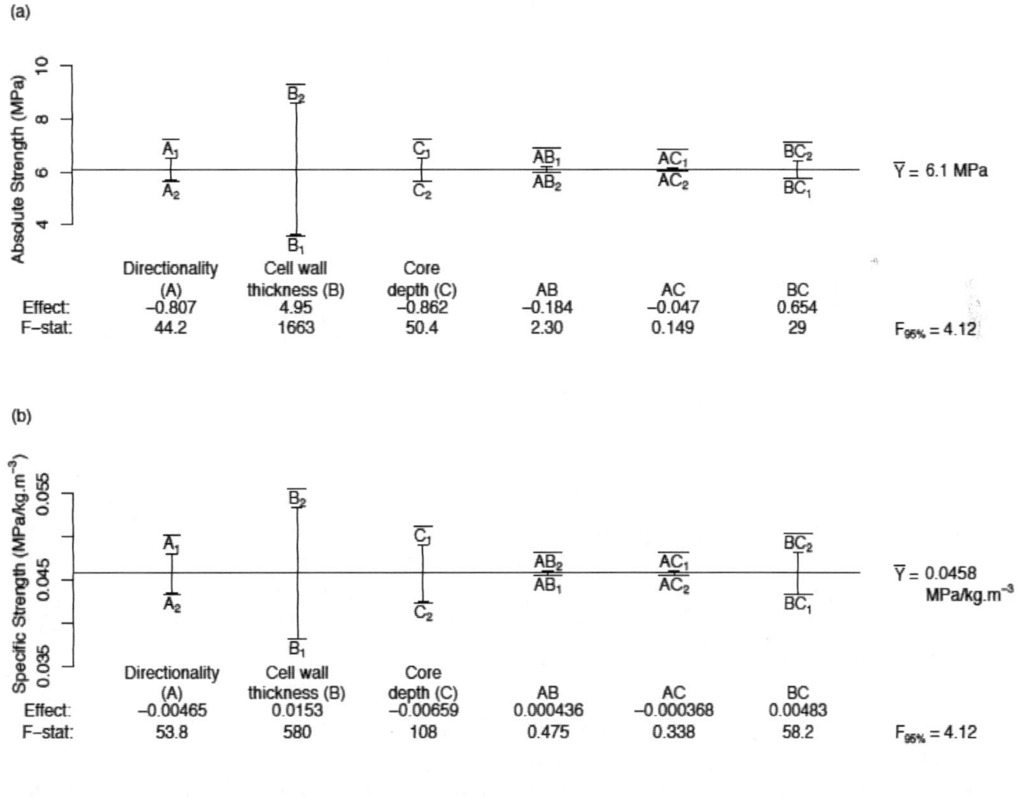

Figure 5. Effects of various factors on (a) absolute and (b) specific (relative to weight) flatwise bare compressive strengths of honeycomb cores.

As illustrated by the parallel lines in Figure 6, it is noted that there is no interaction between directionality and cell wall thickness (AB), and directionality and core depth (AC). Unparallel lines in cell wall thickness and core depth (BC) plot indicate that these factors interact with each other. At a lower cell wall thickness (i.e. 0.6 mm), a lower core depth (i.e. 20 mm) resulted in higher absolute and specific compressive strengths. However, the effects of core depth become insignificant at a higher cell wall thickness (i.e. 1.2 mm). This phenomenon can be explained based on the dominant failure mode for each case. The cores with thick cell walls failed by compressive failure, and this failure is dependent on composites' compressive strength while the cores with thin cell walls failed by buckling, and this failure is limited by core depth. So, it is natural to observe a strong interaction between core depth and cell wall thickness for cores with thin walls and such effect diminishes for cores with thick walls because the compressive strength of the core is bounded by the compressive strength of the parent material.

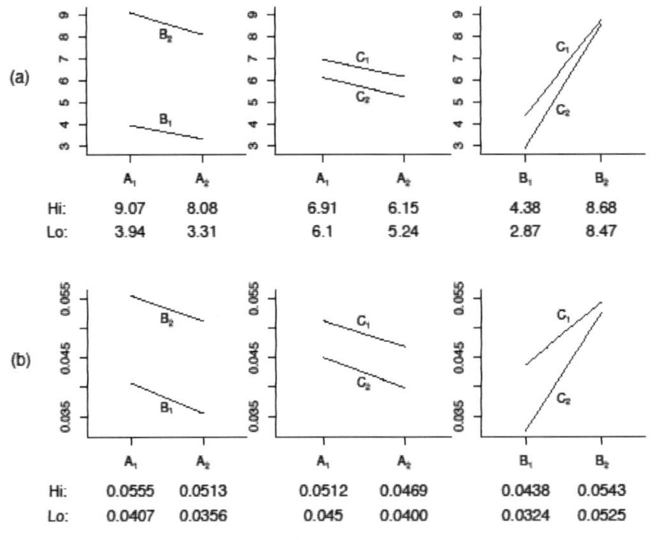

Figure 6. Interaction effects on (a) absolute and (b) specific (relative to weight) flatwise bare compressive strengths of honeycomb cores.

3.2. Compressive Modulus

Figures 7 shows the effects of various factors on both absolute and specific flatwise bare compressive moduli of honeycomb cores. The grand means of absolute and specific moduli are 238 MPa and 1.83 MPa/kg.m^{-3}, respectively. ANOVA shows that directionality (A) and cell wall thickness (B) as well as the interaction between directionality and cell wall thickness (AB), and the interaction between cell wall thickness and core depth (BC) significantly affect the absolute compressive modulus of cores. In terms of specific compressive modulus, however, only directionality (A) and the interaction between cell wall thickness and core depth (BC) affect the property significantly. As expected, the modulus of F-MD honeycomb cores is higher than those of F-TMD due to the alignment of fibres and PP molecular chains in the compressive loading direction. The theoretical equation presented by Gibson and Ashby [22] for predicting the compressive modulus of hexagonal honeycomb cores based on the modulus of parent materials (*E*s) and cell configuration shows that by doubling the cell wall thickness, the compressive modulus is doubled (refer to Equation 1).

$$E_3 = \left(\frac{t}{l}\right)\frac{1+\frac{h}{l}}{\left(\frac{h}{l}+\sin\theta\right)\cos\theta}E_S \tag{1}$$

This explains why absolute compressive modulus for cores with 1.2 mm thick cell walls is almost double that of cores with 0.6 mm thick cell walls. However, by doubling the thickness of the cell walls, the mass of the cores (for the same number of cells) increases proportionately too. As expected, the effects of cell wall thickness on the specific modulus of honeycomb cores would diminish. The effects of core depth appear not to be very important in relation to the compressive modulus, which is inline with the observations made by Meraghni et al. [23].

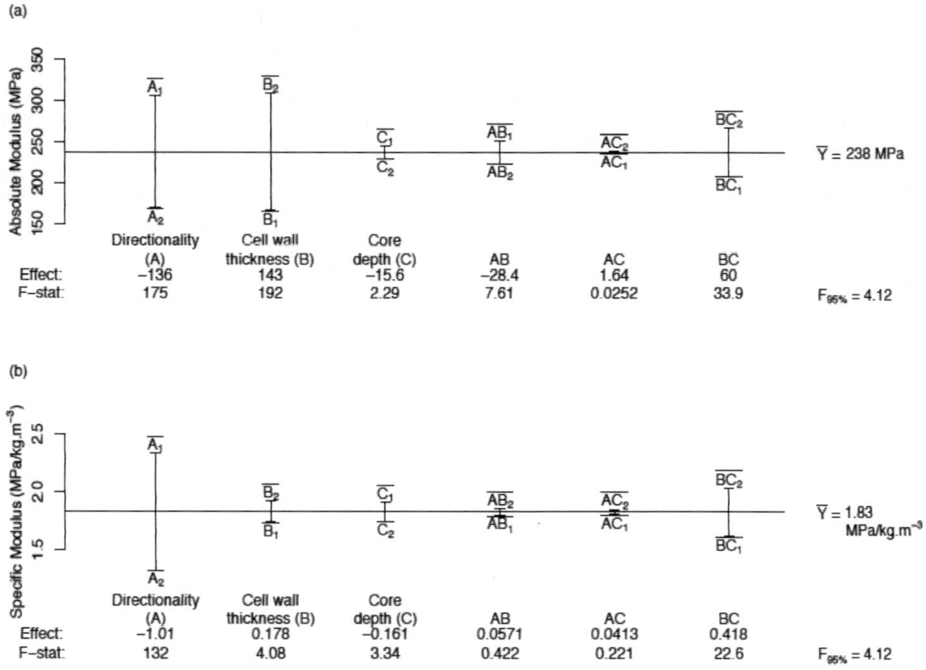

Figure 7. Effects of various factors on (a) absolute and (b) specific (relative to weight) flatwise bare compressive moduli of honeycomb cores.

The distinctive unparallel lines in the interaction plots, shown in Figure 8, indicate the presence of strong interaction between cell wall thickness and core depth for both absolute and specific compressive moduli. At a higher cell wall thickness (1.2 mm), a higher core depth (30 mm) resulted in higher compressive moduli and vice versa.

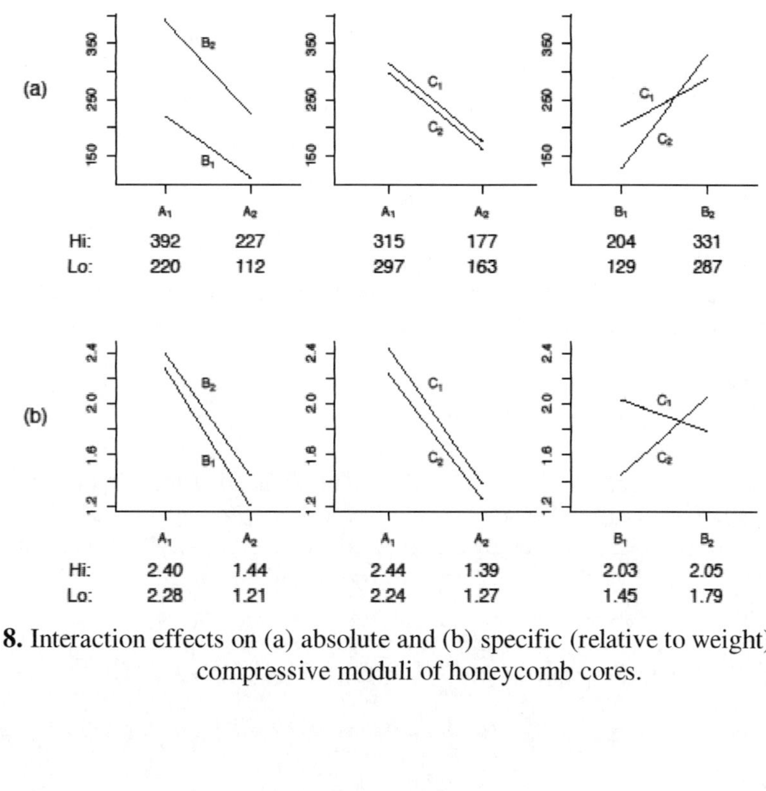

Figure 8. Interaction effects on (a) absolute and (b) specific (relative to weight) flatwise bare compressive moduli of honeycomb cores.

As explained earlier, buckling is dependent on the cell width-to-thickness (l/t or h/t) and the cell width-to-core depth (l/H or h/H) ratios. For cores with thin cell walls, a shallower core has more resistance towards buckling deformation and thus possesses a higher compressive modulus in comparison to a thicker core. On the other hand, cores with thicker cell walls are not susceptible to buckling deformation. Because of this stable structure, it allows for the core to have a higher core depth. Besides this, a mild interaction between directionality and cell wall thickness was observed — both F-MD and 1.2 mm thick cell wall resulted in a higher compressive modulus in comparison to their counterparts, as expected.

4. Conclusion

To date, there has been no extensive study on core materials made of natural fibre/thermoplastic composites, let alone kenaf/polypropylene honeycomb cores. This characterization has shed light on the mechanical performance of this new class of material.

The effects of directionalities, cell wall thickness and core height on flatwise compressive and shear properties of the cores were examined. Cell wall thickness, core depth, directionality, and the interaction between core depth and cell wall thickness influenced both absolute and specific compressive strengths of kenaf/polypropylene honeycomb cores. Directionality and cell wall thickness as well as interaction between cell wall thickness and core depth influenced the absolute compressive modulus of honeycomb cores. However, only directionality and interaction between cell wall thickness and core depth affected specific compressive modulus.

References

[1] Dan Zenkert. An introduction to sandwich construction. Sandwich construction. London : Engineering Materials Advisory Services, London, 1995.

[2] Andre Da Silva and Stelios Kyriakides. Compressive response and failure of balsa wood. *International Journal of Solids and Structures*, 44:8685–8717, 2007.

[3] Gurit. Data sheet for Gurit PVC structural foam core, 2014.

[4] Mark S. Thompson, Ian D. McCarthy, Lars Lidgren, and Leif Ryd. Compressive and shear properties of commercially available polyurethane foams. *Journal of Biomechanical Engineering*, 125(5):732–734, 2003.

[5] EXPOL. Technical product guide: Solid insulation lightweight construction, 2014.

[6] Dyplast. DyTherm phenolic insulation, 2014.

[7] Ultraton. Ultraton P.E. and E.V.A. foams — Data sheets, November 2014.

[8] JSP. Material physical properties for ultra low density polypropylene foam molded from ARPRO expanded polypropylene beads, 2014.

[9] Jochen Pflug. Bio-based panels, November 2014.

[10] Plascore. PP Polypropylene Honeycomb – Data Sheet, 2014.

[11] ATL Composites Pty Ltd. CORE/Paper honeycomb, 2014.

[12] S. Rao, K. Jayaraman, and D. Bhattacharyya. Short fibre reinforced cores and their sandwich panels: Processing and evaluation. *Composites Part A:* Applied Science and Manufacturing, 42(9):1236 – 1246, 2011.

[13] S. Banerjee, M. Battley, and D. Bhattacharyya. Shear strength optimization of reinforced honeycomb core materials. *Mechanics of Advanced Materials and Structures*, 17(7):542–552, 2010.

[14] M.Y.M. Zuhri, Z.W. Guan, and W.J. Cantwell. The mechanical properties of natural fibre based honeycomb core materials. *Composites Part B: Engineering*, 58(0):1 – 9, 2014.

[15] Ariel Stocchi, Lucas Colabella, Adrián Cisilino, and Vera Álvarez. Manufacturing and testing of a sandwich panel honeycomb core reinforced with natural-fiber fabrics. *Materials & Design*, 55(0):394 – 403, 2014.

[16] S. Banerjee and D. Bhattacharyya. Optimal design of sandwich panels made of wood veneer hollow cores. *Composites Science and Technology*, 71(4):425 –432, 2011.

[17] N. Sallih, Manufacturing and Evaluation of Kenaf/Polypropylene Honeycomb Cores, in Department of Mechanical Engineering. 2016, The University of Auckland.

[18] Muhammad Khasif Khan and Wang Qingyuan. Experimental and finite element based investigations of in-plane and out-of-plane properties of aluminium honeycomb. *Applied mechanics and materials*, 275-277:111–116, 2013.

[19] Solace Sam-Brew, Kate Semple, and Gregory D. Smith. Preliminary experiments on the manufacture of hollow core composite panels. *Forest Products Journal*, 61(5):381–389, 2010.

[20] M.K. Khan, T. Baig, and S. Mirza. Experimental investigation of in-plane and out-of-plane crushing of aluminum honeycomb. *Materials Science and Engineering: A*, 539(0):135–142, 2012.

[21] S.P. Timoshenko and J.M. Gere. *Theory of Elastic Stability*. McGraw-Hill, 1961.

[22] Lorna J. Gibson and Michael F. Ashby. *Cellular solids : Structure and Properties*. Cambridge; New York: Cambridge University Press, 1999.

[23] F. Meraghni, F. Desrumaux, and M.L. Benzeggagh. Mechanical behaviour of cellular core for structural sandwich panels. *Composites Part A: Applied Science and Manufacturing*, 30(6):767–779, 1999.

Magnetic attitude control torque generation of a gravity gradient stabilized satellite

N M Suhadis*, M B Salleh and P Rajendran

School of Aerospace Engineering, Universiti Sains Malaysia, 14300 Nibong Tebal, Penang, Malaysia

*E-mail: nurulasikin@usm.my

Abstract. Magnetic torquer is used to generate a magnetic dipole moment onboard satellites whereby a control torque for attitude control purposes is generated when it couples with the geomagnetic field. This technique has been considered very attractive for satellites operated in Low Earth Orbit (LEO) as the strength of the geomagnetic field is relatively high below the altitude of 1000 km. This paper presents the algorithm used to generate required magnetic dipole moment by 3 magnetic torquers mounted onboard a gravity gradient stabilized satellite operated at an altitude of 540 km with nadir pointing mission. As the geomagnetic field cannot be altered and its magnitude and direction vary with respect to the orbit altitude and inclination, a comparison study of attitude control torque generation performance with various orbit inclination is performed where the structured control algorithm is simulated for 13°, 33° and 53° orbit inclinations to see how the variation of the satellite orbit affects the satellite's attitude control torque generation. Results from simulation show that the higher orbit inclination generates optimum magnetic attitude control torque for accurate nadir pointing mission.

1. Introduction

One of the most significant disturbance torques for small satellites operated in Low Earth Orbit (LEO) is the gravity gradient torque. It exists from the variation of the Earth's gravitational force over the asymmetric body of the satellite where the Earth's gravitational force pulls the satellite axis with a least moment of inertia pointing towards the Earth with a relative angular velocity of the axis points to the Earth equal to the mean motion of the orbit. Therefore, this property can be used to passively stabilize the satellite with nadir pointing mission by extending the axis that is required to point to the Earth with a flexible mechanical device called a gravity gradient boom to significantly reduce its moment of inertia. Stability analysis of a gravity gradient stabilized satellite made by [1] shows that the stable condition is achieved with the satellite having moment of inertia values as $I_y > I_x > I_z$ after the deployment of the gravity gradient boom. However, this stabilization technique only allows 2-axis control, which is along the roll and pitch axis with an average attitude accuracies thus leaving the satellite in a libration motion while orbiting. Therefore, having an additional actuator onboard the gravity gradient stabilized satellite is necessary to provide 3-axis control and significantly improve its attitude performance. The actuator should be lightweight, low power consumption and inexpensive for low-cost missions. In this regards, a magnetic torquer is the primary option.

Magnetic torquer is a magnetic induction device consists of a coil and a core where the magnetic dipole moment is generated when an electrical current passing through the coil. It is used to generate a controllable value of magnetic dipole moment needed to be generated onboard the satellite in order to generate the required control torque when couples with the geomagnetic field. This technique has been considered very attractive for satellites operated in LEO as the strength of the geomagnetic field is relatively high below the altitude of 1000 km. Through the literature survey, several control strategies and approaches of the magnetic control technique implemented on the gravity gradient stabilized satellites have been reviewed. It can be summarized here that there is only a work found implemented a single magnetic torquer while the rest implemented three magnetic torquers. The work proposed a magnetic libration control scheme to eliminate the libration motion which likely to occur due to the external disturbance torques acting on the satellite [2]. The magnetic torquer was placed along the z axis where the controller was developed based on the energy dissipation principle. Numerical simulations were performed for a circular polar orbit. Results proved the effectiveness of the developed scheme by improving the yaw attitude oscillation between 0° and 2°.

As for the three magnetic torquers, [3] found that the B-dot control law was not effective in providing 3-axis stabilization during nominal attitude pointing phase. It was due to the non-uniformity of the geomagnetic field and the effect of the orbital rates. Therefore, they proposed to use the magnetic torquer cross-product control law for the attitude control where the control error was developed based on the PD controller. It was found from simulations performed for three different inclinations that the control along the yaw axis was achieved and consequently granting 3-axis stabilization. However, as stated by [4], this control algorithm imposes some control constraints which are the unavailability of the control torque during certain orbital regions and the generation of the unwanted cross control torques between the attitude axes. This is due to its tendency to generate the most favorable magnetic torque by interleaving or simultaneously switching any of the three magnetic torquers that are normal to the geomagnetic field direction at any control instant. Therefore, Steyn proposed a controller based on the fuzzy control theory where the developed controller chooses the magnetic torquer to generate control torques at any control instant. This controller was implemented for libration damping purposes onboard SUNSAT, a gravity gradient stabilized microsatellite developed by graduate students at Stellenbosch University with its primary mission objectives were for imaging, communication and data interchange [5]. Three reaction wheels were also mounted along this satellite's body axis to perform maneuvers for the image pointing and stabilization during imaging. The magnetic torquers were also used for momentum dumping of the reaction wheels.

On the other hand, [6] concern was on the incapability of the magnetic toquer to generate exactly the required action due to the orthogonality rule. Therefore, in their work, two approaches were proposed for generating the actual torque out of the ideal control torque: i) based on the torque closest to the ideally sought and ii) based on the torques with two components equal to the ideal. The ideal control torque law was developed based on the reference modern control theory. According to the proposed methods of generating the actual torque, three different algorithms were formulated. These three formulated algorithms were then compared with the algorithm developed by [3]. It was found that one of the formulated algorithms, which is the one with constant gains and alternate switching for roll and yaw, gives better performances with respect to the others. [7] also carried out a similar research. In their work, two ideal control torques were developed based on the nonlinear control theory. One was based on the energy method and the other one was based on the variable control structure method. These developed algorithms were compared with the algorithm of [6]. Simulation results show that both the energy method and the variable control structure method reached their steady state value faster than that of [6].

On the contrary, [8] proposed a fully autonomous magnetic attitude control system for a small satellite with a remote sensing application operated on a sun-synchronous orbit with an inclination of

97.2° at an altitude of 433 km. The control laws were derived separately for the attitude acquisition and station keeping phases. The former was developed using the B-dot control law while for the latter, the algorithm developed by Martel et al. was implemented. This mission was numerically tested and results show that it fulfills the satellite mission requirement. In other publication, [9] reported the performance of the ACS system of a multi-mission microsatellite developed by a group of Italian universities called UNISAT which employs the same control scheme. In a different way, [10] developed ACS system for a micro-satellite called LEAP. The controller was developed using the well-known B-dot control law for both detumbling and normal modes which managed to detumble the satellite before the gravity boom deployment as well as maintained the satellite attitude accuracy after the deployment. [11] proposed a velocity controller using a cross product between the angular velocity and the geomagnetic field whereby series of control algorithms were developed based on the PD control for recovery from the tumbling motion, nominal operation and recovery from a gravity gradient boom upside-down condition. Numerical simulations were performed for a satellite on a near polar orbit and results showed that global 3-axis stabilization was achieved with a promising attitude performance for all the three axes. These control algorithms were implemented onboard the Danish Orsted satellite and this mission was the first 3-axis stabilized satellite flying with magnetic torquers as the sole actuators onboard.

Another research implementing three magnetic torquers onboard was carried out based on the extension of the H_∞ control technique for controlling the linear time variant system where the periodic controllers for the periodic system can be designed. In this work, [12] have developed the discrete time state-space description for the linear time variant system where its transfer function has many properties analogous to the transfer functions for the linear time invariant system. Thus, allowing the formulation of control law for this system implementing the traditional H_∞ control.

Review of various control laws used for magnetic torquer mentioned here give us options to implement magnetic attitude control technique using three magnetic torquers. All the algorithms developed performed well and fulfill their respective requirements. However, the numerical treatments performed are only for specific orbit inclination. Therefore, in this paper, a comparison study of attitude control torque generation performance with various orbit inclination is performed where the structured control algorithm of a gravity gradient stabilized satellite equipped with three magnetic torquers is simulated for 13°, 33° and 53° to see how the variation of the satellite orbit along these orbits affects the satellite's attitude control torque generation as the geomagnetic field cannot be altered and its magnitude and direction vary with respect to the orbit altitude and inclination. The algorithm used is based on the Proportional-Derivative (PD) controller.

2. Satellite configuration

The configuration of the satellite is as depicted in Figure 1. It is equipped with 3 magnetic torquers that are placed along the $+x$, $+y$ and $+z$ axes and a gravity gradient boom along the $-z$ axis. Each of these magnetic torquers controls any two axes that are normal to them. Note that the z axis points toward the Earth, the y axis is normal to the orbital plane and the x axis points toward the satellite's orbital motion and complete the right-hand orthogonal system [13].

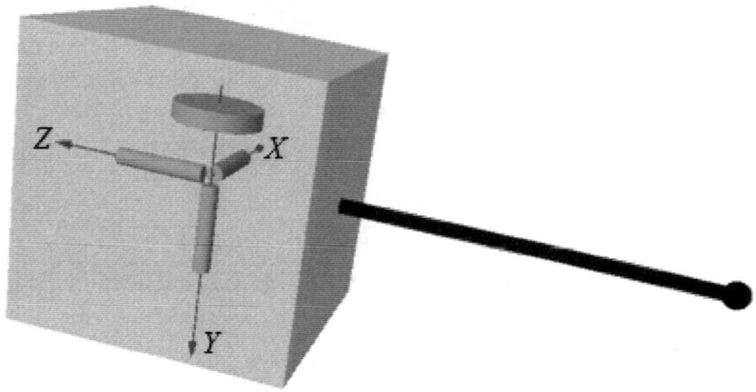

Figure 1. Satellite Configuration.

Based on this configuration, the linearized dynamic equation of motion of the satellite is described in a state-space equation where along the roll/yaw axis the equation becomes

$$\begin{bmatrix} \dot{\phi} \\ \dot{\psi} \\ \ddot{\phi} \\ \ddot{\psi} \end{bmatrix} = \begin{bmatrix} 0 & 0 & 1 & 0 \\ 0 & 0 & 0 & 1 \\ -4\omega_0^2\sigma_x & 0 & 0 & \omega_0(1-\sigma_x) \\ 0 & -\omega_0^2\sigma_z & -\omega_0(1-\sigma_z) & 0 \end{bmatrix} \begin{bmatrix} \phi \\ \psi \\ \dot{\phi} \\ \dot{\psi} \end{bmatrix} + \begin{bmatrix} 0 & 0 \\ 0 & 0 \\ 1/I_x & 0 \\ 0 & 1/I_z \end{bmatrix} \begin{bmatrix} T_x \\ T_z \end{bmatrix} \quad (1)$$

while along the pitch axis is

$$\begin{bmatrix} \dot{\theta} \\ \ddot{\theta} \end{bmatrix} = \begin{bmatrix} 0 & 1 \\ -3\omega_0^2\sigma_y & 0 \end{bmatrix} \begin{bmatrix} \theta \\ \dot{\theta} \end{bmatrix} + \begin{bmatrix} 0 \\ 1/I_y \end{bmatrix} [T_y] \quad (2)$$

The external torque T_x, T_y and T_z in equation (1) and equation (2) consists of two essential parts which are the disturbance torque and the control torque. Their relation can be defined as

$$\begin{bmatrix} T_x \\ T_y \\ T_z \end{bmatrix} = \begin{bmatrix} T_{dx} + T_{mx} \\ T_{dy} + T_{my} \\ T_{dz} + T_{mz} \end{bmatrix} \quad (3)$$

where the disturbance torque consists of the gravity gradient torque, aerodynamic torque, magnetic torque and solar radiation torque, while the control torque is generated by the magnetic torquers.

3. Control strategy

The magnetic control torque is produced when the magnetic dipole moment generated by magnetic torquer interacts with the geomagnetic field. This interaction can be mathematically described as follow

$$\mathbf{T}_m = \mathbf{M} \times \mathbf{B} \quad (4)$$

where $\mathbf{T}_m$ is the magnetic torque vector, **M** is the satellite's magnetic dipole moment vector and **B** is the geomagnetic field vector. This equation clearly explains the generated magnetic control torque direction is normal to both the geomagnetic field vector and the satellite's magnetic dipole moment. Since the control algorithm is for a nominal attitude pointing phase, the magnetic torquer cross-product control law is used and written here as [14]

$$\mathbf{M} = \frac{\mathbf{T}_c \times \mathbf{B}}{|\mathbf{B}|} \tag{5}$$

The PD controller is adopted for controlling this satellite where $\mathbf{T}_c$ in equation (5) is defined as [3]

$$\begin{bmatrix} T_{cx} \\ T_{cy} \\ T_{cz} \end{bmatrix} = \begin{bmatrix} k_{px}\phi + k_{dx}\dot{\phi} \\ k_{py}\theta + k_{dy}\dot{\theta} \\ k_{pz}\psi + \dot{k}_{dz}\psi \end{bmatrix} \tag{6}$$

Therefore, the total generated magnetic control torque along each axis is defined by inserting equation (5) into equation (4), which yielding the following equation

$$T_{mx} = \frac{-(B_y^2 + B_z^2)}{B^2} \cdot [k_{px}\phi + k_{dx}\dot{\phi}] + \frac{B_x B_y}{B^2} \cdot [k_{py}\theta + k_{dy}\dot{\theta}] + \frac{B_x B_z}{B^2} \cdot [k_{pz}\psi + k_{dz}\dot{\psi}] \tag{7}$$

$$T_{my} = \frac{B_x B_y}{B^2} \cdot [k_{px}\phi + k_{dx}\dot{\phi}] - \frac{(B_x^2 + B_z^2)}{B^2} \cdot [k_{py}\theta + k_{dy}\dot{\theta}] + \frac{B_y B_z}{B^2} \cdot [k_{pz}\psi + k_{dz}\dot{\psi}] \tag{8}$$

$$T_{mz} = \frac{B_x B_z}{B^2} \cdot [k_{px}\phi + k_{dx}\dot{\phi}] + \frac{B_y B_z}{B^2} \cdot [k_{py}\theta + k_{dy}\dot{\theta}] - \frac{(B_x^2 + B_y^2)}{B^2} \cdot [k_{pz}\psi + k_{dz}\dot{\psi}] \tag{9}$$

The values of the geomagnetic field magnitude along the orbit can be defined by a series of spherical harmonics as follow [15],

$$(r, \theta, \lambda) = R \sum_{n=1}^{k} \left(\frac{R}{r}\right)^{n+1} \sum_{m=0}^{n} (g_n^m \cos m\lambda + h_n^m \sin m\lambda)\, P_n^m(\theta) \tag{10}$$

where R is the equatorial radius of the earth, (r, θ, λ) are the geocentric distance, co-latitude and east longitude from Greenwich, (g_n^m, h_n^m) are the Schmidt normalized Gaussian coefficients and $P_n^m(\theta)$ is the associated Legendre functions. The general solution of this function expressed in spherical coordinate system can be written as follows

$$B_r = \sum_{n=1}^{k} \left(\frac{R}{r}\right)^{n+2} (n+1) \sum_{m=0}^{n} (g^{n,m} \cos m\lambda + h^{n,m} \sin m\lambda)\, P^{n,m}(\theta) \tag{11}$$

$$B_\theta = -\sum_{n=1}^{k} \left(\frac{R}{r}\right)^{n+2} \sum_{m=0}^{n} (g^{n,m} \cos m\lambda + h^{n,m} \sin m\lambda) \frac{\partial P^{n,m}(\theta)}{\partial \theta} \tag{12}$$

$$B_\lambda = \frac{-1}{\sin\theta} \sum_{n=1}^{k} \left(\frac{R}{r}\right)^{n+2} \sum_{m=0}^{n} m(-g^{n,m} \sin m\lambda + h^{n,m} \cos m\lambda)\, P^{n,m}(\theta) \tag{13}$$

4. Numerical simulation

Series of simulations have been performed using the MATLAB® SIMULINK® software. The components of the geomagnetic field intensity vector expressed in Local Vertical Local Horizontal (LVLH) coordinate system along the 13°, 33° and 53° orbit inclinations with respect to the sampling orbital parameters described in Table 1 have been generated and plotted in **Error! Reference source not found.**.

Table 1. Orbital Parameters.

Parameters	Value
Altitude, h	540 km
Eccentricity, e	≈ 0
Right ascension of ascending node, Ω	0°
Argument of perigee, ω	0°
True anomaly, ν	0°
Simulation time, *t*	20 orbits
Orbital frequency, ω_0	0.0011 rads^{-1}

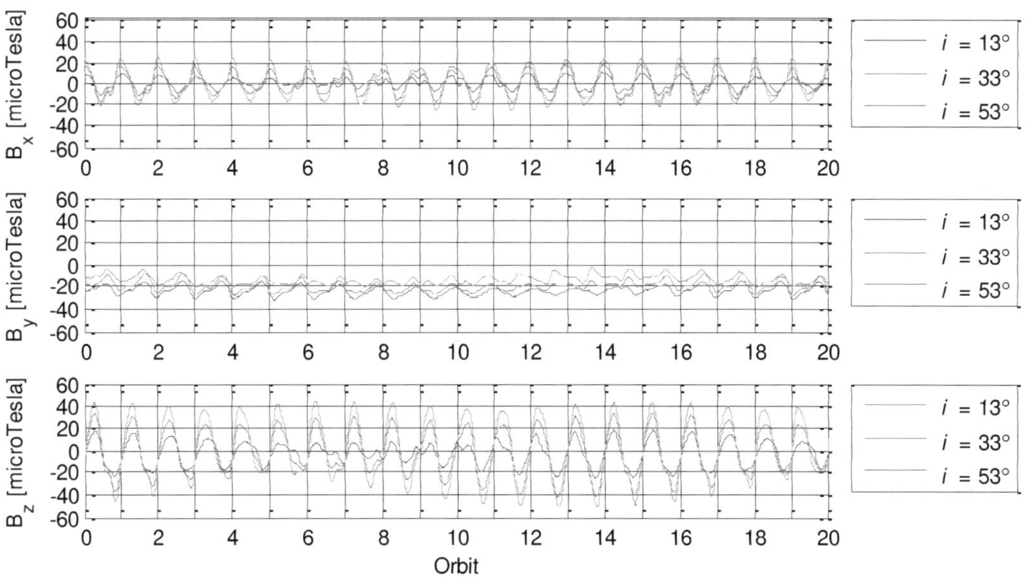

Figure 2. Comparison of the geomagnetic field models along 13°, 33° and 53° orbit inclinations.

The B_x and B_z components vary with respect to the orbital motion as these two components are measured along the $+x$ and $+z$ axis of the LVLH coordinate system which are lying in the orbital plane. Whereas the B_y component is measured along the $+y$ axis which is normal to the orbital plane and only influenced by the rotation of the earth. It is also shown that the magnitude of the geomagnetic field is stronger over the pole and weaker over the equatorial. The tuning of the control gains is based on the geomagnetic field values along the 53° orbit inclination and the same gains are later used to simulate the performance along the 13° and 33° to see the effect of varying the orbit inclination. Details of the satellite parameters and orbital parameters are given in Table 2.

Table 2. Satellite parameters

Parameter	Value
Weight	50 kg
Dimension	$690 \times 366 \times 366$ mm
I_x	178 kgm^2
I_y	181 kgm^2
I_z	4.3 kgm^2
$M_{magnetic\ torquer}$	15 Am^2
Input	$T_{dx} = 12.8 \times 10^{-6} + 8.6 \times 10^{-6}\sin(\omega_0 t)$ Nm $T_{dy} = 55 \times 10^{-6} + 55 \times 10^{-6}\sin(\omega_0 t)$ Nm $T_{dz} = 12.8 \times 10^{-6} + 4.3 \times 10^{-6}\sin(\omega_0 t)$ Nm
Initial conditions	$\phi(0) = 5°, \theta(0) = 5°, \psi(0) = 5°$ $\omega_x(0) = 0, \omega_z(0) = 0$
Gains	$k_{px} = 10, k_{py} = 20, k_{pz} = 100,$ $k_{dx} = 1000, k_{dy} = 1000, k_{dz} = 1000$

The attitude performances of the satellite are depicted in figure 3. The plots show that the roll axis oscillates with an accuracy between -1.5° and 3.4° for 13° orbit inclination, between −2.9° and 3.2° for 33° orbit inclination and between −4.6° and 5.5° for 53° orbit inclination. Meanwhile, the pitch axis oscillates with an accuracy between −4° and 14.5° for 13° orbit inclination, between −3.1° and 12° for 33° orbit inclination and between −3.1° and 8° for 53° orbit inclination. As for the yaw axis, it oscillates with the accuracy between −1.2° and 1.5° for 13° orbit inclination, between −2° and 3° for 33° orbit inclination and between −2.6° and 6° for 53° orbit inclination. It can be clearly seen that the attitude performance of the satellite along three axes improves as the orbit increases. This is due to the geomagnetic field value that is stronger over the pole and weaker over the equatorial.

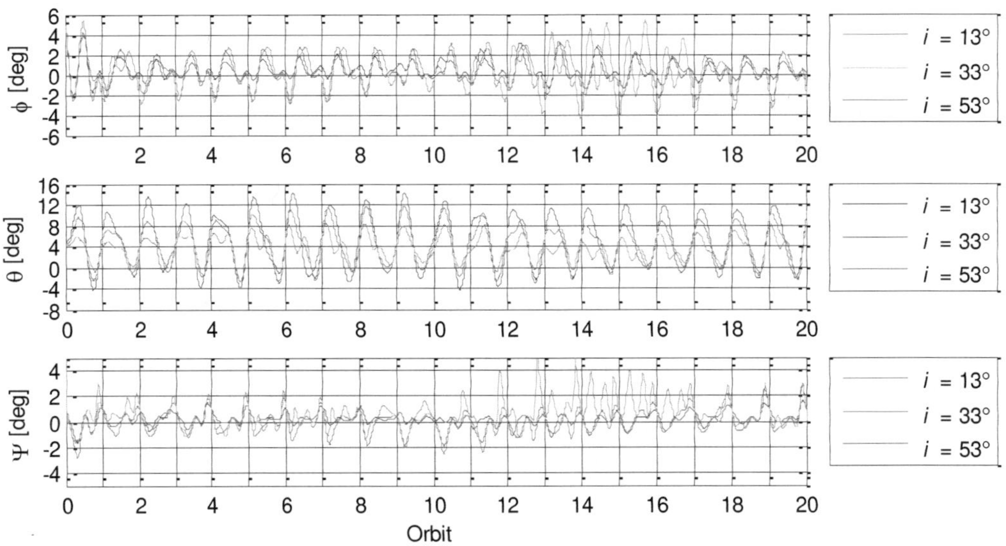

Figure 3. Comparison of the attitude performance.

As for the orbit around the equator, the geomagnetic field is nearly parallel to the pitch axis of the satellite where one of the magnetic torquer is mounted. Thus, the magnitude of B_y is high compared to B_x and B_z. Since the magnetic control torque can only be generated perpendicular to the geomagnetic field and magnetic dipole moment of the magnetic torque, each magnetic torquer controls any two axes that are normal to it. This situation reflects the magnetic dipole moments generated by each magnetic torquers onboard as shown in figure 4 and the torque generated along the roll, pitch and yaw axes as shown in figure 5.

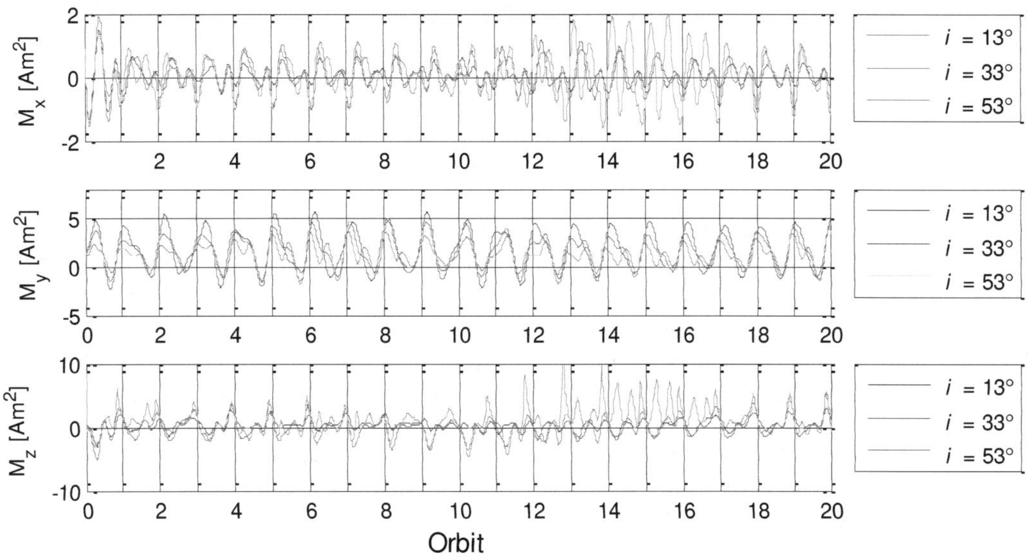

Figure 4. Comparison of the magnetic dipole moment generation.

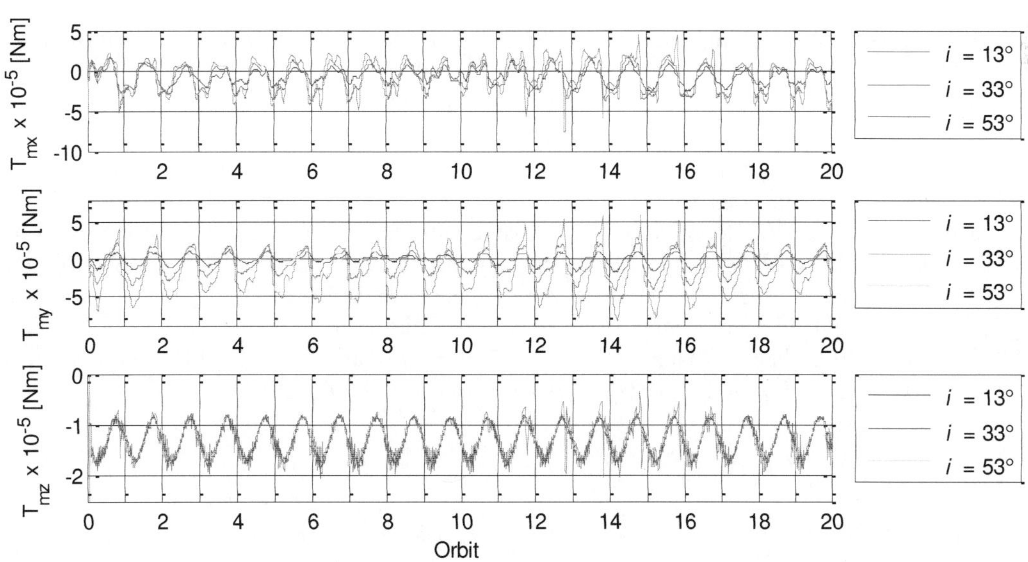

Figure 5. Comparison of the magnetic control torque generation.

5. Conclusion

In this paper, the control torque generation of a gravity gradient stabilized satellite employing 3 magnetic torquers onboard has been simulated for 13°, 33° and 53° orbit inclinations to see how the orbit inclination variation affects the magnetic control torque generation. The control algorithm was structured based on the PD type controller. Simulations were performed using the MATLAB® SIMULINK® codes. Results from the simulations show that the higher orbit inclination generates optimum magnetic attitude control torque for this satellite configuration.

Acknowledgements

The work presented in this paper has been supported by the USM Bridging Grant (304.PAERO.6316112).

References

[1] Kaplan M H 1976 *Modern Spacecraft Dynamics and Control* (New York: John Wiley & Sons)
[2] Maeda K, Hidaka T and Uo M 1997 A magnetic libration control scheme for gravity gradient stabilized small satellite *Adv. in the Astro. Sci.*, 96 (6) pp 887 – 896
[3] Martel, F, Pal P K and Psiaki M 1988 Active magnetic control system for gravity gradient stabilized spacecraft *Proc. of the 2nd AIAA/USU Conf. on Small Satellites* (Utah State University)
[4] Steyn W H 1994 Comparison of low-earth-orbit satellite attitude controllers submitted to controllability constraints J. Guidance, Control and Dynamics 17 (4) pp 795–804
[5] Mostert S and Koekemoer J 1997 The science and engineering payloads and experiments on SUNSAT Acta Astro. 41 (4-10) pp 401–411
[6] Arduini C and Baiocco P 1997 Active magnetic damping attitude control for gravity gradient stabilized spacecraft J. Guidance, and Control 20 (10) pp 117–122
[7] Zhaowei S, Xu Yang and Di Y 2003 Active magnetic control methods for small satellite J. Aero. Eng. 16 (1) pp 38–44
[8] Grassi M 1997a Attitude determination and control for a small remote sensing satellite Acta Astro. 40 (9) pp 675–681
[9] Grassi M 1997b Performance evaluation of the UNISAT attitude control system J. Astro. Sci. 45 p 146
[10] Wang K N, Juang J C and Tsai Y F 2009 Design of the LEAP attitude control system Proc. of the 7th Asian Control Conf. (Hong Kong, China)
[11] Wisnieski R and Blanke M 1999 Fully magnetic attitude control for spacecraft subject to gravity gradient Automatica 35 pp 1201–1214
[12] Kulkarni J and Campbell M 2004 An approach to magnetic torque attitude control of satellites via '*H*∞' control for LTV systems 43rd IEEE Conf. on Decision and Control 1 pp 273–277
[13] Suhadis N M and Varatharajoo R 2012 passive attitude control torque generation performances of a gravity gradient stabilized satellite App. Mech. and Materials, 225 pp 458–463
[14] Stickler A C and Alfriend K T 1974 An elementary magnetic attitude control system AIAA Mech. and Control of Flights Conf. (Anaheim California)
[15] McLean S, Mcmillan S, Maus S, Lesur V Thompson A and Dater D 2004 The US/UK World Magnetic Model for 2005-2010 NOAA Technical Report NESDIS/NGDC-1

Optimum design of a hand-tractor handlebar through metaheuristic algorithms

Apichit Mahachai, Sujin Bureerat, Nantiwat Pholdee*

Sustainable and Infrastructure Research and Development Center, Department of Mechanical Engineering, Faculty of Engineering, Khon Kaen University, Khon Kaen, 40002, Thailand.

*Corresponding author: nantiwat@kku.ac.th

Abstract. This paper presents simultaneous topology shape and sizing design of a walking-tractor handlebar based on using many-objective metaheuristics (MnMHs) in order to reduce vibration transmission, which to some extent causes vibration syndrome disease. The design problem is posed to maximise natural frequencies of the first 5 modes, minimise structure mass and minimise construction cost subjected to stress and displacement constrains with bending and torsion loading conditions. Design variables include topology of side bar stiffeners, shape, and size of all bars. Several MnMHs are used to solve the problems while the results are compared. With this study, the performances of MnMHs for solving optimum design of a walking-tractor handlebar is examined whereas the optimal structures are obtained.

1. Introduction

Nowadays, a walking tractor (also known as a power tiller, a hand tractor, or a behind tractor) is used worldwide particularly in developing countries due to compact, cheap and affordability for small and medium agriculture farms. Moreover, it can be applied to many purposes in the farm such as irrigation, transportation, and tillage [1]. Under working conditions, the walking tractor is subject to several mechanical phenomena such as bending stress, torsion stress, and vibrations [2, 3, 4, and 5] while some of such phenomena e.g. vibration is the cause of an operator injuring issue. Various disorders associated with vibration are vascular, musculoskeletal neurological, and articular which know as hand-arm vibration syndrome (HAVS) [6].

Generally, design of a walking tractor is mostly concerned with improving structural strength based on static analysis [5, 7 and 8]. For the dynamic analysis, the walking tractor handlebar was optimised [2] with two objectives to minimise structural mass and maximise natural frequency. The work was extended to be many-objective optimisation in [9]. This work presented many objective optimisation that minimises structural mass, maximises structural strength in two loading conditions and maximises natural frequencies. Although, many-objective optimisation has been proposed [2 and 9] to serve several requirements of walking tractor handle bar design, it was found that the results obtained in the previous study are difficult to apply in real situations due to expensive and complicated-to-manufacture handlebars being obtained. In this regard, improving design optimisation of a walking tractor handle bar with manufacturing requirements is still a challenge.

This work presents many-objective optimisation of a walking tractor handlebar. The considered design variables include topology of side bar stiffeners, shape, and sizes of all bars. For more practical use, the sizing variables are set to be discrete to fit with the commercialised bar sizes available. The design problem is posed to maximise natural frequencies of the first 5 modes, minimise structure mass and minimise construction cost subjected to stress and displacement constrains with 2 loading conditions, bending and torsion. To deal with the many-objective problem, several MnMHs including, Differential Evolution for Multi-objective Optimisation (DEMO) [10], Multi-objective Harmony Search (MOHS) [11], Hybrid multi-objective real-code population based incremental learning and differential evolution algorithm (MORPBIl) [2], Multi-objective Particle Swarm Optimisation (MOPSO) [12], Unrestricted Population Size Evolutionary Multi-objective Optimisation Algorithm (UPS-EMOA) [13], Non-Dominated Sorting Genetic Algorithm II (NSGA-II) [14] and real-code population-based incremental learning and differential evolution algorithm (RPBILDE) [15] are employed to tackle the problem. The MnMHs performances are investigated and optimal walking tractor handlebar structures are obtained.

2. Design problem

In order to design taking account of ergonomic effect or human factors simultaneously with optimum structural strength, a walking tractor handlebar should have maximised natural frequencies so as to avoid vibration resonance from external excitation that may transmit to a user. Besides, minimising bending and torsion deflection can, to some extent, increase structural reliability. The minimisation of structural mass or volume will affect structural cost. The design variables include topology of side bar stiffeners, shape, and sizes of all bars. Fig. 1 and Fig. 2 show an initial handlebar structure which comprises the main handlebars and stiffeners. The shape of the structure is defined by the parameters; y_1, y_2, y_3, y_4, y_5, z_1 and z_2 while sizing of the structure is defined by the parameter; r_1, r_2, r_3, r_4, r_5, r_6, r_7, r_8, r_9, r_{10} t_1, t_2, t_3 and t_4. The parameters *A, B, C, D, E, F* and $G \in [0, 1]$ are used to define shape of the bars, which can be either a round or a square tubing cross-sections as presented in Table 1, while the parameters *H, I,* and *J* are also used to define shape of bars, which can be either a round or a square bar cross-sections as presented Table 2. Moreover, a certain sections of the structure including *H, I* and *J* can determine a topology as displayed in Fig.3. Material density and young modulus of all bar are set to be 2x1011 N/ m2 and 7800 Kg/m3 respectively.

The many-objective optimisation problem is posed to maximise natural frequencies of the first 5 modes, minimise structure mass and minimise construction cost subjected to stress and displacement constrains with 2 loading conditions, bending and torsion as shown in Fig.4. The optimisation problem can expressed as follow,

Min: $f_1(\mathbf{x}), f_2(\mathbf{x}), f_3(\mathbf{x}), f_4(\mathbf{x})$

$$\text{Subject to}$$
$$\sigma_{\text{bending max}} \leq \sigma_{\text{allow}}$$
$$\sigma_{\text{torsion max}} \leq \sigma_{\text{allow}}$$
$$-0.02 \leq y_1 \leq 0.06, \text{ m}$$
$$-0.025 \leq z_1 \leq 0.025, \text{ m}$$
$$-0.03 \leq y_2 \leq 0.06, \text{ m}$$
$$-0.06 \leq y_3 \leq 0, \text{ m}$$
$$-0.06 \leq y_4 \leq 0, \text{ m}$$
$$-0.06 \leq y_5 \leq 0, \text{ m}$$
$$-0.06 \leq z_2 \leq 0.009, \text{ m}$$
$$0.005 \leq t_4 \leq 0.02, \text{ m}$$
$$A, B, C, D, E, F, G, \in \{0, 1\}$$
$$H, I, J \in \{0, 1, 2\}$$

where **x** is a vector of design variables. f_1 and f_2 are the maximum displacements in the y-direction (vertical direction) due to the bending load and the torsion load, respectively, while f_3 and f_4 are the inverse of the sum of natural frequencies in the first five modes and the structural mass, respectively.

$$\mathbf{x} = \{ y_1, y_2, y_3, y_4, y_5, z_1, z_2\, r_1, r_2, r_3, r_4, r_5, r_6, r_7, r_8, r_9, r_{10}\, t_1, t_2, t_3, t_4, A, B, C, D, E, F, G\, H, I, \text{and } J \}^T$$

f_1 = max displacement due to the bending load
f_2 = max displacement due to the torsion load

$$f_3 = \frac{1}{\omega_1 + \omega_2 + \omega_3 + \omega_4 + \omega_5}$$

f_4 = structure mass
$\sigma_{bending\ max}$ = maximum von Mises stress due to the bending load
$\sigma_{torsion\ max}$ = maximum von Mises stress due to the torsion load
σ_{allow} = allowable stress (450 ×106 N/m2)

Structural analyses are conducted to find the stiffeners matrix, natural frequency, bending and torsion stress of structure. The handlebar structure is modeled as 3D beams and shells. Influence each node in finite element analysis has 6 degrees of freedom.

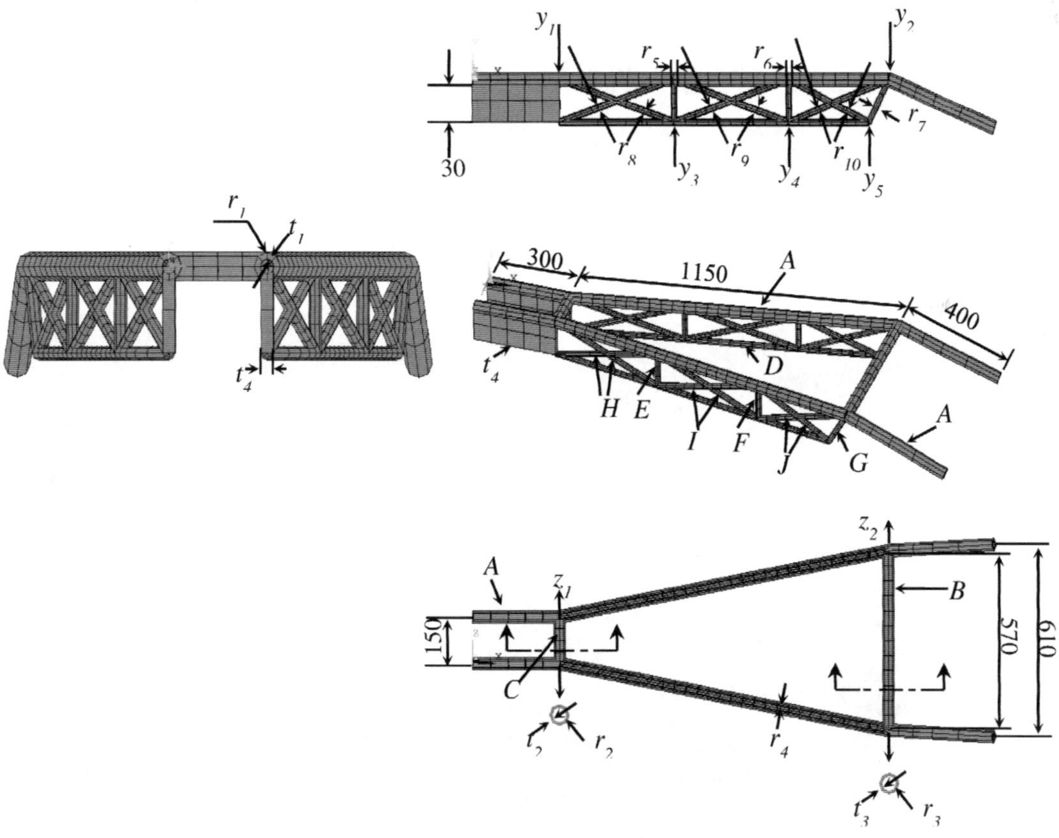

Figure 1. The Design variables.

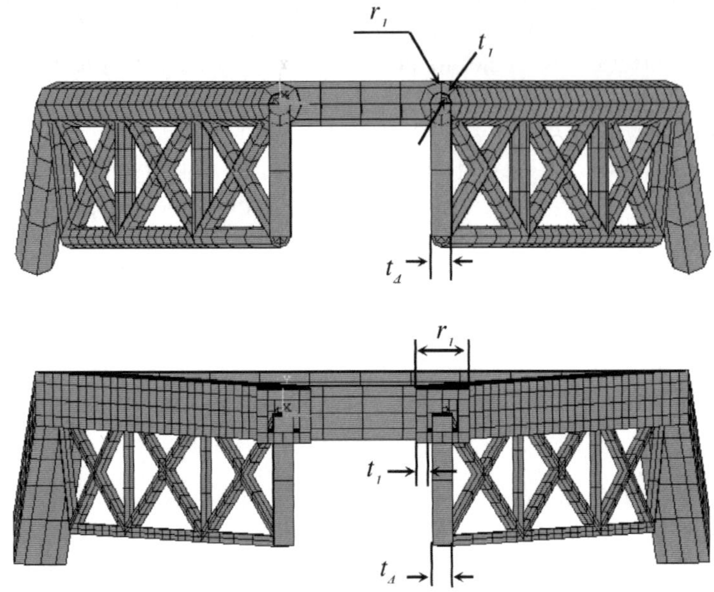

Figure 2. Available cross-sections

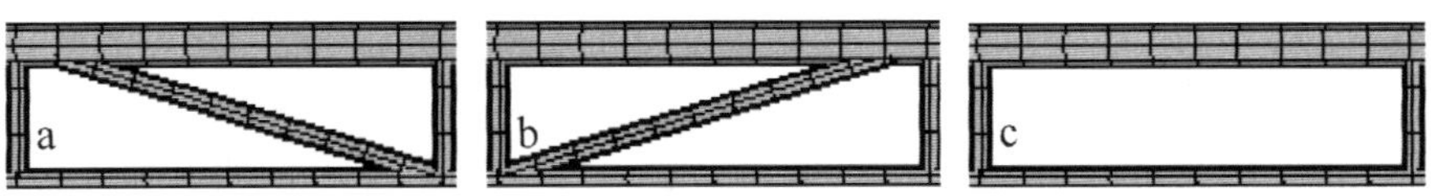

Figure 3. Topology design select of cross-section *H, I* and *J*
(a) diagonal left (b) diagonal right and (c) none.

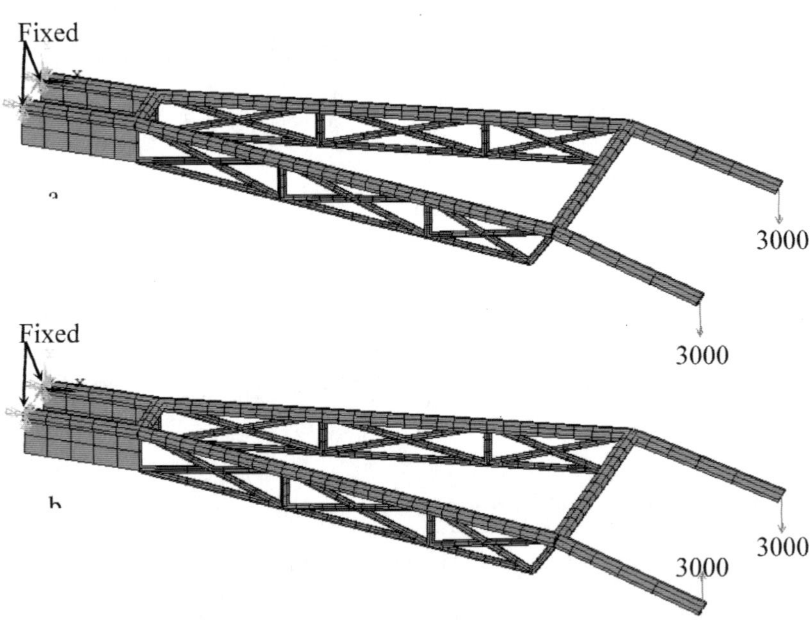

Figure 4. (a) Bending load and (b) Torsion load

IOP Conf. Series: Materials Science and Engineering **370** (2018) 012033 doi:10.1088/1757-899X/370/1/012033

Table 1. Available cross-section *A, B, C, D, E, F* and *G* subject to r_1, r_2 and r_3

a	Round tubing section dimension		note
Set	$\sum_{i=1}^{3} r_i$ (mm.)	$\sum_{i=1}^{3} t_i$ (mm.)	
1	10.85	2	
2	13.6	2.3	
3	17	2.3	
4	21.35	2.3	
5	24.3	2.3 3.2	

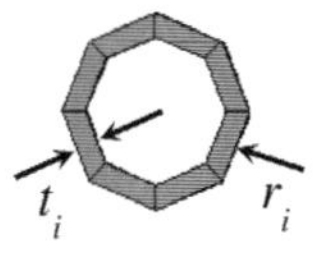

b	Square tubing section dimension		note
set	$\sum_{i=1}^{3} r_i$ (mm.)	$\sum_{i=1}^{3} t_i$ (mm.)	
1	25	2 2.3	
2	32	2.3 3.2	
3	28	2.3 3.2	
4	50	2.3 3.2	

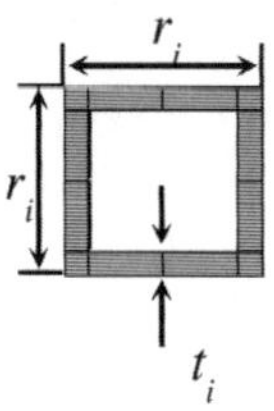

Table 2. Available cross-section *H, I* and *J* subject to r_4, r_5, r_6, r_7, r_8, r_9 and r_{10}

a	Round bar section dimension	note
Set	$\sum_{i=4}^{10} r_i$ (mm.)	
1	3	
2	4	
3	4.5	
4	5	
5	6	
6	7.5	
7	9.5	
8	11	
9	12.5	

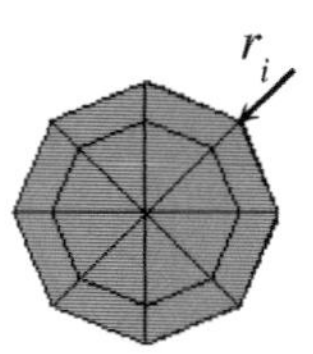

b	Square bar section dimension	note
set	$\sum_{i=4}^{10} r_i$ (mm.)	
1	10	
2	12	
3	16	
4	19	
5	25	
6	32	
7	38	

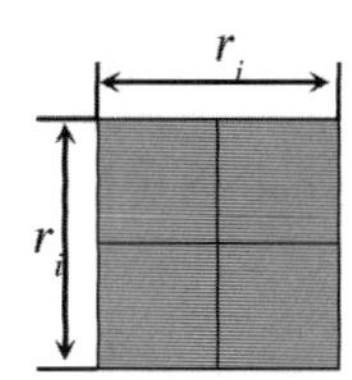

3. Numerical Experiment

The proposed many-objective design problem will be solved by several MOEAs including [9]:

- Differential evolution for multi-objective optimisation (DEMO) [10] using real codes with crossover probability, scaling factor and probability of choosing an element from an offspring in crossover for DE operators being 0.7, 0.8, and 0.5, respectively.
- Multi-objective harmony search (MOHS) [11] using harmony memory considering rate, minimum pitch adjustment rate, maximum pitch adjustment rate and minimum bandwidth rate being 0.5, 0.2, 2, 0.45, and 0.9, respectively.
- Multi-objective population-based incremental learning (MOPBIL) [2] using binary code. The learning rate, mutation probability, and mutation shift are set as 0.25, 0.05, and 0.2, respectively.
- Multi-objective Particle Swarm Optimisation (MOPSO) [12] using starting inertia weight, ending inertia weight, cognitive learning factor and social learning factor being 0.75, 0.1, 0.75 and 0.75, respectively.
- Unrestricted population size evolutionary multi-objective optimisation algorithm (UPS-EMOA) [13] using crossover probability, scaling factor, probability of choosing element from offspring in crossover, minimum population size, and burst size being 0.7, 0.8, 0.5, 10, and 25 respectively.

- Non-dominated sorting genetic algorithm II (NSGA-II) [14] using real codes with crossover mutation probabilities of 1.0 and 0.1 respectively.
- Real-code population-based incremental learning and differential evolution algorithm (RPBILDE) [15] using real codes with NI = 40 where each probability tray produces 5 design solutions. Crossover probability, scaling factor and probability of choosing an element from an offspring in crossover for DE operators are set as 0.7, 0.8, and 0.5 respectively.

Each algorithm is used to solve the problem stated in Section 2 for five optimsiation runs. The population size is set to be 100 while the number of iterations is 250. One optimisation run takes 100×250 function evaluations. It should be noted that the total number of function evaluations used in this study can be considered insufficient for some meta-heuristics according to the literature; nevertheless, this value is set so as to look for only really powerful algorithms. The hypervolume indicator as detailed in [16 and 17] will be used to measure the optimisers' performance. Note that the optimisation parameter settings detailed above are obtained from using several settings for each optimiser and selecting the one that gives the best results.

4. Results and Discussion

After performing 5 optimisation runs of the five MnMHs on solving the proposed walking tractor handlebar optimisation problem, comparison results based on hypervolume indicator is shown in table 3. The mean value of hypervolume is used to measure algorithm's search convergence while the standard deviation (STD) value is used to measure algorithm's search consistency. The higher mean hypervolume values is the better convergence while the lower STD is the better consistency. From Table 3, the best performer based on mean hypervolume value is UPS-EMOA while the second best and the third best are MODE and RPBILDE, respectively. For the measure of search consistency based on STD of hypervolume values, the best performer is MODE while the second best, which obtained slightly higher STD, is UPS-EMOA. Overall, the UPS-EMOA is said to be the best performer for solving the proposed walking tractor handlebar optimisation problem in this study in terms of both search convergence and search consistency. Figure 5 shows some of the optimum structures obtained from using UPS-EMOA.

Table 3. Comparison results based on hypervolume indicator

Algorithm	Runs						
	1	2	3	4	5	Mean	STD
MODE	0.831	0.834	0.833	0.833	0.834	0.833	0.00110
MOHS	0.814	0.815	0.815	0.812	0.818	0.8148	0.00194
MOPBIL	0.807	0.793	0.795	0.799	0.8	0.7988	0.00483
MOPSO	0.819	0.813	0.819	0.816	0.819	0.8172	0.00240
UPS-EMOA	0.835	0.835	0.833	0.833	0.836	0.8344	0.00120
NSGA-II	0.831	0.823	0.831	0.824	0.828	0.8274	0.00338
RPBILDE	0.832	0.835	0.832	0.832	0.831	0.8324	0.00136

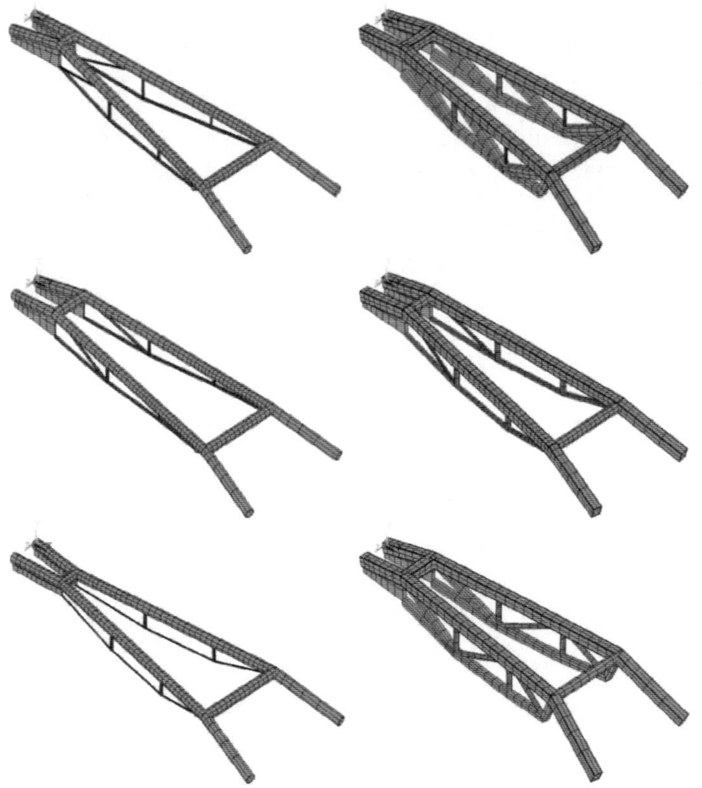

Figure 5. Some of the optimum structures obtained from the best performer UPS-EMOA.

5. Conclusions

In this work, simultaneous topology, shape and size design of a walking-tractor handlebar based on using many-objective metaheuristics (MnMHs) is presented. The design problem is posed to maximise the sum of natural frequencies of the first 5 modes, minimise structure mass and minimise construction cost subject to stress and displacement constrains with 2 loading conditions, bending and torsion. Design variables include the topology of side bar stiffeners, shape, and size of all bars. Several MnMHs are used to solve the problem while the results are compared to based on the hypervolume indicator. With this study, the best algorithm of the five implemented MnMHs is UPS-EMOA. For future work, more important parameters for ergonomic design will be added while a more advanced many-objective optimiser will be developed.

Acknowledgement

The authors are grateful for the support from the Thailand Research Fund (TRF).

References

[1] Kanyakam S, Srisomporn S and Bureerat S 2009 *In: Proceedings of the 23rd Conference of the Mechanical Engineering Network of Thailand, Chiang Mai, Thailand.*
[2] Bureerat S and Sriworamas K 2007 *Advance in Soft Computing.* **39** 223–32
[3] Dewangan K N and Tewari V K 2010 *International Agricultural Engineering Journal.* **19** 48-57
[4] Mojica J M M, Yap M A B, Layaoen H D Z and Amongo R M C 2016 *Formerly the Philippine Agriculturist.* ISSN 0031-7454
[5] Fabbri A, Cevoli C and Cantalupo G 2017 *Journal of Agricultural Engineering.* **48** 2
[6] Bovenzi M 1998 *Int Arch Occup Environ Health.* **71** 509-15

[7] Kanyakam S and Bureerat S 2007 *IEEE Congress on Evolutionary Computation.* **4162** 169
[8] Tewaria V K and Dewanganb K N 2009 *biosystems engineering.* **103** 146-58
[9] Mahachi A, bureeat S and Pholdee N 2017 *Advances in Computational Design*. in progress.
[10] Robic T and Filipic B 2005 *Evolutionary Multi-Criterion Optimization.* **3410** 520-33
[11] Sivasubramani S and Swarup K S 2011 *Electrical Power and Energy System.* **33** 745–52
[12] Reyes-Sierra M and Coello Coello C A 2006 *International Journal of Computational Intelligence Research.* **2** 287–308
[13] Aittokoski T and Miettinen K 2010 *Optimization Methods and Software.* **25** 841-58
[14] Deb K, Pratap A, Agarwal S and Meyarivan T 2002 *IEEE Transactions on Evolutionary Computation* vol **6** 182-97
[15] Pholdee N and Bureerat S 2013 *Information Sciences.* **223** 136-52
[16] Zitzler E and Thiele L 1998 *Lecture Notes in Computer Science 1498: Parallel Problem Solving from Nature – PPSN V* **1498** 292-301
[17] Bandyopadhyay S and Mukherjee A 2014 *IEEE Transactions on Evolutionary Computation* **99** 1

International Conference on Aerospace and Mechanical Engineering (AeroMech17) IOP Publishing
IOP Conf. Series: Materials Science and Engineering **370** (2018) 012034 doi:10.1088/1757-899X/370/1/012034

The Influence of Hoop Diameter on Aerodynamic Performance of O-Ring Paper Plane

N. I. Ismail[1,a], Hazim Sharudin[1], R. J. Talib[1], A. A. Hassan[1] and H. Yusoff

Faculty of Mechanical Engineering, Universiti Teknologi Mara, Kampus Pulau Pinang, 13500 Permatang Pauh, Pulau Pinang, Malaysia.

[a]Corresponding's author: iswadi558@ppinang.uitm.edu.my

Abstract. The O-ring paper plane can be categorized as one of the Micro Air Vehicle (MAV) based on their characteristics and size. However, the aerodynamics performance of the O-ring paper plane was not fully discovered by previous researchers due to its aerodynamics complexity and various hoop diameters. Thus, the objective of this research is to study the influence of hoop diameters towards the aerodynamics performance of O-ring paper plane. In this works, three types of O-ring paper plane known as Design 1, 2 and 3 with different hoop diameter were initially developed by using the ANSYS-Design Modeler. All the design was analyzed based on aerodynamic simulations works executed on ANSYS-CFX solver. The results suggested that Design 3 (with larger hoop size) produced better C_L, C_{Lmax} and AoA_{stall} magnitude compared to other design. In fact, O-ring paper plane with larger hoop size configurations showed potential in providing at least 5.2% and 5.9% better performance in stability ($\Delta C_M/\Delta C_L$) and aerodynamic efficiency (C_L/C_{Dmax}), respectively. Despite the advantages found in lift performances, however, O-ring paper plane with larger hoop size configurations slightly suffered from larger drag increment ($C_{Dincrement}$) compared to smaller hoop size configurations. Based on these results, it can be presumed that O-Ring paper plane with larger hoop sizes contributed into better lift, stability and aerodynamic efficiency performances but slightly suffered from larger drag penalty.

1. Introduction

Micro Air Vehicle (MAV) is defined as a small, portable flying vehicle designed for performing reconnaissance and surveillance missions. MAV also known as a micro scale class of Unmanned Aerial Vehicle (UAV). A basic MAV flight control system includes a miniature receiver with a wire antenna, two miniature servos to drive elevons, an electronic speed controller, and a radio controller with an uplink command transmitter[1]. There are lots of advantages of MAV includes silence, stealth, diminutive and practically undetectable. Due to the MAV characteristics, it has numerous potential applications for both military and civilian[2]. Based on the previous study, MAV can be categorized according to 5 basic types known as flapping, fixed, membrane flexible, rotary and morphing [3]as shown in Figure 1. At maximal dimension of 15 cm and nominal flight speeds of around 10 m/s[4], MAVs can perform missions such as environmental monitoring, surveillance, and assessment in hostile situations. MAV normally operated at Reynolds number range between 10,000 and 100 000, however the flow separation at this Reynolds Number range has led to sudden increases in drag and aerodynamic efficiencies loss.

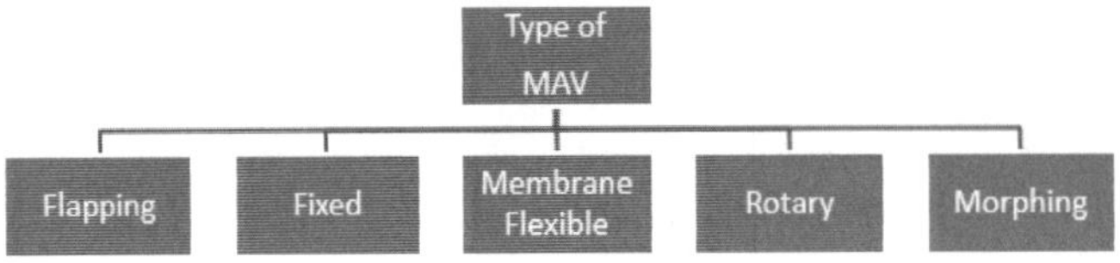

Figure 1. Category of MAV

Through a generation, paper plane has been introduced as an introduction to aeronautical engineering for some many ages. Paper plane is a simple glider that made out from a piece of paper. By using the standard origami technique, the paper plane can be fold into any desired shape and the size of the wingspan also can be adjusted during the folding process[5]. Paper plane can be categorized as one of the MAV because of their sizes. In recent times, paper plane has undergone many changes in terms of their design. O-Ring paper plane (Figure 2) can be categorized as one of the evolution of paper plane designs which is unconventional and complex design. This type of wing is also known as closed wing[6]. Although the O-Ring paper planes are easy to produce, the aerodynamics performances behind this kind of paper plane is still not fully discovered because of its aerodynamics complexity and various hoop sizing. Thus, the objectives of this works to study the influence of hoop diameters towards the aerodynamics performance of O-ring paper plane. In this works, the O-ring CAD designs are initially developed by using the ANSYS-Design Modeler before the aerodynamic analysis was executed based on ANSYS-CFX simulation works. Three types of O-ring paper plane with different hoop sizing known as Design 1, 2 and 3 are used for comparison works.

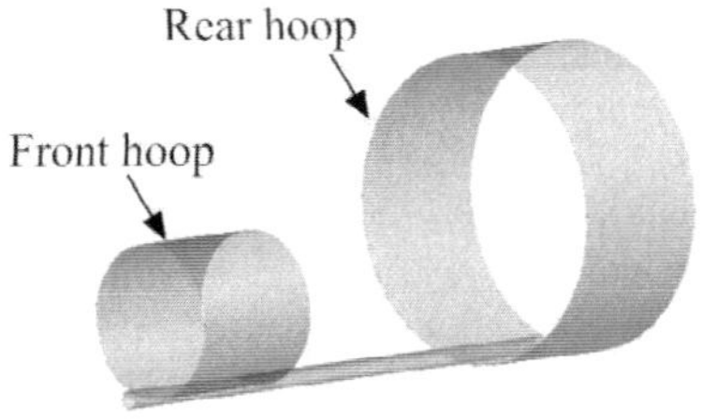

Figure 2. O-Ring paper plane

2. Methodology

In the present research, the O-ring CAD designs are initially developed by using the ANSYS-Design Modeler before the aerodynamic analysis was executed based on ANSYS-CFX simulation works. To solve the turbulent flow issue, 3D RANS equations coupled with SST k-ω turbulent equation are employed under the assumption of a steady, incompressible and turbulent airflow field.

2.1. O-ring paper plane model

The O-Ring paper plane models were develop based on its common configuration in which the front hoop will be smaller than the rear hoop configuration. For comparison works, three selected designs known as is Design 1, Design 2 and Design 3 were used. Each design has different hoop configurations as presented in Table 1 and Figure 3.

2.2. Domain Sizing and Mesh Generation

All the simulation method used in this works were set up in the ANSYS-Workbench framework. The O-Ring CAD design was sliced into half to ease the simulation process and shorten the simulation time. The airflow domain was built surrounding the O-Ring design based on standard sizing as shown in Figure 4. All the O-Ring paper plane part and domain were combined as one-part to ensure that the mesh conforming characteristics achieved in the flow domain. Unstructured hybrid mesh and inflation layers with ANSYS SOLID 187 3D element type was created for all design. The first cell above the hoop surface is set at $y^+ \leq 1$. Results of the grid independent study shows that the optimized grid achieved around 727745 elements as shown in Figure 5.

Table 1. The characteristic of the 3 selected designs

O-Ring Design	Design 1	Design 2	Design 3
Front hoop diameter	50 mm	70 mm	90 mm
Rear hoop diameter	100 mm	120 mm	140 mm
Wing area	7800 mm^2	9880 mm^2	1190 mm^2
Chord length	52 mm	52 mm	52 mm
Length	200 mm	200 mm	200 mm
Reynold number	100, 139	100, 139	100, 139
Velocity	29.1 m/s	29.1 m/s	29.1 m/s

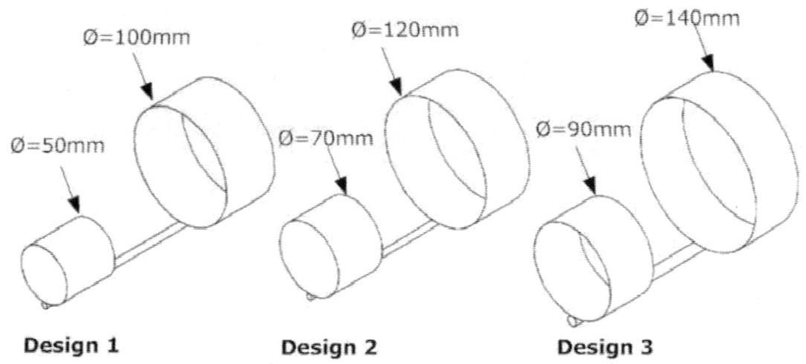

Figure 3. Different hoop sizing on each O-Ring paper plane design

2.3. MAV wing simulation

The boundary conditions imposed on the air domain are also shown in Figure 4. The location of inlet and outlet indicated by flow vectors (Figure 4). The flow velocity was specified at the inlet with velocity of 29.1 m/s. Zero pressure boundary condition is implemented at the outlet to ensure airflow continuities. The symmetrical wall and side walls (opposite the symmetrical wall) imposed as symmetrical and slip surface boundary conditions, respectively. Non-slip boundary surface imposed on hoop surface and automatic wall function is fully employed to solve the flow viscous effect. The angle of attack (AOA) for all O-Ring design was varied between -10° to 40° with 2° interval.

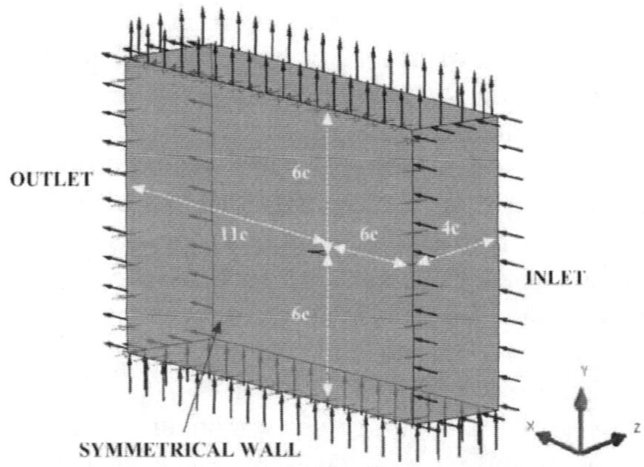

Figure 4. The computational airflow domain sizing and boundary conditions

A steady state, incompressible flow Navier-Stokes equation (RANS) combined with Shear Stress Turbulence (SST) model was used to solve the flow problems over each O-Ring design. The simulation convergence was control based on the magnitude of momentum residual (below 1.0×10^{-5}) and monitoring the stability of of lift coefficient (C_L) and drag coefficient (C_D) value.

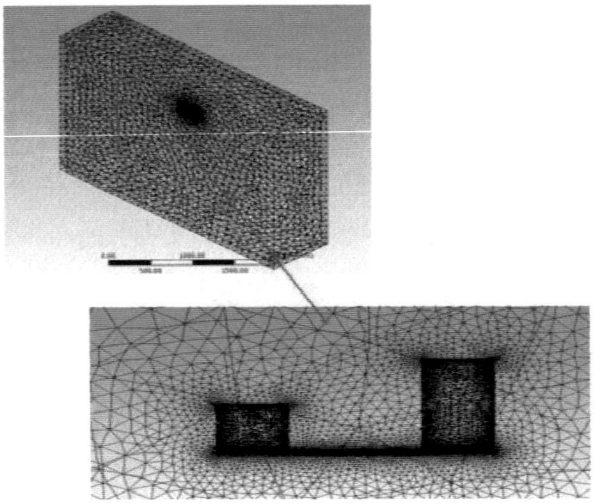

Figure 5. The optimize mesh and inflation layer for ANSYS-CFX solver

3. Results and Discussion

3.1. Aerodynamic performance of O-ring paper plane

Based from the ANSYS-CFX simulation, the aerodynamics performance analysis on each O-Ring design is viewed in detail with particular study given on the aerodynamics performance characteristics such as lift coefficient (C_L), drag coefficient (C_D) and moment coefficient (C_M). The aerodynamic efficiency (C_L/ C_D) was also been analysed based from the of C_L and C_D performances. To show the variation of the aerodynamics performances, each result is presented based on comparison study between the three O-Ring paper plane designs.

3.2. Lift coefficient distribution

Figure 6 shows the C_L performance for all three O-Ring paper plane designs. The graph shows that each design produces a different C_L magnitude but almost similarly in overall trends. In general, the magnitude of C_L for each design have increased proportionally with the angle of attack (AoA) increment particularly in the pre-stall area (AoA area below the stall angle). The C_L magnitude peaks (C_{Lmax}) at its highest magnitude at the stall angle (AoA_{stall}) before it slightly decreases and almost plateau at the post-stall area (AoA>AoAstall).

Based on the C_L analysis taken at AoA =0° to 20 °, the results show that Design 3 produced better C_L magnitude compared to the other designs. In average, Design 3 produce about 16.6% and 23.3% higher C_L magnitude compared to than Design 2 and Design 1, respectively. Based on the C_{Lmax} results, Design 3 exhibited the highest C_{Lmax} at C_{Lmax}=0.613. This C_{Lmax} magnitude is about 6.3% at and 14.2% higher than Design 2 and Design 3 produced, respectively. In AoA_{stall} analysis, the results exhibited that Design 2 and Design 3 produce almost similar AoA_{stall} magnitude at AoA_{stall} =30°. However, Design 1 has induced the earliest stall angle at AoA_{stall} =26° which is 15.4% earlier than Design 2 and Design 3 produced. Based on these results, it obviously shows that Design 3 and Design 2 with larger hoop sizing has promising advantages in producing better C_L magnitude and C_{Lmax}

performance. Thus, one can presume that the O-ring paper plane with larger hoop size which obviously has larger aspect ratio potentially induces the better C_L performances.

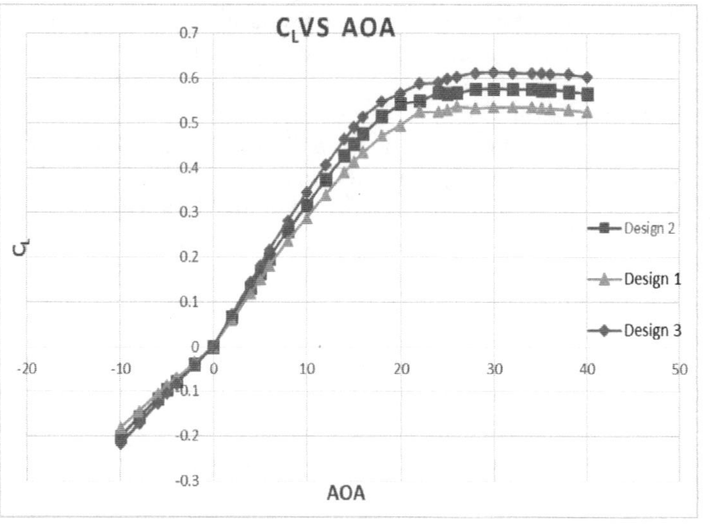

Figure 6. C_L performances for all O-Ring paper plane designs.

3.3. Drag coefficient distribution

The aerodynamic studies of O-Ring paper plane continued on the drag coefficient (C_D) performance as shown in Figure 7. Based on the simulation results, the results exhibited that each designs produced almost consistent and similar C_D trend throughout the AoA range. Generally, the C_D magnitude for each design increase with the AoA increment.

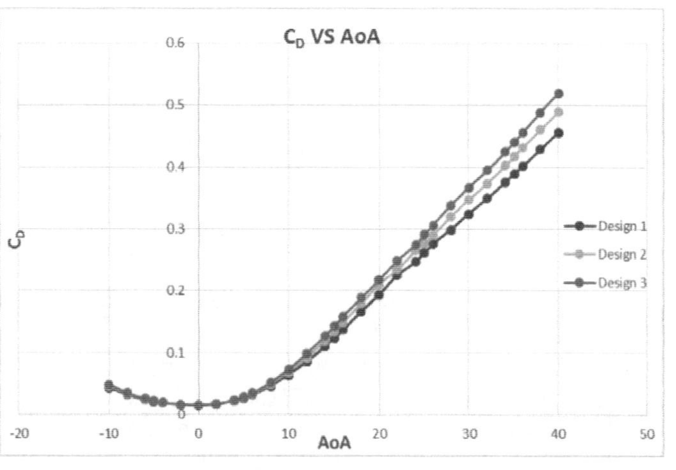

Figure 7. C_D performances for all three O-Ring paper plane design.

Based on the minimum drag magnitude ($C_{D\ min}$) analysis, the results indicated that all three designs exhibited the lowest C_D magnitude at AoA=0°. The detail analysis shows that Design 3 produced the lowest $C_{D\ min}$ among the design at $C_{D\ min}$ =0.01546. Design 2 produced the $C_{D\ min}$ =0.0157 which is 1.4% higher than Design 3 while Design 1 produce the highest $C_{D\ min}$ value of at $C_{D\ min}$ =0.016. This value is about 3.5% higher than the $C_{D\ min}$ magnitude of found in Design 3. Despite the $C_{D\ min}$ advantages, the magnitude of C_D increment ($C_{D\ increment}$) of Design 3 is the highest among the design at

$C_{D\ increment}$ = 22.82% particularly at AoA =0° to 20 °. This is followed by Design 2 and Design 1 at $C_{D\ increment}$ = 22.33% and 21.36%, respectively. However, based on comparative study, the difference in $C_{D\ increment}$ between the designs is below 2%. Thus, one can be presumed that larger hoop size may induced larger drag penalty. This is due to larger surface area found in Design 3 and Design 2 which expose and interact to the airflow which may contribute into the larger skin drag[7].

3.4. Moment coefficient distribution

The graph in the Figure 8 shows the simulation results for the pitching moment coefficient (C_M) performance that has been measured at the wing leading edge. The results indicate that each designs produced almost consistence C_M curve throughout angle of attack range. In general, the magnitude of C_M for all three designs decreased in a non-linear pattern as the value of the C_L increase. The trend clearly shows that the non-linear C_M curve began at the $-C_L$ magnitude up to the C_{Lmax} magnitude before the C_M curve started to induce an irregular pattern. Analysis on C_M results is specifically focused to capture the C_M slope magnitude ($\Delta C_M / \Delta C_L$) which taken at the AoA 5° to 20°. The initial analysis shows that each design produced a negative $\Delta C_M / \Delta C_L$ magnitude which use as an initial indicator for aircraft stability during the flight [8].

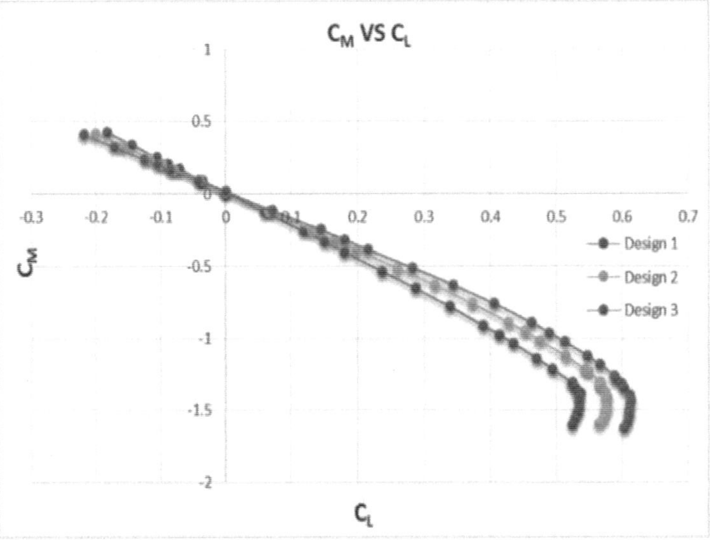

Figure 8. C_M performances on three O-Ring paper plane design.

Based on $\Delta C_M / \Delta C_L$ analysis, Design 3 exhibited the steepest slope magnitude at $\Delta C_M / \Delta C_L$ = -0.444. Design 2 and Design 1 generated lower $\Delta C_M / \Delta C_L$ magnitude at $\Delta C_M / \Delta C_L$ = -0.4222 and -0.396 respectively. This $\Delta C_M / \Delta C_L$ magnitude is about 5.2% (Design 2) and 12.1% (Design 1) lower than Design 3 produced. In aerodynamics study, the $\Delta C_M / \Delta C_L$ magnitude are used to indicate the level of stability for an aircraft. Steeper $\Delta C_M / \Delta C_L$ slope means better stability on MAV wing [8]. Thus, based on this statement, one can presumed that Design 3 is the the most stable among the design. In fact, larger hoop size (found in Design 3 and Design 2) induced better stability O-Ring paper plane.

3.5. Aerodynamic efficiency

The graph in the Figure 9 presents the aerodynamic efficiency (C_L/C_D) for all three design of O-Ring paper plane. In general, the C_L/C_D curves for the all designs increased linearly and peaks approximately at C_L = 0.16 - 0.18. The peak point for the C_L/C_D magnitude is known as the maximum aerodynamic efficiency ($C_L/C_{D\ max}$). Based on the $C_L/C_{D\ max}$ magnitude, the result shows that each design induced the C_L/C_{Dmax} at the low AoA value between 5° to 6° (or equivalent to $C_L = 0.16 \sim 0.18$). Then, C_L/C_D curve starts to decrease after $C_L/C_{D\,max}$ up to the $C_{L\,max}$ point (AoA_{stall}).

The detail study on $C_L/C_{D\,max}$ magnitude has been conducted to explain the maximum aerodynamic efficiency for all designs. The result shows that Design 3 generated the highest C_L/C_{Dmax} magnitude at 6.27403. This is followed by Design 2 and Design 1 at $C_L/C_{D\,max}$ =5.92613 and $C_L/C_{D\,max}$ =5.5559, respectively. Based on these $CL/C_{D\,max}$ results it shows that Design 3 produce about 5.9% and 13.1% better $C_L/C_{D\,max}$ magnitude than Design 2 and Design 1, respectively. This is possibly due to bigger surface area found in Design 3 which induce better C_L distribution and consequently enhance the C_L/C_{Dmax} performance. From this result, one can presume that the O-Ring with bigger hoop size induces the better $C_L/C_{D\,max}$ performances.

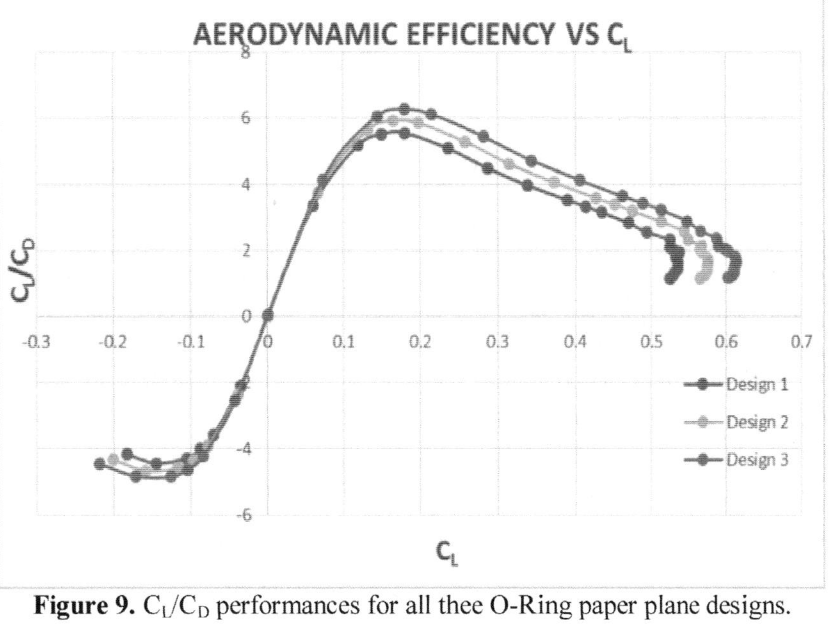

Figure 9. C_L/C_D performances for all thee O-Ring paper plane designs.

4. Conclusion

In this work, the O-Ring paper plane known as Design 1, Design 2 and Design 3 are compared to evaluate the influence of hoop sizing towards the aerodynamic performances. The comparison works have been done by focusing on the common aerodynamic performances such as C_L, C_D, C_M and C_L/C_D.

Based on C_L performance, the result shows that Design 3 produces at least 16.6% higher C_L magnitude compared to other design. In fact, O-Ring paper plane larger hoop configuration (as found in Design 3 and Design 2) produced better C_{Lmax} and AoA_{stall} magnitude compared to smaller hoop configuration.

Based on C_D performance, the result shows that the O-Ring paper plane with larger hoop configuration (Design 3) able to produce at least 1.4% better $C_{D\,min}$ compared to smaller hoop configurations (Design 2 and Design 1). However, Design 3 slightly suffer from larger $C_{D\,increment}$ with at least 2% larger $C_{D\,increment}$ magnitude compared to Design 2 and Design 1.

Based on ΔC_M / ΔC_L analysis, Design 3 promisingly exhibited the steepest slope magnitude at $\Delta C_M/\Delta C_L$ = -0.444 which is about 5.2% and 12.1% better than Design 2 and Design 1 produced, respectively. Steeper ΔC_M / ΔC_L slope means better stability on MAV wing. Thus, the results clearly show that larger hoop size (found in Design 3 and Design 2) induced better stability O-Ring paper plane.

The detail study on C_L/C_{Dmax} magnitude reveals that Design 3 exhibited the highest C_L/C_{Dmax} magnitude at 6.27403. This magnitude is about 5.9% and 13.1% better than Design 2 and Design 1 produced, respectively. This result indicates that larger larger hoop size configuration (found in Design 3 and Design 2) contributes into better the aerodynamic efficiency on O-Ring paper plane.

In conclusions, one can presume that O-Ring paper plane with larger hoop sizes contributes into better lift, stability and aerodynamic efficiency performances. However, it may slightly suffer from larger drag penalty probably due to larger skin drag friction. In future works, an experimental study will be conducted to assess the O-Ring paper plane configurations with view to validate the simulation results.

Acknowledgement

Authors acknowledge technical and financial support from Universiti Teknologi MARA Cawangan Pulau Pinang and the Government of Malaysia.

References

[1] H. Liu, X. Wang, T. Nakata, and K. Yoshida, "Autonomous Control Systems and Vehicles," vol. 65, 2013.
[2] H. Wu, D. Sun, Z. Zhou, S. Xiong, and X. Wang, "Micro Air Vehicle: Architecture and Implementation," in *International Conference 0n Robotics & Automation*, 2003, pp. 1–6.
[3] N. I. Ismail, "Aerodynamic Performances and Flow Structure Investigations On Active Twist Morphing MAV Wing," Universiti Teknologi MARA, 2014.
[4] N. I. Ismail, M. A. Tasin, R. J. Talib, A. H. Zulkifl, M. H. Basri, and M. M. Mahadzir, "A Review on MAV Design Challenges," *Australian Journal of Basic and Applied Sciences*, vol. 10, no. 7, pp. 84–90, 2016.
[5] N. B. Feng, K. Q. Mei, P. Y. Yin, and J. U. Schlüter, "On the Aerodynamics of Paper Airplanes," in *27th AIAA Applied Aerodynamics Conference*, 2009, no. June, pp. 1–16.
[6] P. Roglev, "MDO Framework for Conceptual Design of Closed Wing UAV," *International Journal of Scientific Research Engineering & Technology (IJSRET)*, vol. 2, no. 8, pp. 526–531, 2013.
[7] N. I. Ismail, "Aerodynamic Performances of Twist Morphing MAV Wing," PhD Thesis Dissertation, Universiti Teknologi MARA, 2015.
[8] N. I. Ismail, A. H. Zulkifl, M. Z. Abdullah, M. H. Basri, N. S. Abdullah, A. H. Zulkifli, M. Hisyam Basri, and N. Shah Abdullah, "Computational Aerodynamic Analysis on Perimeter Reinforced (PR)-Compliant Wing," *Chinese Journal of Aeronautics*, vol. 26, no. 5, pp. 1093–1105, Oct. 2013.

IOP Conf. Series: Materials Science and Engineering **370** (2018) 012035 doi:10.1088/1757-899X/370/1/012035

The contributions of human factors on human error in Malaysia aviation maintenance industries

H Padil[1*], M N Said[1], and A Azizan[1]

[1]Universiti Kuala Lumpur-Malaysia Institute of Aviation Technology. Lot 2891 Jalan Jenderam Hulu, 43800 Dengkil, Selangor, Malaysia

*Corresponding author: husnapadil@unikl.edu.my

Abstract. Aviation maintenance is a multitasking activity in which individuals perform varied tasks under constant pressure to meet deadlines as well as challenging work conditions. These situational characteristics combined with human factors can lead to various types of human related errors. The primary objective of this research is to develop a structural relationship model that incorporates human factors, organizational factors, and their impact on human errors in aviation maintenance. Towards that end, a questionnaire was developed which was administered to Malaysian aviation maintenance professionals. Structural Equation Modelling (SEM) approach was used in this study utilizing AMOS software. Results showed that there were a significant relationship of human factors on human errors and were tested in the model. Human factors had a partial effect on organizational factors while organizational factors had a direct and positive impact on human errors. It was also revealed that organizational factors contributed to human errors when coupled with human factors construct. This study has contributed to the advancement of knowledge on human factors effecting safety and has provided guidelines for improving human factors performance relating to aviation maintenance activities and could be used as a reference for improving safety performance in the Malaysian aviation maintenance companies.

1. Introduction

Aviation is one of the transport industries where the change of technology is eminent. However one must understand that maintenance of aircraft does not entirely depend on technology alone. In fact the main contributor to aircraft maintenance is human who the aircraft maintenance personnel are. The work of aircraft maintenance personnel obviously is vital to the safety and economic viability of airlines operations[1]. Aviation safety must not be compromised by inadequate maintenance. To ensure that maintenance is carried out properly, the aircraft maintenance personnel must be properly trained [2]. Aviation maintenance has changed over the years; new aircraft manufactured with new materials and systems operated with the latest technology. However, one aspect of aviation maintenance that has not changed is that most maintenance task is done by human. As the tasks are performed by human, errors will somehow surface and if not properly contained, can cause accidents where lives are lost. Currently, there are numerous hurdles in the human factors pertaining to aviation maintenance. This research is a fundamental proactive study to understand issues relating to human errors in aviation and also to ensure safe and effective maintenance processes so that aircrafts released from maintenance are safe and timely. Emphasis and study on this subject will provide confidence to the aviation authorities and on those aviation industries that has shown commitment in battling human error contribution to aircraft maintenance within Malaysia. Although the human factors are highly emphasized in the human factors management literature, human factors alone may not only be the factors contribute to the human error in aviation industry[3]. The result shows that, if the human factors are coupled with the organization, then the strategy for reducing human error will be better in aviation maintenance industries. This implies that the following constructs of human factors (software, hardware, environment, liveware(I) and liveware(O) and also efforts of organization (quality of support, company policy, workforce, finance strategy, and safety culture) must be managed and controlled within aviation maintenance companies in order to reduce human error. Therefore, managers in aviation maintenance industry should consider human factors and organization as a facilitating initiative for reducing human error in aviation environment.

2. Methodology

2.1. Development of Questionnaire

A survey was conducted to find out the perception of the maintenance personnel on the significant human factors that would contribute to human error in aircraft maintenance environment. The most efficient and productive way to formulate effective questions as well as set meaningful framework for the research is to involve at least some of the respondents in the construction of the questionnaire[4]. Basing on the comments above, the final draft of the questionnaire was sent to the aviation maintenance experts and practitioners from the industries and academician the objectives to evaluate the feasibility of carrying out the research in this environment, to acclimatize with the organizational environment, especially in areas that are not familiar to the researcher, to identify which category of maintenance personnel in aviation industries that would have the greatest impact on human error in aviation maintenance and to communicate with the potential survey respondents so that the research is well supported and can be conducted smoothly with minimum hassle.

2.2. Data Collection

Data are simple facts or recorded measures of certain phenomena, while information refers to a body of facts in a format suitable for decision making or in a context that defies relationship between pieces of data collection and procedure. Collecting data using survey questionnaires methods could be administrated using mail, or delivered personally. Initially, a survey questionnaire has been distributed to the selected 18 aviation maintenance industries in Malaysia. Respondent were selected from the several categories of aviation maintenance personnel such as Supervisor. License Aircraft Engineer (LAE), Instructor and Technician. Some of the questionnaires were personally handed directly to the respondent, but a major portion of the questionnaire was mailed to the leader for distribution. The estimated times taken by respondents to answer the questionnaires are taking about three month. A total of 500 questionnaires distributed on the first stage, however only 146 responded have returned to the researcher within 2 to 5 weeks from the day of the questionnaires being dispatched. The follow up process was carried out with telephone calls, fax transmission, and re-mail the questionnaire to the respective aviation industries with the same amount of questions, and the total of 315 respondents were recorded after the second follow up process. In the research of assessing human factors variables in aviation maintenance, the data will be used to provide quantitative data for the purpose of analysis. The data will be used as the research on assessing human factors is more of cognitive science where by the perception of the people involved in the maintenance environment towards the casual factors of human errors must be considered. Survey conducted on the aviation maintenance personnel by using questionnaire covering elements that normally contributes to human errors in aircraft maintenance industry in Malaysia.

2.3. Structural Equation Modelling

The Structural Equation Modelling (SEM) was used to test and analyse the relationship of the research model. SEM aims to examine the inter-related relationships simultaneously between a set of posited constructs, each of which is measured by one or more observed items.SEM involves the analysis of two models; a measurement or factor analysis model and a structural model[5]. The measurement model specifies the relationships between the observe measures and their underlying constructs, with the constructs allowed to inter-correlate. The structural model specifies the posited causal relationships among the constructs. To study the research model with the measurement model, the structural model with a path diagram is constructed. Single headed arrows represent causal relationships between variables. Goodness of fit tests is conducted with the survey data to examine the efficiency of the structural model. To assess the overall model fit without being affected by sample size, alternative standalone fit indices less sensitive to sample size are use.

3. Results

From the path analysis, it is observed that Human Factors and Organization were significant towards dependent variable Human Error. With reference to the significant importance, independent Organization factors (0.405) were more significant compared to independent Human Error (0.324) which was referring to the estimate value stated in the Table 1.

Table 1. Regression Weight on Human Factor, Human Error and Organization

	Unstandardized Estimate	P-value	Standardized Estimate
HE < ---------- HF	.569	***	.324
ORG < ---------- HF	.501	.008	.252
HE < ---------- ORG	.632	***	.405

Note: *** indicate that the p-value is <0.001

Table 2. Standard Regression Weight of Factors in the Conceptual Model

	Unstandardized Estimate	P-value	Standardized Estimate
SW < ---------- HF	1.000	***	.754
HW < ---------- HF	1.214	***	.871
ENVT < ---------- HF	1.069	***	.714
LW(I) < ---------- HF	1.110	***	.818
LW(O) < ---------- HF	1.120	***	.768
QS < ---------- ORG	1.000	***	.803
CP < ---------- ORG	.920	***	.836
WF < ---------- ORG	.908	***	.818
FS < ---------- ORG	.878	***	.883
SC < ---------- ORG	.867	***	.741
INST < ---------- HE	1.000	***	.781
SR < ---------- HE	.901	***	.848

Note: *** indicate that the p-value is < 0.001

Based on weight and ranking, the order of significance of the five dimensions studied, the result is presented in Table 2 with Human Error as dependent variable and Software. Hardware. Environment, Liveware (I) and Liveware (O) as independent variables, it indicated that variable Software, Hardware, Environment. Liveware (I) and Liveware (O) were significant at 95% confidence level (p-value > 0.001) with variable Software taken as reference group. In this model, variable Hardware (0.871) is the most significant factors, followed by variable Liveware (I) (0.818). Liveware (O) (0.768), Software (0.754) and Environment (0.714).

Furthermore, when independent variable Quality Support is taken as reference factor, it is found that independent variables Quality Support, Company Policy, Workforce. Finance Strategy and Safety

Culture were significant in 95% confidence level (p-value > 0.001). Finance Strategy (0.883) is the most significant factor influencing organization. It continues with other independent variables Company Policy (0.836), Workforce (0.818), Quality Support (0.803) and Safety Culture (0.741).

4. Discussion

This study has shown the significant relationship between human factor based on SHELL model and human error in aviation maintenance[6]. Based on the hypothesis the human factors contribute to human error in aviation maintenance. In the Organization dimension, three items are first in the weight order of significant factors: Finance Strategy, Company Policy and Workforce [7][8]. With respect to Finance Strategy, aviation maintenance personnel perceive that recognition for their professional performance is reflected in the level of their salary and company incentive. Decision makers may need to be reminded that salary and incentive issue directly affect employee mood and indirectly affect employee involvement in an organization, and that they create a potential aviation safety risk factor if the staff feels that they are being underpaid and underappreciated. The factor of company policy refers to the need to establish a policy which encourages all staff to actively report any mistakes or improper condition professionally to avoid creating a blaming culture. In the first dimensions of Human Factors, concerning on aviation personnel and their interactions with hardware such as correct equipment, tools and parts available, is first in the weight order of the Hardware dimension. Confucius stated that it is necessary to have effective tools, equipment, and proper aircraft parts to do a good job[9]. However, in many first-line maintenance units, such as line maintenance or maintenance stations, aviation personnel cannot use or obtain the appropriate tools for their job due to equipment or parts shortages, unfinished checkups, or workplace limitations. On the second dimensions, almost all experts focus on the risks concerning Liveware, confirming that humans are still the major source of risk in aviation maintenance operations[10][11][12]. The research findings indicate that aviation personnel rank second in the survey, in the middle of the five dimensions in Human Factors. Professional personnel with a positive attitude believe that safety standards make the company more competitive, "safety works until we are busy"[12][13] while employees with a negative attitude feel that such rules impede operations and reduce efficiency in this highly competitive business environment. Ahlstrom and Hartman[14] identified a lack of communication and coordination breakdowns, a significant factor generally agreed upon by the maintenance experts is in the weight order of the Liveware dimension. As is true in many other professions, aircraft maintenance involves time pressure and professional competition; therefore, it is easy for aviation maintenance personnel not to freely and frequently communicate with each other about their maintenance failures and successes, and to make unclear statements and practice poor listening skills. While aviation maintenance personnel may listen to supervisors' orders and instructions when doing their jobs, they often refuse outside suggestions for improvement because of feelings of professional superiority and stubbornness. Consequently, it is easy for aviation maintenance personnel to develop blind spots that cause maintenance error. Insufficient details in maintenance manual and poor illustrations are the two factors in the weight order of the Software dimension. A maintenance manual refers to the manual references ensuring that the newest aviation standards, rules, regulations, and skills descriptions are contained in aviation maintenance manuals. Certain workplace norms of inappropriate practices are accepted by company employees, such as when aviation maintenance personnel make no use of their manuals in their routine work schedule. These norms usually violate the standards for maintenance procedures, but they are tolerated and have been in use within companies for a long time. Long term exposure to hazardous environment comes first in the weight order of the Environment dimension. Therefore, aviation maintenance personnel should be informed and trained to handle hazardous materials. In addition to providing aviation maintenance personnel protective gears such as protective clothing, rubber gloves, ear muffs and safety glasses, airline companies should also hold regularly scheduled health and safety seminars. In terms of Workforce, aircraft maintenance operations require professional certificates and rotations. Frequent changes in management team or leadership in organization will disturb also the maintenance team performance. The importance of management attitudes justifying the important attached by other researchers[15]. It is difficult to train aviation personnel. The number of qualified personnel available for job vacancies is usually far less than required by the scheduled flight increases of airline companies. Aircraft deterioration and the additional work hours needed to maintain newly purchased airplanes also present problems on the job. In a situation without sufficient manpower, aviation maintenance personnel must improve their efficiency in response to the increased work load. Therefore this study suggests that human factor alone may not only be the factors contribute to the human error in industry.

5. Conclusion

The results from the study provide an empirical support on the impact of human factors and organization to human error. Results showed that there were significant relationships of human factors on human errors. Human factors had a partial effect on organizational factors while organizational factors had a direct and positive impact on human errors. It was also revealed that organizational factors contributed to human errors when coupled with human factors construct. This study has contributed to the advancement of knowledge on human factors effecting safety and has provided guidelines for improving human factors performance relating to aviation maintenance activities and could be used as a reference for improving safety performance in the Malaysian aviation maintenance companies.

References

[1] K. A. Pettersen and T. Bjørnskau, "Organizational contradictions between safety and security - Perceived challenges and ways of integrating critical infrastructure protection in civil aviation," *Saf. Sci.*, vol. 71, no. PB, pp. 167–177, 2015.
[2] Y. H. Chang and Y. C. Wang, "Significant human risk factors in aircraft maintenance technicians," *Saf. Sci.*, vol. 48, no. 1, pp. 54–62, 2010.
[3] A. Shanmugam and T. Paul Robert, "Ranking of aircraft maintenance organization based on human factor performance," *Comput. Ind. Eng.*, vol. 88, pp. 410–416, 2015.
[4] R. W. McLeod, "Human factors in barrier management: Hard truths and challenges," *Process Saf. Environ. Prot.*, vol. 110, pp. 31–42, 2017.
[5] J. C. J. Anderson and D. D. W. Gerbing, "Structural equation modeling in practice: A review and recommended two-step approach," *Psychol. Bull.*, vol. 103, no. 3, pp. 411–423, 1988.
[6] T. Shell, "a Human Factors Approach for the Analysis and the Encoding of Aviation," *Environment*, pp. 463–468, 2004.
[7] N. McDonald, S. Corrigan, C. Daly, and S. Cromie, "Safety management systems and safety culture in aircraft maintenance organisations," *Saf. Sci.*, vol. 34, no. 1–3, pp. 151–176, 2000.
[8] "Maintenance hangars for the next generation of larger aircraft (2003)," vol. 81, no. 2003, pp. 2017, 2017.
[9] B. Emami-Mehrgani, W. P. Neumann, S. Nadeau, and M. Bazrafshan, "Considering human error in optimizing production and corrective and preventive maintenance policies for manufacturing systems," *Appl. Math. Model.*, vol. 40, no. 3, pp. 2056–2074, 2016.

[10] CAA, "An Introduction to Aircraft Maintenance Engineering Human Factors for JAR 66," *Hum. Factors*, pp. 127, 2002.
[11] Civil Aviation Authority, "Human factors in aircraft maintenance and inspection [CAP 718]," no. 12, pp. 51, 2002.
[12] Civil Aviation Authority, "CAP 716: Aviation Maintenance Human Factors (EASA / JAR145 Approved Organisations): Guidance Material on the UK CAA Interpretation of Part 145 Human Factors and Error Management Requirements," *Hum. Factors*, 2006.
[13] J. R. González Dan, J. Arnaldos, and R. M. Darbra, "Introduction of the human factor in the estimation of accident frequencies through fuzzy logic," *Saf. Sci.*, vol. 97, pp. 134–143, 2017.
[14] V. Ahlstrom and D. G. Hartman, "Human Error in Airway Facilities," Security, no. January, 2001.
[15] K. Mearns, S. N Whitaker, S. M. and R. Flin, "Safety climate, safety management practice and safety performance in offshore environments," Safety Science, vol. 41, pp. 641-680, 2003.

Aviation Gas Turbine Engine Emissions: Drop in Alternative Fuel and its Challenges

H Mohd Noh[1], G A Rodrigo [2], N A Abdul Rahman[1], S Ismail[1], M A Shafie[1], M W Zainal Ariffin[1], A A Ahmad[1], R Basit[1], A Khalid[1], N H R Yahaya[1], M R Yusoff[1] and J Othman[3]

[1]Universiti Kuala Lumpur – MIAT, Lot 2891 Jalan Jenderam Hulu, 43900, Dengkil, Selangor
[2]Air Transport Department, ETSIAE, Universidad Politecnica De Madrid, Pz Cardenal Cisneros, 28040, Madrid, Spain
[3]Universiti Kuala Lumpur - MFI Section 14, Jalan Teras Jernang, 43650 Bandar Baru Bangi, Selangor, Malaysia

Corresponding email: hazariah@unikl.edu.my

Abstract. The growing demand of air travel in aviation industries contributes to increases in carbon emissions. As far as technological, infrastructure and operation is concerned, carbon-neutral growth (CNG) cannot be achieved due to fossil fuel usage. The involvement of biojet fuel is the only measures that can be adapted to reduce up to -50% of Carbon emissions. This paper will discuss the gas emissions from the aircraft engine and the challenges it poses to Airlines and operation as well as adapting the Alternative drop-in fuel as its solutions. Alternative fuel using biomass has been approved by American Society for Testing and Materials (ASTM) to be a drop-in fuel with no alteration on the gas turbine engine, hence will ease and generate cost saving for the Airlines. The challenges remain, as the concern on the fuel properties and characteristics, distributions, environment and economic, have been the focal point for policymakers and researchers. As a conclusion, the authority and the government need to join hand in creating the opportunity for alternative fuel in the aviation industries to ensure the reduction in aviation carbon emissions.

1. Introduction

Aviation industries are growing and seeking for its expansion. The rapid growth resulting in the rise of aviation greenhouse gas emissions (GHG). The serious concern about the growth trajectory of the industry and emissions have led to call for market measures such as emissions trading and carbon levies to be introduced to restrict demand and prompt innovation [1]–[4]. The Aircraft emissions will give the impact surface to air quality through the formation of ozone and Particulate Matter (PM) as shown in figure 1. PM with a diameter less than 2.5 μm (PM2.5) can be generated either by direct emission (e.g. engine soot) or by way of gaseous precursors e.g. Nitrogen Oxide (NOx) and Sulphur Oxide (SO_2), which can be oxidized to form additional PM2.5. Increases in ground-level ozone and PM2.5 and subsequent population exposure can lead to a higher frequency of human health incidences, such as morbidity or mortality [5]. The environmental impacts of Aviation under the

Atmospheric Chemistry and Physics can be seen in figure 1, considering the impacts of noise and emissions and interdependencies of potential mitigation option [5]–[7].

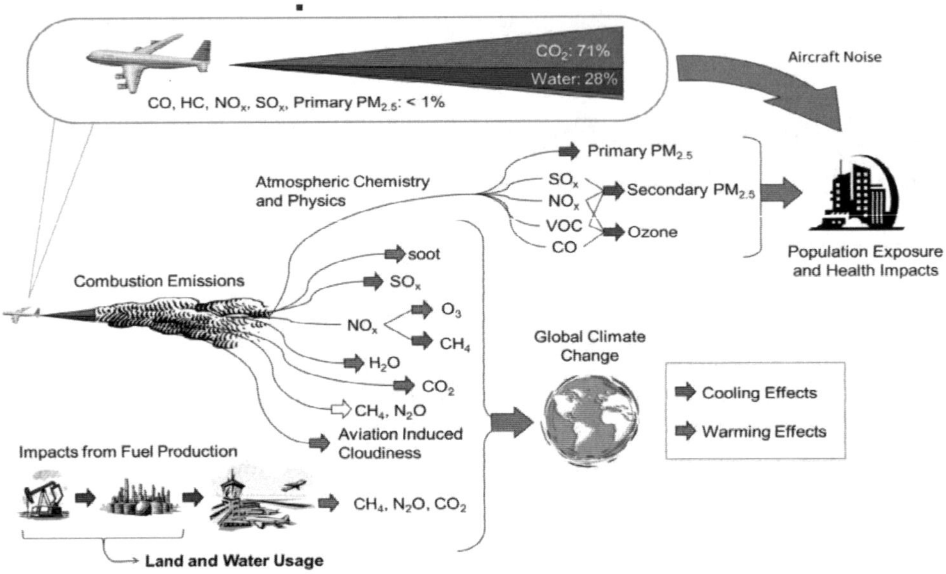

Figure 1. Environmental impact of Aviation considering environmental impacts of noise and emissions and interdependencies of potential mitigation options (source by IATA)

According to [4] International Civil Aviation Organization (ICAO) forecasts in its '2016 Environmental Report' in comparison to 2010, as shown in figure 2, when the annual international aviation emissions stood at 448Mt, the international aviation emissions will increase by 52% to 68%. It is estimated that the annual emissions of Carbon Oxide (CO_2)(682Mt to 755Mt) by 2020. On the other hands, between 169% and 185% (estimated annual emissions of 1205Mt to 1278Mt) by the year 2040 and up to 284% to 300% (estimated annual emissions of 1721Mt to 1794Mt) by 2050 depending on the level of technological and operational improvements.

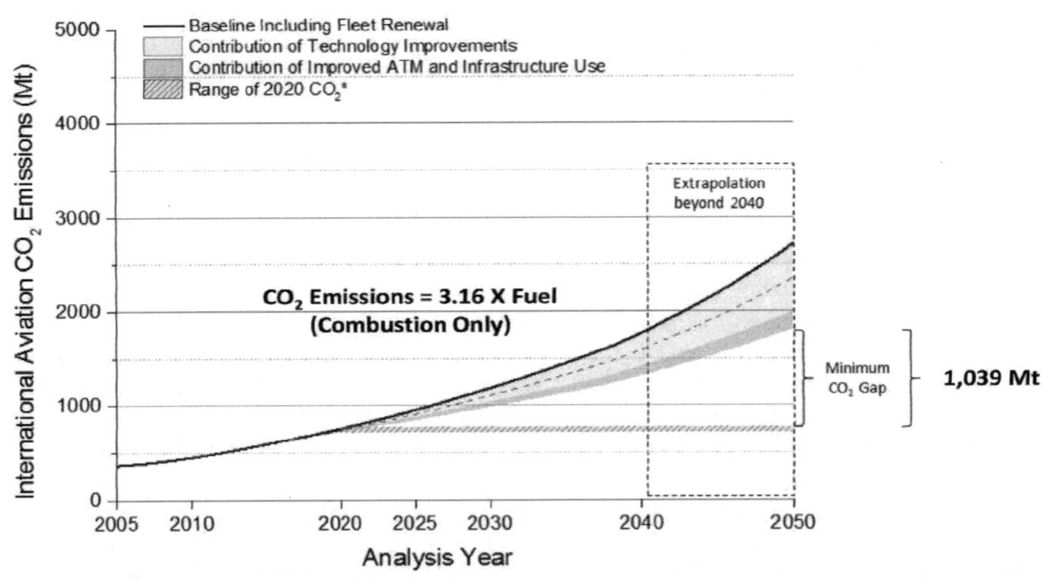

Figure 2. International Aviation emissions and Analysis (Source by ICAO)

IOP Conf. Series: Materials Science and Engineering **370** (2018) 012036 doi:10.1088/1757-899X/370/1/012036

Thus, even under the most optimistic scenario, taken into account the effectiveness of technical and operational measures, the aviation CO_2 emissions in 2050 are still expected to be 3.8 times higher than in 2010 emissions due to the forecasted strong increase in aviation activities. Technical and operational measures are therefore on their own insufficient to achieve CNG from 2020, see Figure 3 [4].

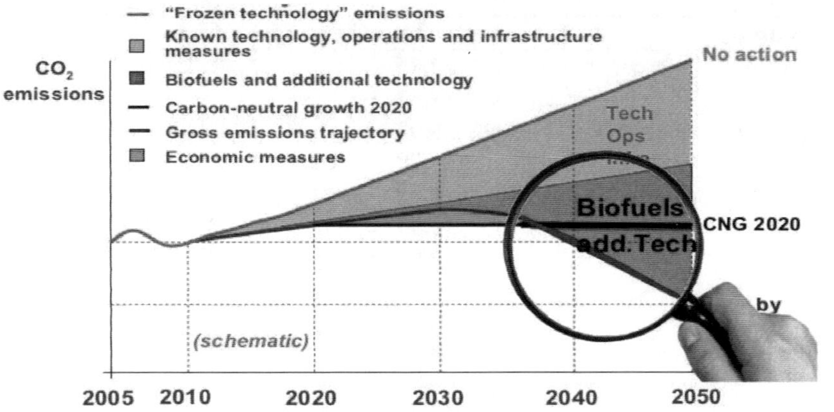

Figure 3. Emissions Reduction roadmap

Aviation Initiative for Renewable Energy in Germany or Aireg has come and joined forces with the policymakers, the scientific community in a plan to halve its carbon emissions by 2050. In mid-June 2011, Aireg has reached a milestone for more renewable energy in aviation using biofuels as an indispensable to achieving this aim [8]. According to [9], CO_2 emissions study continues with the evaluation of CO_2 emissions' data growth from the year 2010 to 2013. This data sources are gathered from the European Commission's Eurostat Air Transport Statistics (Eurostat) and EUROCONTROL flight plans database. The study (Table 1) has shown the descrease of CO_2 emissions in 29 countries in 2013, together with the evolution of CO_2 from 2010. Overall, there is a reduction of 3.1%, 209.1 Mtonnes due to new technologies, Air Traffic Management (ATM) and new renewable energy. The changes in the fuel efficiency are analyzed and the potential reasons for those changes are investigated [10].

The aviation biojet fuel was solely depending on its feasibility studies to compare with the rail and road transport, where electric propulsion systems do not offer an alternative way to reduce the emissions. While batteries remain too weak and too heavy for aircraft. The possibility of using renewable wind, hydro, and solar power in commercial aviation, unfortunately, is ruled out for the foreseeable future. So, the capability to offer climate-friendly mobility from renewable energies inaviation is the only mode of transport that has to rely on the use of biofuel [11].

2. Definition of drop-in biofuels

Safety is paramount to aviation authority, hencethere will be no requirement of any changes to aircraft engine fuel systems, distribution methods, or storage facilities [12]. Therefore, aviation biofuel can be a "drop-in" fuel that meets or exceeds internationally recognised as biojet fuel. As ASTM D4054 is where the approval for biojet fuel specification with unchange operating.

The ASTM D4054 [13] was developed as a guide related to testing and the procedures ensuring certification of the drop in alternative fuels have remained unchanged to the limitation of engine operating and aircraft operations, (Figure 4). Some aviation alternative fuels may require new specialized fuel handling systems, which is why the drop-in approach is the only practical solution for aviation [14]. A series of test flights in 2008 until 2017 demonstrated the viability of a drop-in aviation biofuel where among contributors were Virgin Atlantic, Air New Zealand, Continental Airlines, Japan Airlines and KLM Royal Dutch Airlines in cooperation with Boeing, Airbus, UOP, CFM, Pratt &

Whitney, and GE [15]. Figure 5 below demonstrates the global biojet fuel flights showing the support and initiatives from various airlines and biofuel producers with the number of commercial flight using biojet fuel.

Table 1. CO_2 Emissions in 29 countries.

	Country	CO_2 emissions 2013 (Mtonnes)	Changes from 2010 (%)
1	**UK**	**47.36**	**-2.60%**
2	**Germany**	**37.62**	**-4.20%**
3	**France**	**26.92**	**-4.30%**
4	**Spain**	**17.49**	**-11.50%**
5	**The Netherlands**	**16.12**	**-0.60%**
6	**Italy**	**15.05**	**-4.30%**
7	Switzerland	6.56	9.00%
8	Belgium	5.38	-4.90%
9	Greece	4.69	0.60%
10	Portugal	4.21	0.00%
11	Denmark	3.44	5.50%
12	Norway	3.26	18.50%
13	Sweden	3.16	2.60%
14	Austria	2.99	-9.70%
15	Ireland	2.85	9.20%
16	Finland	2.56	5.80%
17	Poland	2.07	-8.40%
18	Czech Republic	1.3	-4.40%
19	Cyprus	1.14	2.70%
20	Bulgaria	1.03	19.80%
21	Luxembourg	0.9	-30.20%
22	Romania	0.84	-10.60%
23	Hungary	0.59	-32.20%
24	Latvia	0.49	16.70%
25	Malta	0.37	12.10%
26	Lithuania	0.24	20.00%
27	Estonia	0.17	41.70%
28	Slovakia	0.17	-15.00%
29	Slovenia	0.14	-12.50%
Total EU27+2		**209.1**	**-3.10%**

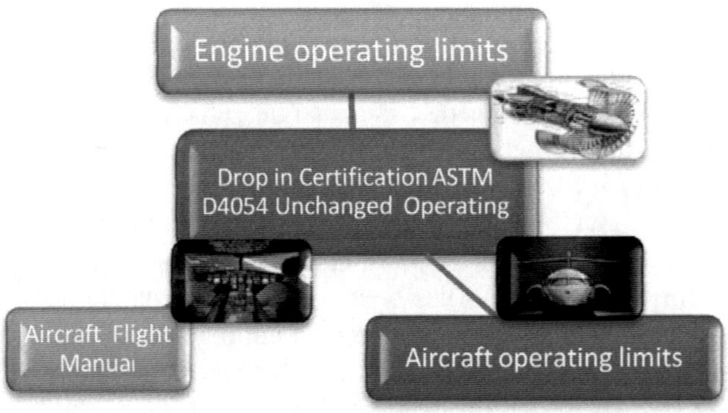

Figure 4. ASTM Approval Specification with no changes in aircraft engine,

Aircraft Flight Manual and Aircraft Operating limits

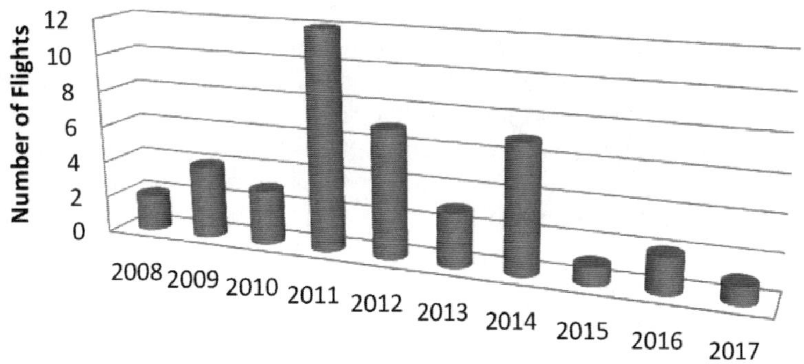

	2008	2009	2010	2011	2012	2013	2014	2015	2016	2017
■ Series1	2	4	3	12	7	3	7	1	2	1

Figure 5. Biojet fuel Commercial Flight from 2009 to 2017.

Considering the diversity of drop-in biofuel processes and product options, the following definition is used throughout this report to provide a functional representation of what is meant by a drop-in biofuel. Conventional biofuels have a distinct chemical nature, and so they can be accurately defined by their chemical composition alone. In contrast, drop-in biofuels consist of a mixture of many different types of hydrocarbons, the properties of which, just like petroleum fuels, is typically characterized by the mixtures' functional characteristics such as distillation profile, viscosity, acidity and more. It should be able to be readily "dropped in" to the existing petroleum infrastructure and be handled in the same way as petroleum fuels, without requiring significant infrastructure adjustments. Future technology in global biojet fuel adaptation will take into consideration all as mentioned in figure 6, where starting from handling safety to Material Compatibility in the overall expect of future Technology.

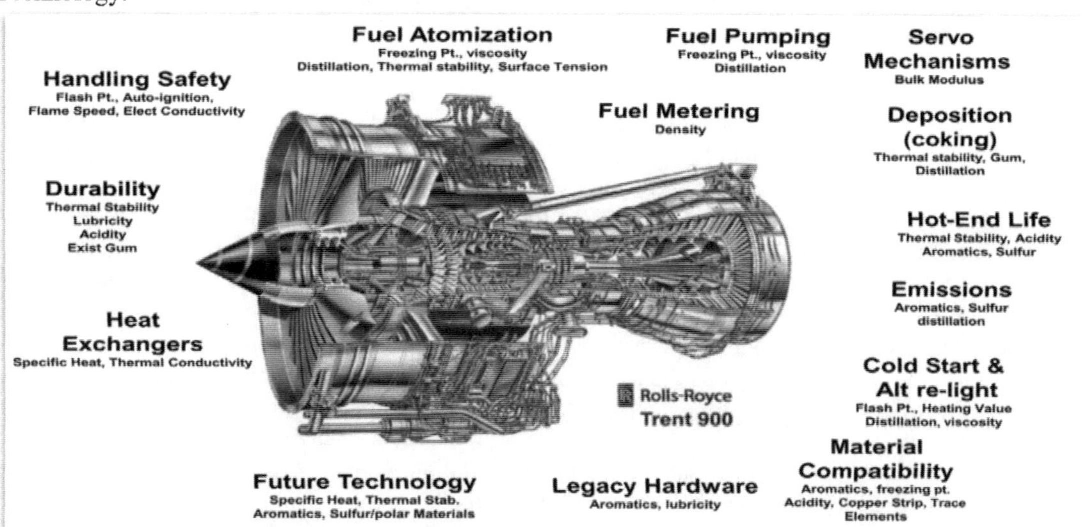

Figure 6. Future technology in global Biojet fuel adaptation (Source IATA)

3. Biojet fuels challenges

The demand and needs of this alternative fuel are exceeding, but there are many challenges need to be tackled. Challenges to overcome and eliminate the drawbacks of the biojet fuel will be crucial such as the technical and production issues, the environmental challenges, distribution problems, feedstock availability and sustainability [16]. In promoting biojet fuel to Airlines industries, the different incentive policies around the world are revised and compared to, and some indications are offered on the most recommendable procedures that might be carried out in the near future [17]. This can be a support for refuelling the future with the new alternative biojet fuel. This paper will place the interest of the main challenges that need to be addressed and provides an overview of the recent developments worldwide toward commercial-scale deployment.

3.1. Technical and production issues

The increase in commercial flights using alternative fuel demonstrated the technical feasibility of these fuels in aviation and the keen interest of airlines. Designer (manufacturer) and maintenance cannot be isolated. The airworthiness of the aircraft fuel systems needs to be ensured to enable the aircraft to fly and to prevent any malfunctions in the system throughout the flight. In this case, when the new alternative energy and the use of biojet fuel by airlines are mandated in European countries, a closer alignment of goals is needed, the new policies and standard among engine manufacturers, the airlines and the MRO, where engine contracts such as total care and other service elements should be revised and be considered priorities [18]. However, the production of these fuels is still in its early phase with only a limited volume currently available. There are still significant challenges to overcome before these fuels can represent a substantial share of the global jet fuel supply[19].

3.2. The environmental challenges

The rising on the demands of air travel will increase the emissions of CO_2. Due to that, the potential of alternative fuels for GHG emissions reductions is a strong motivator for their introduction into commercial aviation operations. Moreover, the aviation community has demonstrated its commitment to the environmental, social and economic pillars of the sustainability of alternative fuels [20]. A large number of aviation stakeholders are represented in organizations such as the Roundtable for Sustainable Biomaterials (RSB) and the Sustainable Aviation Fuel User Group (SAFUG), which aims to promote sustainable practices in agriculture and energy biomass production [15]. The outcomes of the life cycles analysis in measuring the CO_2 emission can be at par for the total emissions contribution in reducing the environmental impact [21].

3.3. Distribution challenges

One of the biggest challenges in enhancing the projection of biojet fuel is the distribution challenges. The lack of certainty of fuel supply availability at an acceptable cost over a long term can be a significant barrier to achieving increased deployment of many bioenergy projects. Most food and crops are grown and sold on an annual basis, often for widely varying commodity prices. So there may be a reticence by growers at committing to continue to grow specific energy crops over an extended period. The developers of a large plant may, therefore, take overall responsibility for obtaining the required biomass, either by securing long-term contracts well in advance or producing their biomass supplies [22]. If using local landowners to grow the biomass crops under contract, they will need to specify the form in which the biomass feedstock is required and even have significant input.

3.4. Economic challenges

Economic can be another challenge that is covering all Aviation industry players. The specific market can be a finite decision for any biofuel producers. As bioenergy developer will not invest in constructing a bioenergy plant without first securing a biomass fuel supply over the long term. A reliable market for the heat, power or biofuels to be produced also needs to be identified and purchase agreements signed. As current, government initiative and act as an investor or banker will need to have confidence that all aspects of the project are well defined, and contracts and warranties are all in place because a plant is likely to require a substantial financial commitment over a relatively extended period before a return on investment results.

Many bioenergy projects are technically feasible, but investments do not proceed because other forms of energy appear to be more cost competitive. A significant barrier results where the relatively high costs for or biofuels as demand is not promising at this instant hence, cannot compete on purely economic terms with fossil fuels used to provide the same amount of useful energy.

4. Conclusions

The mandate of using biojet fuel will be soon to be implied with the EU Clean Sky Initiative in European countries and will be projected to be oblique around the world as one of the step in going green.

4.1. Authority role and responsibilities

IATA and ICAO are playing important roles in the projection of the drop in biojet fuel into Airlines. The research, initiatives, economic measures, the policy will be covered through its role. As IATA's role representing the global community of international airlines is to: Play a leading role in standard setting for alternative drop-in fuels by in the areas of technical certification and logistics, and provide related technical support [23]. Renewable jet fuel is designed to be compliant with Jet a/a-1 fuel specifications and its crucial to make sure the drop-in alternative to conventional petroleum-derived fuel in a range of performance metrics, including fit for purpose, combustion performance and greenhouse gas emission reduction potential, without compromising on performance quality.

As far as jet fuel properties are concerned, Aviation will ensure that safety is the top priority, and all fuels are suited for aviation. While there are about a dozen other properties that have to be met by fuels to qualify for ASTM certification as usable in existing engines, the oxygen content (or H/C ratio) and carbon number range can be considered as the minimum, most basic characteristics need to be monitored. The authorities, together with non-government organisations, will need to join hand and putting a stake in biojet fuel project. In tackling the carbon emissions reduction, future interest could include policy planners, local government agencies, transport authorities, civil aviation authorities, health and safety bodies, environmental and amenity groups, landowners and potential growers of the biomass.

The challenges can lead to opportunities of this new alternative energy which needs to be broadened in various aspects for sustainability. Turbine engine can be focusing in its maintainability in towards the aircraft system, where supporting the demand for engine technology. In addition, biofuels may provide valuable economic opportunities to communities that can develop new sources of income in many developing nations. The opportunity can overcome the challenges, and its measurement of the readiness is crucial in biojet fuel adaptation in the aviation industry.

Acknowledgment

The researchers would like to thank you MOHE for the research funding under the Fundamental Research Grant Scheme FRGS/1/2015/TK10/UNIKL/03/1.

References

[1] IATA 2009 IATA Environment Report.
[2] ICAO 200Environmental Report.
[3] Boeing 2008 Environment Report.
[4] ICAO 2010 ICAO Environment report.
[5] M. R. Withers 2014 *Environmental and economic assessment of alternative transportation fuels* Massachusetts Institute of Technology.
[6] A. W. Schäfer and I. A. Waitz 2014 *Transp. Policy* **34**.

[7] D. S. Lee, D. W. Fahey, P. M. Forster, P. J. Newton, R. C. N. Wit, L. L. Lim, B. Owen, and R. Sausen 2009 *Atmos. Environ.*, vol. **43**, 3520
[8] Lufthansa 2011 *Policy Brief July 2011 Biofuels : Aviation moving at a clip.*
[9] F. Amizadeh, G. Alonso, A. Benito, and G. Morales-Alonso 2016 *J. Air Transp. Manag.* **55** 9
[10] F. Aminzadeh, G. A. Rodrigo, and G. M. Alonso 2017 *Analysis of evolution of commercial air transport CO2 emissions in the European Union.*
[11] Lufthansa 2011 *Policy Brief July 2011.*
[12] C. Dumas 2011 *Airbus Alternative Fuels* pp. 1–22.
[13] C. Aviation, A. Fuels, C. Team, M. Rumizen, and S. T. Specialist, 2013 *ASTM D4054 Users ' Guide* ASTM Int., pp. 1–54.
[14] A. Zschocke and S. Scheuermann 2012 *High Biofuel Blends in Aviation (HBBA) ENER/C2/2012/ 420-1 Interim Report.*
[15] SAFUG 2010 *Governments' Unique Role in Sustainable Aviation Biofuel.*
[16] T. Kandaramath Hari, Z. Yaakob, and N. N. Binitha, 2015 *Renew. Sustain. Energy Rev* **42** 1234
[17] H. M. Noh, A. Benito, and G. Alonso 2016 *Transp. Res. Part D Transp. Environ.* **46**, 298
[18] H. Mohd Noh, G. A. Rodriges, N. A. Abdul Rahman, H. Noh, G. A. Rodriges, and N. A. Abdul Rahman 2015 *Applied Mechanics and Materials* **747** 325
[19] IATA 2013 *Report on Alternative Fuels 8th edition*
[20] C. W. I. A. W. Niven Winchester, Dominic McConnachie 2013 *Market Cost of Renewable Jet Fuel Adoption in the United States.*
[21] D. B. Agusdinata, F. Zhao, and D. A. Delaurentis 2012 *Sustainability of biojet fuels : A multiactor life cycle* 27.
[22] S. Nair and H. Paulose 2014 *Emergence of green business models: The case of algae biofuel for aviation* vol. **65**.
[23] IATA 2009 IATA *Sustainable Alternative Aviation Fuels Strategy.*

IOP Conf. Series: Materials Science and Engineering **370** (2018) 012037 doi:10.1088/1757-899X/370/1/012037

Refuelling the future: Progress towards testing drop-in biofuels in replacing conventional fuel for commercial flights

H Mohd Noh[1], M N Mahamad Taher[1], G A Rodrigo[2], N A Abdul Rahman[1], S Ismail[1], M Mat Rani[1], I Mohd Salleh[1], Y Dahdi[1], W N S Wan Abdul Razak[1], M S Mat Ghani[1], M R Yusoff[1] and A Benito[2]

[1]Mechanical Department, Universiti Kuala Lumpur – MIAT, Lot 2891, Jalan Jenderam Hulu, 43900, Dengkil, Selangor

[2]Air Transport Department, ETSIAE, Universidad Politecnica De Madrid, Pz Cardenal Cisneros, 28040, Madrid, Spain

Corresponding email: hazariah@unikl.edu.my

Abstract. Due to different motivations, including the interest in reducing the dependency on fossil fuel and environmental implications, drop-in biofuels are a reality in today's commercial aviation. This paper summarizes the state-of-the-art of biomass-origin kerosene certification and provides references to the commercial flights performed so far by all airlines around the world. Results prove that the normal operation of the flights using the drop-in biofuel do not experience any repercussion in the performance in both engine and maintenance.

1. Introduction

Air transport is one of the most rapidly grown transport sectors. Forecasts give future air traffic yearly growth rates of 4 % [1, 2, 3, 4, 5]. Despite of the benefits from that growth, there are concerns about the increase in aviation greenhouse gas emissions (GHG). The significant growth suggests aviation could become a significant factor over the coming decades [6, 7, 8, 9]. Carbon Oxide (CO2) is considered the most important greenhouse gas emitted by aircraft, responsible for 3% of global fossil fuel consumption and 12% of transportation-related CO2 emissions [10]. Recent studies suggest that if the global economic growth continues, aviation CO2 emissions are likely to experience a greater than three-fold increase between 2000 and 2050 [11, 12, 13].

In response to concerns over the global environmental impacts of aviation, stakeholders have committed to strategies of mitigation related to efficiency improvement (fleet replacement, use of larger aircraft, increased density seating inside aircraft, improvements in Air Traffic Control (ATC) and navigation procedures). The calculation of the practical consequences of all those elements is rather complicated, but according to International Air Transportation (IATA) the results show an average improvement in efficiency, measuring in ton of fuel per (Revenue Ton per Kilometers) RTK, around 1.9% yearly [14] for the IATA members, although other studies indicate lower fuel per RTK [15]. The International Civil Aviation Organization (ICAO) Council, in its climate change mitigation program, has set an aspirational target for the World Air Transport sector of 2.0% yearly CO2 improvement until 2020.

Sustainable, advanced-generation aviation biofuels will play a large role in reducing CO2 emissions [6, 16]. The European (Emission Trading Scheme) ETS Directive [17]considers a zero emission factor for aviation biofuels, although assessments of life-cycle CO2 emissions typically show a potential reduction compared to fossil jet fuel in the order of 30-90%, depending on feedstock and production processes [18]. As for the biofuel usage in aviation, most of the specification and operational questions have already been answered and no technological show-stopper is seen thus far. However, the economic viability is still far from being made secured [19, 20, 21].

In 2009, data gathered suggest that the proportion of biofuels in total fuel consumption by commercial aviation was 0.5% and will rise to 15.5% in 2024 in a "moderate" scenario, and to 30.5% in an "ambitious" scenario [22]. In the European Union, the European Commission has launched the "European Advanced Biofuels Flightpath", an industry-wide initiative to speed up the market uptake of aviation biofuels in Europe. It provides a roadmap to achieve an annual production of two million tonnes of sustainably produced biofuel for aviation by 2020 [23] This research paper investigates the state-of-the-art of the biojet fuel usage considering biofuel certification and the utilization of drop in biofuels for commercial aviation around the world.

2. Drop-in Biofules Tests on Commercial Flights

In 2008, a ground test was performed on a CFM 56-7B with a new alternative energy using biomass algae and jatropha for the biojet fuel in a 50% mix with the current jet fuel [24, 25]. In September 2009, alternative fuel was produced using the Fischer Tropsh (F-T) process and was certified for aviation usage by American Standard Testing Material (ASTM) International Standard D7566 [26] A 50% blend of F-T synthetic fuel with conventional fuels is currently used by some biojet fuel flight tests in commercial aviation [27]. On July 1st 2011, ASTM approved the jet fuel product slate of Hydroprocessed Esters and Fatty Acids (HEFA) under alternative fuel specification ASTM D7566 [28]. HEFA fuel that meets this specification can be mixed with conventional jet fuel, up to a blend ratio of 50%. [29]. HEFA is currently the leading process for producing renewable jet fuel [27, 30, and 31]. A summary of biofuel tests performed on commercial flights can be seen in the figure 1 below:

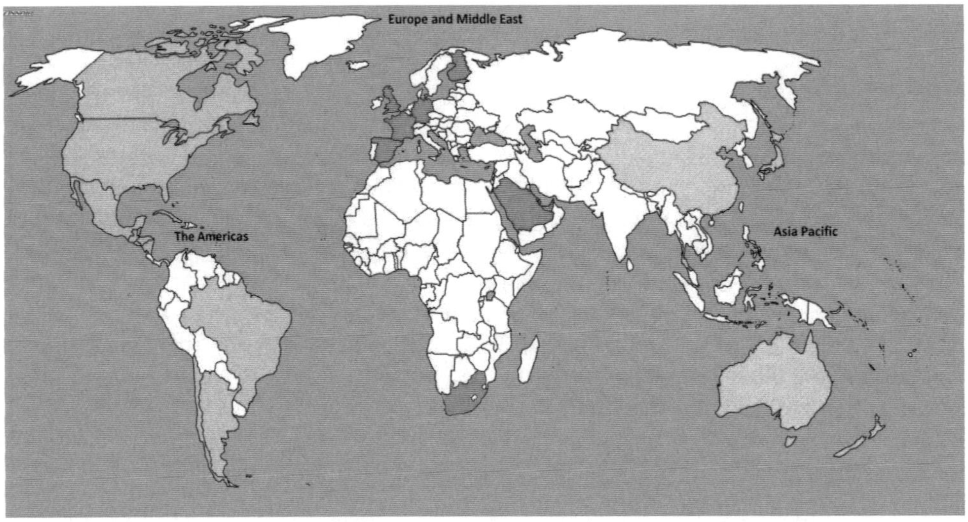

Figure 1. The world biojet fuel map for Commercial Aviation flight

2.1 Asia Pacific

Policy and regulations initiatives on the aviation biofuels and renewable energy are crucial when considering the ICAO global and international policy and strategy. Continuous development and improvement will be needed to accelerate the utilisation of national potential resources, investment and implementation. Table 1 presents the flight test and scheduled flight in Asia Pacific region using biojet fuel.

Table 1. Asia Pacific Flights

CARRIER	AIRCRAFT	PARTNERS	DATE	BIO-FUEL	BLEND	DESTINATION
Virgin Atlantic	B747-400	Boeing, GE Aviation	23.02.2008	Coconut Babassu	20% one engine	Amsterdam- London
Air New Zealand	B747-400	Boeing, Roll Royce	30.12.2008	Jatropha	50% one engine	Two hour flight test
JAL	B 747-300	Boeing, P&W, Honeywell UOP	30.01.2009	Camelina, Jatropha, Alage blend	50% one engine	Demo flight
Air China	B747-400	Boeing, Petro China, Honeywell UOP	June 2011	Jatropha	50% of one engine	Test Beijing airspace
Thai Airways	B777-200	SkyNRG, Roll Royce Boeing	21.12.2011	Castor seed	One engine	Bangkok- Chiang Ma
QANTAS	A330	Airbus	April 2012	Cooking Oil	50%	Sydney- Adelaide
ANA	B787	Boeing	April 2012	Cooking Oil	10% Blend	Evert's Paine Field(KPAE to Hane (HND)
Jetstar	A320	Airbus	April 2013	Cooking Oil	50% Blend	Melbourne -Hobart
Hainan Airline	B737-800	Boeing, Sinopec	21.03.2015	Cooking Oil	50% Blend	Shanghai – Beijing
Singapore Airlines	A350-900	AltAir, Fuels, SkyNRG, NAFCO, CAO	3.5.2017	Cooking Oil	50% Blend	San Francisco-Singapore

Thailand, Singapore and Indonesia are working in ensuring the contribution towards the refueling the future with biojet fuel. This can be seen when Thai Airways becomes one of the earliest to conduct the Biojet fuel flight test back in the year 2011. The details are as below:

- Thai Airlines (22.1.2011): B777-200
- Engine: Roll Royce 877
- Partner: Rolls Royce, Boeing, SkyNRG
- Biofuel use: Castor Seed
- Destination: Bangkok to Chiang Mai

Singapore is committed to the global effort to reduce international aviation Emissions in partnership with the Civil Aviation Authority of Singapore (CAAS). In May 2017, Singapore Airlines conducted biojet fuel flight test , SQ31, where details information as below;

- Singapore Airlines (3 May 2017): (3-month projects A350-900 Engine Rolls-Royce Trent XWB-84.
- Partner: AltAir Fuels, SkyNRG and American Fuel Corporation (NAFCO), China Aviation Oil (Singapore)
- Biofuel use: cooking oils and conventional jet fuel
- Destination: San Francisco to Singapore

2.2 Europe and Middle East

European Countries' contribution cannot be questioned in terms of using biojet fuel in their projects, research and development. Most of the biojet fuel producers and providers were European companies and the details of carrier, aircraft type and biofuel types can be seen in table 2 below.

KLM carrier from Netherland, is the most Airlines that participate in Biojet fuel flight (test schedule flight). This can be seen starting from the early years of Biojet fuel projects in 2009 until 2016. In the eyar 2014, KLM participated in 26 total flights from Amsterdam (AMS) to New York (JFK). Currently KLM is an active Airlines which have the highest collaboration with biojet fuel producer's SkyNRG and Neste which base in Netherland.

Table 2. Europe and Middle East Flights

CARRIER	AIRCRAFT	PARTNERS	DATE	BIO-FUEL	BLEND	DESTINATION
QATAR Airline	A340-600	Airbus , Shell	12.10.2009	Gas to Liquid (Not biofuel)	50% four engi	London - Doha
KLM	B747-400	GE, Honeywell UOP	23.11.2009	Camelina	50% one engi	1.5 hours flight test
Iberia	A320	Airbus, Iberia, SkyNF		Camelina	25% of one engine	Madrid- Barcelona
KLM	B737-800	KLM, SyNRG	29.06.2011	Cooking oil	50% of one engine	Amsterdam - Paris
Finnair	A319	Airbus, SkyNRG	July 2011	Cooking Oil	50% Blend	Amsterdam- Helsinki
Air France	A321	Airbus, SkyNRG	October 2011	Used Cooking Oil	50% blend	Toulouse- Paris
Etihad	B777-300ER	Boeing	Jan 2012	Vegetable cooking o	Fuel blend	Abu Dhabi – Seattle
Lufthansa	A321	Airbus, Lufthansa Technik (LHT) and MTU Aero Engines	15 July – 27 December 2011	Jatrohpa, Camelina, Animal fat	50% of one engine	Frankfurt-Hamburg
KLM	B777-200ER	Boeing, SkyNRG	March 2013	Cooking Oil	50% blend	JFK, New York Weekly flight to Amsterdam (Total 26 flights)
KLM	A330-200	Airbus, SkyNRG, ITAKA	May 2014	Cooking Oil	20% Blend	Amsterdam – Carribean Island (Aruba & Bonaire)
Finnair	A330	Airbus, SkyNRG, Statoil Aviation	Sept 2014	Cooking Oil	Fuel blend	Helsinki- New York
Lufthansa	A320	Amyris Total	September 2014	Renewable Farsane- Sugarcane	10% blend (Max)	Frankfurt- Berlin
Scandinavian Airlines	B737-800	SkyNRG Nordic	07.10.2014	Cooking Oil	10% Blend	Stockholm - Oslo
Scandinavian Airlines	B737-800	SkyNRg Nordic	11.11.2014	Cooking Oil	48% Blend	Trondheim – Oslo
Norwegian	B737-800	SkyNRG Nordic	11.11.2014	Cooking OIl	50% Blend	Bergen – Oslo
KLM	Embraer 190	Neste, Renewable Jet Fuel	31.06.2016	Camelina	50% Blend	Oslo – Amsterdam
Mango Airlines	B737-800	SkyNRg, Boeing	15.07.2016	Nicotine-free tobacc plant Solaris		Johannesburg – Cape Town

2.3 The Americas

The Americas on the other hand, contributes more than 14 flight tests mostly from United States and Brazil. Many of the flight tests have been conducted using biofuel but more to military purposes. Details of the flight test is shown in Table 3.The blend of biojet fuel being used is up to 50%, and most common types being used such Algae, Camelina Jathropa and used cooking oil. In July 2014, when the World cup was held in Brazil, all flight inbound and outbound from Orlando to Brazil were using Biojet fuel produce by Amryis and Total fuel producer.

3. Biojet Fuel Flight

From the total biojet fuel flight test conducted from 2008 until 2017, the type of aircraft that is widely used for flight test is the B737 series which is 11 flights. As B737 series is among the simplest type of fleet to be monitor without creating any conflict with the system itself. Meanwhile, the second higher types of aircraft used in six flight tests are from A320 and the least type used from A340,A 350, Embraer 190, B787, and B757 with only one flight test. See figure 2. As a logical opinion, Airbus and Boeing determined in supporting this biojet fuel flight along together with the biojet fuel producers.

IOP Conf. Series: Materials Science and Engineering **370** (2018) 012037 doi:10.1088/1757-899X/370/1/012037

Table 3. The Americas Flights

CARRIER	AIRCRAFT	PARTNERS	DATE	BIO-FUEL	BLEND	DESTINATION
Continental	B 737-800	Boeing, GE Aviation, CFM, Honeywell UOP	7.01.2009	Algae and Jatropha	50% one engine	Houston- over Gulf of Mexico
United	A319	Rentech	30.04.2010	GTL	40% two engine	90 minute flight test
TAM	A320	Airbus , CFM	23.11.2010	Jatropha	50%	45 minutes flight test
Interjet	A320	CFM, Safran, EADS,	1.04.2011	Jatropha,	50%	Mexico City- Tuxtla Gutierrez
BOEING	747-8F	Boeing new model aircraft	June 2011	Camelina	15% mix all four engines	Paris Airshow
Aeromexico	B777-200	Boeing, ASA	August 2011	Jatropha	30% blend	Mexico- Madrid
Thompson Airways	B757-200	Boeing, SkyNRG	October 2011	Used Cooking Oil	50% blend	Birmingham Airport
Continental	B737-800	Boeing, Solazyme	November 2011	Algae	40% blend	George Bush Intercontinental Airpo t- O'Hare Chicago Airport
Alaska Airline		Boeing, Bombardier, SkyNRG	November 2011	Used Cooking Oil	20% blend	Seattle - Washington
LAN	A320	SkyNRG, Air BP Copec	March 2012	Used cooking oil (SkyNRG)	Fuel blend	Santiago – Concepcio Chile
Aero-mexico	B777	Boeing, Honeywell Green Jet,	June 2012	Jatropha , Camelina	50% blend	Mexico city- Sao Paul Rio + 20
GOL	B737	Boeing, Honeywell Green Jet,	June 2012	Cooking oil, Inedibl Corn Oil	50% blend	Sao Paulo – Rio Janei Rio + 20
Air Canada	A319	SkyNRG	June 2012	Used cooking oil (SkyNRG)	50% blend	Toronto - Mexico City
LAN	A320	Airbus, LATAM Airlines, Terpel	August 2013	Camelina	20% Blend	Bagota- Calli
GOL	B737 Fleet	Boeing, Inter Americ Bank, Amyris, Total	July 2014	Renewable Farsane- Derived Sugarcane	10% Blend	Orlando- Sao Paulo (World Cup carrier)

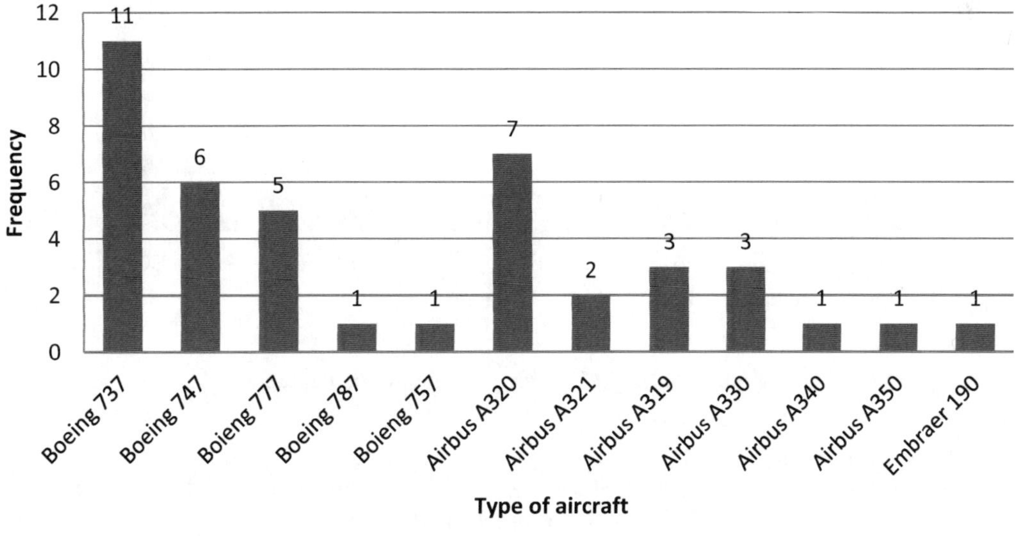

Figure 2. Biojet fuel Flight and aircraft type

In figure 3 below, highlights indicate the types of biofuel frequently used in the test flights. Cooking oil is shown to be most frequently used in 20 flights, followed by, Jathropa (9 flights) , Camelina (8 Flights) and Algae(3 Flight). This will not included the mixture of one or two of biomass.

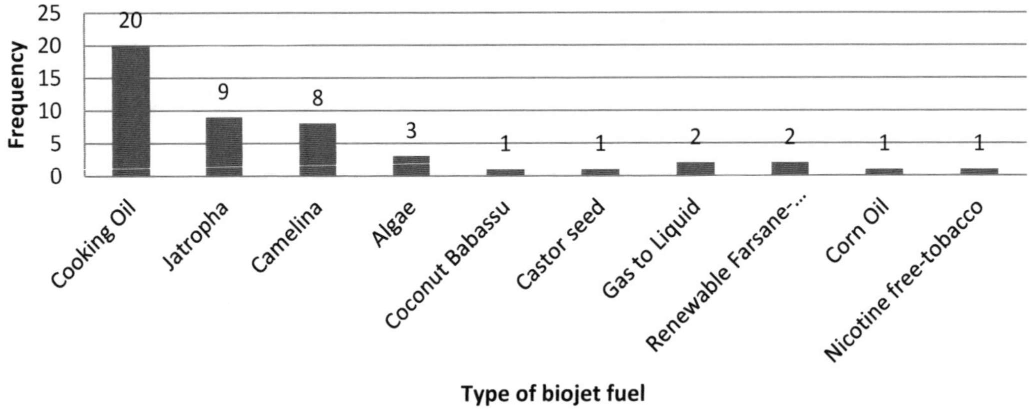

Figure 3. Biojet fuel flight with alternative fuel type

Whereas in Figure 4, the highest fuel blend used for biojet fuel flight is 60.61% with a 50 % blend. The mixture consists of fuel from jatropha, algae, camelina, GTL- Gas to liquid, cooking oil, Brassica, and carnita. Meanwhile, the second higher fuel blend is 18.18% with a 20 % blend which uses alternative fuel from coconut, babassu, jatropha, cooking oil and camelina. The least percentage of fuel blend used is 3.03% with a 30 % blend which is from alternative fuel type, Jatropha. Due to direct conversion from used cooking oil, compares with other types of biomass its contributes more to the flight test without any restriction in high conversion cost.

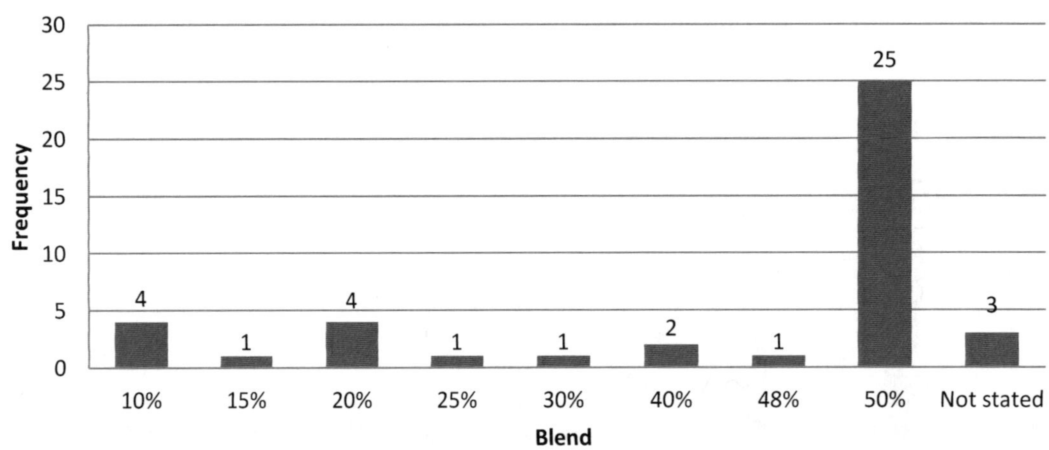

Figure 4. Biojet fuel Flight blending percentage.

4. Global Demand on Biojet Fuel

The longest experience so far using biofuel was the Lufthansa BurnFAIR Project. One engine (V2533) of a Lufthansa Airbus A321 was refueled with a biofuel blend from 15.07.11 until 27.12.11 and performed 1,187 scheduled flights between Hamburg and Frankfurt. The biofuel blend was supplied by Nestle Oil and contained a mix of 50% biosynthetic kerosene and 50% conventional kerosene. Observation of the engine trend analysis showed a reduction in fuel flow of 1.2% which was due to the result of higher fuel heating value of the bio fuel blend [32]. In addition, all components were reported to be in good condition: fuel tanks with no microbial growth, fuel line and fitting with normal conditions, fuel filter with no sediments or unusual material.

The used of Biofuel on flight tests so far has proven that it can be a drop-in fuel without any major modification to the engine [33]. The aromatics (molecule with a carbon ring of unsaturated bonds) remain an impediment [34, 35, 36]. Undergoing research in creating biojet fuel – Fischer Trop (F-T) with aromatics is currently under the (ASTM) D4060 [31]. The main problem encountered about land availability in harvesting biomass for biojet fuel and sustainability means that it is not prudent at this time to assume that in 2050 biofuels could account for more than 10% of global aviation fuel [37]. At the same time, the growing global demand for air travel has led to collective motivated research to obtain more sustainable alternative fuels [38, 39]. The incentives in promoting the biojet fuel throughout the globe need to be measured as an indication to project much more sustainable flight to be conducted [40, 41].

5. Conclusion

The implementation on Biojet Fuel in the aviation industry will need to be measured through Fuel Readiness Level and Technology Readiness Level. As this will contribute to ensure further development on refuelling the future for the commercial aviation using the new alternative fuels. What is more, future collaboration from fuel manufacturer, Airlines and Engine manufacturer around the world such as Shell, GE Company, Airbus and Boeing will strengthen the path for forthcoming progress. The sharing platform from government bodies, policy makers, engine manufacturer, aircraft manufacturer and biofuel manufacturer need to be strengthend in ensuring the projection of the usage of biofuel for commercial aviation can be adapted worldwide.

Acknowledgment

The researchers would like to thank you MOHE for the research funding under the Fundamental Research Grant Scheme FRGS/1/2015/TK10/UNIKL/03/1.

References

[1] EUROCONTROL 2010 Long Term Forecast – Flight movements 2010-2030.
[2] IATA, 2012 Airline Industry Forecast 2012-2016.
[3] Airport Council International 2013 Global Traffic Forecast 2012-2031.
[4] Airbus 2012 Global Market Forecast 2012-2031 Airbus, Blagnac Cedex.
[5] Boeing 2012 Current Market Outlook 2012, Boeing Commercial Airplanes, Seattle.
[6] ATAG (Air Transport Action Group) 2012 Aviation Benefits Beyond Borders (EU), October 2012
[7] European Commission, COM 697.
[8] Kurniawan, J., S., Khardi, S. 2011 *Environmental Impact Assessment Review* **31** 240
[9] Macintosh, A., Wallace, L., 2009 *Energy Policy* **37** 264
[10] Simone, N., Stettler, M., Barrett, S., 2013. *Transportation Research Part D: Transport and Environment* **25** 33
[11] ICAO Information Paper 2009 Global Aviation CO2 Emissions Projections to 2050.
[12] Lee, D. S., Fahey, D. W., Forster, P. M., Newton, P. J., Wit, R. C. N., Lim, L. L., Sausen, R. 2009. *Atmospheric Environment*, **43(22- 23)**, 3520
[13] Alonso, G., Benito, A., Lonza, L., & Kousoulidou, M. 2014 *Journal of Air Transport Management* **36** 85
[14] IATA 2013 Aviation and the environment, ITDI, Montreal.
[15] Peeters, P., 2013 *ICAO CAEP WG3 Information Paper*, Madrid.
[16] Krammer, P., Dray, L., Köhler, M. 2013 *Transportation Research Part D: Transport and Environment* **23** 64
[17] European Commission 2008 Directive 2008/101/EC of 19 November 2008 amending Directive

2003/87/EC so as to include aviation activities in the scheme for greenhouse gas emission allowance trading within the Community.

[18] Stratton, R. W., Wong, H. M. Hileman, J. I. 2010 Life Cycle Greenhouse Gas Emissions from Alternative Jet Fuels, PARTNER Project 28 report

[19] Gorter, H. De, & Just, D. R. 2010 *Energy and Agricultural Policy* **32(1)**, 4

[20] Halog, A., & Manik, Y. 2011 *Sustainability* **3(12)** 469

[21] Demirbas, A. 2009 *Applied Energy*, **86 (1)** 108

[22] S. Sgouridis, P. A. Bonnefoy, and R. J. Hansman 2011 *Transp. Res. Part A Policy Pract.* **45**, no. 10 1077, 2011.

[23] James D. Kinder and Timothy Rahmes 2009 Evaluation of Bio-Derived Synthetic Paraffinic Kerosenes (Bio-SPK)

[24] Jain, S., & Sharma, M. P. 2011 *Renewable and Sustainable Energy Reviews* **15(1)** 438

[25] ASTM D 1655. 2006 Standard specification for aviation Turbine fuels, American Society for Testing and Materials, [Online] http://www.astm.org/Standards/D1655.htm (Accessed: 15May 2013).

[26] IATA 2009 Alternative Fuels Foreword 2009 (pp. 1–92). [online] http://www.iata.org/publications/Documents/2009-report-alternative-fuels.pdf (Accessed: 10 April 2010).

[27] IATA fact fuel sheet 2013 FACT SHEET: Fuel The global airline industry's, (December 2012),[Online]http://www.iata.org/pressroom/facts_figures/fact_sheets/pages/index.aspx (Accessed: 10 January 2013)

[28] Liu, G., Yan, B., & Chen, G. 2013 *Renewable and Sustainable Energy Reviews* **25** 59

[29] IATA 2009 IATA ANNUAL REPORT 2009. [Online] http://www.iata.org/pressroom/Documents/IATAAnnualReport2009.pdf (Assessed: 15th January 2011)

[30] IATA 2013 IATA 2013 Report on Alternative Fuels 8th edition Dec 2013. [Online] http://www.iata.org/publications/documents/2013-report-alternative-fuels.pdf. (Accessed: 29 April 2014).

[31] MTU Aero Engine. Report on Technical Memorandum Project FAIR (2012). [Online] http://www.puresky.de/en#/results-of-the-six-month-long-term-trial/at-a-glance/ (Accessed: 13 March 2013)

[32] Law, C. K. 2012 *AIAA Journal* **50(1)** 19

[33] Agusdinata, D. B., Zhao, F., & Delaurentis, D. A. 2012 *Environmental science technology Journal* **27**

[34] Daggett, D. L., Hendricks, R. C., Walther, R., & Corporan, E. 2007 Alternate Fuels for use in Commercial Aircraft. Most, (April), 1–8. [Online]. http://gltrs.grc.nasa.gov/reports/2008/TM-2008-214833.pdf (accessed date; 15th April 2012)

[35] Dufferwiel, S. 2011 Alternative Aviation Fuels – Aromatics & Thermal Stability, The University of Sheffield-Research Council UK for Low carbon Report-(May 2011).

[36] Rosillo-calle, F., Teelucksingh, S., & Seiffert, M. 2012 The potential and Role of Biofuels in Commercial Air Transport- Biojetfuel IEA UK Task Force 40 http://www.bioenergytrade.org/downloads/T40-Biojetfuel- Report-Sept2012.pdf (Accessed: 2 September 2013)

[37] Gegg, P., Budd, L., & Ison, S. 2014 The market development of aviation biofuel: Drivers and constraints. *Journal of Air Transport Management*, **39**, 34

[38] Nair, S., & Paulose, H. 2014 *Energy Policy*, **65**, 175

[39] Blakey, S., Rye, L., & Wilson, C. W. 2011 *Proceedings of the Combustion Institute*, 33(2), 2863

[40] H M. Noh, A. Benito, and G. Alonso 2016 *Transp. Res. Part D Transp. Environ.***46** 298

[41] H. Mohd Noh, M. N. Mohd Tahir, Y. Nur Halimatun Radhiah, N. A. Abdul Rahman, G. A. Rodrigo, and J. Othman 2017 *Sci. Int.* **29** 1

IOP Conf. Series: Materials Science and Engineering **370** (2018) 012038 doi:10.1088/1757-899X/370/1/012038

Optimisation of flight dynamic control based on many-objectives meta-heuristic: a comparative study

Sujin Bureerat[a], Nantiwat Pholdee[a,*] and Thana Radpukdee[b]

[a]Sustainable and Infrastructure Research and Development Center, Department of Mechanical Engineering, Faculty of Engineering, Khon Kaen University, Khon Kaen, 40002, Thailand.
[b]Department of Industrial Engineering, Faculty of Engineering, Khon Kaen University, Khon Kaen, 40002, Thailand

*E-mail: nantiwat@kku.ac.th

Abstract. Development of many objective meta-heuristics (MnMHs) is a currently interesting topic as they are suitable to real applications of optimisation problems which usually require many ob-jectives. However, most of MnMHs have been mostly developed and tested based on stand-ard testing functions while the use of MnMHs to real applications is rare. Therefore, in this work, MnMHs are applied for optimisation design of flight dynamic control. The design prob-lem is posed to find control gains for minimising; the control effort, the spiral root, the damp-ing in roll root, sideslip angle deviation, and maximising; the damping ratio of the dutch-roll complex pair, the dutch-roll frequency, bank angle at pre-specified times 1 seconds and 2.8 second subjected to several constraints based on Military Specifications (1969) requirement. Several established many-objective meta-heuristics (MnMHs) are used to solve the problem while their performances are compared. With this research work, performance of several MnMHs for flight control is investigated. The results obtained will be the baseline for future development of flight dynamic and control.

1. Introduction

Recently, meta-heuristics (MHs) also known as evolutionary algorithms (EAs) are very popu-lar in the applications to engineering design optimisaiton. The methods mostly developed ac-cording to inspiration of physical law or natural phenomena such as genetic evolutionary, food finding of animal or insect, etc [1-7]. The advantage of MHs is that they are global search optimisation. With the balance of exploitation and exploration, they can avoid local trapping. In addition, due to no requirements of function derivative of MHs, they can deal with almost any kind of objective function and design variables. Also, they can explore a Pa-reto front in a single optimisation run for the multiobjective (usually for 2-3 objectives) or many objective problem (more than 3 objectives). In this regard, MHs are becoming more popular than the classical gradient-based methods nowadays, particularly for multiobjective and many-objective optimisation problems. However, MHs usually have low search conver-gence and consistency due to some random strategies in its search procedure. The develop-ment of a new MH to eliminate this drawback is therefore challenging.

Development of a meta-heuristic for many-objective optimisation problems is a cur-rently interesting topic as most of real engineering design problems usually need to optimise several objective functions at the same time. Several ideas to develop MHs for many-objective problems (many objective meta-heursitic, MnMHs) have been presented, for exam-ple, modifying the selection and/or updating mechanisms [8, 9], reducing the spaces of objec-tive functions [10-12], applying a two-archive technique [13, 14], using a point-preference approach [15-17]. Although several MnMHs have been proposed successfully, researchers have usually focused on implementing them for test functions problem without considering constraints [12]. Therefore, investigation of MnMHs performance for real engineering applica-tions is still wide open.

This paper presents a comparative study of several MnMHs for many objective optimi-sation of flight dynamic control. The design problem is posed to find control gains as design variables. The objective functions are minimising the control effort, the spiral root, the damp-ing in roll root, sideslip angle deviation, and maximising the damping ratio of the Dutch-roll complex pair, the Dutch-roll frequency, bank angle at pre-specified times 1 seconds and 2.8 second subjected to several constrained based on Military Specifications (1969) requirement. A number of many-objective meta-heuristics (MnMHs) including multiobjective evolutionary algorithm based on decomposition (MOEA/D) [18], improved two-archive algorithm (Two_Arch2) [13], hybridisation of real-code population-based incremental learning and dif-ferential evolution (RPBILDE) [19] preference-inspired co-evolutionary algorithm using goal vectors (PICEA-g) [17], and knee point driven evolutionary algorithm for many-objective optimization (KnEA) [16] are used to tackle the problem while C-indicator as detailed in [20] are used for performance assessment.

2. Flight dynamic control optimization problem

In this study, the optimisation problem of a flight control in lateral and directional modes is consider. The flight dynamic model used is the same as those presented in [21-23]. The motion of an aircraft can be described in three main axes; roll axis (longitudinal), Pitch axis (lateral) and Yaw as shown in Figure 1. The state equation represents the lateral/directional motion of an aircraft can be expressed as [21-23]:

$$\dot{\mathbf{x}} = \mathbf{Ax} + \mathbf{Bu} \tag{1}$$

where $\mathbf{x} = [\beta, r, p, \phi]^T$

β = Sideslip
r = yaw rate,
p = roll rate,
ϕ = bank angle,
$\mathbf{A}$ = kinetic energy matrix,
$\mathbf{B}$ = Coriolis matrix,
$\boldsymbol{u}$ = control vector

The control vector, **u,** can be express as;

$$\mathbf{u} = \mathbf{C}\mathbf{u}_p + \mathbf{Kx} \tag{2}$$

where $\mathbf{u}_p$ is a pilot's control input vector while **C** and **K** are the gain matrices expressed as

$$\mathbf{C} = \begin{bmatrix} 1 & 0 \\ k_5 & 1 \end{bmatrix},$$

$$\mathbf{K} = \begin{bmatrix} k_6 & k_1 & k_2 & 0 \\ k_7 & k_3 & k_4 & 0 \end{bmatrix}$$

The parameters k1¬-k7 are control gain coefficients which need to be determined.
Substitution eq. (2) into the eq. (1), the state equation for the lateral/directional motion of the aircraft is expressed as;

$$\dot{\mathbf{x}} = (\mathbf{A} + \mathbf{BK})\mathbf{x} + \mathbf{BCu}_p \tag{3}$$

Design optimisation of the control system of an aircraft is found to be many-objective optimisation as there are several criteria that need to be satisfied such as control stability, ac-curacy, sensitivity, control effort, etc. while the control gain coefficients are set as design variables. In this work, the many-objective problem of the flight lateral/directional control is set to minimise; spiral root (λ_S), roll should damping (λ_R), the control effort ($\sum_{i=1}^{7} k_i$), sideslip angle deviation ($\Delta\beta$) , and maximise; damping ratio of Dutch-roll complex pair (ξ_D), Dutch-roll frequency (ω_D), bank angle at 1 and 2.8 seconds ($\phi(1)$ and $\phi(2.8)$), subjected to several constraints for a fighter aircraft based on the Military Specifications (1969) requirement . The optimisation problem can be expressed as:

$$\text{Min: } \mathbf{f}(\mathbf{x}) = \left\{ \lambda_s, \lambda_R, \sum_{i=1}^{7} k_i, \Delta\beta, \frac{1}{\xi_D}, \frac{1}{\omega_D}, \frac{1}{\varphi(1)}, \frac{1}{\varphi(2.8)} \right\} \tag{4}$$

Subjected to:

$\lambda_S \leq -0.01$

$\lambda_R \leq -3.75$

$\xi_D \geq 0.5$

$\omega_d \geq 1$

$\varphi(1) \geq 90^\circ$

$\varphi(2.8) \geq 360^\circ$

$\Delta\beta \leq 0.75^\circ$

The parameter $\lambda_s, \lambda_R, \xi_D$ and ω_d can be calculated based on the eigenvalues associated with the term of (A+BK) while β and ϕ can be found from dynamic response. The design variables are control gains coefficient in the matrix K (x = {k1, k2, k3, k4, k5, k6, k7}T). The kinetic energy matrix (A) and the Coriolis matrix (B) are defined as;

$$\mathbf{A} = \begin{bmatrix} -0.2842 & -0.9879 & 0.1547 & 0.0204 \\ 10.8574 & -0.5504 & -0.2896 & 0 \\ -199.8942 & -0.4840 & -1.6025 & 0 \\ 0 & 0.1566 & 1 & 0 \end{bmatrix},$$

$$\mathbf{B} = \begin{bmatrix} 0 & 0.0524 \\ 0.4198 & -12.7393 \\ 50.5756 & 21.6753 \\ 0 & 0 \end{bmatrix}$$

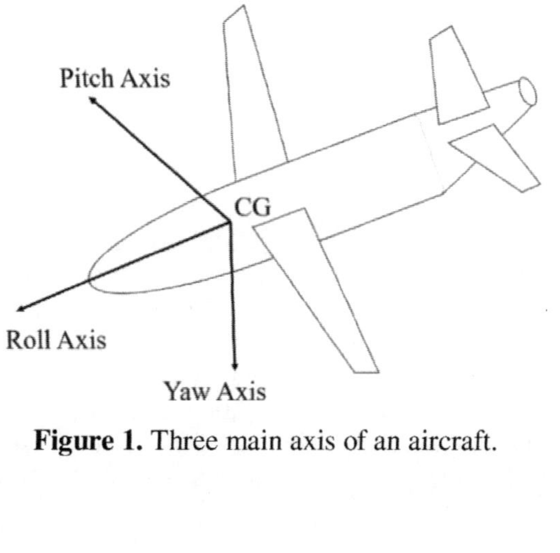

Figure 1. Three main axis of an aircraft.

3. Numerical Experiment

Five established meta-heuristics including, MOEA/D, Two-Arch2, RPBILDE, PICEA-g, and KnEA, are used to solve many-objective problems of the flight dynamic control problem described in Section 2. The MnMHs parameter settings used in this study (details of notations can be found in the corresponding references of each method) are given as [24]:

- MOEA/D: The algorithm is coded by authors of this paper based on the reference [18]. The number of neighbouring weight vectors, crossover and mutation probabilities are set as 6, 1.0, and 0.1 respectively.
- Two-Arch2: Use the code from H.Wang et al. [13]. The crossover probability and mu-tation probability are set to be 1 and 0.1, respectively.
- RPBIDE: The code is provided by the authors of [19]. Scaling factor, crossover prob-ability, and probability of choosing element from offspring in crossover are set to be 0.8, 0.7 and 0.5, respectively.
- PICEA-g: Use the code from R.Wang et al. [17]. All optimization parameters such as SBX parameter, type of crossover, probability of crossover between a pair, probability of in-ternal crossover, etc. are set as default values from [17].
- KnEA: used the default values and the parameter settings from X. Zhang et al. [16].

Each optimiser is used to find a Pareto optimal front of the problems as detailed in Section 2 for 10 optimisation runs. The population size is set to be 50 and the number of gen-erations is set to be 200. For an MnMH using different population size, the search process will be terminated with the total number of function evaluations (FEs) equal to 50×200 FEs. In addition, as most of MnMHs used in this study is box-constrained many-objective optimiser, to deal with constraints, a penalty function is used. The function is expressed as follow;

$$f_p(\mathbf{x}) = (1+\varepsilon_1 \cdot v)^{\varepsilon_2}$$

where f_p is a penalised function value, ε_1 and ε_2 are selected considering the exploration and exploitation rates of the search space.

$$v = \sum_{i-1}^{ng} v_i$$

where n_g and v_i are the number of constraints and constraints violation, respectively. The v_i value can be calculated as:

$$v_i = \begin{cases} \max(\lambda_S + 0.01, 0) \\ \max(\lambda_R + 3.75, 0) \\ \max(0.5 - \xi_D, 0) \\ \max(1 - \omega_d, 0) \\ \max(90 - \varphi(1), 0) \\ \max(360 - \boldsymbol{\varphi}(2.8), 0) \\ \max(\Delta\beta - 0.75, 0) \end{cases}$$

4. Results and discussions

After performing 10 optimisation runs of the five MnMHs on solving the flight dynamic control optimisation problem. The search performances are investigated based on the C-indicator. The C-indicator will compare a pair of optimisers which results in totally 10 pairs.

Figure 2 shows the box-plots of the C-values comparing of all optimisers. In the box-plots, the upper and lower lines show the maximum and minimum C-values while the middle line shows median of the C-values from 10 optimisation runs. The box-plot at row i and column j give the results of C (optimiser_i, optimiser_j) which is required to compare with C (op-timiser_j, optimiser_i) whereas the higher median of C the better. The quantitative assessment is given in Tables 1 where each value on row i and column j is the mean value of C (optimis-er_i, optimiser_j). The comparison is evaluated in the same way as the box-plot comparison. For example, average C (MOEA/D, Two_arch2) = 0 while average C (Two_arch2, MOEA/D) = 0.041256. This means Two_arch2 is better than MOEA/D. From the table, it is found that Two_arch2 is the best optimiser while the others are ranked in the following sequences: (2) KnEA, (3) RPBILDE, (4) PICEA-g and (5) MOEA/D.

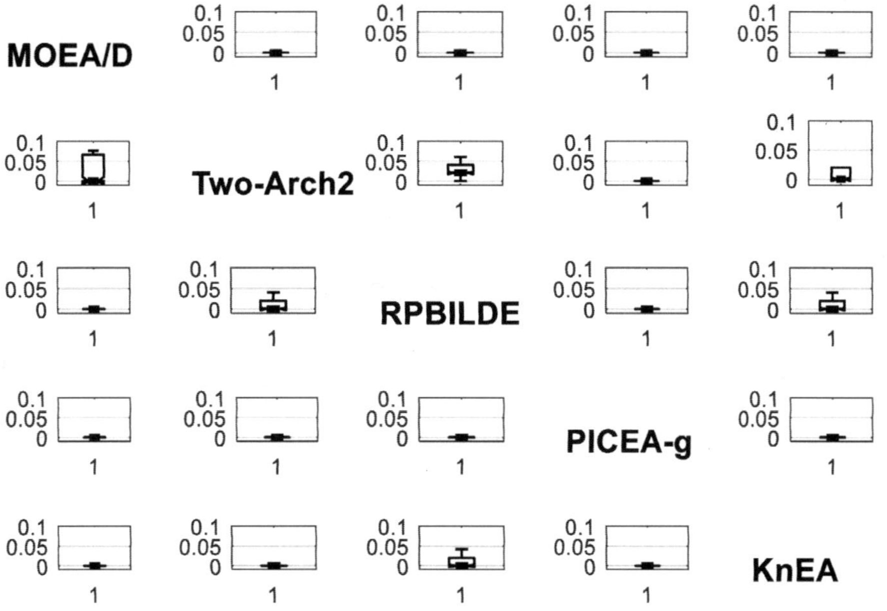

Figure 2. Box plot of C-values

Table 1. Mean C-values of all MnMHs

MOEA/D	0	0.002816	0.0022	0.0018
0.041256	Two-Arch2	0.027603	0.004	0.0094
0.015269	0.0068	RPBILDE	0.0028	0.007
0.0025	0	0.001021	PICEA-g	0.0006
0.010611	0.0014	0.008299	0.0022	KnEA

5. Conclusion

Many-objective constrained optimisation of flight dynamic control has been proposed and five many-objective meta-heuristics are implemented to solve the problem. The comparative results shown that Two_Arch2 is the best performer based on C-indicator values. Future work will be directed to the development of a new powerful meta-heuristic for solving real applica-tion of flight dynamic and control.

Acknowledgement
Technical The authors are grateful for the support from the Thailand Research Fund (TRF).

References

[1] Socha K and Dorigo, M 2008 European Journal of Operational Research **185** 1155

[2] Venter G.and Sobieszczanski-Sobieski J 2003 AIAA Journal **41** 1583
[3] Storn R.and Price K 1997 Journal of Global Optimization **11** 341
[4] Back T 1996 Evolutionary Algorithms in Theory and Practice (Oxford University Press: Oxford)
[5] Lindfield G and Penny J 1995 Numerical Methods Using MATLAB Ellis Horwood
[6] Baluja S 1994 Population-Based Incremental Learning: A Method for Integrating GeneticSearch Based Function Optimization and Competitive Learning In CMU-CS-94-163 t. r. Ed. (Carnagie Mellon University: Pittsburgh).
[7] Hwang C R 1988 Acta Applicandae Mathematicae 12 108.
[8] Deb K and Jain H 2012 In Evolutionary Computation (CEC), 2012 IEEE Congress on pp 1.
[9] Bader J and Zitzler E 2011 Evol. Comput **19** 45.
[10] Bandyopadhyay S and Mukherjee A 2014 Evolutionary Computation, IEEE Transactions on **1**
[11] Murata T and Taki A 2010 In 10th International Conference on Hybrid Intelligent Systems (HIS) 2010 pp 175
[12] Pholdee N Bureerat S and Yıldız A R 2017 International Journal of Vehicle Design **73** 20
[13] Wang H, Jiao L and Yao X 2015 IEEE transactions on evolutionary computation **19** 524
[14] Nuaekaew K, Artrit P, Pholdee N and Bureerat S 2017 Expert Systems with Applications **87** 79
[15] Cheng R, Jin Y, Olhofer M, and Sendhoff B 2016
[16] Zhang X, Tian Y and Jin Y 2015 IEEE transactions on evolutionary computation **19** 761
[17] Wang R, Purshouse R C and Fleming P J 2013 IEEE transactions on evolutionary computation **17** 474
[18] Qingfu Z and Hui L 2007 IEEE Transactions on evolutionary computation **11** 712
[19] Pholdee N, and Bureerat S 2013 Information Sciences **223** 136
[20] Zitzler E, Deb K and Thiele L 2000 Evolutionary Computation **8** 173.
[21] Tabak D, Schy A, Giesy D and Johnson K 1979 Automatica **15** 595
[22] Adra S F, Hamody A I, Griffin I and Fleming P J A 2005 In IEEE Congress on Evolutionary Computation, 2005. vol 1 pp 1
[23] Rostami S and Neri, F 2016 Integrated Computer-Aided Engineering **23** 313
[24] Pholdee N, Bureerat S, Jaroenapibal P and Radpukdee T 2017, In International Symposium on Neural Networks Springer pp 143.

Influence of Heat Treatment on the Corrosion of Carbon Steel in Environment Containing Carbon Dioxide and Acetic Acid

M.Adham Adnan[1], Kok-Eng Kee [1]*, Pandian Bothi Raja[2], Mokhtar Che Ismail[1], Saeid Kakooei[1]

[1]Mechanical Engineering Department, Universiti Teknologi PETRONAS, Bandar Seri Iskandar, 31750 Tronoh, Perak, Malaysia
[2]Center for Corrosion Research, Universiti Teknologi PETRONAS, Bandar Seri Iskandar, 31750 Tronoh, Perak, Malaysia

*Corresponding author: keekokeng@utp.edu.my

ABSTRACT. Experimental investigations were carried out to study the influence of heat treatment on the corrosion behaviour of API 5L X65 pipeline steel in the presence of acetic acid HAc and carbon dioxide CO2 environment. CO2 corrosion has been recognized as one of corrosion threats resulting in a loss of pipeline integrity. In this work, the specimens taken from the X65 steel pipe were subjected to respective heat treatment which was annealing, normalizing and quenching. As-received specimens were included as control. CO2 corrosion experiments using three-electrode glass cell system were performed with and without acetic acid. Corrosion rates were measured by means of weight loss and Linear Polarization Resistance (LPR) electrochemical methods. Results showed that heat treatments were found to affect the corrosion rates. Annealed specimen showed the lowest corrosion rate while quenched specimen appeared to corrode more compared to other specimens. The presence of acetic acid in aqueous CO2 environment generally aggravated the corrosion rate for all heat-treated specimens.

1. Introduction

The structural integrity of transportation pipelines is of paramount importance in the oil and gas industries. Carbon steel remains the preferred choice of pipeline material because it is economical, durable and adequate in strength[1]. Depending on the specific applications of the carbon steel components, different forms of heat treatment process can be applied to alter the mechanical properties, such as the yield strength, tensile strength, hardness, impact resistance and ductility. Normalizing is typically applied to the pressure retaining components: seamless steel pipe, pipe fittings, shell for vessel, annealing heat treatment is applied to parts having uniform wall thickness, while quenching is applied to components subjected to high pressure service such as injection nozzle or high service temperature such as boiler. However, the carbon steel pipeline material is susceptible to internal corrosion when subjected to hydrocarbon products containing corrosive species such as CO_2, H_2S and/or organic acids.

Heat treatment is a process of heating and cooling metals to achieve desired physical and mechanical properties through the modification of the crystalline structure. Heating temperature, length of time and rate of cooling are some of the significant variables that could impact the properties of the heat-treated metals [1,2]. The common forms of heat treatment done to carbon steel are annealing, normalizing and quenching. In annealing, the steels are held above the austenitic temperature and then slowly cooled in a furnace to produce coarse grain structure of pearlite with excess of ferrite or cementite. This results in annealed parts with uniform softness and stress relieved. In normalizing, the steels are heated until it passes the austenitic temperature and then air-cooled to produce refined grain structure. The normalized parts have improved hardness, strength and machinability. This heat treatment is more extensively used since it is more economical. In quenching, the steels are rapidly cooled in water or oil to produce needle-like martensitic structures [3]. It is apparent that the choice of heat treatment cycle and cooling path determine the final microstructure and properties of the steels. Generally, refined grain microstructure in normalizing yields to combined high strength and high ductility; however it tends to increase the difference in potential for galvanic action to occur, thereby leading to higher corrosion rate [3,4]. For instance, the impact strength is improved after heat treatment as a result of residual stress relief [5]. On the other hand, the corrosion and electrical properties can be slightly affected after the heat treatment.

CO_2 corrosion is considered as sweet corrosion that commonly occurs internally in the transportation pipelines carrying hydrocarbon products. CO_2 gas is often produced alongside with hydrodrocarbon phase, hence it is readily present in the produced fluids. Dry CO_2 gas is not corrosive, but becomes corrosive when it dissolves in formation water to form carbonic acid (H_2CO_3):

$$CO_2 + H_2O \leftrightarrow H_2CO_3 \tag{1}$$

Carbonic acid is a weak acid that ionizes partially into bicarbonate (HCO_3^-) ion and carbonate (CO_3^{2-}) ion in two-step reactions [6] shown in equation (2).

$$H_2CO_3 \leftrightarrow H^+ + HCO_3^- \tag{2}$$

$$HCO_3^- \leftrightarrow H^+ + CO_3^{2-} \tag{3}$$

The cathodic reactions are either from the proton hydrogen reduction in Eq. (4) through the buffering effect or the direct reduction of carbonic acid on the metal surface [6,7] in Eq. (5).

$$2H^+ + 2e^- \leftrightarrow H_2 \tag{4}$$

$$2H_2CO_3 + 2e^- \leftrightarrow H_2 + 2HCO_3^- \tag{5}$$

Organic acid is naturally present in the oil field brines. Acetic acid (HAc) appears to be the most common type of organic acid with concentrations ranging from hundreds to thousands of ppm. Field experience considered HAc to be a corrosion threat. A number of laboratory work have been performed on the role of HAc in CO_2 corrosion over wide ranging test conditions as summarized in the literature review [8,9]. Some studies focused on the role of acetic acid in top of line corrosion (TLC) cases, which it accelerates the cathodic reaction and local corrosion rate [10,11]. Others researchers argued that HAc has a negligible effect on the overall corrosion rate, but seems to stimulate localized attack [12]. Similar to carbonic acid, acetic acid is a weak acid that dissociates partially in water into hydrogen ion and acetate ions:

$$HAc \leftrightarrow H^+ + Ac^- \tag{6}$$

However, HAc is a stronger acid compared to carbonic acid (pKa 4.76 versus pKa 6.37 at 25°C). As such, it is generally accepted that HAc affects the cathodic reaction and aggravates the corrosion in CO_2-containing media [13,14]. The main concern is the presence of free acetic species as a buffer which can dissociate partially to provide H^+ source when needed in cathodic reaction [8]. The degree of HAc dissociation is governed by pKa, pH and temperature. The lower the pH, the more likely the HAc dissociates. The concentration of the free acetic species is given in the following mass conservation equation [12]:

$$[HAc_{free}] = {[HAc_{total}]} \Big/ {\left(1 + K_a/[H^+]\right)} \tag{7}$$

where K_a is the dissociation constant for HAc that varies with temperature.

Table 1 shows the calculated dissociation of acetic species at 25°C, pH 5.5, in which approximately 15% of total HAc is undissociated acetic species.

Table 1. Concentration of undissociated HAc and Ac^- at 25°C, pH 5.5

Total Hac (PPM)	Undissociated Hac (PPM)	Acetate, Ac^- (PPM)
100	15	83
500	76	417
1,000	152	834
1,500	228	1251

As demonstrated by several related work [8,13], the self-dissociation of two weak acids in CO_2/HAc containing media can be more corrosive due to the buffering effect. The Fe^{2+} ions are formed based on the following two overall aqueous reactions:

$$2HAC + Fe \leftrightarrow Fe^{2+} + 2Ac^- + H_2 \quad (8)$$

$$2H_2CO_3 + Fe \leftrightarrow Fe^{2+} + 2HCO_3^- + H_2 \quad (9)$$

The charge balance equation for electroneutrality condition (10) of the bulk aqueous solution is expressed as:

$$[H^+] + 2[Fe^{2+}] = [OH^-] + [Ac^-] + [HCO_3^-] + 2[CO_3^{2-}] \quad (10)$$

2. Methodology

2.1. Materials

Eight (8) specimens were sectioned from a carbon steel pipe grade API 5L X65, with the chemical composition shown in Table 2. The specimens were classified into 4 groups: A, B, C and D with each group having two specimens. Group A was the as-received specimen as control. For annealing, group B specimens were furnace heated to 750°C, hold for 30 minutes and allowed to be cooled in the furnace for 24 hours. For annealing, group C specimens were furnace heated to 750°C, hold for 30 minutes and allowed to cool in still air to ambient temperature. In quenching, group D specimens were furnace heated to 750°C, hold for 30 minutes and quenched in water medium. After the heat treatment, the specimens were etched for surface examination using optical microscope. Corrosion experiments were performed by weight loss method and linear polarization resistance (LPR) method in a standard three-electrode glass cell set up. Prior the experiment, the exposed surfaces of the specimens were consecutively polished with silicon carbide sand paper up to 600 grit, rinsed in de-ionised water, cleaned in alcohol in an ultrasonic bath and finally air-dried with blower.

Table 2. Elemental composition (wt%) of API 5L X65 carbon steel

Element	C	Mn	Si	Cr	Ni	Mo	Fe
wt.%	0.13	1.25	0.35	0.12	0.08	0.16	balance

2.2. Test setup

Corrosion experiments were performed in a 1-litre glass cell using three-electrode system as shown in Figure 1. The electrodes comprise silver/silver chloride (Ag/AgCl) as the reference electrode (RE), stainless steel rod as the counter electrode (CE) and mounted sample as the working electrode (WE). The test solution was prepared from 3 wt.% sodium chloride (NaCl) dissolved in de-ionised water. Prior to the test, the solution was bubbled continuously with CO_2 gas for at least 45 minutes to remove the dissolved oxygen. The gas bubbling continued during the experiment to ensure wet CO_2 saturation and to maintain slight positive pressure over atmospheric. The solution pH was adjusted by adding HCl or NaOH.

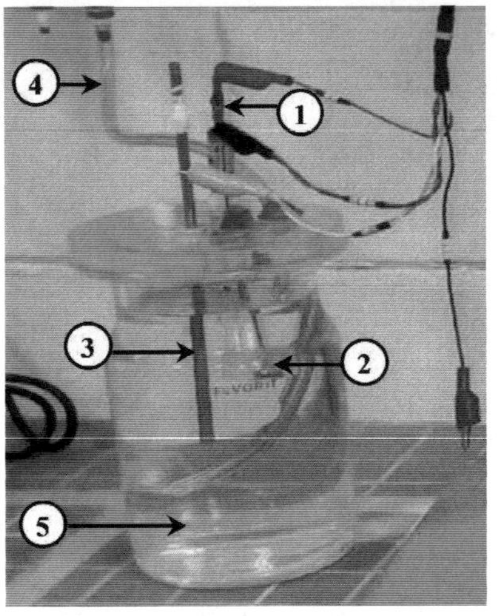

Figure 1. Three electrode glass cell setup: (1) reference electrode, (2) working electrode, (3) counter electrode, (4) CO_2 gas inlet, (5) 3 wt.% NaCl aqueous solution.

2.3. Test Parameters

Table 3 shows the test parameters for the glass cell experiments. The test parameters were chosen so as to simulate the CO_2 corrosion in the presence of acetic acid encountered in flow lines. For comparison, the first set of tests was performed without acidic acid HAc in CO_2 environment at pH 5.5. For the second set of tests, HAc of 1000 ppm was added at the start of test to the solution and the solution pH is adjusted to pH 5.5. Linear polarization resistance (LPR) technique, as per ASTM G59 [15], was used to monitor the in-situ corrosion rate at every half hour throughout the 48-hour duration. In LPR, the working electrode was polarized at ±10 mV from the open circuit potential at a sweep rate of 0.125 mV/s. Weight loss by immersion technique as per ASTM G31 [16] is performed for 50 to 65 hours as a measure of general corrosion rate. The calculation for general corrosion rate is given as:

$$CR = \frac{\Delta m \times 24 \times 365 \times 10}{\rho \times A \times t} \tag{11}$$

where Δm is mass loss (g), ρ is density (g/cm^3), A is exposed surface area (cm^2) , t is exposure time (h).

Table 3. Test matrix.

Element	Value			
Steel sample	API 5L x-65			
Heat treatment	Group A	Group B	Group C	Group D
	Control	Annealing	Normalizing	Quenching
Test solution	3 wt% $NaCl_{(aq)}$, saturated with CO_2			
Solution pH	5.5			
Temperature	25°C			
Total HAc (ppm)	0, 1000			
Measurement techniques	Linear polarization resistance LPR : 48 h Weight loss : 50 to 65 h			

3. Results and Discussion

3.1. Microscopy Analysis

The optical microstructure surface for four groups: (a) as-received control (b) annealed; (c) normalize; and (d) quenched specimens of X65 carbon steel are shown in Figure 2. With heat treatment applied, different microstructure surfaces were observed. Annealed specimen displayed a much coarse pearlite distribution. Normalized specimen showed shorter grains. For quenched sample, the needle shape morphology was identified as martensite structure.

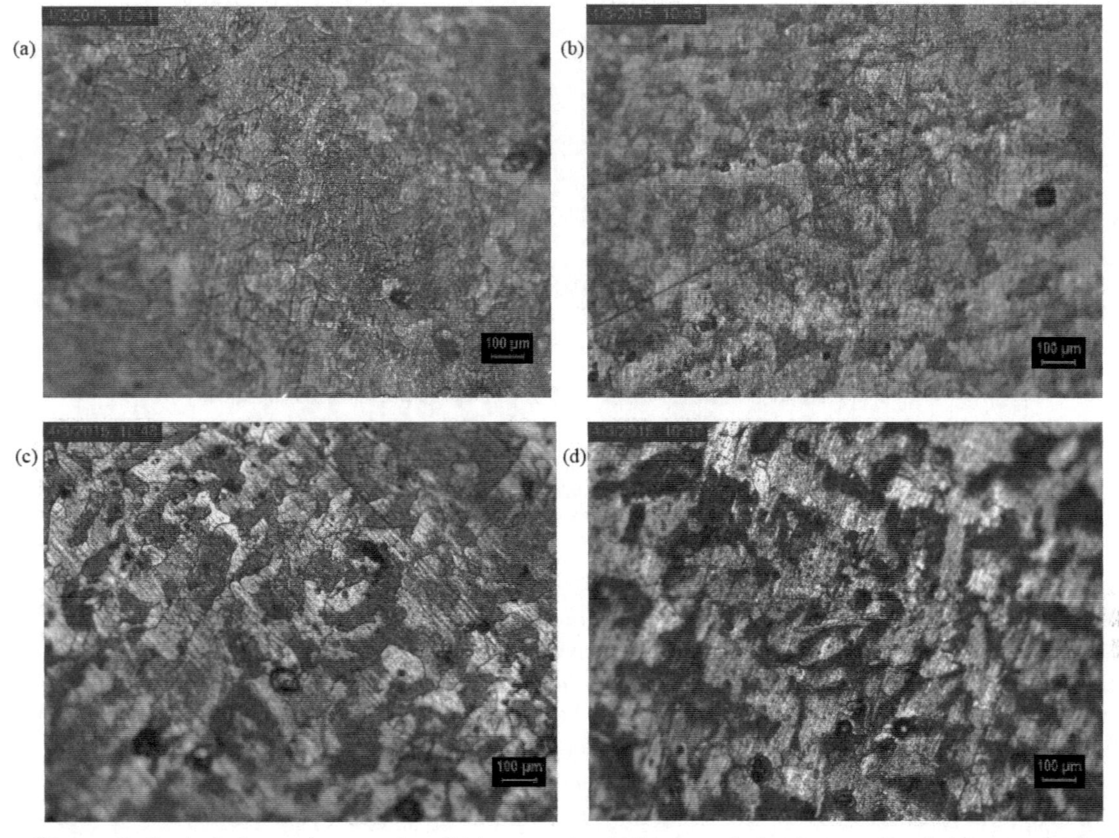

Figure 2. Optical photomicrographs of (a) as-received; (b) annealed; (c) normalized; (d) quenched specimens (400 X).

3.2. Effect of Heat Treatment

Figure 3 shows the variations of LPR corrosion rate with time (48 hours) at pH 5.5 and 25°C for samples subjected to different types of heat treatments in saturated CO_2 environment without the presence of acetic acid. All LPR corrosion rates of the specimens showed increasing trend in the first 5 hours before it remained consistent throughout. For as-received control specimen, the corrosion trend was consistent with the final corrosion rate was 1.3 mm/yr. Upon comparison, the trend of corrosion rate for normalized specimen matched with the control specimen. This showed that the as-received control specimen was similar to the normalized steel. The quenched specimen showed the highest corrosion rate with the final corrosion rate at 1.7 mm/yr while the annealed steel specimen showed the lowest with a final corrosion rate of 1.0 mm/yr. Results showed that annealed specimen with coarser grain structure exhibited better corrosion resistance over other specimens. The findings were comparable to a similar study by Siedu [17], in which the bigger the grain structure the more corrosion resistant it was. This can be explained by larger grains tends to minimize the difference in potential for galvanic cell to occur [3,4].

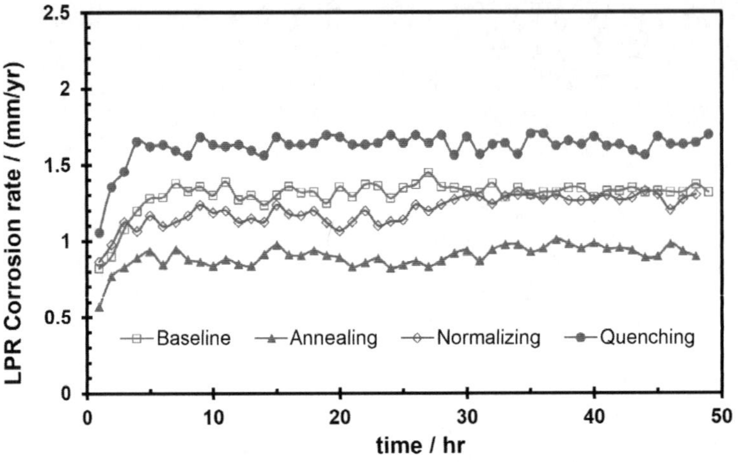

Figure 3. LPR corrosion rate versus time for baseline and heat treated samples at conditions: 25°C, pH 5.5, 1 bar pCO_2, 0 ppm HAc.

3.3. *Effect of Acetic Acid*

Figure 4 shows the variation of LPR corrosion rates with time (48 hours) for baseline and different heat treated specimens in CO_2/HAc containing environment at pH 5.5 and 25°C. In such condition, the undissociated HAc is approximately 152 ppm or 15% of the total 1000 ppm HAc. LPR results show that the overall average corrosion rates for all specimens increase to about 2 mm/yr, except for the annealed specimen at about 1.5 mm/yr. The increment in average corrosion rates ranges from 30% to 55%. This can be explained by the presence of weak acetic acid that acts as a buffer source for hydrogen that contributes to additional cathodic reactions. Similar findings were also reported by others [8].

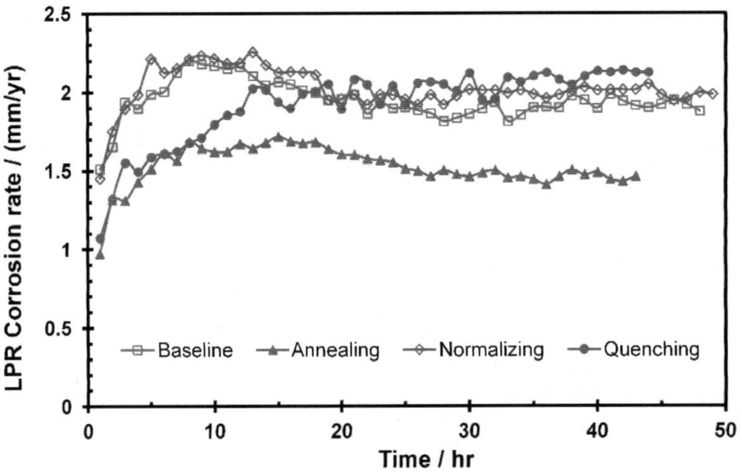

Figure 4. LPR corrosion rates versus time for baseline and 3 types of heat treated samples at condition: 25°C, pH 5.5, 1 bar pCO_2, 1000 ppm HAc.

3.4. *Weight Loss Results*

In order to better capture the long term corrosion effect, weight lost (WL) specimen were submerged in the test fluids for more than 60–90 hours at 25°C in 3 wt% NaCl saturated with CO_2. Without the addition of HAc, the WL results in Figure 5 show that the corrosion rate of the quenched specimen was the highest at 1.7 mm/y which agreed with the LPR results. The annealed sample had the lowest corrosion rate at 1 mm/y and was comparable to the LPR results. With the addition of 1000 ppm HAc, the WL corrosion rates for all specimens generally increased to about 2 mm/yr except for annealed steel specimen at 1.5 mm/y as shown in Figure 6. The results were in agreement with the LPR results [18]. Overall, the corrosion rates for all specimens were higher because of the addition of acetic acid at pH 5.5 that provides additional hydrogen source for the cathodic reactions. The annealed specimen with coarser grains was more corrosion resistant than the other specimens.

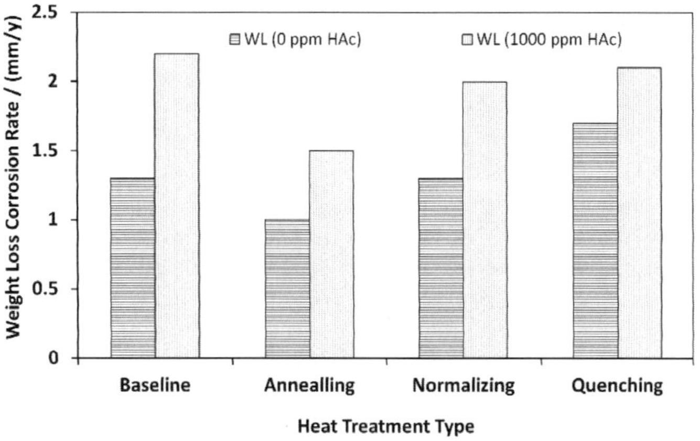

Figure 5. Comparison of weight loss (WL) specimens with and without HAc for heat treated samples.

4. Conclusions

The wok studied the corrosion behaviour of heat treated steel specimens in CO_2/HAc containing media. Based on the results discussed in this work, the summary is as follows:

4.1. Heat treatment was found to affect the corrosion rates. The annealed specimen with coarser grain displayed better corrosion resistance. The corrosion rate results for annealed specimens were 35% lower than the other heat treated specimens.

4.2. The corrosion behavior of as-received specimens was similar to that of normalized steel specimens.

4.3. The presence of acetic acid was found to aggravate the overall corrosion rates by 30% to 55% in CO2/HAc environment. At pH 5.5, there is 15% undissociated HAc in the solution that can readily provide additional hydrogen ion for cathodic reaction.

Acknowledgement

The authors would like to acknowledge the financial assistance received from Universiti Teknologi PETRONAS to carry out this work. Facilities provided by Centre for Corrosion Research are gratefully acknowledged.

References

[1] Gorni A 2004 *Steel Forming and Heat Treating Hand Book* (Brazil: Sao Vicente SP)
[2] Callister JWD 1997 *Material Science and Engineering: An Introduction* (New York: John Wiley & Sons, Inc) p 290
[3] Seikh AH 2013. *Journal of Chemistry* **2013** 1
[4] Oluyemi DO, Oluwole OI, Adewuyi BO 2011 *Materials Research* **14** 135
[5] Chen H-C, Cheng G-H 1989 *Journal of Materials Science* **24** 1991
[6] Nešić S, Nordsveen M, Nyborg R, Stangeland A 2003 *Corrosion Sc.* **59** 489

[7] De Waard C, Lotz U 1993 *CORROSION/1993* paper no 69 (Houston: NACE)
[8] George KS, Nešić S 2007 *Corrosion* **63** 178
[9] Nafday OA 2004 *Film formation and CO2 corrosion in the presence of acetic acid* (M.S. thesis: Ohio University)
[10] Okafor PC, Nesic S 2007 *Chemical Engineering Communications* **194** 141
[11] Gunaltun Y, Larrey D 2000 *NACE CORROSION/2000* paper no 71 (Houston: NACE)
[12] Amri J, Gulbrandsen E, Nogueira R 2011 *NACE CORROSION/2011* paper no 11329 (Houston: NACE)
[13] Crolet J, Thevenot N, Dugstad A 1999 *NACE CORROSION/1999* paper no 24 (Houston: NACE)
[14] Huang L, Brown B, Nesic S 2014 *NACE CORROSION/2014* paper no 4374 (Houston: NACE)
[15] ASTM Standard G59 2014 *Standard Test Method for Conducting Potentiodynmaic Polarization Measurements* (West Conshohocken: ASTM Int)
[16] ASTM Standard G31 2004 *Standard Practice for Laboratory Immersion Corrosion Testing of Metals* (West Conshohocken: ASTM Int.)
[17] Seidu SO, Johnson Kutelu B 2013 *J.Minerals and Materials Characterization and Engineering* **01** 95
[18] Ferreira L, Raquel M et al 2015 *Mat. Res.* **18** 245

Static Structural Analysis of a Variable Span Morphing Wing for Unmanned Aerial Vehicle

M Bashir and P Rajendran

School of Aerospace Engineering, Universiti Sains Malaysia, Engineering Campus, Nibong Tebal, Pulau Pinang, Malaysia.

Email: aeparvathy@usm.my

Abstract. While the primary reason to develop an adaptive wing is the aerodynamic benefits, the primary hindrance is the structural and vibrational considerations due to the unsteady nature of the airflow during the flight. Hence this study forms an important part of the morphable wing technology. In this paper, the design of a moderate aspect ratio variable span wing will be performed. The morphing wing is modeled structurally to observe the effect of spanwise load distribution on the wing structure. For the structural design and analysis of the unmanned aerial vehicle (UAV) under this study, commercial software Solidworks and Ansys/Static Structural/Modal are used. The static structural analyses of the wing are performed under different load conditions. The results of these analyses show that the designed structure is safe within the flight envelope. It is observed that the wing-root bending moment increases drastically due to an increase in the wingspan. Thus, the bending moment along the wingspan of the morphing wing is much larger than that of the conventional wing which results in an increase in the deflection of the free-end. The maximum stress for the un-extended wing configuration increases for the extended wing configuration.

1. Introduction

Capability of changing the wing during flight can be referred to "Morphing". This capability may result in economical fuel consumption, increase in mission adaptability and performance. Fixed wing aircraft are designed for considering the flight envelope and the mission profiles. For instance, fixed wing aircraft which are designed for high altitude level flight is not suitable for a dog fight. The aim of a morphing wing is the ability to adapt the wing for a variety mission phases.

Morphing wings coupled with the availability of light-weight, high-strength materials over the last couple of decades; have exhibit increase in aerodynamic efficiency with increasing aspect ratio. For instance a study has shown the efficiency of high altitude long endurance configurations also increases with increasing wing aspect-ratio and decreasing structural weight [1]. However, the resulting structure exhibits low structural stiffness, leading to significant static aeroelastic deformations at the tip of extended sections.

Several studies have been conducted on using composite materials for the design of aircraft wing. Newman Iii et al [2] demonstrated a computationally efficient with high-fidelity and integrated static aeroelastic analysis procedure. Here, the aerodynamic analysis consists of solving the nonlinear Euler equations by using an upwind cell-centered finite-volume scheme on unstructured tetrahedral meshes.

They presented static aeroelastic analysis results for a flexible wing in low subsonic, high subsonic (subcritical), transonic (supercritical), and supersonic flow conditions.

Bae et al [3] presented aerodynamic and aeroelastic analysis of variable span morphing wing (VSMW). The aeroelastic analysis shows that the flexibility of morphing wing structure increases as the wing span increases. At a given flight condition, the aerodynamic deformation is much larger than that of the conventional wing. Static aeroelastic considerations show that a VSMW requires increased bending stiffness because the wing deformation due to bending is much more significant than the deformation due to twist, which demands that VSMW designers include larger bending stiffness in their formulation.

Also, Santos et al [4] developed a fully functional VSMW system covering areas from aerodynamic optimization and structural design, through prototyping and flight testing. The results showed that both deployment and load tests revealed satisfactory performance for VSMW concept. However, deployment can be improved by increasing the skin stiffness at the inboard fixed wing tip with an internal stiff rib (between sandwich facings). Or else, this can also be achieved using an external lighter rib similar to an end plate around the perimeter of the airfoil.

Andersen et al [5] utilized an evolutionary process to analyze and characterize the dynamic aeroelastic features. It was concluded that NextGen's experience gained during the flutter prediction efforts indicated significant opportunities for aeroelastic modeling and analysis method developments. Aeroelastic characterization of morphing aircraft structures would benefit from both advances in automated structure, aerodynamic, and spline modeling. Elham [6] concluded that using composite materials in designing a wing box will reduce the weight up to 40% compared to aluminum wing box.

Potter et al [7] also investigated the physics-based structural analysis and also found that the optimal rib-spacing in a wing box should be optimized, and the use of composite materials reduced the weight up to 30% compared to the aluminum one. Gürses et al [8] studied the structural design to increase and decrease the camber control surface to match selected airfoil profiles. Schorsch et al [9] developed an elastomer based skin for seamless morphing of a rib based adaptive trailing edge. The skin material development focused on low temperature requirements for cruise flight conditions, resistance to environmental conditions as well as long fatigue life.

Xu [10] focused on the effect of wing skin thickness on the stiffness and strength of a composite wing. Recently, Naidu and Adali [11] designed a wing box structure for Medium Altitude Long Endurance UAV. However, they only considered a sandwich structure, using isotropic and composite materials. Pecora et al [12] worked on the design process to enable the high-fidelity reproduction of target shapes under the action of aerodynamic and inertial loads; and avoid any detrimental impact on flight safety as well as on the overall aircraft maintenance plans.

2. Methodology

The wing baseline geometry in this paper is similar to Phoenix UAV wing that has a moderate aspect ratio UAV. Nevertheless, the methodology presented in this paper can be used to design any arbitrary aircraft wing having moderate aspect ratio. Since we are employing VSMW, the area under the lift distribution is equal to the total lift produced by the morphing wing. When the aircraft would undergo the telescopic extension and increase the wingspan from 1.2 m to 1.8 m, it would achieve a higher aspect ratio. Keeping the chord lengths constant, but increasing the span to 0.6 m we see an increase in the aspect ratio. Additionally, the wing is considered rectangular because the chord of extended section is nearly equal to the chord of fixed wing profile.

Since the internal structure of the given UAV wing was unknown, the current structure was designed according to the standard wing structural design. The current structure consists of 3 main components: 10 ribs (6 in fixed wing part with 10 cm rib spacing, and 4 in extended part with 15 cm spacing), 2 spars at 25% and 75% wing chord, and skin panels. The rib spacing was chosen to have the maximum buckling strength of the panels, while the spars position was chosen to accommodate the maximum torsional stiffness according to the design practices in the wing structural design. Summary of the wing geometry specification is shown in Table 1.

Table 1. Wing Geometry Specification.

Wing Feature	Value
Wing Semi-Span WOE	1.2 m
Wing Semi-Span WE	1.8 m
Fully Extended Wing Area	0.36 m^2
Aspect Ratio	9
Chord	0.2 m
Airfoil	NACA 0018

2.1 Load Analysis of Wing

A static finite element method was used extensively during the design processes. In this study, the moderate-aspect-ratio VSMW is investigated. Lift load is considered as important criteria while designing an aircraft. Fuselage and wing are the two main regions where lift load acting in an aircraft. Here 80% of the lift load is acted on the wings (i.e., maximum lift load is acted on the wings) and remaining 20% in acted on the fuselage. Therefore in wings the maximum load is acted nearer to the wing roots.

Equation 1 below estimates lift, L produced by UAV where n is design load factor and W is the weight of UAV. Table 2 below summarizes the calculated features of investigated wing similar to Phoenix UAV. Considering the weight of the aircraft, design load factor, and factor of safety, the total design load of UAV can be estimated. This force is converted into the pressure load as uniformly distributed load by dividing this force for the semi wing area of 0.4 m^2.Therefore, the total pressure load applied from the bottom of the surface is 200 Pa.

$$L = n \times W \tag{1}$$

Table 2. Summarizes the calculated features of investigated wing.

Details	Calculated Value
Weight of UAV	500N
Design Load Factor	3g
Factor of Safety	1.5
Total Design Load	450 N
Total Lift Load on Fuselage (20%)	50 N
Total Lift Load on Wings (80%)	400N

2.2 Modelling and Grid Generation

The solid modeling of the airfoil was made with the help of Solidworks 16.0 as shown in Figure 1. The materials and their sectional properties of wing components are presented in Table 3. In addition, mechanical properties of an isotropic material, Aluminum 2024-T3 used in the Ansys Engineering model has been presented in Table 4.

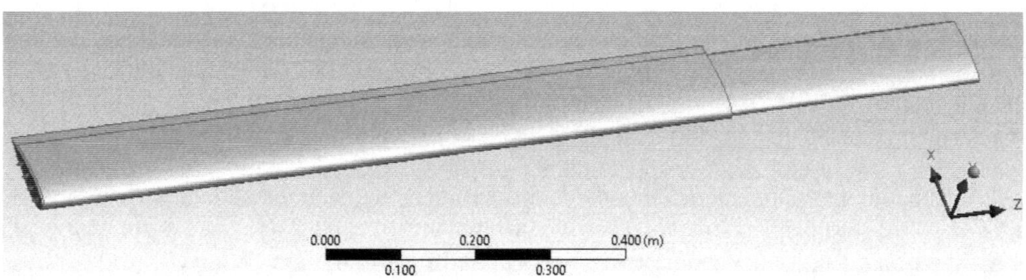

Figure 1. The solid modeling of the airfoil.

Table 3. Materials and their sectional properties of wing components.

Element Property	Material	Thickness (mm)
Spar Webs	Structural Steel	2.54
Ribs	Structural Steel	0.80
Aluminium Skin	Aluminium 2024-T3	1.50

Table 4. Properties of Aluminium 2024-T3.

Property	Value
Density	2780 kg/m3
Young's Modulus, E	73.1 GPa
Shear Modulus, G	28.0 GPa
Poison's Raito, v	0.33
Ultimate Strength	483 MPa
Yield Strength	385 MPa
Shear Strength	283 MPa

The unstructured mesh has been generated for CFD static structural analysis of the moderate-aspect-ratio wing model. Unstructured meshes are usually flexible because they conform to intricate shapes easily but they have drawback for treating the viscous boundary layer region. The wing meshes for both un-extended and extended span are shown in Figure 2.

(a)
(b)

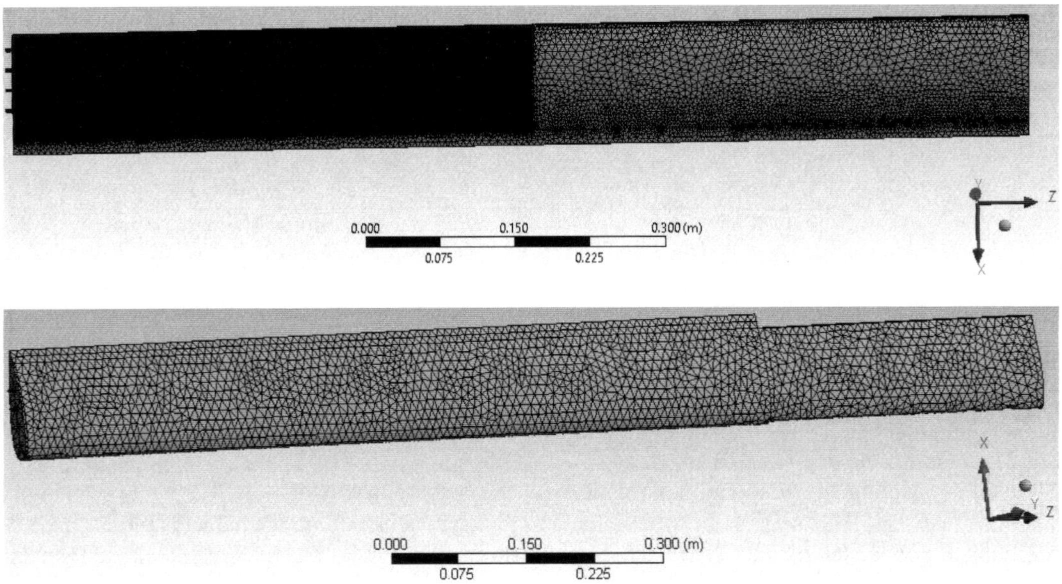

Figure 2. Wing meshes for both (a) un-extended and (b) extended span.

3. Results and Discussion

After the generation of pressure boundary condition, the wing is fixed from its one end and then the static analysis is performed. The result is given as deformation of the wing in model scale along with Von Mises stress and strain plots. The boundary conditions to analyse the wing conditions are developed by creating pressure load on the lower surfaces of wing. Based on the Equation 1, pressure load value of 200 Pa has been estimated. The five different cases shown in Figure 3 are analysed one by one analysed based upon the span change and applied load conditions is given in the following subsections.

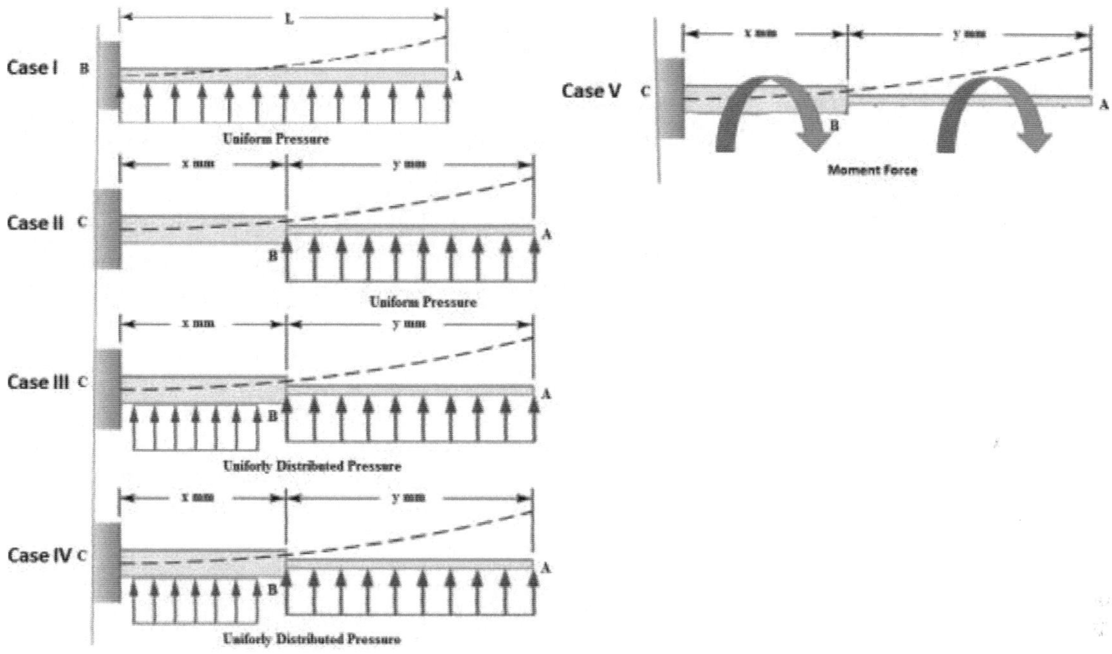

Figure 3. The five case studies based on the span change and applied load condition.

3.1 Case I - Without Extension
Figure 4 depicts the deformation of the un-extended wing along with stress and strain values. From the figure, it is so clear that the tip of the wing under goes maximum deformation and the root region is under tremendous stress, but it is still under control. The maximum deformation value shows 8.7375 x 10^{-5} and also the stresses are below the allowable the value. It is clear that the structure is safe enough to take the loads. This structure is safe enough to withstand this particular manoeuvring with enough factor of safety.

3.2 Case II - With Extension but load on extended Part only

In this case, the pressure is applied on the extended section only. Figure 5 shows the total deformation, Von Mises stress, and Equivalent strain. It is seen that maximum deflection occurs at the tip of extended wing, and the material starts to yield when the Von Mises stress reaches a critical value, yield strength. The stress distribution for the given loads has been observed. Here, the stress is distributed uniformly but maximum stresses are developed nearer to root of wing section. In this case, the Von Mises stress observed in the wing analysis is 72 MPa. The Yield strength of the aluminium alloy AA 2024-T3 is 385 MPa. Thus, the structure is safe because the stress magnitude which was obtained from the analysis is less than the yield strength of the structural material.

3.3 Case III - With Extension but uniformly distributed load over the wing surface

Figure 6 depicts the case of static analysis of the span extended wing with uniformly distributed pressure performed using ANSYS Workbench. As expected, it shows that the maximum deflection increases from 0.00087375 m for the un-extended wing case to 0.0082961 m for the case of telescopic extension of the wing with uniformly distributed pressure throughout the wing. From the figures it is so clear that the root region is under tremendous stress, but it is still under control. The maximum stress value shows 3.8669e 7 Pa and still within the allowable value. It is clear that the structure is safe enough to take the loads.

IOP Conf. Series: Materials Science and Engineering **370** (2018) 012040 doi:10.1088/1757-899X/370/1/012040

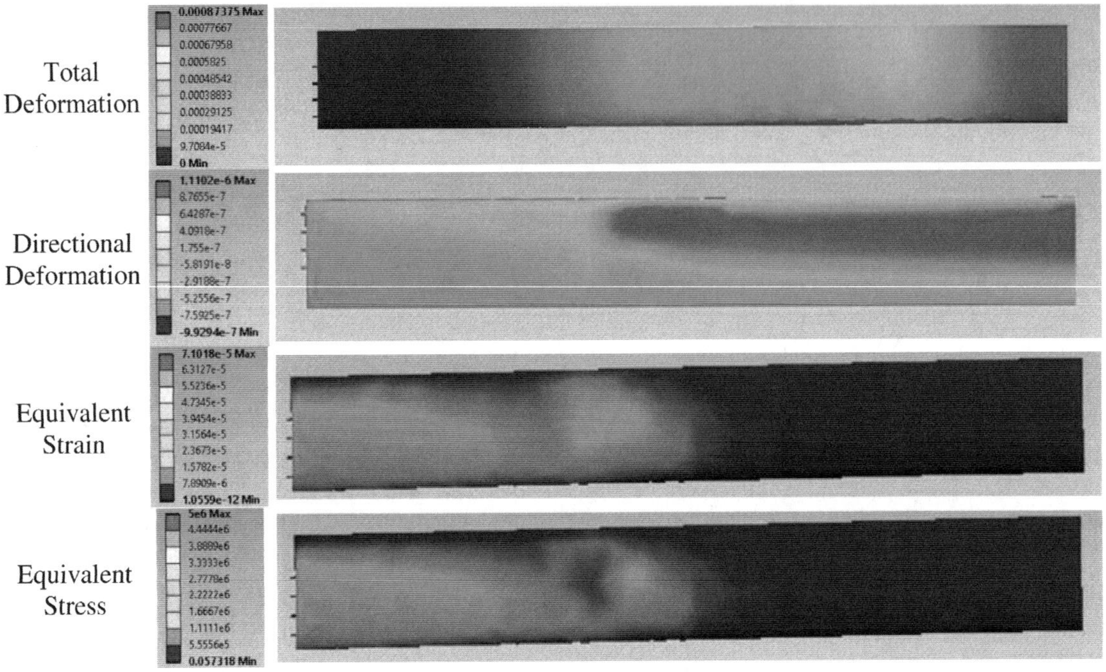

Figure 4. Case I - Without Extension

Figure 5. Case II - With Extension but load on extended Part only

3.4 Case IV – Modal analysis of Wing

In order to determine the effects of inertial loads during manoeuvres and natural frequencies, modal analysis was also performed. Modal analysis comprises of deformation calculations due to loads at higher load factors and estimation of natural frequencies and mode shapes. The mode shapes for corresponding frequencies are given in Figure 7.

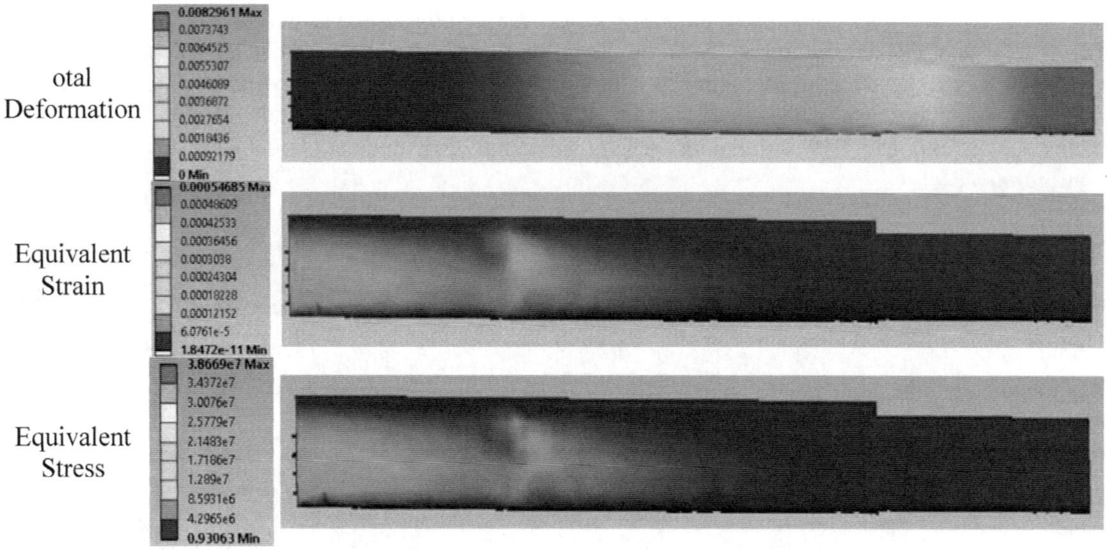

Figure 6. Case III - With Extension but uniformly distributed load over the wing surface

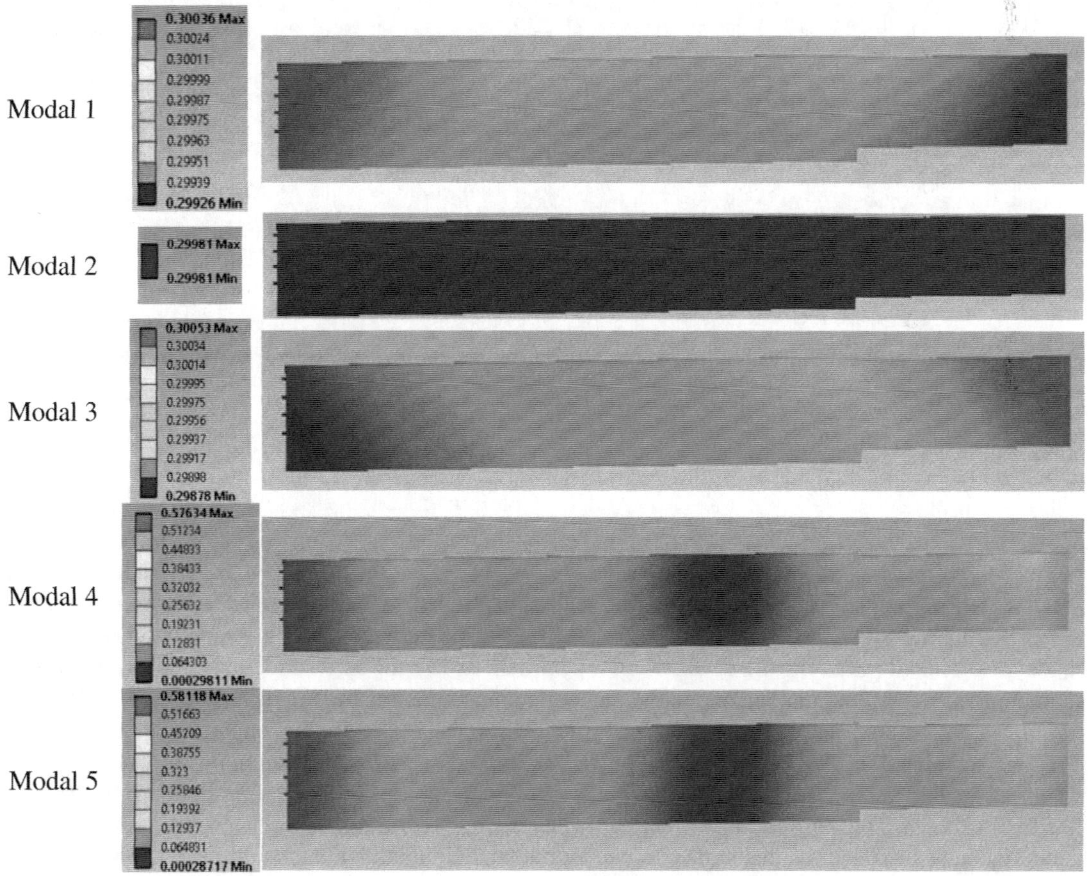

Figure 7. Case IV – Modal analysis of Wing

IOP Conf. Series: Materials Science and Engineering **370** (2018) 012040 doi:10.1088/1757-899X/370/1/012040

3.5 *Case V - Application of Moment Force over the Leading Edge of Wing*

Figure 8 shows that the wing-root bending moment increases drastically due to an increase in the wingspan. Thus, the bending moment along the wingspan of the morphing wing is much larger than that of the conventional wing which results in an increase in the deflection of the free-end. The maximum stress for the un-extended wing configuration increases for the extended wing configuration.

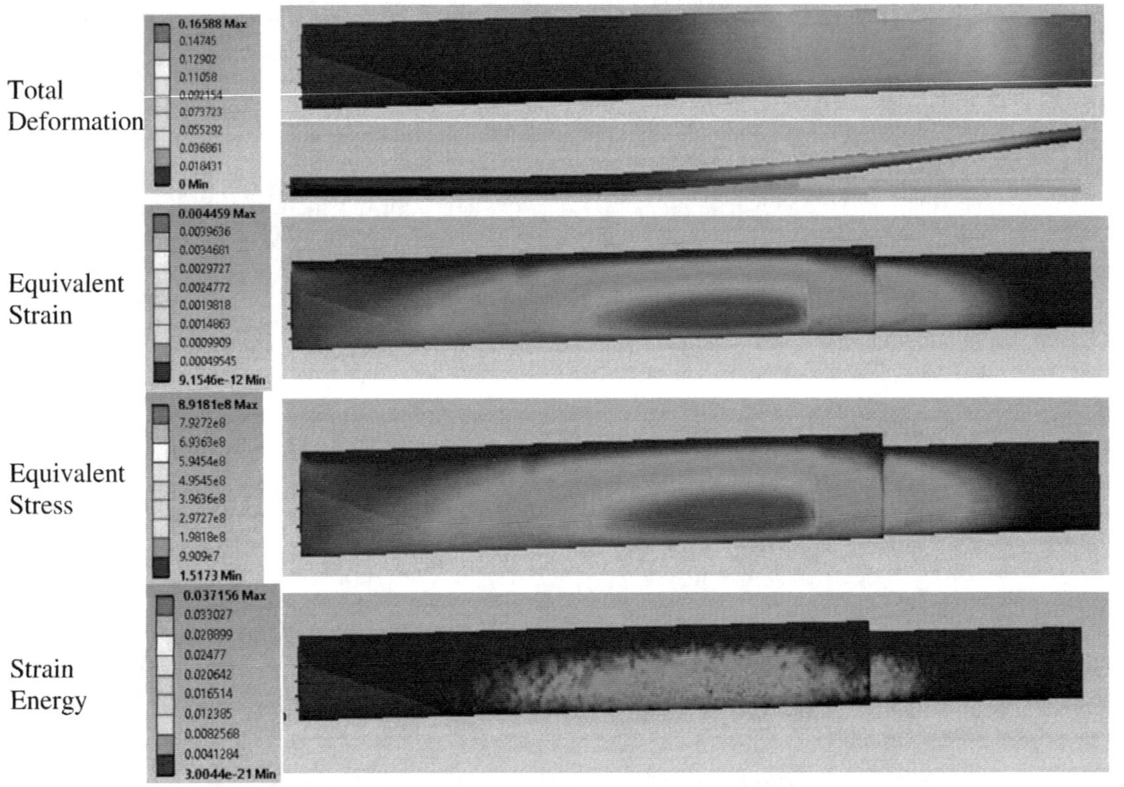

Figure 8. Case V of Moment Force over the Leading Edge.

4. Conclusion

Morphing UAV wing model as per the design considerations is developed in Solidworks and the model is imported into Ansys/Static Structural and subjected to various loading. The loading are calculated according to the aerodynamic lift load analysis by considering various design factors and the deflection over have been estimated. As seen in the results, the wing is severely affected by the loads on extended part, and the deflection is maximum at the tip of extended section. Moreover, the moment loading applied in one case shows the deflection increases further by the moment force. Also, the stress and strain values increased at the hinged joints between two wing sections. Von misses stress is calculated in order to know the maximum stress levels and minimum stress levels on the wing. Their differences are shown clearly with the contour stress levels. The deflection and stress levels are shown from minimum to maximum in the color contours. Their values are given side by side. The wing structure has been optimally analyzed which satisfies the strength and stability criteria.

Acknowledgement

This publication was supported by Universiti Sains Malaysia Grant No. 304/PAERO/6315002.

References

[1] M. J. Patil, D. H. Hodges, and C. E. Cesnik, "Nonlinear aeroelasticity and flight dynamics of high-altitude long-endurance aircraft," *Journal of Aircraft,* vol. 38, pp. 88-94, 2001.

[2] J. N. Iii, P. Newman, A. T. Iii, and G.-W. Hou, "Efficient nonlinear static aeroelastic wing analysis," *Computers & fluids,* vol. 28, pp. 615-628, 1999.

[3] J.-S. Bae, T. M. Seigler, and D. J. Inman, "Aerodynamic and static aeroelastic characteristics of a variable-span morphing wing," *Journal of aircraft,* vol. 42, pp. 528-534, 2005.

[4] P. Santos, J. Sousa, and P. Gamboa, "Variable-span wing development for improved flight performance," *Journal of Intelligent Material Systems and Structures,* vol. 28, pp. 961-978, 2017.

[5] G. R. Andersen, D. L. Cowan, and D. J. Piatak, "Aeroelastic modeling, analysis and testing of a morphing wing structure," in *Proceedings of 48th AIAA/ASME/ASCE/AHS/ASC structures, structural dynamics and materials conference*, 2007, pp. 23-26.

[6] A. Elham, "Effect of wing-box structure on the optimum wing outer shape," *The Aeronautical Journal,* vol. 118, pp. 1-30, 2014.

[7] C. L. Potter, S. G. Russell, V. Kim, Z. Liu, and D. N. Mavris, "Design Space Exploration of Wing Box Substructures Configuration Using SPANDSET," in *16th AIAA/ISSMO Multidisciplinary Analysis and Optimization Conference*, 2015, p. 2789.

[8] E. Gürses, İ. O. Tunçöz, Y. Yang, P. Arslan, U. Kalkan, H. Tıraş, M. Şahin, S. Özgen, and Y. Yaman, "Structural and aerodynamic analyses of a hybrid trailing edge control surface of a fully morphing wing," *Journal of Intelligent Material Systems and Structures,* vol. 28, pp. 979-991, 2017.

[9] O. Schorsch, A. Lühring, C. Nagel, R. Pecora, and I. Dimino, "Polymer based morphing skin for adaptive wings," in *Proceedings of the 7th ECCOMAS thematic conference on smart structures and materials (SMART 2015), Ponta Delgada, Azores, submitted*, 2015.

[10] R. Xu, "Optimal design of a composite wing structure for a flying-wing aircraft subject to multi-constraint," 2012.

[11] Y. Naidu and S. Adali, "Design and Optimization of a Medium Altitude Long Endurance UAV Wingbox Structure," *R & D Journal of the South African Institution of Mechanical Engineering,* vol. 30, pp. 22-29, 2014.

[12] R. Pecora, I. Dimino, F. Amoroso, and M. Ciminello, "Structural design of an adaptive wing trailing edge for enhanced cruise performances," in *Proceedings of 24th AIAA/AHS Adaptive Structures Conference. San Diego*, 2016.

Optical Wear Inspection of Countersink Drill Bit for Drilling Operation in Aircraft Manufacturing and Assembly Industry: A Method

Raiminor Ramzi[1], Elmi Abu Bakar[2]

[1] School of Mechanical Engineering, Universiti Sains Malaysia, 14300, Nibong Tebal, Pulau Pinang, Malaysia.
[2] School of Aerospace Engineering, Universiti Sains Malaysia, 14300, Nibong Tebal, Pulau Pinang, Malaysia.

Corresponding author: meelmi@usm.my

Abstract. Assembling aircraft skin requires thousands of rivets in order to fasten the skin together with rib. Therefore, there will be a lot of drilling operations daily. This could lead to wastage of drill bit usage if the operation is not fully utilized or reduce the hole quality if it is overused. An inspection system is developed to monitor the condition of the countersink drill bit and put these issues under control. Currently, the existing optical wear inspection system is capable to detect and measure the percentage of wear of the drill bit. The theory and application of image processing used in the detection and measurement were already discussed in previous work. In this paper, a method is proposed to inspect the wear of countersink drill bit by using the current existing system. Several key factors that may affect the accuracy of the optical inspection such as relation between illumination and image quality, and subject positioning and alignment are discussed in this paper. An external illumination system is developed to improve the quality of image captured. The drill bit positioning and alignment are corrected using image registration method in image processing. Another relevant precaution is also added into account before finally implement the complete step of performing the inspection.

1. Introduction

Aircraft manufacturing and assembly industry involves several machining operations such as drilling, riveting, sanding and painting. Drilling and riveting are done in the sub-assembly stage while sanding and painting usually come after these process for finishing. Thousands of rivets require to fasten the skin and rib together. Countersunk or flush rivet is preferable to use for assembling this exterior aircraft parts because it is more aerodynamic. It is flush with the aircraft skin and thus reducing drag on it. Countersink drill bit is necessary to create a countersunk hole to fit the rivet in. The shape and diameter of the countersink drill bit is not consistent along its length because it has conical shape in the middle which is used to create the countersunk hole.

Drilling process in aerospace industry become more challenging as modern aircraft structures are made up of composite material. It is a material known for its high strength to weight ratio but worst machinability [1]–[4]. High strength is good for the aircraft but it would be extremely hard to drill and may reduce the life expectancy of the drill bit. Therefore, a stronger and tougher material such as polycrystalline diamond (PCD) and carbide bit are commonly used to drill composite material in order to improve the machinability efficiency [4].

The usage of the drill bit has no indicator that can control the process. An overused of the drill bit may reduce the quality of hole while stopping the usage before the end of the tool life may cause wastage. This is where an inspection system is needed. There are a lot researches that has been done in monitoring drill bit using the optical approach. One of the earliest was already done in 1990 by T. Pfeifer and J. Elzer. They propose a vision system and measure the flank wear of drill bit by applying the knowledge of digital image processing [5]. The research on monitoring the wear continues until nowadays with different approaches, methods and objectives. J.C. Su *et al* use edge detection method to extract the boundary of the cutting lips and divided the region to multiple region in order to measure the flank wear [6]. Yu-Teng Liang and Yih-Chih Chiou proves that a real time flank wear measurement method is possible by measuring the wear on 3 drill bits with different materials [7]. Another group of researchers, Guifang Duan *et al.* developed an automatic optical inspection (AOI) for microdrill bit in PCB manufacturing application. The system is capable to perform phase identification of the drill bit condition [8]. Different approach was done by A. Volkan Atli *et al* as monitoring the tool condition from side view rather than top view. A concept of deviation from linearity (DEFROL) is used to measure the changes of the drill bit point angle [9].

From previous research, a system is developed to inspect and monitor the condition of the drill bit wear [10]. It is capable of measuring the amount of wear in term of percentage. This paper proposed a method using optical approach in inspecting wear of countersink drill bit which is used for drilling operation in aircraft manufacturing and assembly industry. The detection method and process flow will not be discussed in details as it already presented in previous work. This paper will be focused on the methodology of the inspection system which includes the system design, illumination system, inspection precaution and procedure.

2. Method approach

2.1. System design

Figure 1 shows the actual configuration of the system. The method used in the inspection system is by monitoring the cutting lips region from top. The system basically can be divided into two parts which is hardware and software. The hardware of the system consists of a digital microscope as the sensor for capturing images and it has its own lighting. However, light emitting diode (LED) strips are used for additional illumination for a better image quality. Servo motors are used on both sides to control the rotation angle of the LED strip with simple programming using Arduino. Chuck is used to lock the drill bit into position and secure its alignment with the digital microscope.

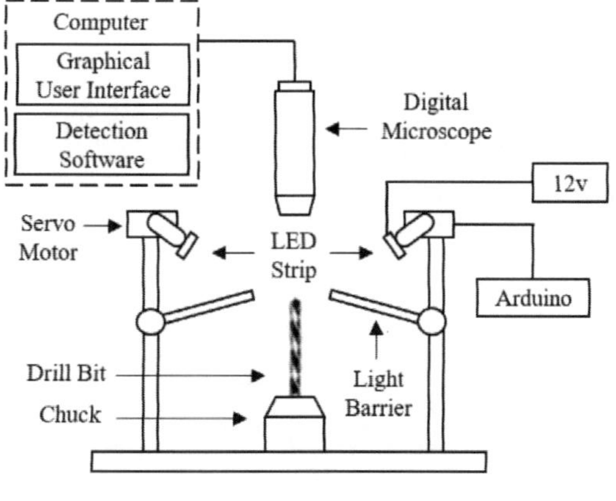

Figure 1. Schematic of the actual inspection system.

IOP Conf. Series: Materials Science and Engineering **370** (2018) 012041 doi:10.1088/1757-899X/370/1/012041

The software consists of a graphical user interface which is developed along with the system using MATLAB. The software is capable of capturing the sample for the inspection with a special guideline which will be explained in the next section. It also includes a detection system to detect and measure the wear in term of percentage. This hardware and software is integrated together to become a complete inspection system

2.2. System illumination

The inspection process is focused on monitoring the cutting lips area of the drill bit. A very good light source required to obtain a high quality and clear image of the cutting lips region. The existing light from the digital microscope is not satisfying the requirement of the inspection. The light basically lightens the entire region of the drill bit which could make the detection process more difficult. Additional light source using LED strip is added on both side and projected to the drill bit with help of the law of reflection. The idea is to project the light at certain angle directly to the cutting lips so that the light enters the lens perfectly and the image captured will only focused or highlighted at the cutting lips only.

The law of reflection defines that when light falls into a smooth surface, the angle of the reflection is equal to the angle of incident and the incident ray, reflected ray and the normal line all lies in the same plane. Figure 2 illustrate the definition of the reflection law where, i is the incident angle of the incident ray and r is the reflection angle of the reflected ray.

The additional illustration is projected at certain angle which denotes as β, angle of the light projection which remains unknown at the moment. The only information available is the drill bit point angle, α, used 130°. The relation between these two angles is defined with help of the diagram shown in figure 3. Therefore, the angle of the light projection, β is can be calculated for any different drill bit point angle, α.

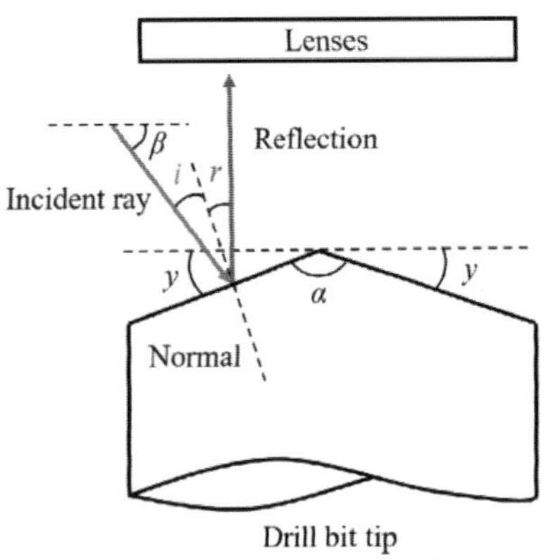

Figure 2. Law of reflection concept used for the additional illumination to ensure light rays enters to the lenses perfectly.

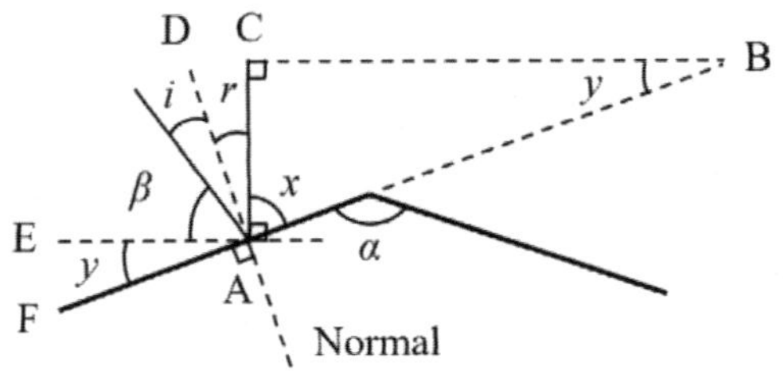

Figure 3. Diagram used to define the relationship between angle α and β.

From figure 2, it is clear that,

$$y = \frac{180° - \alpha}{2} \tag{1}$$

From the perpendicular angle $\llcorner CAE$ in figure 3,

$$\beta = 90° - i - r \tag{2}$$

Using the law of reflection, $i = r$. Therefore, equation (2) become,

$$\beta = 90° - 2r \tag{3}$$

From right triangle ΔACB, the total angle of a triangle is 180°, therefore:

$$180° = 90° + x + y$$
$$x = 90° - y \tag{4}$$

From perpendicular angle $\llcorner DAB$,

$$90° = x + r$$
$$r = 90° - x \tag{5}$$

Substituting equation (4) into (5),

$$r = y \tag{6}$$

Rearranging equation (1) and (6) into (3), therefore the relationship between α, drill bit point angle and β, angle of the light projection can be defined as follow:

$$\beta = 90° - 2\left(\frac{180° - \alpha}{2}\right)$$
$$\beta = \alpha - 90° \tag{7}$$

The LED projection angle is controlled using motor servo on each side. Both servo motors are programmed through Arduino to hold the LED and projected the lights at angle β calculated using equation (7). The brightness of the light source is also an important factor of capturing a clear image. Too much or not enough light could remove an important information from the image. Therefore, dimmer is added to the system in order to control the brightness of the LED in the illumination system.

2.3. Subject Positioning

The detection is done by comparing the image of brand new and worn drill bit and measuring the difference of both images as wear. This comparison requires the image taken each time to be exactly at the same position and size. Therefore, a software is developed to capture the image and included with reference line as a guide as shown in figure 4. This will maintain the size and position of the subject each time the inspection is done. However, the subject positioning is difficult to maintain along the inspection process because there is still a slightly small difference between images taken. This small difference can affect the accuracy of the detection as it increases or decreases the amount of wear detected.

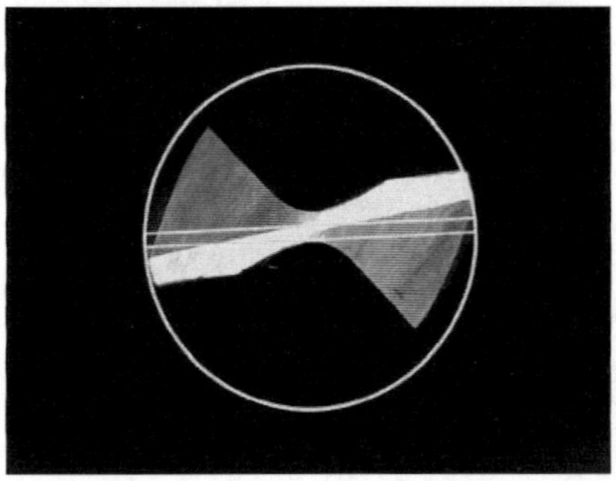

Figure 4. Reference line used to guide and maintain the size, position and orientation of the drill bit during inspection process

3. Result and discussion

3.1. Illumination result

3.1.1. Application of light barrier. Monitoring a countersink drill bit has its own challenge. As mentioned before, countersink drill bit has a conical shape in the middle and this conical region is visible from top view. The region is unwanted as the inspection system is only focused on the cutting lips region to monitor the wear. This is because the conical region is least affected compare to the cutting lips area. The presence of the unwanted region is affecting the inspection result. Therefore, the light barrier as shown in figure 1 is added on both side to eliminate the conical region by blocking lights to the region. It helps the system to focus only to the cutting lips area. Figure 5 proves that the application of the light barrier improves the image acquisition of the inspection system.

Basically, the conical region around the drill bit in figure 5(a) distract the detection algorithm from detecting the region of interest. This image resulted as shown in figure 5(b) which cause error to the detection. A very good example of sample image and its result can be seen as in figure 5(c) and (d) respectively. The detection system clearly detects the region of interest and able to filters all unwanted information. The application of the light barrier to the system has enhanced the inspection result by avoiding the possibility of error in the detection. This enable the system to apply the automatic detection of the cutting lips region which replaced polygon crop, a user defined cropping method to select the region of interest in previous research [10].

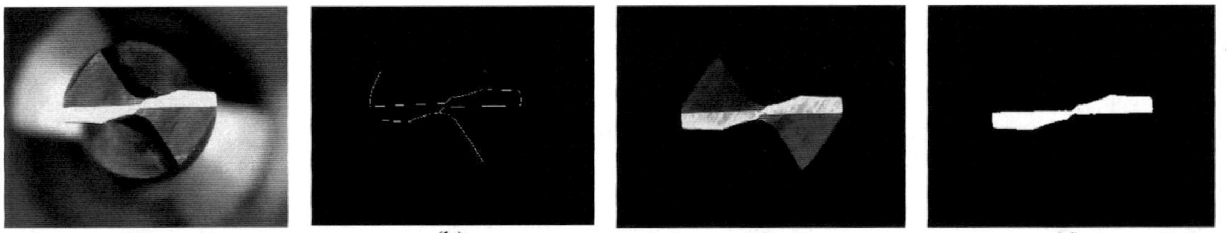

(a) (b) (c) (d)

Figure 5. Comparison of images capture with and without application of light barrier and its results (a) Images captured without using light barrier (b) detection result of image without using light barrier (c) image capture with light barrier (d) detection result of image with light barrier

3.1.2. Angle of light projection. In this research, the illumination is important as it affects the accuracy of the inspection system. The additional LED lighting system is included with servo motors on both side to control the projection angle of both LED. A set of data is calculated for various drill bit point angle, α from 125.0° to 140.0° based on the equation (7) derived in previous part and tabulated in Table 1. The current countersink drill bit used has a drill bit point angle of 130°. Therefore, from Table 1 the best angle of light projection, β for the current system is 40.0°. The value is programmed through Arduino to control the servo motors at the calculated angle. This will ensure that the projection is accurate and fixed along the inspection process.

Table 1. Angle of light projection, β calculated for various drill bit point angle, α using equation (7).

α	y	i	r	β
125.0°	27.5°	27.5°	27.5°	35.0°
130.0°	25.0°	25.0°	25.0°	40.0°
135.0°	22.5°	22.5°	22.5°	45.0°
140.0°	20.0°	20.0°	20.0°	50.0°

3.1.3. Lighting and brightness. The illumination system is tested with several different lighting sources and conditions. Three lighting sources are from ambient light, digital microscope light and the additional LED light with projection angle of 40.0°. Figure 6 shows the difference of the results from different lighting sources and conditions. Figure 6(a) is the result taken using ambient light only without digital microscope light and LED light. This condition is inconsistence as the ambient light may varies in brightness and direction of light source in different situation. Capturing the image with digital microscope light resulting a very clear image as show in figure 6(b) but it does not meet the requirement for the inspection. As explained earlier, the inspection system focus on monitoring the wear on the cutting lips area only. However, the digital microscope light lightens all the entire region of the drill bit which make the detection become harder and resulting the similar result as in figure 5(b). Applying the light barrier to this situation does not make any difference as the lights from the digital microscope is projected directly to the top of the drill bit.

Projecting the LED lights at 40.0° from both side resulting a clear image that focusing at the cutting lips region as shown in figure 6(c) and (d). The difference between the two images is the brightness of the LED light. Figure 6(c) shows the result of image captured using the LED light with high brightness at 1100 lux while figure 6(d) with low brightness at 400 lux. These two images are the acceptable range of the brightness used in the inspection. This limit is necessary as it will be a guidance to control the LED brightness. Over brightness will affect the accuracy of the inspection as the boundary of the cutting lips is extra glowing and cause the system to detect a larger area of the cutting lips. Low brightness is good in differentiating the boundary of the cutting lips but if the brightness is too low, it may cause the detection system harder to detect the boundary itself and resulting a similar result as in figure 5(b). The light barrier enhanced the quality of the image acquisition by removing the unwanted region from the image. This setup combination is the best at acquiring an image that is required by the inspection system compare with image taken using ambient light and digital microscope light.

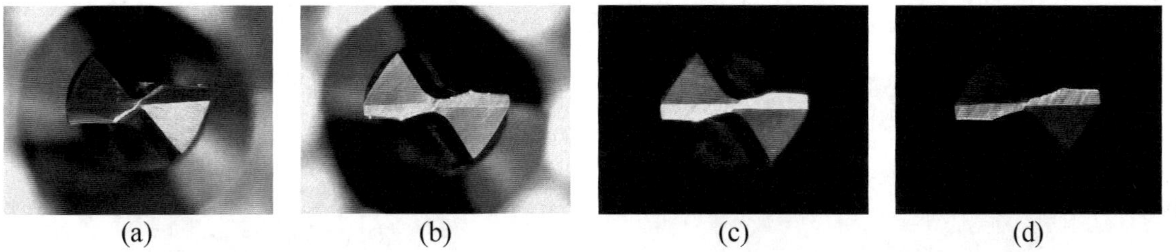

(a) (b) (c) (d)

Figure 6. Images captured from various illumination brightness and light sources (a) ambient light (b) digital microscope light (c) high brightness LED strip (d) low brightness LED strip

3.2. Positioning correction

Besides the illumination of the system, the positioning of the drill bit is also important. The inspection is a repetitive process and the consistency in image acquisition is very important because this will affect the output results. As described before, there is still a small difference between images taken although the reference line is used as guide in maintaining the position, orientation and size of the subject. The problem is solved by using image registration method in image processing. It is used to align the images based on a reference image. This method is included in the detection system in order to eliminate the small difference aforementioned. It is used to perform the correction to the position, orientation and size of the drill bit based on the reference image. Figure 7(a) shows a detection result without using image registration method. The orientation of the drill bits (brand new and worn) are intentionally misaligned to show the effectiveness of this method. This is proven in figure 7(b) as both images are aligned according to the reference image (brand new drill bit). This will make the images to be comparable and enabling the system to monitor and accurately measure the changes of the cutting lips region as wear.

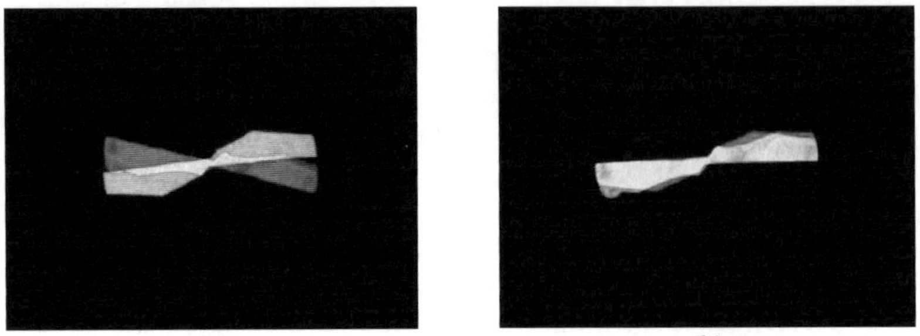

(a) (b)

Figure 7. The comparison of the detection (a) without image registration (b) with image registration

3.3. Inspection precaution and procedure

The material of the workpiece drilled is a combination of composite layers and aluminium, since this system is applied in the drilling process of aircraft manufacturing and assembly. The composite material will produce dust when drilling while the aluminium will produce swaft as a result from the machining operation. During the drilling process, the adhesion of the swaft is possible due to the heat produce by the friction between the drill bit and the workpiece. This adhesion may distract the important information of the image which is the region of the cutting lips. The presence of dust also will prevent the lights from reflecting well to the lenses causing the image taken is not clear enough as shown in figure 8. Figure 9 shows the detection result of an inspection using the same image in figure 8. It can be clearly seen that this precaution is a compulsory step for the inspection. Therefore, it is important to clean the drill bit tip with wire brush before performing the inspection to remove the dust and adhesion of swarf around the cutting lips. This precaution is crucial for this inspection to ensure the images taken is clear and qualified to be used for wear detection.

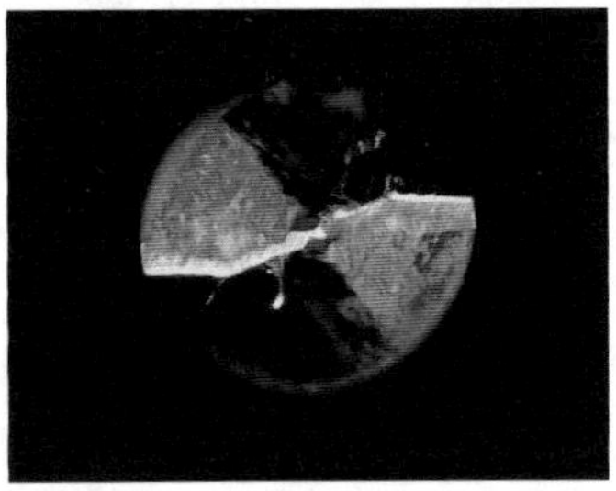

Figure 8. An example of unclean drill bit image

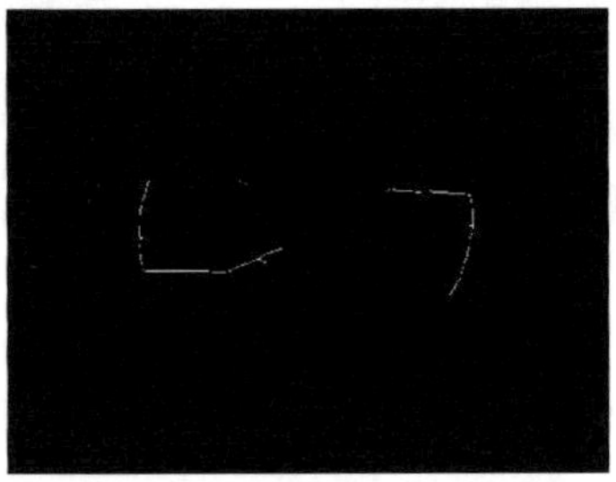

Figure 9. Detection result from the unclean drill bit image

The inspection process is done by following a standard procedure based on the output of the studies.

It starts with preparing the setup of the full system as the configuration shown in figure 1. The digital microscope is connected to the software in computer via USB cable. Then, the tip of the drill bit is cleaned with wire brush to remove dust and swarf before clamp into the chuck. The servo motors is started and the LED lights is on. The servo motors are ensured to be in the correct angle. The drill bit is adjusted into the correct position and orientation with help of the reference line built in the software. The LED brightness is adjusted into the acceptable range. The final step before capturing the image is adjusting the light barrier until the conical region of the countersink drill bit is not visible in the image. After the image is captured, it is ready to be fed to the detection system for wear measurement. The taken steps and precaution are proposed for the other similar monitoring system.

4. Conclusion

This paper proposed a method of performing the wear inspection on a countersink drill bit using an optical approach. The illumination system used is able to capture a good and clear images that is required for the inspection system. A range of brightness is set up to 400 lux to 1100 lux based on the effectiveness of the detection. Image registration method is added to the detection system to ensure the position and alignment of the drill bit for every images captured. Finally, a complete inspection procedure and precaution taken is proposed.

Acknowledgement

The author would like to thanks to Ministry of Higher Education (MOHE) Malaysia for funding support through FRGS research grant (203/PMEKANIK/6071296) and awarding him MyBrain 15 (MyMaster) scholarships.

References

[1] C. R. Dandekar and Y. C. Shin 2012 Modeling of machining of composite materials: A review *Int. J. Mach. Tools Manuf.* **57** 102–121.

[2] D. Iliescu, D. Gehin, M. E. Gutierrez, and F. Girot 2010 Modeling and tool wear in drilling of CFRP *Int. J. Mach. Tools Manuf.* **50** 204–213.

[3] A. M. Abrão, P. E. Faria, J. C. C. Rubio, P. Reis, and J. P. Davim 2007 Drilling of fiber reinforced plastics: A review *J. Mater. Process. Technol.* **186** 1–7.

[4] R. Teti 2002 Machining of Composite Materials *CIRP Ann. - Manuf. Technol.* **51** 611–634.

[5] T. Pfeifer and J. Elzer 1990 Measuring drill wear with digital image processing *Measurement* **8** 132–136.

[6] J. C. Su, C. K. Huang, and Y. S. Tarng 2006 An automated flank wear measurement of microdrills using machine vision *J. of Mater. Process. Technol.* **180** 328–335.

[7] Y. Liang and Y. Chiou 2006 An Effective Drilling Wear Measurement based on Visual Inspection Technique Proc. of the 9th Joint Conf. on Information Science (kaohsiung, Taiwan) pp 2–5.

[8] G. Duan, H. Wang, Z. Liu, J. Tan, and Y.-W. Chen 2014 Automatic optical phase identification of micro-drill bits based on improved ASM and bag of shape segment in PCB production *Mach. Vis. Appl.* **25** 1411–1422.

[9] A. V. Atli, O. Urhan, S. Ertürk, and M. Sönmez 2006 A computer vision-based fast approach to drilling tool condition monitoring Proc. Inst. Mech. Eng. Part B J. Eng. Manuf. vol 220 pp 1409–1415.

[10] R. Ramzi and E. A. Bakar 2016 Wear Detection on Twist Drill Cutting Lips using Digital Microscope Proc. of The 7th TSME Int. Conf. on Mechanical Engineering (Chiang Mai, Thailand).

IOP Conf. Series: Materials Science and Engineering **370** (2018) 012042 doi:10.1088/1757-899X/370/1/012042

Prediction of Trace Element based Energizing Sensor Control System using PWM

Mohammad Nizar Bin Mohamed Zukri[1], Elmi Bin Abu Bakar[1,*], Naoki Uchiyama[2], Mohamad Nazir Bin Abdullah[3]

[1]School of Aerospace Engineering, Universiti Sains Malaysia, Penang, MALAYSIA.
[2]Department of Mechanical Engineering, Toyohashi University of Technology, Aichi, JAPAN.
School of Electrical and Electronic Engineering, Universiti Sains Malaysia, Penang, MALAYSIA.

*E-mail: meelmi@usm.my

Abstract. A real-time system for field-work monitoring wastewater laden with heavy metal in industrial discharge through wireless communication network was developed. The monitoring system poses an interesting challenge in order to determine existing metal ion in the solution whereas the previous result only consider total dissolve ion. This paper aims to distinguish the metal ion based on reaction determination in solution. The control algorithm was implemented as generating voltage input for energize conductivity sensor since the voltage corresponding to oxidation and reaction based on standard reduction potential. Implementation of ATmega2560 microcontroller for control voltage fed on sensor equivalent to controlling the PWM duty cycle. PID controller was designed uses a microcontroller (Arduino) platform with manual tuning for identify reaction process and sufficient voltage input. From the experimental result, is found that the proposed PI controller has excellent tracking and measurement performance. Low-pass filter was applied in programming to make the system understand that signal has achieved stable. The development of hardware and software of the closed loop system has an enhancement of measurement performance and high feasibility for SME's company in economic point of view. The desired objective is to achieve a system with the stable measurement and sufficient voltage supply. This system will provide an accurate and precise control efficiently without using costly component and complicated circuit.

1. Introduction

Industrial wastewater discharge is a one of major that contribute to water pollution. Water pollution can be characterized as pH level, total suspended solid (TSS), total dissolve solid (TDS), amount of heavy metals, chemical oxygen demand (COD) and biological oxygen demand (BOD) [1]. Recently, many researchers in Malaysia conducted research consider heavy metal existing in order to pertain the environment for sustainable [2].

In a few decades, there are many methods established for determine existing of heavy metal ion in industrial wastewater since the pollutant has an endanger effect to aquatic life and harm human [3]. In general practice, sample from effluent discharge was collected manually and transported to the central laboratory to analyses heavy metal contents. Unfortunately, there will give an irrelevant result for heavy metal contents during samples transportation due to light exposure, dissolved oxygen and precipitation, thus the sampling will give inefficient data [4]. Due to environmental threat from wastewater associate with particular risk heavy metal even at low concentration, the development of analytical method with low cost and fast detection method is become interest for most researcher [5].

A variety of inorganic techniques approaches are described in literature for trace elements of heavy metal. The method is difference based on principle and type of analysis. This paper demonstrated simple instrument to determine metal concentration based on electroanalytical method. Electroanalytical is the technique that concern on analytic solution by measure the potential (volt) and current (ampere) in interfacial process of chemical or physics species [6]. From this fundamental concept, this technique can be divided to the various others method based on aspects of the cell controlled and measured. This method is suitable for develop the field work, real-time and on-line environmental analysis due to high performances such as low detection limits, good stability and wide linear response range.

2. Related Works

For nowadays, it will be significantly to SME's industrial consider to measure heavy metal contents to remove and control the toxicants before to release directly into environment through final discharge point. Thus, simplified technique with fast detection device must be develop to monitor the discharge regularly and build a pollution prevention system. The aim of this work is to install the device into final discharge point whereas the data can be quickly and accurately assess the toxic contaminant in the real-field work.

2.1. Block Diagram of System

Figure 1 illustrates the schematic of a PID controller with a feedback loop. The sensor reading block measures the conductivity value for the solution through the conductivity sensor. At initial state, the sensor will read an analogue value from the physical phenomenon, then the analogue value will converted into ion concentration values (ppm). The process variable for the system will represent an absolute value (reading after – reading before) by using low pass filter. Low pass filter is implement in the system to obtain a stable reading. From the stable reading, we can identify and predict the reaction process for electron flow and concentration from a particular solution. Thus, in order to make the system realize the stable reading is achieved, the set point is set for zero. The process variable is aim to approaches the zero reading through the voltage supplied. PID controller uses a feedback loop to control the voltage supply and energize the sensor. The aim of this controller is ensure the process variable close to zero reading by energize the sensor with various voltage. The controller uses an error (the difference between Absolute value and Set Point) in each loop to control the PWM. It will give automatically determine value of various voltage in the system whereas measures the error in each loop and gives the necessary voltage signal to the system.

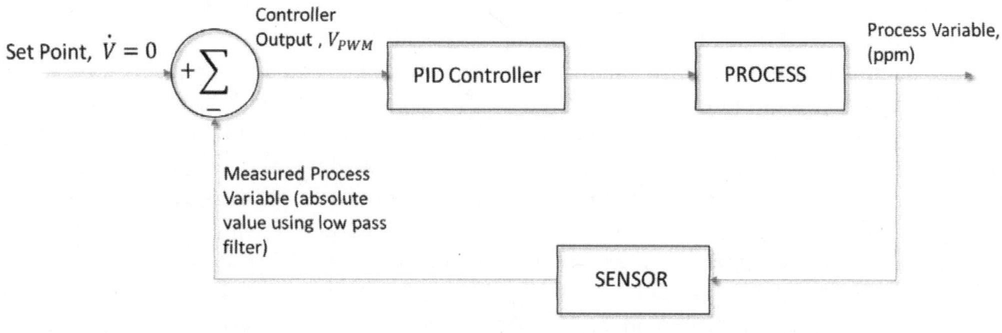

Figure 1. Block Diagram of the System

2.2. System Hardware

The system includes a set of submersible water pumps, piping network system, sensoring unit (electrical conductivity, pH and temperature) from dfrobot, RF95W LoRa wireless module, controller unit (Arduino mega 2560 and Arduino nano), amplifier circuit unit, power supply SPP1272 battery 12V 7.2AH, electronic GPI flow rate digital meter and hydraulic bench. This experimental model was carried out by control of the Arduino based due to the development of wireless instrument on final discharge point. This microcontroller as Table 1 was consider in the system since it is user-oriented dominated such as size and configuration, easy to manipulate, appropriate time constants, broad spectrum tasks to be solved and easier to access through Matlab/Simulink environment [7].

Table 1. Arduino microcontroller model

Specification	**Description**
Processor	ATmega2560
Processor Speed	16MHz
Architecture	8 bit
RAM Memory	256KB
Programming Language	C Variant
Digital Pins	54 (15PWM)
Analog Pins	16 (0-5V)

2.3. Theoretical Summary of consider method

For establish equipment nowadays has a great standards to determine heavy metal ions in laboratory [8]. The most dominant techniques are Inductive Coupled Plasma Mass Spectrometry (ICP-MS), Inductive Coupled Plasma with Atomic Emission Spectrometry (ICP-AES) [9][10], Atomic Absorption Spectrometry (AAS), Atomic Emission Spectrometry (AES), and Cold Vapor Atomic Absorption Spectrometer (CVAAS) [11][12]. These modern methods use optical detection are more selective and sensitive compare to conventional method [13], however there are not convenient for fields work such as batik factories because it require specifically techniques, long time storage, high cost and bulky. Bioanalytical techniques such as nuclei acid through DNA probes also used to trace heavy metal ions [14]. However, this development techniques recently useful for food and clinical application [15]. In general, most literature emphasizes electroanalytical technique for field work instrument development, real-time and online wastewater status due to high performances such as low detection limit, good stability and wide linear response [16][5].

3. Methodology

3.1. Flow Chart

The figure 2 illustrates the project flow chart. It mainly included 3 phase namely perception layer, signal conditioning layer and terminal layer. At perception layer, system will be initiate from a low voltage until the process variable given a stable reading. The system will compute the error, from an error, PID controller will work for necessary voltage signal. At signal conditioning layer, first reading is providing as an analog value. Then, the value will apply smoothing to get an average analog value. Then, the signal will filter using kalman filter and obtain an average kalman voltage signal. These value will give information for terminal display and process variable of the system. At third phase, terminal layer will display the desired information in graphical user interface GUI which is the status of the solution. In this phase, there will be several mathematical formulations, transmission and receiving through wireless and display monitor. The parameter such as existence metal, treatment efficiency, fixation rate and water quality is decided for output display. A wireless system is adding on the system with aided of graphical user interface. At last, an identification, verification and validation is conducted.

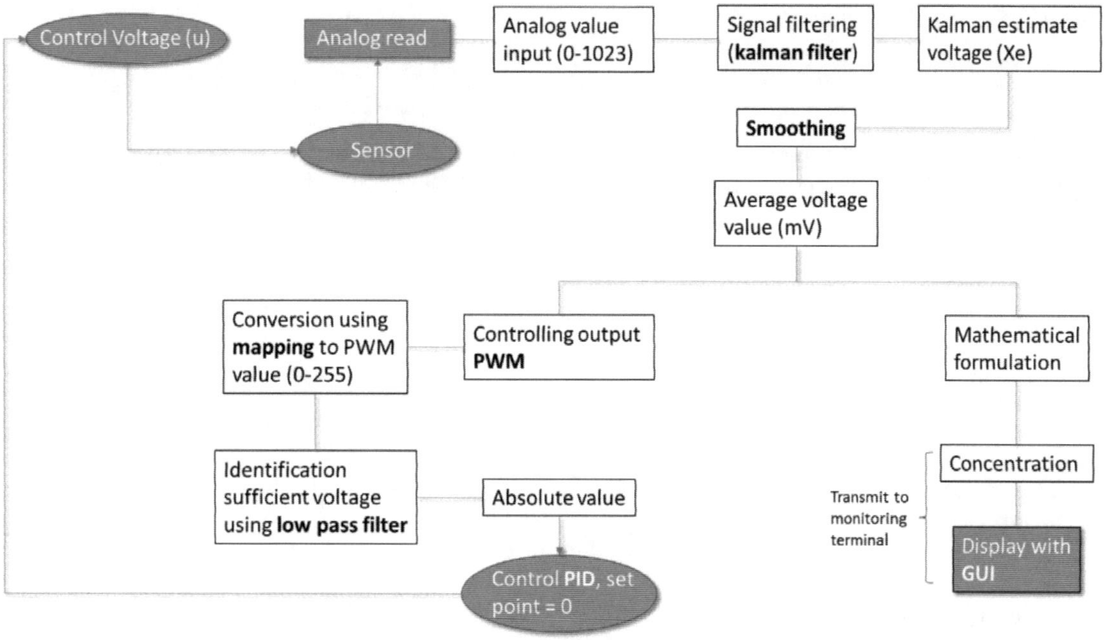

Figure 2. An illustration of whole system

3.2. System Overview

The system uses conductivity sensor. As figure 3, an alternating current (I) was applied at an optimal frequency (typically 94Hz) between two electrodes. The potential voltage between two electrodes will be measured according to ion concentration the solution (V). The resulting voltage is referring also the resistance in the solution. From the measurement of current source and potential voltage, the conductance will be calculated (I/V). Since, cell constant is equal to value of 1, the conductivity value will directly obtain from conductance measurement as written at equation 5.

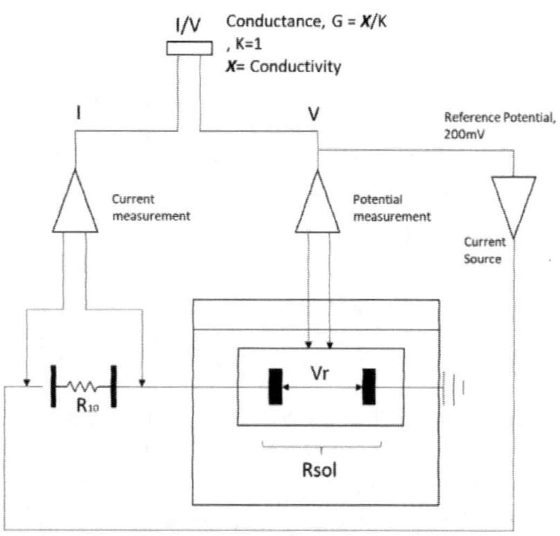

Figure 3. Diagram of the system

3.3. Mathematical Formulation

A current produced in external circuit that connected to the electrodes. This principle follow the ohm's law [17]:

$$E = IR \tag{1}$$

Where E is the applied potential (volt), I is the current measured and R is the resistance of the electrolyte solution between two cell electrodes.

The electrical conductance, G is a reciprocal of solution resistance (S.I units: S Siemens), order literature expressed as mho. However, the measurement of conductance or resistance depends on distance, L between cell electrode and microscopic surface area (geometric area x roughness factor), A of the electrode (A for both electrodes are identical). These parameter can be relate in conductivity formula [18];

$$conductance, G = \frac{1}{R} \tag{2}$$

$$conductivity, \chi = G\frac{L}{A} \tag{3}$$

$$G = \frac{1}{R} = \chi\frac{A}{L} \tag{4}$$

Conductivity is directly proportional to the ratio of distance and area, *L/A*. Therefore, in particular measurement and instrumentation, the ratio *L/A* is usually constant and rigidly fixed called as cell Constant, (K_c) [19]. Thus,

$$\chi = G\frac{L}{A} = GK_c \tag{5}$$

The transfer function of the measurement circuit will be;

$$V_{output} = \frac{R_{10}(820\Omega)}{R_{solution}} x \left|V_{input}\right| \tag{6}$$

Where $R_{solution}$ is the measurement of resistance in aqueous solution. To conclude, equation is written as;

$$conductivity, \chi = \frac{K_c}{R_{(820\Omega)} x \left|V_{input}\right|} x \, V_{output} \tag{7}$$

K_c is the cell constant where the value is equal to 1. V_{input} is depend on the signal generating on PWM. Thus, the ion concentration is linear with the output conductivity and given equation as;

$$Ion\ concentration, C\ (ppm) = 6.9887\chi + 1.6167 \tag{8}$$

3.4. Pulse Width Modulation (PWM)

The technique for Pulse Width Modulation is uses an average analog output signal from digital output. The supply signal consists of a train voltages pulses where the width of individual pulses will control the effective voltage level to the system. From the figure 4, it shows the PWM duty cycle where the voltage signal comprised of pulses with duration, τ_0 and it repeat for every τ_c units of time. The digital output from arduino will run maximum at V_S= 5V. Thus, on a PWM pin, the analogWrite an 8-

bit value function will be used to modulate and set the duty cycle. Hence, the effective voltage from analogWrite the value (0-255) that corresponding to signal voltage in range of 0V to 5V. The output voltage are linearly related according to equation 1 [20].

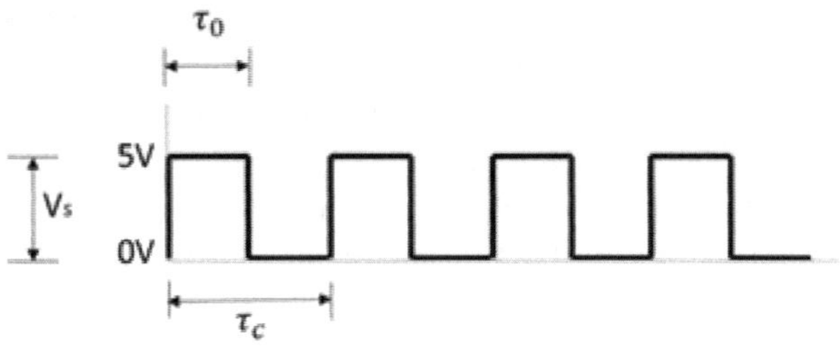

Figure 4. Nomenclature of PWM Duty Cycle

$$PWM\ output = 255\, x \frac{\tau_0}{\tau_c} = 255\, x \frac{V_{eff}}{V_s}\ , V_{eff} = V_s \frac{\tau_0}{\tau_c} \tag{9}$$

3.5. Sensor Design

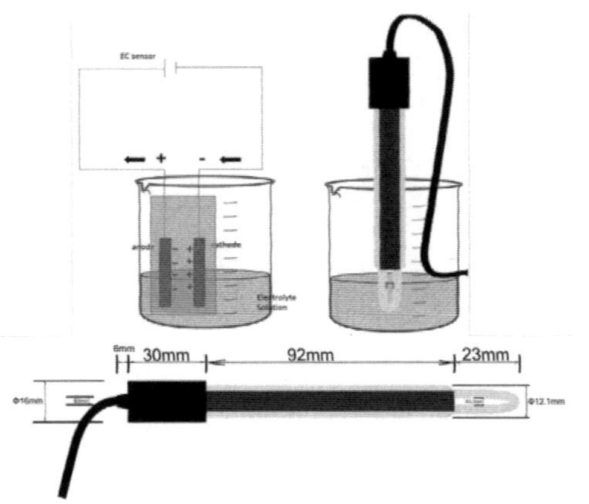

Figure 5. Sensor configuration

Table 2. Sensor specification

Specification	**Description**
Operating Voltage	5V
PCB Size	45mm x 32mm
Operating Temperature	5-40 degree Celsius
Electrode	Electrode Constant K=1
Temperature Sensor	DS18B20 waterproof
Cable length	60cm
PH2 interface	3-pin SMD

4. Experimental Test

To calibrate the conductivity sensor, the calibration plot is used and compare with the calibrate solution. One set measurement data obtained will establish an equation based on least square method. Then, an experimental is set up and the aim for this work is to remove noise from digitization of ADC and random fluctuation signal from the sensor. In this experiment, a synthetic solution with known concentration is prepared as a sample. The sensor (conductivity and temperature) is placed in the container of the sample. The solution will remain static at constant room temperature.

4.1. Signal Filtering

For data acquisition, the performance characteristics figure 6 clearly demonstrates that introduction of Kalman Filter enables to decrease random fluctuation signal. The signal is acquired from the analog value of the sensor. At this moment, smoothing Kalman voltage will give better plot average for particular measurement. With respect to the random character of the process, a comparison running average of Kalman voltage and unfiltered voltage is plotted. When observe the plotting, running average of Kalman Filter represent noise reduction is achieved and influential measurement readings.

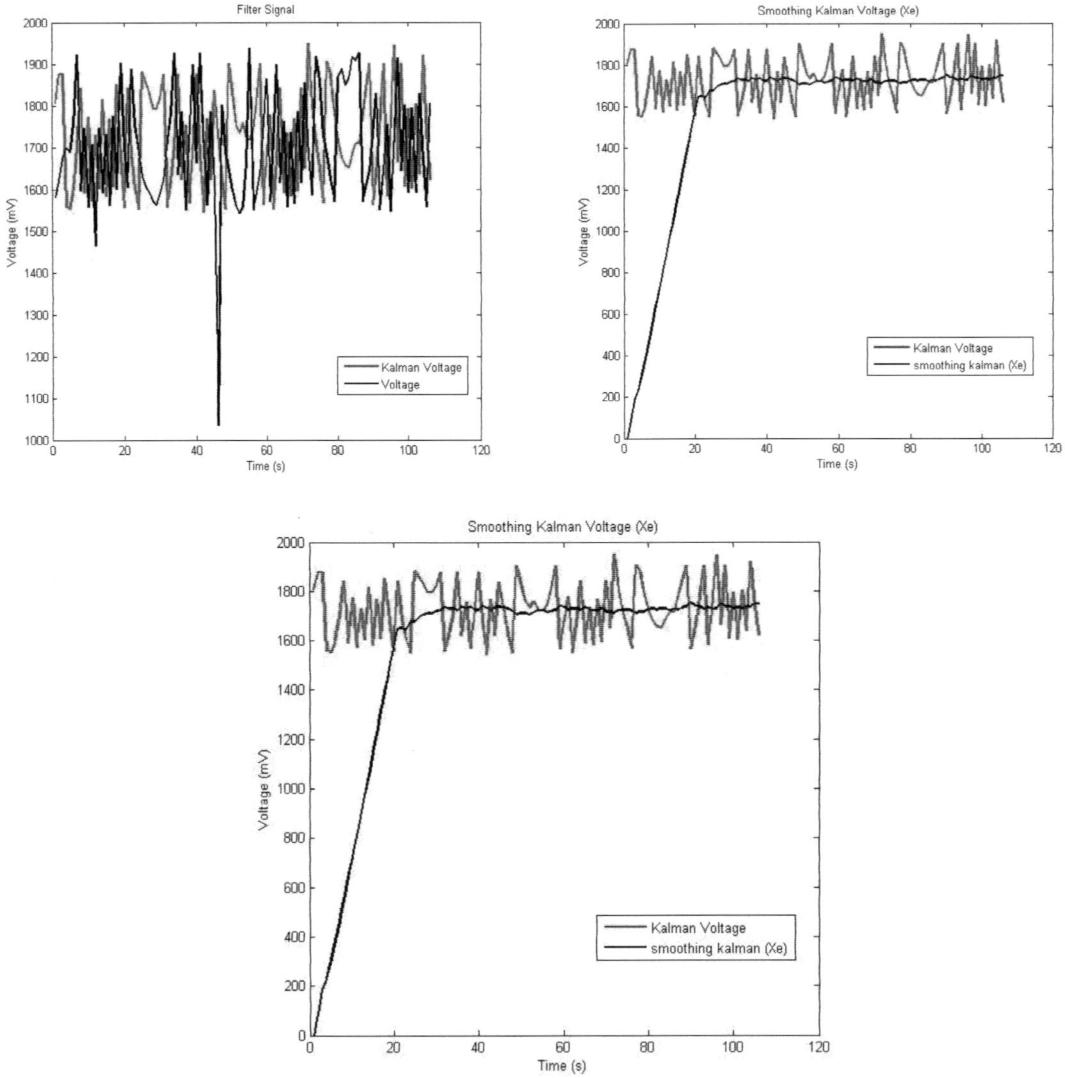

Figure 6. A set of Signal Filtering

4.2. Plot PID Controller

In this section, the PID control is discussed. This controller was applied for making the system enables to select the voltage signal and energize the sensor. As shown in figure 7, three modes for P, PI and PID controller are compared. Set point is set to zero value. The signal will be consider as good performance when approximating to zero value of absolute reading. Since the signal is always fluctuated, it is hard to get zero absolute reading for those signal. However, it is clear in those figures that response PI is better than P and PID mode.

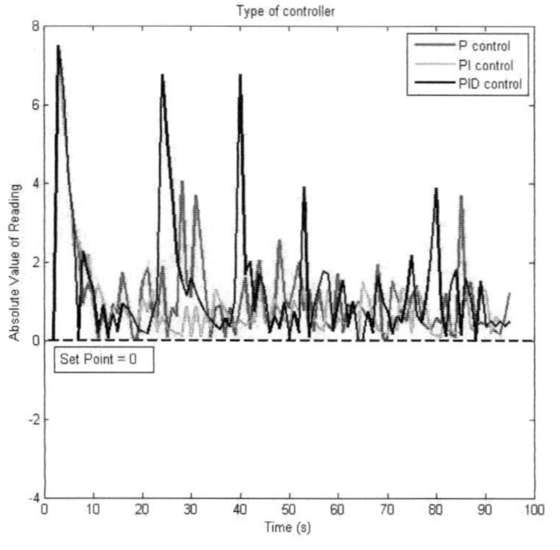

Figure 7. Plots of P, PI, PID modes controller

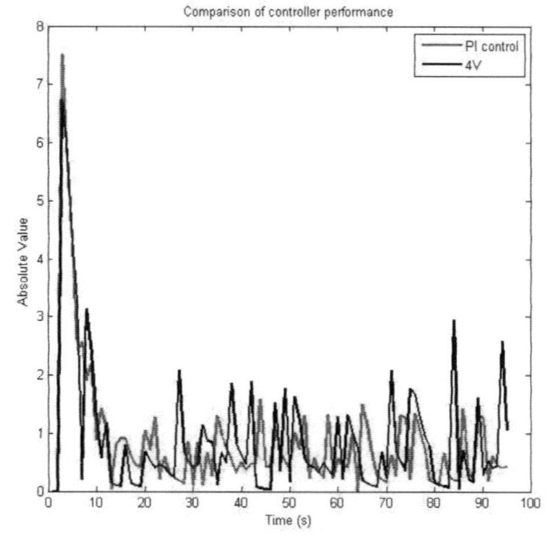

Figure 8. Plots of PI mode controller and Set Voltage

The signals were obtained using the specific instruments. A set number of signals were obtained in static solution and keeping same position of probe. These signals has been through signal filtering by kalman, smoothing and low pass filter where the results led to reduce random signal. The improved signal is then processed using PI control. As shown in figure 8 shows the response of PI control voltage and set 4V voltage. It shows the slightly improvement in controlling the voltage. A series of control measurement of the PI control produce good response with a standard deviation 1.0 and standard deviation for set voltage gave 1.135. The differentiation for two set of measurements has a slightly improvement due to fluctuation voltage energize the sensor. However, the contribution for PI control can be propose at such the voltage supply can be automatically defined sufficiently for reaction occurred.

In figure 9 shows the reading from the sensor response with the time. From the set voltage, Steady State Error SSE is existing for this mode. The systems become unstable with oscillation. When PI controller is applied, ample overshoot is observed, the SSE is reducing and give the voltage signal close to the set point.

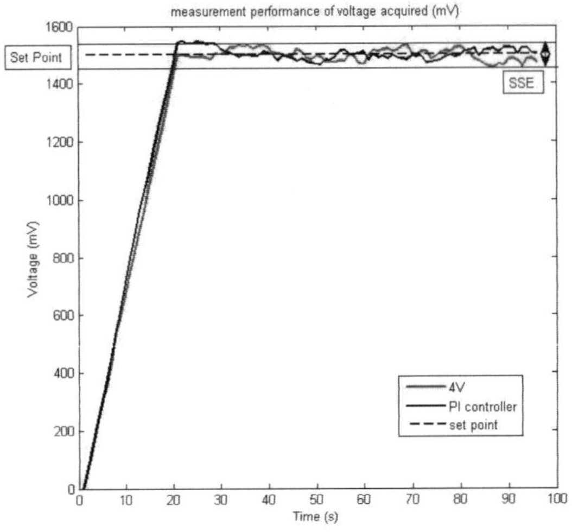

Figure 9. Measurement response of voltage signal from the sensor

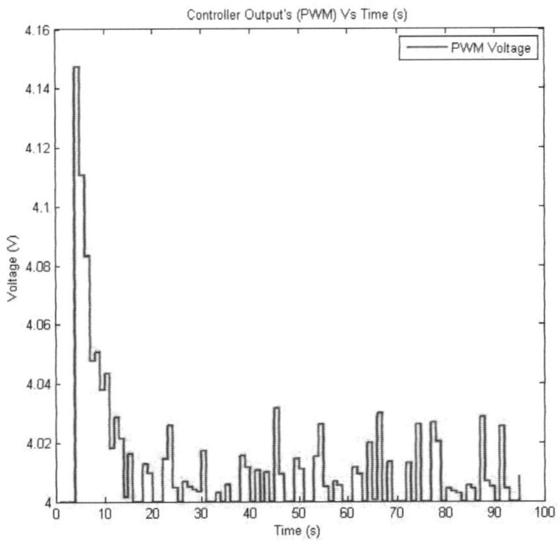

Figure 10. Controller's output VS Time of PI Controller

5. Conclusions

PID controller is designed using manual tuning in experimental work and their performance is compared. It is shown graphically, the controller is essential part and can be implement for the system. With this implement, the system can give voltage sufficiently for a particular reaction. To be conclude, we found that an exact reading for the concentration is difficult due to the digitizing error.

Acknowledgements

This paper was carried out as a part of collaboration program between USM and TUT with transdisciplinary research name "Tools for Heavy Metal Mitigation: Detection and Removal from the environment". The author would like to thank the USM for funding support through RU Top-Down research grant (1001/PAERO/6740041).

IOP Conf. Series: Materials Science and Engineering **370** (2018) 012042 doi:10.1088/1757-899X/370/1/012042

References

[1] F. Al-badaii, M. Shuhaimi-othman, and M. B. Gasim, "Water Quality Assessment of the Semenyih River , Selangor , Malaysia," vol. 2013, 2013.

[2] H. a. Aziz, M. N. Adlan, and K. S. Ariffin, "Heavy metals (Cd, Pb, Zn, Ni, Cu and Cr(III)) removal from water in Malaysia: Post treatment by high quality limestone," Bioresour. Technol., vol. 99, no. 6, pp. 1578–1583, 2008.

[3] S. K. Gunatilake, "Methods of Removing Heavy Metals from Industrial Wastewater," J. Multidiscip. Eng. Sci. Stud., vol. 1, no. 1, pp. 12–18, 2015.

[4] H. Li, A. Shi, M. Li, and X. Zhang, "Effect of pH , Temperature , Dissolved Oxygen , and Flow Rate of Overlying Water on Heavy Metals Release from Storm Sewer Sediments," vol. 2013, 2013.

[5] M. D. Tutulea, I. Cretescu, D. Sibiescu, and C. Stan, "Electrochemical sensors for heavy metal ions detection from aqueous solutions," Environ. Eng. Manag. J., vol. 11, no. 2, 2012.

[6] O. a Farghaly, R. S. A. Hameed, and A.-A. H. Abu-Nawwas, "Analytical Application Using Modern Electrochemical Techniques," Int. J. Electrochem. Sci., vol. 9, no. 2014, pp. 3287–3318, 2014.

[7] M. Huba, P. Bisták, and T. Huba, "Filtered PI and PID control of an Arduino based thermal plant," IFAC-PapersOnLine, vol. 49, no. 25, pp. 336–341, 2016.

[8] Z.-L. Lv et al., "A simplified electrochemical instrument equipped with automated flow-injection system and network communication technology for remote online monitoring of heavy metal ions," J. Electroanal. Chem., vol. 791, pp. 49–55, 2017.

[9] A. Note, "Monitoring Heavy Metals by ICP- OES for Compliance with RoHS and," vol. 2003, 2006.

[10] E. Physics, "ICP-MS DETERMINATIONS OF HEAVY METALS IN SURFACE WATERS Environment is an essential element of human existence . It is a result of interference of natural elements – earth , air , water , climate , biosphere – with elements created by human activity [1]," vol. 57, pp. 1184–1193, 2012.

[11] J. Sanders, "Monitoring Heavy Metals by Atomic Absorption Spectroscopy for Compliance with RoHS and WEEE Directives," Semicond. Anal. Environ. Introd., vol. 1, no. 40, pp. 1–6, 2012.

[12] B. J. Farey and L. a. Nelson, "Atomic Absorption Spectrometry," Tech. Instrum. Anal. Chem., vol. 5, pp. 67–94, 1982.

[13] V. Gayla, "Inductively Coupled Plasma : the Future of Heavy Metals Testing," Life Sci. Tech. Bulletub, no. 17/April, 2009.

[14] J. Wang, G. Liu, and A. Merkoç i, "Electrochemical coding technology for simultaneous detection of multiple DNA targets," J. Am. Chem. Soc., vol. 125, no. 11, pp. 3214–3215, 2003.

[15] M. R. Knecht and M. Sethi, "Bio-inspired colorimetric detection of Hg2+ and Pb2+ heavy metal ions using Au nanoparticles," Anal. Bioanal. Chem., vol. 394, no. 1, pp. 33–46, 2009.

[16] X. Pei, W. Kang, W. Yue, A. Bange, W. R. Heineman, and I. Papautsky, "Disposable copper-based electrochemical sensor for anodic stripping voltammetry," Anal. Chem., vol. 86, no. 10, pp. 4893–4900, 2014.

[17] Sas, "Conductivity-theory and practice," Anal. Radiom., vol. D61M002, pp. 1–50, 2004.

[18] F. Prieto García, E. Barrado Esteban, M. Vega, and L. Debán, "a Rapid Estimation of Metal Contents in Wastewater Treatment for Conductivity Measurements," J. Chil. Chem. Soc., vol. 50, no. 3, pp. 547–551, 2005.

[19] "A Guide to Conductivity Measurement," pp. 6–16.

[20] G. Recktenwald, "1 Basic PWM Properties," 2011.

Real-time Collision Avoidance and Path Optimizer for Semi-autonomous UAVs.

A.F Hawary[a], N.A Razak[b]

School of Aerospace Engineering, Engineering Campus, Universiti Sains Malaysia, 14300, Nibong Tebal, Penang, Malaysia

Corresponding author: [a]aefaizul@usm.my

Abstract. Whilst UAV offers a potentially cheaper and more localized observation platform than current satellite or land-based approaches, it requires an advance path planner to reveal its true potential, particularly in real-time missions. Manual control by human will have limited line-of-sights and prone to errors due to careless and fatigue. A good alternative solution is to equip the UAV with semi-autonomous capabilities that able to navigate via a pre-planned route in real-time fashion. In this paper, we propose an easy-and-practical path optimizer based on the classical Travelling Salesman Problem and adopts a brute force search method to re-optimize the route in the event of collisions using range finder sensor. The former utilizes a Simple Genetic Algorithm and the latter uses Nearest Neighbour algorithm. Both algorithms are combined to optimize the route and avoid collision at once. Although many researchers proposed various path planning algorithms, we find that it is difficult to integrate on a basic UAV model and often lacks of real-time collision detection optimizer. Therefore, we explore a practical benefit from this approach using on-board Arduino and Ardupilot controllers by manually emulating the motion of an actual UAV model prior to test on the flying site. The result showed that the range finder sensor provides a real-time data to the algorithm to find a collision-free path and eventually optimized the route successfully.

1. Introduction

The path planner is considered a key element of the unmanned aerial vehicle (UAV) navigation system. Basically, its task is to compute the optimal path from a start point to an end point. Unlike the trajectories for commercial airlines, UAVs trajectories are constantly changing depending on the terrain and conditions prevailing at the time of their flight. A simple trajectory optimization for typical UAVs is to find a shortest path using search algorithms as in [1], though other objectives, e.g., distance travelled, the average altitude, the fuel consumptions are equally important but not covered within the scope of this work. As an example, the authors of [2] proposed the use of a genetic algorithm (GA) to improve the TSP exploration and better avoid local minima when searching for an optimal path. The authors of [3] also use a SGA, but the collision is not included.

In this paper, we study the effectiveness and the practicality of UAV path planner and collision detection using an on-board controller for fixed-wing UAVs. Essentially, we combine brute force and non-deterministic algorithms to develop an operational path planner modified from [4], and the avoidance based on [5] but we use a laser range finder instead. We present two important contributions. First, we propose a simple integration between Ardupilot (APM) and Arduino Mega, which mutually work to optimize the route as well as avoiding obstacles. This allows us to use a generic optimization algorithm (without major modification) as the search algorithm. In our case, we use the Nearest Neighbour (NN) and the Simple Genetic Algorithm (SGA), but these could be replaced by other algorithms. Second, we present a technique to combine both the SGA and the NN while minimizing the communication between the processes in order to achieve a near real-time performance. Both algorithms have been widely used for robot path planning. However, to our knowledge, there exists no rigorous comparison between the two algorithms when applied to this particular problem. The results we present in this paper provide a clear insight how to develop a simple algorithm for UAV path planning in real complex environments. The remainder of this paper is organized into sections. Section 2 provides the details of the UAV specification. We present in Section 3 the methodology to combine and integrate the algorithms. And finally we include our result and conclusion in Sections 4 and 5, respectively.

2. UAV Specification

This study utilizes a fixed-wing model aircraft (model X-UAV Sky surfer X8) with 1.4 m wing span, 0.95 m length and weighs about 0.6 kg. The body is made of Expanded PolyOlefin (EPO) foam and powered by a brush-less motor. The flap mechanisms are activated by four servos using an Ardupilot (APM) as an on-board flight controller. The built-in proportional-integral-derivative (PID) controller uses various input parameters (e.g. airspeeds, roll rate, yaw rate, etc.) through a 6 degree-of-freedoms (6-DOF) inertial measurement unit (IMU) to achieve in-flight stability. A Global Positioning System (GPS) unit is also installed to provide its position (latitude, longitude) during navigation. In addition, a separate Arduino Mega is installed as a path-planner and collision module where the algorithm proposes in this paper executes. A picture of the UAV is shown in Figure 1.

Figure 1. X-UAV Sky surfer X8 model aircraft.

This UAV features a typical 4-channel setup with a single propeller where its stability depends on the deflection angle of the control surfaces such as aileron, rudder, and elevator as shown in Figure 2 below.

3. Methodology

It is worth mentioning here that, the UAV can fly remotely with good stability using APM controller even without a path planner. Although APM has its own mission planner, it does not have the optimizing and re-routing capabilities. Hence, we propose a separate module to handle the path-planning and collision avoidance algorithm to speed up the computation. In general, the UAV route is made from a series of waypoints called target points (TPs). We define that TP is a set of points where the UAV has to visit or reach in order to cover certain areas. TPs can be a random or specified waypoints defined by users within the area of interest. Then, the path optimizer would compute the best route that covers all TPs in a way that it minimizes the distance and the number of crossing paths. In real-time UAV mission, the pre-planned path is considered valid, unless there is a potential collision that requires real time re-routing. In order to do this, the optimizer and the obstacle avoidance module has to work together to optimize a return journey that begins from a specific point. Therefore, we relate this problem to a classical Travelling Salesman Problem (TSP) [6] with a slightly modified algorithm to embed collision avoiding capabilities.

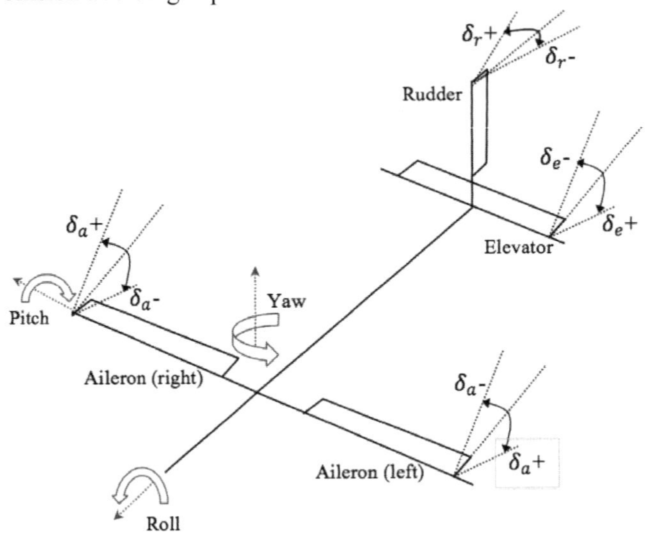

Figure 2. Control surface specification ($\delta_a = \pm 40°, \delta_e = \pm 30°, \delta_r = \pm 30°$).

3.1. *Travelling Salesman Problem*

TSP simply rules the salesman to plan a return journey through all cities so that it makes the most profit without visiting any cities twice. The profit can be any functions that contribute to his profit, e.g., distance, flight time. From graph theory, TSP can be described as follows:

TSP = *(G, f, t)*

where, $G = (V, E)$ is a complete graph, f is a function $V \times V \rightarrow Z$, and $t \in Z$, G is a graph that contains a travelling salesman tour with cost that does not exceed t in case t is restricted.

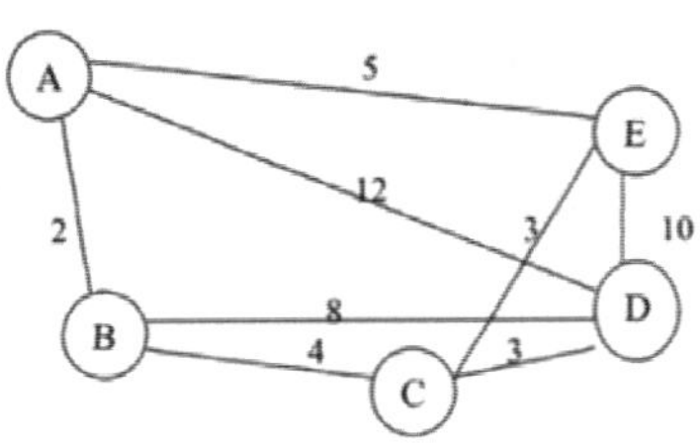

Figure 3. A simple TSP routing problem.

Figure 3 illustrates how a shortest return path from point A is obtained using TSP. Let us route Path1 from A, B, C, D, E and A and route Path2 from A, B, C, E, D and A. Computing the length of Path1 results in 24 units and Path2 has 31 units respectively. Obviously, Path2 is favourable due to shorter in length. Solving this kind of problems over hundreds of points are complex; therefore, a systematic algorithm is needed. Mathematically, TSP generalizes the question of Hamiltonian circuit in a graph as shown in Eq.(1).

$$F = min\sum_{i=1}^{n-1} dist(x_i, x_{i+1}) + dist(x_{n-1}, x_1) \tag{1}$$

Where, x is the city and i =1,2,3,...n represents the number of cities. There are numerous methods available to solve TSP. Among those methods, we favour a brute force method due to its simplicity and fast i.e., Nearest Neighbour. In addition, we combine the brute force with a meta-heuristic evolutionary algorithm SGA proposed in [7] by adopting Lin-Kernighan tour.

3.2. Routing Algorithm

SGA belongs to the larger class of evolutionary algorithms (EAs), which generate solutions using techniques inspired by the evolution theory. Its heuristic search mimics the process of natural evolution through selection, crossover and mutation within the *gene* in the chromosome. This method provides a wider search spectrum to locate the global solutions in optimization problem. Due to its nondeterministic nature, SGA lacks of real-time capability and time consuming. This is critical in the operation that requires instant decisions such as disturbance, collision or lost tracked from its original route. Moreover, the delay could prolong the UAV in an 'unguided state' before the algorithm produces a new solution. Therefore, we explore a simple brute force method using NN to take over the 'unguided state' and only kick in when the path collides with obstacles. As opposed to SGA, NN is rather simple in both constructions and operations where it always favours the nearest point. Thus, the decision can compute almost instantly and simple to process on any on-board controllers.

3.3. Collision Avoidance

Collision avoidance measures time of flight between the UAV and the obstacles to obtain the distance if the beams collide with an object. Using the speed of light as a constant, the laser rangefinder (LIDAR) sensor calculates the duration between the time a pulse is sent and the time it is received and then determines the distance to the object. In this project, a tiny rangefinder Lite 3 from Roboshop.com is used as it provides reliable and extremely precise measurements with great repeatable accuracy and short response time up to 500 measurements per second and can measure up to 40 m measuring range. Moreover, the communication to the module is simplified via I2C protocol and highly compatible with Arduino Mega. Basically, the distance, D between the two points is given by Eq.(2);

$$D = \frac{ct}{2} \tag{2}$$

where c is the speed of light ($3 \times 10^8\ m/s$) in the atmosphere and t is the amount of time for the rount-trip between two points given by Eq.(3).

$$t = \frac{\phi}{\omega} \tag{3}$$

where ϕ is the phase delay made by the light travelling and ω is the angular frequency of the optical wave.
Substituting Eq.(2) into Eq.(3) yields,

$$D = \frac{1}{2}ct = \frac{1}{2}c\frac{\phi}{\omega} = \frac{c}{4\pi f}(N\pi + \Delta\phi) = \frac{\lambda}{4}(N + \Delta N) \tag{4}$$

where, λ is the wave length of $\frac{c}{f}$, $\Delta\phi$ is the part of phase delay that does not fulfil π, N is the integer number of wave half-cycles of the round trip and ΔN the remaining fraction part.
The collision mechanisms built on a single laser rangefinder which is pointing to five different directions by rotating (motorized) the sensors as shown in Figure 4 below.
The pointing directions of the LIDAR are prefixed with an *RF* for rangefinder followed a specific direction, i.e., head, up, down, left and right. The state of each direction of the rangefinders is used to control the deflection of rolling angles (Aileron), yawing angles (rudder) and pitching angles (elevator). This setup provides narrow angle collision detection within 20 m at nominal glider cruising speed of 10 m/s. That means; the controller has two

seconds to compute a new avoiding route before collision with approximately 1000 readings. Depending on the state of the sensors, a decision is based on a look-up table to command a necessary control strategy to the APM for every condition as shown in Table 1.

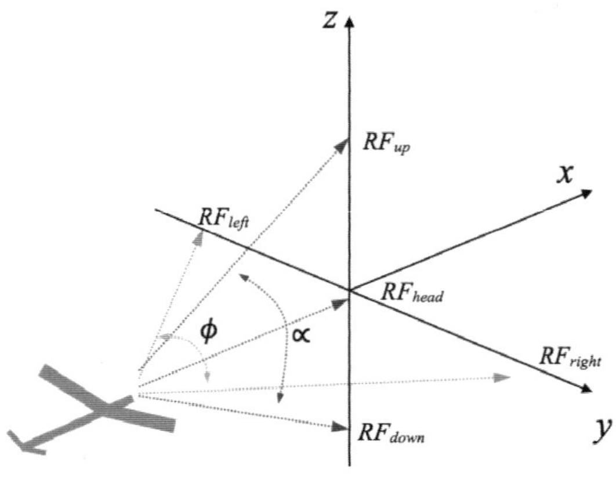

Figure 4. LIDAR setup based on motorized pointing direction.

Table 1. A look-up decision table for collision avoidance.

Case	Rangefinder Status (1 = block, 0 = clear)					Decision	Surface control			
	RF_{left}	RF_{right}	RF_{up}	RF_{down}	RF_{head}		Aileron (δ_a)	Rudder (δ_r)	Elevator(δ_e)	Throttle (%)
1	0	0	0	0	0	Proceed	0	0	0	0
2	1	1	1	0	0	Proceed	0	0	0	0
3	0	1	1	1	1	Left turn	-20	0	-10	+5
4	0	0	1	1	1	Left turn	-20	0	-10	+5
5	0	0	0	1	1	Left turn	-20	0	-10	+5
6	0	0	0	0	1	Left turn	-20	0	-10	+5
7	1	0	1	1	1	Right turn	+20	0	-10	+5
8	1	0	0	1	1	Right turn	+20	0	-10	+5
9	1	0	1	0	1	Right turn	+20	0	-10	+5
10	1	1	1	1	1	Sharp Climb	0	0	-30	+20
11	1	1	1	0	1	Sharp Climb	0	0	-30	+20
12	1	1	0	1	1	Climb	0	0	-15	+10
13	1	1	0	0	1	Climb	0	0	-15	+10

The signals of the control surfaces are operating based on pulsing and monitor. The pulse's interval is between 1 to 2 milliseconds. Note that, regardless of the state of other sensors, for as long as the RF_{head} is 0 or unblocked, the decision is remained. Otherwise, the decision depends on the positioning of the obstacles detected by the other sensors. As for the case of 10-13, the UAV has to climb up or sharply climb, but it remains on the same original route with elevated trajectories. For the case of 3, 7, 8 and 9, the decision is chosen based on the direction which is clear from obstacles. However, for the case of 4, 5 and 6 where both left and right sensors are clear the decision, whether to take left turn or right turn are not defined. For this case, a simple rule is applied considering the distance of the next point to the original point on the route which is taken care by NN algorithm and then re-route by SGA. In order to do this, NN will search and calculate the nearest point during collision and assign a set of temporary

TPs to navigate to a new TP. While NN diverting the route through TP, S will take its current position and calculate for a new optimal route and then terminate NN routine. This process repeat if the UAV faces the obstacles.

3.4. Algorithm Configuration

Within SGA, the selection of the parents is based on Roulette Wheel Selection (RWS) and a single-point crossover and flip-bit mutation. We set the probability of crossover P_c and P_m to be 0.8 and 0.01 respectively. The population size is set to 10 and max generation to 10. As for the NN, there is no additional setting other than computing the nearest trajectory point. The combined algorithm for both methods is shown in the algorithm listing below.

```
//Genetic Algorithm main routine
Initialization : Assign SGA parameters, such as pc, pm, popsize, max_generation, home, etc.
Generate chromosome of [popsize],
   Assign [nearest point] as a starting point
  Permutate randomly
  Evaluate their fitness values.
  gen = 0.
Main()
While generation < max_generation.
     Select individuals [popsize] from current gen using RWS to generate the mating pool.
     While population < popsize
         Select two individuals from the mating pool randomly
         Perform single-point crossover with probability pc,
         Perform bit-flip mutation for every gene of the offspring with probability pm.
         Then insert the mutant into the new population.
         Evaluate the fitness value for every new individual in the new population.
         Replace the current population with the new population. Generation = generation+1.
Submit the fittest individual as the results of GA to APM controller to execute.
End
//Nearest Neighbour Interrupt routine
Read sensor states
  Refer decision look-up table
  Determine the nearest point
  Assign temporary target point
  Send command to APM to navigate to a temporary target point
  Assign [nearest point] to GA as the next point
Return
```

Ideally, the original solution produced by SGA algorithm alone would continue valid for as long as the route does not interfere with any obstacles.

4. Simulations

A simple functional test performed to compare the solution of three different algorithms namely Random, NN and SGA (based on Lin-Kernighan Tour) and is comparable with Concorde TSP Solver. The test involves 100 random points within a defined space. The simulation result for every algorithm as shown in Figure 5 below.

All algorithms have been run using 100 random points shown in Figure 5(a). Obviously, the solution from Random algorithm is not appropriate for flight path planning (too many crossing paths). A slightly better solution produced by NN algorithm is shown in Figure 5(c) with several zigzag paths. As discussed earlier, NN is good in determining the nearest point quickly without having to rely on expensive processing time and yet the route is acceptable. Interestingly, the solution in Figure 5(d) has shown a smooth path without any path-crossing, and it is solved using SGA with slightly longer computing time with significant improvement on the length. So, SGA could produce a smooth-and-shorter path, but it takes slightly long as compare to the NN. Therefore, we take advantage of the instantaneous decision by NN and the solution optimality from SGA to provide the best from both worlds to embed a simple-and-practical algorithm for UAV.

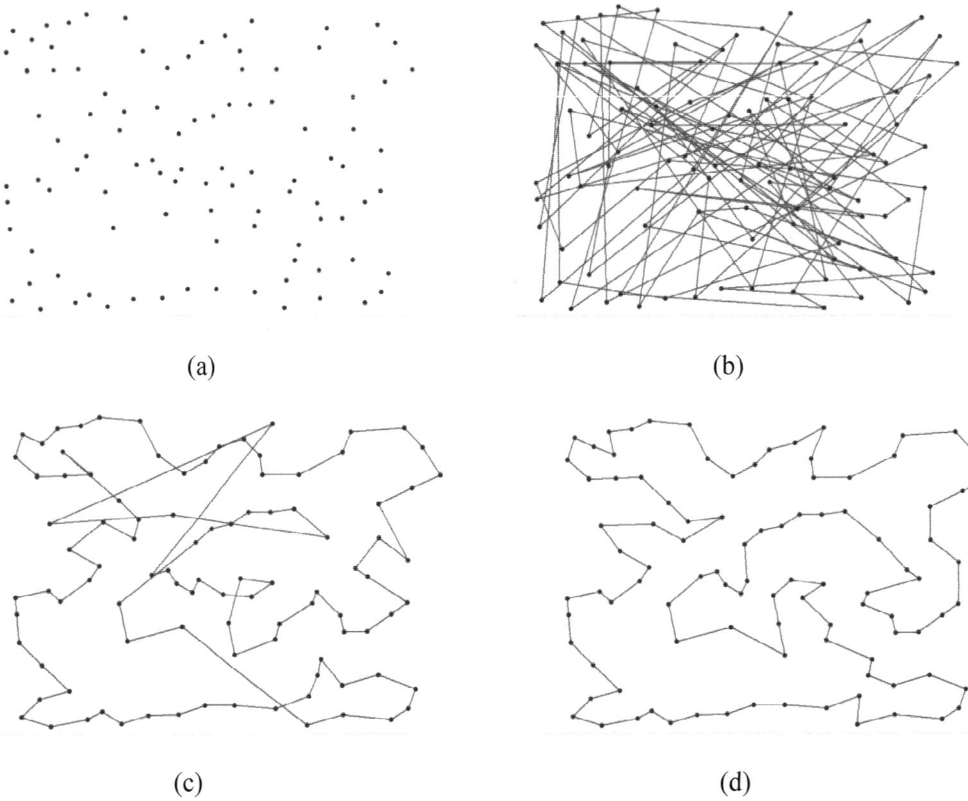

Figure 5. (a) 100 random points, (b) A solution using Random algorithm (t = 0.57s , l =5283), (c) A solution using Nearest Neighbor (t = 0.75s, l =962), (d) A solution using Genetic Algorithm (t = 0.91s, l =767).

5. Result and Discussion

The algorithm was tested with a simple obstacle starting from a point called Home. If no obstacle detected along the route, it would follow the pre-planned path which begins from 'Home' all the way to 1, 2, 3…11 and returns to 'Home' as shown in Figure 6(a). Ideally, the UAV would follow the pre-planned route from the beginning till the end. However, if the pre-planned route intersects with an obstacle on the heading direction (RF_{head}=1), the flight path is diverted from Home to point 11 through two temporary points derived from NN algorithm that seeks a nearest point and eventually diverts its journey to point 11 instead and continue to 10, 9…2, 1 and return Home (Figure 6(b)). This is because at the point of collision, NN sees that TP 11 is nearer that TP 1 hence the decision is to turn right instead. It is observed that while NN is calculating a nearest point, SGA runs its permutation and rearranges the way point from 1, 2, 3…10 and 11 previously to a new way point beginning from 11, 10, 10…2, 1 and returns Home. Since NN algorithm runs in the interrupt mode, it provides a sufficient time for the on-board processor to compute S hence reduces the 'unguided state' delay.

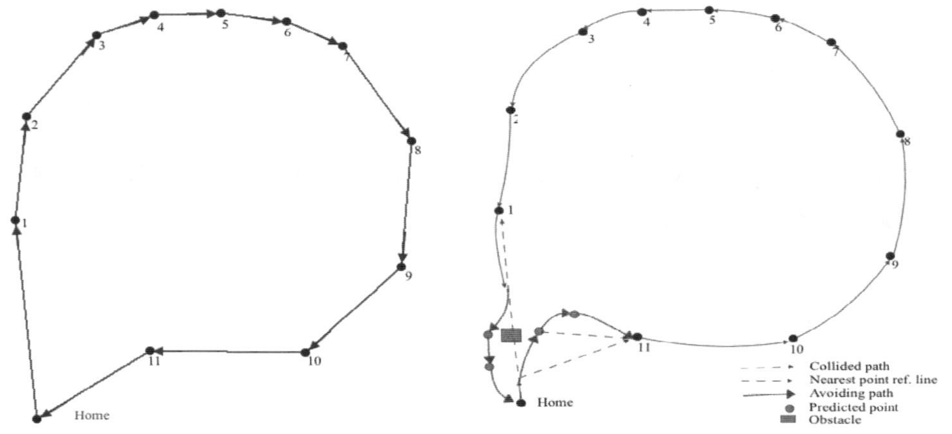

Figure 6. (a) Pre-planned route without obstacles. (b) Simulated real-time route with an obstacle.

Figure 6 shows the combined approaches used here could provide a simple real-time path planning and obstacle avoidance solution for the UAV researchers and hobbies out there. Since the scope of this work is to simulate the behavior of the UAV along the route, we only show a typical motion by moving the UAV manually. However, the position tracking data is not presented al it can be easily extracted from a GPS module. Moreover, due to the lack of the actual and real-time flight testing, the detail response of the control surfaces is not presented here. We reserve the detail observation of the flight performance during the real-time performance test.

6. Conclusion

In conclusion, we observe that the result from this study and the initial experimental provide a practical solution to embed the complete algorithm into the Arduino Mega and APM. Not only, it provides a simple solution; it is also affordable to be implemented especially for the beginner with minimal efforts. Hence, for the scope of this paper, we consider the objective of this study is achieved.

References

[1] X. Z. Gao, Z. X. Hou, X. F. Zhu, J. T. Zhang, and X. Q. Chen, "The shortest path planning for manoeuvres of UAV," *Acta Polytech. Hungarica*, vol. 10, no. 1, pp. 221–239, 2013.

[2] Y. Yu, Y. Chen, and T. Li, "A New Design of Genetic Algorithm for Solving TSP," in *Computational Sciences and Optimization (CSO), 2011 Fourth International Joint Conference on*, 2011, pp. 309–313.

[3] J. Y. Potvin, "Genetic algorithms for the traveling salesman problem," *Ann. Oper. Res.*, vol. 63, no. 3, pp. 337–370, 1996.

[4] J. Tisdale, Z. K. Z. Kim, and J. Hedrick, "Autonomous UAV path planning and estimation," *IEEE Robot. Autom. Mag.*, vol. 16, no. 2, pp. 35–42, 2009.

[5] N. Gageik, P. Benz, and S. Montenegro, "Obstacle detection and collision avoidance for a UAV with complementary low-cost sensors," *IEEE Access*, vol. 3, pp. 599–609, 2015.

[6] D. L. Applegate, R. E. Bixby, V. Chvátal, and W. J. Cook, "The Traveling Salesman Problem: A Computational Study," *Princet. Univ. Press*, p. 593, 2006.

[7] Y. Deng, Y. Liu, and D. Zhou, "An Improved Genetic Algorithm with Initial Population Strategy for Symmetric TSP," *Math. Probl. Eng.*, vol. 2015, 2015.

Evaluation of Particle Image Velocimetry Measurement Using Multi-wavelength Illumination

HC Lai[1], TF Chew[1] and NA Razak[1]

[1] School of Aerospace Engineering, Engineering Campus, Universiti Sains Malaysia, 14300 Nibong Tebal, Penang, Malaysia

Email: norizham@usm.my

Abstract. In past decades, particle image velocimetry (PIV) has been widely used in measuring fluid flow and a lot of researches have been done to improve the PIV technique. Many researches are conducted on high power light emitting diode (HPLED) to replace the traditional laser illumination system in PIV. As an extended work to the research in PIV illumination system, two high power light emitting diodes (HPLED) with different wavelength are introduced as PIV illumination system. The objective of this research is using dual colours LED to directly replace laser as illumination system in order for a single frame to be captured by a normal camera instead of a high speed camera. Dual colours HPLEDs PIV are capable with single frame double pulses mode which able to plot the velocity vector of the particles after correlation. An illumination system is designed and fabricated and evaluated by measuring water flow in a small tank. The results indicates that HPLEDs promises a few advantages in terms of cost, safety and performance. It has a high potential to be develop into an alternative for PIV in the near future.

1. Introduction

Particle image velocimetry (PIV) is a technology that has benefitted the field of fluid flow measurement for decades. The name PIV was proposed to distinguish the term itself from the laser speckle mode in 1984, as there was an argument that illumination of particles in fluid flows by a light sheet would seldom create a speckle pattern in the image plane [1]. It is undeniable that the contribution of PIV invention towards the fluid flow measurement field is essential and irreplaceable from the past till today. As the name of PIV itself suggest, the idea is to trace the particles in fluid flow by capturing two or more images with very small time interval. With this, reliable fluid flow properties such as the direction, velocity, vorticity and shear strain of flow can be obtained.

A PIV setup is the integration of a few systems with respective function to perform as desired. Firstly is the illumination system. A common PIV system uses laser as the light source. The Neodyme-YAG (Nd:YAG) laser is the most widely used laser source for PIV [2]. It is pulsed in the form of a thin sheet into the test section for the tracing particles to reflect the light and to be captured by the recording system. Today, the development of solid state illumination resulted in mass production of a wide range of HPLEDs available in the market. It provides alternatives that can be used to replace the laser illumination in the traditional PIV setup as laser source's is costly in price and might cause safety issue if it is not handle with extra care during operation.

The system that works closely together with the illumination system is the recording system. The recording system basically refers to a camera or video recorder. Both systems are required to be synchronised in order to achieve the correct timing for light pulsing and image acquisition. The quality and the capability of the recording system will basically determine the capability of the entire PIV system. For example, by fitting a camera pinhole model to the two cameras using single or multiple views of a 3D calibration plate, it is capable to perform a stereo-PIV measurement for 3D flow analysis [3].

Another essential part of a PIV is the computing system or a computer. The computer with the aid of applicable software will analysed the acquired images and results the flow characteristic of the fluid. There are many software available that can be used to perform PIV result analysis such as PIVView, OpenPIV and PIVLab, which is a Matlab-based application. Photo editing software are also sometimes required to perform colour filter on the raw images, or convert the raw images to files that are compatible for the PIV software.

Although a test section is an individual system from the PIV, it must not be neglected as it provides a platform to hold the test model with desired flow condition. A test section must be optically transparent in order for the light sheet to pass through and shone into the flow across the test model. The test section can part of wind tunnel, water channel, a transparent 3D-printed model to study the internal flow or even as simple as just a large space with desired fluid flow condition, depending on the objectives of the experiment.

Seeding particles or tracing particles are added into the fluid flow in the test section in order to visualise the flow. The main idea is to get the suitable seeding particles that have about the same density with the fluid so that it could mix well without floating or settling in the fluid. It is also worth to study the characteristics of seeding particles in order to determine its suitability for a respective experiment. Zhang [4] had performed a study on several seeding particles such as hollow glass spheres, polymer micro-spheres and solidified particles to determine their availability to perform PIV measurement in liquid helium. On the other hand, a seeding particles choice evaluation test is performed by Rafika [5] in order to choose suitable seeding particles to study the acoustic streaming flow generated by High Intensity Focused Ultrasound (HIFU). In short, the selection of seeding particles is crucial for PIV system is able to generate reliable results.

Improvements and exploring availabilities of PIV in order to maximize its functions and capabilities are widely conducted. Finding alternatives to replace the laser illumination system for PIV are no longer a new research topic in flow measurement field. It can be trace back to 1993, where Hshen, Ma and Zhuang [6] from Beijing, China developed a type of new PIV, which is known as the white light bubble image velocimetry (WLBIV) by modifying the traditional PIV illumination system. As the name suggest, the illumination system of the WLBIV used a flash set as the multi-pulse light sources for the PIV film while the particle tracer in the flow field are oxygen bubble generated by the electrolyzed water. High-power laser is also not required in WLBIV but as a result, the system has high signal to noise ratio.

Willert and colleagues [7]-[8] performed PIV experiment using of High Power Light Emitting Diode (HPLED) as an alternative to laser-base illumination system in PIV. In the research, it was found that HPLEDs promise some attractive advantages in comparison to laser. LEDs are able to operate beyond their continuous damage threshold when operating in short duration or pulsed. The light pulses generated by the HPLEDs are sufficient to illuminate and image micron-sized particles in flow velocimetry.

A research on colour-sequence enhanced particle streak velocimetry (CSPSV) which is a new approached introduced by Wang [9] and colleagues from China. CSPSV consists of three types of controlled flash lights of different colours and one type of white light. The change of illumination colour of the test zone in a given sequence added the time sequence information onto a single frame, which enables easier execution of the algorithm for three-dimensional reconstruction and streak pair matching. CSPSV also combines the advantages of particle tracking velocimetry and particle streak velocimetry, which reduces the requirement of high-speed cameras. CSPSV also can be easily extended and used to measure the air flow in large space.

2. PIV illumination system with HPLEDs

As high power LED (HPLED) has been chosen as the alternative to replace the traditional laser PIV illumination source, an illumination system has to be constructed in order to support the illumination of HPLEDs in desired way. The selected HPLEDs for this research project are green and red colours as shown in Figure 1 which are capable to perform in both continuous and pulsing mode. Both HPLEDs has maximum power of 100W and 6.25cm^2 size of illumination area. Other main components of the illumination system include the HPLEDs electronic driving circuit, optic fibre guide, cylindrical and convex lenses, power supply and microcontroller.

Red and green HPLEDs are selected to replace laser as illumination source. Green HPLEDs is chosen because it has the highest quantum efficiency and human eyes are most sensitive to the middle of the spectral range, which is around 550nm and is in the range of green visible light wavelength. A traditional laser-based PIV also uses laser that illuminates green light. Red colour HPLED is chosen as the second colour instead of other available colour HPLEDs such as blue and yellow because red colour has a higher eye sensitivity function compare to blue colour. Yellow colour is better than red in terms of wavelength because it is more sensitive to the eyes. However, the wavelength range of green and yellow is very near which might have the tendency to cause difficulties in differentiating both green and yellow light.

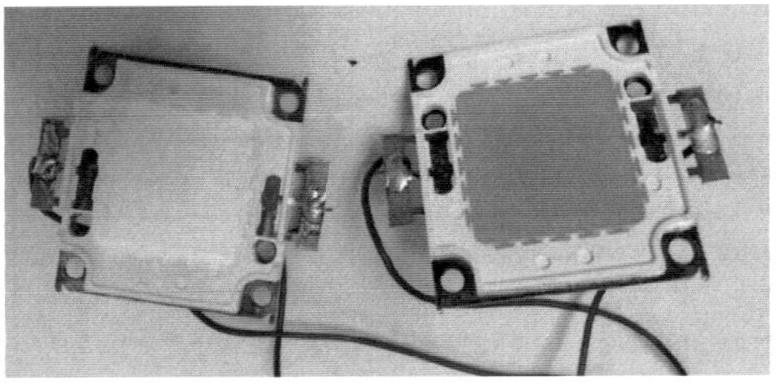

Figure 1. Green (left) and red (right) HPLEDs with maximum power of 100W.

It would be simpler to connect the HPLEDs directly to power supply to lighten it but it might cause damage to the HPLEDs and the desirable pulsing are not able to be control. Therefore, an electrical control circuit is essential to be developed in order to light up the HPLEDs with desirable pulsing. The electrical components required and respective specification to construct the driving circuit of the HPLEDs is presented in Table 1 below.

Table 1. Components in the LED driver control circuit

Symbol in Circuit	Components
R1,R3	1 Ω resistors
C1,C3	10 μF capacitors
C2,C4	2200 μF capacitors
L1	Red Colour HPLED
L2	Green Colour HPLED
Q1, Q2	IRF540N MOSFET
S1, S2	Switch
R2	0.165Ω resistor
R4	0.110Ω resistor

IOP Conf. Series: Materials Science and Engineering **370** (2018) 012044 doi:10.1088/1757-899X/370/1/012044

Both HPLEDs are driven using two different circuits which draw power from different power supply but are controlled using the same microcontroller as shown in Figure 2. The MOSFET in the circuits are triggered by a microcontroller, Arduino Uno. A pulsing program is uploaded into Arduino in order to trigger the circuit as desired. In order to stabilize the electrical energy in the circuit, the presence of capacitors store extra electrical energy and discharge electrical energy when voltage is lower than usual especially during pulsing.

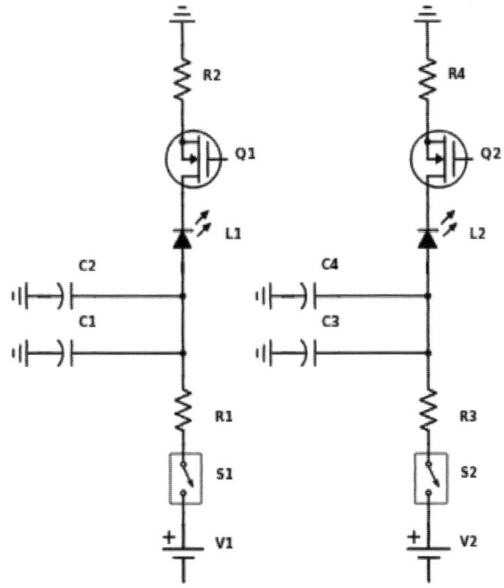

Figure 2. Schematic diagram of HPLED electrical driving circuit for both red (left) and green (right).

The entire illumination system with the circuits and optics are integrated into a box with the size of one CPU unit as shown in **Figure** *3*. A beam splitter is place directly in front of both HPLEDs to reflect the light illuminated by each HPLED in two different directions. In order to minimize the light loss from the HPLEDs, convex lenses are used to focus the light that passes through the beam splitter into the optic fiber guide inlet. As the HPLEDs radiate large amount of heat during operation, cooling fans are installed to cool down the HPLEDs and circuits to avoid the components from overheating damage.

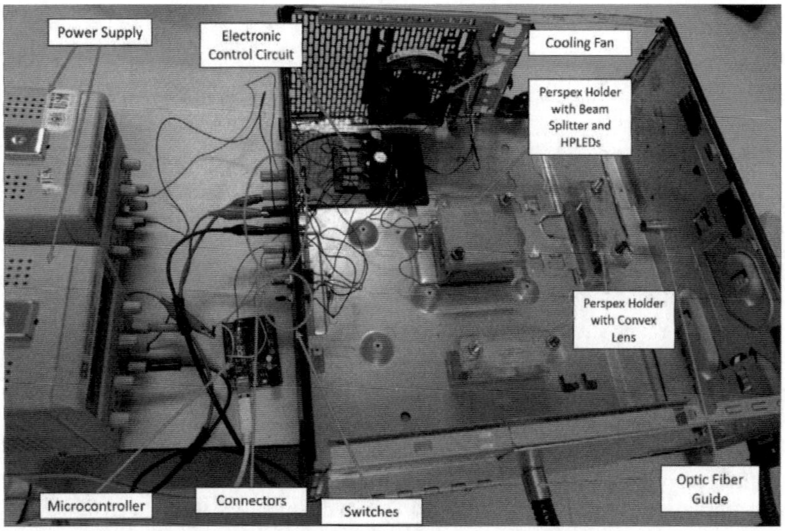

Figure 3. Final design of the illumination system platform.

2.1. HPLEDs characterization and pulsing operation

After the setup is done, the characteristics of both HPLEDs have to be studied. The light intensity of both HPLEDs are measured in terms of illuminance which its SI units is lux (lx) by varying the power supply to the HPLEDs. It is important to measure the illuminance of both HPLEDs in order to get them illuminate light with same intensity while operating as both HPLEDs illuminate light with different intensity under the same power. The measurement is performed using a digital photo sensor, BH1750 which is controlled by using a microcontroller. The microcontroller is connected to a computer for programming purpose. The coding to operate BH1750 digital light sensor is uploaded to the microcontroller [10]. The illuminated lights are shone into the photo sensor through the optic fiber guide.

The measurement is conducted in a totally dark condition to avoid stray light to introduce noise. Both HPLEDs are characterized separately under the same condition. The power towards the HPLEDs is controlled by manipulating the voltage of the power supply. The voltage is started with maximum value and slowly decreased to the point where the illuminance reading is zero. At each voltage, the average illuminance reading for ten seconds will be taken and recorded. Before taking the next reading, the HPLED will be cooled down to avoid the temperature factor which may affect the performance of the HPLED.

After the completion of characterizing the HPLEDs, it is also essential to make sure both HPLEDs pulsed appropriately with the aid of microcontroller to fulfill the operation mode of a PIV system. The microcontroller was programmed to pulse multiple LED in sequence. Both HPLEDs will be pulsed for 150 ms and then switched off while the interval between pulses is also 150 ms. Green HPLED are pulsed after the red HPLED. The interval between pulses can be set based on the flow velocity to achieve an optimum measurement condition, i.e. usually the higher the velocity, the lower the interval between pulses and vice versa. The pulsing of the HPLEDs will restart whenever the microcontroller is reset.

2.2. HPLEDs illuminance against power

Even though both HPLEDs are from the same manufacturer, both HPLEDs do not share the same properties in terms of capability to draw power from the power supply and also the intensity. The light intensity against the power of both HPLEDs are plotted and presented in Figure 4. It can be clearly seen that red HPLEDs draw more power from the power supply compare to the green HPLED, i.e. the maximum power to pulse red HPLED are higher than green HPLED. Besides, red HPLED started to emit light at higher power supply compare to green HPLED. However, green HPLED emits light with much higher intensity compare to red HPLED. When both HPLEDs are driven at maximum power available, green HPLEDs illuminate light with almost ten times the illuminance of the red HPLED. This might be caused by the sensor as it is not wavelength corrected.

From Figure 5.1, it also can be observed that the illuminance of both HPLEDs increase together with the power. It generally shows a linear trend at the beginning which fulfills the theory which states that illuminance is directly proportional to power [11]. However, deviation in illuminance is spotted when the power approaches maximum for both HPLEDs. This is because when the HPLEDs are operating in high power, large amount of heat generated by the HPLEDs increases the operating temperature which causes the internal resistance of the HPLEDs to increase. This phenomenon indirectly affects the power across the HPLEDs, causing the illuminance to be slightly reduced. The illuminance threshold of the HPLEDs might also affect the HPLEDs performance when the illuminance is approaching maximum.

From the illuminance test, the respective power required by each HPLED to provide light with the same intensity can be determined. It is crucial to achieve the same intensity for both HPLEDs while operating in PIV system. If both HPLEDs do not emit light with the same illuminance, more particles can be observed while the HPLED that illuminate light with higher intensity is pulsed. This will cause one of the images in the image set acquired to have higher seeding particles density which lead to higher noise ratio and reduce the accuracy of the PIV measurement.

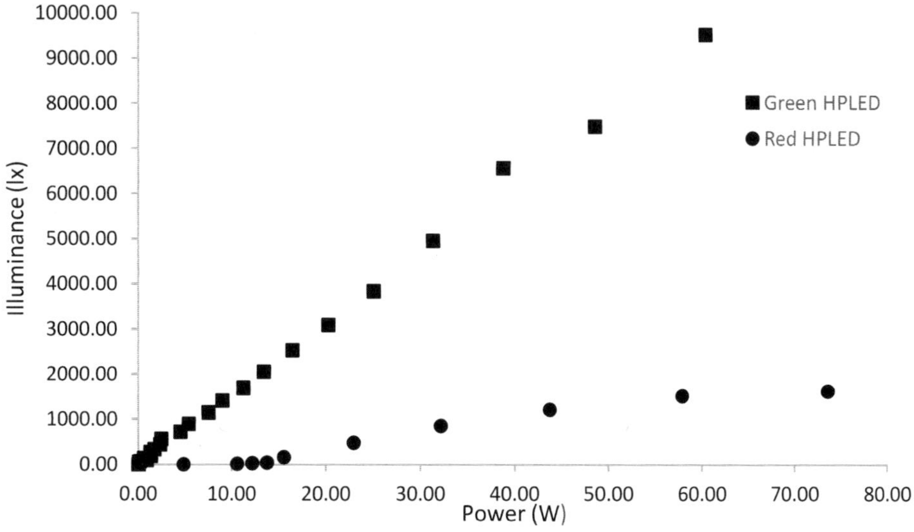

Figure 4. Illuminance against power for both red and green HPLEDs.

3. Illumination system evaluation by non-uniform flow measurement

After the illumination system is ready, a simple experiment is conducted in order to evaluate the reliability of the designed PIV illumination system. **Error! Reference source not found.** illustrates the setup of the HPLEDs illuminated PIV. A cylindrical lens is responsible to focus the light that emitted from optics fiber guide into a thinner sheet. A tank that contains tap water acts as the test section of the experiment. It is seeded using polyamide with a mean diameter of 50 μm. The light is shone to the water tank while the camera is located perpendicularly to the light sheet. The camera used in this experiment is Canon EOS 1300D. The lens of camera is set to focus on the particles in the plane of the light sheet. The HPLEDs are pulsed as mentioned while the camera captured the image of particles inside the tank with 1 s exposure.

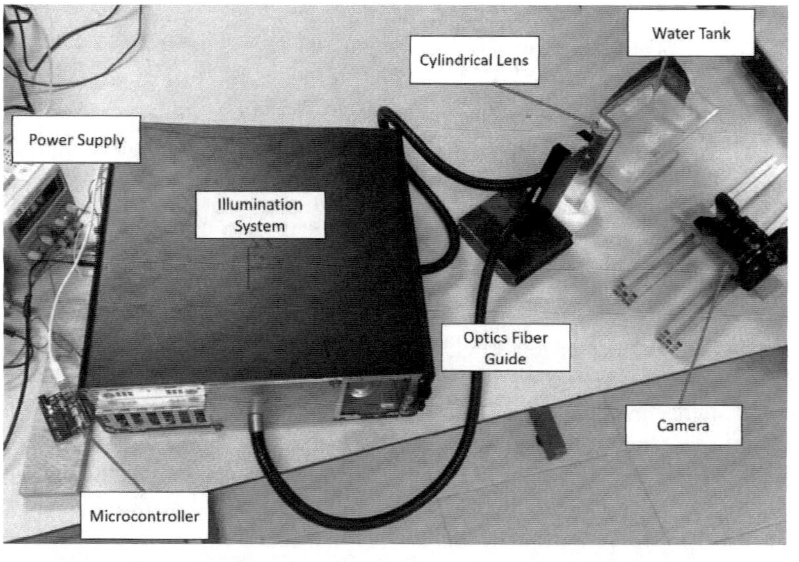

Figure 5. HPLEDs PIV setup to conduct flow measurement.

To evaluate the capability and reliability of the new designed PIV illumination system, the experiment is carried out by simply measuring the flow inside the tank. The mixture of water and seeding particles inside the test section are randomly disturbed using a wooden stick before any images are acquired. As the intention of this research is to prove the concept of multi-wavelength illumination system in PIV, no measurement was conducted on any known flow model for the time being.

The acquired images are required to undergo an image post-process before it is able to be analysed by any PIV analysis software available. Colour filtering function in the graphic editor, IrfanView are used to filter the acquired image into two separated red and green images. Both filtered images are converted into greyscale images which are then imported to PIVLab, a MATLAB based PIV analysis software for analysis. To reduce the time of analysis, only a section at the centre of the tank where the flow is significant known as region of interest (ROI) are selected to perform the analysis. PIVLab performed cross-correlation on the acquired images and plot the velocity flow field which will be further discuss in the next session.

3.1. Image Analysis

The acquired raw image before any post process was performed is shown in **Figure *6***. It can be observed that streaks are present and the particles can also be seen at both ends of the streaks. Each particle appears in both red and green colour in the streak, as it scattered the red and green light illuminated from the HPLEDs. The present of streaks aid to validate the accuracy of the velocity vector plotted by PIVLab. The streaks are not present after the raw image acquired undergoes colour filter process and each particle can be clearly seen. It has no doubt that correlation can be performed by PIV analysis software.

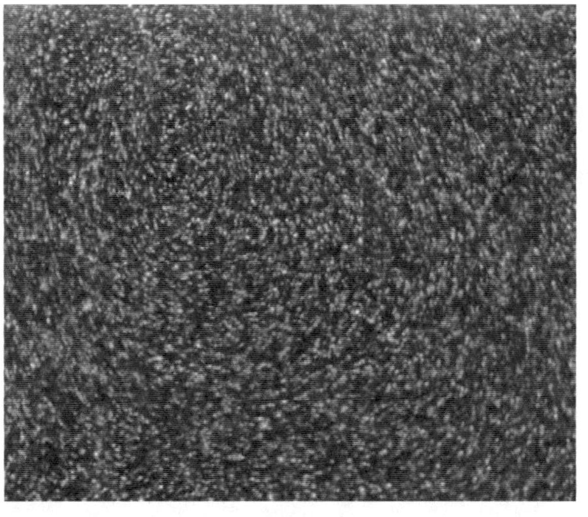

Figure 6. Raw image of random flow in the water tank.

Cross-correlation was performed by PIVLab and the result is presented in the form of velocity vector as shown in Figure 7. The result represents the instantaneous flow condition inside the water tank. Vectors with higher magnitude indicate higher velocity of the flow. As the flow is randomly disturbed so it is reasonable that the velocity vector plotted appear in random direction. The plotted velocity vector can be verified by comparing with the streak present in the raw image acquired. By comparing the plotted velocity vector with the streak in the raw image acquired, it can be said that the result obtained is reliable and promising. This proves that multi-wavelength HPLEDs are capable as the illumination system of a PIV system.

Figure 7. Velocity flow field of a random flow in the water tank.

3.2. Multi-wavelength HPLEDs illumination advantages

The evaluation of multi-wavelength HPLEDs illumination system in PIV does not only rely on the results. Attractive advantages by the multi-wavelength illumination system also added value to it. First of all, HPLEDs' cost is relatively lower compare to laser-based illumination system. Using multi-wavelength HPLEDs illumination system also allows the high-speed camera to be replaced by a normal DSLR camera, which also reduce the cost of the entire PIV system. A multi-wavelength illumination PIV system is might be quarter the cost of the traditional laser based PIV.

Besides, from the previous findings, HPLEDs provide incoherent light source with a wider wavelength range, which alleviates several issues to speckle artifacts found in laser-based illumination system. HPLEDs are also safer to operate and maintain compare to lasers. This is because PIV usually used a Class IV laser which is eye hazard if direct, reflected or diffusely-reflected beam is viewed and also possible to cause skin and fire hazard. Although HPLEDs is not hazardous, it does not mean that it is eye-friendly especially when operating at extended period.

The only disadvatage of HPLEDs is large amount of heat is generated in operation. Even though the maximum power available for the HPLEDs is 100W and it will be damaged due to over power during operation, the increase in temperature causes the internal resistance of the HPLEDs to increase which lead to a drastic decrease in the illuminated light intensity. This might affect the results obtained from the PIV measurement. Hence, effective cooling measurements are required to increase the effective operation time of the HPLEDs at maximum power and avoid the heat generated from damaging the components in the driving circuit.

4. Conclusion

In conclusion, the idea of replacing traditional illumination source with HPLEDs is possible according to the reliable results shown in previous chapter. Its advantages over lasers source which are reduction in cost, longer lifespan and acceptable performance are attractive. The only disadvantage of HPLEDs is that they will dissipate large amount of heat with high operating power and caused unsteady illuminance. The introduction of dual colours HPLEDs also successfully reduces the cost of the experiment by replacing high speed camera with normal recording system. Correlations are able to perform and velocity vectors of fluid flow are able to be extracted for low speed flow applications. Thus, the feasibility of dual colours PIV is ensured.

Acknowledgement
This work is partly supported by School of Aerospace Engineering, Universiti Sains Malaysia via 814238 RU and 60313017 research grant.

5. References

[1] Adrian RJ 2005 Twenty years of particle image velocimetry *Exp. in Fluids* **39**(**2**) p 159-169
[2] Brossard C et al. 2009 Principles and applications of particle image velocimetry, *AerospaceLab* (**1**)
p 1-11
[3] Wieneke B 2005 Stereo-PIV using self-calibration on particle images *Exp. in Fluids* **39**(**2**) p 267-280
[4] Zhang T, Celik D, and Sciver SV 2004 Tracer particles for application to PIV studies of liquid helium *Journal of low temperature physics* **134**(**3**) p 985-1000
[5] Ben HSR et al. 2017 PIV for the characterization of focused field induced acoustic streaming: seeding particle choice evaluation *Ultrasonics* **76** p 217-226
[6] Hshen G, Ma G and Zhuang F 1993 White Light Bubble Image Velocimetry *Chinese Journal of Aeronautics* **14**
[7] Willert C, Stasicki B, Klinner J and Moessner S 2010 Pulsed operation of high-power light emitting diodes for imaging flow velocimetry *Meas. Sci. Technol.* 075402-075402.
[8] Willert C, Stasicki B, Klinner J and Moessner S 2009 Pulsed Operation of High Power Light Emitting Diodes for Flow Velocimetry *8th Int. Symp. on Particle Image Velocimetry*
[9] Wang H, Li X, Shao X, Wang B and Lin Y. 2017 A colour-sequence enhanced particle streak velocimetry method for air flow measurement in a ventilated space *Building and Environment* ***112*** p 77-87. doi:10.1016/j.buildenv.2016.11.015
[10] Rawashdeh M 2016 *BH1750 Digital Light Sensor*
[11] Song J 2010 What is Illuminance ?

IOP Conf. Series: Materials Science and Engineering **370** (2018) 012045 doi:10.1088/1757-899X/370/1/012045

Control torque generation of a CMG-based small satellite with MTGAC system: a trade-off study

M B Salleh, N M Suhadis*, P Rajendran, N M Mazlan

School of Aerospace Engineering, Universiti Sains Malaysia, 14300 Nibong Tebal, Penang, Malaysia

*E-mail: nurulasikin@usm.my

Abstract. In this paper, the gimbal angle compensation method using magnetic control law has been adopted for a small satellite operating in low earth orbit under disturbance toques influence. Three light weight magnetic torquers have been used to generate the magnetic compensation torque to bring diverge gimbals at preferable angle. The magnetic control torque required to compensate the gimbal angle is based on the gimbal error rate which depends on the gimbal angle converging time. A simulation study has been performed without and with the MTGAC system to investigate the amount of generated control torque as a trade-off between the power consumption, attitude control performance and CMG dynamic performance. Numerical simulations show that the satellite with the MTGAC system generates more control torques which leads to the additional power requirement but in return results in a favorable attitude control performance and gimbal angle management.

1. Introduction

Control moment gyroscope (CMG) is a momentum exchange device that controls satellite attitude based on the conservation of angular momentum of the satellite system [1]. It has a torque amplification characteristic which makes it as a perfect actuator for fast slew maneuver and precise attitude pointing control over other actuators such as momentum wheels and reaction wheels. These features increase the operational envelop of the satellites and subsequently increase the return of mission data and reduce payload complexity [2,3]. Control torque is generated by rotating the CMG's gimbal at certain gimbal angle rate. This causes the direction of the flywheel's angular momentum vector to change and results in a generation of control torque, orthogonal to both flywheel's spin and gimbal axes [4]. There are three types of CMG i.e. single gimbal CMG (SGCMG), double gimbal CMG (DGCMG) and variable speed CMG (VSCMG). Among these three, SGCMG is less complex and cheaper than VSCMG and it has more significant torque amplification over DGCMG [5-7]. However, SGCMG has drawback in term of singularity where at certain gimbal angle orientations, the control torque cannot be generated by the CMG cluster thus resulting in a loss of satellite attitude controllability. The singularity problems have lead towards the development of singularity avoidance or steering laws that can avoid or escape the CMG cluster from singular states [8-12].

Another concern about the CMG system is the gimbal angle offset due to disturbance torques. The constant disturbance torques cause the gimbals to drift away from preferred or reference angles. This situation reduces the repeatability of slew maneuvers and can also drive the CMG cluster into saturation singularity state which most of the steering laws fail to avoid or escape. Since the CMG system lacks ability to compensate the drifted gimbals, an auxiliary controller to maintain the CMG gimbal at preferable angle is required. In principle, another external control torque is applied on the satellite to counteract the disturbance torques.

Magnetic control torque offers simple, cheap and more sustainable option to counter this problem. Therefore, it is more preferable for small satellite application, especially for low Earth orbit (LEO) missions. The technique utilizes the coupling action between the geomagnetic field and the magnetic dipole moment induced either by three or two magnetic torquers on-board the satellite. McElvain [13] and Stickler and Alfriend [14] proposed a control equation which was later became the standard for wheel unloading control law that uses magnetic control torque. The momentum unloading strategy was designed such that the produced counter torque is a function of the excess momentum to be removed. Meanwhile, Camillo and Markley [15] derived analytical formulae based on the bang-bang and linear controllers in order to optimize the performance of the reaction wheel angular momentum unloading after considering the limitations of using magnetic torquers. Then, an optimal magnetic momentum control for inertially pointing spacecraft with reaction wheel was studied by Lovera [16]. The optimal approaches based on periodic linear quadratic (LQ) and H_∞ control were proposed and their effectiveness was compared with the classical control law. The simulation results showed that the periodic control was more efficient than the classical control algorithm in achieving the desired momentum unloading performances. Suhadis and Varatharajoo [17] studied the active magnetic control technique to unload the excess wheel angular momentum of a small biased-momentum satellite in a nominal operation. They adopted the proportional-integral (PI) controller to keep the wheel's angular momentum within the bias value. Two cases were considered where two and three magnetic torquers were considered for each case, respectively. The results from the numerical studies showed that the momentum unloading performance with three magnetic torquers was better than with two magnetic torquers where a faster unloading control can be achieved.

On the other hand, the implementation of magnetic control torque to compensate drifted CMG gimbals can be found from Lappas et al. work [18]. The control strategy was designed such that the required magnetic control torque generated by the magnetic torquers proportional to the gimbal angle error rate. A controller gain known as gimbal angle converging time was introduced which need to be properly selected in order to obtain sufficient compensation torque. In the numerical study, three magnetic torquers aligned with the satellite's principle axis were considered to compensate the drifted gimbal of 4-SGCMGs system in the conventional pyramid configuration. It was found that, the proposed magnetic torque controller was able to bring back the drifted gimbals to the preferred angles within acceptable error band. However, the effect of the gimbal angle compensation controller on generated control torques and attitude control performances were not covered in their study.

In this paper, the gimbal angle compensation system coined as magnetic torque gimbal angle compensation (MTGAC) system is adopted to compensate the gimbal angle of a 4-SGCMGs cluster. The system is implemented onboard a small satellite platform operating in low earth orbit (LEO) under disturbances torques. This magnetic torque controller is cooperated with the standard PD (proportional-derivative) controller that controls the satellite attitude. Meanwhile, the Singularity Robust (SR) inverse steering law is utilized as the singularity avoidance law for the CMG cluster. The study compares the generated control torques during attitude pointing mode without and with the MTGAC system and then discuss in terms of trade-off between power consumption, CMG dynamics and attitude control performance. All presented algorithms are numerically treated for simulation studies via simulation tools.

2. Satellite attitude control structure

2.1 Satellite dynamics and kinematics

The MTGAC system onboard a CMG-based controlled small satellite uses three magnetic torquer coils to generate magnetic compensation torque. These magnetic torquers are placed along the satellite's body-fixed axes (X_B, Y_B, Z_B) where the Z_B axis points toward the earth, the Y_B axis is normal to the orbital plane and the X_B axis points toward the satellite's orbital motion and complete the right-hand orthogonal system. In addition, a cluster of 4-SGCMGs is also configured such that the axes of the system are along the satellite's body-fixed axes, respectively. Figure 1 shows the configuration of the MTGAC magnetic torquers as well as 4-SGCMGs system.

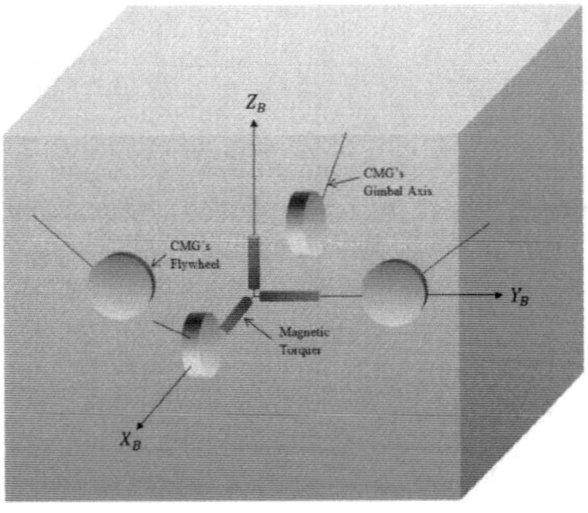

Figure 1. Satellite configuration with four SGCMGs and three magnetic torquers.

Based on the figure 1, the equation of motion of a satellite with a CMG cluster and three magnetic torquers is governed by the following equation of motion [7]

$$\dot{\mathbf{H}}_\mathbf{s} + \boldsymbol{\omega} \times \mathbf{H}_\mathbf{s} = \mathbf{T}_\mathbf{ext} \tag{1}$$

where $\mathbf{H}_\mathbf{s}$ is the angular momentum vector of the total system expressed in the spacecraft body-fixed control axes, $\boldsymbol{\omega}$ is the satellite angular velocity vector, and $\mathbf{T}_\mathbf{ext}$ is the external torques vector. The external torque, $\mathbf{T}_\mathbf{ext}$ is the disturbance torque from the environment $\mathbf{T}_\mathbf{d}$ (i.e. aerodynamic torque, magnetic torque and solar radiation torque) for case without the MTGAC system whereas for case with the MTGAC system it is the total of disturbance torque and magnetic control torque from the gimbal angle compensation system, $\mathbf{T}_\mathbf{ext} = \mathbf{T}_\mathbf{d} + \mathbf{T}_\mathbf{M}$.

The total angular momentum vector of the system consists of the angular momentum of the satellite and the CMG cluster and it is given as

$$\mathbf{H}_\mathbf{s} = \mathbf{I}\boldsymbol{\omega} + \mathbf{h} \tag{2}$$

Substituting equation (2) into equation (1) gives

$$\left(\mathbf{I}\dot{\boldsymbol{\omega}} + \dot{\mathbf{h}}\right) + \boldsymbol{\omega} \times (\mathbf{I}\boldsymbol{\omega} + \mathbf{h}) = \mathbf{T}_\mathbf{ext} \tag{3}$$

and by introducing the internal control torque vector, $\mathbf{u}$ generated by the CMG system, equation (3) can be written as

$$\mathbf{I}\dot{\boldsymbol{\omega}} + \boldsymbol{\omega} \times \mathbf{I}\boldsymbol{\omega} = \mathbf{u} + \mathbf{T}_\mathbf{ext} \tag{4}$$

$$\dot{\mathbf{h}} + \boldsymbol{\omega} \times \mathbf{h} = -\mathbf{u} \tag{5}$$

where $\mathbf{I}$ is the moment of inertia of the satellite and $\mathbf{h}$ is the angular momentum vector of the CMG cluster expressed in the satellite body-fixed axes. For the commanded control torque input, $\mathbf{u}$ the CMG torque command is chosen as

$$\dot{\mathbf{h}} = -\mathbf{u} - \boldsymbol{\omega} \times \mathbf{h} \tag{6}$$

in which equation (6) is incorporated with the steering law to determine the gimbal angle rate command.

2.2 CMG dynamics

In this study, a cluster of four identical SGCMGs in pyramid configuration is selected to be employed onboard small satellite as the attitude controller. The configuration has a skew angle of $\beta = 54.73$ degree and it is commonly considered due to uniformity of the momentum envelop and minimum redundancy [7]. The control torque generated by the CMG cluster is the time derivative of the CMG angular momentum. Based on the SGCMGs configuration it can be obtained as follow

$$\dot{\mathbf{h}} = \mathbf{A}(\delta)\dot{\boldsymbol{\delta}} \tag{7}$$

where $\mathbf{A}(\delta)$ is the Jacobian matrix defined as

$$\mathbf{A}(\delta) = \frac{\partial \mathbf{h}}{\partial \boldsymbol{\delta}} = h_0 \begin{bmatrix} -cos\beta \cos\delta_1 & \sin\delta_2 & cos\beta \cos\delta_3 & -\sin\delta_4 \\ -\sin\delta_1 & -cos\beta\cos\delta_2 & \sin\delta_3 & cos\beta\cos\delta_4 \\ sin\beta\cos\delta_1 & sin\beta\cos\delta_2 & sin\beta\cos\delta_3 & sin\beta\cos\delta_4 \end{bmatrix} \tag{8}$$

where h_0 is the magnitude of flywheel's angular momentum. From equation (7), the gimbal angle rate is determined by considering the pseudoinverse steering law given by

$$\dot{\boldsymbol{\delta}} = \mathbf{A}^{+}\dot{\mathbf{h}} = \mathbf{A}^{\mathrm{T}}(\mathbf{A}\mathbf{A}^{\mathrm{T}})\dot{\mathbf{h}} \tag{9}$$

Since singularity is the drawback of SGCMG, the Singularity Robust (SR) inverse steering law has been used in this study due to its simplicity to aid the CMG system avoiding or escaping singular states [8]. Thus, the gimbal angle rate command becomes $\dot{\boldsymbol{\delta}} = \mathbf{A}^{\mathrm{T}}(\mathbf{A}\mathbf{A}^{\mathrm{T}} + \lambda\mathbf{I})^{-1}\dot{\mathbf{h}}$ where λ is the SR invers steering law constant that need to be properly selected. A singularity index $m = \det(\mathbf{A}\mathbf{A}^{\mathrm{T}})$ is used to indicate the state of the CMG system where for $m = 0$ it indicates that the CMG system is in singularity state. It is worthwhile to mention that, the SR steering inverse does not always avoid or escape the CMG system from singular states, especially saturation singularity.

2.3 Attitude Controller

The attitude of the satellite is controlled based on the required control torque commanded by the attitude controller. For a real time implementation, the standard PD (proportional-derivative) controller is adopted. The quaternion error and angular rate vectors are fed to quaternion feedback controller and it is modelled as [19]

$$\mathbf{u} = -\mathbf{K_p}\mathbf{q_e} - \mathbf{K_d}\boldsymbol{\omega} \tag{10}$$

where $\mathbf{q_e} = [q_{1e} \quad q_{2e} \quad q_{3e}]^{\mathrm{T}}$ is the attitude quaternion error vector. By considering the one-to-one equivalence between the direction cosine matrix elements and the quaternion vector elements, the required command control torque can be expressed as [20]

$$\begin{bmatrix} u_x \\ u_y \\ u_z \end{bmatrix} = \begin{bmatrix} -K_{p_x} q_{1e} q_{4e} - K_{d_x}\omega_x \\ -K_{p_y} q_{2e} q_{4e} - K_{d_y}\omega_y \\ -K_{p_z} q_{3e} q_{4e} - K_{d_z}\omega_z \end{bmatrix} \tag{11}$$

The proportional and derivative gains are given as $\mathbf{K_p} = 2{\omega_n}^2[\mathbf{I}]$ and $\mathbf{K_d} = 2\zeta\omega_n[\mathbf{I}]$ respectively, where ζ is the damping ratio and ω_n is the natural frequency. For better attitude performance, the gains need to be properly selected by chosen suitable value of damping ratio and natural frequency.

3. Magnetic torque gimbal angle compensation (MTGAC) system

The gimbal angle offset due to disturbance torques is compensated using magnetic control torque. This torque is generated when the generated magnetic dipole moment of the magnetic torquers inside the satellite couples with the Earth's magnetic field vector [21]. This interaction can be described as [20]

$$\mathbf{T_M} = \mathbf{M} \times \mathbf{B} \tag{12}$$

where $\mathbf{T_M}$ is the generated magnetic torque vector, $\mathbf{M}$ is the magnetic dipole moment vector generated by the magnetic torquers inside the satellite and $\mathbf{B}$ is the Earth's magnetic field vector.

The Earth's magnetic field or geomagnetic field is mathematically expressed by a spherical harmonic model known as Complex Geomagnetic Field Model (CGFM). In this study, however, the Simplified Geomagnetic Field Model (SGFM) is considered to reduce computational burden and it is modelled in the local vertical and local horizontal (LVLH) coordinate system as [22]

$$\begin{bmatrix} B_x \\ B_y \\ B_z \end{bmatrix}_{LVLH} = \frac{M_e}{a^3}\begin{bmatrix} \sin i \cos(\omega_0 t) \\ -\cos i \\ 2\sin(\omega_0 t)\sin i \end{bmatrix} \tag{13}$$

where $M_e = 7.9 \times 10^{15}\text{Wbm}^{-1}$ is the Earth's magnetic strength of the dipole vector, a is the orbit's semi-major axis, ω_0 is the orbital frequency and t is the time measured from $t = 0$ s at the ascending node of the geomagnetic equator. Based on equation (13), it is obvious that the geomagnetic field experienced by the satellite is depends by the orbit inclination and semi-major axis or altitude of the satellite. Satellite orbiting at high orbit inclination with low altitude will experience more geomagnetic than satellite at low orbit inclination of the same altitude.

By having the variation of geomagnetic field along the satellite's orbit, the magnetic dipole moment, $\mathbf{M}$ need to be induced by the magnetic torquers can be obtained from the cross product of the commanded magnetic torque and the geomagnetic field, $\mathbf{B}$ and it is given as [23]

$$\mathbf{M} = \frac{\mathbf{B} \times \mathbf{T_c}}{B^2} \tag{14}$$

where $B = \|\mathbf{B}\|$ is the magnitude of the geomagnetic field and $\mathbf{T_c}$ is the magnetic torque required to compensate the gimbals angle offset. The required magnetic torque is designed to be dependent on the gimbal angle error rate and it is defined as [19]

$$\mathbf{T_c} = -\mathbf{A}(\delta)\frac{\boldsymbol{\delta}^* - \boldsymbol{\delta}}{t_c} \tag{15}$$

where $\mathbf{A}(\boldsymbol{\delta})$ is the Jacobian matrix, $\boldsymbol{\delta}^*$ is the preferred gimbals angle vector, $\boldsymbol{\delta}$ is the nominal gimbals angle vector and $\boldsymbol{t_c}$ is the time for the current gimbal angles to converge to the preferred gimbal angles. By substituting equation (14) and (15) into equation (12), the magnetic control torque generated by the magnetic torquers is expressed as follow

$$\mathbf{T_M} = -\frac{\mathbf{B} \times \mathbf{A}(\delta)\frac{\boldsymbol{\delta}^* - \boldsymbol{\delta}}{t_c}}{B^2} \times \mathbf{B} \tag{16}$$

The expression in equation (16) tells that the gimbal angle converging time t_c influences the amount of magnetic torque that needs to be generated and the required magnetic torque can be achieved whenever the magnetic dipole moment vector is perpendicular to the geomagnetic field vector.

4. Simulation studies

All mathematical models in the previous chapters were made amenable for numerical simulations in the Matlab®-Simulink™. The mission of the satellite was to perform attitude pointing mode i.e. roll (R) = 30°, pitch (P) = 0° and yaw (Y) = 0° for 25 orbital cycles. First, the simulation was performed without the MTGAC system (Case I). Then, it was followed by a simulation with the MTGAC system (Case II) at orbit inclination of 65°. All simulation parameters used in this study are summarized in table 1. The PD-controller gains are obtained through fine tuning using heuristic method whereas the value of t_c was selected to be 100s for optimal gimbal angle compensation performance [24].

Table 1. Simulation Parameters.

Parameter	**Value**
Orbital Parameters	
Altitude, h	500 km
Orbit inclination, i	65 deg
Orbital frequency, ω_0	0.0011 rad/s
Orbital cycle	10 orbits
Disturbance Torques	
T_{dx}	$8 \times 10^{-5} \sin(\omega_o t)$ Nm
T_{dy}	$8 \times 10^{-6} + 5 \times 10^{-5} \cos(\omega_o t) + 8 \times 10^{-5} \sin(\omega_o t)$ Nm
T_{dz}	$8 \times 10^{-6} + 5 \times 10^{-5} \cos(\omega_o t)$ Nm
Satellite Parameters	
Satellite's moment of inertia, $\mathbf{I}$	$\begin{bmatrix} 4.5 & 0 & 0 \\ 0 & 4.5 & 0 \\ 0 & 0 & 4.5 \end{bmatrix}$ kgm^2
Initial attitude, $\mathbf{RPY}_{\text{initial}}$	$[0 \quad 0 \quad 0]^{\mathrm{T}}$ deg
Desired attitude, $\mathbf{RPY}_{\text{desired}}$	$[30 \quad 0 \quad 0]^{\mathrm{T}}$ deg
Initial gimbal angle, $\boldsymbol{\delta}_{\text{initial}}$	$[0 \quad 0 \quad 0 \quad 0]^{\mathrm{T}}$ deg
Preferred gimbal angle, $\boldsymbol{\delta}^*$	$[0 \quad 0 \quad 0 \quad 0]^{\mathrm{T}}$ deg
Maximum gimbal angle rate, $\dot{\boldsymbol{\delta}}_{\mathbf{max}}$	$\pm[10 \quad 10 \quad 10 \quad 10]^{\mathrm{T}}$ deg/s
Flywheel angular momentum, h_0	0.47 Nms
Gimbal angle converging time, t_c	100 s
Magnetic dipole moment limit, $\mathbf{M}_{\mathbf{limit}}$	$\pm[12 \quad 12 \quad 12]^{\mathrm{T}}$ Am^2
Proportional gain, $\mathbf{K_p}$	$[3.86 \quad 3.86 \quad 3.86]^{\mathrm{T}}$ Nm/rad
Derivative gain, $\mathbf{K_d}$	$[5.84 \quad 5.84 \quad 5.84]^{\mathrm{T}}$ Nms/rad
SR-inverse steering law constant, λ	0.01

The generated CMG torques without and with the MTGAC system during attitude pointing mode are shown in figure 2. For Case I, the maximum control torque along the x-axis is 8.5×10^{-5} Nm in positive direction and -8.5×10^{-5} Nm in negative direction. Along the y-axis, the maximum control torque is 1.1×10^{-4} Nm in positive direction and -9.0×10^{-5} Nm in negative direction. Meanwhile, along the z-axis the maximum control torque generated by the CMG system is 6.0×10^{-5} Nm in positive direction and -4.5×10^{-5} Nm in negative direction. On the other hand, for Case II, the maximum control torque along the x-axis is 5.0×10^{-5} Nm in positive direction and -8.5×10^{-5} Nm in negative direction, 3.5×10^{-5} Nm in positive direction and -4.5×10^{-5} Nm in negative direction along the y-axis whereas along the z-axis the maximum control torque is 7.5×10^{-5} Nm in positive direction and -6.5×10^{-5} Nm in negative direction. By comparing the torques between these two cases, the CMG system generates more attitude control torque in total for Case I since it needs to counter for constant disturbance torques that act on the satellite during attitude pointing mode. In contrast, for Case II the constant disturbance torques that caused the gimbal to drift from its preferred angle are countered by the magnetic control torque from the MTGAC system.

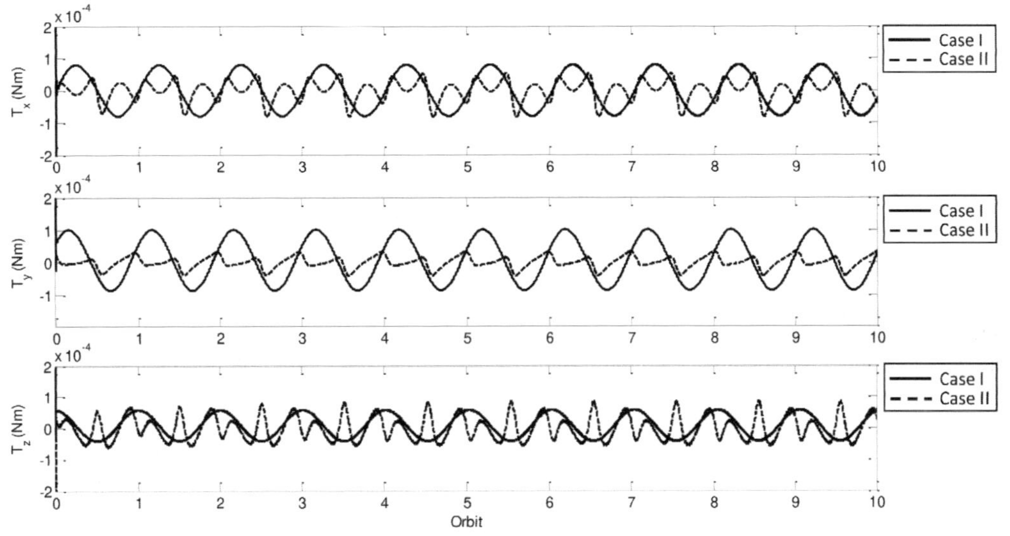

Figure 2. CMG torque during attitude pointing mode.

Figure 3 shows the magnetic control torques generated by the MTGAC system. The maximum magnetic torque along the x-axis is 1.0×10^{-4} Nm and -1.0×10^{-4} Nm in positive and negative direction, respectively. Along the y-axis, the maximum magnetic torque is 1.1×10^{-4} Nm in positive direction and -8.0×10^{-5} Nm in negative direction whereas along the z-axis the maximum magnetic torque is 1.1×10^{-4} Nm in positive direction and -1.3×10^{-4} Nm in negative direction. Summing the generated torques tor each cases, the total control torque for Case II is in fact larger than for the Case I. Total control torque for Case I is found to be 2.3×10^{-4} Nm and -2.2×10^{-4} Nm, in positive and negative direction, respectively whereas for Case II, the total control torque is 4.8×10^{-4} Nm in positive direction and -5.1×10^{-4} Nm in negative direction. Since the total generated control torque has direct influence to the amount of power consumption, the implementation of the MTGAC system to compensate gimbal angle offset leads to additional power requirement for the CMG-based controlled satellite. For small satellite platform where power is one of the main design constraints, this seems to be less favorable. However, the implementation of gimbal angle management system is crucial since the CMG system requires auxiliary control torque to compensate drift gimbals in which if not properly treated can lead to saturation singularity where sometime singularity avoidance law fails to avoid or escape.

The advantages of utilizing the MTGAC system are depicted in figure 4 and figure 5 respectively where a similar simulation is performed for 20 orbital cycles. Based on figure 4, the satellite fails to maintain at the desired attitude through the simulation time for Case I whereas the attitude of the satellite is successfully maintained about the desired attitude with certain accuracies for Case II. The loss in the attitude control of satellite for Case I is due to the gimbal angle offset where the gimbal for all SGCMGs gradually diverge from the preferred angle, as shown in figure 5. In fact, the #2 SGCMG

IOP Conf. Series: Materials Science and Engineering **370** (2018) 012045 doi:10.1088/1757-899X/370/1/012045

suddenly diverges towards unfavorable negative angle after 18th orbital cycle which causes the satellite to loss its attitude controllability. In contrast, for Case II the gimbals are successfully compensated about the preferred angle thus allowing the satellite to fulfil the attitude pointing mission.

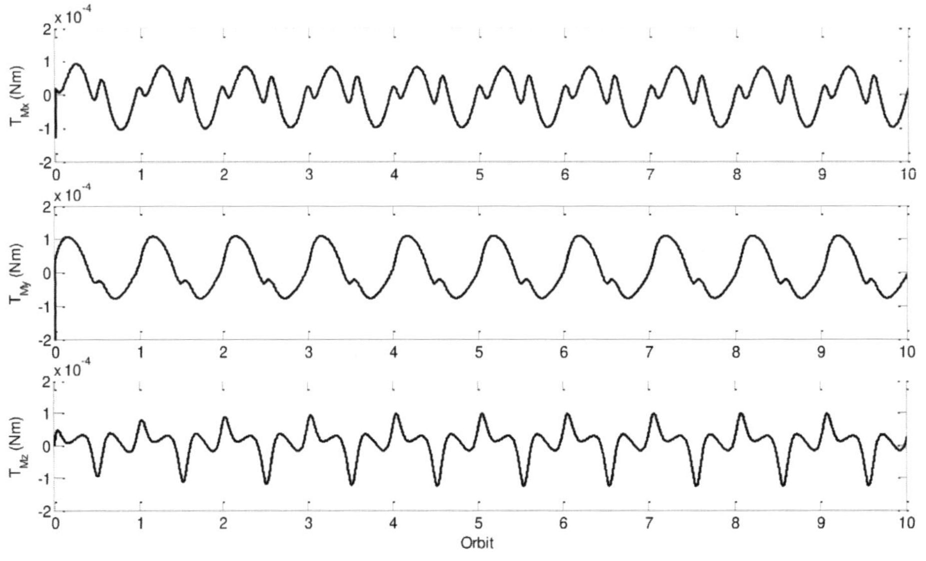

Figure 3. Magnetic torque generated by the MTGAC system during attitude pointing mode.

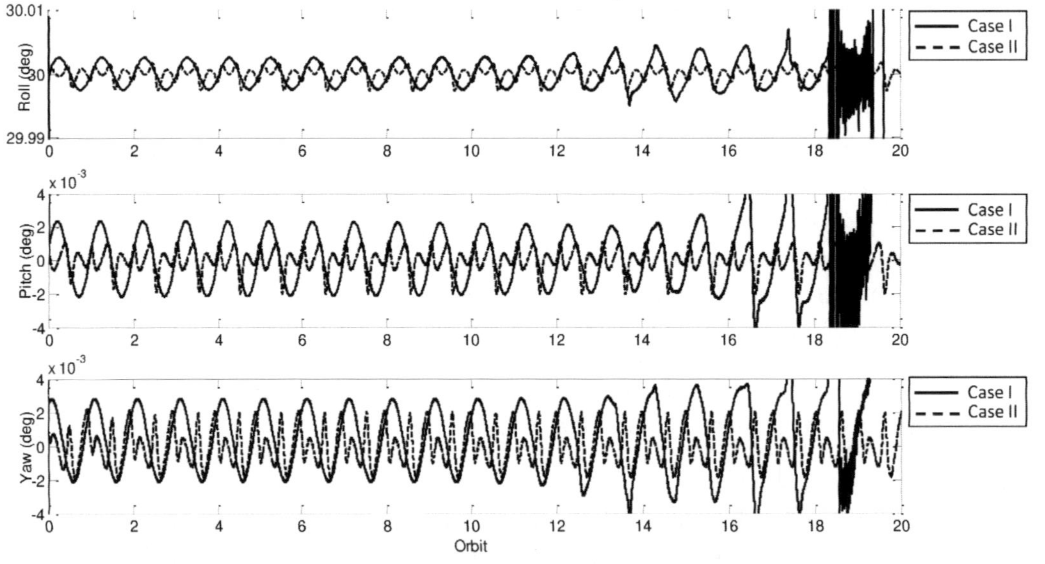

Figure 4. Satellite attitude during attitude pointing mode for 20 orbits.

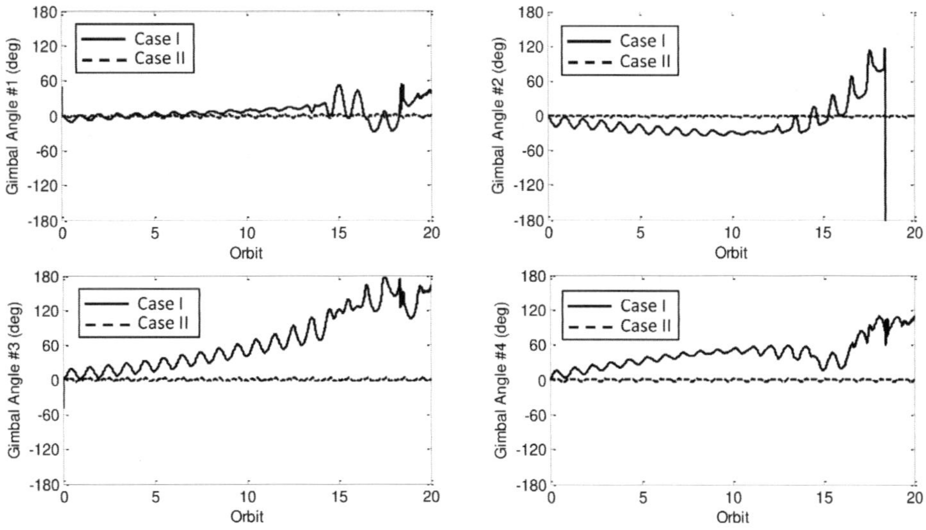

Figure 5. Gimbal angle of the CMG system during attitude pointing mode for 20 orbits.

5. Conclusion

In this paper, the control torques generated by CMG-based small satellite without and with the MTGAC system has been studied. The MTGAC system which is meant to compensate gimbal angle offset is cooperated with the PD-based attitude controller. All algorithms were implanted on MATLAB®-SIMULINK™ for simulation studies where the generated control torques are discussed as a trade-off between power consumption, attitude control performance and CMG dynamics. Results from the simulations show that the implementation of MTGAC system leads to additional power consumption due to addition of generated magnetic control torque. However, this addition leads to a better attitude control of the satellite as well as gimbal angle management of the CMG system which allows the satellite to fulfil its mission. The optimal power consumption of the satellite with the MTGAC system can be achieved if the software and hardware of system is carefully designed.

Acknowledgement

This research and publication was supported by Universiti Sains Malaysia Grant No. 304/PAERO/ 60315002.

References

[1] Omagari K, Fujihashi K and Matunaga S 2008 CMG configuration and control for rapid attitude maneuver of small spacecraft, *Int. Symp. on Artificial Intelligence, Robotics and Automation in Space*-iSAIRAS

[2] Lappas V, Steyn W H and Underwood C I 2002 Attitude control for small satellites using control moment gyros, *Acta Astronautica*, Vol. 51, No. 1-9, pp 101-11

[3] Defendini A, Lagadec K, Guay P, Blais T and Griseri G 1999 Low cost CMG-based AOCS design, *Proc. 4th ESA Int. Conf. on Spacecrft Guidance, Navigation and Control System, ESTEC*, Noodwijk, The Netherlands

[4] Omagari K, Usuda T and Matunaga S 2005 Research of control momentum gyros for micro-satellites and 3-DOF attitude dynamics simulator experiments, *Proc. of The 8th Int. Symp. on Artificial Intelligence, Robotics and Automation in Space-iSAIRAS*, Munich, Germany

[5] Bhat S P and Tiwari P K 2006 Controllability of spacecraft attitude using control moment

gyroscopes, *Proc. of the 2006 American Control Conf.*, Minneapolis, Minnesota, USA

[6] Lappas V J, Steyn W H and Underwood C I 2002 Practical results on the development of a control moment gyro based attitude control system for agile small satellites, *17th AIAA/USU Small Satellite Conf. Proc., SSC02-VIII-8*

[7] Wie, B. 1998 *Space vehicle dynamics and control*, AIAA Education Series, pp. 667-708

[8] Bedrossian, N S et al 1990 Steering law design for redundant single gimbal moment gyroscope, *J. of Guidance, Control and Dynamics*, 13(6) pp 1083-89

[9] Asghar S, Palmer P L and Roberts 2006 Exact steering law for pyramid-type four control moment gyro systems, *AIAA/ASS Astrodynamics Specialist Conf. and Exhibit*, Colorado

[10] Vadali S R, Oh H –S and Walker S R 1990 Preferred gimbal angles for single gimbal control moment gyros, *J. of Guidance, Control and Dynamics*, 13(6) pp 1090-95

[11] Leve F A and Fitz-Coy N G 2010 Hybrid steering logic for single-gimbal control moment gyroscopes, *J. of Guidance, Control, and Dynamics,* Vol. 33, No. 4 pp 1202-12

[12] Takada K, Kojima H and Matsuda H 2010 Control moment gyro singularity-avoidance steering control based on singular-surface cost function, *J. of Guidance, Control, and Dynamics,* Vol. 33, No. 5 pp 1442-50

[13] McElvain R J 1962 *Satellite Angular Momentum Removal Utilizing the Earth's Magnetic Field*, Torques and Attitude Sensing in Earth Satellites, Academic Press, New York pp 137-157

[14] Stickler A C and Alfriend K 1976 Elementary magnetic attitude control system, *J. of Spacecraft and Rockets*, Vol. 13, No. 5 pp 282-287

[15] Camillo P J and Markley F L 1980 Orbit-average behavior of magnetic control laws for momentum unloading, *J. of Guidance and Control*, Vol. 3, No. 6 p. 563-568

[16] Lovera M (2001) Optimal Magnetic Momentum Control for Inertially Pointing Spacecraft, European Journal of Control, vol. 7, No. 1, pp 30-39

[17] Suhadis N M and Varatharajoo R 2009 Satellite attitude performance during the momentum dumping mode, *Int. Rev. of Aerospace Engineering*, Vol. 2, No. 3, pp 133-138

[18] Lappas V, Steyn W H and Underwood C I 2001 Control moment gyro gimbal angle compensation using magnetic control during external disturbances, *AIAA Guidance, Navigations, and Control and Exhibit*, Canada

[19] Lappas V J 2002 A control moment gyro (CMG) based attitude control system (ACS) for agile small satellites, *Doctoral Thesis*, University of Surrey

[20] Sidi M J 1997 *Spacecraft Dynamics and Control*, Cambridge University Press, pp 322-327,

[21] Suhadis N M and Varatharajoo R 2012 Passive attitude control torque generation performances of a gravity gradient stabilization satellite, *Appl. Mechanics and Materials*, Vol. 225, pp 458-463

[22] Suhadis N M and Varatharajoo R 2009 Satellite attitude performance during the momentum dumping mode, *Int. Rev. of Aerospace Engineering (I.RE.AS.E)*, Vol. 2, N. 3 pp 133-138

[23] Ismail Z and Varatharajoo R 2010 A study of reaction wheel configuration for a 3-axis satellite attitude control, *Adv. in Space Research 45*, pp 750-759

[24] Salleh M B and Suhadis M N 2015 Magnetic gimbal angle compensator of CMG-based controlled small satellite, Int. Rev. of Aerospace Engineering (I.RE.AS.E), Vol. 8, N. 1 pp 10-15

IOP Conf. Series: Materials Science and Engineering **370** (2018) 012046 doi:10.1088/1757-899X/370/1/012046

Mechanical performance of laminated natural fibre bio-composites and hybrids

M. Shukur Zainol Abidin [1,2,*], Liang Yeng Fei [1], Pang Chin Wee [1], A.Anjang [1,2], I. Lister Daungkil [1]

[1] School of Aerospace Engineering, Universiti Sains Malaysia, 14300 Penang, Malaysia.
[2] Cluster for Polymer Composites (CPC), Universiti Sains Malaysia, 14300 Penang, Malaysia.

*Corresponding author: aeshukur@usm.my

Abstract. Composite materials are highly utilized in the current era due to their tailored specified properties, such as high strength-to-weight, corrosion resistance, inflammable properties, wear resistance and etc. Bio-composites are materials that consisted of matrix (resin) and reinforcements from natural fibres. The mechanical properties of bio composite materials, such as coir, and palm oil fibre, were analysed and evaluated. Fabrication method via combination of laminating process and compression moulding were developed and utilised to manufacture the specimens. Impact, tensile, compression and bending test were carried to evaluate the mechanical characterisation of the bio-fibre composites. However, the mechanical performance of the natural composite were found to be inferior to that of conventional glass fibre composites. Hybridization of natural fibre with glass fibres composite shows a significant increase in specific strength and specific modulus.

1. Introduction

The demand for composite material has been increasing over the years. More industries, especially aerospace related, are searching for cheaper and higher quality material alternatives. Composite consist of two or more material constituents commonly known as reinforcing phase and a matrix. Rayon, nylon, carbon and glass fibres are common synthetic reinforcing phase [1]. Increasing environmental concerns has led to more attention and focus on development of bio-composite. The great potential for using natural fibre reinforced composite lies in their eco-friendliness and wide potential applications for automotive, mobile phone and biomedical industries. Natural fibres such as jute, sisal, coir, hemps, as well as banana and pineapple leaves are generally extracted from renewable sources, fully biodegradable, non-toxic, and can be easily recycled to reduce the materials' carbon footprint [2].

In the recent years, there has been a growing interest in the application of natural fibres as reinforcements for polymer matrices [3]. Natural fibre has good potential as reinforcement in thermoplastic and thermoset polymer composites mainly due to low density and high specific properties of fibres [4]. Natural fibres have the properties, composition, structures and features that are suitable to be used as reinforcements or fillers in polymer composites [5]. Additionally, the synergy between plant-based natural fibre with conventional synthetic fibre into polymeric matrices leads to a new class of fibre-reinforced composite known as hybrid composites. It has been widely reported that hybrid composite materials shows improved mechanical properties like tensile, flexural and impact strength comparable to glass fibre reinforced composites [6-11].

Glass fibre reinforced polymers (GFRPs) are a fibre reinforced polymer made of a plastic matrix reinforced by fine fibres of glass. Fibre glass is a lightweight, strong, and robust material used in different industries due to their excellent properties. GFRP composites are largely used mainly due to a combination of low cost and good mechanical properties. Although strength properties are somewhat lower than carbon fibre and it is less stiff, the material is typically far less brittle[12]. The incorporation of natural fibre with GFRP improves the tensile, flexural and impact strength of the materials [8] and placing the GFRP layers at the ends possess good mechanical strength [13]. The strength properties of natural fibre composites are somewhat lower, because of less stiff and typically less brittle. Reinforcing glass fibre into the natural composites enhanced tensile and flexural properties without any effect on tensile and flexural module. In addition, adding natural fibre with glass fibre improves thermal properties and water resistance of the hybrid composites [14].

2. Methodology

2.1 Material

Palm oil fibres and coir fibres were used as the natural fibre reinforcement for the composites. The coir fibres and oil palm fibres were supplied by Green Tree Garden & Landscape Nursery and United Palm Oil Industries Sdn. Bhd. respectively. The density of coir fibre is 1.2 g/cm^3, whereas for oil palm fibres is 1.0 g/cm^3. The fibre glass used throughout this project is chopped strand mat (CSM) with a density of 2.5 g/cm^3. CSM is utilized in this project because of its randomly distributed characteristics, which is similar to that of natural fibres. A generic commercially available polyester resin (FRP 101) was used for this study. The polyester resin was mixed with hardener (Luperox DDM-F) at a ratio of 1:100 to obtain the matrix for the composites.

2.2 Fabrication process

Clean natural fibres were handpicked and separated from impurities, such as sand and leaves. These impurities may cause inconsistent density of the specimen, as well as affecting the results of the tests. Separated fibres were then weighted using electronic weighing balance before being allocated on designated plate with area of 160 mm × 160 mm. The total mass of the natural fibre was divided by 4 as 4 plies of fibre sheets will be used. To evenly distribute the mass of the fibre, the mass of each ply of fibre was divided by 9, to be allocated on the designated squares as shown Figure 1. For instance, the total mass of coir to be used in CFCR is 27.65 g. This amount will be divided by 4, which makes it 6.92 g per ply, then divided by 9 making it 0.768 g per square grid.

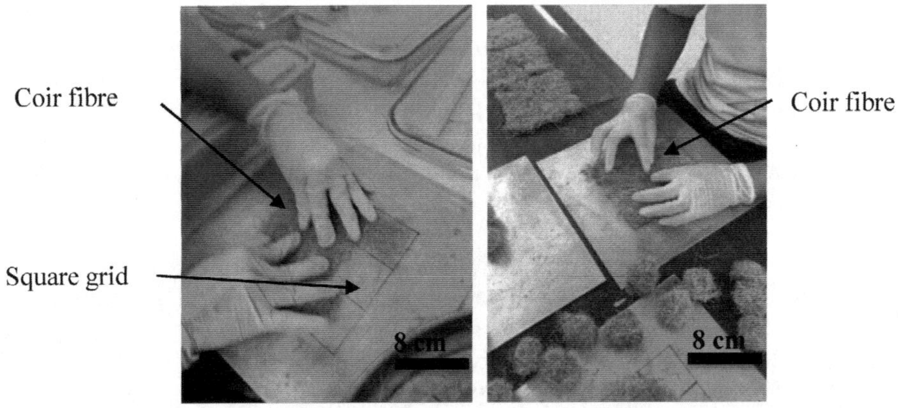

Figure 1. Fibre allocated on aluminium plate with area of 150mm × 150mm

Upon fibre allocation, the fibres are dried by placing in the oven at approximately 50 °C for at least 5 hours. The purpose of drying the fibre is to extract moisture from the fibre to prevent the water from affecting the results, as well as enhancing the curing process. The fibre sheet was placed on a plastic paper, then being rolled into a much thinner sheet of fibre. Compression moulding method was used to fabricate the composite plates that will be used for testing. Compression moulding provides good surface finish of composite and it is easy to perform. Aluminium plate moulds were used to compress the composite into shape.

The aluminium bases and mould were waxed thoroughly to prevent the composite from attaching onto the surface during curing process. The first layer of the fibre sheet was aligned on the mould after applying the polyester resin. A roller was used to spread the resin evenly. This step is repeated until all fibre sheets were stacked up. Another aluminium base was used to cover the specimen. A heavy metal plate was placed on the aluminium mould to ensure uniform distribution of force during compression. The mould was then being compressed with hydraulic jack. Excessive resins were removed during this process, and then the specimen was left to cure for at least 12 hours. The fibre sheets stacking configuration is summarised in Table 1. After curing, the specimen was demoulded and trimmed in accordance to the specific test standards using a wet diamond cutter.

Table 1. Fibre sheets stacking configuration

Type	Ply Configuration	Ply Sequence
GFRC	8 plies of Glass Fibres	G/G/G/G/G/G/G/G
CFRC	8 plies of Coir Fibres	C/C/C/C/C/C/C/C
PFRC	8 plies of Palm Fibres	P/P/P/P/P/P/P/P
HCRC	4 plies of Glass Fibre + 4 plies of Coir Fibres	G/G/C/C/C/C/G/G
HPRC	4 plies of Glass Fibre + 4 plies of Palm Fibres	G/G/P/P/P/P/G/G

2.3 Characterisation method

2.3.1 Specimen density measurement

Each sample was cut into a square with both width and length at 20 mm. The length, width, and thickness of the specimens were measured again using a Vernier calliper to obtain a more accurate volume. The mass of each sample was measured using a digital weighing scale. The average volume and mass were obtained from three samples of each specimen type. The mass of each sample was divided by the volume to obtain the density.

2.3.2 Tensile testing

A tensile test was carried out based on ASTM D3039 on the composite samples to determine the tensile strength, elastic modulus, tensile strain and modulus of toughness. The composite plate was cut into five pieces at 160 mm × 25 mm × 3 mm using a cutting machine. The test section length of 100 mm was marked by a permanent marker. The test was conducted using an Instron 3367 Universal Testing Machine. An extensometer was place on the middle of test section to measure the local strain. The crosshead movement was set at 2 mm/min.

2.3.3 Bending testing

A bending test was carried out based on ASTM D790 on the composite samples to determine the flexural strength, flexural modulus, and flexural strain. The composite plate was cut into five pieces at

160 mm × 25 mm × 3 mm using a cutting machine similarly to tensile test specimen. The test section length of 100 mm was marked by a permanent marker. The crosshead movement was set at 2 mm/min

2.3.4 Compression testing
The compression test was carried out based on ASTM D695 on the composite samples to obtain the compressive strength, average value and standard deviation. The composite plate was cut into 10 specimens. 5 longer specimens were cut to be at dimensions of 150 mm × 12.7 mm × 3 mm while another 5 shorter specimens were cut to be at dimensions of 122mm × 12.7 mm × 3 mm using a cutting machine. The gauge length of longer specimen is 38 mm while the gauge length for shorter specimen is 10 mm. The crosshead movement was set at 1.2 mm/min.

2.3.5 Impact testing
Impact testing was carried out in accordance with ASTM D256 on the composite samples to obtain the impact energy and specific impact energy. The composite plate was cut into 5 specimens with the dimension of 64 mm × 12.7 mm × 3.2 mm. The impact resistance of the specimen involves flexural shock using standardized pendulum-type hammer when breaking specimens under one swing. A 15 Joule hammer was selected as the energy required to break the specimen. After impact, the energy required to break the specimen was indicated at the impact gauge.

3. Result and Discussion

Compression moulding method would give a good surface finish and easier to obtain many specimens with fixed dimensions depending on the mold used. The thickness of the mold was at 3 mm. However, composite specimens that contains natural fibre has irregular thicknesses of 3-6 mm. The pure natural fibre (CFRP and PFRP) composite plate was the thickest followed by hybrid (CGFRP and PGFRP) and pure glass fibres (GFRP) specimens. This is because the natural fibre sheets expands more during the layup process with polyester resin. The car jack used to compress the mold also does not provide sufficient force to control the thickness. As a result, the CFRP, PFRP, CGFRP, and PGFRP may contain lower fibre content per unit volume due to increased thickness. Hence, the mechanical testing results for each fabrication batch of the same type of fibres used may be different due to the variations of thickness.

3.1 Density Measurement
As expected, the density of the GFRC were higher than that of both natural fibre composites and hybrid composites. The density of the composites are tabulated in Table 2. Furthermore, the densities measured are in agreement with the expected densities as estimated via Rule of Mixture. Therefore the voids in these samples should be sufficiently low and does not degrade the mechanical performance of the composites.

Table 2. Measured density of the composites.

	Density, kg/m^3
GFRC	1656.38 ± 13.02
CFRC	1260.26 ± 6.43
PFRC	1217.88 ± 10.11
HCRC	1466.27 ± 9.82
HPRC	1420.69 ± 9.06

3.2 Tensile properties
GFRC (pure glass composite) has the highest ultimate strength, strain-to-failure and elastic modulus followed by HPRC (hybrid palm), HCRC (hybrid-coir), CFRC (pure coir), and PFRC (pure palm) as tabulated in Table 3. It was expected and shown that pure glass composite would give the highest

tensile properties. Hybrid composite were superior in terms of tensile properties in comparison to pure natural fibre composite. This shows that the added glass fibre on the hybrid composite greatly improves the tensile property of natural fibre reinforced composites.

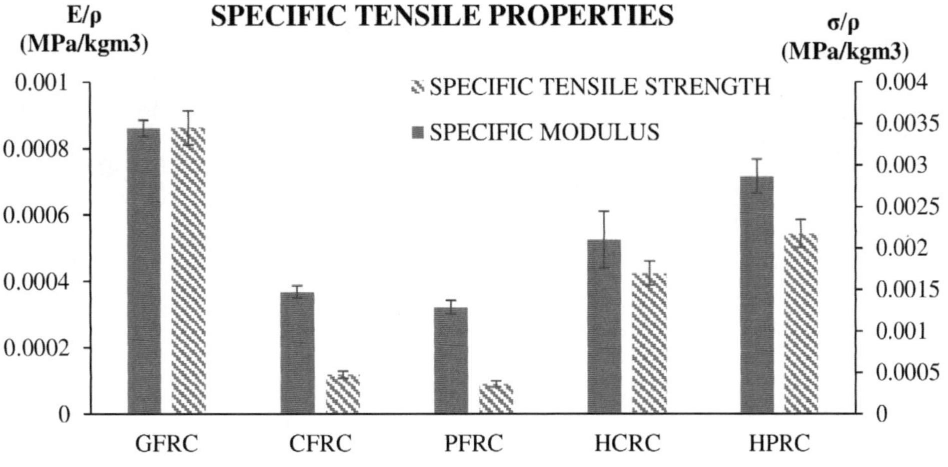

Figure 2. Specific tensile properties of the composites.

3.3 Bending properties

GFRC has the highest flexural strength followed by HCRC, HPRC, CFRC, and PFRC. Although the hybrid composite can withstand higher maximum load than pure glass composite, the actual flexural strength of hybrid composite is still inferior compared to pure glass composite. Table 3 also clearly shows that the hybridization causes the flexural strength of NFCs to increase significantly, in this case is nearly 5 times the strength of pure NFCs. The elastic modulus of HCRC is lower than HPRC and GFRC but higher than CFRC and PFRC. This result shows that hybrid composites are nearly as good as pure glass composite in terms of flexural stiffness. Pure NFCs were still far inferior in flexural stiffness.

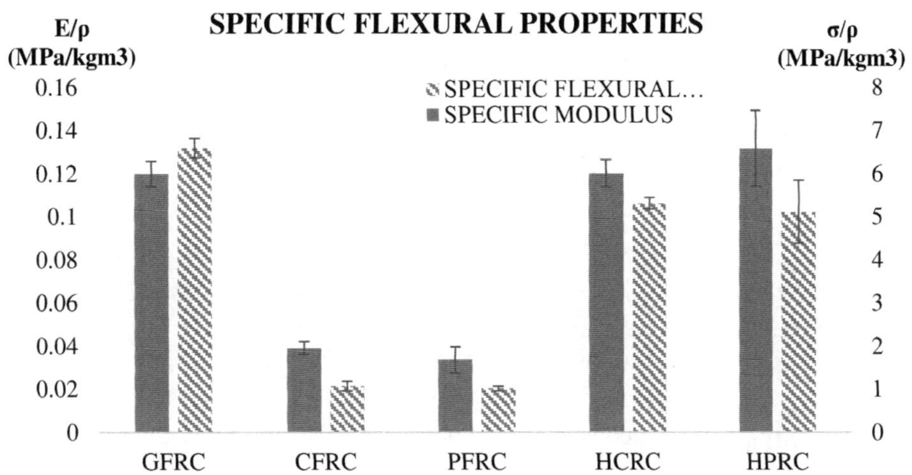

Figure 3. Specific flexural properties of the composites.

3.4 Compression properties

GFRC has the highest modulus while there is not much difference between CFRC and PFRC as well as between HCRC and HPRC. Glass composite has the highest modulus here while hybrid composite comes in second.

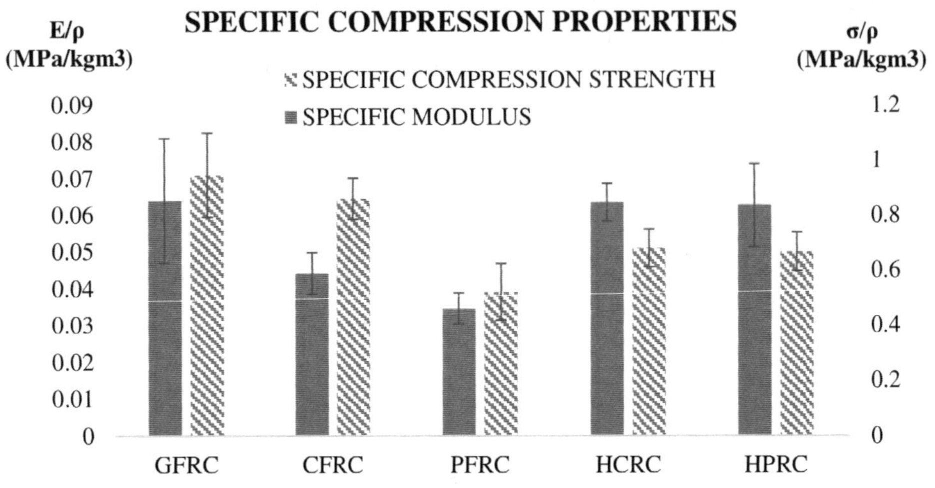

Figure 4. Specific compression properties of the composites.

3.5 Impact properties

GFRC has the highest impact energy and specific impact energy compared to the other hybrid composite with about 50% difference in impact resistance. Pure coconut and pure oil palm recorded the lowest impact resistance with 10% and 20% respectively less than the GFRC. On the other hand, the hybrid composites of HCRC and HPRC indicated an increase in impact properties as much as 12.8% and 34.8% respectively. These results shows that hybridization of the composite contribute to the improved mechanical properties of the natural fibre composites.

Table 3. Mechanical properties of the composites.

	COMPOSITE SYSTEMS		GFRC	CFRC	PFRC	HCRC	HPRC
MATERIAL PROPERTIES	Tensile	**σ_{max} (MPa)**	1.430 ± 0.085	0.150 ± 0.014	0.109 ± 0.011	0.622 ± 0.054	0.772 ± 0.060
		E (MPa)	5.71 ± 0.16	1.86 ± 0.09	1.57 ± 0.10	3.08 ± 0.50	4.07 ± 0.29
		Strain to Failure (mm/mm)	0.053 ± 0.005	0.011 ± 0.002	0.011 ± 0.003	0.045 ± 0.007	0.049 ± 0.005
	Compression	**σ_{max} (MPa)**	117.3 ± 19	81.15 ± 7.1	47.56 ± 9.38	74.75 ± 7.49	70.99 ± 7.47
		E (MPa)	1410.82 ± 375.04	741.44 ± 95.06	560.08 ± 69.86	1240.01 ± 99.78	1185.96 ± 213.29
		Strain to Failure (mm/mm)	0.075 ± 0.045	0.16 ± 0.032	0.092 ± 0.052	0.075 ± 0.01	0.077 ± 0.01
	Bending	**σ_{fm} (MPa)**	218.42 ± 7.39	26.95 ± 2.80	24.53 ± 1.34	155.64 ± 3.93	145.35 ± 20.79
		E (MPa)	9928 ± 477	2464 ± 189	2039 ± 374	8808 ± 452	9355 ± 1248
	Impact	**Impact Energy, (kJ/m^2)**	103 ± 7	8.45± 1.71	17.57 ± 1.62	43.60 ± 3.89	44.8 ± 6.37
		Specific Impact Energy, (kJ/m^2)	0.063± 0.004	0.0067 ± 0.0014	0.014 ± 0.0013	0.030 ± 0.0026	0.032 ± 0.0045

4. Conclusion

From fibre alignment, composite fabrication, and mechanical testing to wind turbine blade fabrication, natural fibres has been used as a one of the reinforcement material. The better way to produce natural fibre sheet is through rolling as the procedure were simpler than hydraulic press which also provides less uniform flattening. It is known that NFCs have poor mechanical properties. However, through hybridization of natural fibre with glass fibre, the tensile and flexural properties had been improved significantly. In flexural test, the specific strength and specific modulus of hybrid bio-composite are higher than that of pure glass composite. In compression test, the effects of hybridization on compressive properties were negligible and only offer slight differences. The idea of hybridization has widened the possible applications of natural fibre in mechanical structures. Due to the abundance of natural fibres, the cost of making strong and reliable structures such and the wind turbine blade can be reduced significantly.

Acknowledgement

The authors would like to acknowledge and thank Mr Hasfizan bin Hashim and Mr Mohd Shahar bin Che Had for their technical expertise and generous assistance in the fabrication process of the composites. This study was conducted under USM Short Term Grant 304/PAERO/6315057.

References

[1] Hull, D. and T.W. Clyne, *An Introduction to Composite Materials.* 2 ed1996, Cambridge: Cambridge University Press.
[2] Dong, Y., et al., Polylactic acid (PLA) biocomposites reinforced with coir fibres: Evaluation of mechanical performance and multifunctional properties. *Composites Part A: Applied Science and Manufacturing,* 2014. **63**: p. 76-84.
[3] Ahmad, F., H.S. Choi, and M.K. Park, A Review: Natural Fiber Composites Selection in View of Mechanical, Light Weight, and Economic Properties. *Macromolecular Materials and Engineering,* 2015. **300**(1): p. 10-24.
[4] Wambua, P., J. Ivens, and I. Verpoest, Natural fibres: can they replace glass in fibre reinforced plastics? *Composites Science and Technology*, 2003. **63**(9): p. 1259-1264.
[5] Faruk, O., et al., Progress Report on Natural Fiber Reinforced Composites. *Macromolecular Materials and Engineering*, 2014. **299**(1): p. 9-26.
[6] Salleh, Z., et al., Fracture toughness investigation on long kenaf/woven glass hybrid composite due to water absorption effect. *Procedia Engineering*, 2012. **41**: p. 1667-1673.
[7] Sharba, M.J., et al., Effects of kenaf fiber orientation on mechanical properties and fatigue life of glass/kenaf hybrid composites. *BioResources*, 2015. **11**(1): p. 1448-1465.
[8] Sanjay, M.R., G.R. Arpitha, and B. Yogesha, Study on Mechanical Properties of Natural - Glass Fibre Reinforced Polymer Hybrid Composites: A Review. *Materials Today: Proceedings*, 2015. **2**(4): p. 2959-2967.
[9] Pickering, K.L., M.A. Efendy, and T.M. Le, A review of recent developments in natural fibre composites and their mechanical performance. *Composites Part A: Applied Science and Manufacturing*, 2016. **83**: p. 98-112.
[10] Yan, L. and N. Chouw, Crashworthiness characteristics of flax fibre reinforced epoxy tubes for energy absorption application. *Materials & Design*, 2013. **51**: p. 629-640.
[11] Kinloch, A., et al., Tough, natural-fibre composites based upon epoxy matrices. *Journal of materials science*, 2015. **50**(21): p. 6947-6960.
[12] Chawla, K.K., *Composite Materials : Science and Engineering*. 3 rd ed2012, New York: Springer-Verlag New York.
[13] K, S.A. and V. S., Tensile, flexural and interlaminar shear properties of woven jute and jute-glass fabric reinforced polyester composites. *Journal of Materials Processing Technology*, 2008. **207**: p. 330-5.
[14] Jarukumjorn, K. and N. Suppakarn, Effect of glass fiber hybridization on properties of sisal fiber polypropylene composites. *Composites: Part B*, 2009. **40**: p. 623-627.

IOP Conf. Series: Materials Science and Engineering **370** (2018) 012047 doi:10.1088/1757-899X/370/1/012047

Impact Resistance of Bamboo Fabric Reinforced Polypropylene Composites and Their Hybrids

N Z M Zuhudi[1*], K Jayaraman[2], R J T Lin[2], N M Nur[1]

[1]Aerospace Section, Universiti Kuala Lumpur – Malaysian Institute of Aviation Technology (UNIKL MIAT), Lot 2891, Jalan Jenderam Hulu, 43800 Dengkil, Selangor, Malaysia

[2]Center for Advanced Composite Materials (CACM), Department of Mechanical Engineering, The University of Auckland, Private Bag 92019, Auckland, New Zealand

*Corresponding author: zuhairah@unikl.edu.my

Abstract. Bamboo in woven fabric form embedded in the polymer results in easier material handling during production and reduction in the manufacturing cost of the composites. In the current study, the effect of hybridization of bamboo fabric with glass fibres on the impact properties of the composites made from them has been studied. The composites were fabricated using a compression moulding method. Bamboo fibres were in the form of twill-weave fabric. Impact tests were conducted using an instrumented drop weight impact test system. It can be seen that the peak load and energy absorbed increase with the increase in the applied impact energy. For the BPP composites, the applied energy of the impact was increased until penetration was occurred at 55 J. While for the BGPP hybrid composites, they were penetrated at the applied energy of only 35 J. However, The hybrid composites exhibit a greater peak load resistance than the BPP composites. The crack damage in the composites was also reduced with the presence of fabric reinforcement. These preliminary results indicate that bamboo fabric is truly a new contender for developing excellent and balanced properties of composites such as for the interior components.

1. Introduction

1.1. Impact properties of composites and their hybrids

Impact resistance of a structure can be investigated using several test methods such as Charpy, Izod and instrumented falling weight testing. Drop weight impact (DWI) testing can provide more information than Charpy and Izod impact testing. The test output gives more relevant data regarding toughness ability and damage behaviour of composites compared to Charpy and Izod test results. According to Richardson et al. [1], impact can be divided into two categories; high velocity and low velocity impact. High velocity impact response is dominated by stress wave propagation through the materials, in which localised damage occurs rapidly. The impact event is over before the stress waves reached the edge of the structure, so boundary condition effects can be neglected. In contrast, in low velocity impact, boundary conditions are important because the impact duration is long enough for the structure to respond to impact and more energy is absorbed. Most of the impact test work reported in the literature has been conducted using low velocity impact testing in order to be able to observe the damage mode initiation and propagation.

The nature and extent of impact damage is dependent on parameters such as impact velocity, impact angle, shape of the impactor, laminate material (including fibre volume fraction and lay-up), and laminate geometry [2]. It is important to better understand the impact responses of such materials especially hybrid composites, because it has been already proved that incorporation of glass with natural fibres will lead to an enhancement in the mechanical properties of the resulting hybrid composites.

Low velocity impact of glass fibre reinforced composites in thermoset matrices has been the subject of many experimental studies [3,4]. GPP behaviour and damage tolerance under low velocity impact also has been discussed and well-reviewed [5-7]. However, there has not much work on the impact resistance of natural fibre reinforced composites (NFRCs) [8] particularly in PP based matrices for example the work by Bledzki et al. [9]. Interestingly, more researchers are focussing on the effects natural fabric on impact resistance in PP based matrices [10-12]. A few studies have reported the behaviour of NFRCs under drop weight impact testing; however, no such research has been carried out with natural fabric-glass hybrid composites in PP matrices. Most of work on natural fabric hybrid composites used thermoset resin such as polyester [9, 13-14]. For example, Ahmed et al. [2] explored the effects of hybridisation on low velocity impact damage of woven jute fabric reinforced isothalic polyester composites. Their results indicated that jute composites have better impact energy absorption capacity than jute-glass hybrid composites but their damage tolerance capability is lower. Benevolenski et al. [13] performed work on the transverse perforation impact behaviour of flax mat-reinforced PP composites with additional discontinuous cellulose (Lyocell) and discontinuous glass fibre mat at 50 wt.% fibre content. They found the resistance of flax-PP to perforation on impact was strongly improved by the hybridisation with Lyocell and glass fibres. More recently, Pandita et al. [15] carried out drop weight impact tests on jute-glass polyester composites. They found that the impact resistance of their hybrid composites was higher than that of jute woven composites. They concluded that the woven glass at the outer surface of a composite can act as a strong skin.

1.2. Impact behavior study

A typical load-displacement curve derived from drop impact test on the composites is shown in figure 1. The peak load (F_m) is the maximum load that the specimen can sustain on fracture, which indicates the beginning of significant damage. The associated energy absorbed up to this point is symbolized by E_m and it represents the energy to initiate crack. After F_m , the dropping off in force indicates crack propagation and E_p represents the energy absorption in this phase. The total energy absorption is defined as:

$$E_t = E_m + E_p \tag{1}$$

where E_t is the area under the load-deflection curve is [16].

Figure 1 shows the two typical load-displacement curves obtained in an impact event. Shaded areas in the figure represent the energies absorbed by the composites during impact tests resulting in closed and open type curves. A closed type curve as shown in figure 2 (a) shows a rebounding phenomenon happens in the specimen. An open type load-displacement curve has a horizontal section at the very end, a post perforation frictional section. In order to identify the true energy absorption due to damage formation in the specimens, the post-perforation friction section needs to be removed from the curve. For this purpose, the initial part of the descending section of the curve may be extended to the displacement axis, shown as an estimated point in figure 2 (b). The absorbed energy can be determined from the initial kinetic energy minus the rebound kinetic energy using the initial and rebound velocities. The difference between absorbed energies calculated from these curves and that calculated with kinetics energy methods is negligible.

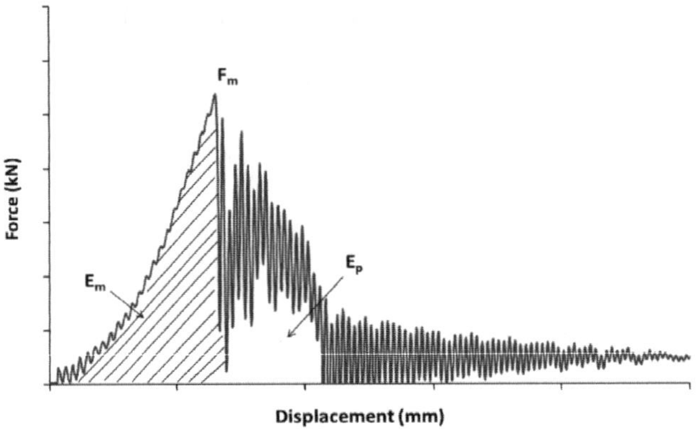

Figure 1. Schematic diagram of force versus displacement from a drop weight impact test showing the area under curve

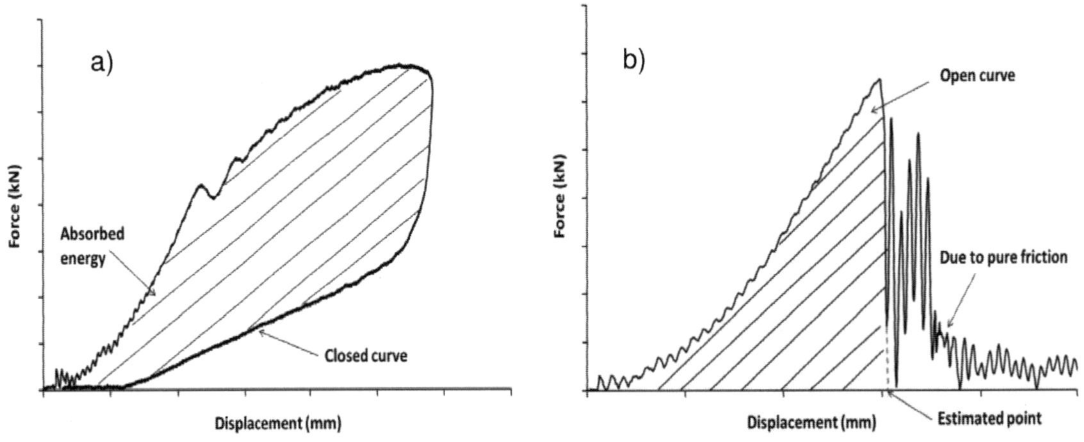

Figure 2. Schematic diagram of force versus displacement from a drop weight impact test for a) non-perforated specimen and b) perforated specimen

The penetration threshold is defined as the impact energy at which the impactor does not rebound from the specimens [16]. The perforation threshold is defined as the absorbed energy when the tup penetrates the specimen. In the case of a thick specimen, the hemispherical nose may completely bury into the specimen, resulting in penetration without reaching perforation, as shown in figure 3 (a). However, it should be noted that it is hard to observe such penetration when the specimen is much thinner than the radius of the impactor nose. The impactor nose used in this study had a hemispherical shape with a radius of 8 mm while the composites were in the range of 1.8 to 3.0 mm nominal thickness. Accordingly, penetration of specimens is generally followed by perforation. Figure 3 (b) shows the impactor at the onset of perforation and figure 3 (c) illustrates the perforation through the specimen.

It is important to study in detail the behaviour of the composite under impact in order to estimate not only the consequences of the energy absorption, but also how the impact damage can be minimized in some way. In this research, the objective of this paper is to carry out a preliminary study of the effect of hybridization of bamboo fabric with glass fibres on the impact behaviour of the composites impacted at different applied energies. The result reported here is associated with the previous results reported elsewhere [17].

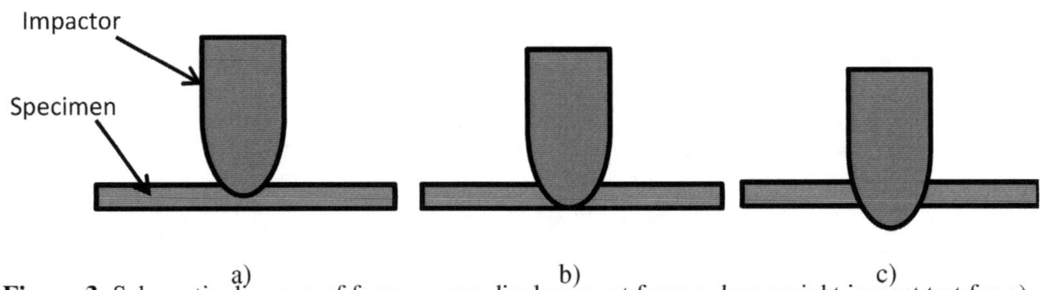

Figure 3. Schematic diagram of force versus displacement from a drop weight impact test for a) penetration, b) impactor is at the onset of perforation and c) perforation [16]

2. Materials and Methods

The raw materials used to fabricate the BPP composites in this work were polypropylene and woven bamboo fabric. The matrix used was polypropylene (PP) random copolymer, Moplen RP241G, manufactured by Lyondell Basell Industries and supplied by Field International Ltd., Auckland, New Zealand. The PP sheets have a nominal thickness of 0.38 and 0.58 mm. 100% bamboo fabric twill-woven was obtained from Xinchang Textiles Co.Ltd., Guangzhou, China. The bamboo fabrics with a width of 1500 mm and weight of 220 gsm were used, having specification of 20*20 tex and 108*58 per square inch for yarn count and density, respectively. The glass pre-preg supplied by Plytron ICI Ltd. UK with nominal glass volume fraction and density of 35% and 1480 kg/m^3 . The nominal thickness and sheet width are 0.47 mm and 240 mm, respectively.

2.1. Methods

2.1.1. Fabrication of composites.

The compression moulding process was used in this research to produce composite laminates. The closed mould was heated until the required temperature of 185°C was reached. The ply stack that had been dried earlier was placed in the mould cavity for pre heating, and the loaded mould was continuously heated for about five minutes without pressure to allow the polypropylene to start melting and percolating through the fibres. At this point, the consolidation pressure of 0.80 MPa was applied and held steady for five minutes. During this impregnation stage, pressure was applied to force the molten polymer into the fabrics while removing the excess air and volatiles. During the cooling period, the pressure applied was maintained until the temperature of the mould cavity dropped down to 40°C or lower when the laminate could be removed from the mould. The bamboo polypropylene (BPP) composites and their bamboo-glass polypropylene (BGPP) hybrid composites were fabricated in the range of about 55-65% fibre weight fraction, all in warp direction. The hybrid composites were fabricated with about 30% of bamboo and 20% of glass fibre weight fraction.

2.1.2. Instrumented drop weight impact test.

Impact test were conducted using an instrumented drop weight impact test system (DYNATUP 8250) on samples size of 150 x 100 mm. The test was performed using a drop weight with a steel (Thyrodur 2550) hemispherical striker tip of 16 mm diameter. Different impact energies of 20, 25, 35, 45 and 55 J were applied to the composites. The energy was obtained by changing the drop height, with constant impactor mass of 9.745 kg. For every type of laminate five specimens were tested. The specimen was tested using a test rig with a toggle clamp at all edges. The impact data was generated by the data system in terms of force-time, energy-time, force-displacement graphs which can be displayed from the output, along with the computed values of input energy, impact velocity, peak load, total energy, maximum energy at peak load, etc.

3. Results and Discussion

3.1. Effects of applied impact energy

A summary of the impact properties computed for the BPP composites for different energy levels when tested using the toggle clamps is shown in table 1. The BPP composite was not penetrated with the applied energy of 25 J. The applied energy of the impact was then increased until penetration was achieved at 55 J. It can be seen that the peak load and energy absorbed increase with the increase in the applied impact energy. Typical impact force versus displacement of BPP composites as a function of applied impact energy is shown in figure 4. A closed-loop type behaviour for applied energy lower than 55 J shows that complete perforation did not take place. The impactor rebounded in such cases. The force increased to a peak before dropping off in several phases. The gradual dropping off led to more displacement before failure and it is associated with higher energy absorption.

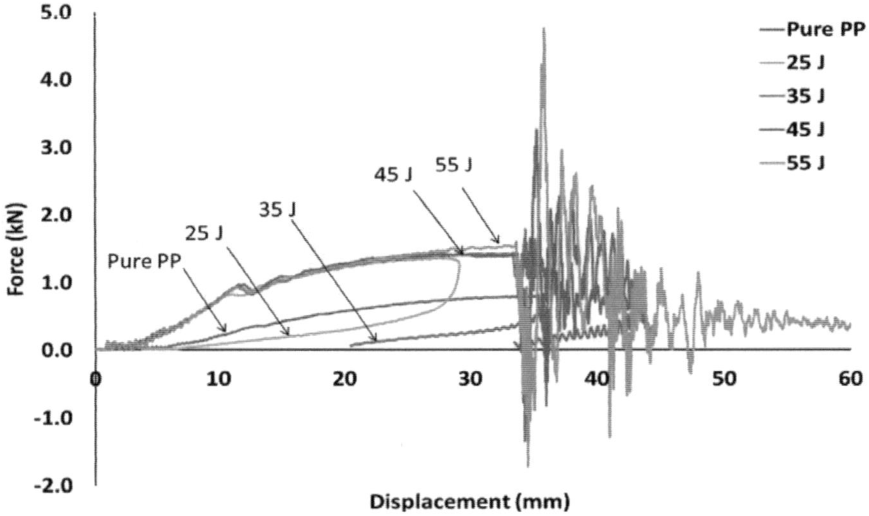

Figure 4. Typical impact force versus displacement of the BPP composites at different applied impact energies

Table 1. Impact properties of the BPP composites at different applied energies

Applied Energy (J)	Peak Load (kN)	Energy absorbed (J)	Displacement to max. load (mm)	Impact responses
25	1.365±0.3	21.93±0.2	27.12±0.8	Indentation
35	1.422±0.2	30.52±0.5	32.89±0.2	Indentation
45	1.452±0.8	32.07±0.8	33.64±1.1	Indentation
55	1.599±0.6	32.40±0.1	33.65±1.5	Perforation

Damage patterns in BPP composites impacted at various applied energies are shown in figure 5. The BPP composites were not penetrated or perforated with impact energies of 25 J, 35 J and 45 J. The impactor bounced back. At 25 J up to 45 J, the damage in the composites was limited to small indentations at the point of impact and became obvious on the other side of composites as the applied energy increased. However, as can be seen, the impact damage at 55 J is more profound at the back surface. Fibre failure and matrix cracks were obvious at the back surface. The damage propagates from the back surface (tension side) to the front surface (compression side) until the maximum force is reached.

The study of these properties using drop weight impact testing provided insight into the capability of the BPP composites in term of impact resistance and energy absorption properties. The force in the BPP composites increased before dropping off. The gradual dropping off indicated crack propagation occurred and energy was then, absorbed by the bamboo fabric. The impactor bounced back at energies lower than 55 J, as indicated by the load-displacement curves for the BPP composites. Damage in the composites impacted at 25 J to 45 J was limited to small indentations at the point of impact. 55 J was

IOP Conf. Series: Materials Science and Engineering **370** (2018) 012047 doi:10.1088/1757-899X/370/1/012047

required to perforate the specimens. These results indicate the greater impact load carrying ability of the BPP composites. This is because of greater stiffness and load bearing ability in the bamboo fabric compared to the pure PP. A larger damage area was associated with the increase in the applied energy. This is consistent with findings elsewhere in the literature [2].

The higher applied impact energy required to perforate the BPP composite can be explained in terms of the fibre toughness. Woven fabric inhibits the initiation of interlaminar cracks. The fabric composites offer excellent resistance to impact damage through the cross-over points, which act as stress distributors. The impact loading applied on the fabric composites beyond the perforation threshold energy level produces crack initiation within the ply in the fabric composites, which begins to propagate through the thickness, but has to cut through the fibre. Unless the energy available is high enough to fracture the fibre tow, the crack is arrested [18].

It is widely accepted that an improvement in the impact properties of polymer composites with high stiffness reinforcements, such as glass, can be obtained by mixing them with tougher and higher energy absorption reinforcements such as natural fibres. There have been no studies reported in the literature on the impact response of bamboo fabric and glass unidirectional reinforced PP hybrid composites.

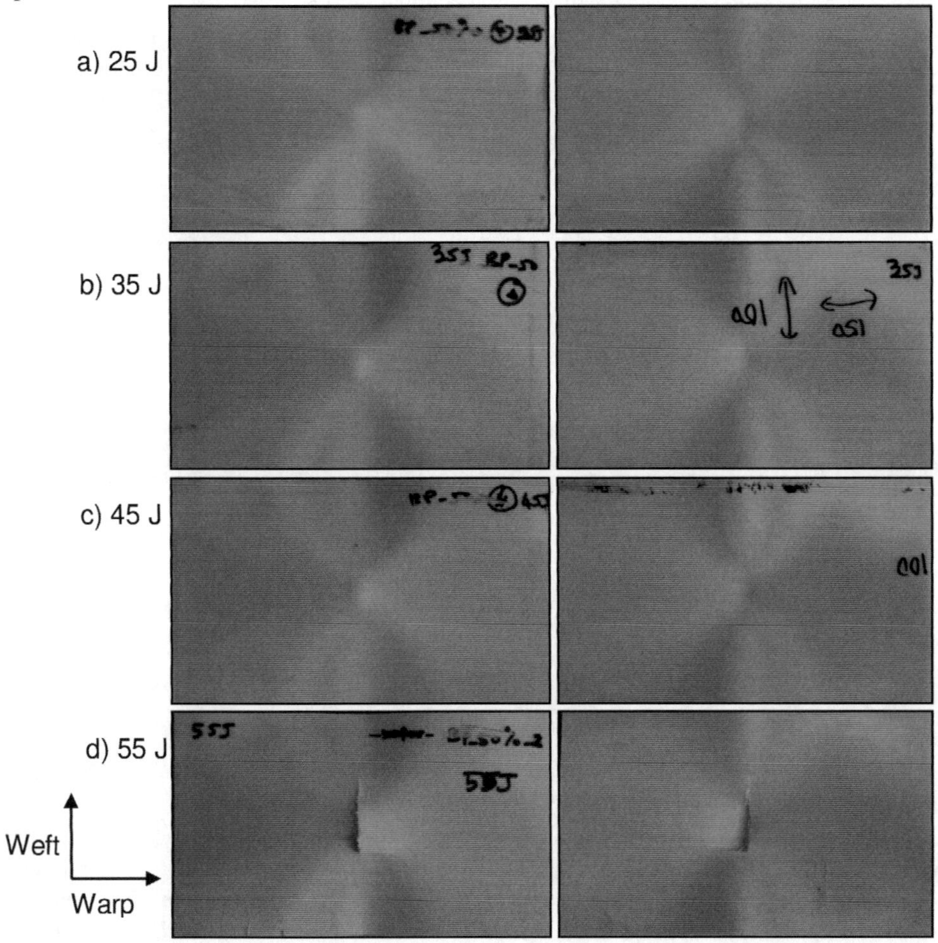

Figure 5. Photos of damage patterns (front and back view) in the BPP composites impacted at different applied energies when tested using the toggle clamp

3.2. Effect of hybridization

A summary of the impact properties computed for the BGPP hybrid composites for different energy levels when tested using the toggle clamps is summarised in table 2. The BGPP hybrid composites were not penetrated with the applied energy of 25 J. The hybrid composites started to be penetrated at the impact energy of 35 J. It can be seen that the peak load and energy absorbed increase with the increase in the applied impact energy. Typical impact force versus displacement of the hybrid composites as a function of applied impact energy is shown in figure 6. A higher impact force and energy was absorbed in the composites for higher applied energy. No significant difference observed as the applied energy increased.

Photos of damage patterns in the hybrid composites impacted at different applied energies are shown in Figure 7. It can be seen that perforation of the composites occurred at 35 J. The applied energy was then increased to 45 J and 55 J to observe the damage area. The impact damage in the hybrid composites at 35 J is mainly in the form of delamination, matrix cracking and fibre breakage. There was no difference in the length of cracks and the size of the damage area with the increase in applied energy. The crack was initiated at the bamboo-glass interface.

Table 2. Impact properties of the BGPP hybrid composites at different applied energies

Applied Energy (J)	Peak Load (kN)	Energy absorbed (J)	Displacement to max. load (mm)	Impact responses
35	2.015±0.1	10.08±0.1	33.68±0.5	Perforation
45	2.216±0.2	12.78±0.1	34.78±0.3	Perforation
55	2.366±0.8	13.44±0.2	35.99±0.1	Perforation

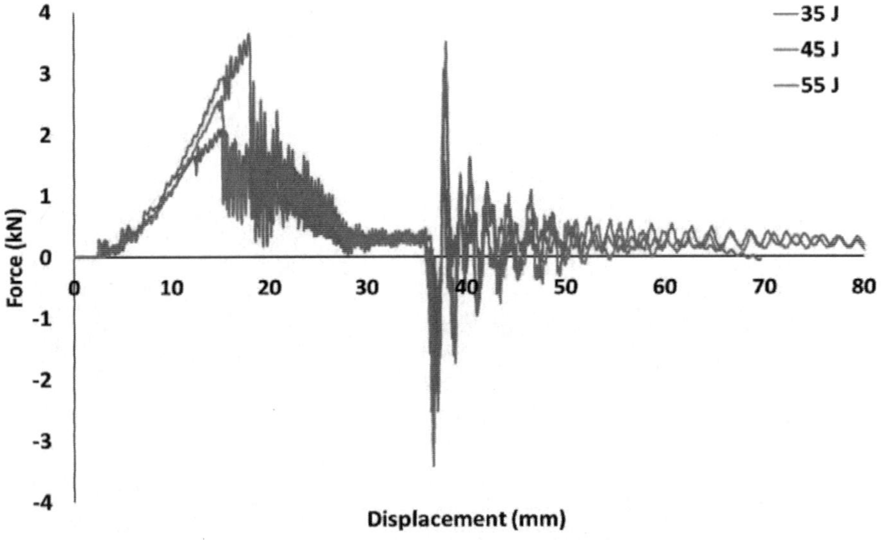

Figure 6. Typical impact force versus displacement of the BGPP hybrid composites at different applied impact energies when tested using the toggle clamps

For the effect of hybridisation, the force versus displacement for the hybrid composites was similar with that observed in the BPP composites. A gradual dropping off of force led to much higher deflection values. The bamboo fabric offers excellent resistance to impact damage. The peak force increased with the increase in the applied impact energy. When compared with BPP composites, the hybrid composites withstand a higher peak force but lower energy absorbed. This increase in peak force is due to the greater stiffness of the glass plies. The higher energy absorbed in the BPP composites than that in the hybrid composites, indicates that the BPP composites have better energy absorption than the hybrid composites. The displacements in the BPP and hybrid composites are comparable and it depends mainly on the stiffness and thickness of the composites and applied impact energy. The hybrid composites were perforated at 35 J applied energy, compared to 55 J for the BPP composites. The damage was greater for the BPP composites, indicating the better damage resistance capability of the hybrid composites. The hybrid composites exhibit a greater peak load resistance than the BPP composites. The pattern of impact damage in the hybrid composites was strongly affected by the mechanical properties of the glass. Because of the fibre alignment in the glass, the delamination tended to occur when the crack and fibre failures were found along the glass fibre direction. The presence of bamboo fabric in the hybrid composites reduced the amount of delamination, although the hybrid composites displayed a lower energy absorbed than the BPP composites. From the results obtained, the hybrid composites give better impact resistance in terms of much higher peak force. The glass layer in the hybrid composites effectively withstood greater impact peak load than the BPP composites. This result is consistent with previous hybridisation work by Ahmed et al. [2] and Sarasini et al. [19].

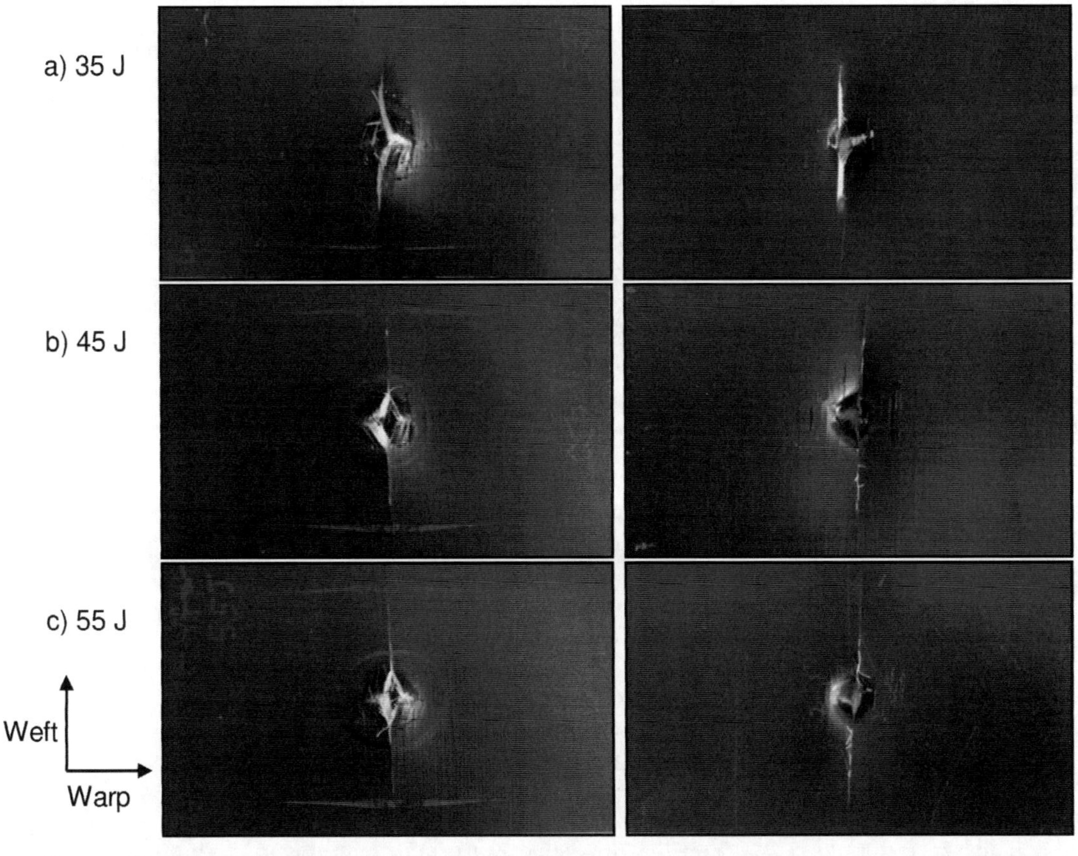

Figure 7. Photos of damage patterns (front and back view) in the hybrid composites impacted at applied energies and tested using the toggle clamps different

Hybridisation of high performance fibres has been reported to be an efficient way to improve the impact performance of composites. According to Santulli et al. [20], a hybrid laminate had much more impact resistance (up to four times for the same laminate thickness) when impacted on the glass side. It was thus suggested that the sandwich configurations with glass fibre in the skin and natural fibre in the core can be considered as the most suitable configuration for high impact resistance. They identified the following factors for development of natural fibre/glass hybrid bio-composites: larger fibre volume fraction, improved effectiveness of the interface in dissipating impact damage, and modification of the configuration to improve impact properties. This phenomenon occurred because the interlaced yarns resulted in an uneven and crimped surface of the woven fabric, which restrained of the growth of delamination caused by mismatching of the bending stiffness. The delamination propagated along the glass fibre direction. According to Acha et al. [21], the radial cracking mechanism is less effective in terms of toughness (at least when the cracks are isolated), however, in this study, the toughness of the bamboo fabric was higher than that of the glass fibres. This may increase the energy absorbed during impact. Another reason which contributed to the higher energy absorption was the deformation mechanism such as fibre pull-out and debonding, which influenced the damage behaviour and led to higher energy absorbed.

4. Conclusion

The study was undertaken to evaluate the effect of hybridisation on impact resistance of the composites. The applied energy of the impact was then increased until penetration was achieved at 55 J for the BPP composite. It can be seen that the peak load and energy absorbed increase with the increase in the applied impact energy. The perforation impact energy was at 55 J for the BPP composite, compared to that of neat PP, at 20 J. For the BGPP hybrid composites, they were not penetrated with the applied energy of 25 J, however, were penetrated at the applied energy of 35 J. The crack damage in the BPP composites was also reduced with the presence of fabric reinforcement. The damage was greater for the BPP composites, indicating the better damage resistance capability of the hybrid composites. The hybrid composites exhibit a greater peak load resistance than the BPP composites. The hybrid composites exhibit a greater peak load resistance than the BPP composites. This evidence suggests that the possibility that hybridisation can provides the potential of achieving a desired combination of properties. This research will serve as a base for future studies and makes several noteworthy contributions to the usage of bamboo fabric composites as alternative natural reinforcement and a more cost-effective usage of synthetic fibres can be produced by replacing the partially with the bamboo fabric.

Acknowledgement

This research is supported by the Universiti Kuala Lumpur Malaysia Institute of Aviation Technology (UniKL MIAT), through a doctoral scholarship provided to the first author. Thanks also go to the staff at the Center for Advanced Composite Material (CACM), The University of Auckland.

References

[1] Richardson M O W and Wisheart M J 1996 *Compos. Part A: Appl Sci Manu.* **27**(12): p. 1123-1131.
[2] Ahmed K S, Vijayarangan S, Kumar 2007 *J Reinf Plast Comp.* **26**(10): p. 959-976.
[3] Aslan Z, Karakuzu R and Okutan B 2003 *Compos Struct.* **59**(1): p. 119-127.
[4] Trudel B D, Bureau M N, Denault J and Fisa B 2003 *Polym Composite.* **24**(4): p. 499-511.
[5] Jacob G C, Fellers J F, Simunovic S and Starbuck J M 2002 *J Compos Mater.* **36**(7): p. 813-850.
[6] Santulli C, Brooks R, Long A C, Warrior N A and Rudd C D 2002 *Plast Rubber Compos.* **31**(6): p. 270-277.
[7] Alcock B, Cabrera N O, Barkoula N M and Peijs T 2006 *Compos Sci Technol.* **66**(11): p. 1724-1737.
[8] Dhakal H N, Skrifvars M, Adekunle K and Zhang Z Y 2014 *Compos Sci Technol.* **92**(0): p. 134-141.
[9] Bledzki A and Faruk O 2003 *Appl Compos Mater.* **10**(6): p. 365-379.
[10] Kannan T G, Wu C M, Cheng K B and Wang C Y 2013 *J Ind Tex.* **42**(4): p. 417-433.
[11] Garkhail S K, Heijenrath R W H and Peijs T 2000 *Appl Compos Mater* **7**(5-6): p. 351-372.
[13] Benevolenski O I, Karger-Kocsis J, Mieck K P and Reubmann T 2000 *J Thermoplast Compos Mater.* **13**(6): p. 481-496.

[14] Morye S S and Wool R P 2005 *Polym Composites.* **26**(4): p. 407-416.
[15] Pandita S D, Yuan X, Manan M A, Lau C H, Subramanian A S and Wei J 2014 *J Reinf Plast Comp*. **33**(1): p. 14-25.
[16] Atas C and Sayman O 2008 *Compos Struct.* **82**(3): p. 336-345.
[17] Zuhudi M N Z, Jayaraman K and Lin R T J 2016 *Polym Polym Compos.* **24** (9)
[18] Padaki N V, Alagirusamy R, Deopura B L, Sugun B S and Fangueiro R 2008 *Ind J Fibre Tex Res.*
[19] Sarasini F, Tirillò J, Valente M, Ferrante L, Cioffi S, Iannace S and Sorrentino L 2013 *Mater Des*. **49**(0): p. 290-302.
[20] Santulli C, Janssen M and Jeronimidis G 2005 *J Mater Sci.* **40**(13): p. 3581-3585.
[21] Acha B A, Marcovich N E and Reboredo M M 2005 *J Appl Polym Sci*. **98**(2): p. 639-650.

Mechanical and Thermal Characterization of Alkali Treated Kenaf Fibers

S A S Abdullah[1*], N Z M Zuhudi[1], NIS Anuar[2], M D Isa[1]

[1]Universiti Kuala Lumpur, Malaysian Institute of Aviation Technology, 43800 Dengkil, Selangor, Malaysia
[2]Malaysian Timber Industry Board, 56100 Cheras, Kuala Lumpur, Malaysia

*Corresponding author: sfathinadlina@unikl.edu.my

Abstract. Research on bio composite for automotive and aerospace application has been extensive with the advancement of natural fiber yarn and woven technology. Malaysia has marked kenaf as its main crop commodity by 2020. Surface modification of natural fibers is one of the significant areas in current biocomposite research. Alkali treatment removes certain amount of lignin, hemicellulose, and wax on the surface of fiber, besides depolymerizing cellulose structure and increasing percentage of crystallinity. Surface modification with NaOH of 3%, 6% and 9% concentration with various lengths of immersion time was conducted. The effect of alkali treatment on the mechanical strength and thermal degradation of kenaf fibre were investigated by means of fiber bundle tensile test and thermogravimetric analyser (TGA). Alkali treatment strongly modifies the thermal behaviour of the fibers, being particularly effective in the removal of noncellulosic matter. In addition, the mechanical properties of kenaf fibers revealed higher tensile strength for NaOH treated fibers.

1. Introduction

Manufacturing high performance composites from natural fibres is actively pursued by researchers across the globe. Automotive and aerospace industry has started on natural fiber composite since 1999, with Kenaf fibers as seat backs, packaging trays, door inserts, load floors and pillars, in vehicle interiors [1]. Boeing developed sidewall panel for a 737 aircraft made of flax-thermoset composite in 2013[2]. BMW i3 with its lightweight design produced side panel trim and instrument panel made of hemp in 2013, and Kenaf in random mat form in 2014 [3].

Kenaf is the name given to fiber from the bast of stems of plants of the genus Hibiscus, family Malvaceae and species H. cannabinus L. In Malaysia the government is heavily promoting the development of kenaf, as the next major industrial crop for the country in line with its policy in developing new sources of economic growth. Kenaf has a short gestation period of only four months and is a high carbon dioxide absorbent [4]. Kenaf growth does not need pest control and absorbs chemical and heavy metals from the soil. It also has a wider range of adaptation to climates and soils than any other fiber plant in commercial production [4].

Not unlike other natural fibers, kenaf as plant fiber offers the undisputable advantage of renewability for which production requires little energy and involves carbon dioxide absorption whilst returning oxygen to the environment. It also has the advantage of low hazard manufacturing processes and less abrasion to processing equipment, besides involving much less emission of toxic fumes when subjected to heat as compared to current synthetic materials [5]. However, the main disadvantages of natural fiber as composite reinforcements are the poor compatibility between fiber and matrix, flammability and high moisture sorption [6].

This leads to an important area in natural fiber composites which is pre-treatment of fibers. Alkaline treatment commonly involves sodium hydroxyde (NaOH) reacting with the hydroxyl group of the natural fibers, removing hemicellulose, minimal amount of lignin, wax and oil that surround the external surface of the fiber, leading to increase in surface roughness [7]. This facilitates mechanical interlocking which leads to an improvement in fiber-matrix adhesion. [8]. In addition, alkalisation caused fiber fibrillation i.e. axial splitting of the elementary fibers, leading to a decrease in fiber diameter, increasing the aspect ratio and the effective surface area available for wetting by a matrix in composite [9]. High moisture content in fibers is attributed to presence of hemicellulose, and removal of hemicellulose by alkalization has been reported to reduce moisture absorption in natural fiber [10]. Alkalization also improved the strength of individual fibers. Removal of hemicelluloses cause the inter fibrillar regions of the fiber to becomes less dense and rigid, hence fibrils are more capable of rearranging themselves along the direction of tensile deformation, promoting even load distribution in the fibers and reduces stress concentration [11]. Edeerozey et al immersed kenaf in NaOH solution with different concentrations (3%, 6% and 9% NaOH) for 3 hours at room temperature and fiber bundle tensile test showed NaOH 6% giving the highest strength[12]. Changes in chemical constituents of natural fiber due to alkalization improved the strength and moisture content of fiber and composites fabricated from.

Natural fibers are combustible in nature. When exposed to fire or high intensity heat source, polymer matrix and fibers will thermally decompose to yield flammable and non-flammable volatile gases [13]. The released flammable volatile gas will react with surrounding oxygen and form the final combustion product accompanied by heat release, which will continue the combustion cycle [14]. Literature have reported that variations in the flammability of different plant fibers can be attributed in part to the chemical composition of the fibers [15]. Cellulose is known to have high flammability due to its high crystallinity, whereas hemicellulose and lignin improves fire resistance; hemicellulose produced non-combustible gas whilst lignin produced char that reduces the amount of flammable volatiles released, slowing the chain branching reactions in the combustion cycle [15]. Hence alkalization affects flammability with the change in chemical composition it caused. Thermal degradation condition is crucial in the understanding of flammability properties of material. Thermal stability is one of the means by which natural fiber develop their inherent flame resistant [28].

Thermal degradation of natural fibers has always been a crucial factor for the development of natural fiber composites. The majority of natural fibers have low degradation temperatures of below 200°C due to its cellulosic nature [16]. Alkalization has been reported to increase thermal stability as reported by Guduri et al. on flax fibers [17]. Liu et al. reported that the temperature at the maximum rate of decomposition obtained through weight percentage and derivative weight percentage for the Indian grass fibres increased with the increase in alkali concentration, thereby resulting in the improvement of the thermal stability of the fibre after alkali treatment [18]. However there has been contradicting results, as reported by Puglia et al. who indicated that thermal stability of the alkali-treated flax fibres (Phormium tenax) decreased slightly as shown by the earlier degradation temperature as compared to untreated flax fibres [19]. Similar results have been reported by Zhu et al. based on their experiments on also flax fibres [20]. Research in thermal degradation of alkali treated kenaf composite is scarce, even more so on the kenaf fiber itself.

This research work aims to study the effect of alkali treatment on the thermal degradation and mechanical strength of kenaf fibers. This preliminary study will lead to an in depth understanding of fire and mechanical behaviour of alkali treated kenaf composites, with regards to the main aim of this research which is the manufacture-ability, durability and other functional properties of kenaf composites.

2. Samples preparation and experimental procedure

2.1. Materials selection

IOP Conf. Series: Materials Science and Engineering **370** (2018) 012048 doi:10.1088/1757-899X/370/1/012048

Kenaf fibre with enzyme retting extraction grade was obtained from Dynamic Agrofarm Sdn Bhd (Pahang, Malaysia) at the harvesting age of 4 months. The extraction method was conducted by first spraying the kenaf stem with pectinase-rich enzyme and letting it rest for 24 hours. This was followed by manually peeling the kenaf bast from the core, and soaking it in a tank of water with the enzyme for 3 days, and finally dried under sunlight for 24 hours. The kenaf fibers were cut into an average length of 100 mm.

2.2. Alkali treatment
Sodium hydroxide (analytical grade) pellets purchased from Merck were prepared of 3%, 6% and 9% concentration. The kenaf fibers of 100 mm length were separated into ten groups of 20 grams each and each group (except one) was immersed into the three different NaOH concentration for duration of 1, 3 and 24 hours. The fibers were then washed with running water mixed with 0.1% acetic acid until a neutral pH is achieved, dried in a hot air oven at 60◦C for a period of 24 hours to remove the excess moisture.

Table 1. Designation of kenaf fibers.

Designation	Treatment	Soaking Time
Raw	-	-
NaOH3%1hr	NaOH 3%	1 hour
NaOH3%3hr	NaOH 3%	3 hours
NaOH3%24hr	NaOH 3%	24 hours
NaOH6%1hr	NaOH 6%	1 hour
NaOH6%3hr	NaOH 6%	3 hours
NaOH6%24hr	NaOH 6%	24 hours
NaOH9%1hr	NaOH 9%	1 hour
NaOH9%3hr	NaOH 9%	3 hours
NaOH9%24hr	NaOH 9%	24 hours

2.3. Fiber bundle tensile test
Kenaf fiber exists in fiber bundles. In accordance with ASTM D 3822, the tensile properties of fiber were determined with Shimadzu Ez-L10NX fiber tensile strength testing machine. The sample was set at fiber gauge length of 10 mm at the crosshead speed of 5 mm/min, and tested as in figure 4. For each of the ten sets of sample as in table 1, 20 fiber bundles with diameter less than 60 μm were chosen and tested individually to determine the average fiber bundle strength. Prior to the testing, average fiber bundle diameter were obtained from biological microscope.

Figure 1. Fiber tensile strength testing machine.

2.4. Thermogravimetric analysis (TGA)

Thermogravimetric analysis is a thermal analysis techniques used to measure the mass change, thermal decomposition and thermal stability of material. Thermal gravimetric analysis was conducted using TGA Q500 machine. Samples weighing approximately 5 mg were subjected to pyrolysis in nitrogen environment to a maximum temperature of 650 °C at a heating rate of 20 °C/min. The weight loss was recorded in response to increasing temperature, with final residue yield on set of degradation temperature and number of degradation steps reported.

3. Results and discussion

3.1. Physical properties

Kenaf fibers exist in a fiber bundle form. This is observed when the fiber was partially pulled apart. Figure 1 shows a visual representation and the magnified image of untreated kenaf fiber in its fiber bundle form obtained from a microbiological microscope of 100 μm magnification. From literature, micro scale fiber bundle exists in 50-100μm diameter [21]. A single kenaf fiber can only be obtained by breaking down the kenaf fiber bundles into elementary fibers, with diameter of 10-20 μm [21].

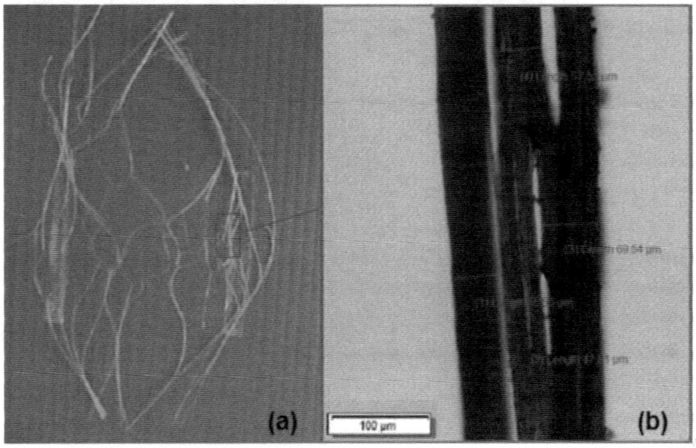

Figure 2. (a) kenaf fiber bundle (b) kenaf fiber bundle magnified at 100x.

For the purpose of studying the fiber bundle strength, only those with diameter between 40 to 65 μm were chosen. The diameter of each fiber was measured at three different random locations to obtain the average value, as shown in figure 2. Fibers of more than 65 μm were observed to be observably intertwined and were separated for tensile strength test.

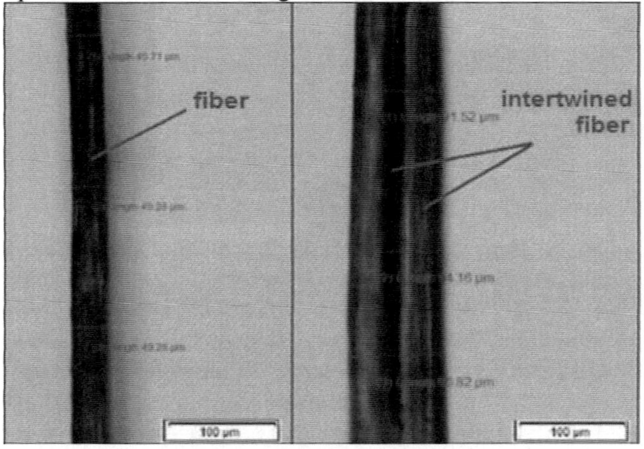

Figure 3. Magnified fiber bundle image.

It was observed that as the alkali concentration increases, the fiber bundles were easier to separate, demonstrating partial removal of hemicellulose and lignin as binding agent had occurred at the lower NaOH concentration and a more complete removal had occurred at the higher NaOH concentration. Fibers treated with higher concentration also showed a more severe physical change. Figure 3 shows a comparison of raw and kenaf fiber treated with concentration of 3%, 6% and 9% treated for 24 hours. During alkalization, almost all of hemicellulose and a certain amount of lignin, wax and oil covering the external surface of the fiber cell wall are removed, thus causing the fibers to appear dry and bristly. Hemicellulose is a semi crystalline polymer of glucose that acts as a binding agent to the cellulose strand. Hemicellulose is covered with lignin, an amorphous polymer which also acts as a binder. Removal of hemicellulose and partial removal of lignin caused the kenaf fibers to loose most of its binding agent [22]. Figure 3 showed a comparison of physical change shown by kenaf fibers soaked for duration of 24 hours. Samples soaked at the lesser time showed the same rough and bristly condition but a lesser severity.

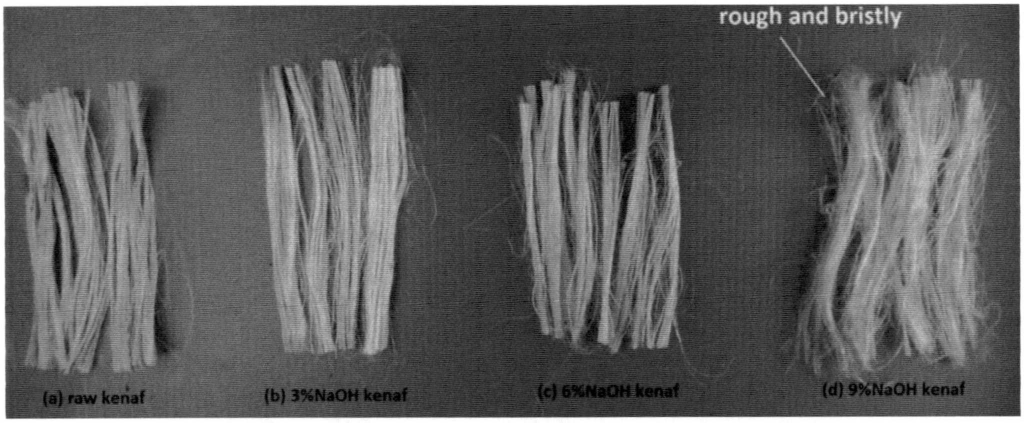

Figure 4. Untreated and NaOH treated kenaf fibers (24 hours).

3.2. Fiber bundle strength test

The overall results for kenaf fibers treated for soaking time of 1 hour, 3 hours and 24 hours were summarized in table 2. All three soaking time shows a similar pattern, which is an increase of average maximum tensile strength of the fibers with increasing concentration of NaOH, up to 6% NaOH. However, NaOH concentration of 9% show a dramatic drop of strength, with value recorded even lower than that of the untreated fibers. Figure 5 shows tensile strength of untreated and treated kenaf fibers of various NaOH concentrations treated for 1 hour.

Table 2. Average tensile strength (MPa).

Soaking Time	Raw	NaOH 3%	NaOH 6%	NaOH 9%
0 hour	330.42	-	-	-
1 hour	-	358.89	407.72	309.84
3 hours	-	367.39	412.29	297.65
24 hours	-	330.25	386.37	267.54

This increased tensile strength with the increase of NaOH concentration may be explained by the improved uniformity of cellulose structure after the removal of the amorphous and semi-crystalline hemicellulose and lignin [18]. The aligned fibrils promotes an even load distribution in the fibers, and a higher crystallinity index leads to a better packing of cellulose chains [11]. The dramatic drop of strength could be the result of the damage of fiber cell wall caused by high concentration of NaOH, as a similar scenario was observed by P. Methacanon [10]. As reported by Mwaikambo and Ansell [23], a very high concentration of NaOH would certainly damage the fiber and consequently reduce the tensile strength of the fiber.

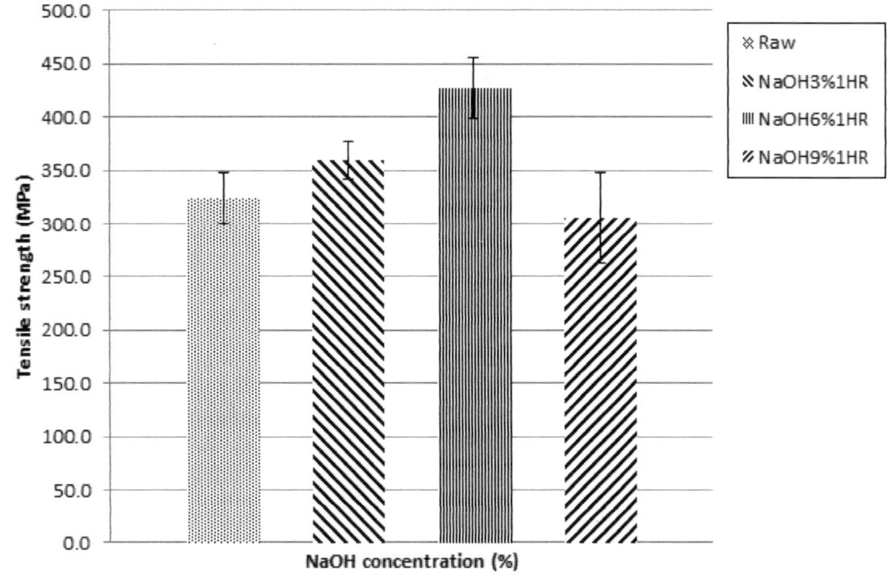

Figure 5. Tensile strength of untreated and treated kenaf fibers of various NaOH concentrations treated for 1 hour.

3.3. Thermogravimetric analysis (TGA)

Results from the TGA run is as shown in figure 6, and a clearer presentation of degradation is as presented in the derivative thermogravimetric graph (DTG) in figure 7. Peaks represent the rate of weight degradation (%/min). Untreated kenaf fibers started to lose weight earlier than the other samples as shown by the curve below 100°C. This is attributed to the higher moisture content on untreated fibers due to the presence of hemicellulose.

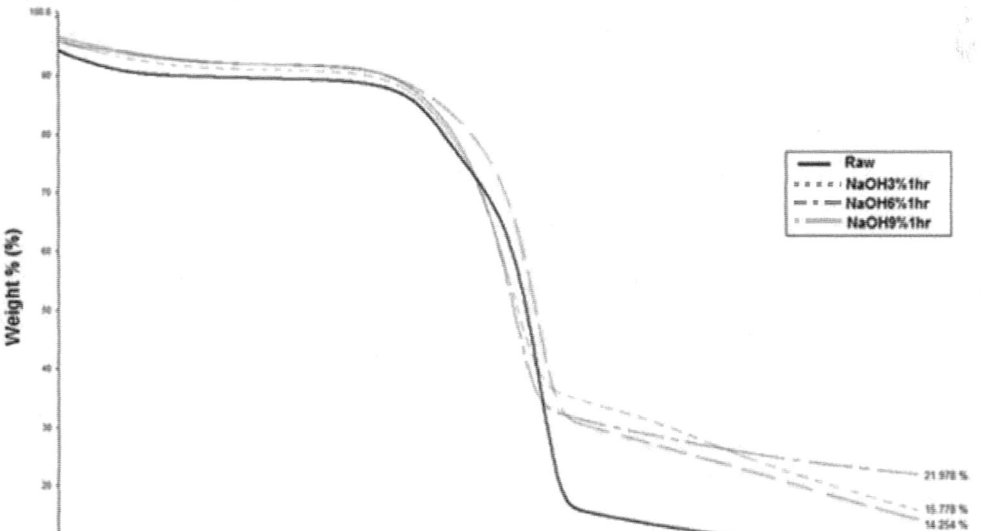

Figure 6. TGA of raw and treated kenaf fibers.

Initial result on thermal stability is characterized by the temperature at which 15% weight loss occurred, referred to as $T_{15\%}$. The $T_{15\%}$ for raw kenaf is 292.78°C as compared to 308.57°C for kenaf treated with NaOH 9%, with NaOH 3% and NaOH 6% having $T_{15\%}$ values in between. The higher temperature for 15% weight loss is an early indication of thermal stability of treated as compared to the untreated kenaf fibers [24].

Table 3. Data analysis from TGA and DTG.

Designation	Treatment	$T_{15\%}$ (°C)	Peak height (maximum weight loss)(wt%/min)	Char residue at 636.4°C (%)	Total weight decomposition (%)
Raw	NIL	292.78	30.653	7.526	65.52
NaOH3%1hr	NaOH 3%	297.82	18.195	15.778	48.18
NaOH6%1hr	NaOH 6%	301.3	22.275	21.978	59.98
NaOH9%1hr	NaOH 9%	308.57	25.724	14.254	59.55

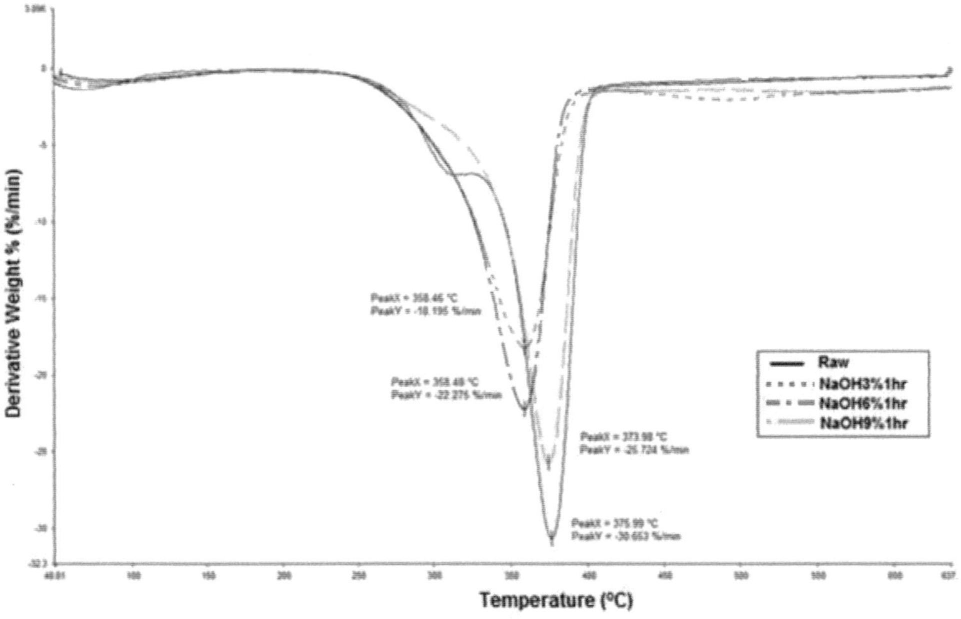

Figure 7. DTG of raw and treated kenaf fibers.

DTG curve showed that whilst treated kenaf fiber exhibits single step degradation, raw kenaf fibers displayed twostep degradation, with the first peak occurring at 270 °C. In this temperature range, degradation is associated with hemicellulose and some parts of lignin. Treated kenaf fibers did not experience loss of weight at 270°C, which is explained by the removal of hemicellulose during alkalization [25]. DTG graph of fibers treated with 9% NaOH showed a complete removal of hemicellulose as no peak is observed.

The second stage is associated with degradation of cellulose and lignin. Peaks at the range of 320 to 375°C exhibits decomposition of cellulose, and a wider range of up to 420°C is for the decomposition of lignin [23]. DTG curve showed NaOH treated fibers (3% and 6%) with an earlier degradation temperature than untreated fibers. This indication of thermal instability might be due to the removal of fiber cell walls, which reduce the thermal protection of the treated fibres. The exposed cellulose fibrils experienced direct heat without any layering from the cell wall [26]. Cellulose will degrade through a process called pyrolisis, forming levoglucosan which will decompose to produce flammable volatiles [14]. Another factor causing thermal instability in treated fiber is the crystallinity. For cellulosic fibers in particular, higher levels of crystallinity results in higher levels of levoglucosan during pyrolysis and consequently increased flammability [15]. This is in accordance to reports of increased flammability for fibers with higher cellulose content [15].

The peaks on the DTG represent the maximum rate of weight loss. Treated fibers have a much lesser rate of degradation due to the removal of most cementing components. By treating the fibers with NaOH 3%, the maximum rate of weight loss is reduced by almost half from 30.653 wt%/min to 18.196 wt.%/min, indicating thermal stability [27]. Similarly, total decomposition weight, as obtained from the area under the DTG curve, also showed a lesser decomposition prior to pre-treatment of fibers. However, the degradation rate is observed to increase as NaOH concentration used increased. This may be due to the removal of lignin, which leads to lesser char in the fibers to retard degradation.

4. Conclusion

Kenaf fibers were treated at various concentrations of sodium hydroxide and soaking time to study the effect of alkali treatment on the physical, mechanical and thermal properties of kenaf fibers. From this study, it can be concluded that:

1. Tensile properties of fiber bundle increase with NaOH treatment of up to NaOH concentration of 6%. Removal of impurities improves the uniformity of the fibers resulting in an increase in strength. However a high concentration of 9% will degrade the strength of fiber cell walls. Duration of soaking of 1 to 3 hours is sufficient for a good treatment.
2. Absence of hemicellulose in treated fibers is observed by the one step degradation instead of a two-step degradation depicted by raw fibers. The peak at 270°C correlates with the decomposition temperature of hemicellulose.
3. Treated fiber showed a slower rate of degradation and reduced total weight decomposition due to the absence of hemicellulose and lignin.
4. Exposed cellulose structures and diminishing fiber cell wall of alkali treated fibers accelerated degradation process. High crystallinity of treated fiber also caused formation of levoglucosan, decreasing thermal stability in fibers. This showed that fire resistance of fibres may not be improved with alkali treatment. Therefore, it is suggested that further study on the effect of fire retardant on thermal stability of the kenaf fibres is conducted.

References

[1] Hassan F, Zulkifli R, Ghazali MJ and Azhari CH 2007 *Kenaf Fiber Composite in Automotive Industry : An Overview* **7** 315–21
[2] Eldridge D. 2013 *Eur Plast News* 45–7
[3] Schmiedel I and Barfuss GSN 2013 *ATZ Worldwide* **116** 20–3
[4] Karimi S, Tahir P, Karimi A, Dufresne A and Abdulkhani A 2014 *Carbohydrate Polymers* **101** 878–85
[5] Pickering KL, Efendy MGA and Le TM 2016 *Compos Part A Appl Sci Manuf* **83** 98–112
[6] Li X, Tabil ÆLG and Panigrahi ÆS 2007 *J Polym Env* **15** 25–33
[7] Chandrasekar M, Ishak MR, Sapuan SM, Leman Z and Jawaid M. 2017 *Plast Rubber Compos* **46** 119–36
[8] Amel BA, Paridah MT, Sudin R, Anwar UMK and Hussein AS. 2013 *Industrial Crops and Products* **46** 117–23
[9] Exequiel S. Rodríguez and Analía Vázquez 2006 *8th Int Conf Flow Process Compos Mater* 105–12
[10] Methacanon P, Weerawatsophon U, Sumransin N, Prahsarn C and Bergado DT. 2010 Carbohydr Polym **82** 1090–6
[11] Symington MC, Banks WM, West OD and Pethrick RA. 2009 *J Compos Mater* **43** 1083–108
[12] Edeerozey AMM, Akil H, Azhar AB and Ariffin MIZ 2007 *Materials Letters* **61** 2023–5
[13] Mngomezulu ME, John MJ, Jacobs V and Luyt AS. 2014 *Carbohydr Polym* **111** 149–82
[14] Bhattacharyya D, Subasinghe A and Kim NK 2015 C*omposites Part A: Applied Science and*

Manufacturing **65** 91-9
[15] Chapple S and Anandjiwala R 2010 *J Thermoplast Compos Mater* **23** 871–93
[16] Araújo JR, Waldman WR and De Paoli M 2008 *Polym Degrad Stab* **93** 1770–5
[17] Guduri B, Semosa H and Meng Y 2009 *Proc 17th ICCM Int Conf Compos Mater Edinburgh* 927–31
[18] Liu W, Mohanty AK, Drzal LT, Askel P and Misra M. 2004 *J Mater Sci* **39** 1051–4
[19] Puglia D, Monti M, Santulli C, Sarasini F, De Rosa IM and Kenny JM 2013 *Fibers Polym* **4** 423–7
[20] Zhu J, Brington J, Zhu H and Abhyankar H 2015 *J Sci Res Reports* **4** 1–11
[21] Jiang W, Han G, Zhang Y, Liu S, Zhou C and Song Y 2017 Sci Rep **7** 1240
[22] Araújo JR, Waldman WR and De Paoli MA 2008 *Polymer Degradation and Stability* **93** 1770-75
[23] Mwaikambo LY and Ansell MP 2002 *J Appl Polym Sci* **84** 2222–34
[24] Faseha S, MS Islam, M Hasan and Hassan MM 2014 *Mater Des* **54** 425–9
[25] Razali N, Salit MS, Jawaid M, Ishak MR and Lazim Y 2015 *BioResources* **10** 1803–23
[26] Azwa ZN and Yousif BF 2013 *Malaysian Postgrad Conf* 256–64
[27] Azwa ZN and Yousif BF 2013 P*olymer Degradation and Stability* **98** 2752–9
[28] Harrocks A R 1983 *JSDC* **99** 191-7

IOP Conf. Series: Materials Science and Engineering **370** (2018) 012049 doi:10.1088/1757-899X/370/1/012049

Investigation of micro burner performance during porous media combustion for surface and submerged flames

Ayub Ahmed Janvekar[1*], M. Z. Abdullah[2], Z. A. Ahmad[3], Aizat Abas[1], Ahmed A. Hussien[1], Pramod S. Kataraki[1], Mazlan Mohamed[1,4], Azmi Husin[1,5], Khairil Fadzli[2,6]

[1]School of Mechanical Engineering, Universiti Sains Malaysia, Engineering Campus, 14300 Nibong Tebal, Penang, Malaysia.
[2]School of Aerospace Engineering, Universiti Sains Malaysia, Engineering Campus, 14300 Nibong Tebal, Penang, Malaysia
[3]School of Materials & Mineral Resources Engineering, Engineering Campus, 14300 Nibong Tebal, Penang, Malaysia
[4]Advances Material Research Cluster, Faculty of Bioengineering and Technology, Universiti Malaysia Kelantan, Jeli Campus, 17600 Jeli, Kelantan, Malaysia
[5]Faculty of Mechanical Engineering, Universiti Teknologi MARA, 13500, Penang, Malaysia
[6]Department of Mechanical and Manufacturing Engineering, Universiti Putra Malaysia, 43400, Serdang, Selangor, Malaysia

*Corresponding author: ayubjanvekar@gmail.com

Abstract. Porous media combustion is considered to be one of the popular choice due to its tremendous advantages. Such type of combustion liberates not only super stable flame but also maintains emissions parameters below thresholds level. Present study incorporates reaction and preheat layer with discrete and foam type of materials respectively. Burner was made to run in ultra-lean mode. Optimum equivalence ratio was found out to be 0.7 for surface flame, while 0.6 during submerged flame condition. Maximum thermal efficiency was noted to be 81%. Finally, emissions parameters where recorded continuously to measure NOx and CO, which were under global limits.

1. Introduction

The natural gas industry is booming across the world for more than a decade. Saving combustible liquid fuel has become a key factor to both domestic and industrial market [1]. Additionally, major impact to any nation across the world is due to the super-fast depletion of fossil fuel. So, every possible attempts are made to save fossil fuels in some or the other way. One of the most popular area which consumes large amount of fuels is rigorous usage of burner which fulfills domestic needs [2-4]. Thus, porous medium burners (PMBs) can be taken a better option to adopt so that control over superfast depleting of fossils fuels is retained below threshold level. The major advantage of using porous materials inside burner housing leads to high temperature profiles and lower count in CO and Nox. Recent literature has repeatedly confirmed the PMB are it's the first choice to developed efficient burners [5-7]. Apart of being showing high thermal efficacy they do play vital role in lean flammability limits at higher burning rates [8-11]. Zhou, Huang [12] made at combustion performance and emission behaviors of a typical atmospheric induction gas stove and testing in both plain and plateau regions to evaluate the impact of altitude on combustion characteristics. While Wan, Gao [13] focused on temperature and flame behaviors under the tunnel ceiling for interacting flames. Interestingly, Zhong and Yu [14] and Singh, Hui [15] took butane as source of fuel to perform combustions and noted the surface and gas-phase reactions in microscale catalytic partial oxidation and soot formation in non-premixed counterflow flames respectively. Generally, two types of porous media are used inside the combustion zone namely, discrete and foam. Enabling porous media shall boost the origin of surface and submerged flame in more stable manner. A flame can be so called as submerged if there it is not visible to naked eye. While surface flame can be regarded as convention domestic stove flame.

2. Experimental setup

Fig. 1 indicates burner housing by highlighting major components. A complete layout of the system is as shown in Fig. 2 and actual setup in Fig. 3. Burner was divided into two layer namely reaction and preheat layer with each thickness of 10mm. Reaction zone was made with alumina spheres of 5mm radius. Supply of air-butane was fed from the bottom of the housing. Preheat layer was made of porcelain foam with 8 ppcm and a porosity of 84%. Thermocouple (K-type) was used to measure surface temperature (T1) while surface probe was used to for wall temperature at suitable locations (namely T2, T3, and T4). T1 was placed at mid of reaction zone, T2 at the junction between reaction and preheat layer while T3 at the mid of preheat layer. Thermocouples are connected to DAQ. Thermal images where taken by FLUKE thermal imager. Flow of air and butane was controlled by digital flow meter was measure L/min while emissions were captures using KANE-9106 QUINTOX Combustion Analyzer in ppm (parts per million).

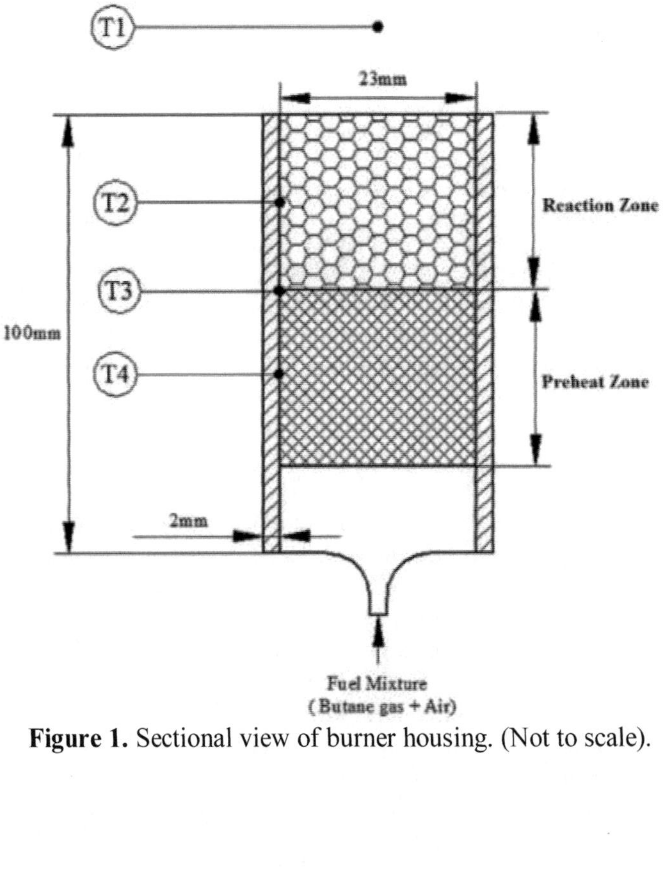

Figure 1. Sectional view of burner housing. (Not to scale).

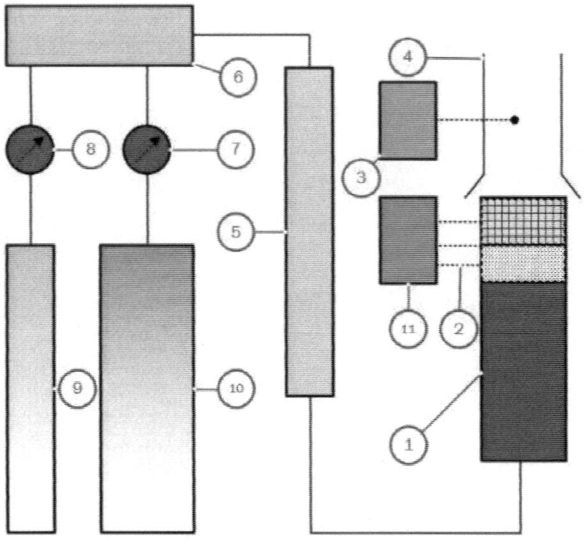

Figure 2. Layout of combustion system. (1) Housing, (2) Thermocouple, (3) Emission gas analyzer, (4) Gas conduit, (5) Mixing unit, (6) Pre-mix, (7) Air flow meter, (8) Butane flow meter, (9) Butane supply, (10) Air inject, (11) DAQ (Data acquisition system).

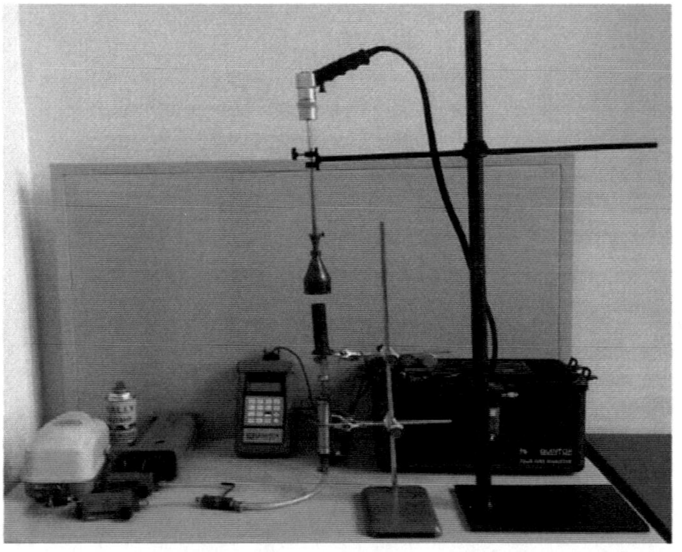

Figure 3. Experimental burner setup.

3. Analysis

Equivalence ratio (ER) is as shown in Equation (1), where the actual air–fuel ratio (Ma) and while stoichiometric air fuel ratio (Ms) for butane gas, was can be calculated by considering Avogadro's and Dalton's laws [16], which was computed a value of 30.95 with air at room temperature as shown in Table 1. Stable flame with minimum fuel consumption was obtained at 0.1 Lpm (litre per min). Trails were performed at room temperature. Maximum surface temperature was noted as 620°C at 0.7 ER, while maximum average temperature was recorded 165oC at 0.6 ER as shown in Fig 4 and 5. The possible reason of burner to undergoes surface or submerged flame mainly depends on availability of oxygen and surface area to perform combustion. Furthermore Fig. 6 indicate the actual photographic images at these respective ER. Thermal images are used to capture the temperature distribution at

IOP Conf. Series: Materials Science and Engineering **370** (2018) 012049 doi:10.1088/1757-899X/370/1/012049

critical ER. With the help thermal imager generated photographic images indicate the sniped overview of temperature variation as shown in Fig. 7 and 8. Thermal efficiency was calculated at ER = 0.6 where in maximum surface flame the was recorded as 81%. Finally, with the help of certified gas analyzer, emitted emissions from the burner surface where fed to gas probe to measure NOx and CO. It was found out that emission where less than 1ppm since the minimum amount to detect the ppm for the utilized gas analyzer is 1pmm.

$$\mathrm{ER} = \frac{M_s}{M_a} \quad (1)$$

Table 1. ER for butane and air flow rate.

Butane (Lpm)			0.1		
Air (Lpm)	3.1	3.4	3.8	4.4	5.1
ER	1.0	0.9	0.8	0.7	0.6

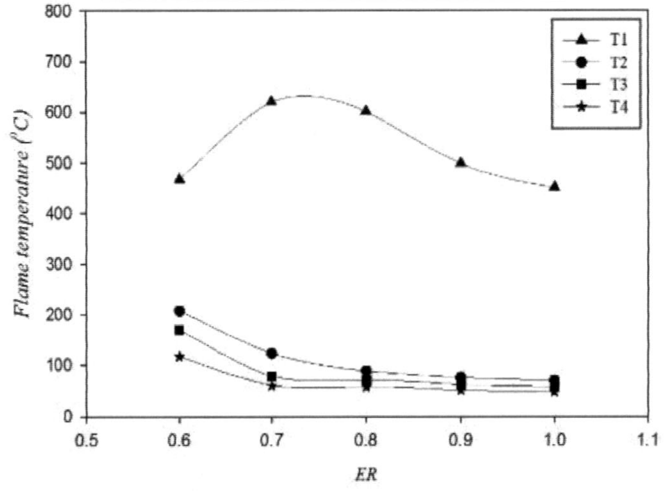

Figure 4. Flame temperature

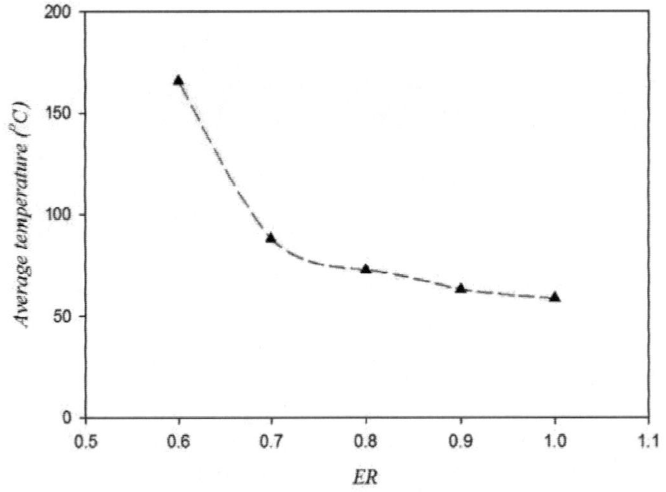

Figure 5. Average temperature

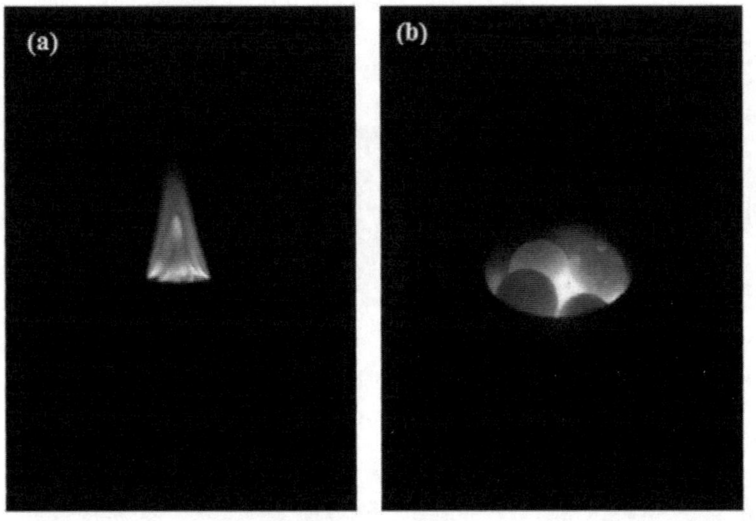

Figure 6. Actual photo

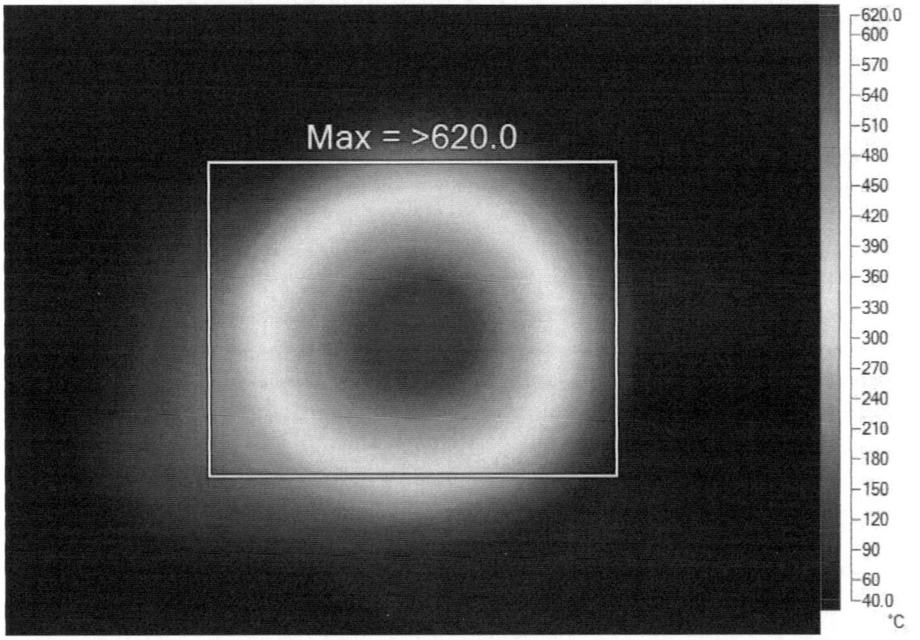

Figure 7. Thermal image (Top view).

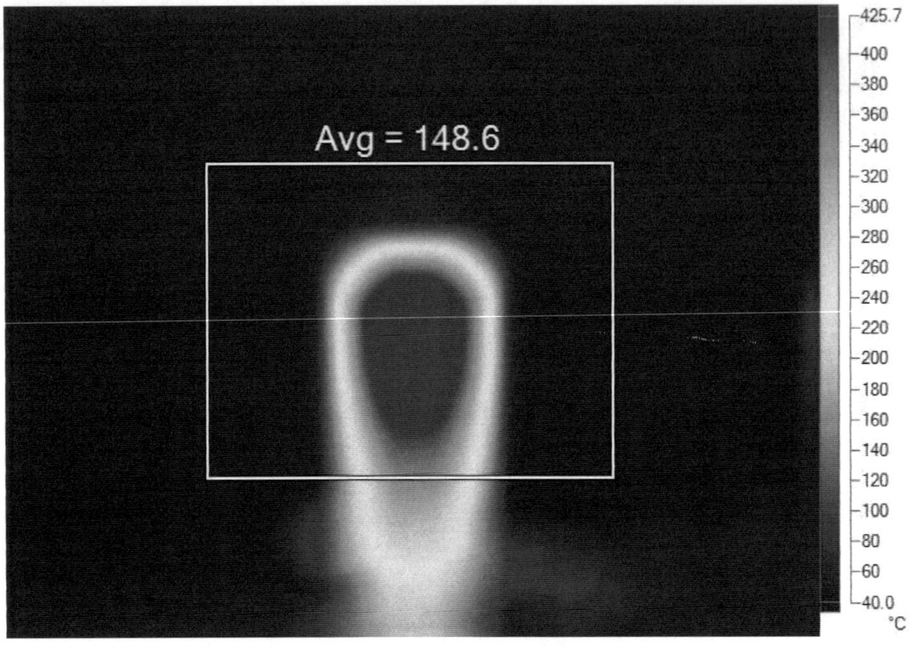

Figure 8. Thermal image (Side view).

4. Conclusions

Presented experimental work deal with clear bifurcation between surface and submerged flame based on ER. Predefined thickness of reaction and preheat layer was used to performed combustion. Optimum ER for surface flames and submerged flame are 0.7 and 0.6 respectively. Maximum temperature recorded at best performing ER was 620°C leading to thermal efficiency of 81%. Emission parameters such NOx and CO where less than 1ppm.

References

[1] Dmitriev, A.M., et al., The effect of methyl pentanoate addition on the structure of premixed fuel-rich n-heptane/toluene flame at atmospheric pressure. Combustion and Flame, 2015. 162(5): p. 1964-1975.

[2] Udayraj, et al., Heat and mass transfer through thermal protective clothing – A review. International Journal of Thermal Sciences, 2016. 106: p. 32-56.

[3] Jiang, X., D. Mira, and D.L. Cluff, The combustion mitigation of methane as a non-CO2 greenhouse gas. Progress in Energy and Combustion Science, 2016.

[4] Arya, P.K., et al., DME blended LPG as a cooking fuel option for Indian household: A review. Renewable & Sustainable Energy Reviews, 2016. 53: p. 1591-1601.

[5] Janvekar, A.A., et al., Effects of the preheat layer thickness on surface/submerged flame during porous media combustion of micro burner. Energy, 2017. 122: p. 103-110.

[6] Janvekar, A.A., et al. Assessment of porous media burner for surface/submerged flame during porous media combustion. in AIP Conference Proceedings. 2017. AIP Publishing.

[7] Janvekar, A.A., et al., Assessment of porous media combustion with foam porous media for surface/submerged flame. Materials Today: Proceedings, 2017.

[8] Daneshvar, H., R. Prinja, and N.P. Kherani, Thermophotovoltaics: Fundamentals, challenges and prospects. Applied Energy, 2015. 159: p. 560-575.

[9] Al-Abdeli, Y.M. and A.R. Masri, Review of laboratory swirl burners and experiments for model validation. Experimental Thermal and Fluid Science, 2015. 69: p. 178-196.

[10] Zanganeh, J., B. Moghtaderi, and H. Ishida, Combustion and flame spread on fuel-soaked porous solids. Progress in Energy and Combustion Science, 2013. 39(4): p. 320-339.

[11] Mujeebu, M.A., et al., Trends in modeling of porous media combustion. Progress in Energy and Combustion Science, 2010. 36(6): p. 627-650.
[12] Zhou, Y., et al., Comparative study on the combustion characteristics of an atmospheric induction stove in the plateau and plain regions of China. Applied Thermal Engineering, 2017. 111: p. 301-307.
[13] Wan, H.X., et al., Experimental study on ceiling gas temperature and flame performances of two buoyancy-controlled propane burners located in a tunnel. Applied Energy, 2017. 185: p. 573-581.
[14] Zhong, B.-J. and Y.-W. Yu, The surface and gas-phase reactions in microscale catalytic partial oxidation of n-butane/air mixture. Fuel, 2016. 186: p. 623-628.
[15] Singh, P., X. Hui, and C.J. Sung, Soot formation in non-premixed counterflow flames of butane and butanol isomers. Combustion and Flame, 2016. 164: p. 167-182.
[16] Ismail, A.K., et al., Application of porous medium burner with micro cogeneration system. Energy, 2013. 50: p. 131-142.

IOP Conf. Series: Materials Science and Engineering **370** (2018) 012050 doi:10.1088/1757-899X/370/1/012050

Passive control of base pressure with static cylinder at supersonic flow

Mohammed Asadullah*, S A Khan, Waqar Asrar, and E Sulaeman

Department of Mechanical Engineering, International Islamic University Malaysia, Kuala Lumpur, Malaysia

*Email: shiblibhai@gmail.com

Abstract. An experimental investigation for over expanded, perfectly expanded and under expanded supersonic jets is presented to study the effectiveness of a static cylinder to reduce base drag at Mach 2 through a converging diverging nozzle for a cross-sectional duct nozzle area ratio of 9. A static cylinder of 2 mm diameter at 2 mm from the side wall of a square duct and 8 mm from the square nozzle exit in the base region is installed as a passive control device. Base pressures in the wake flow after sudden expansion of jets in to a square nozzle have been measured. The length width ratio of the duct is 10. The jets were operated at different nozzle pressure ratios in the range from 2 to 9. The flow field in the square duct was also observed for all cases. Static cylinder as a passive controller was found to reduce the base drag as high as 59 percent at NPR = 9 and 14 percent at NPR = 6. The base pressure depends on Mach number, area ratio, length to width ratio, and nozzle pressure ratio (NPR). When flow from the nozzle is over expanded, the static control is ineffective till NPR is 6. The level of expansion plays a vital role. The flow flux in the square duct remains almost identical with and without control for most of the cases. However, at higher NPRs namely 6, 7.8, and 9 the control results in an increase as well as a decrease in the wall pressure along the duct. Passive control of the base flow is effective for higher NPRs. The flow field for perfectly expanded nozzle is dominated by the shock waves.

Key words: Wall pressure - base pressure - nozzle pressure ratio - passive control

Nomenclature

Abbreviation

M	Mach number
NPR	Nozzle pressure ratio

Roman Symbols

A_1	Exit area of nozzle [mm^2]
A_2	Cross-sectional area of duct [mm^2]
L	Duct length [mm]
P_a	Atmospheric pressure [cm of Hg]
P_b	Base pressure [cm of Hg]
P_w	Wall pressure [cm of Hg]
P_0	Settling chamber pressure [cm of Hg]
W_n	Width of nozzle exit [mm]

W Duct width [mm]
X Axial distance along x-axis [mm]

1. Introduction

The major augmentation to the aerodynamic drag of a bluff body, or an abrupt change in cross-section is due to the low pressure in the base area, inside the recirculation zone. Decreasing the depression on the base would then be of considerable importance in many engineering applications. The physical mechanism influencing the drag at the base is still not well understood [1]. Passive flow control due to its simplicity has been largely reported by several researchers in the literature. A cylinder is a bluff body, which becomes an effective controller as its diameter is reduced. A stationary cylinder placed normal to a flow produces a repeating pattern of swirling vortices caused by separation of flow behind the cylinder. As the diameter of the cylinder is reduced the shedding behind the cylinder also reduces. Flow control of separation and/or the structure of shear layer(s) in the depression zone can be achieved by various methods like blowing, suction, surface roughness etc. These have been studied by researchers in the past such as [2]. Review articles by [3], [4], and [5] present a comprehensive bird's eye view of the diverse ways for subduing vortex shedding. According to [2] passive flow control is effectual for low Reynolds number only and is ineffective at high Reynolds number. So, it needs to be studied. In the studies reported so far, interaction of cylinders with each other, effect of different cylinder diameter on flow field have been studied for incompressible flows only.

2. Literature Review

2.1. The existing passive and active controls.

The base pressure in high speed flow has been divided in to passive and active control [6]. [7] studied flows sudden expansion for high speed compressible subsonic flows in circular pipes. They came to conclusion that the non- dimensional base pressure and hence the base drag is deeply affected by the geometric parameters. [8] studied active flow induced cavity oscillations of incidence. They observed that a base cavity acts as an effective control to decrease base drag by increasing the base pressure in axis-symmetric high speed compressible flows. [9] did experiments to study the effect of ribs on an abruptly enlarged flow field. He concluded that the ribs effect on the flow field and on base pressure was very significant. [10] reviewed different techniques for reduction of base, after body drag and turbulent base flows problems. He explored the developments that have taken place on the use of passive techniques to control the base drag by locked vortex after-bodies, ventilated cavities, simple cavities, multi-step after-bodies and simple after-bodies employing a non-axis-symmetric boat-tailing in different speed regimes. [11] studied subsonic flow near wake and the effect of base cavities on it. They studied the result of the base cavity on the wake behind thin 2D bluff body in the subsonic regime. Mach 2.5 flow is studied for base bleed control of near wake flow field by [12]. Their results across the base plate are comparatively uniform radial pressure gradients. [13] reviewed the theory of base pressure in incompressible steady base flow. [14] performed computational study of base drag reduction for a projectile in different flight phases. DNS of turbulent flow was studied over a backward facing step to get a better understanding of reattachment length, wall shear stress and oscillations [15]. An investigation of base flow control by wall pressure analysis was studied by [16]. The methods for the passive control of base pressure studied so far have been through grooves, cavities, ribs, tabs etc. Also, a lot of work using numerical simulation has been done to understand the theory of this complicated flow and obtain correlation between the physics of the boundary layer, recirculation zone and the wake behind reattachment, [1].

2.2. The proposed controller

Detailed investigation on the interaction of two circular cylinder in various arrangements in a flow field has been performed and its physics somewhat understood [17], but the cylinders were not placed in a recirculation zone nor to reduce the base drag. Suppressing vortex shedding of a cylinder at low

IOP Conf. Series: Materials Science and Engineering **370** (2018) 012050 doi:10.1088/1757-899X/370/1/012050

Reynolds numbers was achieved by using a smaller cylinder outside the recirculation area in the vicinity of the main cylinder [18]. From the review of flow control using a cylinder, it is evident that the passive cylinder itself being a bluff body should have small diameter so that its own vortex shedding is minimized. Detailed numerical study of the near wake area of a circular cylinder was done by [19]. Experimentally and numerically flow physics of incompressible flow about fixed cylinder pairs was performed for various cylinder arrangements. [20]. Thus, from the literature we see the urge to control base drag but there is not a single case to the best of authors knowledge where cylinder is used to control base pressure in compressible flows for different high-speed regimes. It is proposed to use a cylinder as a passive controller to reduce the base drag from a converging diverging nozzle flow at Mach 2.

3. Experimental Setup and procedure

3.1. High speed flow facility

The experiments were done in B.I.T research centre at supersonic research laboratory, Mangalore, Karnataka, India. The test facility consists of internal flow apparatus. Compressed dry air at a very high pressure from the storage tank is ducted to the settling chamber through a pressure regulatory valve. Once the flow attains equilibrium in the settling chamber it is abruptly expanded through the square nozzle into the suddenly expanded square duct as shown in figure 1. Pressure measurement is done by transducer 9205 (NI) and DAQ. Measurement were done by 16 channel DAQ ranging from approximately 0-150 psi. It takes 250 samples per seconds, takes average and writes on the hard disk. The LabVIEW software as virtual instrumentation along with DAQ acquires data and displays the pressure readings from all the 16 channels, concurrently in the computer screen.

3.2. Fabrication process

By using analytical approach, we calculated the design Mach number and then fabricated supersonic nozzle of Mach 2 having semi-divergence angle of 5 degree and nozzle exit area as 10*10 mm square, based on the design. The material used was brass. Once fabricated, the nozzle was calibrated for Mach 2. Square duct of area ratio 10 with taps for pressure measuring and sliding glasses for flow visualization were also fabricated as shown in figure 1 ,2 and 3. A 2- mm diameter steel control cylinder was used to

control the base pressure. It is located at 8 mm from the base and 2 mm from the wall. Base pressure was measured at 4 places at 11.5mm from centre. Flange of square nozzle has 8 holes to fasten it with the square duct. The sudden expansion square duct was also fabricated from brass with its length 10 times its width i.e. L/W= 10.

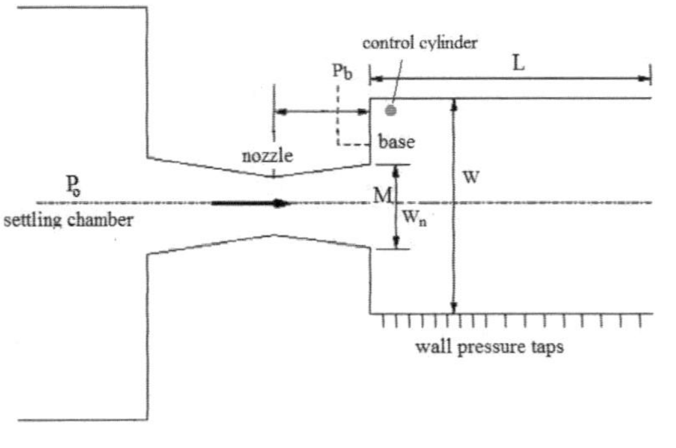

Figure 1. Experimental setup

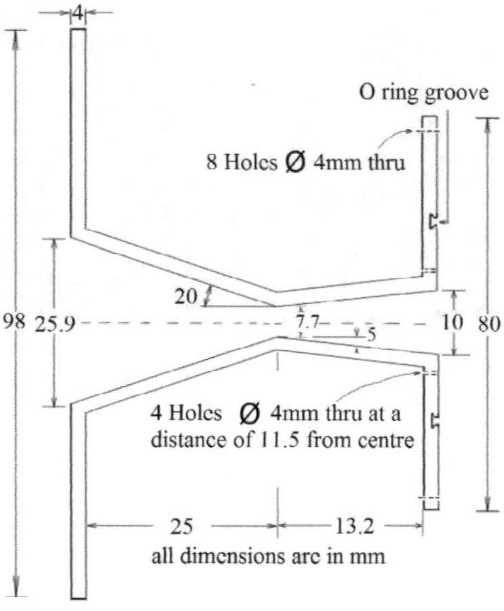

Figure 2. Schematic drawing of the Nozzle.

Duct wall pressure taps of 1 mm ID steel tubes were used. Wall pressure taps began at 7 mm from the nozzle exit. These were 3 mm apart initially and the distance between them in the axial direction progressively increased until maximum duct length of 300 mm as shown in figure 3(a). The NPR used for this study were 2, 3, 6, 7.8, and 9. The L/W ratio was 10, Area ratio (A2/A1) was 9 for all NPRs and the nozzle exit Mach number 2. The settling chamber pressure was recorded under steady state conditions. At steady flow conditions the pressure at the base and the on the wall in the square duct were recorded by pressure transducers.

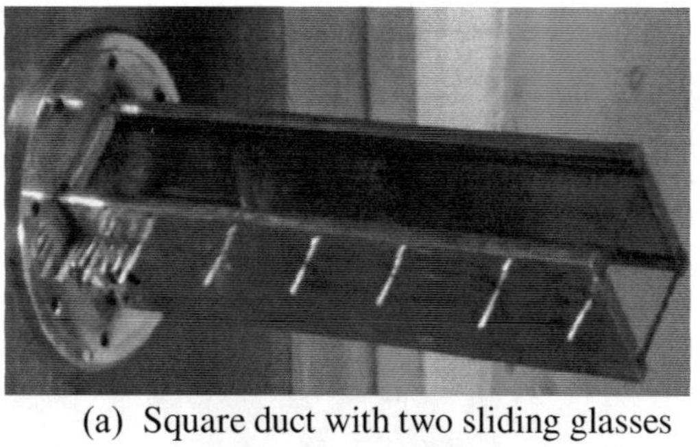

(a) Square duct with two sliding glasses

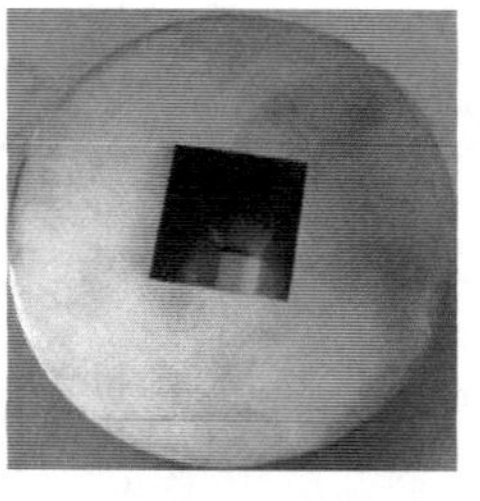

(b) Nozzle inlet,

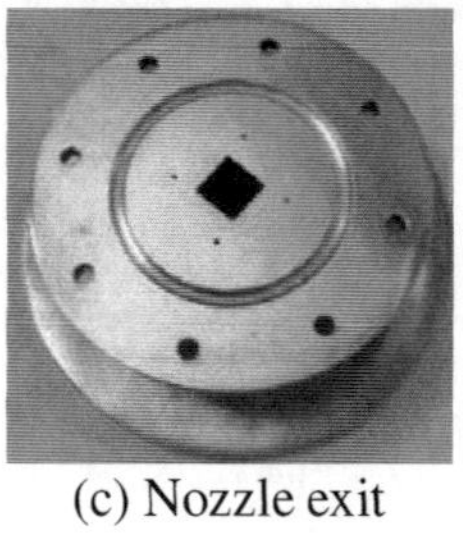

(c) Nozzle exit

(d) Cylinder

Figure 3. Photographic view of fabricated parts

Figure 4. Photographic view of nozzle, duct and cylinder attached to the settling chamber

Measurements for all NPRs with and without control were completed for L/W =10 for the square duct of 30*30 mm square cross- sectional area and length 300 mm. The main aim of this investigation is to study the workability of a static cylinder as a passive control device placed normal to the flow for controlling base pressure. To see the increase and its effectiveness in base pressure at different NPR with and with no control, the base pressure results are also presented in terms of percentage.

4. Results and discussions

4.1. Base pressure

The investigation done experimentally concentrates on the effectiveness of cylinder as passive control, located in the base region of an abrupt expanded square duct, to control the pressure at the base. The parameters
considered are the area ratio, L/W ratio, and the nozzle pressure ratio (NPR) at Mach 2. The percentage change in base pressure is given by,

$$P_b = [(P_{b\ control}-P_{b\ no\ control})/P_{b\ no\ control}] \times 100$$

The dependence of base pressure on the NPR for an area ratio of 9 at L/W = 10, for NPRs from 2 to 9 are presented in figures 5 and 6. It is well known from the literature that whenever the jets are over expanded the control either by active or passive means are not very effective. Similar trends are seen in figure 5. The percentage increase in the base pressure up to NPR 6 is less than 20 percent, whereas, when the jets are perfectly expanded or under expanded the control mechanism results in substantial increase in the base pressure value and as high as 59 % increase is recorded.

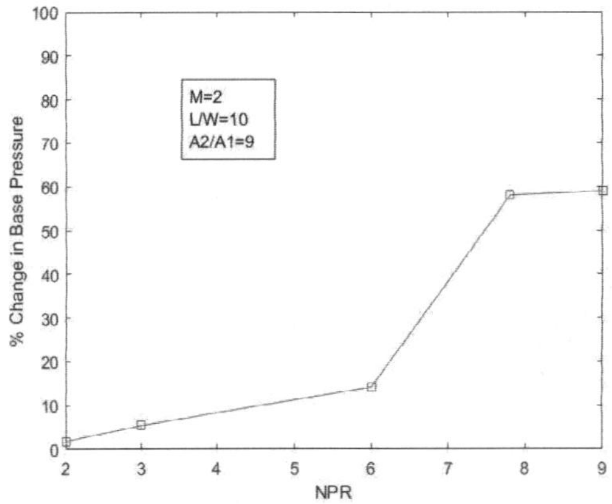

Figure 5. Percentage Change in Base Pressure with NPR

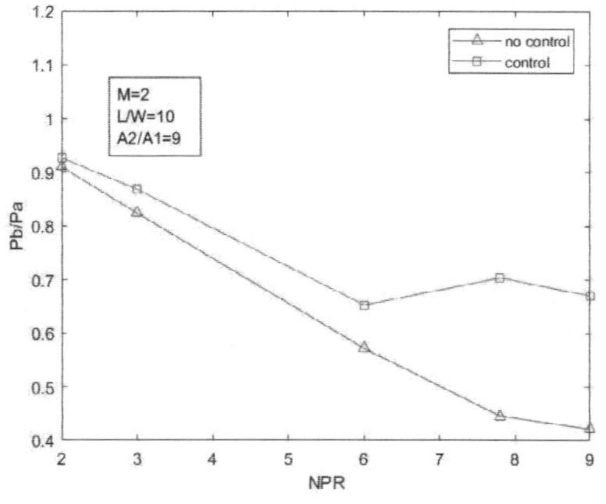

Figure 6. Base pressure variation with NPR

The practical reason for the enhancement of base pressure may be the impression of the shock at square nozzle exit which flushes the flow away out of the base region, thereby making the vortex positioned at the base weak. Further the weak vortex confronts with the presence of passive control in the form of static cylinder. This may be due to increase in NPR, leading to reduction in the level of over expansion; thus, at the nozzle exit weak oblique shock is present as compared at lower NPRs.

4.2. Wall pressure

The variation of base pressure with NPR is shown in figure 7. At low NPRs, the base pressure is very high with and without control mechanism. This trend is due to the level of expansion. With an increase in NPR the base pressure continues to decrease till the flow becomes perfectly expanded. It is also seen that as long as the flow from the nozzle is over expanded the control effectiveness is marginal. When the jet is perfectly expanded or under expanded the control mechanism is effective. This trend matches with the findings of earlier studies [21]. When a passive control mechanism is introduced to manage the base pressure, there is a possibility that it might adversely affect the nature of the flow field in the square duct. To observe this undesirable effect, the wall pressure along the square duct was measured. Wall pressure distribution along the length of the duct for NPR = 2 is shown in Figure 7. At NPR=1.89 the ideal critical condition is reached. Hence at NPR 2, flow at the exit of the nozzle will be over expanded. Accordingly, the wall pressure in the enlarged duct begins with a very high value Pw/Pa = 0.9, decreases up to X/W = 1, which seems to be the reattachment length, then further downstream due to the presence of the shock wave a jump in the wall pressure rendering it to be equal to atmospheric pressure is detected. Further, due to the presence of the expansion wave wall pressure marginally decreases and then further downstream it reaches atmospheric pressure. Results shown in Figure 8 are for NPR = 3 and they show similar trends as discussed for Fig. 6 with the exception the NPR has slightly increased and hence the level of over expansion has gone down marginally, and the

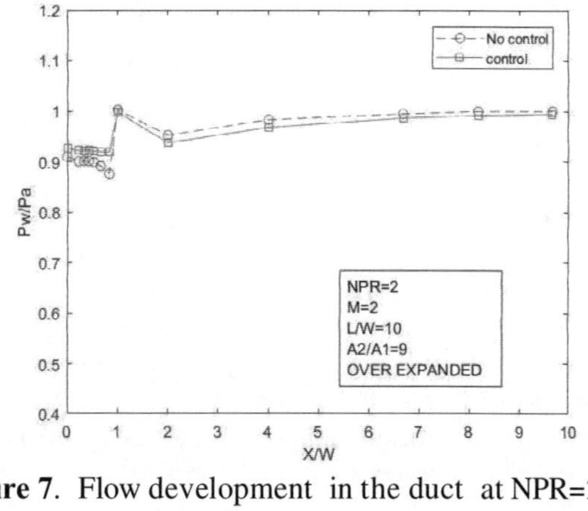

Figure 7. Flow development in the duct at NPR=2

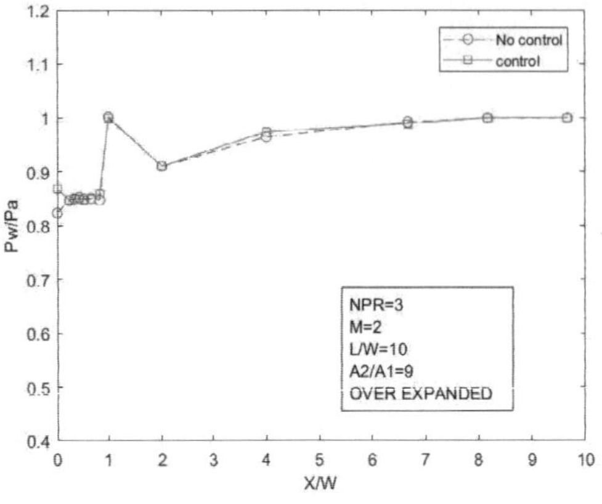

Figure 8. Flow development in the duct at NPR=3

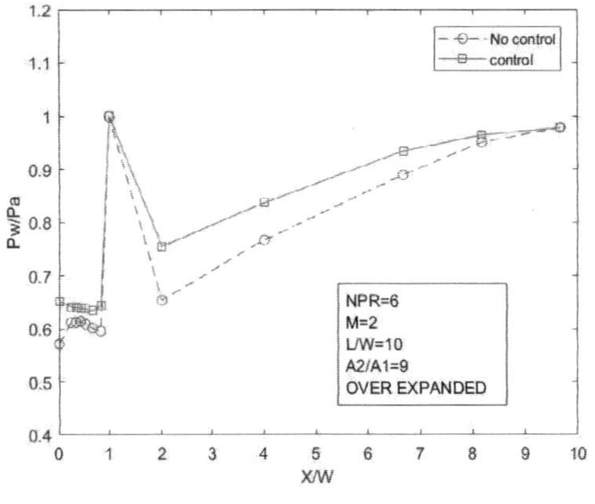

Figure 9. Flow development in the duct at NPR=6

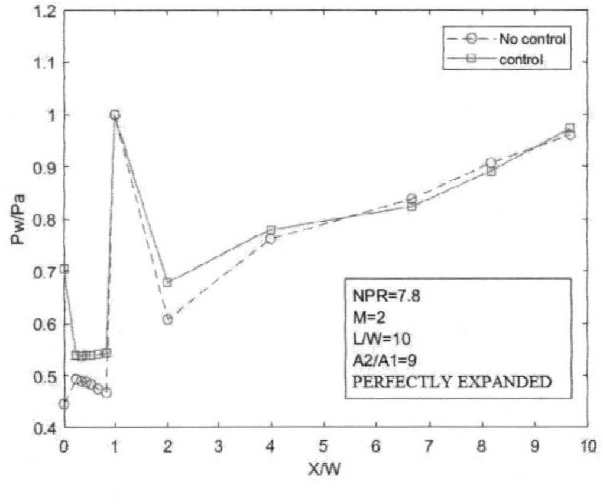

Figure 10. Flow development in the duct at NPR=7.8

Wall pressures for perfectly expanded nozzle flow is presented in figure 10. The difference in the results for NPR 6 and 7.8 is that for NPR 7.8 at X/W = 0, the pressure at the base region has further decreased and the control mechanism in the form of static cylinder in the wake region results in substantial increase in the base pressure and this increase is 59% which will be useful in reducing the base drag in the case of external flows in rocket and missile aerodynamics. In figure 10., at X/W =1, an oblique shock is present which results in a sudden increase in wall pressure taking it to the level of atmospheric pressure. In the downstream at X/W = 5 onwards the control mechanism results in a marginal decrease of wall pressure. This figure also proves that the flow from perfectly expanded nozzles is not shock free instead it is dominated by shock waves. Figure 11. presents the results for the highest NPR tested. For this NPR = 9, the flow is under expanded (i. e. Pe/Pa = 1.15). For this case the results show similar trends as was seen when NPR was 7.8 with the exception that the magnitude of base pressure has marginally decreased for both cases of flow with and without the control mechanism. In this situation the flow is under expanded, here again when the control was employed results in decrease of wall pressure from x/w = 4 onwards. This trend could be due to presence of the shock train in the duct and its strength, reflection from the wall as well as the shock interactions.

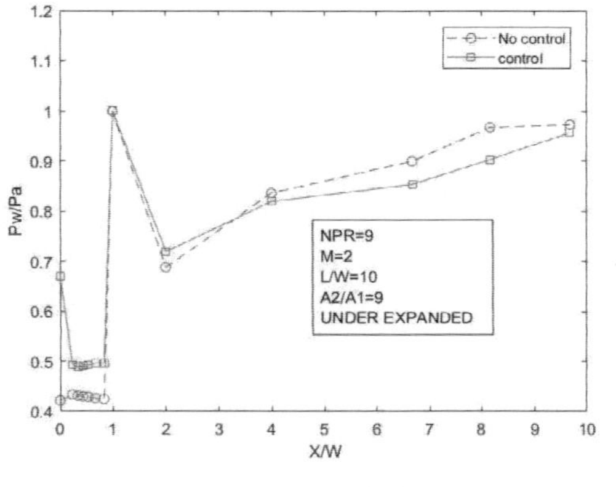

Figure 11 Flow development in the duct at NPR=9

5. Conclusions

Based on the above discussions we conclude that the base pressure is a function of Area ratio, L/W ratio, and NPR, for a particular Mach number. When flow from the nozzle is over expanded the passive control, mechanism is less effective till NPR is 6. This indicates that expansion level plays a vital role. Once the flow is perfectly expanded or under expanded the control mechanism becomes effective and base suction in the base area is reduced and 59 percent increase in the base pressure is achieved. The flow field in the duct remains almost identical with and without the control for most of the cases. However, at higher NPR namely, 6, 7.8, and 9 the control device results in an increase as well as decrease in the wall pressure along the duct. Passive control of the base flow seems to be effective for higher NPRs. The flow field for perfectly expanded nozzle is dominated by shock waves.

Acknowledgments

Authors wishing to acknowledge and thank Bearys Institute of Technology for utilizing research facility in supersonic laboratory.

References

[1] Mariotti A, Buresti G and Salvetti M 2015 Connection between base drag, separating boundary layer characteristics and wake mean recirculation length of an axisymmetric blunt-based body, *Journal of Fluids and Structures* 55 191-203

[2] Mittal S and Raghuvanshi A 2001 Control of vortex shedding behind circular cylinder for flows at low Reynolds numbers *International journal for numerical methods in fluids* 35 421-47

[3] Gad-el-Hak M and Bushnell D M 1991 Separation control *Journal of fluids engineering* 113 5-30

[4] Griffin O M and Hall M 1991 Vortex shedding lock-on and flow control in bluff body wakes *Journal of fluids engineering* 113 526-37

[5] Zdravkovich M 1981 Review and classification of various aerodynamic and hydrodynamic means for suppressing vortex shedding *Journal of Wind Engineering and Industrial Aerodynamics* 7 145-89
[6] Jahanmiri M 2011 Aircraft Drag Reduction: An Overview. *Chalmers University of Technology)*
[7] Vijayaraja K, Elangovan S and Rathakrishnan E 2008 Re-Attachment in Sudden Enlargements *International Journal of Turbo and Jet Engines* 25 95-104
[8] Cattafesta L, Williams D, Rowley C and Alvi F 2003 Review of active control of flow-induced cavity resonance *AIAA paper* 3567 2003
[9] Rathakrishnan E 2001 Effect of ribs on suddenly expanded flows *AIAA journal* 39 1402-4
[10] Viswanath P 1996 Flow management techniques for base and afterbody drag reduction *Progress in Aerospace Sciences* 32 79-129
[11] Kruiswyk R and Dutton J 1990 Effects of a base cavity on subsonic near-wake flow *AIAA journal* 28 1885-95
[12] Mathur T and Dutton J C 1996 Basebleed experiments with a cylindrical afterbody in supersonic flow *Journal of Spacecraft and Rockets* 33 30-7
[13] Tanner M 1998 Theories for base pressure in incompressible steady base flow *Progress in aerospace sciences* 34 423-80
[14] Suliman M, Mahmoud O, AlSanabawy M and Abdel-Hamid O 2009 Computational Investigation of base drag reduction for a projectile at different flight regimes. In: Paper: ASAT-13-FM-05, *13th International Conference on Aerospace Science and Aviation Technology*, Military Technical College, Kobry Elkobbah, Cairo, Egypt,
[15] Le H, Moin P and Kim J 1997 Direct numerical simulation of turbulent flow over a backward-facing step *Journal of fluid mechanics* 330 349-74
[16] Khan S, Bashir M and Asadullah M 2016 An investigation of base flow control by wall pressure analysis in a suddenly expansion nozzle *Journal of Scientific Research and Development*
[17] Zdravkovich M 1977 Review of flow interference between two circular cylinders in various arrangements *Journal of Fluids Engineering* 99 618-33
[18] Strykowski P and Sreenivasan K 1990 On the formation and suppression of vortex 'shedding'at low Reynolds numbers *Journal of Fluid Mechanics* 218 71-107
[19] Wissink J and Rodi W 2008 Numerical study of the near wake of a circular cylinder *International journal of heat and fluid flow* 29 1060-70
[20] Jester W and Kallinderis Y 2003 Numerical study of incompressible flow about fixed cylinder pairs *journal of Fluids and Structures* 17 561-77
[21] Samimy M, Zaman K and Reeder M 1993 Effect of tabs on the flow and noise field of an axisymmetric jet *AIAA journal* 31 609-19

IOP Conf. Series: Materials Science and Engineering **370** (2018) 012051 doi:10.1088/1757-899X/370/1/012051

Stability derivatives of a oscillating wedges in viscous hypersonic flow

Pavitra S[1*], Lavanya S[2] and S A Khan[3]

[1]Assistant Professor, Department of Mathematics, A.J.I.E.T, Mangalore, Karnataka, India
[2]Professor, Department of Mathematics, P.A.C.E, Mangalore, Karnataka, India
[3]Professor, Department of Mechanical Engineering, Faculty of Engineering, IIUM, Gombak Campus, Kuala Lumpur, Malaysia

*E-mail: pavitrashanbag@gmail.com

Abstract. In this paper an oscillating wedge has been considered, and the fluid slabs are kept at 900 to the wedge surface. The solutions of the continuity, momentum,and energy equations are obtained. By using the Rankine-Hugoniot relations for shockwaves, we can find the conditions behind the shock.This theory is unsteady one because of the consideration of effect of secondary wave reflections.Solutions are obtained for hypersonic flow over the wedge by varying different wedge semi vertex angles.These results shows extremely good consistency with Hui's predictions. When the effects of unsteadiness are considered then there is considerable change in the magnitude of the damping derivatives near the leading edge or initial 40 percent of the pivot positions and this difference is only marginal when we further down towards the trailing edge. However, this effect of unsteadiness is not visible in case of the stiffness derivatives.It is observed that the stiffness derivative increases with the increase in the wedge angle due to the increase in the plan form area of the wedge, resulting in the variation in the surface pressure distribution of the wedge.Further, due to the increment in the wedge angle the centre of pressure shifts towards the trailing edge. **Keywords:** Angle of attack, hypersonic flow, Mach number, oscillating wedge

1. Introduction

The knowledge of hypersonic flow over an oscillating wedge is a key to investigate the results related to spacecraft (shuttle) and re-entry body. Analytical study of unsteady flow with a small disturbance approach shows excellent results with the test done in the laboratory. So that we can avoid wastage of time and high investment for the experiment.

Appleton included secondary wave effects to analyse the problems connected to 2-D oscillating wedges. To improve an unsteady piston theory of an oscillating wedge, Appleton theory collaborated with Ghosh's large deflection similitude.

2. Literature Survey

Investigation of Tsien's [1] work was limited to conservative motion for unsteady hypersonic flow. Hayes and probstein [5, 10] invented that, the equivalence principle proved by Hayes is applicable to both steady and unsteady flow. Hamaker and wong[3] showed complete dynamic similarity for symmetric body. Light hill developed a theory on oscillating airfoils in hypersonic flow, which includes pitching oscillations. Light hill [4] introduced a parameter δ to measure maximum inclination angle of Mach wave in the flow field. He assumed $M1\delta \leq 1$, which is of the order of maximum deflection of a streamline. Light hill [4] combined the two –dimensional unsteady flow problem to that of a gas flow in a tube driven by a piston which leads to "Piston Analogy". In Lighthill piston theory, the concept of secondary wave effects has not been considered. McIntosh [8] and Appleton [7] have included the effects of secondary wave reflected from the shock and showed independently that, the neglect of these effects provides a reduction to the strong shock piston theory. Ghosh and Mistry [11] studied the flow over an oscillating two-dimensional wedge and developed a theory on quasi steady flow. Same concept has been used by Ghosh [12] to solve problems related to axis-symmetric flow. Rasta and Khan [13] extended Ghosh's theory for attached shock.

3. Analysis

In the isentropic process, expression for the pressure on a piston as a power series in its velocity is given by,

$$\frac{P_2}{P_1} = 1 + \gamma\frac{U_p}{a_1} + \frac{\gamma(\gamma+1)}{4}\left(\frac{U_p}{a_1}\right)^2 + \frac{\gamma(\gamma+1)}{12}\left(\frac{U_p}{a_1}\right)^3 + ------$$

where P_1& P_2 represents pressure ahead and after the shock in the fluid slab respectively.

Consider the governing equations:

Continuity equation: $\frac{\partial \rho}{\partial \tau} + \frac{\partial}{\partial y}(\rho\, v) = 0$

Momentum equation: $\rho\left(\frac{\partial v}{\partial \tau} + v\frac{\partial v}{\partial y}\right) + \frac{\partial p}{\partial y} = 0$

Energy equation: $\left(\frac{\partial}{\partial \tau} + v\frac{\partial}{\partial y}\right)\ln\left(\frac{p}{\rho^{\gamma}}\right) = 0$

where ρ is the density, p is the pressure and, v is the velocity component in the y direction and τ is the time which is measured from the instant the fluid slab reaches the leading edge of the wedge. The piston velocity of the wedge is given by

$$v_p = U_\infty \sin(\theta + \delta) + \dot{\theta}(x - hL)$$

where L is the length of the flat plate, hL is the position of the pivot point and θ is the angular displacement, which is positive for the nose down movement of the plate.

Pressure perturbation over the piston is given by,

$$\bar{P_p} = \rho_2 a_2 U_\infty \left[\left(\frac{1+w}{1-w}\right)\cos\delta + \left\{\left(2\frac{1+w\delta'}{1-w\delta'} - \frac{1+w}{1-w}\right)\frac{x}{L} - \left(\frac{1+w}{1-w}\right)h\right\}\frac{\dot{\theta}_L}{U_\infty}\right]$$

Restoring moment over the plate is given by

$$-M_r = \int_0^L (x - hL)\,\bar{P_p}\,dx$$

The stiffness derivative and damping derivatives of the plates in the transferred axes,

$$-C_{m_\theta} = \frac{1}{\frac{1}{2}\rho_\infty U_\infty^2 C^2}\left(-\frac{\partial M_r}{\partial \theta}\right)_\theta$$

$$-C_{m_{\dot\theta}} = \frac{1}{\frac{1}{2}\rho_\infty U_\infty^2 C^2}\left(-\frac{\partial M_r}{\partial \dot\theta}\right)_{\dot\theta}$$

Non-dimensionalizing must be done with respect to $\frac{1}{2}\rho_\infty U_\infty^2 C^2$ and $\frac{1}{2}\rho_\infty U_\infty^2 C^3$.

The stiffness derivative and damping derivatives of the plates reduces to the following expressions.

$$-C_{m_\theta} = \frac{2\rho_2 a_2}{\rho_\infty U_\infty \cos\delta}\left[\frac{1}{2}\left(\frac{1+w}{1-w}\right) - \left(\frac{1+w}{1-w}\right)h' \cos^2\delta\right]$$

$$-C_{m_{\dot\theta}} = \frac{2\rho_2 a_2}{\rho_\infty U_\infty \cos^3\delta}\left[\frac{1}{3}\left(2\frac{1+w\delta'}{1-w\delta'} - \frac{1+w}{1-w}\right) - \left(\frac{1+w\delta'}{1-w\delta'}\right)h' \cos^2\delta + \left(\frac{1+w}{1-w}\right)h'^2 \cos^4\delta\right]$$

Since the stiffness and damping derivatives for the wedges are equivalent to those of two flat plates, the RHS of above two equations are multiplied by 2. Therefore, the stiffness derivative of the wedge is

$$-C_{m_\theta} = \frac{4\rho_2 a_2}{\rho_\infty U_\infty \cos\delta}\left(\frac{1+w}{1-w}\right)\left[\frac{1}{2} - h' \cos^2\delta\right]$$

And the damping derivative of the wedge is

$$-C_{m_{\dot\theta}} = \frac{4\rho_2 a_2}{\rho_\infty U_\infty \cos^3\delta}\left[\frac{1}{3}\left(2\frac{1+w\delta'}{1-w\delta'} - \frac{1+w}{1-w}\right) - \left(\frac{1+w\delta'}{1-w\delta'}\right)h' \cos^2\delta + \left(\frac{1+w}{1-w}\right)h'^2 \cos^4\delta\right]$$

4. Results and Discussions

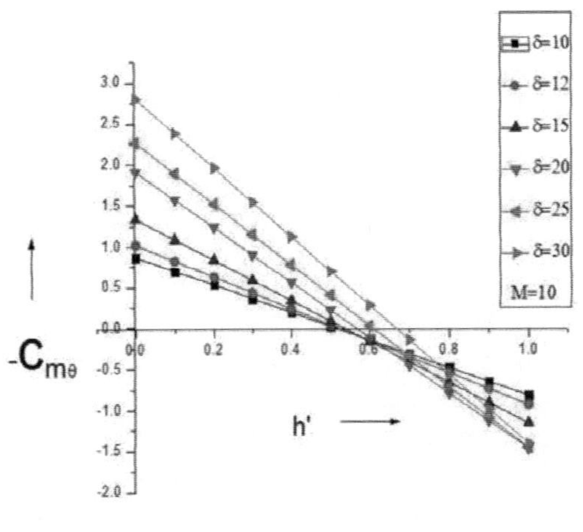

Figure 1. Stiffness Derivative Vs pivot position at M=10.

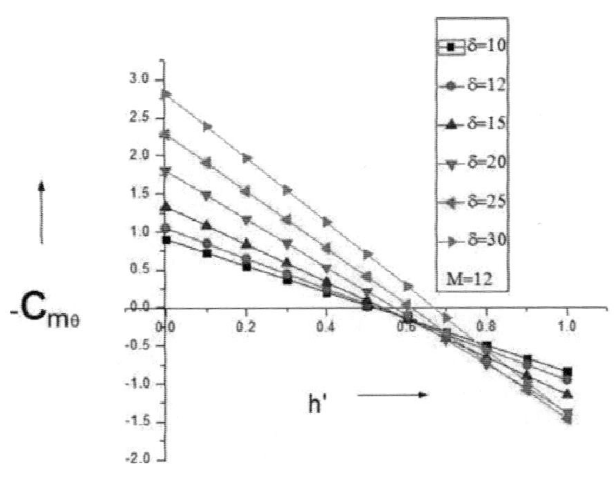

Figure 2. Stiffness Derivative Vs pivot position at M=12.

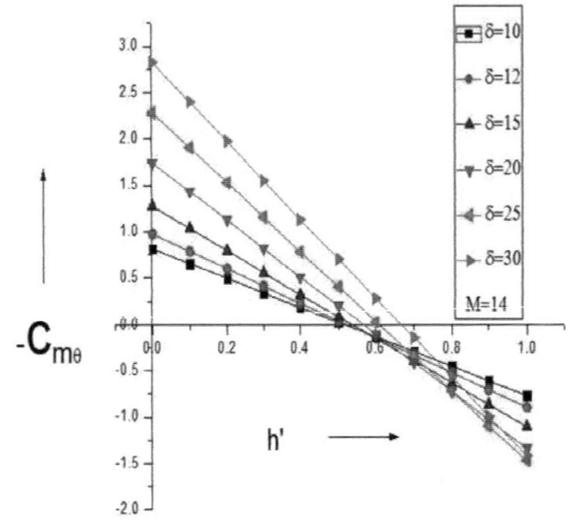

Figure 3. Stiffness Derivative Vs pivot position at M=14.

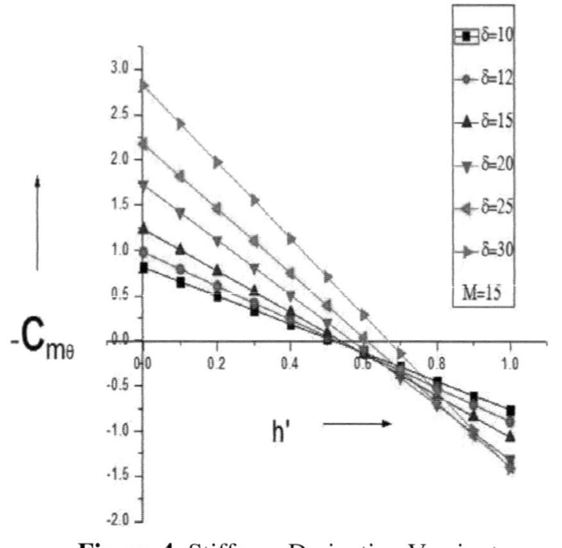

Figure 4. Stiffness Derivative Vs pivot position at M=15.

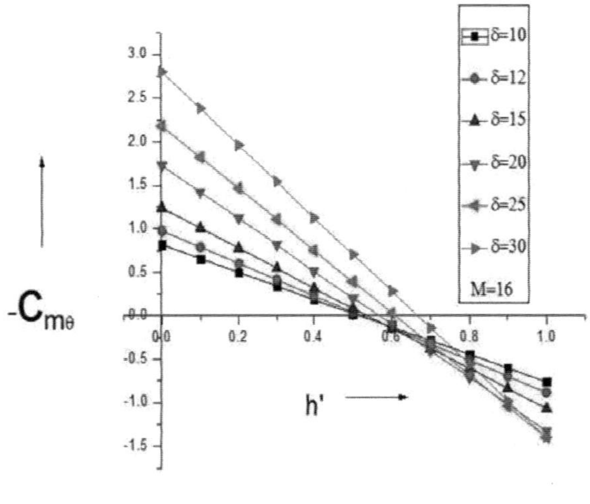

Figure 5. Stiffness Derivative Vs pivot position at M=16.

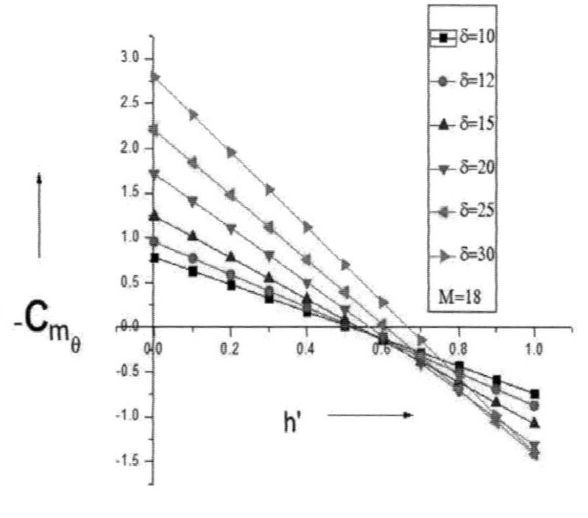

Figure 6. Stiffness Derivative Vs pivot position at M=18.

The stiffness derivatives of the oscillating wedge of different wedge semi vertex angles have been calculated by using Ghosh and Mistry (1980) and the present theory. Figures 1 to 3 shows the change in stiffness derivatives of oscillating wedge for various semi vertex angles at δ = 100, 120, 150, 200, 250, 300 with respect to pivot point position by varying Mach numbers from M = 10, M = 12, M = 14. It is seen that increase in the value of semi vertex angle results in increase in the value of stiffness derivative up to M = 14. From Figures 4 to 7 it is observed that the magnitude of stiffness starts decreasing for M = 15, M = 16, M = 18 and M = 20, but there is no much variation. Thus, it satisfies Mach number independence principle. The stiffness value is maximum at the nose of the wedge in the initial stage and then it starts decrease linearly with respect to the pivot position.

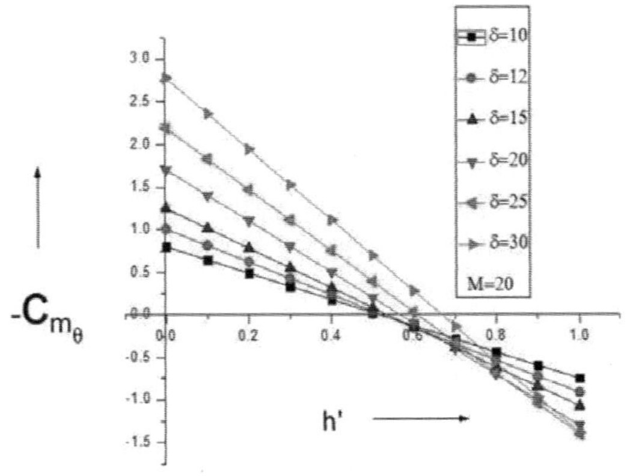

Figure 7. Stiffness Derivative Vs pivot position at M=20.

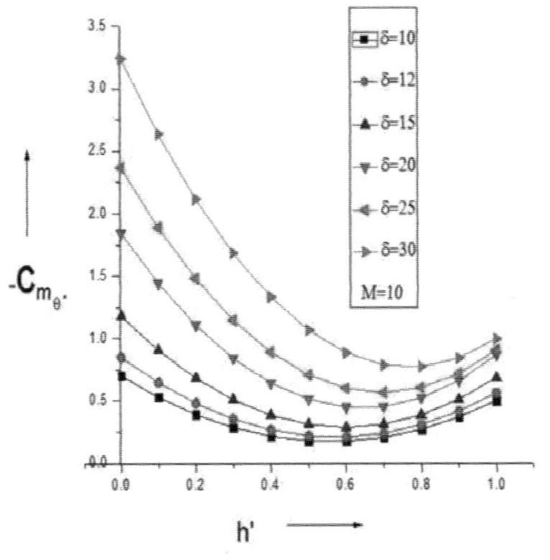

Figure 8 .Damping derivative Vs pivot position at M=10.

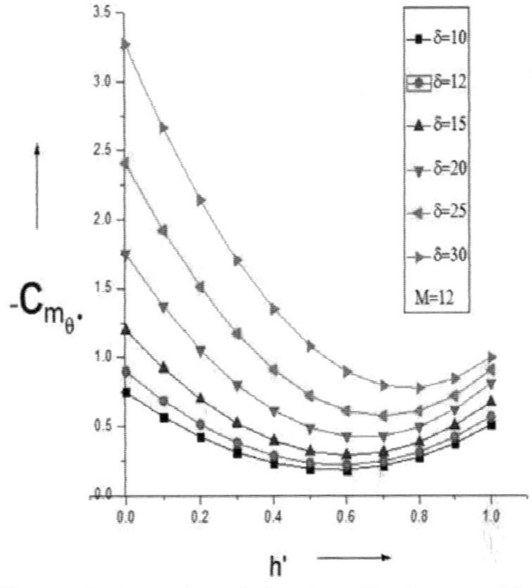

Figure 9. Damping derivative Vs pivot position at M=12.

Figures 8 to 14 shows the variation of damping derivatives with respect to pivot positions from 0 to 1 for Mach numbers M =10, 12,14,16,18 and 20 respectively for different semi-vertex angles from δ =10 to 30. In damping case, we can observe that magnitude increase with increase in wedge semi vertex angle. There is no much difference in the magnitude of damping derivative with pivot point position for Mach numbers M=15 to 20,thus it holds Mach number independence principle. Figure 15compares the results for planar wedge with Ghosh and Mistry (1980) theory for M=100, δ=150.Discrepancies in the present theory with that of Ghosh and Mistry (1980) theory for the plane wedge are attributed to the considerations of secondary wave reflections and the unsteady effects in the present theory. Also, the stiffness and damping derivatives estimated by the Ghosh and Mistry (1980) were quasi-steady whereas, the present theory evaluates unsteady stiffness and damping derivatives. One of the advantage of the present theory is that by subtracting the damping derivative by quasi-steady method from unsteady damping derivative to obtain damping derivatives due rate of change of angle of attack, which is the only way to evaluate the contribution to damping derivative in pitch due to rate of change of angle of attack. The effect of secondary waves used to increase the damping for forward pivot position and decrease it for rearward position. This trend is attributed due to the variation in the surface pressure distribution of the wedge.

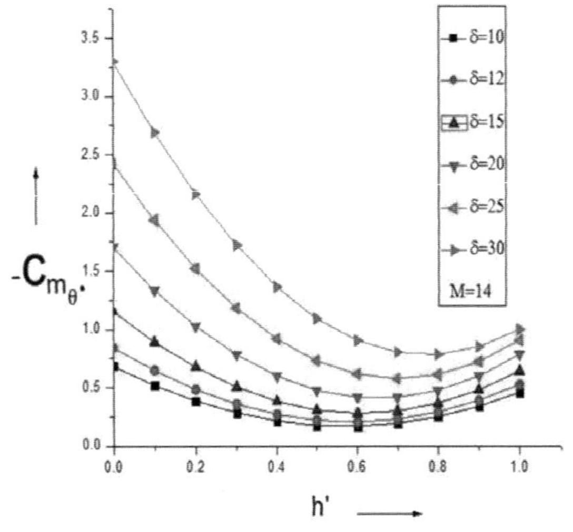

Figure 10. Damping derivative Vs pivot position at M=14.

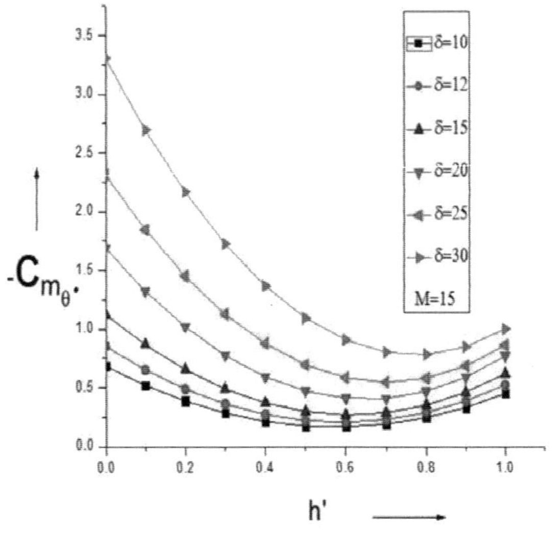

Figure 11. Damping derivative Vs pivot position at M=15.

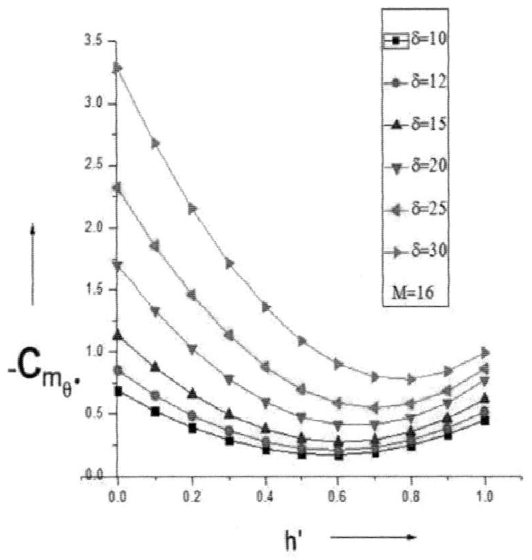

Figure 12. Damping derivative Vs pivot position at M=16.

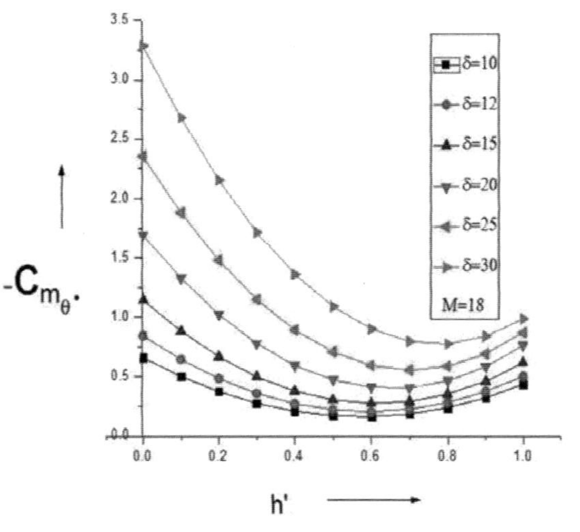

Figure 13. Damping derivative Vs pivot position at M=18.

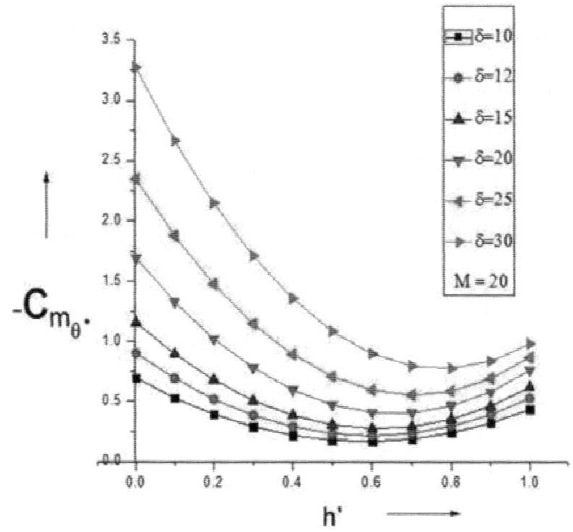

Figure 14. Damping derivative Vs pivot position at M=20.

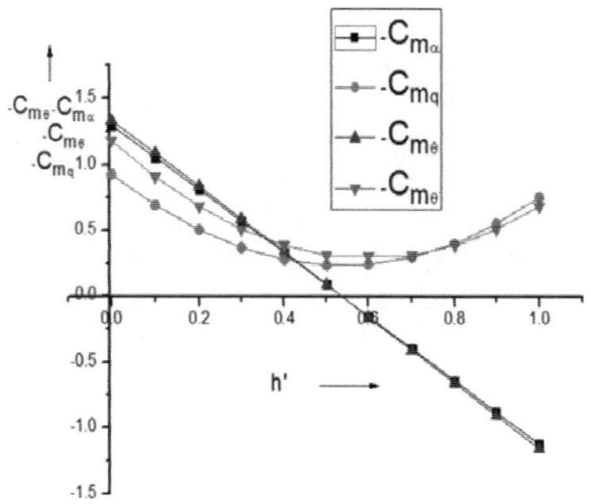

Figure 15. Stiffness and Damping of plane wedge for M = 10, δ = 15

5. Conclusions

- The unsteady effect and secondary wave reflections has been considered in the present theory. When the effects of unsteadiness are considered then there is considerable change in the magnitude of the damping derivatives near the leading edge or initial 40 percent of the pivot positions and this difference is only marginal when we further down towards the trailing edge. However, this effect of unsteadiness is not visible in case of the stiffness derivatives.
- From the results it is observed that the stiffness derivative increases with the increase in the wedge angle due to the increase in the plan form area of the wedge, resulting in the variation in the surface pressure distribution of the wedge.
- Due to the increment in the wedge angle the center of pressure shifts towards the trailing edge.

References

[1] Tsien, H. S, Similarity laws of hypersonic flow, *Journal of Mathematical Physics*, 1946, Vol.25, Pp.247-251.

[2] Carrier, G. F, The Oscillating Wedge in Supersonic Stream, *Journal of Aeronautical Sciences*, 1949, Vol.16, pp. 150-152

[3] Hamaker,F.M.,and Wong,T.J., The similarity law for non-steady hypersonic flows and Requirements for dynamic similarity of related bodies in free flight,1952,*NACA Technical Note* no.2631.

[4] Lighthill, M. J, Oscillating air foil at high Mach numbers, *Journal of Aeronautical Sciences*, 1953,
Vol. 20, pp. 402-406.

[5] Hayes, W.D. and Probstein,R.F, Viscous hypersonic similitude, *Journal of Aero Space Science*, 1959, Vol.26, pp.815–825.

[6] Miles, J. W, Unsteady flow of hypersonic speeds, Hypersonic flow, *Butterworths scientific publication*, London, England,1960, pp.185-197.

[7] Appleton, J.P., Aerodynamic pitching derivatives of a wedge in hypersonic flow,1964, *AIAA Journal*, Vol.2, pp.2034-2036.

[8] McInthosh, S.C., Hypersonic flow over an oscillating wedge,1965, *AIAA journal*,Vol.3,pp.433-440.

[9] Hayes, W.D. and Probstein, R.F. Hypersonic flow theory, *Academic Press*, 1966, New York.

IOP Conf. Series: Materials Science and Engineering **370** (2018) 012051 doi:10.1088/1757-899X/370/1/012051

[10] Hui, W.H., Stability of Oscillating Wedges and Caret Wings in Hypersonic and Supersonic Flows, *AIAA Journal*, 1969,Vol. 7, pp. 1524-1530.

[11] Ghosh K.and Mistry B.K., Large incidence hypersonic similitude and oscillating non planar wedges, *AIAA Journal*, 1980,Vol. 18, pp. 1004-1006.

[12] Ghosh K, Hypersonic large deflection similitude for oscillating delta wings, *The Aeronautical journal*, 1984, pp.357-361.

[13] Crasta A, Khan S.A, Hypersonic similitude for planar wedges, *International Journal of Advanced Research in Engineering and Technology*, 2014, Vol.5 (2), pp.16–31.

Experimental Investigation of the Base Flow and Base Pressure of Sudden Expansion Nozzle

Ahmed Saleel[1*], M. A. Ali Baig[2], S. A. Khan[3]

[1]Assistant Professor, King Khalid University, Abha, Saudi Arabia.
[2]School of Mechanical Engineering, CMR Technical Campus, Hyderabad, Telangana.
[3]Professor, Department of Mechanical Engineering, IIUM, Kuala Lumpur, Malaysia.

*E-mail: sakhan@iium.edu.my

Abstract. This paper presents an experimental investigation of an airflow from convergent-divergent axisymmetric nozzles expanded suddenly into circular duct of larger cross- sectional area than that of nozzle exit area, focusing attention on the base pressure and the flow development in the duct. To investigate the influence of area ratios and nozzle pressure ratios on the flow field developed in the duct, the micro jets of 1 mm orifice diameter located at 900 interval along a pitch circle diameter 1.3 times the nozzle exit diameter were employed as the controller of the base pressure. The Mach number investigated in the present study was 1.87, 2.2 and 2.58. The area ratios of the present study are 2.56, 3.24, 4.84 and 6.25. The nozzle pressure ratio (NPR) used were 3 and 5. The length-to-diameter ratio of the enlarged duct was varied from 10 to 1.The level of expansion at the nozzle exit (i.e. before sudden expansion) influences the wall pressure very strongly. From the results it is observed that for NPRs 3 there is no appreciable gain in the base pressure, and hence control employed in the form of micro jets is not effective for this NPR, however, at NPR 5, there is significant change in the base pressure values for all the area ratios. This clearly indicates that the level of expansion plays an important role to dictate the value of the base pressure and ultimately the control effectiveness by the micro jets

1. Introduction

THE flow field generated by the infringement of high-speed flows usually results in a very unsteady flow field. When such flows are generated at the rear end of aerospace vehicles, this flow can lead to a host of adverse effects that can diminish aircraft performance. One of the important related issues is the occurrence of base flow aerodynamic side loads due to vortex shedding and asymmetric flow separation inside a rocket nozzle. This problem has been well studied through experiments as well as computations. The unsteady separation phenomenon inside the nozzle can lead to steady/unsteady forcing of the thrusting nozzle, as well as associated mechanisms due to fluid–structure interactions [1-5]. The presence of such unsteady side loads can also result in adverse control systems of the vehicle in pitch, roll, and yaw [6]. Since the nozzle is located in the wake of the main body, response of the nozzle and the associated mechanical systems to the oscillating outer flow in the presence of asymmetric loading due to internal flow has been studied [4, 7, 8]. The investigations have demonstrated the origin of these unsteady phenomena at the base of the nozzle are 1) asymmetric separation line, 2) pressure pulsations at the separation and reattachment locations, 3) aeroelastic coupling, 4) transition of separation pattern between restricted shock separation (RSS) and full shock separation (FSS), and 5) external flow instabilities such as buffeting [9]. A computational approach to model the fluid– structure interactions of the one such nozzle using detached-eddy simulations coupled with second-order structural computations for different nozzle configurations has been studied [10].

In fluid dynamics research community, the efficient and effective control of turbulent flow has turned an upcoming goal. Innumerous applications of turbulence flow control include viscous drag reduction in numerous aerodynamic and hydrodynamic applications, regulation of heat transfer, reduction of wall pressure fluctuations and flow-generated noise, reduced oscillation and unsteadiness in resonance-dominated flows, and so on. These control techniques are also used to delay the transition to turbulence, and prevention/delay of boundary-layer separation in various internal and external flows leading to performance loss. The interaction of pressure distribution in the expansion corner with the boundary layer and thickness of upstream flow was studied by Wick [11]. Boundary layer is a cause of fluid for the corner flow and it was found that air expands abruptly after passing through a convergent nozzle. The under expanded gas jets from the blunt bodies was seen to produce a shock structure by applying numerical studies by Menon and Skews [12]. This shock structure was affected by the corners of the nozzle and barrel shocks were observed in the nozzle exit by changing nozzle orientation. Also Muller examined the effect of initial flow direction on the base pressure of nozzle [13]. The effect of base cavities on the base pressure at various angles was studied by Tanner [14]. He found an increase in base pressure by applying cavities, and hence reduction in base drags. The experimental investigation to study the effects of micro jets under the influence of over, under and correct expansion to control the base drag was studied by Rathakrishnan and others [15-22]. The result was very effective in terms of percentage, as micro jets reduced the base drag without affecting the wall pressure distribution. It is found that many techniques can be used to reduce or even suppress the flow separation. The experimental investigation to study the effects of micro jets under the influence of over, under and correct expansion to control the base drag was studied by Khan and others [20-27]. The result was very effective in terms of percentage, as micro jets reduced the base drag without affecting the wall pressure distribution. It is found that many techniques can be used to reduce or even suppress the flow separation.

Therefore in this study, we have employed micro jets as active control devices in our examination, the correlation of base pressure change with flow Mach number has been demonstrated at different L/D ratios.

2. Experimental Procedure

The fringed nozzle-enlarged duct is fixed at the end of the settling chamber by a slot holder arrangement. The base pressure taps and wall pressure taps are measured matching the channels of pressure sensors connected on bread board to a data logger at a time in each run for different Mach numbers and expansion levels, as shown in figure 1. Pressure sensors used for this experiment are Honeywell TruStability® Board Mount Pressure Sensors, HSC Series - High Accuracy, Low Pressure Sensors - HSCDANN015PAAA5 (Absolute) and HSCDANN010BGAA5 (Gauge). Data acquisition is done with the help of Graphtec MIDI LOGGER GL820, with a 20- channel input. The model area ratio (D2/d2) is varied in the range from 1.5 to 4. The L/D ratio is another variable parameter 4, 5, 6 for pipe diameters 19 mm, 16 mm, and 13 mm respectively. NPR can be calculated accurately, while Mach number can be estimated from isentropic relationships. The measurements include the stagnation pressure of the settling chamber, the base pressure, and the wall pressure distribution along the length of the duct. All of the pressures are measured using absolute pressure sensors, except for the settling chamber pressure, which is measured using gauge pressure sensor. A data logger is used for data acquisition. Tests done using pressure sensors and data logger for a particular case with repeated testing for the same, showed that it was repeatable within ±2~3% accuracy with the previous results of the same case.

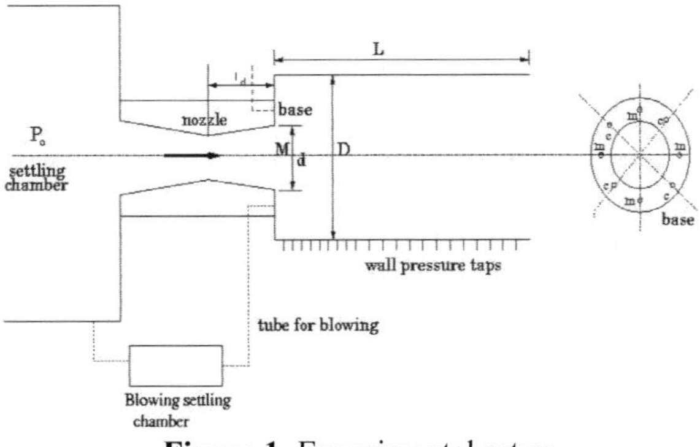

Figure 1. Experimental setup

3. Results and Discussion

The obtained values consist of base pressure (Pb); along the length of the duct and the NPR i.e. stagnation pressure (P0) to back pressure (Patm) ratio. The obtained pressures were made non-dimensional by dividing them with the atmospheric pressure/back pressure. This investigation focuses attention on the effectiveness of active control in the form of micro jets, located at the base region of suddenly expanded axi-symmetric ducts, to modify the base pressure at supersonic Mach number, which is more prone to screech. The parameters considered in the present study are the area ratio of the enlarged duct, L/D ratio of the suddenly expanded duct, the jet Mach number and the level of expansion (NPR).

The dependence of base pressure on Mach number, four area ratios namely 2.56, 3.24, 4.84, and 6.25 for NPRs from 3 and 5, and L/D ranging from 1 to 10, are given in Figure 3 and Figure 4 respectively. Results of the base pressure with and without control are compared. It is seen from these results that in supersonic regime the Mach number has got very strong influence on the base pressure. For a given Mach number the nozzle pressure ratio (NPR) which dictates the level of expansion has a strong role to play on the control effectiveness of the micro jets. Also, it is seen that with increase of NPR the control becomes more effective in increasing the base pressure for Mach numbers in the range from 1.87 to 2.58. The physical reason for this may be the influence of the shock at nozzle exit which turns the flow away from the base region, thereby weakening the vortex positioned at the base. This results in increase of base pressure since the weakened vortex at the base encounters the mass flow injected by the micro-jets. Therefore, the turning away tendency of incoming flow comes down leaving the vortex almost intact. At this situation when the micro jets are introduced they may propagate without any deflecting tendency, thereby entraining some mass from the standing vortex and convecting it away from the base causing the base pressure to assume higher values than those for without control. It is well known from literature that passive controls perform better in the presence of favourable pressure gradient. In the present study the combined effect of favourable pressure gradient and the relief due to area ratio on the active control effectiveness is investigated. The percentage change in base pressure as function of length to diameter ratio and area ratio are presented for NPR =3 and 5.

Figure 3 (d) represent the results of base pressure dependence on area ratio for L/D= 4. For NPR = 3 results do not show any change from the previous L/D. Later, results for NPR = 5 are shown in igure 4 (d), it is seen that for Mach 1.87 the results are similar to that of for L/D = 5. But for other Mach numbers namely 2.2 and 2.58 the values of base pressure are on the higher side. The physical reasons for this behavior may be the strength of the oblique shock and the reattachment length at these Mach numbers. Figure 3 (e) epresent the change in base pressure with respect to Area ratio for L/D = 5. Results for NPR = 3 are shown in Figure 3 (e). As this NPR belongs to the over expanded case for the Mach number tested and Pb assumes very high value. The control effectiveness is only marginal for all the area ratio and Mach number tested except for Mach 2.58 for area ratio 6.25, Mach 1.87 for area ratio 2.56. Results for NPR = 5 are shown in Figure 4 (e), and they exhibit similar trend as discussed earlier except the magnitude of the base pressure as the level of over expansion has come down.

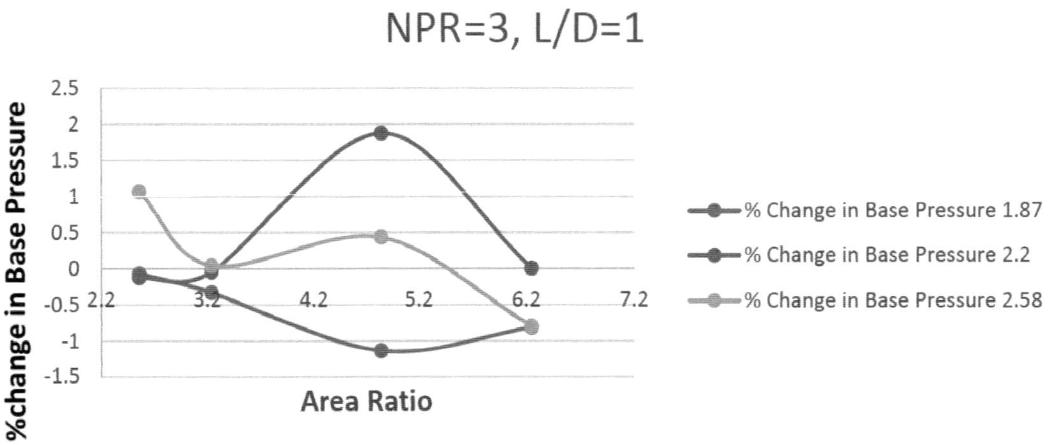

(a)

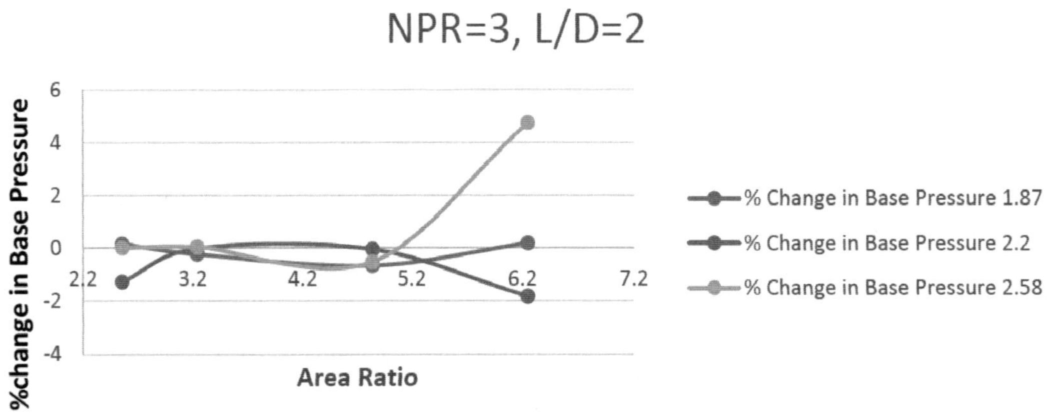

(b)

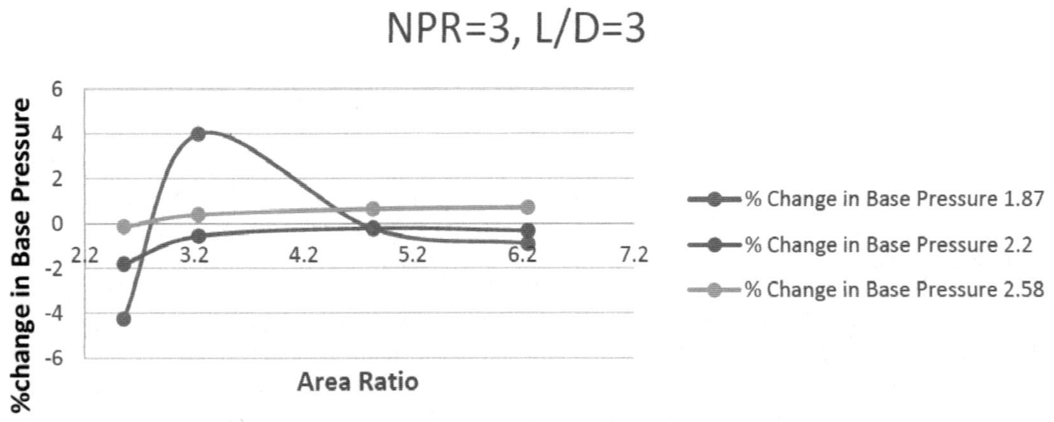

(c)

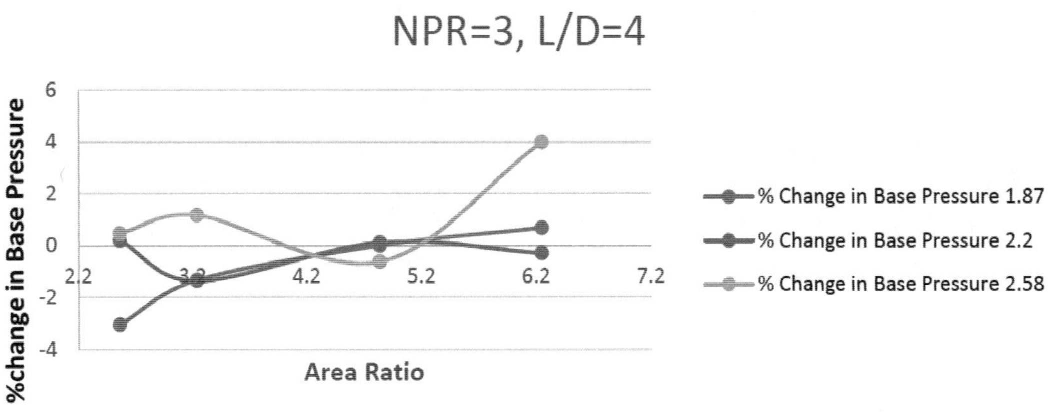
NPR=3, L/D=4
%change in Base Pressure
6
4
2
0
-2
-4
2.2
3.2
4.2
5.2
6.2
7.2
Area Ratio
% Change in Base Pressure 1.87
% Change in Base Pressure 2.2
% Change in Base Pressure 2.58

(d)

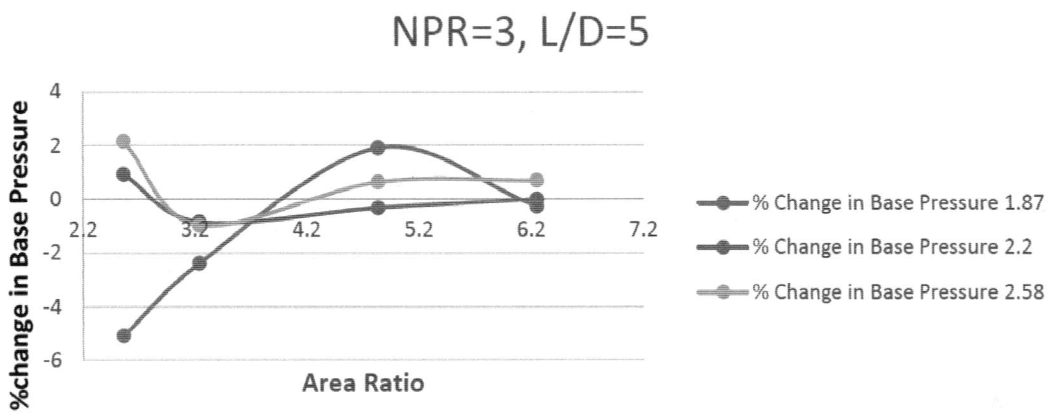
NPR=3, L/D=5
%change in Base Pressure
4
2
0
-2
-4
-6
2.2
3.2
4.2
5.2
6.2
7.2
Area Ratio
% Change in Base Pressure 1.87
% Change in Base Pressure 2.2
% Change in Base Pressure 2.58

(e)

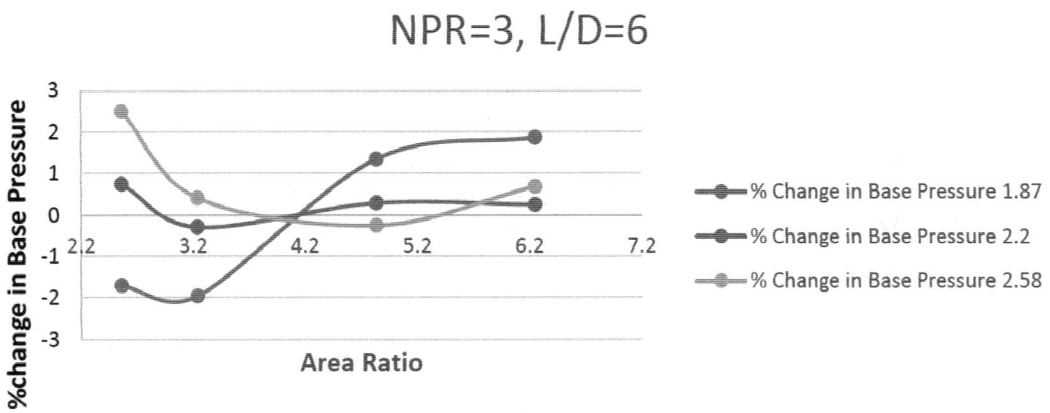
NPR=3, L/D=6
%change in Base Pressure
3
2
1
0
-1
-2
-3
2.2
3.2
4.2
5.2
6.2
7.2
Area Ratio
% Change in Base Pressure 1.87
% Change in Base Pressure 2.2
% Change in Base Pressure 2.58

(f)

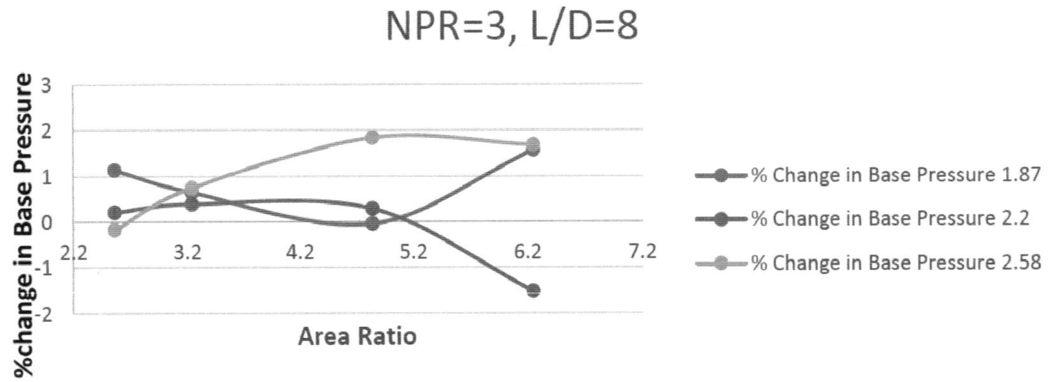

(g)

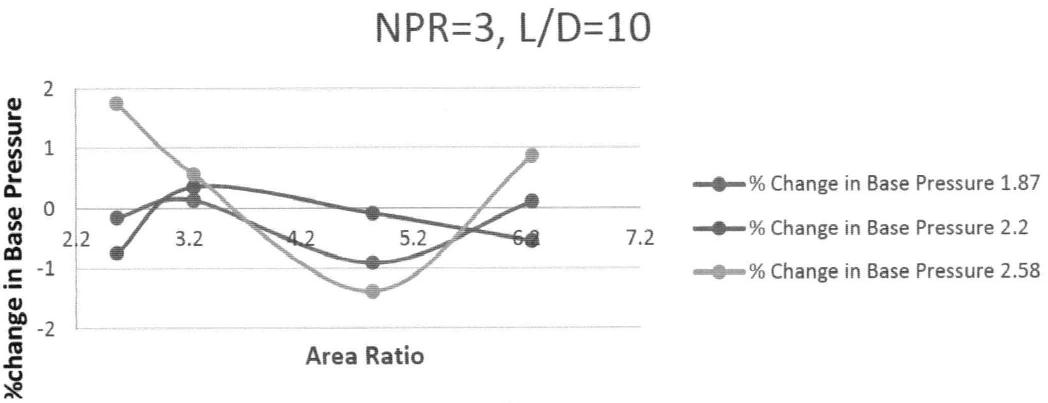

(h)

Figure 2: Base pressure variation with Area ratio for different Mach numbers at NPR 3.

Results are shown in Figure 3 (a) to (h) represents the plots of dependence of % change in base pressure on A2/A1 for NPR 5. In the case of NPR = 5, it can be seen that the behaviour is the same as that of at L/D = 4 excepting that the magnitude of base pressure has slightly increased. The reasons for this increase may be due the decrease in the duct length and the base pressure is influenced by the back pressure. Further, it is seen that the minimum Pb is at L/D = 6 which is in agreement with the results of Rathakrishnan and Sreekanth [28], for subsonic and transonic flow. The control is only of marginal influence on the base pressure for all values of L/D for highest level of area ratio. Further, it is seen that when the area ratio decreases the trend is different, microjets tend to reduce Pb for all the area ratios. It becomes independent from L/D > 6 onwards. However, for area ratio 4.84 it is found that at L/D = 10 there is a drastic change in the behaviour. The base pressure decreases drastically with and without control. The physical reason for this behaviour may be that for this area ratio the relief enjoyed by the suddenly expanded flow is much more than what it was for other area ratios. Hence, the level of expansion causing the formation of shock or expansion fan at the nozzle lip will have a strong effect on Pb as well as its control ability. It happens when the free shear layer expanding into the suddenly expanded passage finds sufficient relief, it re-attaches downstream of the reattachment point for lower area ratios. This increase in reattachment length helps the formation of a powerful vortex at the base causing low base pressure. But for Mach number 1.87 and above even lower NPRs are unable to dictate the reattachment as they do at lower Mach numbers since the shock wave at the nozzle lip becomes stronger with the increase of overexpansion level causing the flow to deflect towards the shock. This causes hindrance to the formation of a strong vortex at the base. Because of this the base pressure shoots up with increase of Mach number. Also, the control becomes ineffective since the shock strength dominates the flow process. These results imply that the flow field becomes sensitive to the relief effect at the expanded plane. However, it should be realized that increase of area ratio beyond some limiting value will not ensure the effects mentioned above for suddenly expanded flows both in subsonic and supersonic flows.

A physical explanation for this interesting phenomenon calls for a deeper investigation around these parameters. In regards of this, in the present investigations even the measurement of sound for Mach number 1.87, L/D = 10. Further, this is well known from the literature that there is a possibility of screech tone in the Mach number range 1.6 -1.9. This may be the reason for oscillatory behaviour of the wall pressure and drastic reductions in the overall sound pressure level and hence very low noise.

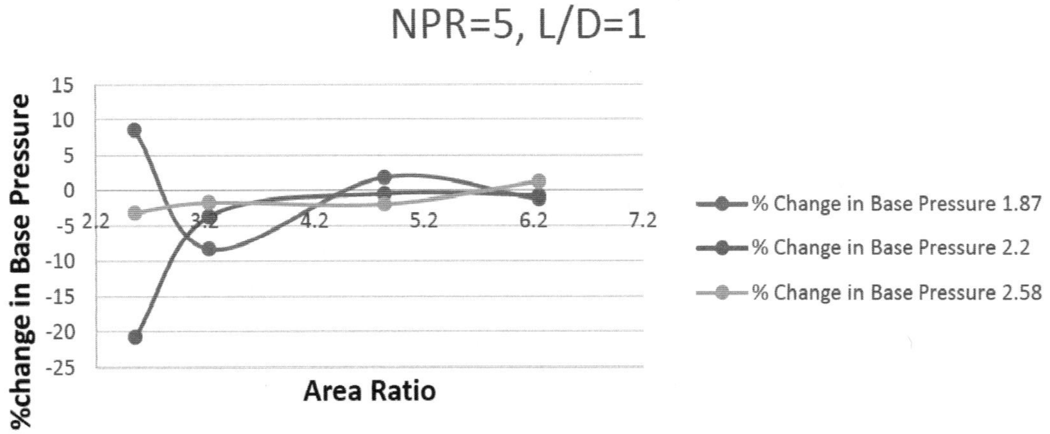

(a)

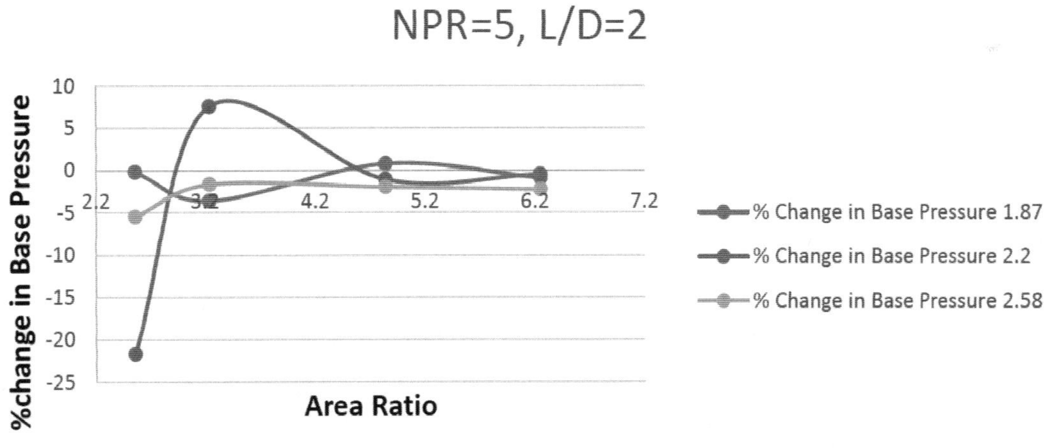

(b)

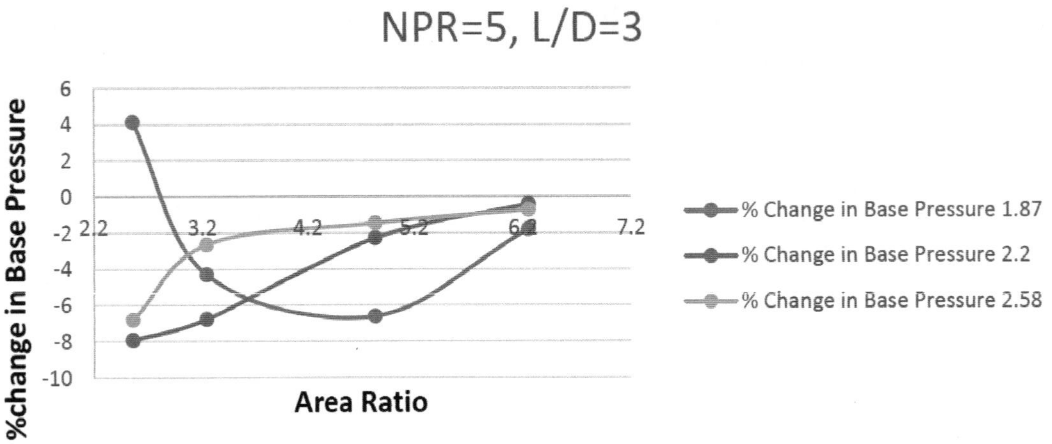
NPR=5, L/D=3
%change in Base Pressure
Area Ratio
6
4
2
0
-2
-4
-6
-8
-10
2.2
3.2
4.2
5.2
6.2
7.2
% Change in Base Pressure 1.87
% Change in Base Pressure 2.2
% Change in Base Pressure 2.58

(c)

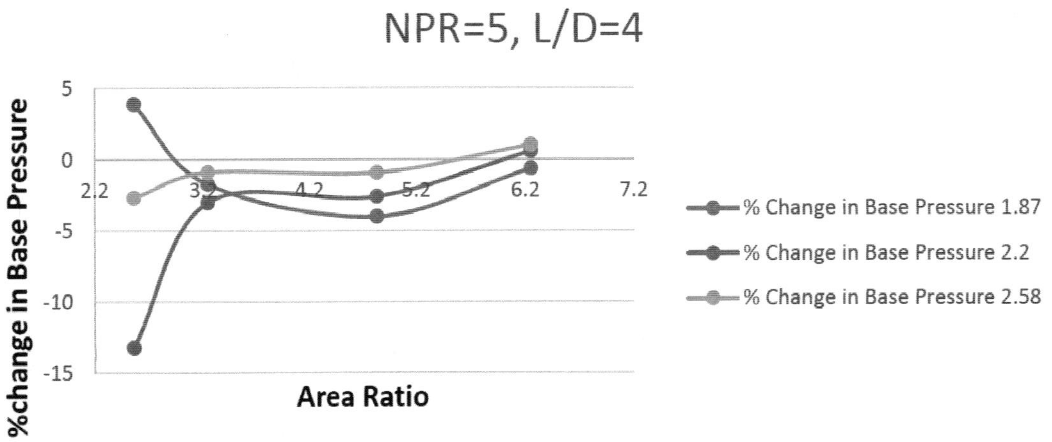
NPR=5, L/D=4
%change in Base Pressure
Area Ratio
5
0
-5
-10
-15
2.2
3.2
4.2
5.2
6.2
7.2
% Change in Base Pressure 1.87
% Change in Base Pressure 2.2
% Change in Base Pressure 2.58

(d)

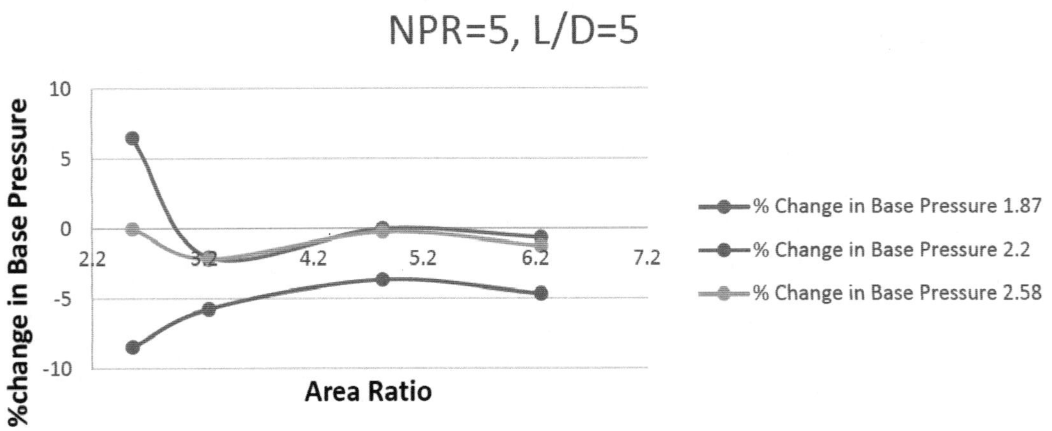
NPR=5, L/D=5
%change in Base Pressure
Area Ratio
10
5
0
-5
-10
2.2
3.2
4.2
5.2
6.2
7.2
% Change in Base Pressure 1.87
% Change in Base Pressure 2.2
% Change in Base Pressure 2.58

(f)

IOP Conf. Series: Materials Science and Engineering **370** (2018) 012052 doi:10.1088/1757-899X/370/1/012052

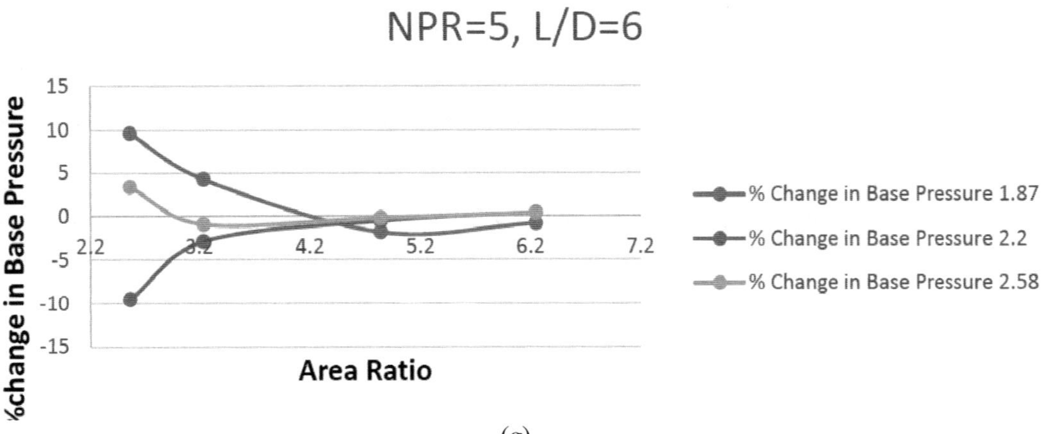

(g)

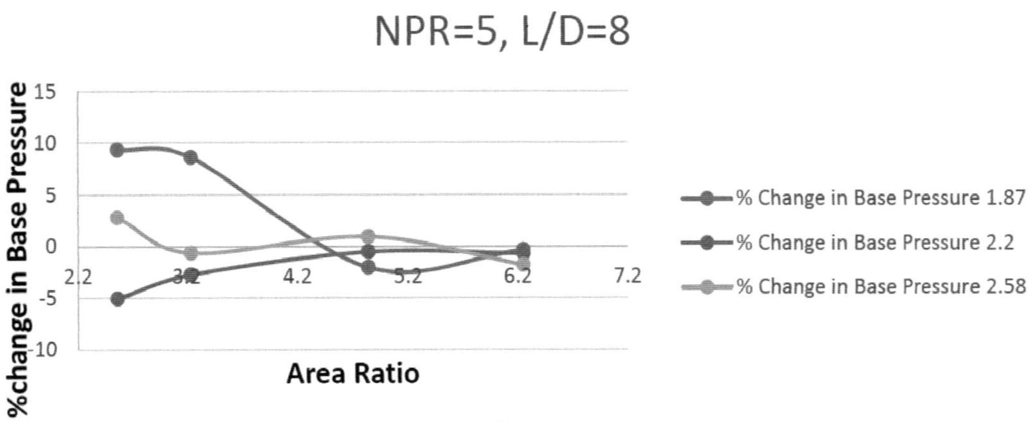

(h)

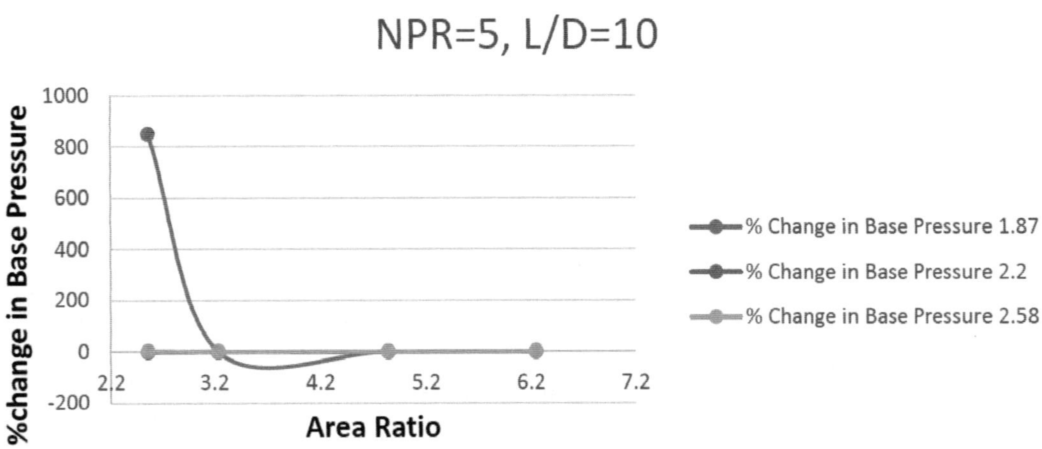

(i)

Figure 3: Base pressure variation with Area ratio for different Mach numbers at NPR 5.

4. Conclusions

From the above results Control with the help of micro jets to control base pressure has been established. The flow field in the duct remains unaltered when jets are highly over expanded. With the increase in NPR the control in the form of micro jets becomes for effective for all NPR 5 as compared to NPR 3. The flow field is dominated by the presence of the waves both strong as well as the weak waves or acoustic waves. The reflection of the waves from the wall, recompression and recombination's are taking place in the base region as well as partially in the duct wall, thereby making the flow oscillatory. The micro jets can be used as effective controllers, increasing the base suction to appreciable level for some combination of parameters. The nozzle pressure ratio plays a key role in deciding the magnitude of base pressure with and without control, in the high supersonic jet Mach number regime too.

All the non-dimensional wall pressure values exhibited in this paper are within an uncertainty band of ± 2.6%. All the investigation results are within the range of ± 3 per cent.

References

[1] Lefrancois E, Dhatt G, Vandromme D. Fluid–structural interaction with application to rocket engines. *International journal for numerical methods in fluids*. 1999;30(7):865-95.

[2] Blades EL, Luke EA, Ruf J. Fully Coupled Fluid-Structure Interaction Simulations of Rocket Engine Side Loads. *AIAA paper*. 2012;3969.

[3] Brown AM, Ruf J, Reed D, D'Agostino M, Keanini R, editors. Characterization of side load phenomena using measurement of fluid/structure interaction. *38th Joint Propulsion Conference and Exhibit, AIAA Paper*; 2002.

[4] Ostlund J, Muhammad-Klingmann B. Supersonic flow separation with application to rocket engine nozzles. *Applied Mechanics Reviews*. 2005;58(3):143-77.

[5] Dumnov G, editor. Unsteady side-loads acting on the nozzle with developed separation zone. *32nd Joint Propulsion Conference and Exhibit, AIAA Paper*; 1996.

[6] Srivastava N, Tkacik PT, Keanini RG. Influence of nozzle random side loads on launch vehicle dynamics. *Journal of Applied Physics*. 2010;108(4):044911.

[7] Frey M, Hagemann G, editors. Flow separation and side-loads in rocket nozzles. *35thAIAA/ASME/SAE/ASEE Joint Propulsion Conference and Exhibit*; 1999.

[8] Damgaard T, Östlund J, Frey M. Side-load phenomena in highly overexpanded rocket nozzles. *Journal of Propulsion and Power*. 2004;20(4):695-704.

[9] Verma SB, Stark R, Haidn O. Relation between shock unsteadiness and the origin of side-loads inside a thrust optimized parabolic rocket nozzle. *Aerospace Science and Technology*. 2006;10(6):474-83.

[10] Lüdeke H, Calvo J. A Fluid Structure Coupling of the Ariane-5 During Start Phase by DES. 2008.

[11] Wick RS. The effect of boundary layer on sonic flow through an abrupt cross-sectional area change. *Journal of the Aeronautical Sciences*. 2012.

[12] Menon N, Skews B. 3-D shock structure in underexpanded supersonic jets from elliptical and rectangular exits. *Shock Waves: Springer*; 2005. p. 529-34.

[13] HALL J, CR, Mueller T, Roache PJ. Influence of initial flow direction on the turbulent base pressure in supersonic axisymmetric flow. *Journal of Spacecraft and Rockets*. 1970;7(12):1484-8.

[14] Tanner M. Base cavity at angles of incidence. *AIAA Journal*. 1988;26(3):376-7.

[15] Khan SA, Rathakrishnan E. Active control of suddenly expanded flows from overexpanded nozzles. *International Journal of Turbo and Jet Engines*. 2002;19(1-2):119-26.

[16] Khan SA, Rathakrishnan E. Control of suddenly expanded flows with micro-jets. *International Journal of Turbo and Jet Engines*. 2003;20(1):63-82.

[17] Khan S, Rathakrishnan E. Control of suddenly expanded flow. *Aircraft Engineering and Aerospace Technology*. 2006;78(4):293-309.

[18] Khan SA, Rathakrishnan E. Active Control of Suddenly Expanded Flows from Underexpanded Nozzles. *International Journal of Turbo and Jet Engines*. 2004;21(4):233-54.

[19] Ashfaq S, Khan SA, Rathakrishnan E. Active Control of Flow through the Nozzles at Sonic Mach Number. *International Journal of Emerging trends in Engineering and Development*. 2013;2(3):73-82.

[20] Chaudhary ZI, Shinde VB, Bashir M, Khan SA. Experimental Investigation on the Effectiveness of Active Control Mechanism on Base Pressure at Low Supersonic Mach Numbers. *Innovative Design and Development Practices in Aerospace and Automotive Engineering: Springer*; 2017. p. 197-209.

[21] Baig MAA, Al-Mufadi F, Khan SA, Rathakrishnan E. Control of base flows with micro jets. *International Journal of Turbo and Jet Engines*. 2011;28(1):59-69.

[22] Baig MAA, Khan SA, Rathakrishnan E. Active Control Of Base Pressure In Suddenly Expanded Flow For Area Ratio 4.84. *International Journal of Engineering Science and Technology*. 2012;4(5):1892-902.

[23] Khan S, Bashir M, Ullah MA. An investigation of base flow control by wall pressure analysis in a suddenly expansion nozzle. 2016.

[24] Chaudhary Z, Shinde VB, Bashir M, Khan S. Experimental investigation of the base flow from the nozzles with sudden expansion. International Journal of Applied Engineering Research.10(92):2015.

[25] Baig MAA, Khan SA, Rathakrishnan E. Effect of Mach number In a Suddenly Expanded Flow for Area Ratio.

[26] Baig MAA, Khan SA, Rathakrishnan E. Control of Nozzle Flow in Suddenly Expanded Duct with
Micro Jets. *International Journal of Engineering Science & Advanced Technology* [IJESAT].2:789-95.

[27] Ashfaq S, Khan S, Rathakrishnan E. Control of suddenly expanded flow for area ratio 3.61. *International Journal of Advanced Scientific and Technical Research*. 2013(3):798-807.

IOP Conf. Series: Materials Science and Engineering **370** (2018) 012053 doi:10.1088/1757-899X/370/1/012053

Micro-hydrokinetic turbine potential for sustainable power generation in Malaysia

M B Salleh[a)], N M Kamaruddin[b)] and Z Mohamed-Kassim[c)]

School of Aerospace Engineering, Engineering Campus, Universiti Sains Malaysia, 14300 Nibong Tebal, Penang, Malaysia

[b)] Corresponding author: fazreena@usm.my

Abstract. Micro-hydrokinetic turbine (μ-HKT) technology is considered a viable option for sustainable, green and low cost power production. In recent years, there is growing number of research and development on this technology to replace conventional power production systems such as fossil fuel as well as to provide off-grid electrification to communities in remote areas. This paper provides an overview of μ-HKT system, the implementation of the technology and the potential of using μ-HKT in Malaysia. A review on the climate in Malaysia shows that its average annual rainfall is higher than the world's average annual rainfall. It contributes to the total hydropower resource of about 29,000 MW which is available all year-round. Currently, hydropower resource contributes only 7.4% of the total electrical power production in Malaysia but is expected to increase with the main contribution coming from μ-HKT. However, the μ-HKT technology has not been adopted in Malaysia due to some challenges that hinder the development of the system. This paper reviews the μ-HKT technology and its potential for application in Malaysia, particularly in remote areas.

1. Introduction

Utilization of electrical energy plays an important role in economic growth and contributes to the improvement of standard of living. Fossil fuel-based power generation is found to be the cheapest available alternative and has been the most common option for metropolitan and rural applications [1]. For example, diesel-power generator is used in many remote areas due to its simplicity, durability and requires less maintenance. However, it is observed that carbon dioxide (CO_2) equivalent emission of greenhouse gases produced by coal is 1689 g CO_2eq/kWh whereas natural gas contributes up to 930 g CO_2eq/kWh [2]. Eventually, it contributes to the climate change and global warming crisis. Due to the depletion of fossil fuel reserves, the rising of its prices put burden on electrical power usage for those who live in rural areas as most of the residents are underprevileged.

Current scenarios have encouraged the exploitation of renewable energy resources. An ideal renewable energy production technology should not only be able to sustain for future usage but must also have minimum negative impact on the environment and society [3]. Among the available renewable energy resources such as biomass, solar, wind, geothermal; hydropower energy holds the prime position in terms of contribution to the world's electricity generation under the renewable energy resource category [4]. Since water is almost 800 times denser than air, the hydropower system can extract energy of about 61.32% higher than wind turbine, even at low speed [5] [6]. In addition, it is found that small hydropower systems either at micro or nano-scale are the most reliable options to be considered particularly to power remote areas situated in close proximity to flowing water such as rivers and waterfalls [7]. This technology has capability to provide a cost-effective off-grid source of electricity less than 100kW in rural areas where populations are small and demand for power consumption is low [8].

Micro-hydrokinetic turbine (μ-HKT) system is a hydropower technology that generates electricity by harnessing kinetic energy of flowing water in a river or a stream with low elevation. The free-flowing water rotates turbine blades which turns a generator via drive shaft connection. The generator then converts the mechanical energy into electrical energy to power electrical loads. In principle, a μ-HKT system consists of five subsystems i.e. water turbine, generator, power control unit, support structure and transmission system. The water turbine of μ-HKT system can be categorized into two main groups, depending on the rotor axis with respect to the river flow, namely horizontal axis (or axial flow) turbines and vertical axis (or cross-flow) turbines [9][10]. Horizontal turbines have rotor parallel to the river flow and usually employ propeller-type rotors whereas vertical axis turbines have rotors orthogonal to the river flow. Examples of vertical axis turbines are paddle wheel, Darrieus, Gorlov and Savonius turbines [11]. Both horizontal and vertical axis turbines have their own technical advantages and disadvantages. For instance, horizontal axis turbines have self-starting capability but require high generator coupling cost due to underwater placement. Meanwhile, vertical axis turbines have lower efficiency but emit less noise due to reduced blade tip losses [9][10]. In general, the selection of the turbine type depends on the factors affecting the design and development such as hydrology and topology of a potential site, complexity, availability of technology and pertinent costs.

Unlike a conventional hydropower system, a μ-HKT does not require a construction of water reservoir or dam at strategic site to store water or canal to divert water from the main stream via penstock. Therefore, it offers minimum construction cost and minimum environmental impact which could disturb biodiversity within the potential site parameter. There are huge numbers of potential sites for the implementation of μ-HKT as it requires free-flowing stream [12].

Due to this fact, there is growing number of study on the development of μ-HKT technology recently. Researchers have done different works on various aspects which includes system design optimization, placement of turbine, augmentation application, reliability analysis, environmental monitoring and techno-economic feasibility [2]. The research on this technology is not only opted for in-land river but also extended for ocean current applications. At the same time, there are even field trial studies that have been performed all around the world to measure the practicality and feasibility of using μ-HKT system particularly for remote area application.

In this paper, several case studies on utilizing small-scale hydrokinetic turbine particularly for remote area application are discussed, respectively. The emphasis is given on the technology, system performance, reliability and problems encountered by the previous field trial studies. Based on the case study findings, the potential application of μ-HKT and its challenges in Malaysia are discussed. This paper provides an overview and potential of μ-HKT application which can be used for sustainable power production in Malaysia.

2. Case Study

Field test studies intended for remote electrification from Australia, Brazil and the United Kingdom (UK) are considered in this paper. These field tests are presented chronologically based on the publication year of the reports. Practicality aspects of small hydrokinetic turbine, technology implemented, performance, efficiency, reliability and pertinent problems are analysed and discussed for each case comprehensively.

2.1 Field trial in Australia

The evaluation of small axial flow turbines performance intended to generate electricity for Nguiu community in Apsley Strait, to replace the use of diesel fuel, was performed by Tuckey et al. [13] and

Swenson [14]. In their work, the performance of two different types of turbines was compared. In the first test, the Tyson turbine with its original high solidity rotor as shown in Figure 1(a) was utilized but the drive train was modified to fit for two car alternators. Then, in the second test, the original rotor was replaced with a lower solidity propeller turbine as shown in Figure 1(b). The second turbine was further modified to accommodate a low speed permanent magnet generator. Both turbines were mounted on pontoon which served as a structural support to keep the turbines submerged in the water. The configuration of small axial flow turbine by using Tyson turbine and low solidity turbine are summarized in Table 1.

(a) (b)

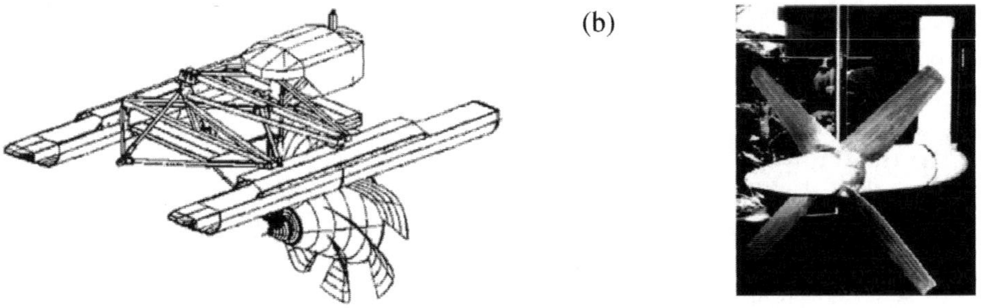

Figure 1. (a) High solidity turbine (Tyson turbine) [15] and (b) low solidity turbine [14].

Table 1. Configurations of small axial flow turbine for field study in Apsley Strait.

Type of turbine	Configuration
High solidity turbine (Tyson turbine)	• 2 m diameter rotor • 7 propeller blades, high solidity type • 4-stage drive train, turbine to alternator ratio of 182:1 (gear ratio) • 2 car alternators • Mounted on pontoon
Low solidity turbine	• 2 m diameter rotor • 4 propeller blades, low solidity type, 70 rpm at 2.1 m/s • 14:1 gear ratio • Low speed permanent magnet generator (rated 600 rpm) • Mounted on pontoon

A ten months of field trial showed that the maximum power coefficient C_p of the Tyson turbine was 0.17 at 1.6 m/s of river flow. The efficiency of the drive train and maximum efficiency of the alternators was estimated at 74% and 44%, respectively. Meanwhile, after a field trial of about 18 months, the maximum C_p of the second turbine was 0.32, obtained at water velocity of 1.1 m/s and 3.5 tip speed ratio. The drive train efficiency of 87.5% and generator efficiency of 89% were achieved thus resulting in the overall efficiency of 25%, higher than the first turbine with only 5.5%. However, both turbines were found to have problems with debris in which marine growth like seaweeds were found attached to the turbines. In addition, the gearbox was damaged by floating mangrove trees in the first field trial.

2.2 Field trial in Brazil

Tiago [16] demonstrated the capability of generating AC power directly using a small water turbine developed by the University of Brasilia. Some innovations were presented in the project as depicted in Figure 2 such as the use of cone-shaped grid in the turbine entrance for debris protection and a stator or guide blades fitted in front of the rotor. The purpose of the stator is to direct the water flow in the turbine in such a way to increase the angle of attack of the propeller thus optimizing the transformation of hydraulic energy. In addition, a suction tube is used at the outlet of the turbine and a cone was used at the center of the turbine to minimize turbulence in the water stream.

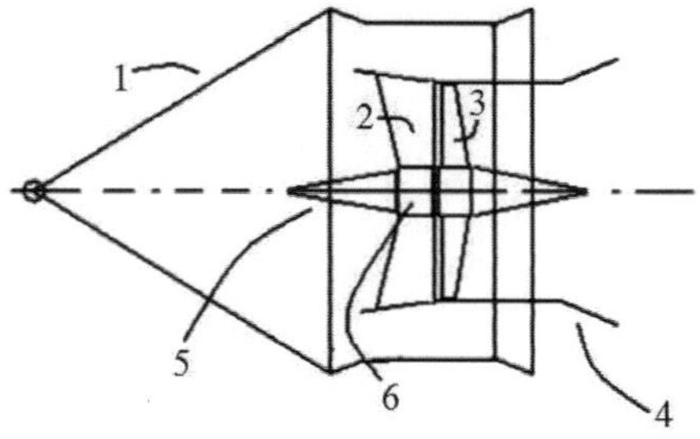

Figure 2. Turbine components: (1) cone-shape protecting grid, (2) stator or guide vanes, (3) propeller, (4) suction tube, (5) center cone and (6) transmission box [17].

The turbine was installed by simply suspending it from the river bank using a lever where the generator was mounted above the river as shown in Figure 3. This practical method allows the turbine to be adjusted according to the water level and lifted up for maintenance without sending men down into the river. Meanwhile, to counter for varying load due to varying river current, an electronic control system was designed to maintain the electrical load on the transmission grid.

Figure 3. Fully submerged turbine in the river was hold by a long lever at the river bank [17].

Both vertical flow and axial flow turbines with different configurations were tested empirically in the field. It was found that the implementation of suction tube significantly increased the overall performance of the turbine. The best results were obtained with a six-blade turbine having 30% solidity at 2 m/s river current. Overall, the turbines were able to generate electrical power of about 1kW using a 2kVA and 220 V AC generator enough to power a remote medical station in the state of Bahia, Brazil with loads including a refrigerator, a freezer and some lightbulbs.

2.3 Field trial in the UK

A turbine having similar design as previously studied by Bahaj et al. [18] in a cavitation tunnel and a towing tank at the University of Southampton was placed on a field test at Yarmouth Pier by Lowe [19]. The turbine consists of three propeller blades of low solidity with a rotor diameter of 2.75 m as shown in Figure 4, suspended from jetty. Instead of using metal shafts or gears, the kinetic energy of the turbine was transferred to a generator through mechanical power transmission by using pressurized water.

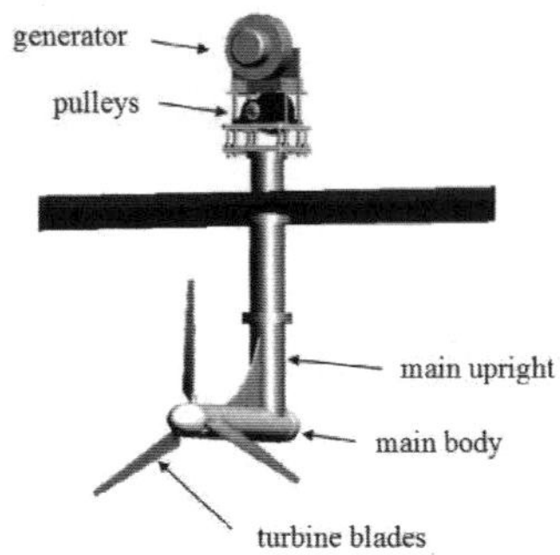

Figure 4. Design of the turbine adopted for the field trial study [18].

During field trial, the turbine was found to be able to generate about 1kW of electrical power at 1.25 m/s water velocity. The overall efficiency of the system is calculated to be about 27%. Similar to other axial flow turbines discussed earlier, the turbine was also found to encounter debris problem. It was reported that seaweed was found attached to the blades and the hub of the turbine shortly after being put in the water as shown in Figure 5. The performance of the turbine was found to be adversely affected by the debris problem. The presence of seaweed on turbine blades caused an increase in drag which slowed down the rotation of the turbine thus reducing the efficiency of the turbine. This problem has led to suggestions on avoiding or shedding debris automatically.

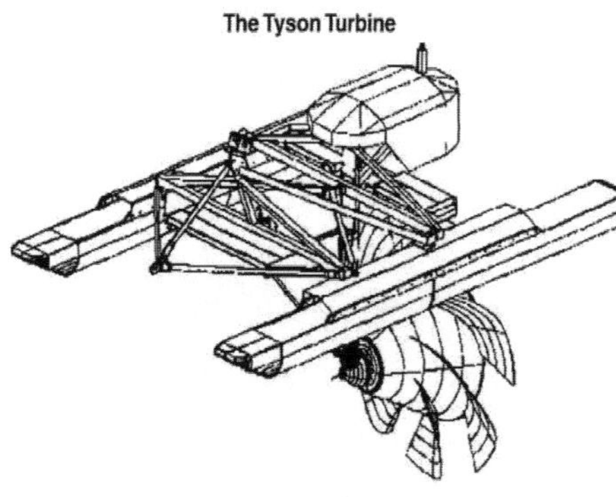

Figure 5. Seaweed on turbine [19].

3. Potential Application Of M-Hkt In Malaysia

Until December 2013, Malaysia has the maximum power generation capacity of 29, 748 MW with maximum load demand of 18, 902 MW [20]. According to the government forecast, the average annual primary energy consumption growth from the year 2004 to 2030 is 4.3%. In 2030, it is projected that the primary energy consumption will be tripled that of 2004 [21]. As the demand for energy consumption is increasing, with the increase of fossil-fuel prices as the primary energy resources, Malaysia is prioritizing renewable energy resources for electrical power production [22].

Currently, more than half of electrical power is generated from coal and natural gas which share the same percentage of 43.70%. This is followed by the renewable hydropower which contributes 8.70%, of total electrical power production, diesel (2%), fuel oil (1.2%) and other renewable resources (solar, wind and biomass) as shown in Figure 6. The percentages indicate that the electrical power production from renewable energy in total is so far less than 10%. Nevertheless, the energy generated through hydropower was found to increase year by year from 1990 to 2013 and Malaysia is targeting to increase its renewable energy production from 217 MW in 2011 to 11.5 GW by 2050 [20][23]. It is predicted that the main contribution to the increment of renewable energy resources will be from the

IOP Conf. Series: Materials Science and Engineering **370** (2018) 012053 doi:10.1088/1757-899X/370/1/012053

utilization of hydropower resource specifically small hydropower system because it is most economical, simple, low operation and maintenance costs with shorter duration of construction, environmental friendly and can give direct off-grid supply of electrical power.

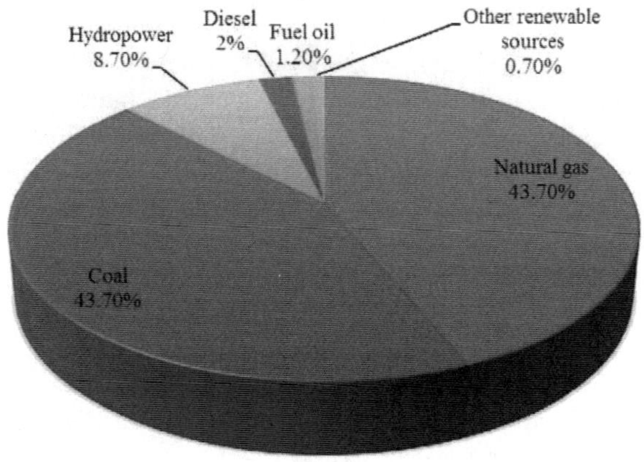

Figure 6. Electrical power production resources in Malaysia [24].

Malaysia is drained by a dense network of rivers and streams originated from highlands which act as catchment areas as shown in Figure 7. In Peninsular Malaysia, there are about 150 major river basins that flow downstream to the sea, the longest being the Pahang river (459 km) in the East Coast. Meanwhile, there are 50 major river basins in East Malaysia which mostly are larger than those in West Malaysia [25][26] while the Rajang river across the state of Sarawak being the longest river (563 km) in Malaysia and the Kinabatangan river in Sabah (560 km) being the second longest.

Due to the latitude of Malaysia that lies in the equatorial zone, the country has a tropical climate which allows reception of precipitation throughout the year. The rainfall is governed by the northeast and southwest monsoons and Malaysia receives rainfall the most during the period between these two monsoons [26]. It is estimated that the average annual rainfall in Malaysia is about 2450 mm in Peninsular Malaysia, 2630 mm in Sabah and 3850 mm in Sarawak. In fact, these values are higher than the world's average annual rainfall which is about 750 mm [8]. Unlike non-tropical countries which are prone to face drought crisis during dry season or frozen river during winter, the warm and wet tropical climate allows rivers in Malaysia to flow continuously every year.

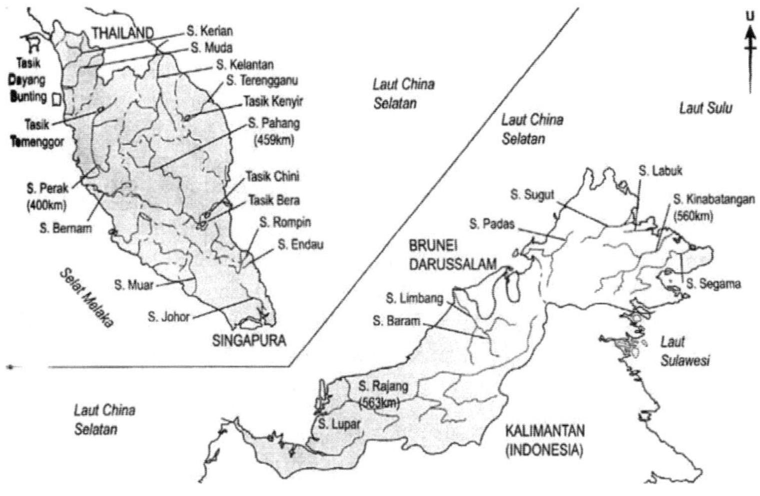

Figure 7. River networks in Malaysia [27].

Based on the vast river networks and the amount of rainfall per year, the hydropower resource is estimated to be 29,000 MW with 500 MW from small hydropower systems [8]. Therefore, Malaysia has a significant potential for the utilization of small scale hydropower especially μ-HKT where a continuous 24 hours of electrical power can be generated all year-round. Since μ-HKT generates electrical power directly from run-off-rivers, there are many potential sites available along the rivers that can opt for off-grid electrical power in remote areas where common grid power supply is less viable and non-economical due to some topological and financial challenges.

Most remote communities in rural areas live near rivers for source of water to perform their daily chores, agricultures and for drinking. These communities live under poverty with lack of access to basic needs including power supply. In Malaysia, there are roughly 8.4% of people in rural areas live below the poverty line and this is mostly found in remote areas in Sabah and Sarawak [28]. Statistical data on poverty in 2014 shows that the incidence of poverty level in Sabah is 3.9 which is the highest followed by Sarawak i.e. 0.9 [28]. The poverty level in remote areas is associated with energy poverty and lack of electricity which hinder positive economic growth and social development of the communities. In 2000, the electrical supply in rural areas in Sabah and Sarawak is found to be 67.05% and 66.91%, respectively. These values are far lower than in Peninsular Malaysia where the percentage of electrical supply in rural areas is above 90% as shown in Table 2.

Table 2. Electricity supply in urban and rural areas in Malaysia in year 2000 [25].

State	Urban	Rural	State	Urban	Rural
Johor	99.53	98.22	Perlis	99.63	99.17
Kedah	99.84	98.58	P. Pinang	99.84	99.16
Kelantan	99.52	97.50	Sabah	89.65	67.05
Melaka	99.90	99.28	Sarawak	93.96	66.91
N. Sembilan	99.61	98.60	Selangor	99.39	97.92
Pahang	99.63	93.96	Terengganu	99.65	98.24
Perak	99.64	96.11	W.P. Kuala Lumpur	99.76	-

For instance, due to the lack of grid electrical infrastructure, indigenous communities who live in thick rainforests in Kapit, Sarawak only rely on diesel generators to power their traditional longhouses [22]. The dependency on diesel-based electrical power contributes various problems to the surrounding environment. Fuel leakage can cause river pollution and reduce soil fertility which are used for agriculture whereas gas emissions from power generation increase greenhouse gases in the atmosphere. At the same time, the increase of global fossil fuel prices put huge burden to the local communities who live under poverty to get fuel supply. The situation becomes even worse as the price could be doubled due to difficulties to transport diesel supply to remote locations. Therefore, the utilization of low cost μ-HKT system is considered as a viable option to replace the conventional electrical power generation using diesel generator among remote communities.

Besides, μ-HKT has huge potential to provide off-grid electrical power at recreational sites located near the river. Some locations along the rivers in Malaysia are commonly developed to be recreational parks or camp sites due to their eco-tourism topological features such as Hutan Lipur Bukit Hijau in Baling, Kedah as shown in Figure 8. The utilization of electrical power from a low cost and environmental friendly small hydrokinetic system can minimize the operational cost of such sites thus encourage more visits from local and foreign visitors. In turn, it helps boost the growth in tourism sector as well as the socio-economics of nearby communities. Besides, the generated electricity can also be distributed to the surrounding communities where many of them still live in poverty with lack of access to the basic needs and limited power supply.

Figure 8. Hutan Lipur Bukit Hijau in Baling, Kedah [29].

4. Challenges

Although the entry of renewable μ-HKT system into the mainstream electrical power production in Malaysia is very promising, its full utilization in the rivers has yet to be realized. It is due to constraints in the current state of technology, debris threats and economical aspects which need to be addressed.

4.1 Technological aspects

Until now, the technology of μ-HKT is not well established yet where some are still under research and development phase. The μ-HKT technology is mostly adopted from the conventional large hydropower system and it does not really fit for small-scale applications. The field trial results from the case studies in the previous section showed that the maximum efficiency of the turbine was 0.27 from the theoretical maximum efficiency of 0.592 [12,18]. Various design aspects of μ-HKT still need to be investigated for optimal operation of the system.

4.2 Debris threat

Like most case studies, the utilization of μ-HKT in Malaysia has to address with the debris problem. Since Malaysian's tropical rivers flow through thick jungle, this threat can be more problematic. Sometimes huge amount of debris such as vines, leaves, logs and other jungle trash may float down the river at high speed during flash flood which can cause serious damages to μ-HKT installed in the river. This situation demand a better robust debris management strategies without reducing turbine overall efficiency and adding complexity to the systems [11].

4.3 Economical barrier

As the μ-HKT system is particularly opted to power electricity for poor communities in remote areas, cost is critical. Since the μ-HKT technology is still new, some components are not readily available in the rural areas, thus development, maintenance and replacement cost could be higher. Techno-economic analysis is required to evaluate the viability of the system design with respect to the cost without burdening the socio-economic status of the remote communities at potential sites.

5. Conclusion

A review on three case studies on the utilization of μ-HKT system has been provided. Each case study illustrated the feasibility and practicality of the system for low cost and low power demand. However, several aspects need to be investigated further for optimal and reliable operation of the system. These aspects are associated with the pertinent problems faced by the system during the field trials such as debris problems and overall efficiency enhancement. The review on the current status of μ-HKT system in Malaysia indicates a significant potential for off-grid power production in remote areas where power supply is very limited. This is strongly supported by the country's vast river networks with estimated small hydropower resource of 500 MW and Malaysia's vision in prioritizing the use of renewable energy. In order to fully utilize its potential, the challenges described here in terms of technological, potential site induced problems and economic issue need to be investigated and analyzed before implementing the μ-HKT system in Malaysia.

References

[1] Cust J, Singh A and Neuhoff K 2007 Rural electrification in India: economic and institutional aspects of renewable [Online]. [Accessed 26th July 2017]. Available from World Wide Web: http://www.eprg.group.cam.ac.uk/wp-content/uploads/2008/11/eprg0730.pdf

[2] Kumar D and Sarkar S 2016 A review on the technology, performance design optimization, reliability, techno-economics and environmental impacts of hydrokinetic energy conversion system, *Renew. and Sustain. Energy Rev.* **58**, pp 796-813

[3] Güney M S and Kaygusuz K 2010 Hydrokinetic energy conversion systems: a technology status review, *Renew. and Sustain. Energy Rev.* **14**(9), pp 2996–3004

[4] Williams G G and Jain P 2011 Renewable energy strategies. Sustain: *a J. of Environ. and Sustain. Issues*, **23**, pp 29–42

[5] Grabbe M, Yuen K, Goude A, Lalander E and Leijon M 2009 Design of an experimental setup for hydro-kinetic energy conversion. *The Int. J. on Hydropower & Dams*, **16**(5), pp 112–116

[6] Sarma N K Biswas A and Misra, R D 2014Experimental and computational evaluation of Savonius hydrokinetic turbine for low velocity condition with comparison to Savonius wind turbine at the same input power, *Energy Convers Manag.*, **83**, pp 88–98

[7] Paish O 2002 Small hydropower: technology and current status, *Renew. and Sustain. Energy Rev.*, **6**(6), pp 537–56

[8] Yah N F, Oumer A N and Idris M S 2017 Small scale hydro-power as a source of renewable energy in Malaysia: a review, *Renew. and Sustain. Energy Rev.,* 72, pp 228-239

[9] Khan, M. J., Iqbal, M. T. and Quaicoe, J. E. (2008) River current energy conversion systems: progress, prospects and challenges, *Renew. and Sustain. Energy Rev.*, **12**, pp 2177–2193

[10] Sørnes, K. (2010) Small-scale water current turbines for river applications. Zero Emission Resource Organization, January 2010. [Online]. [Accessed 23rd August 2017]. Available from Word Wide Web: http://www. zero.no/publikasjoner/small-scale-water-current-turbines-for-river-applica tions.pdf

[11] Anyi M and Kirke B 2010 Evaluation of small axial flow hydrokinetic turbines for remote communities, *Energy for Sustainable Development* **14**, pp 110-116

[12] Vermaak H J, Kusakana K and Koko S P 2014 Status of micro-hydrokinetic river technology in rural applications: a review of literature, *Renew. and Sustain. Energy Rev.,***29**, pp 625-633

[13] Tuckey A M, Patterson D J and Swenson W J 1997 A kinetic energy tidal generator in the Northern Territory—results, *IEEE 23rd International Conference on Industrial Electronics, Control and Instrumentation (IECON)*, New Orleans, pp 937-942

[14] Swenson W J 1999 The evaluation of an axial flow, lift type turbine for harnessing the kinetic energy in tidal flow, *Australian Universities Power Engineering Conference AUPEC*, Darwin, Australia

[15] Shannon R 1996 Waterwheel Engineering, *Proc. of 6th International Permaculture Conference & Convergence*, Perth, Australia

[16] Tiago G L 2003 The state of art of hydrokinetic power in Brazil, *Waterpower XIII*, Buffalo, New York

[17] Els R H and Junior A C P B 2015 The Brazilian experience with hydrokinetic turbines, *The 7th*

Int. Conf. on Applied Energy – ICAE, *Energy Procedia*, **75**, pp 259-264

[18] Bahaj A S, Molland A F, Chaplin J R and Batten W M J 2007 Power and thrust measurements of marine current turbines under various hydrodynamic flow conditions in a cavitation tunnel and a towing tank, *Renew. Energy*, **32**, pp 407-426

[19] Lowe W T 2007 Tidal Turbine, *Master Thesis*, University of Southampton

[20] Commission STE 2015 Malaysia energy statistics 2015. [Online]. [Accessed 24th August 2017]. Available from World Wide Web: http://meih.st.gov.my/documents/10620/f3b9119e-e139-4527-9da6-d77e2eab1c34

[21] Gan P Y and Li Z 2008 An econometric study on long-term energy outlook and the implications of renewable energy utilization in Malaysia, *Energy Policy*, **36**, pp 890–899

[22] Das H S, Tan C W, Yatim A H M and Lau K Y (2017) Feasibility analysis of hybrid photovoltaic/battery/fuel cell energy system for an indigenous residence in East Malaysia, *Renew. And Sustain. Energy Rev.*, **76**, pp 1332-1347

[23] Bekhet H A and Sahid E J M 2014 Highlighting energy policies and strategies for the residential sector in Malaysia, *Int. J. of Energy Economics and Policy*, **4**(3), pp 448-456

[24] Commision ME (2013) National energy balance. [Online]. [Accessed 24th August 2017]. Available from World Wide Web: http://meih.st.gov.my/documents/10620/167a0433-510c-4a4e-81cd-fb178dcb156f

[25] Borhanazad H, Mekhilef S, Saidur R and Boroumandjazi G 2013 Potential application of renewable energy for rural electrification in Malaysia, *Renew. Energy*, **59**, pp 210-219

[26] AQUASTAT 2016 Food and Agriculture Organization of the United Nations [Online]. [Accessed 14th September 2017]. Available from World Wide Web: http://www.fao.org/nr/water/aquastat/countries_regions/MYS/

[27] Geografi dan Anda 2012 2.0. Sungai di Malaysia [Online]. [Accessed 14th September 2017]. Available from World Wide Web: http://gcgeografismkkn.blogspot.my/2012/07/20-sungai-di-malaysia.html

[28] Household income & poverty statistics 2017 [Online]. [Accessed 14th September 2017]. Available from World Wide Web: http://www.epu.gov.my/en/socio-economic/household-income-poverty

[29] Yap Z Y 2016 Top 10 waterfalls in Kedah. [Online]. [Accessed 27th July 2017]. Available from World Wide Web: http://malaysiasaya.my/top-10-waterfalls-kedah

A qualitative study of vortex trapping capability for lift enhancement on unconventional wing

M B Salleh[a)], N M Kamaruddin[b)] and Z Mohamed-Kassim[c)]

School of Aerospace Engineering, Engineering Campus, Universiti Sains Malaysia, 14300 Nibong Tebal, Penang, Malaysia

[b)] Corresponding author: fazreena@usm.my

Abstract. Lift enhancement by using passive vortex trapping technique offers great advantage in small aircraft design as it can improve aerodynamics performance and reduce weight of the wing. To achieve this aim, a qualitative study on the flow structures across wing models with cavities has been performed using smoke wire visualisation technique. An experiment has been conducted at low Reynolds number of 26,000 with angle of attack (α) = 0°, 5°, 10° and 15° to investigate the vortex trapping capability of semi-circular leading edge (SCLE) flat-plate wing model and elliptical leading edge (ELE) flat-plate wing model with cavities, respectively. Results from the qualitative study indicated unique characteristics in the flow structures between the tested wing models. The SCLE wing models were able to trap stable rotating vortices for $\alpha \leq 10°$ whereas the ability of ELE wing models to suppress flow separation allowed stable clockwise vortices to be trapped inside the cavities even at $\alpha > 10°$. The trapped vortices found to have the potential to increase lift on the unconventional wing models.

1. Introduction

The capability to enhance lift generation and to delay flow separation on a wing can improve the aircraft aerodynamic performance. Many studies have been conducted to increase the performance particularly its lift-to-drag ratio by using a flow control technique across the wing which can be classified as active or passive flow control techniques. For instance, the active flow control techniques use high lift devices such as flaps or slats which can be installed on the wing. During take-off, the aircraft would extend these control surface devices to achieve an additional camber effect which would augment the lift [1]. The drawback of this active lift augmentation system, however, it adds substantial weight, complexity and manufacturing cost. Another alternative active control technique is by using power driven streamwise, tangential blowing and suction system [2]-[4]. This technique is mostly effective in the post-stall regime where a slot suction is introduced inside the trailing edge vortex cell of a thick airfoil [5][6]. However, the presence of the slot suction complicates the wing structure design.

Another reliable method to increase the lift is through the usage of passive trapped vortex. Ringleb [7] discovered this technique by introducing vortex cavities on the wing. The technique has proven to potentially delay the flow separation and promote the flow reattachment which can be associated to lift enhancement. Unlike active flow control, passive flow control technique does not require auxiliary power to operate any active mechanisms. The presence of cavities on the wing surface helps to reduce the weight of the wing. Therefore, this technique is favorable to be implemented on a small size aircraft such as an Unmanned Aerial Vehicles (UAVs) and Micro Aerial Vehicles (MAVs). Considering the requirements of having high performance aircraft under power and weight constraints, there is a growing number of research that focus on the study of vortex trapping for lift enhancement.

Yeung [8] has studied vortex trapping technique through numerical approach by using a conformal mapping sequence to explore the ability of an airfoil to trap a single and multiple vortices inside the cavities. In his study, the Joukowsky airfoil with a single cavity, two cavities and three cavities were investigated. Based on the numerical analysis, he concluded that the Joukowsky airfoil with the triple cavities can generate lift 10% higher than the conventional Joukowsky airfoil with the same thickness, camber and angle of attack. The trapped vortices were found to be rotated in a closed elliptical path. Meanwhile, Olsman and Colinius [9] conducted two dimensional direct numerical simulation of the flow across NACA 0018 airfoil without and with cavity at Reynolds number, Re = 20,000. The results of the study showed that the airfoil with a cavity produced a smaller wake as compared to the airfoil without any cavity at angle of attack $\alpha = 10°$.

Meanwhile, Fertis et al. [10] conducted an experimental study on a new airfoil design with a backward-facing steps. They found that the airfoil can increase the lift-to-drag ratio by 28% at $\alpha = 10°$ due to the trapped vortex. Then, Riddle et al. [5] performed similar flow visualization experiment by using a water channel to study the vortex trapping techniques. In their study, two airfoil designs were considered, one with a backward-facing step configuration and the other one with a dual-fence configuration. The dual-fence configuration was found to be more effective in trapping vortex than the backward-facing step as it requires low suction rates. Araque and Nožicka [11] conducted flow visualization study on an NACA 2415 airfoil with a backward-facing step using smoke wire technique. The double-coiled Nichrome wire coated with liquid glycerine produced a clear and distinct smoke streaklines and the results agreed well with the numerical simulation conducted by Araque [12].In addition, a three dimensional instability was also identified through experimental investigations. For instance, Savelsberg and Castro [13] studied the flow inside a large aspect ratio cylindrical vortex cell. They found that the flow in the cavity which was driven by a shear layer forming in the opening showed a nearly solid body rotation, with a low turbulence layer forming in the vortex core. Tee [14] adopted smoke wire technique of Araque and Nožicka [11] to collect a physical evidence of trapped vortex inside the cavity of NACA 0018 wing model. A flat plate wing models with different number of cavity were also tested as a fundamental study. From the experiment, it was found that the blunt semi-circular leading edge nose of the flat plate wing model caused the flow to separate earlier at the leading edge thus resulting in an unstable vortex formation inside the wing model with three cavities. It has been suggested that slender elliptical leading edge nose could potentially delay the flow separation and hence promote stable vortices inside the cavities.

In this paper, the qualitative study on vortex trapping capability technique for unconventional wing models conducted by Tee [14] is further extended by considering two different types of flat plate wing model with cavities i.e. (i) semi-circular leading edge (SCLE) wing model and (ii) elliptical leading edge (ELE) wing model. To simulate a small size aircraft (MAV and UAV) flying condition, wing models were tested under the same low Reynolds number of 26,000 inside a closed-circuit wind tunnel facility. A qualitative approach by using the smoke wire visualization technique was utilized to study the flow structure across the wing models and the dynamic of trapped vortex inside the cavity. At first, the study focuses at the vortex trapping capability of the SCLE wing model at angle of attack $\alpha = 0°$, 5°, 10° and 15° then followed by the ELE wing models for the same angle of attacks. The results obtained for all tested wing models are discussed in term of the capability of the wing models to trap the vortex inside the cavities at different angle of attack with respect to the type of the wing model.

2. Experimental Setup

2.1 Wind tunnel facility

The experiments have been conducted in a closed-circuit wind tunnel at the Wind Tunnel Laboratory, Science and Engineering Research Centre (SERC), Universiti Sains Malaysia. The wind tunnel has a rectangular test section of 1.0 m (width) × 0.8 m (height) × 1.8 m (length), a contraction ratio of 10:1 and a turbulence level of 0.1%. Airflow inside the test section can achieve a maximum speed of 80 m/s and it is recirculated by an axial fan motor and a diffuser downstream the test section. In addition, the test section is equipped with an internal balance mounted on a quadrant mechanism which has the capability to change the angle of attack of the wing models.

2.2 Smoke wire visualization technique

The airflow structures across the wing models have been visualized by using a smoke wire technique, as shown in Figure 1. For this setup, a single Nichrome wire with a diameter of 0.25 mm has been used to generate clear and distinct smoke lines. This wire has been placed vertically upstream with respect to the tested wing models, 0.57 m from the leading edge of SCLE wing model (equivalent to 0.47 m from the leading edge of ELE wing model) and 4.5 cm inbound of the wingtip to provide clear images of the flow structures across the wing models. Electric current were used to heat up the wire by connecting the positive and negative terminal at both end of the wire by using some crocodile clips. A direct current (DC) power supply of 31 V and 1.2 A were used to supply the electrical current across the wire. The smokes were generated when the oil dripping down the heated Nichrome wire evaporated. For the source of smoke, Safex oil has been used due to its non-toxicity. The oil was dripped down through the wire manually from the top of the wind tunnel test section by using a syringe. Since the smokes are white, the top and rear backgrounds of test section have been covered with a black sugar paper to increase the contrast and to avoid any reflection of light source from the surrounding. In order to capture clear images, the smoke streak lines have been illuminated with four parabolic white reflector halogen lamps, with each two were installed at the top and bottom of the test section, respectively.

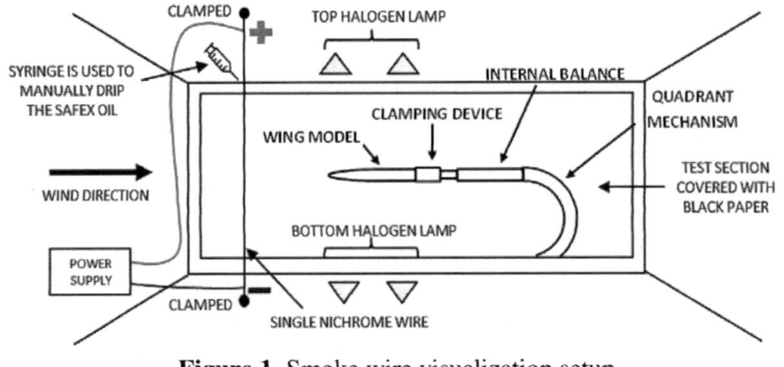

Figure 1. Smoke wire visualization setup.

2.3 Wing models

Two types of wing models were considered in this study, distinguished by the shape of the leading edge nose (i) semi-circular leading edge (SCLE) flat-plate wing model and (ii) elliptical leading edge (ELE) flat-plate wing model, as shown in Figure 2. These wing models have the same thickness of 0.03 m and wing span of 0.5 m. The SCLE wing model has a chord length of 0.25 m which gives the aspect ratio of the wing AR = 2 whereas the ELE wing model has a chord length of 0.35 m which give the aspect ratio AR = 1.43. Both wing models have three cavities of the same size on the upper surface where the dimension of the cavities is adopted from Yeung [8]. In addition, the wing models have holes at the mid-section of the trailing edge to accommodate for the clamping device. All wing models were painted in matte black to reduce the light reflection during experiment.

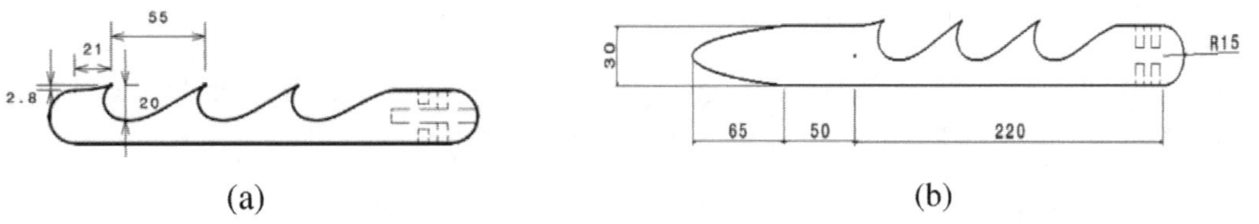

Figure 2. Cross section views of (a) SCLE wing model with three cavities and (b) ELE wing model with three cavities. Units are shown in millimeters (mm).

2.4 Experimental procedure

The wing model was mounted on the internal balance by using a clamping device and its initial angle of attack was set at 0° by using a digital inclinometer. Then, the qualitative study on the flow structures across each wing model was performed at Reynolds number Re = 26,000 for an angle of attack α = 0°, 5°, 10° and 15°, respectively. First, the experiment was performed for the SCLE wing models then followed by the ELE wing models. Since the SCLE and ELE wing models were different in chord length, the airspeed was set at 1.5 m/s for the SCLE wing models whereas for the ELE wing models, the air speed was set at 1.1 m/s corresponding to the similar Reynolds number. A video camera with a resolution of 1920 × 1080i and 30 frames per second (fps) was employed to acquire the images of the airflow structure across the wing models. The camera was placed outside of the test section and located perpendicular to the smoke streak lines, as shown in Figure 3. A black cloth was used to cover the camera during recording process to reduce any external light disturbance. The images of the airflow structure across the wing models were recorded and all acquired images were then analyzed.

Figure 3. A camera was mounted on the tripod stand outside the test section. A black cloth was used to cover the camera to reduce light disturbance from the surrounding.

3. Results

This section presents and discusses the smoke flow visualization of the flow structures across the wing models at different angle of attack, respectively. The experimental results are discussed in terms of the flow structures across the tested wing models and the dynamics of trapped vortices inside the cavities.

In addition, the comparison on the flow structures and capability to trap vortex in the cavity between SCLE and ELE wing models is also discussed. Note that the flow direction of the smoke streak lines presented in the following figures is from left to right.

The airflow dynamics across the SCLE wing model with triple cavities at different angle of attack are shown in Figure 4. At $\alpha = 0°$ as depicted in Figure 4(a), the flow remains attached across the wing model with a stagnation point located at the leading edge nose. The flow right above the first and the second cavity was captured into the cavities to form a clockwise vortex. There was also a vortex trapped inside the third cavity which was formed when the flow structures inside the second cavity escaped and then entered the third cavity. The trajectory momentum and intensity of the vortex flow inside the second cavity was observed to be more distinct than the vortex inside the first cavity which suggests the capability of the trapped vortex to escape the second cavity. Meanwhile, vortex formation was observed in all cavities at $\alpha = 5°$, as shown in Figure 4(b). The trapped vortex inside the first and second cavity was circulating in a clockwise direction and formed as the freestream flow entered the cavities. The intensity of the vortex inside the first cavity was much more distinct than the inside of the second cavity and sufficient to induce flow reattachment downstream the leading edge nose. As the freestream flow above the cavities was sucked in at the cavity trailing edge, it can be deduced that all trapped vortex inside the cavities rotated in a clockwise direction. At $\alpha = 0°$ and 5°, however, the vortex inside the cavities only trapped at the top portion of the cavity. As the angle of attack increased to $\alpha = 10°$ in Figure 4(c), the flow separated earlier at the trailing edge of the wing model and no distinct vortex was found to be trapped inside the first cavity. Moreover, unstable vortex formations were observed inside the second and third cavity which came from the wingtip vortex. These trapped vortex were observed to be turbulent in motion which hindered the flow reattachment to occur. At $\alpha = 15°$, the airflow across the wing model was completely separated, as depicted in Figure 4(d). There was no trapped vortex from the freestream in any cavities except the wingtip vortex that entered the second and third cavity. Based on the results, it can be deduced that the tested SCLE wing model with three cavities was more reliable to trap vortex at a small range of angle of attack. In this range, a stable clockwise rotating vortex can only be captured inside the first and the second cavity. Inside the third cavity, however, the flow structure was turbulent that formed from the wingtip vortex which becomes more prominent as the angle of attack increases.

Figure 5 shows the airflow structures across the ELE wing model with three cavities. At $\alpha = 0°$ the stagnation point was located exactly at the tip of the leading edge nose of the wing model and the flow remained attached on the wing surface. There was a vortex trapped inside the first cavity and it was rotated in a clockwise motion, as shown in Figure 5(a). Unlike the vortex inside the cavities of the SCLE wing model which was trapped only at the top portion of the cavity, the vortex inside the first cavity of the ELE wing model had covered the whole space of cavity thus forming a stable flow structure. Moreover, the intensity of the vortex was distinct than the vortex in the SCLE wing model which suggests that a stronger vortex with a higher momentum was trapped inside the cavity. The clockwise circulation of the vortex inside the first cavity also kept the freestream flow to remain attached towards the trailing edge of the wing model. Detailed observation found that there was a flow injected into the second cavity but the trapped vortex inside this cavity was very weak whereas there was no vortex observed inside the third cavity. Similar flow structures were observed at $\alpha = 5°$ where there was a clear and distinct clockwise vortex trapped inside the first cavity, as shown in Figure 5(b). A faint and weak vortex can also be observed inside the third cavity and it was formed as the wingtip vortex entered the cavity.

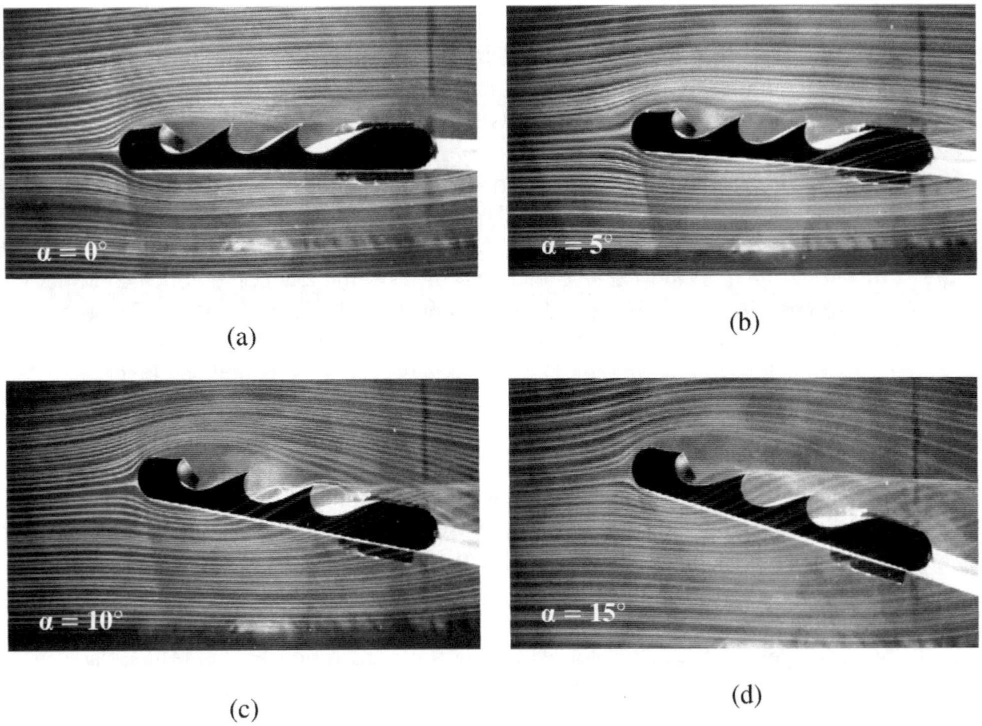

Figure 4. Airflow structures across the SCLE wing model with three cavities.

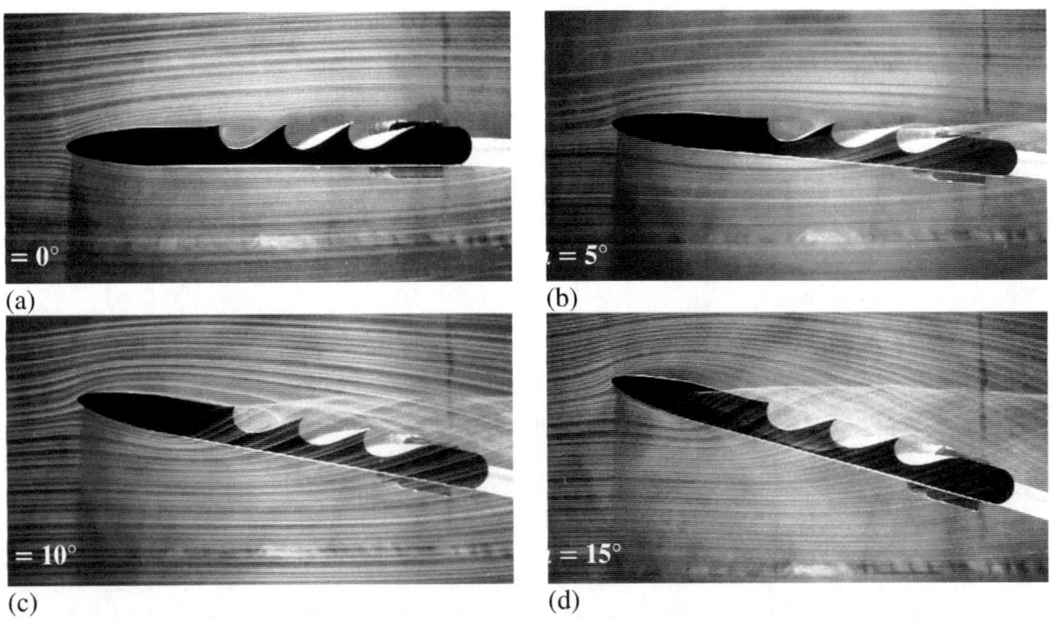

Figure 5. Flow structures at tested angle of attacks for ELE wing model with triple cavities.

For α = 10° a vortex was successfully trapped by the first cavity, as shown in Figure 5(c). The trapped vortex has the similar trajectory strength and as distinct as the case of α = 0° and 5°. The trapped vortex was observed to circulate in a clockwise motion and promoted consistent flow reattachment after the trailing edge of first cavity. In addition, there was a trapped vortex inside the second cavity.

Unlike for the case of SCLE wing model, this vortex was formed as the freestream flow engulfed into the second cavity. However, it has less prominent flow structure as compared to the vortex inside the first cavity thus failed to induce a proper flow reattachment at the trailing edge of the second cavity. This situation caused the freestream flow to slightly fluctuate over the third cavity downstream of the trailing edge of the wing model thus no vortex was observed inside the third cavity. When the wing model was pitched to $\alpha = 15°$, the airflow separated after the leading edge nose but reattached again as it flowed pass the leading edge nose and across the wing model, as depicted in Figure 5(d). It was observed that the shear layer entered the cavities to form a clockwise vortex. Although the intensity of the wingtip vortex grew greater and became more prominent, the trapped vortex inside the cavities was observed to be dominated by the incoming freestream flow above the cavities. The trapped vortex has the capability to successfully promote flow reattachment at the trailing edge of each cavity thus suppressing the flow separation downstream the wing model even at high angle of attack.

4. Conclusion

The vortex trapping capability across two different types of wing models with three cavities have been qualitatively studied using the smoke wire visualization technique. The experiments have been conducted to investigate the characteristic of the flow structures on the upper surface of the wing with the present of cavities. The results have revealed that the flow structures across the wing models were affected by the shape of the leading edge nose. The blunt leading edge nose of SCLE wing model caused the freestream flow to separate at high angle of attack. In contrast, with a slender leading edge nose, the ELE wing model successfully suppressed the flow separation for the tested angle of attacks thus increased the stall angle of the wing. As a result, the SCLE wing models with cavities only able to trap stable clockwise vortex for $\alpha \leq 10°$. These stable vortex were normally found to be trapped inside the first and second cavity whereas the vortex trapped inside the third cavity was unstable and formed as the wingtip vortex entered the cavity. Meanwhile, the ELE wing model was able to trap a stable clockwise vortex inside all cavities. The vortex trapped inside the cavities were found to be more stable and distinct than that of SCLE wing model with less influence from the wing tip vortex even at higher angle of attack $\alpha > 10°$. The trapped clockwise vortex inside the cavities for both wing models have the capability to enhance lift proven by the clockwise motion configuration.

Acknowledgements

The authors acknowledge financial support from Universiti Sains Malaysia, wing model designs and experimental setup assisted by Mr Tham Weng Hong, Ms. Tey Shen Xi, Ms. Tee Yi Hui, Mr. Anuar M Nazri and technical supports from Mr Mohamad Najhan Awang and Mr Mahmud Isa from School of Aerospace Engineering USM.

References

[1] Yeung W W H 2009 Vortex trapping on a surface with an indentation and corrugations *Mathematics and Computers in Simulation*, **79**, pp 3243–57.

[2] Baranov P A, Isaev S A, Prigorodov Y S and Sudakov A G 1999 Numerical modelling of the effect of improving the aerodynamic efficiency of profiles due to the suction in vortex cells *J. of Eng. Physics and Thermophysics* **72** pp 550–53

[3] Baranov P A, Isaev S A, Prigorodov Y S and Sudakov A G 2003 Controlling the turbulent flow past a thick airfoil by means of flow enhancement in a vortex cell using suction from central body surfaces *Fluid Dynamics* **38** pp 387–96

[4] Isaev S A, Miau J J, Sudakov A G and Usachov A E 2015 Analysis of extremal lift behavior of a semicircular airfoil in a turbulent flow at a near-zero angle of attack *Technical Physics Letters* **41** pp 737–39

[5] Riddle T W, Wadcock A J, Tso J and Cummings R M 1999 An Experimental Analysis of Vortex Trapping Techniques *J.of Fluids Engineering* **121** pp 555-59

IOP Conf. Series: Materials Science and Engineering **370** (2018) 012054 doi:10.1088/1757-899X/370/1/012054

[6] Isaev S, Baranov P, Popov I, Sudakov A and Usachov 2016 A Improvement of Aerodynamic Characteristics of a Thick Airfoil with a Vortex Cell in Sub- and Transonic Flow *Acta Astronautica* **132** pp 204-20
[7] Ringleb F O 1961 *Boundary Layer Flow Control Device* US Patent 3,000,40
[8] Yeung W W H 2006 Lift Enhancement of Unconventional Airfoils *J. Mekanikal*, **79**, pp 17-25
[9] Olsman W F J and Colonius T 2011 Numerical Simulation of Flow over an Airfoil with a Cavity *AIAA JOURNAL* **49** No. 1 pp 143-149
[10] Fertis D G 1994 New Airfoil Design Concept with Improved Aerodynamic Characteristics," *J. of Aerospace Eng.* **7** No. 3 pp 328-39
[11] Araque L. V- and Nožicka J 2011 Oil And Smoke Flow Visualization Past Two-Dimensional Airfoils For An Unmanned Aerial Vehicle *Visualization of Mechanical Processes* **1** (3) pp 1-8
[12] Araque L V- 2010 Development of an Airfoil for an Unmanned Aerial Vehicle with Internal Blowing Propulsion System *Ph.D Thesis* Czech Technical University in Prague
[13] Savelsberg R and Castro I 2008 Vortex Flows in Open Cylindrical-Section Cavities *Exp. Fluids* **46** (3) pp 485-97
[14] Tee H Y 2015 Development of Smoke Wire System for Vortex Trapping Application, *Final Year Project Thesis* Universiti Sains Malaysia

Effect of fuselage diameter on aerodynamic characteristics for straight wing at low and high aspect ratio

H C Lai[a)] and N M Kamaruddin[b)]

School of Aerospace Engineering, Engineering Campus, Universiti Sains Malaysia, 14300 Nibong Tebal, Penang, Malaysia

[b)] Corresponding author: fazreena@usm.my

Abstract. A computational code has been developed to investigate the effect of fuselage diameter with a combination of a straight wing on the aerodynamic characteristics. The case studies were performed on both wing-body combination at high wing aspect ratio (AR>4) and low wing aspect ratio (AR<4). The Prándtl's Lifting Line Theory (LLT) and Helmbold's Panel Method (HPM) were used to determine the wing lift and drag characteristics while Roskam's Drag Prediction Method (RDPM) was used to compute the fuselage drag. The computed lift and drag coefficients obtained were then compared to verify the reliability of the computational results. It is found that a larger fuselage diameter contributes to the increment of the total drag of the wing-body combination without significant effect on the lift. The lift is mainly contributed by the wing for both low and high aspect ratio straight wing for low subsonic case.

1. Introduction

In an aircraft design process, each component of the aircraft must be optimized to achieve maximum performance. The most ideal condition in the design process in terms of aerodynamic is the aircraft should have a maximum production of lift and minimum production of drag to improve the efficiency. Hence, it is important to take into account of every surfaces, parts or components of the aircraft that contributes to the aerodynamic performance. The importance of lift and drag for an aircraft are significant to the overall flight performance. The wing generally contributes the most in terms of lift of the aircraft. However, every single part of the aircraft such as fuselage also contributes to the overall lift, albeit small. Fuselage is the main body structure of an aircraft which holds all the smaller parts such as wings, canards, empennage and landing gears together. It also consists of the cockpit, passenger compartment (cabin) and cargo compartment. Thus, designing a fuselage is neither easy nor simple task, especially for a civil aircraft as the passengers comfort must also be taken into account while sizing it and the fuselage covers a large area of an aircraft which most likely contributes to most of the drag. To minimize the effect of this problem, fuselage is usually designed to be streamlined shape to minimize the pressure drag.

In this article, the effect of wing-body combination on aerodynamic characteristics is studied and discussed, where the wing-body combination is the combination of fuselage and wing. The lift of the wing-body combination is not simply the addition of the lift of wing and lift of body alone, but the flow field across the whole wing-body has to be taken into consideration. The main objective of this project is to investigate the effect of fuselage diameter for both straight wing for high AR and low AR on the aerodynamic characteristics. A MATLAB program is constructed to aid the study performed. The program is capable to compute the lift and drag for the wing-body combination and provides the comparison for different fuselage diameter. The program is also capable to assist an aircraft designer to perform the sensitivity analysis on aerodynamics characteristics for a wing-body combination when sizing a fuselage.

2. Literature Review

As the fuselage is a very important component in an aircraft and it poses a huge effect on the overall aircraft aerodynamic characteristics, many studies and researches are performed to study the fuselage design and also the wing-body combination of the aircraft on its aerodynamics characteristics. Some of the studies are carried out on the in-service aircraft in order to predict and improve its aerodynamic performance. These researches are essential as any findings will help in the overall aircraft aerodynamics characteristics and its performance.

The Commercial Aeronautics Sector is well aware that there is a need in fulfilling the society's needs meanwhile reducing the environmental impact and global warming. A balancing between the public's requirement of cheaper fares and the competition of economic infrastructure is hard to achieve without any significant improvements. Abbas and colleagues [1] had developed new aircraft combinations with laminar and turbulent drag reduction technologies and flow control devices to improve the aircraft performance under separated flow conditions of unsteady nature, also to reduce the complexity of the high-lift devices. The main objective of his research is to reduce the induced drag and noise.

The two basic parameters, namely lift and drag are crucial for the structural design of an aircraft. Before building a real aircraft, it is desirable to know the approximation of the aircraft's aerodynamics properties. There are many software that are available in the market to perform simulation on the designed aircraft. However, to achieve a more reliable results, wind tunnel test is usually recommended. Ramon Lopez Pereira [2] from Linköpings University, Sweden had conducted the validation of four different aerodynamic modelling software (Tornado, AVL, PANAIR and a handbook-type preliminary method) by comparing the results obtained with experimental results using wind tunnel. He found that the shape of the wing will caused some of the methods to have difficulties in converging to valid results.

F. Nicolosi et.al. [3] had also performed a study on fuselage aerodynamics prediction methods. The study is performed to develop a CFD-based method to predict longitudinal and lateral-directional aerodynamic coefficients (drag, pitch and yaw) of and aircraft. This study carried out shows that about 30% of and aircraft zero lift drag source is attained to the fuselage. As aerodynamics characteristics are strictly related and will affect the stability of the aircraft, it is vital to obtain a reliable aerodynamic characteristic estimation to carry out a well-designed aircraft. The other group of researchers led by Welstead [4] suggest that using vortex lattice method in aerodynamic characteristics prediction as it produce a reasonably accurate estimation for the lift distribution, induced drag, and the moments for a given planform.

As mentioned above, aerodynamics characteristics will have a direct impact on the aircraft stability, thus, there are researches that are conducted in order to understand the effect of the aerodynamics characteristics on the aircraft's stability behavior. Walter and David [5] had performed the study on the static longitudinal and lateral stability characteristics at low speed of unsweep mid-wing with models of different AR. The research is done by conducting experiments in the Langley stability wind tunnel where both aerodynamics and stability characteristics can be studied on the same time.

In addition to that, Charles F. Hall [6] has also presented the results obtained from a wind tunnel test experimenting on the thin, low aspect ratio wings in combination with a body at subsonic and supersonic speed. The data obtained are lift, drag and pitching moment coefficients for varies Mach number range from 0.25 to 1.9. Furthermore, a series of wing with different planform, aspect ratio, thickness, thickness distribution, wing camber and twist in combination with the fuselage to perform study on the wing-body combination. His research shows that wings with a sharp leading edge had a lower value of minimum drag at supersonic speed.

3. Theory

To determine the lift curve slope for a straight finite wing, lifting line theory (LLT) is the most common method, where lift curve slope of the finite wing, a is given as:

$$a = \frac{a_0}{1+(a_0/\pi e AR)} \tag{1}$$

Where a_0 is the lift curve slope of the airfoil used for the finite wing while e refers to the Oswald efficiency factor which is a correction factor that is proportional to the square of wing lift, C_L^2, taking into account the non-elliptical lift distribution on wings general shape [7, 8].
According to McCormick [9],

$$e = \frac{1}{1+\delta} \tag{2}$$

The induced drag factor, δ is a function of aspect ratio (AR) and taper ratio (TR) which can be obtained graphically. However, the value of δ available is limited up to AR of 10 only. TR is the ratio of the tip chord to the root chord. It will affect the lift distribution and structural weight of the wing [10]. A tapered wing helps to create lift distribution that is more similar to an elliptical lift distribution.

However, LLT is not capable to determine the lift curve slope for the wing in some cases where AR < 4. Hence, modification is done on equation (1), making it into the Helmbold equation which is applicable to low AR straight wing as shown below: [11]

$$a = \frac{a_o}{[1+(a_0/\pi e AR)^2]^{1/2} + a_o/\pi e AR} \tag{3}$$

Induced drag coefficient, $CD0$ for the wing can be determined directly using lift curve slope because $CD0$is directly proportional to the $CL2$.

When the wing is attached to the fuselage, the fuselage contributes a large amount of drag to the entire wing-body combination. Flow separation due the geometry of the fuselage is one of the source of drag. Fuselage will contribute few types of drag which includes friction drag, interference drag, form drag, base drag and induced drag [12].

There are several methods that are capable to compute the drag of the fuselage. The most common method used is the Roskam's Drag Prediction method where it suggest that the fuselage drag coefficient is the summation of the fuselage zero-lift drag and drag due to lift. It can be represented using the following equation [13]:

$$C_{D_f} = C_{D_{0_f}} + C_{D_{L_f}} \tag{4}$$

where:
$C_{D_{0_f}}$ = zero-lift drag coefficient of the fuselage
$C_{D_{L_f}}$ = fuselage drag coefficient due to lift

As the fuselage drag is highly dependent on the fuselage geometry, its diameter and length will be the main parameters that influence the fuselage drag. Hence, the equation below shows the equation that used to model the zero-lift drag which is also the function of the fuselage geometry.

$$C_{D_{0_f}} = R_{wf} C_{f_f} \left[1 + \frac{60}{(l_f/d_f)^3} + 0.0025\left(\frac{l_f}{d_f}\right)\right] \frac{S_{wet}}{S} + C_{D_{b_f}} \tag{5}$$

where:
R_{wf} : Wing-fuselage interference factor

C_{f_f}: Turbulent flat plate skin-friction factor as a function of Mach number and the Re.
l_f: Fuselage length
d_f: Maximum fuselage diameter
S_{wet} : Wetted area of the fuselage
$C_{D_{b_f}}$: Fuselage base drag coefficient

Based on equation (5), in order to obtain a minimal fuselage zero-lift drag coefficient, its wetted area should be minimized and fuselage fineness ratio should be maximized. Another term that will affects the fuselage zero-lift drag coefficient is the base drag which is produce due to the presence of base area of the fuselage. Figure 1 show the fuselage without base area and with base area. The presence of base will leads to flow separation, where the larger the base area, the greater the base drag.

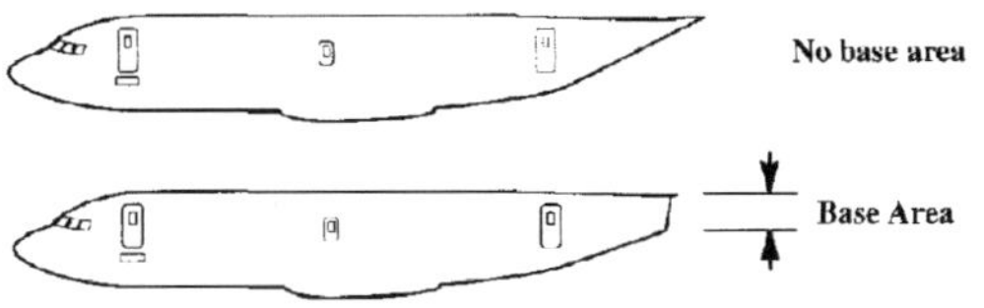

Figure 1. Example of fuselage without and with base area.

The base drag coefficient of the fuselage can be determined by using the equation below [13]:

$$C_{D_{b_f}} = \left\{ \frac{0.029(d_b/d_f)^3}{C_{D_{0_{fus-base}}}(S/S_{fus})^{1/2}} \right\} \{S_{fus}/S\} \tag{6}$$

where,
d_b: Fuselage-base diameter
$C_{D_{0_{fus-base}}}$: Zero-lift drag coefficient of fuselage exclusive of the base from equation (5)
S_{fus}: Fuselage maximum frontal area

From equation (6), the higher base diameter will leads to a greater base drag coefficient as it yields a higher base area.

Similar to the wing, a fuselage will also experience drag that are due to lift. The drag due to lift can be computed using the following equation:

$$C_{D_{L_f}} = \frac{\left(2\alpha^2 S_{b_{fuse}}\right)}{S} + \frac{\left(\eta C_{d_C} \alpha^3 S_{plf_{fuse}}\right)}{S} \tag{7}$$

where,
α: Fuselage angle of attack in radian
$S_{plf_{fuse}}$: Fuselage planform area
C_{d_C} : Experimental steady state cross-flow drag coefficient of a circular cylinder
η: Ratio of the drag of a finite cylinder to the drag of an infinite cylinder

Fuselage drag due to lift coefficient is relatively very small compare to the fuselage zero-lift drag. Thus, in some cases, it is neglected as its value is not significant to the total drag of the fuselage. However, it is taken into account for this project to achieve a more accurate computational result.

Although the individual lift of the fuselage is not calculated as its value is very small, the lift generated by the entire wing-body is still been calculated. The lift curve slope of the wing-body combination can be obtained by the equation below:

$$C_{L_{\alpha_{wf}}} = K_{wf} C_{L_{\alpha_w}} \tag{8}$$

where,
$C_{L_{\alpha_{wf}}}$: Wing-fuselage (wing-body) lift curve
$C_{L_{\alpha_w}}$: Wing lift curve slope
K_{wf}: Wing-fuselage interference factor

Lift curve slope of the wing-fuselage is the function of the lift curve slope of wing with the wing-fuselage interference factor. This is because the flow across the wing and fuselage will interfere at the position where wing is attached to the fuselage. The wing-fuselage interference factor can be obtained by the equation:

$$K_{wf} = 1 + 0.025\left(\frac{d_f}{b}\right) - 0.25\left(\frac{d_f}{b}\right)^2$$

From equation 9, it can be agreed that diameter of fuselage and wingspan are the parameters that will directly affects the lift-curve slope of wing-body.

4. Methodology

This article focuses on the effect of fuselage diameter on aerodynamic characteristics for straight wing at both low and high AR. The chosen airfoil, NACA 65209 is analyzed for both low and high AR in order to establish a consistent comparison. The AR that is used to perform the computation is 2 for low AR case and 6 for high AR case while the Reynolds number (Re) are 100000 and 590000 respectively. Vary in Re is due to different chord length of the model for both cases. Other parameters such as TR, wing span, Mach number, fuselage diameter, fuselage base diameter and fuselage length are consistent for both cases, which are 0.6, 1.1025m, 0.13, 0.15m, 0.075m and 1.125m respectively.

The computation of wing-body combination's aerodynamics characteristics are performed using MATLAB, by constructing relevant coding to perform the simulation. LLT is applied to determine the aerodynamic characteristics of the wing. Meanwhile, fuselage drag is computed by using Roskam's Drag Polar Prediction Method. The general idea of this method is basically summing up the aircraft components that contributes to the drag. Wing-body lift is also calculated by using Roskam's Lift Prediction Method and is compare with lift produced by wing alone.

After obtaining all the results, it is compared with established journal results to verify the accuracy of computed result. Similar to all the values of coefficients that are graphically presented, the established journal result data was extracted by using WebPlotDigitizer and plotted in the same graph with analytical result to the ease of comparison.

5. Results

In this article, two case study are performed. The first case study is performed on low AR NACA 65209 wing (AR=2) and wing-body combination at Re=10^6. The outcome of the program is as shown as below.

Referring to figure 2, it shows that lift coefficient for wing and wing-body combination have almost the same lift slope. However, zooming into the graph, it can be seen that the wing body lift curve slope is slightly lower than the wing. This is because the lift of the wing is calculated separately using LLT which considering only the wing itself. Thus, in the wing-body combination where part of the wing is embedded in the fuselage as attachment, the area of the lifting surfaces decrease leading to the decrease in lift produced by the wing, and the fuselage does not produced any lift in this case, causing the lift generated by the wing-body combination is less than the wing. It can be agreed that the lift of wing-body combination is simply from the wing.

Fuselage contributes significant amount of zero-lift drag as it consists of skin friction drag and pressure drag which due to flow separation. Hence, it can be clearly seen in figure 3 that wing body combination has a higher drag coefficient compare to wing alone. The difference of drag coefficient of the wing-body combination and the wing indicates the amount of drag produced by the fuselage.

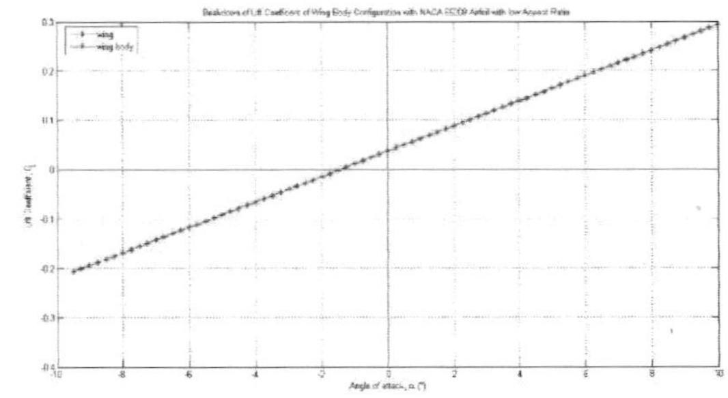

Figure 2. Breakdown of lift coefficient of wing body combination for low AR.

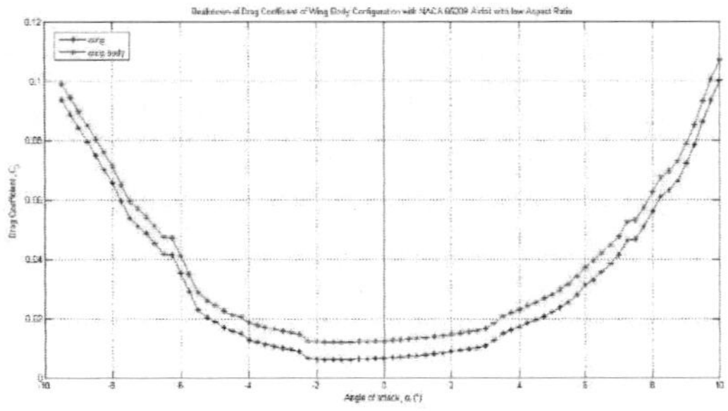

Figure 3. Breakdown of drag coefficient of wing body combination for low AR.

Since the effect of fuselage diameter on lift and drag coefficients is the main objective, the results are clearly presented in figure 4 and figure 5. Figure 4 shows that the lift produced by wing-body combination with different fuselage diameter is essentially constant. This is because the fuselage does not produce any lift in this case, thus changing the fuselage diameter does not affect amount of lift generated. Besides, the varying value is only the fuselage diameter while the wing parameter are kept constant, thus the amount of lift generated by the wing at each wing-body combination is the same, yielding the same results.

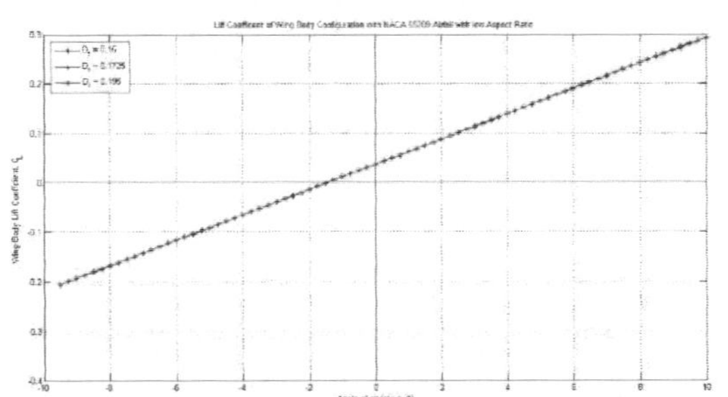

Figure 4. Lift coefficient of wing body combination with different fuselage diameter at low AR.

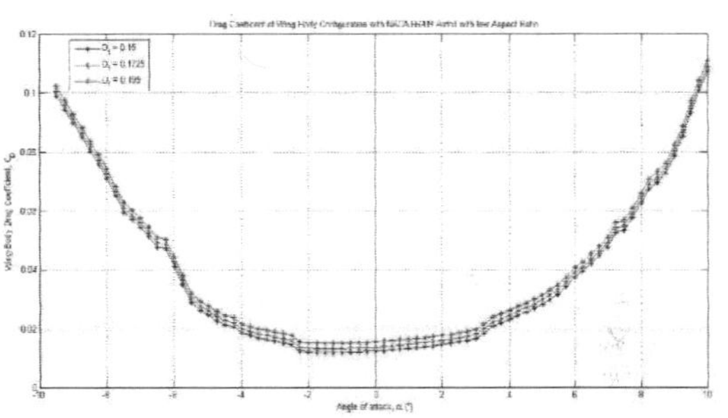

Figure 5. Drag coefficient of wing body combination with different fuselage diameter at low AR.

The effect of fuselage diameter toward the drag coefficient can be seen in figure 5, where it shows that the larger fuselage diameter experience higher drag coefficient. This is due to larger fuselage diameter will has a greater surface and cross-sectional area, resulting in greater skin friction and pressure drag. According to equation (5) and (6), fuselage drag coefficient is the function of fineness ratio which is inversely proportional to the fuselage diameter. Fuselage drag coefficient will increases as fineness ratio decreases. This indicates that the fuselage drag coefficient increases with fuselage diameter.

In order to verify the results obtained from the program, comparison between the computational results and published technical journal are done. The published journal results were experimental data obtained by Walter D. W and David F. T., Jr. [5] are plotted together with the theoretical data and presented in figure 6 and 7 for low AR case. However, the available data obtained from the journal is only started from angle of attack at -4°.

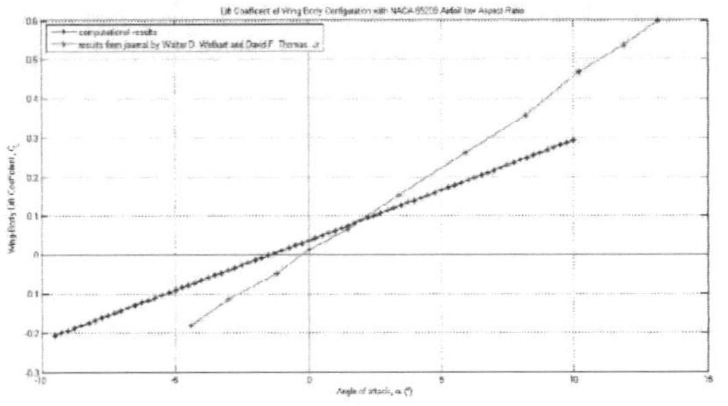

Figure 6. Comparison of theoretical results with journal for lift coefficient against angle of attack with low aspect ratio.

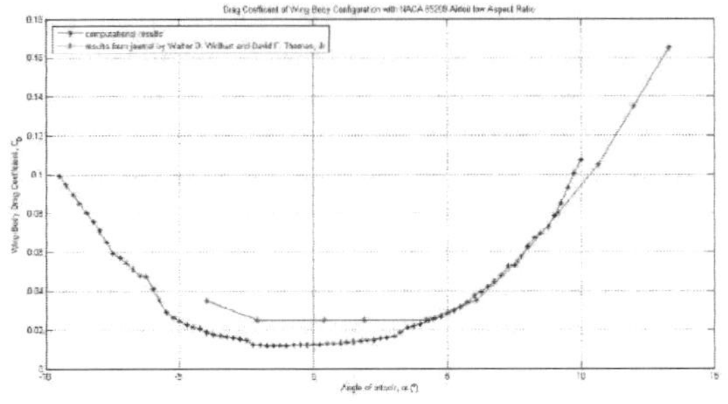

Figure 7. Comparison of theoretical results with journal for drag coefficient against angle of attack with low aspect ratio.

Referring to figure 6, the percentage difference of the experimental data with the computational result is approximately 38.75%. The experimental data has a higher lift curve slope. One of the reasons might be due to the airfoil data used in this study is NACA 65209 while the airfoil used in the experiment is NACA 65A008 as there is no available data for the latter in the airfoil database. The difference in thickness might influence the flow across the airfoil might delay the flow separation. The other reason might due to the difference in Re. The Re used in the experiment is 1.02×10^6 while due the limitations of available airfoil data, the Re used to perform the computation is 1.0×10^6. This might cause the computational results to lose some accuracy.

From figure 7, experimental result for wing-body drag coefficient is almost the same with the computational result when the angle of attack is greater than 4°. It both results has a mean percentage difference of 29.4%. This indicates the program is capable to compute drag for the wing-body combination for low AR case. The difference of both results might due to the experiment is carried out in the wind tunnel to simulate the real case of the flow. However, a lot of undesired disturbance that is unpredictable might occur which has the tendency to affect the accuracy of the result. Besides, the surface roughness of the prototype might contribute more skin friction drag as estimated.

The second case study is performed on high AR (AR = 6) of NACA 65209 wing and wing body combination at Re=5.9×10^5. Similarly, the lift slope for wing-body combination is almost the same lift slope for wing as shown in figure 8. The zoom in view of the graphs shows that the wing-body lift curve slope is lower than the wing lift. The reason of this phenomenon is discussed had been discussed in previous case.

Comparing figure 8 to figure 2 which is the low AR case, the lift curve slope for the high AR case is higher. Based on LLT, high AR wing contributes to a higher finite wing lift slope. The physics behind this phenomena is a longer wing span (higher AR) will generate more lift with larger span-wise lifting surface. Although the wing lift slope varies with AR, angle of attack at zero lift will remains, which is equal to the airfoil zero lift angle of attack. This can be observed from figure 2, 4, 6 and 8.

As the same theory applies for both case, the drag coefficient of wing-body combination in figure 9 is higher than the wing alone. However, comparing figure 3 and figure 9, it is shown that the fuselage drag in high AR is more significant. The fuselage contribute an average of 31% and 12.5% of drag to the wing-body combination for high AR and low AR respectively. This is due to the reference area, S for high AR is smaller, which contributes to higher fuselage base drag coefficient, fuselage zero lift drag coefficient and fuselage drag due to lift coefficient. Thus, it results in higher total drag coefficient.

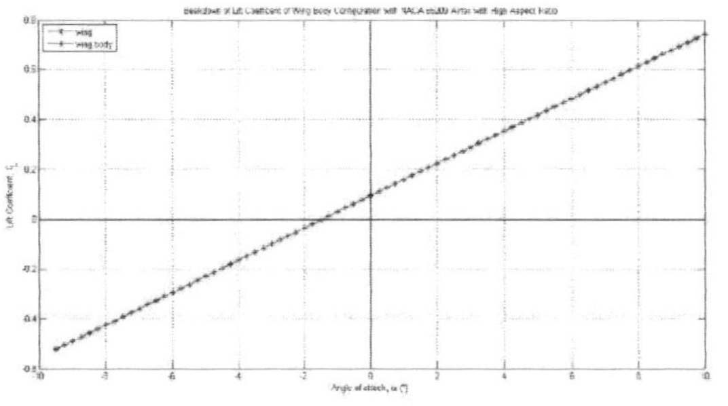

Figure 8. Breakdown of lift coefficient of wing body combination for high AR.

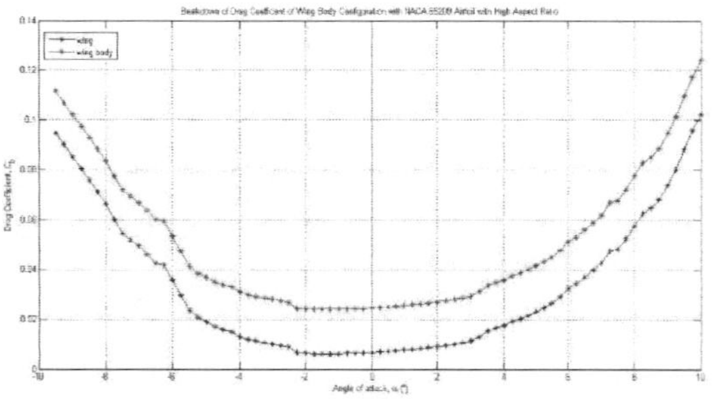

Figure 9. Breakdown of drag coefficient of wing body combination for high AR.

Likewise, the graphs in figure 10 has the same pattern with the low AR case. The difference in fuselage diameter for high AR wing does not affect the lift curve slope, i.e. the three fuselage with different diameter share the same lift curve slope. The only difference between both high AR and low AR case is the lift curve slope which depends on the AR.

Observing figure 5 and 11, drag coefficient increases with fuselage diameter. The increment of drag coefficient with increment of 15% fuselage diameter is compared and the increment are 8% and 14% for low AR and high AR respectively. The reason of more increment in high AR case is basically the same concept as discussed while comparing figure 3 and 9. Ratio of fuselage wetted area to reference area is the main reason on contributing different increment in drag coefficient between different AR.

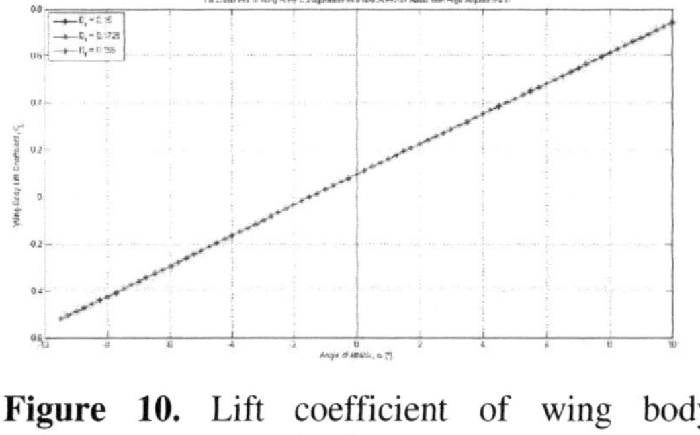

Figure 10. Lift coefficient of wing body combination with different fuselage diameter at high AR.

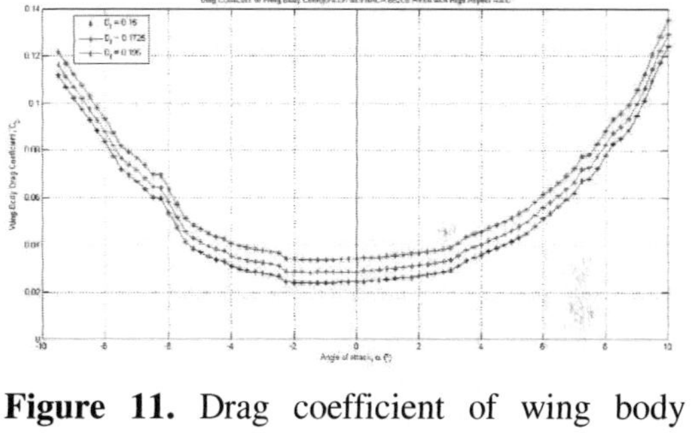

Figure 11. Drag coefficient of wing body combination with different fuselage diameter at high AR.

In order to verify the computational results obtained for high AR wing case, comparison with published journal results is crucial. Figure 12 shows the comparison of result from both computational and journal by by Walter D. W and David F. T., Jr. [5]. Both results shows a percentage difference of approximately 13.85%. This difference might due to theoretical results are always obtained base on the ideal case such as the surrounding temperature and pressure are set to remain constant for the entire simulation performed. However, in reality, temperature and pressure might fluctuates during the experiment. At the same time, some assumptions are also made to simplify the mathematic equation and also to apply some principle or theory in some cases.

Figure 13 shows the comparison of journal results with the computational results of the drag coefficient for wing-body combination. Both results are about the same visually, but it has a difference of approximately 12%. This indicates that the program is capable and reliable to compute drag of the wing-body combination for high AR.

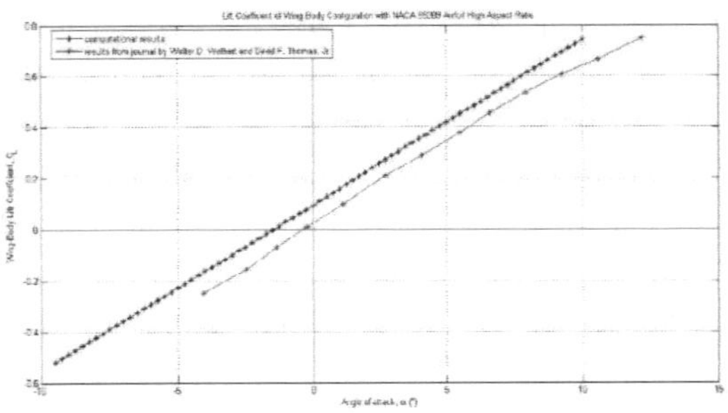

Figure 12. Comparison of theoretical results with journal for lift coefficient against angle of attack with high aspect ratio.

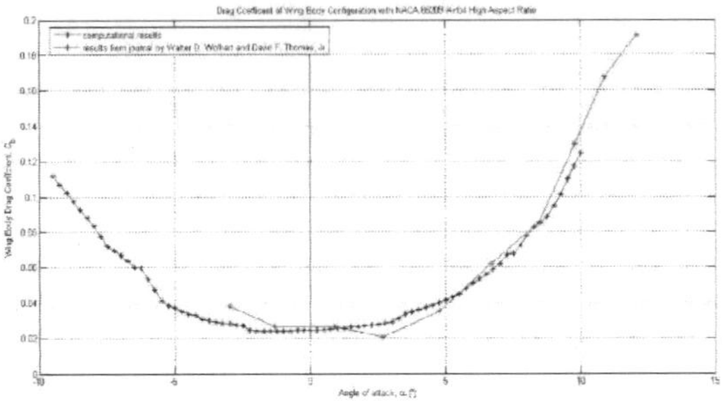

Figure 13. Comparison of theoretical results with journal for drag coefficient against angle of attack with high aspect ratio.

6. Conclusion

The fuselage diameter does not contribute to the increment of the lift coefficient of the wing-body combination. The main contribution of the lift in a wing-body combination is mainly from the wing. Fuselage drag is found to be significant in the combination and highly affected by the diameter with larger drag produced by a larger fuselage diameter. The Prandtl's Lifting Line Theory and Helmbold's Panel Method have managed to successfully compute the lift and drag for the wing while Roskam's Drag Prediction Method only capable to compute the drag of the fuselage the wing-body combination. The computational results agree well with the experimental results based on the trend and this suggests that the program is reliable in computing the lift and drag for a wing-body combination for low and high aspect ratio case.

Acknowledgments

The authors acknowledge Chew Tian Feng, Foo Kar Yen and Khong Jun Yong from School of Aerospace Engineering for their assistance and support during the development of the computational code and analysis of this work.

References

[1] Abbas A, Vicente J and Valero E 2013 Aerodynamic technologies to improve aircraft performance *Aerosp. Sci. Technol.* **100-132**.

[2] López P R 2010 Validation of software for the calculation of aerodynamic coefficients with a focus on the software package Tornado. Linköpings Universitet.

[3] Nicolosi F, Vecchia P, Ciliberti D, Cusati V and Attanasio L 2015 Fuselage aerodynamic prediction methods *33rd AIAA Applied Aerodynamics Conference.*

[4] Welstead J, Reitz B, and Crouse G 2012 Modeling fuselage aerodynamic effects in aircraft design optimization *50th AIAA Aerospace Sciences Meeting* including the New Horizons Forum and Aerospace Exposition

[5] Walter D W and David F T 1956 Static longitudinal and lateral stability characteristics at low speed of unswept-midwing models having wings with an aspect ratio of 2, 4, or 6 Washington: National Advisory Committee for Aeronautics, *Technical Note 3649.*

[6] F Hall C 1953 Lift, drag and pitching moment of low-aspect-ratio wings at subsonic and supersonic speed *NACA Research Memorandum*

[7] Anderson J 2012 *Introduction to flight* (7th ed.) New York: McGraw-Hill

[8] Oswald W 1932 Report No. 408. General formulas and charts for the Ccalculation of airplane performance *Journal of the Franklin Institute* **693-693**

[9] McCormick B 1995 *Aerodynamics, aeronautics and flight mechanics* New York: John Wiley and Sons

[10] Bertin J, and Smith M 1979 *Aerodynamics for engineers* (5th ed.) Englewood Cliffs, N.J.: Prentice-Hall
[11] Anderson J 2001 *Fundamentals of aerodynamics* (5th ed.) Boston: McGraw-Hill
[12] Lan C, and Roskam J 1997 *Airplane aerodynamics and performance* (pp. 157-161). Ottawa, Kan.: Roskam Aviation and Engineering
[13] Roskam J 2004 *Airplane design* Lawrence (Kansas): DARcorporation

IOP Conf. Series: Materials Science and Engineering **370** (2018) 012056 doi:10.1088/1757-899X/370/1/012056

Solving Inverse Kinematics of Robot Manipulators by Means of Meta-Heuristic Optimisation

Kritsada Wichapong, Sujin Bureerat and Nantiwat Pholdee *

Sustainable and Infrastructure Research and Development Center, Department of Mechancial Engineering, Faculty of Engineering, Khon Kaen University, Khon Kaen, 40002, Thailand.

*Corresponding author: nantiwat@kku.ac.th

Abstract. This paper presents the use of meta-heuristic algorithms (MHs) for solving inverse kinematics of robot manipulators based on using forward kinematic. Design variables are joint angular displacements used to move a robot end-effector to the target in the Cartesian space while the design problem is posed to minimize error between target points and the positions of the robot end-effector. The problem is said to be a dynamic problem as the target points always changed by a robot user. Several well established MHs are used to solve the problem and the results obtained from using different meta-heuristics are compared based on the end-effector error and searching speed of the algorithms. From the study, the best performer will be obtained for setting as the baseline for future development of MH-based inverse kinematic solving.

1. Introduction

Robot manipulators are widely used in production lines for the purpose of non-stop lines, high accuracy lines, etc. To control a robot position, a mapping function from the Cartesian space to joint space (also known as inverse kinematic problem) is necessary. Solving such an inverse kinematic problem for a redundant robot is said to be the issue in the field of robotics as singularity can occur in robot position analysis when using conventional methods such as a geometric method [1], and an algebraic method [2-3]. In addition, a classical iterative method [4] is time consuming and the solution is not guaranteed as it depends on an initial solution. In this regards, many researchers have studied alternative ways to deal with the kinematic problem such as using artificial neural networks (ANN) [5-7], using the combination of artificial neural networks and meta-heuristics [8-10]. However, using ANN still has some disadvantages in long training time and less accuracy. As a result, seeking for an alternative strategy to solve the inverse kinematic problem of a redundant robot is interesting and challenging. Using MHs alone for tackling the inverse kinematic problem based on forward kinematics seems to be reasonable as the singularity problem can be avoided. It is flexible without requirements of training data.

The MH is a global search optimization method which has advantages in its derivative-free feature; therefore, they can deal with all kind of objective functions and design variables. It can explore a Pareto front within a single run for multiobjective optimization. As a results, the MHs is widely applied to various applications such as structural optimization [11], flow shop scheduling [12-13], a traveling sale man problem [14], network design [15], damage detection [16], etc. They are also applied to solve the inverse kinematic problem of redundant robots based on solving forward kinematics by defining the optimization problem to find the joint variables attaining Cartesian points of the end-effector closest to the target points [9]. However, due to disadvantages of search convergence rate and consistency of MHs, using MHs for solving the inverse kinematic problem of redundant robots is rare. Development of MHs with acceptable convergence rate and consistency for solving the inverse kinematic problem is always useful. Also, comparative studies of various well established MHs for solving such an inverse problem should be investigated since it will initiate the baseline for MH performance in solving the problem.

In this work, a comparative study of several well established MHs for solving the inverse kinematic problem of a 4D redundant robot based on using forward kinematics as proposed in [9] is conducted. Design variables determine robot joint angular positions which can attain Cartesian points at the target points after solving the forward kinematics. The objective is posed to minimize a distance error between the target points and the calculated points obtained from the given joint angles. Several well established MHs including Differential evolution (DE) [17], Artificial Bee Colony (ABC) [18], Adaptive Differential evolution (JADE) [19], Teaching–learning-based optimization (TLBO) [20], Real-code ant colony optimization (ACOR) [21] and Population-based incremental learning (PBIL) [22 -23] are used to solve the problem for a verity of target points. The results obtained are compared in terms of search speed and accuracy.

2. Kinematic analysis of robotic manipulators

Robot manipulators are constructed by connecting joints and links which are modelled as an open-loop chain mechanism. To control them, kinematic analysis to understand the motion of the robot with respect to a fixed reference is necessary. Normally, kinematic analysis of a robot can be divided into forward and inverse kinematics. The forward kinematic is a problem in which joint angular positions are pre-specified. The end-effectors positions in Cartesian space are then found to meet the given values of the joint angles. The problem is expressed in Equation (1)

$$F_{\text{forward kinematic}}(\theta_1,\theta_2,\theta_3,\theta_4,\theta_5)=(X,Y,Z,\phi_x,\phi_y,\phi_z) \quad (1)$$

where θ_i are joint angles and X, Y, Z, are position of end-effectors in X, Y, and Z, direction. $\emptyset_x, \emptyset_y, \emptyset_z$ are end-effectors orientations in X, Y and Z, direction, respectively.

The inverse kinematic problem on the other hand is the reverse of the forward kinematic which can be defined as Equation (2).

$$F_{\text{inverse kinematic}}(X,Y,Z,\phi_x,\phi_y,\phi_z)=(\theta_1,\theta_2,\theta_3,\theta_4,\theta_5) \quad (2)$$

To solve the forward kinematic problem, the Denavit-Hartenberg function (DH) is usually applied with ease while the inverse kinematic problem still has an issue as mentioned earlier.

In this study, MHs are applied to solve the inverse kinematic problem based on using the forward kinematics. A four-DOF robot model as presented in [9] is employed. The schematic diagram of the robot is shown in Figure 1. To solve the forward kinematic of the robot, DH is applied. The DH transformation matrix can be expressed as

$$ {}^{i-1}_{\ \ i}T = (\propto_{i-1})D_x(a_{i-1})R_Z(\theta_i)Q_i(d_i) = $$

$$\begin{bmatrix} 1 & 0 & 0 & 0 \\ 0 & c\alpha_{(i-1)} & -s\alpha_{(i-1)} & 0 \\ 0 & -s\alpha_{(i-1)} & c\alpha_{(i-1)} & 0 \\ 0 & 0 & 0 & 1 \end{bmatrix} \begin{bmatrix} 1 & 0 & 0 & a_{(i-1)} \\ 0 & 1 & 0 & 0 \\ 0 & 0 & 1 & 0 \\ 0 & 0 & 0 & 1 \end{bmatrix} \begin{bmatrix} c_i & -s_i & 0 & 0 \\ -s_i & c_i & 0 & 0 \\ 0 & 0 & 1 & 0 \\ 0 & 0 & 0 & 1 \end{bmatrix} \begin{bmatrix} 1 & 0 & 0 & 0 \\ 0 & 1 & 0 & 0 \\ 0 & 0 & 1 & d_i \\ 0 & 0 & 0 & 1 \end{bmatrix} \quad (3)$$

$$= \begin{bmatrix} c_i & -s_i & 0 & a_{(i-1)} \\ s_i c\alpha_{(i-1)} & c_i c\alpha_{(i-1)} & -s\alpha_{(i-1)} & -s\alpha_{(i-1)} d_i \\ s_i s\alpha_{(i-1)} & c_i s\alpha_{(i-1)} & c\alpha_{(i-1)} & c\alpha_{(i-1)} d_i \\ 0 & 0 & 0 & 1 \end{bmatrix} \quad (4)$$

where a_i is distance from z_{i-1} to z_i along x_i , α_i is angle between z_{i-1} to z_i measure in plane normal to x_i , d_i is the distance from the origin of the (i-1) coordinate frame to the intersection of the x_{i-1} axis along the z_{i-1} axis, θ_i. is the joint from x_{i-1} to x_i

For the four-DOF robot, the DH variables are shown in Table 1 while transformation matrices of each link can be expressed as the following equations (5-8).

$${}^0_1T = \begin{bmatrix} c_1 & -s_1 & 0 & 0 \\ s_1 & c_1 & 0 & 0 \\ 0 & 0 & 1 & L_1 \\ 0 & 0 & 0 & 1 \end{bmatrix} \quad (5)$$

$${}^1_2T = \begin{bmatrix} c_2 & -s_2 & 0 & 0 \\ 0 & 0 & 1 & 0 \\ -s_2 & -c_2 & 0 & 0 \\ 0 & 0 & 0 & 1 \end{bmatrix} \quad (6)$$

$${}^2_3T = \begin{bmatrix} c_3 & -s_3 & 0 & L_2 \\ s_3 & c_3 & 0 & 0 \\ 0 & 0 & 1 & 0 \\ 0 & 0 & 0 & 1 \end{bmatrix} \quad (7)$$

$${}^3_4T = \begin{bmatrix} c_4 & -s_4 & 0 & L_3 \\ s_4 & c_4 & 0 & 0 \\ 0 & 0 & 1 & 0 \\ 0 & 0 & 0 & 1 \end{bmatrix} \quad (8)$$

$${}^4_5T = \begin{bmatrix} 1 & 0 & 0 & L_4 \\ 0 & 0 & -1 & -L_5 \\ 0 & 1 & 0 & 0 \\ 0 & 0 & 0 & 1 \end{bmatrix} \quad (9)$$

As a consequence, the transformation matrices of the end-effector can be determined as

$${}^e_bT = {}^0_1T\,{}^1_2T\,{}^2_3T \ldots {}^{i-2}_{i-2}T\,{}^{i-1}_{i}T = \begin{bmatrix} n_x & s_x & a_x & p_x \\ n_y & s_y & a_y & p_y \\ n_z & s_z & a_z & p_z \\ 0 & 0 & 0 & 1 \end{bmatrix} \quad (10)$$

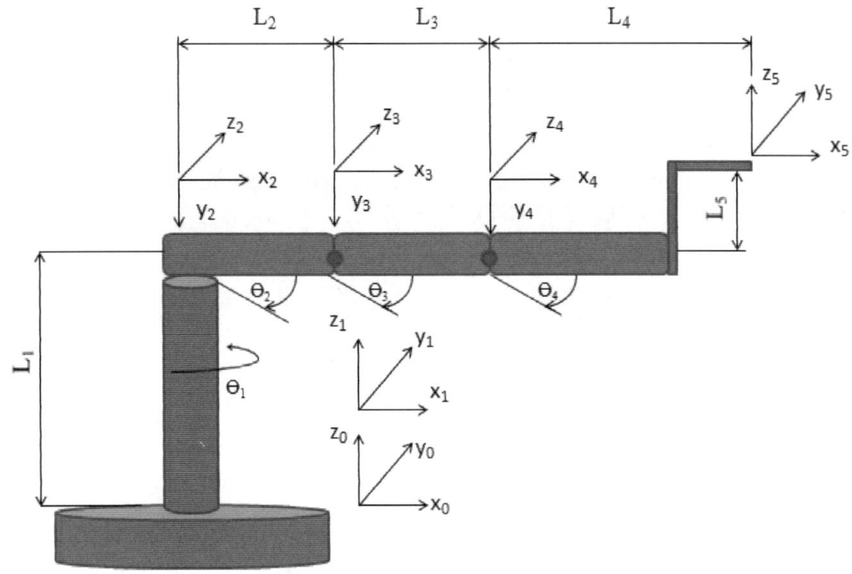

Figure 1. The schematic diagram of the 4D robot

Table 1. D-H parameter of the 4-D robot [9]

i	a_{i-1} (mm)	α_{i-1} (°)	d_i (mm)	θ_{i-1}. (°)
1	0	0	80	θ_1
2	0	-90	0	θ_2
3	120	0	0	θ_3
4	65	0	0	θ_4
5	85	90	25	0

The positions and orientations of the end-effector are shown in Equation (10). P_x, P_y, and P_z, are position of end-effectors in X, Y, and Z, direction while n_x, s_x, a_x, n_y, s_y, a_y, n_z, s_z, a_z, are rotation component of $\emptyset_x$,$\emptyset_y$,$\emptyset_z$. To solve the inverse kinematic problem by using a meta-heuristic optimizer, the objective function can be formulated as

$$\text{Min : Error} = \sqrt{(x_2 - y_x)^2 + (y_2 - y_1)^2 + (z_2 - z_1)^2} \quad (11)$$

Subject to

$$\begin{array}{lll} -90 & < \theta_1 & < 90 \\ -180 & < \theta_2 & < 0 \\ 0 & < \theta_3 & < 145 \\ -90 & < \theta_4 & < 90 \end{array} \quad (12)$$

3. Meta-heuristic

In this work, six well established MHs are compared based on solving the inverse kinematic optimization problem detailed in the previous section. Given that n_P is a population size (unless otherwise specified) and n is a number of design variables, those MHs and their optimization parameter settings (details of notations can be found in the corresponding reference of each method) are detailed as:

- Differential evolution (DE) (1): The DE/best/2/bin strategy was used. A scaling factor, crossover rate and probability of choosing elements of mutant vectors are 0.5, 0.7, and 0.8 respectively.

- Artificial bee colony algorithm (ABC) (2): The number of food sources for employed bees is set to be $n_P/2$. A trial counter to discard a food source is 150.
- Adaptive differential evolution with optional external archive (JADE) (3): Differential evolution with composite trial vector generation strategies and control parameters
- Teaching-learning-based optimization (TLBO) (4): Parameter settings are not required.
- Real-code ant colony optimization (ACOR) (5): The parameter settings are $q = 0.2$, and $\xi = 1$.
- Population-based incremental learning (BPBIL) (6): The learning rate, mutation shift, and mutation rate are set as 0.5, 0.7, and 0.2 respectively.

To verify the search performance in terms of speed and accuracy over the Cartesian space of the robot, 30 testing points for the inverse kinematic problem are assigned leading to be a dynamic problem. The testing points are generated randomly by means of a Latin Hyper Cube sampling technique (LHS). The population size is set to be $n_P = 80$ whereas the number of iterations is 250 for all test problems. The termination criteria are set to stop the search when the objective function value (Error) is lower than 1×10^{-5} or the total number of function evaluations is equal to 80×250 for all optimizers. The penalty function used in this study is a fuzzy set penalty function as detailed in [8].

4. Result

After performing optimization of the 30 points inverse kinematic problems, the results are shown in Tables 2-3. The accuracy of the MHs is measured based on position error of end-effectors as reported in Table 2 while the search speed of the MHs is measured based on number of function evaluations required to reach the target position as reported in Table 3. The lower position error is the more accurate optimizer while the lower number of function evaluations is the faster optimizer. In addition, the total number of successful runs from 30 points inverse kinematic problems is used to measure the search consistency, the more number of successful runs mean the corresponding method is more consistent. Note that, the algorithm that is terminated by the total number of function evaluations reaching the predetermined maximum evaluation number without giving acceptable position error is considered unusable.

From Table 2, the best optimizer based on average position error is TLBO while the second best is DE. When looking at the search consistency based on the number of successful runs, the most consistent optimizer is TLBO while the second best is DE similar to the measure of accuracy. For the search speed as reported in Table 3, the fastest optimizer on solving robot inverse kinematic problems is TLBO which require average 10405 function evaluations to meet the target position. The second best search speed is DE which average 11027 function evaluations to meet the target position. From the table, it was found that only TLBO and DE is usable for solving robot inverse kinematic problems since ABC, JADE, ACOR and BPBIL cannot meet target positions until maximum number of function evaluations (20,000) is reached.

Table 2. Position error (Each cell is the average value from 30 runs)

Point	DE	ABC	JADE	TLBO	ACOR	BPBIL
no.1	$8.1893\text{x}10^{-6}$	$2.0890\ \text{x}10^{-1}$	$3.0283\ \text{x}10^{-4}$	$7.0879\ \text{x}10^{-6}$	$1.5728\ \text{x}10^{-1}$	9.9300
no.2	$8.5071\text{x}10^{-6}$	$1.6049\ \text{x}10^{-1}$	$1.0078\ \text{x}10^{-3}$	$9.4869\ \text{x}10^{-6}$	$5.8601\ \text{x}10^{-4}$	$9.3025\ \text{x}10^{-1}$
no.3	$6.2927\text{x}10^{-1}$	$1.3235\ \text{x}10^{-1}$	$1.3056\ \text{x}10^{-2}$	$7.4932\ \text{x}10^{-6}$	$6.3606\ \text{x}10^{-1}$	3.5103
no.4	$4.0340\text{x}10^{-6}$	$7.2377\ \text{x}10^{-2}$	$2.7022\ \text{x}10^{-5}$	$6.7992\ \text{x}10^{-6}$	$2.0240\ \text{x}10^{-2}$	5.7221
no.5	$9.1439\text{x}10^{-6}$	$2.0680\ \text{x}10^{-2}$	$7.0457\ \text{x}10^{-4}$	$8.3479\ \text{x}10^{-6}$	$4.9800\ \text{x}10^{-2}$	9.8531
no.6	1.7055	$2.1507\ \text{x}10^{-1}$	1.7058	1.7055	1.7055	2.5967
no.7	$7.4033\text{x}10^{-6}$	$3.9507\ \text{x}10^{-2}$	$3.4879\ \text{x}10^{-2}$	$2.2785\ \text{x}10^{-6}$	$1.5208\ \text{x}10^{-1}$	6.1441
no.8	1.4250 x10	$3.9924\ \text{x}10^{-1}$	$7.0118\ \text{x}10^{-2}$	$4.8653\ \text{x}10^{-6}$	$8.3730\ \text{x}10^{-3}$	4.1008
no.9	$2.7551\text{x}10^{-6}$	$7.2261\ \text{x}10^{-1}$	$2.4393\ \text{x}10^{-2}$	$9.1132\ \text{x}10^{-6}$	$1.3526\ \text{x}10^{-1}$	$1.2252\ \text{x}10^{-1}$

no.10	8.7870×10^{-6}	3.4217×10^{-2}	8.9966×10^{-6}	8.9422×10^{-6}	2.9442×10^{-3}	3.3364×10^{-1}
no.11	5.6897×10^{-6}	7.1530×10^{-3}	5.4563×10^{-3}	8.8970×10^{-6}	1.3577×10^{-2}	6.0659
no.12	4.7130×10^{-6}	4.3810×10^{-2}	7.7431×10^{-4}	7.8681×10^{-6}	1.7404×10^{-2}	2.9549
no.13	4.5586×10^{-6}	1.3946	2.9295×10^{-2}	24.643	1.5963×10^{-1}	7.0586
no.14	6.9137×10^{-6}	5.7306×10^{-2}	6.8700×10^{-3}	4.9569×10^{-6}	7.2939×10^{-3}	5.5568
no.15	7.8151×10^{-6}	2.8844×10^{-2}	1.8016×10^{-2}	8.7153×10^{-6}	1.6142×10^{-3}	3.1614
no.16	7.5721×10^{-6}	1.7741×10^{-2}	6.4435×10^{-3}	8.0604×10^{-6}	4.5435×10^{-3}	8.1782
no.17	5.6053×10^{-6}	4.6575×10^{-3}	2.9285×10^{-5}	5.7245×10^{-6}	6.9194×10^{-1}	1.6997
no.18	4.8387×10^{-6}	9.2044×10^{-2}	1.3600×10^{-2}	7.5878×10^{-6}	1.8936×10^{-3}	6.6229
no.19	9.9127×10^{-6}	2.3986×10^{-2}	4.0274×10^{-4}	6.3496×10^{-6}	4.9425×10^{-3}	9.2016
no.20	2.9212	3.9713×10^{-2}	6.8828×10^{-3}	9.8722×10^{-6}	1.7626×10^{-3}	4.3303
no.21	2.5713	7.9942×10^{-2}	1.1605×10^{-4}	2.5713	2.5713	2.3830
no.22	4.4790×10^{-6}	2.0231×10^{-1}	4.7193×10^{-3}	7.4160×10^{-6}	9.5935×10^{-3}	1.7075×10^{-1}
no.23	8.4898×10^{-6}	4.8710×10^{-1}	1.1151×10^{-3}	9.5864×10^{-6}	8.7998×10^{-2}	7.9155
no.24	5.0665×10^{-6}	4.8499×10^{-1}	5.5589×10^{-4}	2.6341×10^{-6}	1.0187×10^{-1}	7.6200
no.25	8.9349×10^{-6}	1.8075	8.8020×10^{-3}	9.5663×10^{-6}	5.5350×10^{-2}	3.6582
no.26	7.2905×10^{-6}	2.4548×10^{-3}	1.1779×10^{-2}	9.8821×10^{-6}	6.2614×10^{-4}	4.5973
no.27	8.8776×10^{-6}	1.0950	1.3468×10^{-4}	5.8790×10^{-6}	1.8359×10^{-1}	1.6161×10^{-1}
no.28	8.9406×10^{-6}	6.1771×10^{-3}	4.0250×10^{-2}	9.8434×10^{-6}	5.1502×10^{-4}	2.0495×10
no.29	8.8427×10^{-6}	4.8073×10^{-1}	8.6287×10^{-4}	6.0543×10^{-6}	1.3084×10^{-1}	2.4813
no.30	6.9763×10^{-6}	4.3567×10^{-1}	1.1489×10^{-2}	4.7662×10^{-6}	1.4399×10^{-1}	1.8845×10
Average	7.3593×10^{-1}	2.9324×10^{-1}	6.7263×10^{-2}	9.6400×10^{-1}	2.3528×10^{-1}	8.1488
Number of Successful Run	24	0	1	27	0	0

Table 3. Average number of function evaluations

Point	DE	ABC	JADE	TLBO	ACOR	BPBIL
no.1	9280	20000	20000	11040	20000	20000
no.2	8960	20000	20000	9440	20000	20000
no.3	20000	20000	20000	8960	20000	20000
no.4	9360	20000	20000	12800	20000	20000
no.5	8800	20000	20000	7520	20000	20000
no.6	20000	20000	20000	20000	20000	20000
no.7	7360	20000	20000	11200	20000	20000
no.8	20000	20000	20000	10400	20000	20000
no.9	9120	20000	20000	10240	20000	20000
no.10	10240	20000	20000	9120	20000	20000
no.11	9440	20000	20000	10080	20000	20000
no.12	8880	20000	20000	10880	20000	20000
no.13	9920	20000	20000	20000	20000	20000
no.14	8320	20000	20000	9280	20000	20000
no.15	9520	20000	20000	8000	20000	20000

no.16	9120	20000	20000	7360	20000	20000
no.17	7280	20000	20000	7840	20000	20000
no.18	10240	20000	20000	7360	20000	20000
no.19	10160	20000	20000	10560	20000	20000
no.20	20000	20000	20000	10400	20000	20000
no.21	20000	20000	20000	20000	20000	20000
no.22	8080	20000	20000	8320	20000	20000
no.23	10160	20000	20000	9440	20000	20000
no.24	8880	20000	20000	9920	20000	20000
no.25	8320	20000	20000	10720	20000	20000
no.26	10400	20000	20000	7520	20000	20000
no.27	8800	20000	20000	8960	20000	20000
no.28	9440	20000	20000	7200	20000	20000
no.29	9440	20000	20000	8800	20000	20000
no.30	11280	20000	20000	8800	20000	20000
Average	11027	20000	20000	10405	20000	20000

Figure 2 shows the search history for the best performer of all algorithms while Figure 3 shows the zoom-in of Figure 2. Form these figures, it was found that, although all algorithms have fast convergence in the beginning, most of them are trapped at some points except for the DE and TLBO. The DE can converge to the optimum result by using lower than 1000 function evaluations whereas the TLBO can converge to the optimum result at about 3500 function evaluations.

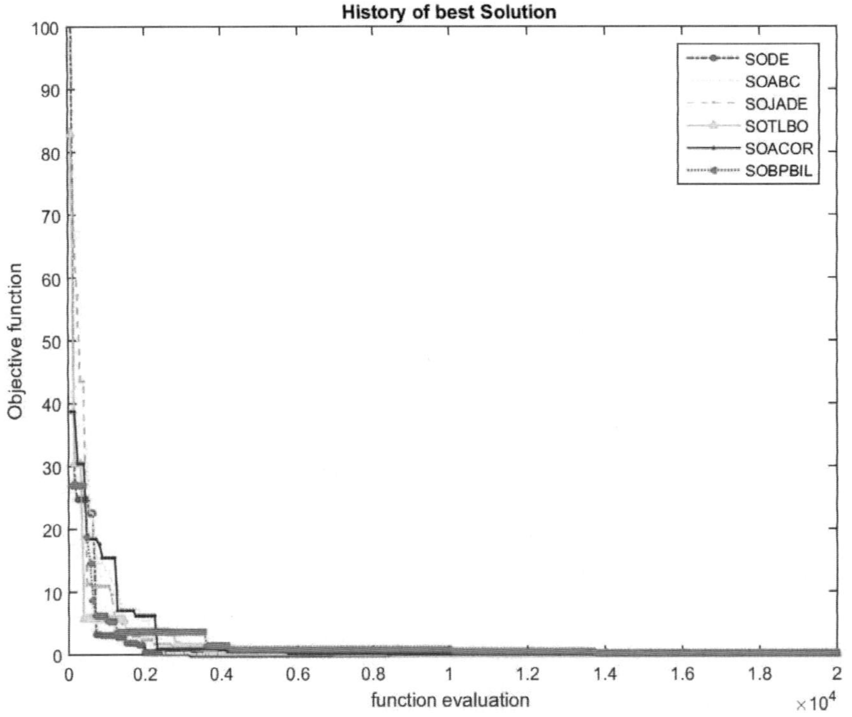

Figure 2. Search history of the best run of all algorithm

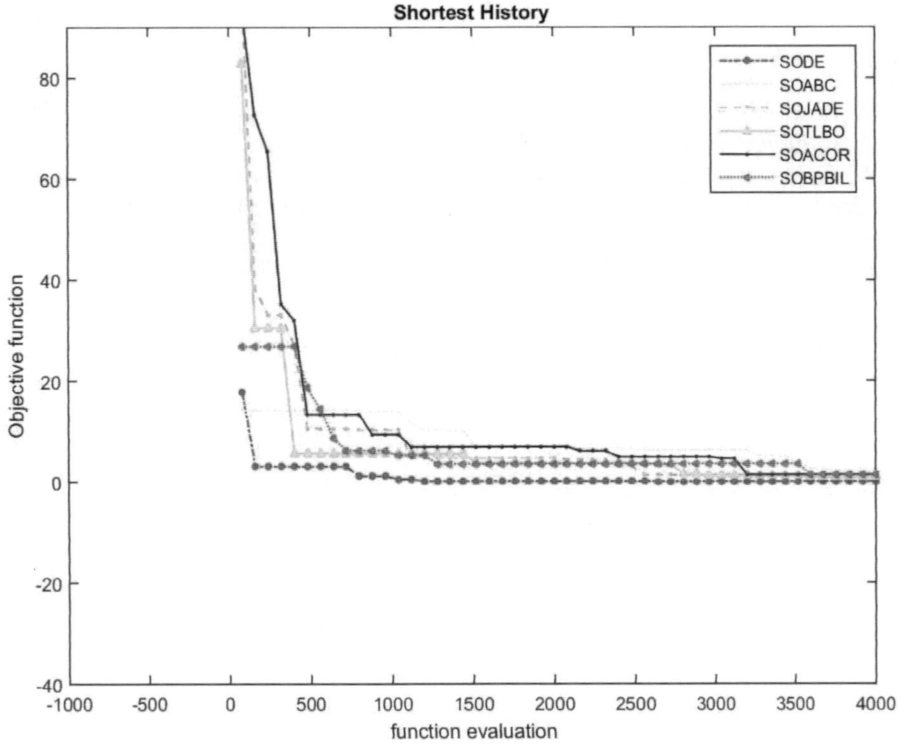

Figure 3. Zoom in of the figure 2

5. Conclusions

In this work, a comparative study of meta-heuristic algorithms (MHs) for solving inverse kinematics of robot manipulators based on using forward kinematic is successfully investigated. The design variables are joint angular displacements used to move the robot end-effector to the target in the Cartesian space while the design problem is posed to minimize error between target points and the robot end-effector position. Six well established MHs are used to solve 30 testing points of inverse kinematic and the results are compared in terms of speed and accuracy. It was found that, the best performer in terms of both speed and accuracy is TLBO while the second best is DE. The others are said to lack on searching results within limitation of the total number of function evaluations as they cannot obtained the satisfy position error. It should be noted that this study is yet to be the real dynamic or on-line optimization problem as the target points are pre-defined before running an optimizer from scratch. With the use of additional numerical schemes such as an archiving technique, the performance of the two best optimizers could still be enhanced.

Acknowledgements

The research described in this paper was financially supported by the Thailand Research Funds (TRF).

References

[1] Ellwood R J, RaatZ J and Hesselbach A 2010 *In: Precision Assembly Technologies and Systems, Chamonix, France* 123-130
[2] Duffy J 1980 *In: Analysis of Mechanisms and Robot Manipulators, Wiley, New York.*
[3] Manocha D and Canny J F 1994 *IEEE Transactions on Robotics and Automation.* **10** 648-57
[4] Korein J U and Badler N I 1982 *IEEE Computer Graphics and Applications*. **2** 71-81
[5] Köker R 2004 *Robotics and Autonomous Systems.* **39** 227-34

[6] Mao Z and Hsia T 1997 *Robotica* **15** 3-10
[7] Guo J and Cherkassky V 1989 *In: International 1989 Joint Conference on Neural Networks, Washington DC, USA* **2** 299-304
[8] Köker R 2013 *Information Sciences.* **222** 528-43
[9] Saka M P, Hasançebi O and Geem Z W 2016 *Swarm and Evolutionary Computation.* **28** 88-97
[10] Köker R 2016 *Engineering with Computers.* **32** 1-13
[11] Bureerat S and Pholdee N *Journal of Computing in Civil Engineering* **30** 04015019
[12] Liu B, Wang L and Jin Y H 2007 *IEEE Transactions on Systems, Man, and Cybernetics, Part B (Cybernetics)* **37** 18-27
[13] Sangsawang C and Sethanan K 2016 *KKU ENGINEERING JOURNAL* **43** 55-61
[14] Nakkaew P, Kantanantha N and Wongthatsanekorn W 2016 *KKU ENGINEERING JOURNAL,* **43** 62-8
[15] Pepper J W and Golden B L 2002 *IEEE Transactions on Systems Man, and Cybernetics, Part A: Systems and Humans* **32** 72-77
[16] Pattnaik S B, Mohan S and Tom V M 1998 *Journal of Transportation Engineering* **124** 368-75
[17] Miguel L F F *Expert Systems with Applications* **39** 9704-14
[18] Storn R and Price K *Journal of Global Optimization* **11** 341-59
[19] Omran M G H, Salman A and Engelbrecht A P 2005 *In:Computational Intelligence and Security: International Conference, China* 15-19
[20] Rao R V, Savsani V J and Vakharia D P 2011 *Computer-Aided Design* **43** 303-15
[21] Socha K and Dorigo M 2008 *Journal of Operational Research* **185** 1155-73
[22] Baluja S 1994 *School of Computer Science, Carnagie Mellon University, Pittsburgh*
[23] Kunakote T and Bureerat S *KKU Research Journal* **19** 62-8

IOP Conf. Series: Materials Science and Engineering **370** (2018) 012057 doi:10.1088/1757-899X/370/1/012057

Experiential study on temperature and emission performance of micro burner during porous media combustion

Ayub Ahmed Janvekar[1*], M Z Abdullah[2], Z A Ahmad[3], A Abas[1], A K Ismail[4], A A Hussien[1], P S Kataraki[1], M H H Ishak[1], M Mazlan[5], A F Zubair[1]

[1]School of Mechanical Engineering, Universiti Sains Malaysia, Engineering Campus, 14300 Nibong Tebal, Penang, Malaysia
[2]School of Aerospace Engineering, Universiti Sains Malaysia, Engineering Campus, 14300 Nibong Tebal, Penang, Malaysia
[3]School of Materials & Mineral Resources Engineering, Engineering Campus, 14300 Nibong Tebal, Penang, Malaysia
[4]Mechanical Section, Universiti Kuala Lumpur Malaysian Spanish Institute, 09000 Kulim Hi-Tech Park, Kedah, Malaysia
[5]Advances Material Research Cluster, Faculty of Bioengineering and Technology, Universiti Malaysia Kelantan, Jeli Campus, 17600 Jeli, Kelantan, Malaysia

*Corresponding author: ayubjanvekar@gmail.com

Abstract. Addition of porous materials in reaction zone give rise to significant improvements in combustion performance. In this work, a dual layered micro porous media burner was tested for stable flame and emissions. Reaction and preheat layer was made up of discrete (zirconia) and foam (porcelain) type of materials respectively. Three different thickness of reaction zone was tested, each with 10, 20 and 30mm. Interestingly, only 20mm thick layer can able to show better thermal efficiency of 72% as compared to 10 and 30mm. Best equivalence ratio came out to be 0.7 for surface and 0.6 for submerged flame conditions. Moreover, emission was continuously monitored to detect presence of NOx and CO, which were under controlled limits.

1. Introduction

Saving of fossil fuels in any best possible way is prime factor for any nation to maintain its growth stable, hence utilization of fuel plays vital role. Consumption of byproducts of fossil fuels like LPG, butane, gasoline so on not only affect large scale industries but do interrupt domestic devices like burner and stove [1-3]. So both developed and developing nations prime moto is to come up with novel idea for improving the systems running of depleting fuels. In present scenarios super-fast demand has seen in liquid/gas fuels since large position of mankind depend on them to run their routine life like burner and stoves [4, 5]. Hence one of the advanced and promising agent can be used so called porous media (PM). Invent these materials in reaction zone of burner shall enhances show significant changes in combustion performances thus making porous media burner (PMB) efficient. The main two prime advantages of enabling PM inside the burner is to generate high temperature profiles (surface and walls) and lower the counts for emission[6-8]. Recent study form a decade has clearly indicated the importance of porous media in burners and their key role has proven to significant [9, 10]. Apart for these benefits, they do effect lean flammability limits at higher burning rates. Zhou, Huang [11] made study on gas supply pressure of the appliance and implies significant influence on actual heat input. Song, Wen [12] made an experimental work on premixed combustion of ultra-low calorific gas (LCG) in an axial and radial gradually-varied porous media burner with annular for heat recirculation and conveyed its importance. Pourhoseini [13] suggested a novel method to on pyrolysis technique for increasing intermediate soot particles concentration and enhancing radiative characteristics of natural gas flame. Park and Kim [14] worked on optimization of the Compact Submerged Combustion Vaporizers by first modeling the heat exchanger unit cell as porous media, then using the entropy minimization method. The porosity and equivalent particle diameter of the unit cell were independently controlled by varying the fin density and fin thickness or fin diameter, respectively and the entropy generation number was calculated numerically. In a typical bi layer porous media consist of top layer know was reaction layer where in combustions occur while lower region is treated as preheat layer. In the present study three different sizes of reaction layer are studied and optimum size thickness is suggested for the desired fuel mixture. In addition, best suited equivalences ratio (ER) was also shortlisted.

2. Experimental setup

Fig. 1 highlights typical insights of burner housing. While Fig. 2 and 3 indicates complete system layout and actual one to one photographic image of setup respectively. A micro burner was made with a cylindrical diameter of 22mm, thickness was 2mm and light was restricted to 100mm. Reaction zone was made of discrete PM of zirconia, with each ball was in spherical shape of diameter 10mm. The layer was tested for three sizes namely 10, 20 and 30 mm. While preheat layer was fixed at 10mm thick size of porcelain PM (foam type of 8 ppcm). A calculative amount of fuel mixture (butane and air) was fed from the base of burner housing to ensure ease flow inside preheat layer. In total four K type of thermocouple where enabled to capture temperature readings. Only one thermocouple places at surface of reaction zone (T1), mid of reaction zone (T2), junction of preheat zone and reaction zone (T3) and finally T4 was kept at mid of preheat layer. All the four thermocouples where wired to data accusation system to get digital readings of temperature profiles. To generate thermal images at various critical ER FLUKE thermal imager was adopted. Digital flow meters where connected across fluid flow tubes to monitor exact flow rates. A unit in terms of L/min was maintained for easy comparison of presented data with previous researchers. KANE-9106 QUINTOX Combustion Analyzer in ppm (parts per million) was used to sense emissions from burner chimney.

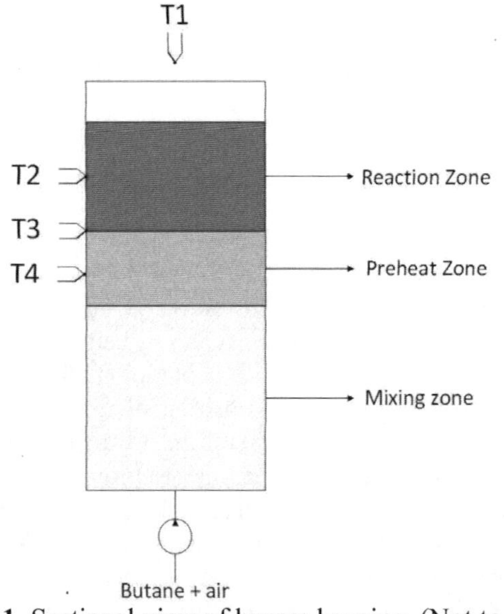

Figure 1. Sectional view of burner housing. (Not to scale).

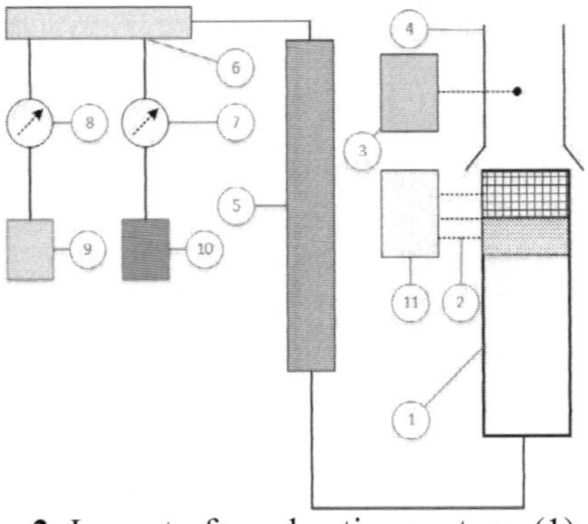

Figure 2. Layout of combustion system. (1) Housing, (2) Thermocouple, (3) Emission gas analyzer, (4) Gas conduit, (5) Mixing unit, (6) Pre-mix, (7) Air flow meter, (8) Butane flow meter, (9) Butane supply, (10) Air inject, (11) DAQ (Data acquisition system).

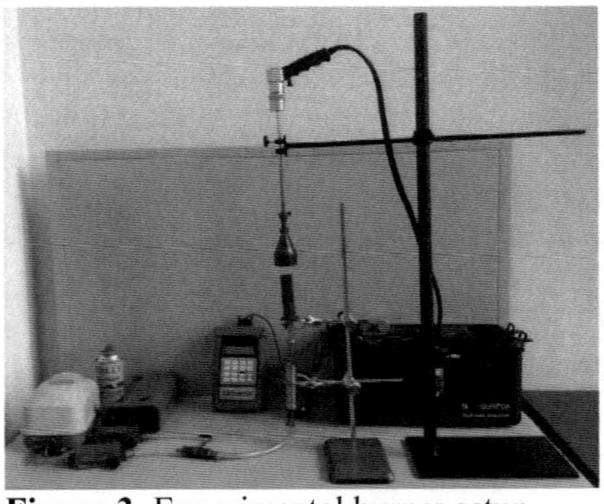

Figure 3. Experimental burner setup.

3. Analysis

In this experimental work a concept of ER is used since the combustion occurs is premixed type. Equation (1) indicates highlights the formulation used to calculate ER, where in actual air–fuel ratio (M_{act}) and while stoichiometric air fuel ratio (M_{sto}) for butane gas, was can be calculated by considering Avogadro's and Dalton's laws [5]. Table 1 indicates actual combination of air and fuel ratio used to get desired fuel mixture. Initially burner was made purposefully to run at low fuel contact so that burner can be made more efficient.

$$\mathrm{ER} = \frac{M_{sto}}{M_{\mathrm{act}}} \qquad (1)$$

Table 1. ER for butane and air flow rate.

Butane (Lpm)			0.1		
Air (Lpm)	3.1	3.4	3.8	4.4	5.1
ER	1.0	0.9	0.8	0.7	0.6

A process of trial and error has shown a minimum amount of fuel used to run the burner at safe and stable flame come out to be 0.1 Lpm (litre per min). Burner was made to run from stoichiometric ER to lean region. Interesting results were seen when all the three cases where plotted with a help of a graph as shown in Fig. 4. A maximum temperature value of 601°C was noted with reaction layer of 20mm thick while 505°C and 556°C where recorded at 10 and 30mm thick reactions zone. The possible reasons of getting such a differences is due to the availability of surface area to perform combustion, the pressure drop across reaction zone and preheat zone, and allowed ER [15, 16]. Burner was under surface flame condition when the ER was in the range between 1.0 to 0.6 and the it gives magnificent profile at ER=0.6 as indicate in Fig. 4. Incidence of submerged flame imitated soon ER cross 0.6 and tends to move towards 0.7. A completely submerged flame was seen with naked eye at ER=0.6. Furthermore, increase in ER led to back flash error thus made burner to shut down, its due to back force created in fuel mixture due to lack of oxygen supply to perfume combustion process[17, 18].

Fig. 5 indicates averaged wall temperature recorded over the wall of the burner, it can be seen that maximum recorded temperature was with 20mm thick layer. Thus optimum choice of selection would be 20mm rather than 10 and 30mm. Once the ideal choice of layer is made a thermal imager was used to get a temperature distribution on the wall of the burner as shown in Fig. 6. Value shown in Fig. 6

for average temperature from imager was 161.1°C, while actual value from DAQ was 165°C which can be acceptable. Thermal efficacy was also measured for critical ER for all the three cases. Maximum efficiency was calculated for 20mm thick layer, reaching up to 72%, while 59% and 66% were noted for 10 and 20mm thick reaction layer respectively. All the thermal efficiency was carried out at ER=0.6, since burner was at optimum performance at this ER. Finally, certified gas analyzer was added to setup to measure emitted emissions from the burner. Injected gasses where fed to gas tip to measure NOx and CO. It was found out that emission where less than 1ppm since the minimum amount the device can detect was 1ppm.

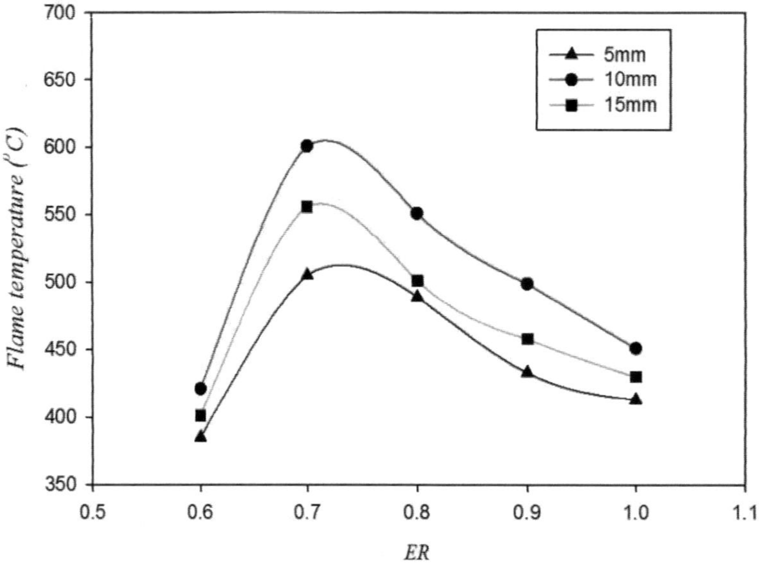

Figure 4. Surface flame temperature above the surface of reaction layer.

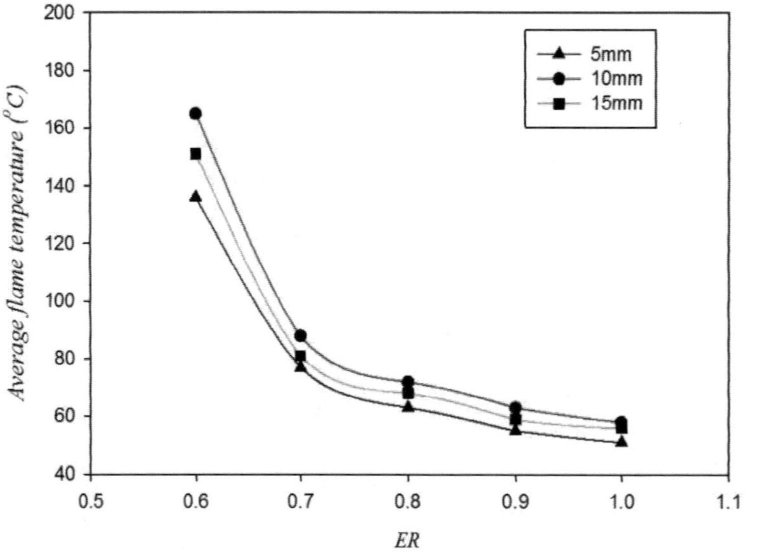

Figure 5. Average flame temperature at walls of the burner.

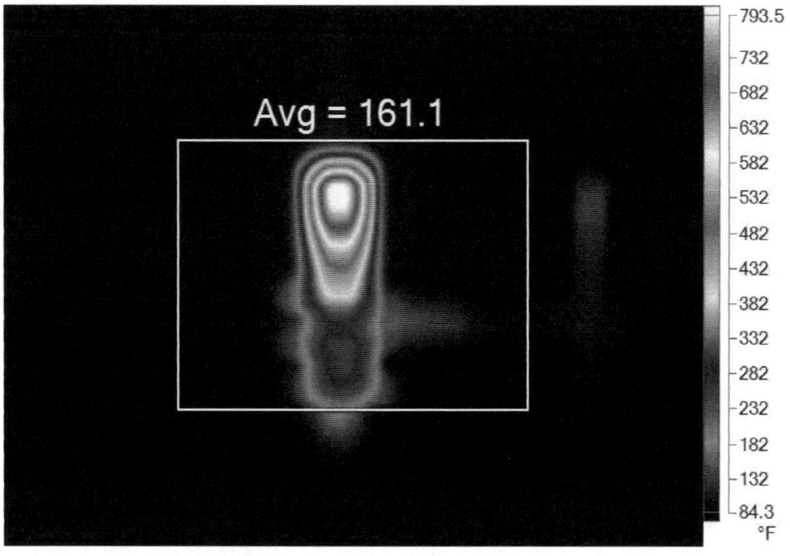

Figure 6. Temperature distribution image from thermal imager.

4. Conclusions

In this experimental work, a micro burner was made to undergo both surface and submerged flames. Burner was made to reach lean ER. Out of three predefined thickness of reaction zone, only 20mm could able to perform better results. Ideal ER for surface flames and submerged flame are 0.7 and 0.6 respectively. Maximum temperature that can be generated from this micro burner could reach up to 601°C, thereby giving a thermal efficiency of 72%. Emitted emission results in low NOx and CO values.

References

[1] Mujeebu, M.A., M.Z. Abdullah, and A.A. Mohamad, Development of energy efficient porous medium burners on surface and submerged combustion modes. *Energy*, 2011. **36**(8): p. 5132-5139.

[2] Mujeebu, M.A., et al., Development of Premixed Burner Based on Stabilized Combustion within Discrete Porous Medium. *Journal of Porous Media*, 2011. **14**(10): p. 909-917.

[3] Mujeebu, M.A., et al., A Mesoscale Premixed LPG Burner with Surface Combustion in Porous Ceramic Foam. *Energy Sources, Part A: Recovery, Utilization, and Environmental Effects*, 2011. **34**(1): p. 9-18.

[4] Ismail, A.K., et al., Effect of ceramic coating in combustion and cogeneration performance of Al2O3 porous medium. *Journal of the Energy Institute*, 2016. **89**(1): p. 81-93.

[5] Ismail, A.K., et al., Application of porous medium burner with micro cogeneration system. *Energy*, 2013. **50**: p. 131-142.

[6] Mustafa, K.F., et al., A review of combustion-driven thermoelectric (TE) and thermophotovoltaic (TPV) power systems. *Renewable and Sustainable Energy Reviews*, 2016.

[7] Mustafa, K.F., et al., Comparative assessment of a porous burner using vegetable cooking oil–kerosene fuel blends for thermoelectric and thermophotovoltaic power generation. *Fuel*, 2016. **180**: p. 137-147.

[8] Mustafa, K.F., et al., Experimental investigation of the performance of a liquid fuel-fired porous burner operating on kerosene-vegetable cooking oil (VCO) blends for micro-cogeneration of thermoelectric power. *Renewable Energy,* 2015. **74**: p. 505-516.

[9] Janvekar, A.A., et al., Effects of the preheat layer thickness on surface/submerged flame during porous media combustion of micro burner. *Energy*, 2017. **122**: p. 103-110.

[10] .Janvekar, A.A., et al. Assessment of porous media burner for surface/submerged flame during porous media combustion. in *AIP Conference Proceedings*. 2017. AIP Publishing.

[11] Zhou, Y., et al., Comparative study on the combustion characteristics of an atmospheric induction stove in the plateau and plain regions of China. *Applied Thermal Engineering*, 2017. **111**: p. 301-307.

[12] .Song, F., et al., Ultra-low calorific gas combustion in a gradually-varied porous burner with annular heat recirculation. *Energy*, 2017. **119**: p. 497-503.

[13] Pourhoseini, S.H., A novel configuration of natural gas diffusion burners to enhance optical, thermal and radiative characteristics of flame and reduce NOx emission. *Energy*, 2017. **132**: p. 41-48.

[14] Park, Y.C. and J. Kim, Submerged combustion vaporizer optimization using Entropy Minimization Method. *Applied Thermal Engineering*, 2016. **103**: p. 1071-1076.

[15] Mujeebu, M.A., et al., Applications of porous media combustion technology - A review. *Applied Energy*, 2009. **86**(9): p. 1365-1375.

[16] Mujeebu, M.A., et al., A review of investigations on liquid fuel combustion in porous inert media. *Progress in Energy and Combustion Science*, 2009. **35**(2): p. 216-230.

[17] Mujeebu, M.A., et al., Corrigendum to "Combustion in porous media and its applications – A comprehensive survey" [Journal of Environmental Management 90 (2009) 2287–2312]. *Journal of Environmental Management*, 2009. **91**(2): p. 550.

[18] Mujeebu, M.A., et al., Combustion in porous media and its applications--a comprehensive survey. *J Environ Manage,* 2009. **90**(8): p. 2287-312.

IOP Conf. Series: Materials Science and Engineering **370** (2018) 012058 doi:10.1088/1757-899X/370/1/012058

Counter Clockwise Rotation of Cylinder with Variable Position to Control Base Flows

Mohammed Asadullah*, S A Khan, Waqar Asrar, and E Sulaeman

Department of Mechanical Engineering, International Islamic University Malaysia, Kuala Lumpur, Malaysia.

*E-mail: shiblibhai@gmail.com

Abstract. Experimental study of supersonic base flow at Mach 2 has been carried out to see the effect of cylinder when rotated counter clockwise inside the dead zone at variable locations near its base to control base pressure for different level of expansion for area ratio 9. Active cylinder of 2 mm diameter rotating counter clockwise when seen from top, is mounted as a controller. Three locations are chosen from the side wall of square duct namely at 2, 4, 6 mm respectively and 8 mm from square nozzle exit in the base region to mount the controller. Base pressure in recirculation zone and wall pressure along the square duct length has been measured with and without control. The experiments were carried out for NPR 2, 3, 6, 7.8 and 8.5. Cylinder when rotated counter clockwise as an active controller were found to reduce the base drag as high as 62 percent at NPR 8.5 when located near to duct wall and 50 percent when located away from duct wall for the same NPR. For perfectly expanded flows at NPR 7.8 the reduction in base drag was 53 percent near duct wall and 44 percent near duct wall. The active controller was up to 19 percentage effective for over expanded flows near to duct wall and up to 12 percent when located away from duct wall. Also, the control did not adversely affect the flow field. **Key words**: Wall pressure - base pressure - nozzle pressure ratio - active control

Abbreviation

A2/A1	Area ratio
M	Mach number
NPR	Nozzle pressure ratio
L/W	Length to width ratio
X/W	Duct axial distance to width ratio

Roman Symbols

A_1	Exit area of nozzle [mm^2]
A_2	Cross-sectional area of duct [mm^2]
L	Duct length [mm]
P_a	Atmospheric pressure [cm of Hg]
P_b	Base pressure [cm of Hg]
P_w	Wall pressure [cm of Hg]
P_0	Settling chamber pressure [cm of Hg]
w	Width of nozzle exit [mm]
W	Duct width [mm]
X	Axial distance along x-axis [mm]

1. Introduction

Abrupt expansion leads to is a complex structure of flow field consisting of separation, recirculation and reattachment [1]. Further flow separation causes low pressure zone at the base, lift reduction and noise generation [2]. In high speed vehicles, such as ballistic missiles, projectiles and rockets the depression at base in jet off condition amounts to base drag as high as 50 percent of the total drag [3]. The base pressure acting at the base of bluff body moving at supersonic speeds is vital in prediction of its base drag [4]. Thus, managing the base pressure with an active controller will be focus of our study. Past researchers have developed tools such as periodic suction and injection [5], micro jets [6], plasma actuator [7] etc. to actively control flow field [8]. Active control techniques require the input energy, making it bit expensive but very effective [9]. Researchers have worked on control of flows for a very long era and a large amount of data is available in literature. So only relevant data to the present study is explored here. To manage base pressure in a desired manner a robust control devise is needed. We proposed a rotating cylinder in counter clockwise direction when viewed from top to modify the base field.

2. Literature Review

An active device for controlling the base pressure in high speed compressible flow regimes, near ambient shock waves, by modifying flow pattern inside the dead zone to authors knowledge is being studied for the first time. The inherent instability problem due to vortex shedding behind the cylinder viewed as bluff body [10] is being studied by ample number of researchers. The problem of flow past a rotating cylinder at low Reynolds number was studied extensively by [11] but he only concentrated on lift due to magnus effect and boundary layer control on an aerofoil. An experimental study of vortex-induced vibration in the crossflow direction of a rotating circular cylinder was carried out by [12]. Suppressing vortex shedding of a cylinder at low Reynolds numbers was achieved by using a smaller cylinder outside the recirculation area in the wake of the main cylinder for incompressible flows [13]. [14] studied the stability analysis of flow past a circular cylinder and concluded rotation helps stability but there is a critical rotation when instability is unavoidable. [15] studied effect of high rotation rates on the laminar flow around a circular cylinder, for higher Reynolds number in the steady flow regime. He concluded that the inherent drag force of cylinder decreases with increasing rotational velocities. [16] investigated on turbulent flow past a rotating cylinder at high Reynolds number. His results revealed stabilization of the acting forces at high spin rates, thus indicating a flow field with suppressed vortex-shedding activity [17]. studied uniform shear flow past a rotating cylinder. [18] investigated vortex shedding for a slow rotating cylinder at different speeds using a stepper motor. [19] studied wake of rotating circular cylinder in an incompressible flow. Rapidly rotating circular cylinder and oscillations were investigated in detail by [20]. Physical mechanism of transition in bluff bodies like rotating cylinder having 3D wake was elaborated in detail by [21]. From review of cylinder study, the flow physics for a single rotating cylinder, and the effect of rotation on cylinder and its interaction with surrounding has been well recorded. Most of the study are done for incompressible low speed flows to reduce the vortex shedding behind the cylinder by various means. One of the ways to get rid of its vortex shedding leading to instability is by reducing its diameter and/or rotating it. Thus, physics of rotating cylinder as a bluff body is well supported by vast literature but there is not a single case to the authors knowledge where cylinder is used to control base drag in compressible high-speed flows. We propose to use cylinder as an active controller rotating in counter clockwise direction at different locations along the base line 'Y-Y' as shown in figure 4.

3. Experimental Setup and procedure

3.1. Supersonic speed facility

The experiments were conducted in B.I.T research centre at supersonic laboratory, Mangalore, Karnataka, India. The supersonic speed facility consists of a settling chamber, with a provision to mount the jet nozzle on its flange. Compressed dry air is stored at a pressure of 200 to a maximum of 300 psi in two high pressure vessels namely storage tank. Compressed air from the storage tank is

IOP Conf. Series: Materials Science and Engineering **370** (2018) 012058 doi:10.1088/1757-899X/370/1/012058

supplied to the settling chamber at a high pressure of 150 psi through a pressure regulatory valve (PRV) as shown in the figure 1 and 2. Once the flow attains equilibrium in the settling chamber, it is expanded through the square nozzle to the suddenly expanded square duct. The desired stagnation pressure at the settling chamber is achieved by controlling the pressure regulatory valve (PRV).

Figure 1. Photographic view of storage tanks

Pressure transducer 9205 is used for measuring base pressure behind exit of nozzle, wall pressure along the duct and stagnation pressure in the settling chamber. It has 16 channels and pressure range is approximately 0-150 psi. It averages 250 samples per seconds and displays the reading. The user-friendly menu driven LabVIEW software along with DAQ acquires data and displays the readings from all the channels simultaneously on the computer screen.

Figure 2. Photographic view of experimental setup

3.2. Fabrication process

The supersonic nozzle using analytical methods was designed for Mach 2 and then fabricated from brass for the same Mach number. Before starting an experiment, it is mandatory for people working in high speed flows to calibrate it. After calibration, the nozzle was attached with the flange of settling chamber at one end and expansion square duct at the other. The exit area of the square nozzle is 10*10 mm square. The Control cylinder of 2 mm diameter shown in figure 3 attached with the motor, positioned on 'YY' base line at 8 mm from nozzle exit was initially mounted at 2 mm from side wall through two roof adjustable table vertically in to the square duct. Flange of square nozzle has 8 holes to fasten it with the square duct. The sudden expansion square duct was also fabricated from brass with its length 10 times its width i.e. L/W = 10. Pressure taps along the duct wall of 1 mm ID steel tubes were used to record wall pressure.

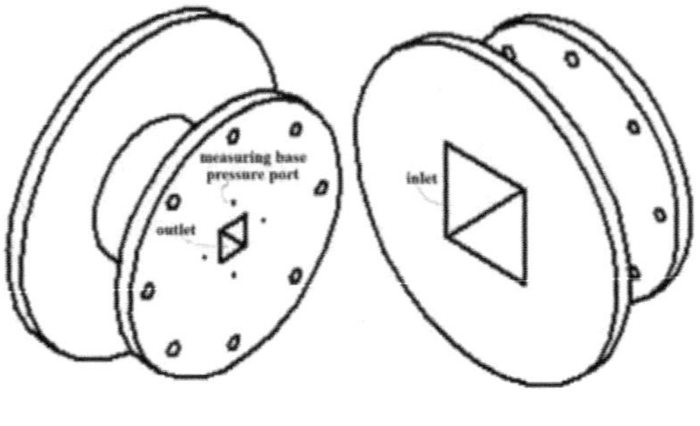

(a) Square nozzle

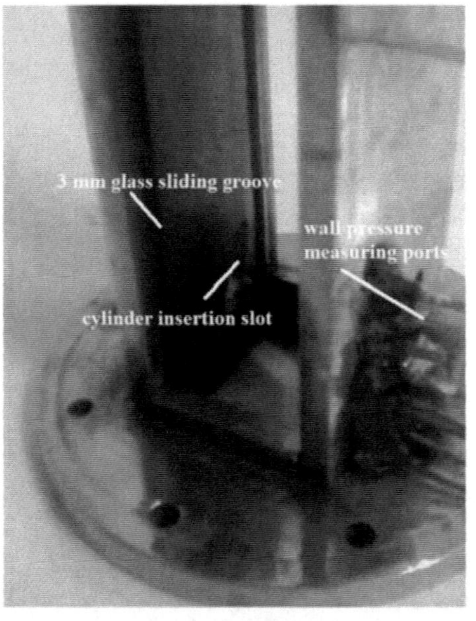

(b) Square duct

Figure 3. Photographic view of fabricated square duct and square nozzle

Subsequently its position is varied along 'YY' at 4 mm and 6mm respectively as shown by green dots in figure 4. Base pressure was measured at 4 places by nozzle ports at 11.5 mm from centre. Taps begins at 7 mm from the nozzle exit and initially these were 3 mm apart and the distance between them in the axial direction progressively increased until maximum duct length of 300 mm as shown in Fig. 5. The NPR used for this study were 2, 3, 6, and 7.8. The area ratio A2/A1 and length width ratio L/W were 9 and 10 for all the NPRs.

Measurements were made for all NPRs with and without control. The settling chamber was provided with wire meshes to attain the stagnation state. When the flow becomes steady, the settling chamber pressure was recorded. At steady flow conditions the base pressure and the wall pressure in the duct were recorded by pressure transducers. Measurements for all NPRs for no control and control were completed for the square duct of 30*30 square mm cross-sectional area. The base pressure and wall pressure were made non-dimensional with atmospheric pressure and were presented as Pb/Pa and Pw/Pa after converting all reading from gauge pressure to absolute pressure.

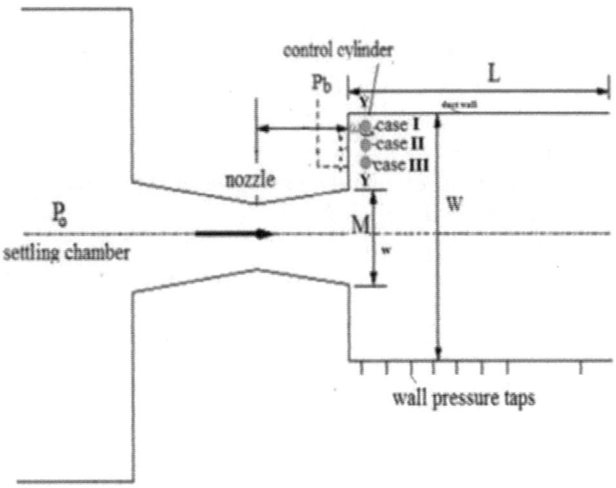

Figure 4. Schematic drawing of the nozzle and duct with different position of control cylinder shown in green dots

Figure 5. Photographic view of control cylinder, adjustable table and square duct

4. Results and discussions

The main aim of this investigation is to study the effect of the active control as well as the position of the cylinder placed normal to the flow and its influence on the base pressure when it is rotated in the counter clockwise direction along the base. The dynamic cylinder is located at 2 mm, 4 mm, and 6 mm from the base line sequentially and the base pressure as well as the wall pressure were recoded with and without control. The percentage change in base pressure is given by,

% Pb = [(Pb control-Pb no control)/Pb no control] × 100

Fig. 5 presents the percentage change in base pressure for various NPRs. From results, it is evident that as long as flow remains over expanded the control effectiveness is marginal and this trend continues till the NPR approaches close to the nozzle pressure ratio required for perfectly expanded case. At NPR

2 the flow is highly over expanded for Mach 2 and hence there is no change the base pressure even when the location of the cylinder was varied from 2 mm to 6 mm from the base line 3and was rotating clockwise as well. However, at NPR 3 the gain is 8 % when active cylinder was at 2 mm from the base

as compared to the other two locations where gain is 5 %. This difference seems to be due to its location in base corner. Further, it is found that control reversal takes place at NPR = 4 and 6.5. For the NPRs in the range 2 to 5 there is increase in the base pressure, and there is decrease in the base pressure for NPRs in the range 5 to 6.5, then for higher NPRs there is considerable increase in the base pressure which is more than sixty percent. The physics behind this phenomenon seems to be the strength of the oblique shock waves as long as the jets are over expanded, location of the rotating cylinder and the strength of the base vortex. From the results, it is evident that when the cylinder is located at 2mm from the base will give maximum increase in the base pressure. When this cylinder was shifted towards main jet there is progressive increase in the base pressure in the range 45 to 48 %.

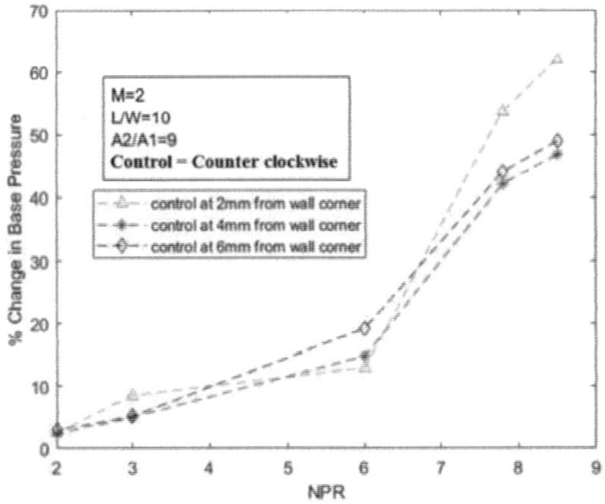

Figure 6. Percentage Change in Base Pressure with NPR

Non-dimensional base pressure variation with NPR for three different locations of the rotating cylinder. As discussed earlier when the dynamic control was located at 2 mm from the base line was most effective, here also we observe the similar results. It is seen that right from NPR 2 till 8.5 there is continuous decrease in the base pressure values for without control case. When the control was placed at 2 mm from the base as the rotating cylinder, it shows mixed trend, control results in increase of pressure for all the NPRs tested, but when compare the control effectiveness with reference to the other two locations of the control namely 4 mm and 6mm from the base, this location of the control results in maximum gain at NPR 3, lowest at NPR 6 and then increase for the NPRs 7.8 and 8.5. The control cylinder located at 4mm results lowest gain in comparison with 6 mm location. When control is located at 6 mm from the base it will be closed to the main jet and as well as the shear layer. Hence due to the combined effect of the interaction of the main jet, boundary layer, dividing stream line, base vortex and the rotating cylinder has resulted in this phenomenon.

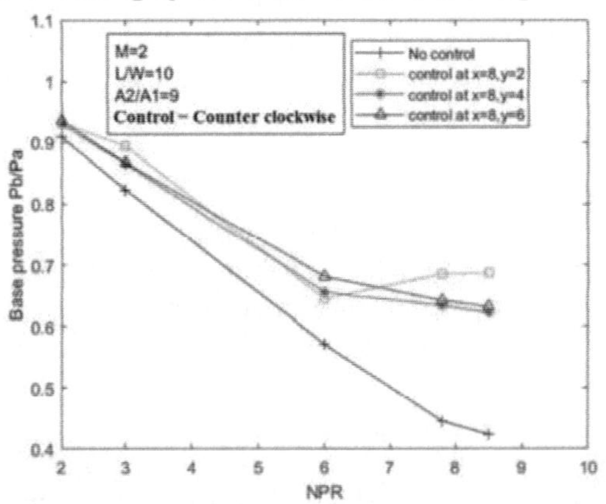

Figure 7. Base pressure variation with NPR

IOP Conf. Series: Materials Science and Engineering **370** (2018) 012058 doi:10.1088/1757-899X/370/1/012058

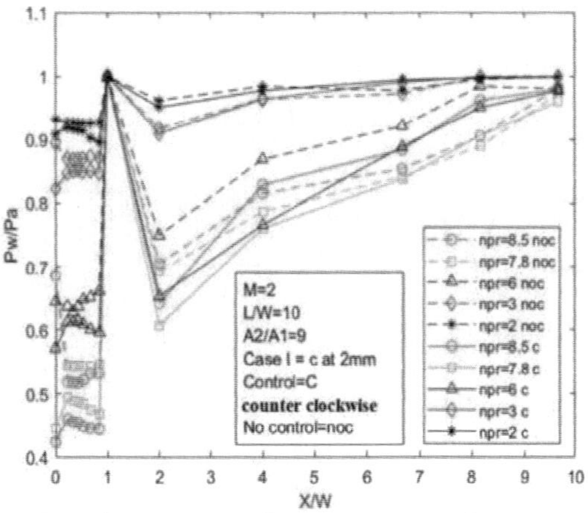

Figure 8. Flow development in the duct at different NPR for Case I

Wall Pressure flow pattern is shown in Fig. 8 for all the NPRs tested with and without control. At NPRs 2 and 3 flows with and without control are identical. Here at NPR 2 and 3 the control results in increase and decrease of the wall pressure. The same is the trend for NPR 6 is seen except the magnitude of control effectiveness has increased considerably due to the reduction in the level of over expansion but at x/w = 8 this increase is negligible due to the back-pressure effect. At NPR 7.8 the control shows mixed trend, due to the presence of weak wave at the nozzle exit, interaction with the dividing stream line, and active control and its location in the base corner.

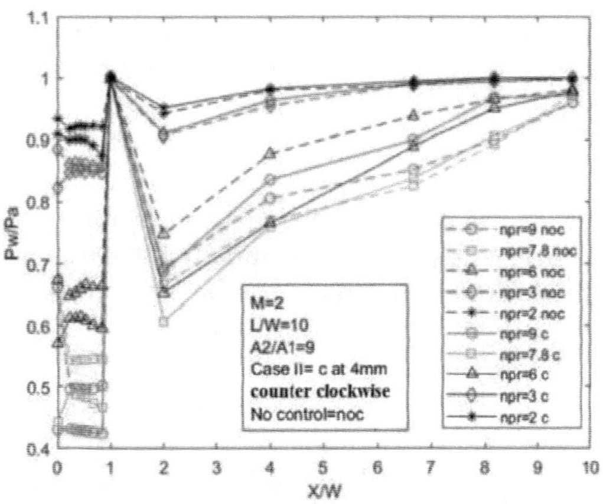

Figure 9. Flow development in the duct at different NPR for Case II

Figure 9 shows wall pressure results for the case II when control cylinder was at 4 mm from the base line. It is seen that wall pressure distribution in the duct with and without control are identical for NPRs 2 and 3, control results in decrease of wall pressure NPR 2 and increase for NPR 3 and this change is marginal and this trend is due to the level of over expansion at the exit of the nozzle. When we look wall pressure results for NPR 6, the initial values have reduced considerably due to the reduction in the level of over exapnsion and control results in appreciable decrease of wall pressure, but the magnitude of decrease in the down stream has reduced considerably due to the back pressure effect. At NPR 7.8 which is required for correct expansion, the results in decrease of wall pressure, at x/w = 1 there is strong wave in form resulting in sudden jump in pressure which is equal to the ambient pressure, then again flow expands to value around 0.65 and 0.6 without and with control and the control reversal takes place at x/w = 6.5 and 9. We see this fluctuating trend due to the jet being

IOP Conf. Series: Materials Science and Engineering **370** (2018) 012058 doi:10.1088/1757-899X/370/1/012058

correctly expanded at NPR = 7.8, where Mach waves will be present, rotating cylinder is entraining some mass towards the base corner.

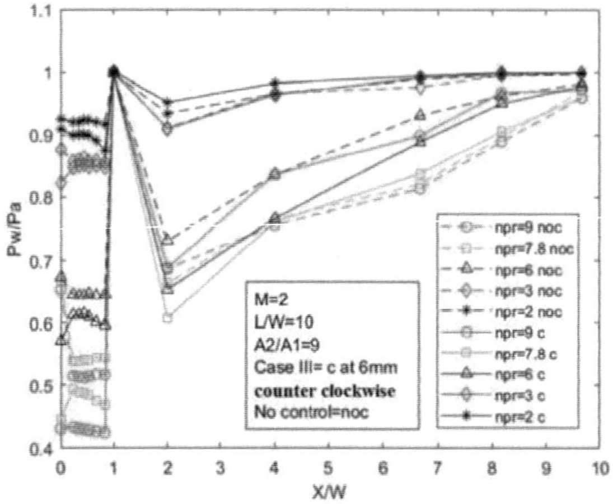

Figure 10. Flow development in the duct at different NPR for Case III

Figure 10 shows the results for case III where rotating cylinder is placed at 6 mm from the base line. This case also presents similar results as discussed earlier with the exception that control mechanism has been placed at 6 mm from the base line. We see a different trend for NPR 7.8 where the effectiveness is marginal as compare to the previous cases. This modification in the wall pressure flow field is due to the shock wave and boundary layer interaction, effect of the location of control mechanism near the main jet as well as the dividing stream line, which is not the case for the other two cases I and II. Hence, we can conclude that overall trend remains the same with minor variations in the values.

5. Conclusions

Base on the above discussions we can draw the following conclusions:

1. The case I when the control was placed at 2 mm from the base is most effective for controlling the base pressure resulting in maximum gain in the base pressure then comes the case III when the control was placed 6 mm from the base line. However, the control results in increasing the base pressure for all the three locations with minor changes in the control effectiveness.
2. The Case II seems to give average results due to its proximity from the base as well as the main jet.
3. When jets are perfectly expanded, they give mixed trend and control reversal takes placed along the enlarged duct wall many times due to the presence of weak wave.
4. Over all the flow field and the trend in the duct remains the same for all the cases. Hence, the control does not adversely affect the wall pressure flow field in the duct.

Acknowledgements

Authors wishing to acknowledge and thank Bearys Institute of Technology for utilizing research facility in supersonic laboratory.

References

[1] Pandey K and Rathakrishnan E 2006 Annular cavities for Base flow control *International Journal of Turbo and Jet Engines* **23** 113-28.

[2] Uruba V, Jonáš P and Mazur O 2007 Control of a channel-flow behind a backward-facing step by suction/blowing *International Journal of Heat and Fluid Flow* **28** 665-72

[3] Viswanath P 1996 Flow management techniques for base and afterbody drag reduction *Progress in Aerospace Sciences* **32** 79-129.

[4] Badrinarayanan M 1961 An experimental investigation of base flows at supersonic speeds *The Aeronautical Journal* **65** 475-82.

[5] Williams D and Amato C 1989 *Frontiers in experimental fluid mechanics*: Springer) pp 337-64.

[6] Khan S A and Rathakrishnan E 2003 Control of suddenly expanded flows with micro-jets International *journal of Turbo and Jet engines* **20** 63-82.

[7] Kühn M, Kühn-Kauffeldt M, Schein J and Belinger A 2017 Plasma actuators for active flow control based on a glow discharge. In: *Journal of Physics: Conference Series*: IOP Publishing) p 012007.

[8] Gad-el-Hak M Flow Control: Passive, Active, and Reactive Flow Management. 2000. Cambridge University Press).

[9] Beaudoin J-F, Cadot O, Aider J-L and Wesfreid J-E 2006 Drag reduction of a bluff body using adaptive control methods *Physics of Fluids* **18** 085107.

[10] Baek S-J and Sung H J 1998 Numerical simulation of the flow behind a rotary oscillating circular cylinder *Physics of Fluids* **10** 869-76.

[11] Badr H, Dennis S and Young P 1989 Steady and unsteady flow past a rotating circular cylinder at low Reynolds numbers Computers *& Fluids* **17** 579-609.

[12] Seyed-Aghazadeh B and Modarres-Sadeghi Y 2015 An experimental investigation of vortex-induced vibration of a rotating circular cylinder in the crossflow direction *Physics of Fluids* **27** 067101.

[13] Strykowski P and Sreenivasan K 1990 On the formation and suppression of vortex 'shedding'at low Reynolds numbers *Journal of Fluid Mechanics* **218** 71-107.

[14] Mittal S and Kumar B 2003 Flow past a rotating cylinder *Journal of Fluid Mechanics* **476** 303-34.

[15] Stojković D, Breuer M and Durst F 2002 Effect of high rotation rates on the laminar flow around a circular cylinder *Physics of fluids* **14** 3160-78.

[16] Karabelas S, Koumroglou B, Argyropoulos C and Markatos N 2012 High Reynolds number turbulent flow past a rotating cylinder *Applied mathematical modelling* **36** 379-98.

[17] Sung H J, Chun C K and Hyun J M 1995 Experimental study of uniform-shear flow past a rotating cylinder *Journal of fluids engineering* **117** 62-7.

[18] Lam K 2009 Vortex shedding flow behind a slowly rotating circular cylinder *Journal of Fluids and Structures* **25** 245-62.

[19] Kimura T, Tsutahara M and Wang Z-1992 Wake of a rotating circular cylinder *AIAA journal* **30** 555-6.

[20] Glauert M 1957 The flow past a rapidly rotating circular cylinder. In: *Proceedings of the Royal Society of London A: Mathematical, Physical and Engineering Sciences*: The Royal Society) pp 108-15.

[21] Thompson M, Leweke T and Williamson C 2001 The physical mechanism of transition in bluff body wakes Journal of Fluids and Structures 15 607-16.

[22] Samimy M, Zaman K and Reeder M 1993 Effect of tabs on the flow and noise field of an axisymmetric jet AIAA journal 31 609-19.

Computation of Stability Derivatives of an oscillating cone for specific heat ratio = 1.66

Aysha Shabana[1*], Renita Sharon Monis[2], Asha Crasta[3], S. A. Khan[4]

[1] Research Scholar, Department of Mathematics, M.I.T.E, Moodabidri and Assistant Professor, Sahyadri College of Engineering and Management, Mangaluru, Karnataka, India.
[2] Research Scholar, Department of Mathematics, M.I.T.E, Moodabidri and Assistant Professor (Sr), SMVITM, Bantakal, Karnataka, India.
[3] Associate Professor, Department of Mathematics, M.I.T.E, Moodabidri, Karnataka, India.
[4]Professor, Department of Mechanical Engineering, Faculty of Engineering, IIUM, Gombak Campus, Kuala Lumpur, Malaysia.

*E-mail: aysha.nawshad@gmail.com

Abstract. In this paper the expressions for stiffness and Damping derivatives are obtained in a closed form for perfect gas where the flow is quasi-steady and axi-axisymmetric, and the nose semi angle of the cone is such that the Mach number M_2 behind the shock $M_2 \geq 2.5$. Results are presented for an oscillating cone for gas with $= 1.666$, at different Mach numbers and semi cone angles. The Stiffness derivative decreases with pivot position and also with semi vertex angle, there is substantial change in the stiffness derivative when semi-vertex has been increased from 5 degrees to ten degrees, further increase in the semi-vertex angle results in marginal change in the stiffness derivative. Due the marginal change in the Mach number level there is marginal increase in the magnitude of the stability and with further increase in the inertia level the stability derivative conform to the Mach number independence principle. The present theory for Oscillating cone is restricted to quasi-steady case. Viscous effects have been neglected. The expressions so obtained for stability derivative in pitch are valid for a slender ogive which often approximates to the whole fuselage of an aircraft. **Keywords:** High Speed Flow, Hypersonic Flow, Oscillating cone, Stiffness derivative

1. Introduction

The present work estimates stability derivatives in pitch for non-slender axi-symmetric ogives oscillating in hypersonic flow. At hypersonic speeds the "nose cones" are often non-slender and blunt. The reason for such a configuration, is the problem of aerodynamic heating and hence the heat transfer at such speeds. Although the present work is not for blunt bodies with detached shocks, once a theory is developed for the ogives with sharp nose, it can then be extended to more practical shapes to incorporate the bluntness.

It is fascinating to note that the study of hypersonic flows, which was limited to slender bodies and low angles of attack, should attain a stage of non-slender shapes and at large angle of attack flows; promising an emphasized development of efficient future hypersonic systems.

Ghosh [1] developed a new hypersonic similitude with the assumptions of attached bow- shock and Mach number behind the shock being greater than 2.5. This similitude was valid for the windward surface of an aero-foil with large flow deflection. His work was extended to oscillating non-planar wedges by Crasta and Khan to calculate the aerodynamic pitching moment derivatives in both Supersonic [2] and Hypersonic flows [3, 4 and 5].

The large deflection similitude of Ghosh [1] has been extended by Ghosh, K., [2] to axisymmetric bodies with attached shock. Equivalence with a new piston-motion with axial symmetry has been established. The cone (or quasi-cone) results from the revolution of a wedge (or quasi-wedge) round the stream-wise axis and a similar revolution of the independent fluid slab [1] produces an axially symmetric conico-annular space. It is shown by Ghosh, K., [2] that the flow past a cone or quasi-cone is equivalent to a piston-motion in this conico-annular space, which is called the similitudinal slab. Although Ghosh, K., [2] gives similitude for cones and quasi-cones, he gives a solution based on the similitude for a cone only. This solution gives a constant density shock layer. Hence the constant density form of the unsteady Bernoulli's equation is used to find pressure on the cone surface in terms of the piston Mach number. Results are obtained for hypersonic flow for a perfect gas over oscillating cones and ogives of different Mach numbers and semi- angles.

2. Analysis

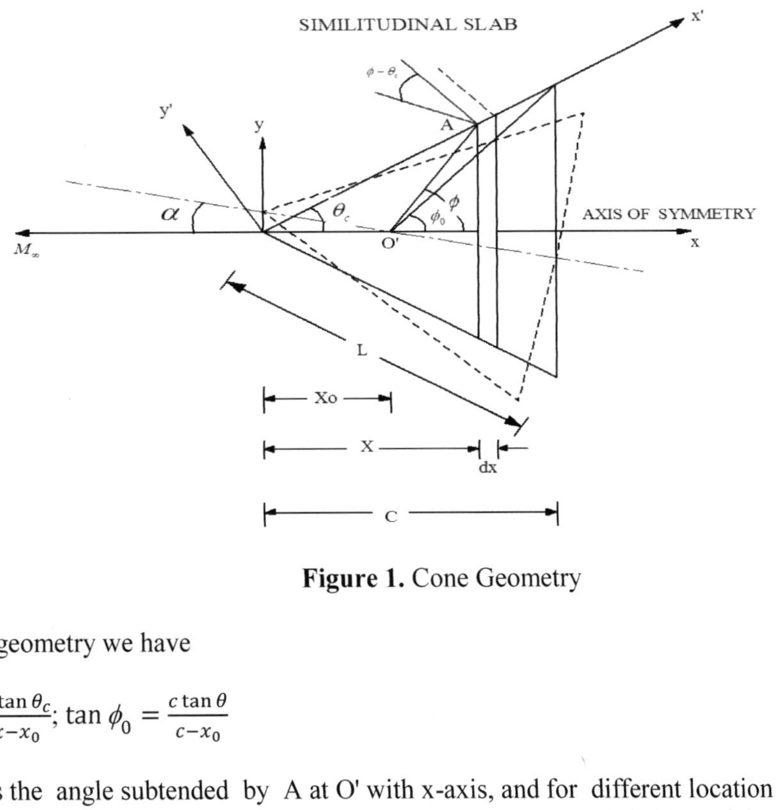

Figure 1. Cone Geometry

From the geometry we have

$$\tan\phi = \frac{x\tan\theta_c}{x-x_0};\ \tan\phi_0 = \frac{c\tan\theta}{c-x_0}$$

where ϕ is the angle subtended by A at O' with x-axis, and for different location of A, ϕ varies from π to ϕ_c, θ_c is the cone semi-angle and c is the chord length. The definition of the Stiffness coefficient denoted by C_{m_α}, is

$$C_{m_\alpha} = \left[\frac{\partial M}{\partial \alpha}\right]_{\alpha,q\to 0} \frac{1}{\frac{1}{2}P_\infty U_\infty^2 S_b c}$$

where S_b = base area of the cone = $\pi(c\tan\theta_c)^2$, c = chord length of the cone. On solving we obtain the following,

The expression for the pressure ratio of a steady cone at zero incidence (Ghosh, K., 1984), provided the bow shock is attached, is

$$\frac{P_{bo}}{P_\infty} = 1 + \gamma M_{po}^2\left(1 + \frac{1}{4}\varepsilon\right) \quad (1)$$

where the density ratio is

$$\varepsilon = \frac{2 + (\gamma - 1)M_{po}^2}{2 + (\gamma + 1)M_{po}^2} \quad (2)$$

and M_{po} is the piston Mach number of the equivalent piston, operating in a conico-annular space; P_{bo} is the pressure on the body (cone) surface at zero incidences.

$$M_{po} = M_\infty \sin\theta_c$$

where θ_c is the cone semi angle.

Now

$$\frac{dP_{bo}}{dM_{po}} = 2P_\infty M_{po}\left[1 + \frac{1}{4}\left(1 + \frac{1}{2}M_{po}\frac{d\varepsilon}{dM_{po}}\right)\right] \quad (3)$$

where

$$\frac{d\varepsilon}{dM_{po}} = \frac{-8M_{po}}{\left[2 + (\gamma + 1)M_{po}^2\right]^2} \quad (4)$$

Thus using the definition of Stiffness and damping derivative and using the above, the expression for Stiffness derivative is obtained as

$$C_{m_\alpha} = D[h^3(1 - 2n^2) - (1 - h)\{H(2 + h) + n^2h(1 + 2h)\}] \quad (5)$$

And the expression for damping derivative is obtained as

$$C_{m_q} = \frac{D}{2}[h^4(2n^2 - 3n^4 - 1) - (1 - h)\{H(3H + h(H + 2n^2) + 2h^2n^2) + n^4h^2(1 + 3h)\}] \quad (6)$$

where

$$D = \frac{2}{3(1 + n^2)}\left[1 + \frac{1}{4}\left(\varepsilon + \frac{1}{2}K\frac{d\varepsilon}{dM_{po}}\right)\right] \quad (7)$$

and

$$H = (1 - h + n^2)$$

3. Results and Discussion

The results were obtained for various flow and geometrical parameters & discussed in this section. The variation of Stiffness derivative with pivot position for Mach number M = 5 is depicted in Figure 2. It is seen that Stiffness derivative decreases with pivot position and also with semi vertex angle. From the figure it is seen that at h = 0 there is substantial change in the stiffness derivative is only when semi-vertex has been increased from 5 degrees to ten degrees, further increase in the semi-vertex angle results in marginal change in the stiffness derivative. However, when we move downstream towards the base of the cone there is substantial increment in the stiffness derivative and this trend could be attributed to the flow field and hence the pressure distribution on the surface of the cone. We also should keep in mind that this is the case of body of revolution and not a simple 2-D case. Here 3-D effect does play an important role.

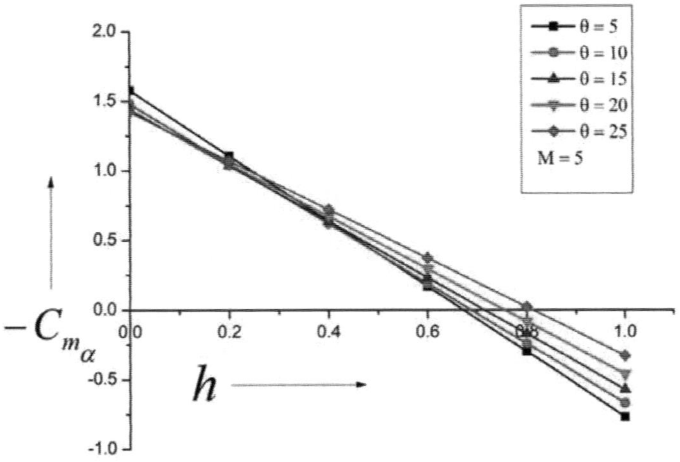

Figure 2. Variation of Stiffness derivative with pivot position for Mach number M = 5

Fig. 3 shows the results for stiffness derivatives for Mach number M = 7. Since all the parameters are same only the Mach number has been increased from M = 5 to 7. Due this marginal change in the inertia level there is marginal increase in the magnitude of the stability derivative and remaining trends are on the similar lines.

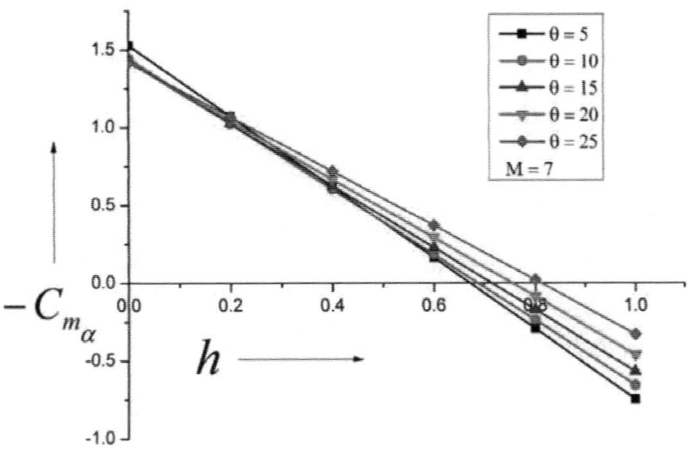

Figure 3. Variation of Stiffness derivative with pivot position for Mach number M = 7

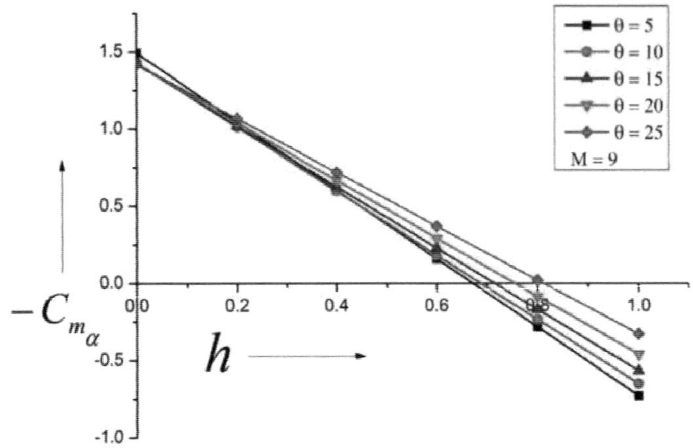

Figure 4. Variation of Stiffness derivative with pivot position for Mach number M = 7

Fig. 4 presents the results of stiffness derivative for Mach number M = 9. Here it is seen that since the Mach number M has further been increased from M = 7 to 9, this has resulted further in the reduction of Stiffness derivatives. However, when we consider the results downstream and more so beyond centre of pressure there is appreciable increase in the magnitude of the stiffness derivatives.

Similar results are shown in figures 5 and 6 for Mach number M = 10 and 15. In figure 6 there is marginal change in the stiffness derivative near the vertex of the cone for the pivot position h = 0 to 0.1, but this change disappears for M = 15. This indicates that Mach number independence principle hold for this Mach number. Further, it is seen that the centre of pressure is independent of Mach number, it remains in the same range, and this may be due to the 3-D effect.

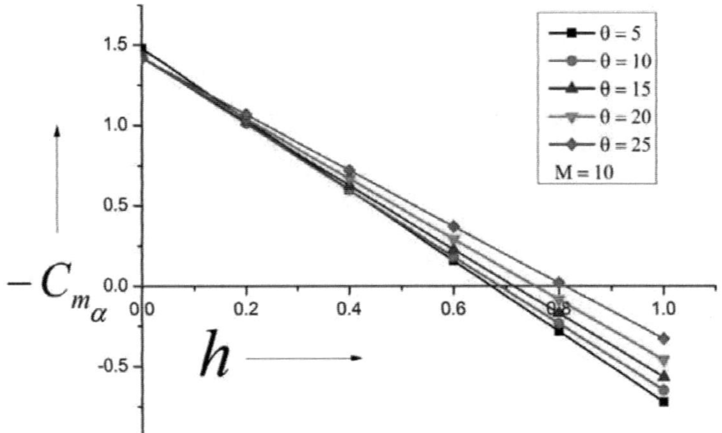

Figure 5. Variation of Stiffness derivative with pivot position for Mach number M = 10

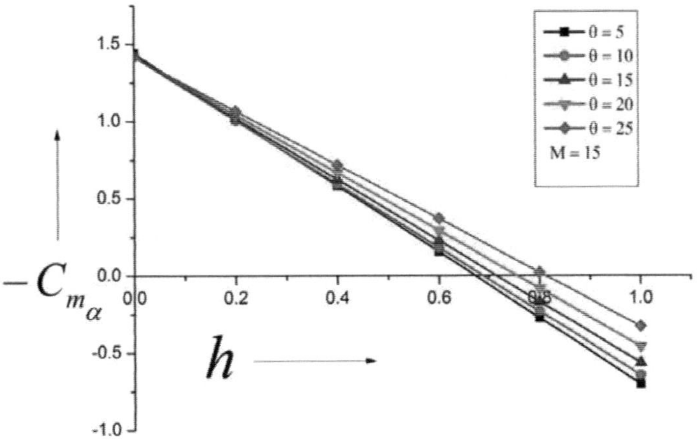

Figure 6. Variation of Stiffness derivative with pivot position for Mach number M =15

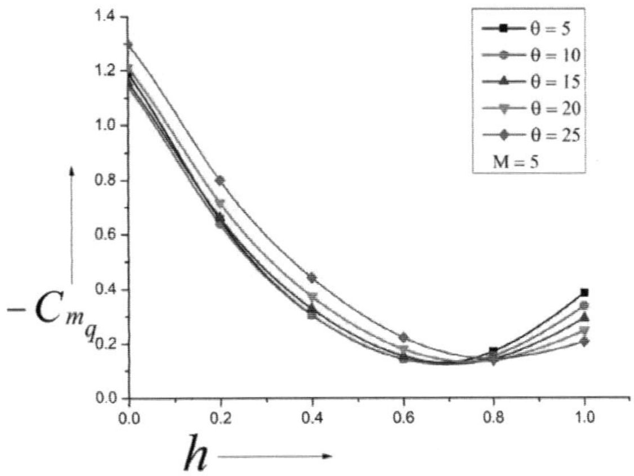

Figure 7. Damping derivative variation with pivot position for Mach number M = 5.

Figs. 7 to 11 present the results of damping derivatives for Mach M = 5, 7, 9, 10 and 15. From figure 7 it is found that when angle theta was increased from 5 to 10 degrees there is 33 % increase in the stability derivative and the same trend continues till it attains minimum value and then there is reversal in the trend. This is due to the location of pivot position. As long as the values of the pivot positions are ahead of centre of pressure it will have specific behaviour and later just opposite behaviour will start. It is also seen that there is forward shift in the minima point as compared to the oscillating wedge, this may due to the 3-D effect and due to the oscillating cone. Further the position of centre of pressure remains in the range of h = 0.7 to 0.8. With further increase in the value of angle theta keeping Mach number same the increase in the damping derivatives is marginal. This may be due to the combined effect of Mach number, angle theta, and the pivot position.

Similar results are seen in Figs. 8 to 11. In all these figures it is observed that with further increase in the Mach number there is marginal decrease in the magnitude of the damping derivative as expected that with the increase in the Mach number the damping derivatives will decrease. Another important observation is that the band of centre of pressure which was between (h = 0.7 to 0.8) is narrowed down with the increase in Mach number. Once the Mach number is M = 10 and above there is change in the values of the damping derivatives due to the attainment of Mach number independence principle.

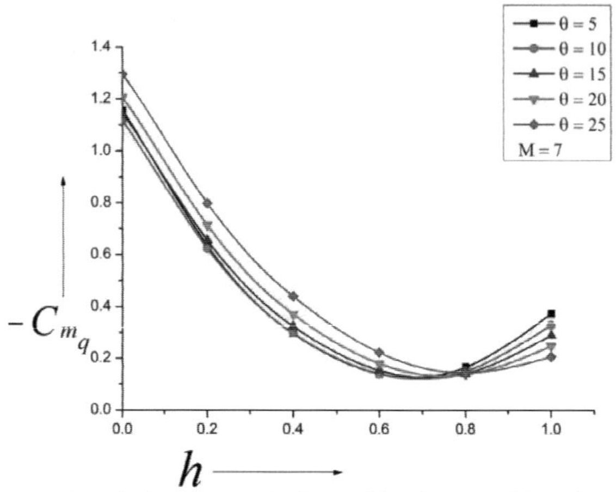

Figure 8. Damping derivative variation with pivot position for Mach number M = 7

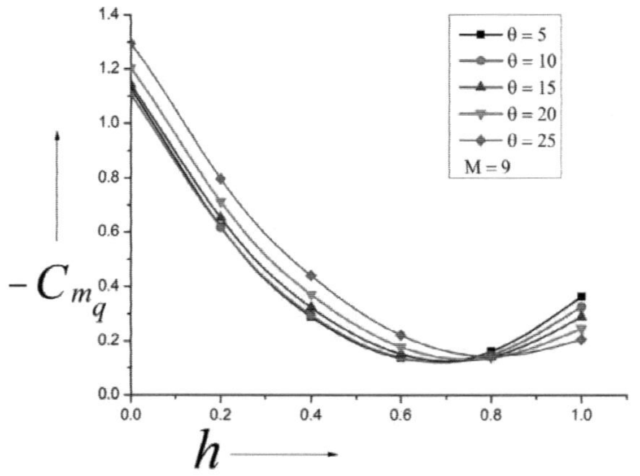

Figure 9. Damping derivative variation with pivot position for Mach number M =9.

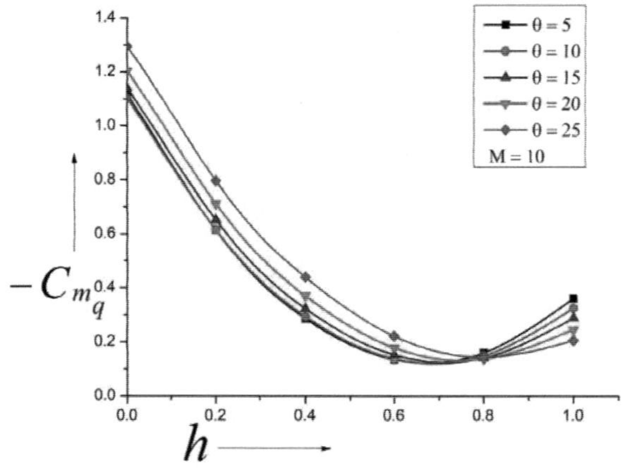

Figure 10. Damping derivative variation with pivot position for Mach number M =10.

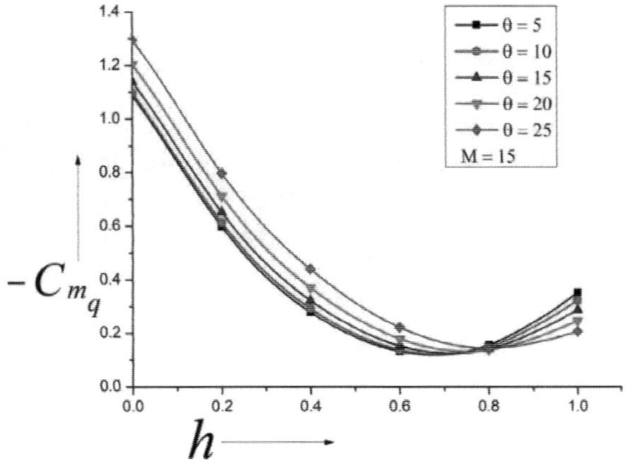

Figure 11. Damping derivative variation with pivot position for Mach number M =15.

4. Conclusions

Based on the above discussion we can draw the following conclusions:

- The Stiffness derivative decreases with pivot position and also with semi vertex angle, there is substantial change in the stiffness derivative when semi-vertex has been increased from 5 degrees to ten degrees, further increase in the semi-vertex angle results in marginal change in the stiffness derivative, this is due to the case of body of revolution and not a simple 2-D case. Here 3-D effect does play an important role.
- Due the marginal change in the inertia level there is marginal increase in the magnitude of the stability.
- It is found that when angle theta was increased from 5 to 10 degrees there is 33 % increase in the stability derivative and the same trend continues till it attains minimum value and then there is reversal in the trend. This is due to the location of pivot position. It is also seen that there is forward shift in the minima point as compared to the oscillating wedge.
- The present theory for Oscillating cone is restricted to quasi-steady case. The main difficulty encountered in treating the unsteady case would be due to the flow being unsteady as well as non-uniform in the conical– annular space even for a steady piston.
- These results are likely to find wide applications in high speed flow problems. Viscous effects have been neglected. The expressions so obtained for stability derivative in pitch are valid for a slender ogive which often approximates to the fuselage of an aircraft.

References

[1] Ghosh, K. (1977), "A New Similitude for Aerofoils in Hypersonic Flow", *Proceedings of the 6th Canadian Congress of Applied Mechanics, Vancouver, Canada,* May 29-June 3, pp. 685-686.

[2] Ghosh K., (1984), *Hypersonic large deflection similitude for oscillating delta wings*, the Aeronautical Journal, pp. 357-361.

[3] Asha Crasta and Khan S. A., (2012), *High Incidence Supersonic similitude for Planar wedge*, International Journal of Engineering research and Applications, 2(5), pp. 468-471.

[4] Asha Crasta and S. A. Khan, (2012), *Estimation of Stability derivatives of an Oscillating Hypersonic delta wings with curved leading edges*, IJMET, 3(3), pp. 483-492.

[5] Asha Crasta, M. A. Baig, S. A. Khan, (2012), *Estimation of Stability derivatives of a Delta wing in Hypersonic flow*, International Journal of Emerging trends in Engineering and Developments 6(2), pp. 505-516.

[6] Asha Crasta and S. A. Khan, (2012*), Oscillating Supersonic delta wing with Straight Leading Edges*, International Journal of Computational Engineering Research, 2(5), pp. 1226-1233.

[7] Asha Crasta and S. A. Khan, (2014), *Effect of Angle of Incidence on Stability derivatives of a wing*, International Journal for Research in Applied Science and Engineering Technology, 2(V), pp. 411-422.

[8] Asha Crasta and S. A. Khan, (2014), *Hypersonic Similitude for Planar Wedges*, International Journal of Advanced Research in Engineering and Technology, 5(2), pp. 16-31.

[9] Asha Crasta and S. A. Khan, (2014), *Estimation of Stability Derivatives for a Planar Wedge in the Newtonian Limit*, IOSR Journal of Mathematics, 10(2), pp. 01-06.

[10] Asha Crasta and S. A. Khan, (2014), *Effect of Mach number on Stiffness and Damping Derivatives for Oscillating Hypersonic Non-Planar Wedge*,IOSR Journal of Mechanical and Civil Engineering, 11(2), pp. 04-11.

[11] Asha Crasta and S.A. Khan, (2014), *Effect of Angle of Incidence on Stiffness and Damping derivatives for Oscillating Hypersonic Non-planar Wedge*, International Journal for Research in Applied Science and Engineering Technology, 2(IV),pp. 229-242.

[12] Asha Crasta and S. A. Khan,(2014*), Supersonic Similitude for Oscillating Non-Planar wedge*, IOSR Journal of Mathematics, 10(2), pp. 15-24.

[13] Khan S. A. and Asha Crasta, (2010), *Oscillating Supersonic delta wings with curved leading edges*, Advanced Studies in Contemporary Mathematics, 20(3), pp. 359-372.

[14] Asha Crasta and S. A. Khan (2015), "*Effect of Angle of attack on Stiffness derivative of an oscillating supersonic delta wing withcurved leading edges*" IOSR-JMCE issue1, Volume12, December,pp12-25.

[15] Asha Crasta and S. A. Khan, "*Effect of Angle of attack on Damping derivative of a delta wing with full sine curved leading edges*" IJETED Journal issue5, Volume1, December- January 2015.

[16] Asha Crasta and S. A. Khan,(2015) "*Estimation of Damping derivative of a delta wing with half sine curved leading edges*" IOSR Journal of Mechanical and civil engineering, Vol.12, issue1,February ,pp40-44.

[17] Asha Crasta and S. A. Khan (2015), "*Estimation of Damping derivative in pitch of a Supersonic Delta wing with curved leading edges*" IOSR Journal of Journal of Mathematics , Vol. 11,issue 1, Jan-feb ,pp.07-15.

[18] Asha Crasta and S.A.Khan (2015),"*Effect of Aspect ratio with angle of attack of an oscillating Hypersonic delta wing with Straight Leading edges*", Mathematical Sciences International research Journal",volume 4,issue2, pp.28-33., ISSN2278-8697.

[19] Aysha Shabana, Renita Monis, Asha Crasta and S.A.Khan , "*Pressure and its derivative with respect to piston Mach number for an oscillating cone* " IOSR Journal of Mechanical and Civil Engineering , volume 2 ,pp. 74-78.

[20] Asha Crasta, Aysha Shabana, Renita Monis, and S.A.Khan , "*Analysis of Variation of Stiffness Derivative with Mach number and angle of attack for a supersonic flow*", IOSR Journal of Mechanical and Civil Engineering , volume 2 ,pp. 97- 104.

[21] Renita Monis, Aysha Shabana, Asha Crasta and S.A.Khan ,"*Supersonic Flow Analysis and Evaluation of Damping Derivative*", IOSR Journal of Mechanical and Civil Engineering, volume 3 ,pp. 29- 36.

[22] Asha Crasta, S. A. Khan , " *Determination of Surface Pressure of an axisymmetric ogive in Hypersonic Flow*" in Mathematical sciences International Research Journal, Vol. 2, Issue2,August 2013,pp.333-335,ISSN:2278-8697.

[23] Asha Crasta and Khan S. A., 2014, '*Stability derivatives of a Delta Wing with Straight leading edge in the Newtonian Limit'*, International Journal of Engineering research and Applications, Vol. 4, Issue 3, Version 2, March, pp. 32-38.

[24] Asha Crasta and S. A. Khan, 2013, '*Stability Derivatives in the Newtonian Limit',* International Journal of Advanced Research in Engineering and Technology, Volume 4, Issue 7, Nov-Dec pp. 276-289.

[25] Asha Crasta, S. A. Khan , 2014, '*Effect of angle of incidence on Roll damping derivative of a delta Wing'*, International Journal of Emerging trends in Engineering and Developments Vol. 2, Issue4, March 2014, pp. 343-356.

[26] Asha Crasta and S. A. Khan, 2014 ,'*Effect of Angle of attack on Stability derivatives of a Delta wing with Straight leading edge in Supersonic Flow'*, International Journal of Mathematics,Vol.10,issue5,Sep-oct 2014. Pp. 01-08.

[27] Asha Crasta, S. A. Khan , 2014, '*Effect of sweep angle on Roll damping derivative of a delta Wing',* International Journal of Emerging trends in Engineering and Developments Vol15, Issue4, Aug-Sep, pp. 45-55.

[28] Asha Crasta and S. A. Khan, 2014 ,'*Effect of Angle of attack on Stability derivatives of a Delta wing in Hypersonic flow with Straight leading edge'*, International Journal of Mechanical and civil engineering, Vol.11,issue5,Sep-oct (ver.2). Pp. 21-29.

[29] Asha Crasta and S.A.Khan 2014,"*Effect of sweep angle on rolling moment derivative of an oscillating supersonic/hypersonic delta wing*", International journal of modern Engg. Research, Vol.4, issue 10,ver3, pp.15-22.

[30] Asha Crasta and S.A. Khan,2016,"Effect of Aspect ratio with roll moment derivative of a delta wing in supersonic flow", International Journal of Advances in Engineering Research, (IJAER) 2016, Vol. No. 12, Issue No. I, July,pp.10-15, e-ISSN: 2231-5152, p-ISSN: 2454-1796.

Design and Fabrication of Flying Saucer Utilizing Coanda Effect

Abdul Aabid, and S. A. Khan

Department of Mechanical Engineering, Kulliyyah of Engineering, International Islamic University Malaysia

*E-mail: aabidhussain.ae@gmail.com

Abstract. Coanda effect is used in several engineering applications with distinctive designs and structures. It is also applied in aircrafts flying at low speeds for a comfortable ride. In this paper, we have designed and modelled Coanda effect in terms of a flying saucer. The fabrication was done by means of structural and electronic components. Electrical motor was used as a propeller to take off and land vertically (VTOL) along with hovering capability. The rotor disc diameter is smaller than the bulbous body unlike a helicopter which makes to fly very stable. Control flaps were used to regulate the path by altering the flow over the streamlined body. The model was then tested with a remote control. Numerical Simulation of the tesla turbine was done using ANSYS 14.5 software and displacements were obtained by applying different forces on designed model. CATIA V5 was used to analyse the shaft of the model to get minimum value of torque at which the shaft starts to deform. **Keywords:** AMB, PMB, RPM, Regular Ball Bearing, VTOL, STOL, VSTOL, USAF, UAV.

1. Introduction

An un-conventional design was patented by Nikola Tesla in 1913 with the objective to develop a turbine with lower complexity, cost, and maintenance requirements than conventional mechanisms. Within the literature, the Tesla turbine has been commonly referred as a boundary layer, multiple-disk, friction, or shear force turbine [1]. The design consists of several, closely-spaced rigid disks set in parallel on a shaft. The co-rotating disks are centred and locked to the shaft. Located near the centre of the disks are orifices that allow for fluid exhaust in the axial direction. The disk-shaft assembly is set on bearings and enclosed within a cylindrical casing. An inlet into the casing is directed approximately tangentially [2].

The idea of the magnetic bearing and its applications has been conceptualized over the years. Today, magnetic bearing technology has become viable because of advances in microprocessing controllers that allow for confident and robust active control. It is advance in following areas: rotor and stator materials and designs which maximize flux, minimize energy losses, and minimize stress limitations; wire materials and coatings for high temperature operation; high-speed micro processing for advanced controller designs and extremely robust capabilities; back-up bearing technology for providing a viable touchdown surface; and precision sensor technology; have put magnetic bearings on the forefront of advanced, lubrication free support systems and The magnetic bearing technology is distant from young, didn't become practical application and the high-speed micro processing enables active magnetic bearing with stable control and dynamic stiffness and damping such application is the most of one gas turbine engine [3]. Turbomachines are machines that transfer energy between a rotor and a fluid, including both turbines and compressors. While a turbine transfers energy from a fluid to a rotor, a compressor transfers energy from a rotor to a fluid. A Tesla turbomachine utilizes the viscous shear forces of a fluid (boundary layer effect) passing near a disk on an axle to transmit torque to and from the fluid. It can be designed to efficiently pump highly viscous fluids as well as low viscous fluids [4].

Once aircraft flying on air, at high Mach number, the thin layer of air that sticks to the wing. This layer of air energies the same speed as the aircraft. Then, shear action or shear plane between that boundary layer and the close inactive air around the aircraft. In aerodynamics, boundary layer drag is completely unwanted precept. But, Tesla could turn that precept, it supposed that boundary layer drag used to do something useful in its applications [5]. Conventional turbines are mostly reaction and impulse type or both. Often technical challenge faced by conventional turbines. Financial feasibility of power plants is depended upon innovations to prevent erosion of mechanical equipment's or alternatives which better handle these conditions. Tesla turbine is an unconventional turbine that uses fluid properties such as boundary layer and adhesion of fluid on series of smooth disks keyed to a shaft [6]. A preliminary design of a Tesla turbine for pico hydro applications has been undertaken employing Rice's analytical method. The efficiency of the preliminary turbine design was near 80% but it is believed that the bulk of efficiency losses will be found in inlet and exhaust flows, which are not considered in this analysis. Challenges remain for the application of Tesla turbines to pico hydro generation [7].

Powered lift rotary wing, vertical takeoff and landing (VTOL), and short takeoff and landing (STOL) aircraft, and aircraft capable of both vertical and short field operations (VSTOL) are the different designed implemented in the literature. The concentration is on the USAF (X-designated), US Army (VZ-designated), and European experimental VSTOL aircraft, some research aircraft supposed to be prototypes for production aircraft into their mission [8].

Unmanned Aerial Vehicles (UAVs) play a prime role in the modern-day warfare where emphasis on surveillance, intelligence-gathering and dissemination of information. Some have a fixed wing design, which bear a resemblance to the one of outmoded achieving plane and require runway to take off and land. Others are using rotors just as helicopter, thus better result as far as hovering is concerned. This is important because the Vertical Take Off and Landing (VTOL) capability plays also a key part for UAV. It may come to the idea of searching for a modern design based on Coanda Effect (Coanda UAV). These have been evolved to generate lift and maneuverability force in a more efficient manner, which have rudder just as air plane, and have two types of flap. It can Take-Off and Land just like helicopter but move faster than helicopter [9], [10], figure 1 shows the designed and fabricated model.

Some more advanced flying saucer's capable of spaceflight have been proposed, often as black projects by aeronautics companies. The Lenticular Reentry Vehicle was a secret project run by Convair for a saucer device which could carry both astronauts and nuclear weapons into orbit; the nuclear-powered system was planned in depth, but is not believed to have ever flown. More exotically, British Rail worked on plans for the British Rail "Space Vehicle" a proposed, saucer-shaped craft based on so far undiscovered technologies such as nuclear fusion and superconductivity, which was supposed to have been able to transport multiple passenger between planets, but never went beyond the patent stage [11].

2. Design Process

From the literature, to fabricate the flying saucer it begins with the assortment of apparatus's and their quantities according to the existing model. Selection of components based on the range and endurance of the flying saucer and it is very important to observe the weight because of payload should be in the certain limit. Choice of Styrofoam is based on the its weight and strength, the selected Styrofoam is similar to a sheet form. Then, Styrofoam was brake into some fragments based on design dimensions to build an assembly of model by using adhesive bond into carbon fiber roads. Later, electronic components enhanced to operate the fabricated model. While assembling several attempts it was performed due to inherent in adhesive bond, the quantity of adhesive is very important to glue into the Styrofoam to reinforced to provide structural strength. The below figure 1 shows that fabricated model after all the components enhanced in to the flying saucer and developed with slightly change of design and dimensions to improve the performance of the existing model.

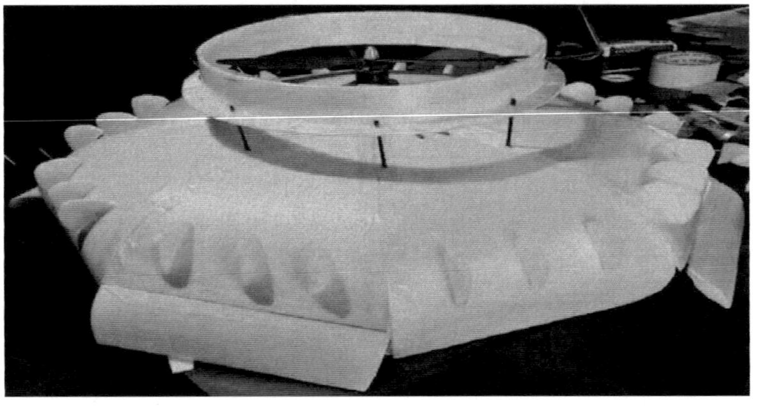

Figure 1. Fabricated model

The description is used for the design as listed below to fabricate the model. Following approach is performed to construct the design and accomplish the Coanda effect flying saucer as shown in figure 2.

Table 1. Description (From existing model)

Specification	
Size	50.5mmx50.5mmx12mm
Wight	21 grams (Inc Piezo buzzer)
IC	Atmega664 PA
Gyro	6050MPU InvenSense Inc.
Auto level	Yes
Input Voltage	4.8-6.0 V
AVR Interface	Standard 6 pin.
Signal from Receiver	1520us (5 channel)
Signal to ESC	1520us

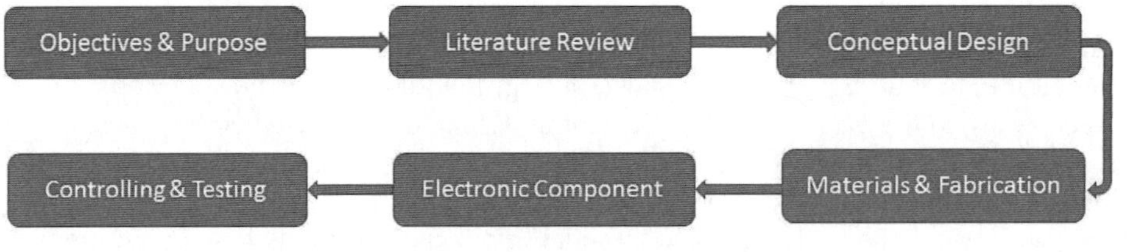

Figure 2. Methodology

3. Components
The consideration of components is hired by the two-major discipline and enhanced for making flying saucer Coanda effect. These components are selected from the literature of existing model and to improve its performance to some extent by changing components.

3.1 Structural Components
The selected materials for the fabrication of the body is Styrofoam, along with which Carbon fibre rods are used as reinforcement to provide structural strength.

- Styrofoam
- Carbon fiber rods
- Electronic Components

3.2 Electronic Components
The electronic components consist of the whole entity from the power supplying battery to the flap actuating serve. The listed below shows all the electronic components used for model.

- Transmitter
- The Receiver
- Motors
- Electronic Speed Controller (ESC)
- Battery
- Flight Control Board

4. Result and Analysis
4.1 Analysis
Analysis of Tesla Turbine: The turbine disc was analysed using ANSYS 14.5 software. The disc structural analysed by applying the force in the direction at which the viscous force act on the disc. The analysis showed the displacement of the disc at an increasing rate from the centre of disc to the edge. The displacement proves that the force acting on the disc will provide the necessary torque to drive the turbine.

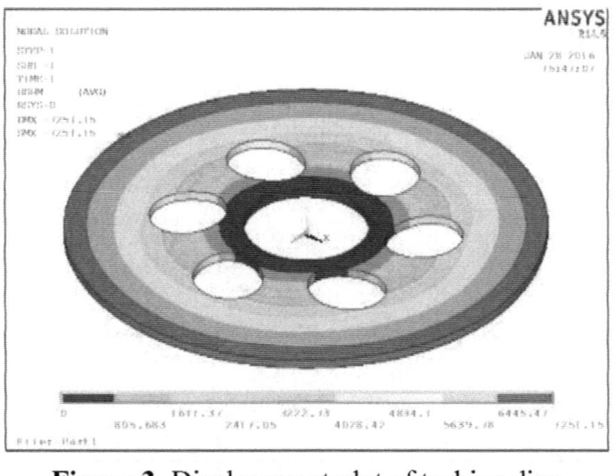

Figure 3. Displacement plot of turbine disc

Furthermore, the shaft was analysed using CATIA v5. The structural analysis of the shaft was done by fixing the pointed end of the shaft and applying the torque where the disc will be fitted. Then we checked for the min torque value at which the shaft started to deform. The value was found to be around 0.225 Nm when the shaft started to deform.

Figure 4. Von-Misses Stress plot of Turbine shaft

MATLAB analysis of AMB: The MATLAB simulation of the AMB system provided the following graph showing the stability and working of the proposed model for the following values of parameters.

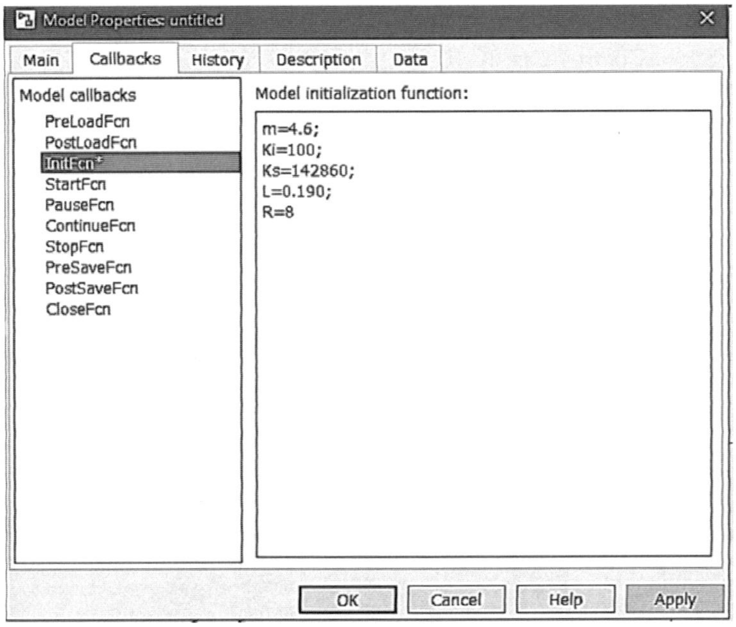

Figure 5. Parameters assigned to linearized AMB and feedback system

From the above input data, the following graphs were obtained. The below graph is gives us the displacement vs. time plot of magnetic bearing. At the start of the setup there will be no displacement as the surface of the rotor does not have an eddy current created on it. As the rotor rotates the eddy currents are formed on the surface and hence Lorentz force starts to act on the surface hence we find a gradual increase in x i.e., the position of the rotor with respect to the coil. After few seconds, the system is stabilized and a constant displacement t line is obtained.

When an external force is introduced to the system the disturbance caused by the force will displace the position of the rotor such vibrations will be dampened and the system will be stabilized by the forces acting on the rotor by the electromagnetic coils. The below graph shows the stabilization of the system for external excitation.

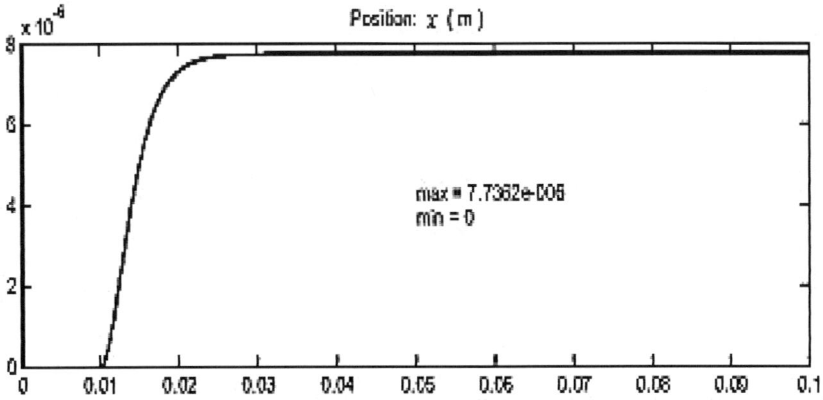

Figure 6. Displacement vs. time plot for AMB feedback system without external force

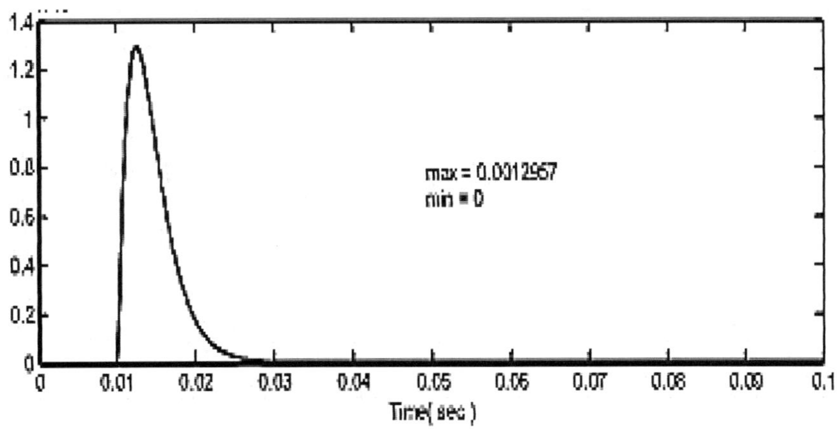

Figure 7. Displacement vs. time plot for AMB feedback system when an external force is applied.

4.2 Experimental Result

After the fabrication and assembly of the model, it was tested for performance parameters by using two different bearings i.e., regular ball bearing and PMB. The results obtained by those tests are listed below.

The first test was to check the max. RPM attained for different pressure of compressed air passed into the turbine. Air pressure was measured using a pressure gauge near the inlet and RPM was found using an RPM meter. As expected the PMB had higher RPM value compared to ball bearing as shown in to the below figure 8.

The setup was made to attain the max. RPM and then passing of compressed air through the inlet was stopped. The RPM was noted at regular intervals of time till the step stopped to run. This process was first conducted using PMB and then with ball bearing. The noted results are as follows.

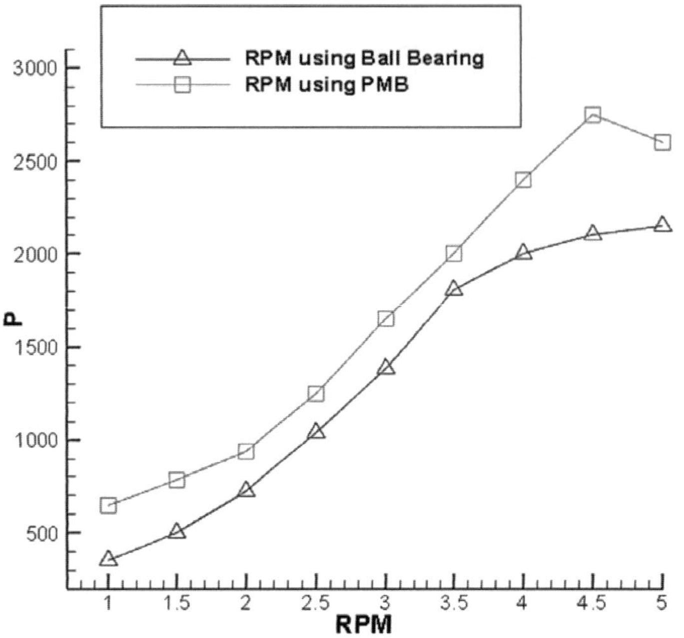

Figure 8. Graphical Representation of Pressure vs. RPM

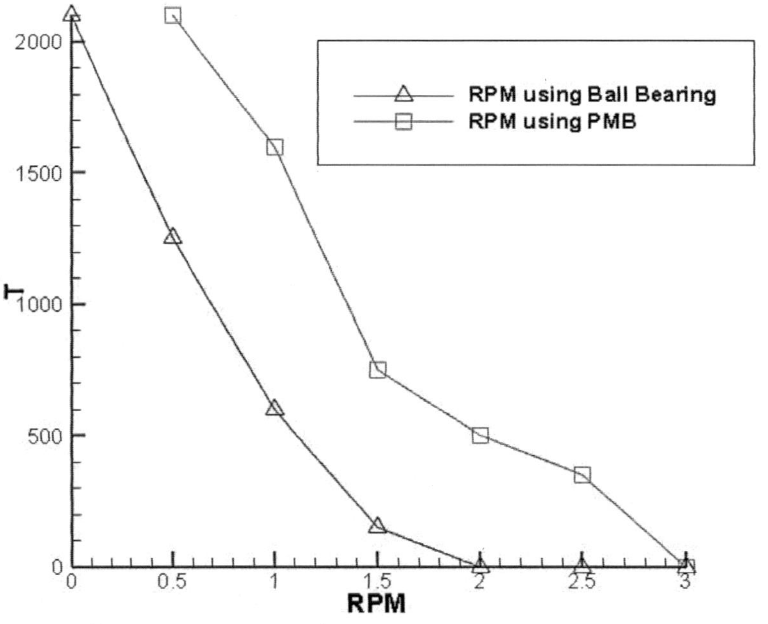

Figure 9. Graphical Representation of Time vs. RPM

5. Conclusion

Flying saucer using Coanda effect was fabricated by using structural and electronic components. The dimensions of the model were selected from the existing literature. Sorties were performed with the help of remote control.

- The results of the ball bearing and PMB comparisons were plotted. Result shows that PMB had higher RPM value compared to ball bearing.

- ANSYS simulation shows that displacement proves the force acting on the disc will provide the necessary torque to drive the turbine.
- CATIA V5 simulation shows that shaft starts to deform at 0.225 Nm.

References

[1] W. Rice, "An Analytica Investigation if Turbines rimental," *Jounal Eng. power*, pp. 29–36, 2013.
[2] N. Tesla, "Fluid Propulsion," 523832, 1913.
[3] D. J. Clark, M. J. Jansen, and G. T. Montague, "An Overview of Magnetic Bearing Technology for Gas Turbine Engines," *NASA Cent. Aerosp. Inf.*, no. 8, 2004.
[4] S. Gayathri, K. Subakaran, B. Somshekar, S. K. Rehman, R. V Angadi, and Y. Vijayakumar, "State and Art of the Active Magnetic Bearing Design and Anlysis by Using MATLAB," *Int. J. Adv. Res. Electr. Electron. Instrum. Eng.*, vol. 3, no. 11, pp. 13168–13175, 2014.
[5] H. P. Borate, N. D. Misal, and P. Lam, "An Effect of Surface Finish and Spacing between Discs on the Performance of Disc Turbine," *Int. J. Appl. Res. Mech. Eng.*, vol. 2, no. 1, pp. 30–36, 2012.
[6] R. J. Pandey, S. Pudasaini, S. Dhakal, and R. B. Uprety, "Design and Computational Analysis of 1 kW Tesla," *Int. J. Sci. Res. Publ.*, vol. 4, no. 11, pp. 1–5, 2014.
[7] B. P. Ho-yan, "Tesla Turbine for Pico Hydro Applications," *Guelph Eng. J.*, no. 4, pp. 1–8, 2011.
[8] R. Ransone, F. Chief, F. Test, A. Force, and F. Test, "An Overview of VSTOL Aircraft and Their Contributions," in *AIAA 2002 Biennial International Powered Lift Conference and Exhibit*, 2002, no. 11, pp. 1–21.
[9] C. Paper, C. Effect, and C. View, "Design and Construction of an Unmanned Aerial Vehicle Based on Coanda Effect," in *International Conference on Mechanical Engineering and Renewable Energy*, 2015, no. 11, pp. 1–5.
[10] B. J. Naudin, "How to build a RC ' Coanda Effect Saucer ,'" 2010.
[11] T. L. Wahl, "Design Guideance for Coanda-Effect Screens," 2003.

Estimation of Stability Derivatives in Newtonian Limit for Oscillating Cone

Aysha Shabana[1*], Renita Sharon Monis[2], Asha Crasta[3], S .A. Khan[4]

[1]Research Scholar, Department of Mathematics, M.I.T.E, Moodabidri and Assistant Professor, Sahyadri College of Engineering and Management, Mangaluru, Karnataka, India

[2]Research Scholar, Department of Mathematics, M.I.T.E, Moodabidri and Assistant Professor(Sr), SMVITM, Bantakal, Karnataka, India

[3]Associate Professor, Department of Mathematics, M.I.T.E, Moodabidri, Karnataka, India

[4]Professor, Department of Mechanical Engineering, Faculty of Engineering, IIUM, Gombak Campus, Kuala Lumpur, Malaysia

*E-mail: aysha.nawshad@gmail.com

Abstract. Stability derivatives in Newtonian limit for an oscillating cone are obtained. Stiffness derivative decreases with pivot position for the entire range of cone angles. For cone angles in the range 20 to 30 degrees there is substantial increase in the stiffness derivative. Damping derivative decreases with pivot position for various cone angles and attains a minima at h = 0.75 and then again with increase in pivot position there is non-linear increment in damping derivatives. There is considerable change in the magnitude, for higher cone angles in the range of 20 degrees and above as the centre of pressure has further moved towards the trailing edge of the cone at pivot position around h = 0.88. Stiffness derivative increases with cone angle for various pivot positions. Damping derivative with cone angle for various fixed pivot positions is seen to increase linearly with cone angle, it is also observed that this trend of linear increment tend to become non-linear for cone angles in the range 20 degrees and beyond. Damping derivative decreases with pivot position h = 0.8 for the entire range of cone angles, however, for pivot position h = 1.0, shows almost constant values up to cone angle of 20 degrees. For cone angle 30 degrees value of damping derivative coincides irrespective of pivot position being at h = 0.8 or 1.0 and this trend is attributed to the 3-D effect of the flow field.
Keywords: Hypersonic flow, Newtonian limit, Oscillating cone, Mach number, Stability derivative.

1. Introduction

The present work evaluates stability derivatives in Newtonian limit for an oscillating cone in hypersonic flow. At hypersonic speeds the "nose cones" are often non-slender and blunt. The reason for such a configuration, is the problem of aerodynamic heating and hence the heat transfer at such speeds. Although the present work is not for blunt bodies with detached shocks, once a theory is developed for the ogives with sharp nose, it can then be extended to more practical shapes to incorporate the bluntness. It is really interesting to note that the study of hypersonic flows, which was restricted to slender bodies and low angles of attack, should attain a stage of non-slender shapes and at large angle of attack flows; promising an emphasized development of efficient future hypersonic systems.

Ghosh [1] developed a new hypersonic similitude with the assumptions of attached bow-shock and Mach number behind the shock being greater than 2.5. This similitude was valid for the windward surface of an aero-foil with large flow deflection. His work was extended to oscillating non-planar wedges by Crasta and Khan to calculate the aerodynamic pitching moment derivatives in both Supersonic [2] and Hypersonic flows [7, 8, 9, 10, 11 and 12].

The large deflection similitude of Ghosh [1] has been extended by Ghosh, K., [2] to axisymmetric bodies with attached shock. Equivalence with a new piston-motion with axial symmetry has been established. The cone (or quasi-cone) results from the revolution of a wedge (or quasi-wedge) round the streamwise axis and a similar revolution of the independent fluid slab [1] produces an axially symmetric conico-annular space. It is shown by Ghosh, K., [2] that the flow past a cone or quasi-cone is equivalent to a piston-motion in this conico-annular space, which is called the similitudinal slab. Although Ghosh, K., [2] gives similitude for cones and quasi-cones, he gives a solution based on the similitude for a cone only. This solution gives a constant density shock layer. Hence the constant density form of the unsteady Bernoulli's equation is used to find pressure on the cone surface in terms of the piston Mach number. Results are obtained for hypersonic flow for a perfect gas over oscillating cones and ogives of different Mach numbers and semi- angles.

2. Analysis

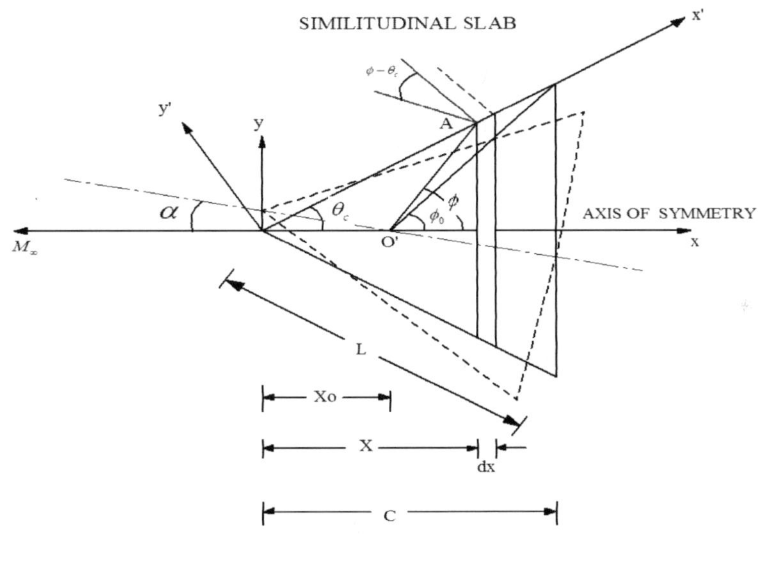

Figure 1. Cone Geometry

From the geometry we have

$$\tan\phi = \frac{x\tan\theta_c}{x-x_0};\ \tan\phi_0 = \frac{c\tan\theta}{c-x_0}$$

where ϕ is the angle subtended by A at O' with x-axis, and for different location of A, ϕ varies from π to ϕ_c, θ_c is the cone semi-angle and c is the chord length. The definition of the Stiffness coefficient denoted by C_{m_α}, is

$$C_{m_\alpha} = \left[\frac{\partial M}{\partial \alpha}\right]_{\alpha,q\to 0} \frac{1}{\frac{1}{2}P_\infty U_\infty^2 S_b c}$$

where S_b = base area of the cone = $\pi(c\tan\theta_c)^2$, c = chord length of the cone. On solving we obtain the following,

The Stiffness derivative for an oscillating cone is given by

$$C_{m_\alpha} = D[h^3(1-2n^2-(1-h)\{H(2+h)+n^2h(1+2h)\})]$$

where

$$D = \frac{2}{3(1+n^2)}\left[1+\frac{1}{4}\left(\varepsilon+\frac{1}{2}K\frac{d\varepsilon}{dM_{po}}\right)\right]$$

$$D = \frac{2}{3(1+n^2)}\left[1+\frac{1}{4}\left\{\left(\frac{2+(\gamma-1)K^2}{2+(\gamma+1)K^2}\right)+\frac{1}{2}K\left(\frac{-8K}{[2+(\gamma+1)K^2]^2}\right)\right\}\right]$$

$$D = \frac{2}{3(1+n^2)}\left[1+\frac{1}{4}\left\{\left(\frac{2+(\gamma-1)M_\infty^2\sin^2\theta_c}{2+(\gamma+1)M_\infty^2\sin^2\theta_c}\right)+\frac{1}{2}M_\infty\sin\theta_c\left(\frac{-8M_\infty\sin\theta_c}{[2+(\gamma+1)M_\infty^2\sin^2\theta_c]^2}\right)\right\}\right]$$

$$D = \frac{2}{3(1+n^2)}\left[1+\frac{1}{4}\left\{\left(\frac{2+(\gamma-1)M_\infty^2\sin^2\theta_c}{2+(\gamma+1)M_\infty^2\sin^2\theta_c}\right)+\frac{8}{2}\left(\frac{M_\infty^2\sin^2\theta_c}{[2+(\gamma+1)M_\infty^2\sin^2\theta_c]^2}\right)\right\}\right]$$

In the Newtonian limit M_∞ tends to infinity and γ tends to unity,

$$D = \frac{2}{3(1+n^2)}\lim_{\substack{M_\infty\to\infty\\ \gamma\to 1}}\left[1+\frac{1}{4}\left\{\left(\frac{2+(\gamma-1)M_\infty^2\sin^2\theta_c}{2+(\gamma+1)M_\infty^2\sin^2\theta_c}\right)+\frac{8}{2}\left(\frac{M_\infty^2\sin^2\theta_c}{[2+(\gamma+1)M_\infty^2\sin^2\theta_c]^2}\right)\right\}\right]$$

By applying limit

$$D = \frac{2}{3(1+n^2)}$$

∴ The Stiffness derivative in the Newtonian limit is given by,

$$C_{m_\alpha} = \frac{2}{3(1+n^2)}[h^3(1-2n^2)-(1-h)\{H(2+h)+n^2h(1+2h)\}]$$

∴ The Damping derivative for an oscillating cone is given by

$$C_{m_q} = \frac{D}{2}[h^4(2n^2-3n^4-1)-(1-h)\{H(3H+h(H+2n^2)+2h^2n^2)+n^4h^2(1+3h)\}]$$

where

$$D = \frac{2}{3(1+n^2)}\left[1+\frac{1}{4}\left(\varepsilon+\frac{1}{2}K\frac{d\varepsilon}{dM_{po}}\right)\right]$$

In the Newtonian limit M_∞ tends to infinity and γ tends to unity,

Following the same steps as in stiffness derivative we get

$$D = \frac{2}{3(1+n^2)}$$

∴ The Damping derivative in the Newtonian limit is given by,

$$C_{m_q} = \frac{D}{2}[h^4(2n^2 - 3n^4 - 1) - (1-h)\{H(3H + h(H + 2n^2) + 2h^2n^2) + n^4h^2(1 + 3h)\}]$$

Various results have been plotted and discussed.

3. Result and Discussion

Before analysing the results, we have to keep in mind that we are evaluating stability derivatives in the Newtonian limit. In the Newtonian limit the specific heat ratio γ of air, normally its value is 1.4 but in this case it will tend to one. Another variable is the Mach number which has a wide range but in view of the Newtonian limit it will tend to infinity. Which implies that the inertia of the flow will no more be a variable. This means that in the present case results will reflect the effect of the geometric parameters alone. Keeping these facts in mind we need to discuss the results.

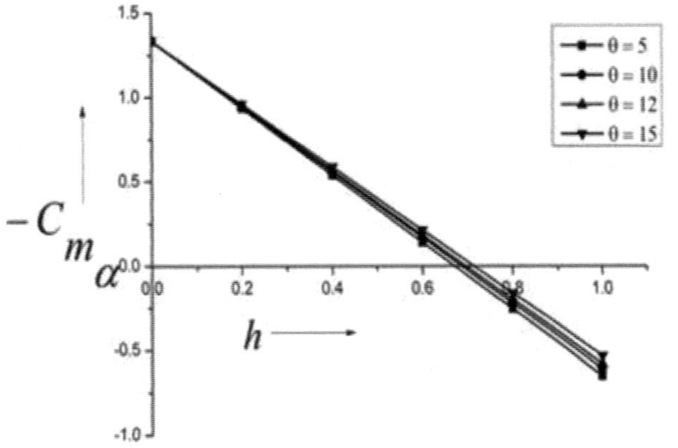

Figure 2. Stiffness derivative versus pivot position

Figure 2 shows the dependence of stiffness derivative with pivot position for various cone angles in the range from 5 to 15 degrees. Stiffness derivative decreases with pivot position for various cone angles and further it is noticed that up to 20% from the nose of the cone there is no variation between stiffness derivatives with various cone angles due to the negligible increase in the surface area of the cone and hence the surface pressure value. Slight variation appears when pivot position moves beyond 40% and substantial change at h = 1 which is 100% from the nose of the cone.

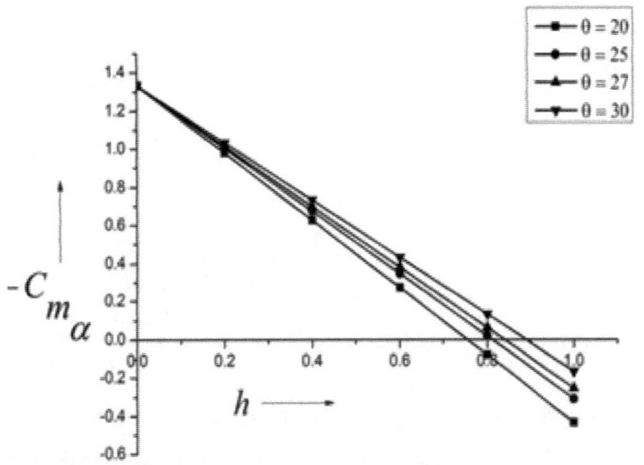

Figure 3. Stiffness derivative versus pivot position

Figure 3 shows the dependence of stiffness derivative with pivot position for various cone angles in the range from 20 to 30 degrees. Stiffness derivative decreases with pivot position for various cone angles as was seen in the previous case in figure 2, further it is seen that up till 10% from the nose of the cone there is no variation between stiffness derivative with cone angles for the entire range of the cone angle. As the cone angle increases the difference in stiffness derivative values for pivot position 20% onwards from the nose of the cone is clearly visible. This may be due to increase in the cone angle and hence the surface area and the resultant surface pressure for the range of the geometrical parameters of the present study. From the figure it is found that the centre of pressure is at h = 0.75, 0.82, 0.85, and 0.9. This shift in the centre of pressure due to the progressive increase in the cone angle will result in increased static margin as the centre of gravity remains fixed. However, we have to take final call on the maximum value of the static margin depending on the mission requirement and the type of application. For instance in case of fighter planes we are interested to have very low static margin as it will be helpful for Pilot in manoeuvring the fighter plane, whereas, in case of spinning shells there is a limiting value of the static margin for the operational requirement to enable the shell to avoid the base landing. For shells the fuzzes attached to the nose they function only due to the impact hence, base landing of the shells from the Guns must be avoided.

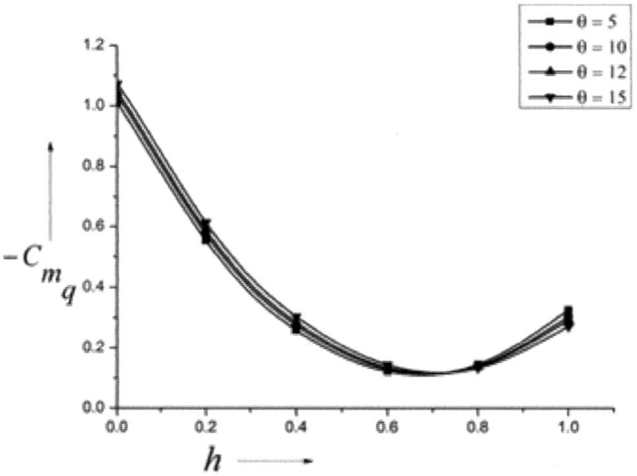

Figure 4. Damping derivative versus pivot position

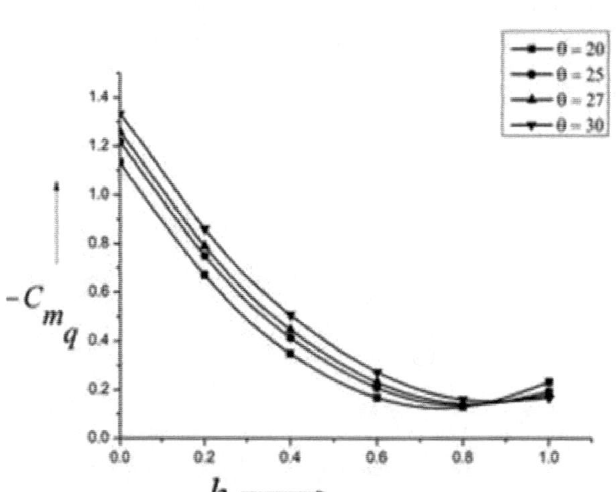

Figure 5. Damping derivative versus pivot position

Figure 4 and 5 show the dependence of damping derivative with pivot position for various cone angles. From the figure it is seen that damping derivative decreases with pivot position with cone angles attains a minima at h = 0.75 and then increases. There is no much variation in the values of damping derivative for cone angles from 5 to 15 degrees (Fig. 4). However, figure 5 shows considerable change in the magnitude, for higher cone angles the centre of pressure has further moved towards the trailing edge of the cone at around h = 0.88. As observed in case of stiffness derivative the trends are on the similar lines and the reasons for this trend has been discussed earlier and the same discussion holds here as well.

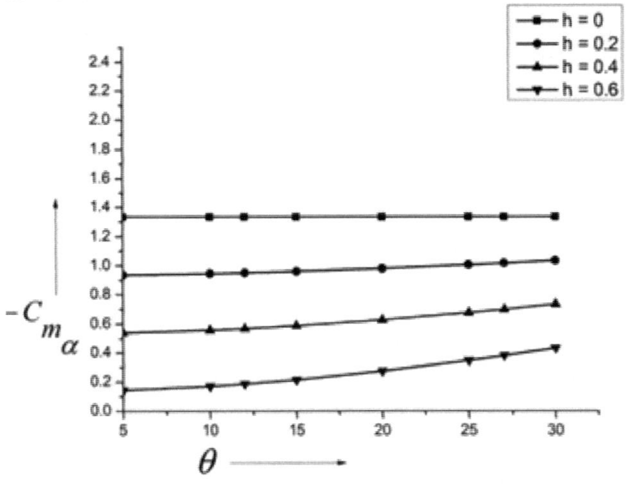

Figure 6. Stiffness derivative versus cone angle

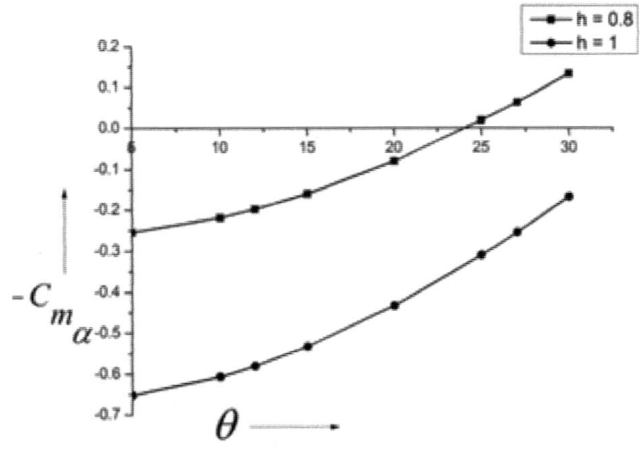

Figure 7. Stiffness derivative versus cone angle

Figs. 6 and 7 show the variation of stiffness derivative with cone angle for various pivot positions (i. e. h = 0, 0.2, 0.4, 0.6, 0.8, and 1.0). As expected the stiffness derivative decreases with pivot position for a given cone angle as discussed earlier (Figs. 2 and 3). Further, it is seen that with the increase in the cone angles the stiffness derivative increases linearly for entire range of cone angles and all the pivot positions except for h = 0.6, 0.8, and 1.0. It is also seen that the non-linearity has crept in for h = 0.6, 0.8 and 1.0. This non-linear trend is also seen for cone angle 20 degrees and above for lower pivot positions namely 0.2 and above. This trend could be attributed to the pressure distribution on the surface of the cone.

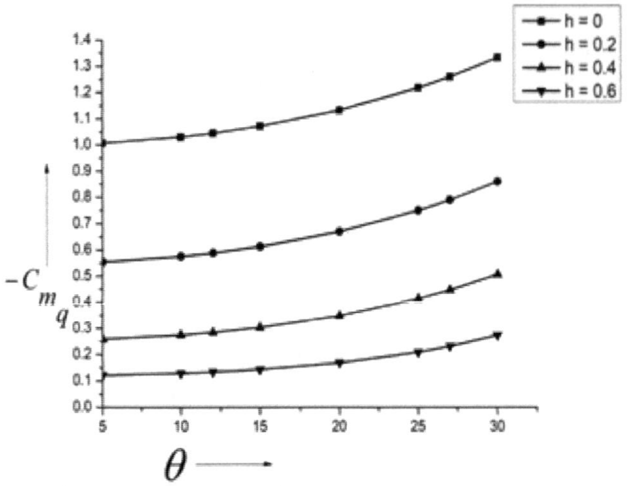

Figure 8. Damping derivative versus cone angle

Figure 8 shows the dependence of damping derivative with cone angle for various fixed pivot positions. Damping derivative increases linearly with cone angle, it is also observed that this trend of linear increment tend to become non-linear for cone angles in the range 20 degrees and beyond.

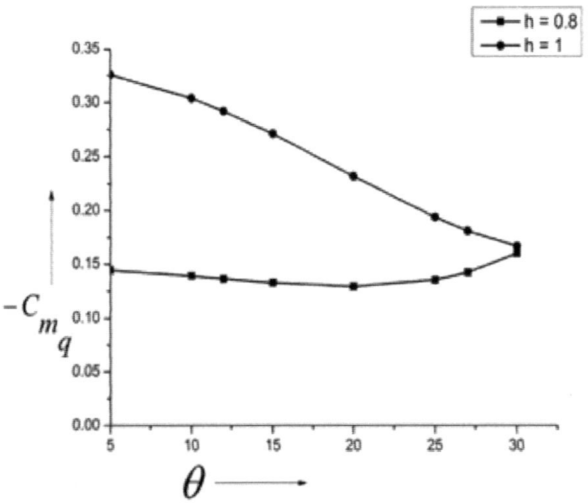

Figure 9. Damping derivative versus cone angle

Fig. 9 shows the variation of damping derivative with cone angle for pivot position h = 0.8 and 1.0. Damping derivative decreases with pivot position h = 0.8 for the entire range of cone angles. When we consider the results for pivot position h = 1.0, shows almost constant values up to cone angle of 20 degrees. For cone angle more than 20 degrees the damping derivative starts increasing. It is observed that at cone angle of 30 degrees value of damping derivative coincides irrespective of pivot position being at h = 0.8 or 1.0. This may be due to the 3-D effect of the flow field, cone surface pressure and the effect of body of revolution.

4. Conclusion

Based on the above discussion we can draw the following conclusions:

- Stiffness derivative decreases with pivot position for the entire range of cone angles. For lower range of pivot positions and cone angle up to 15 degrees this increment is not seen.

- For cone angles in the range 20 to 30 degrees there is substantial increase in the stiffness derivative and the trend of decrement continues.
- Damping derivative decreases with pivot position for various cone angles and attains a minima at h = 0.75 and then again with increase in pivot position there is non-linear increment in damping derivatives. There is considerable change in the magnitude, for higher cone angles in the range of 20 degrees and above as the centre of pressure has further moved towards the trailing edge of the cone at pivot position around h = 0.88.
- Stiffness derivative increases with cone angle for various pivot positions. Further, it is seen that with the increase in the cone angles the stiffness derivative increases linearly for entire range of cone angles. Non-linearity is observed for h = 0.6, 0.8 and 1.0.
- Damping derivative with cone angle for various fixed pivot positions is seen to increase linearly with cone angle, it is also observed that this trend of linear increment tend to become non-linear for cone angles in the range 20 degrees and beyond.
- Damping derivative decreases with pivot position h = 0.8 for the entire range of cone angles, however, for pivot position h = 1.0, shows almost constant values up to cone angle of 20 degrees. For cone angle 30 degrees value of damping derivative coincides irrespective of pivot position being at h = 0.8 or 1.0 and this trend is attributed to the 3-D effect of the flow field.

References

[1] Ghosh, K. (1977), "A New Similitude for Aerofoils in Hypersonic Flow*", Proceedings of the 6th Canadian Congress of Applied Mechanics, Vancouver, Canada,* May 29-June 3, pp. 685-686.

[2] Ghosh K., (1984), *Hypersonic large deflection similitude for oscillating delta wings*, the Aeronautical journal, pp. 357-361.

[3] Asha Crasta and Khan S. A., (2012), *High Incidence Supersonic similitude for Planar wedge*, International Journal of Engineering research and Applications, 2(5), pp. 468-471.

[4] Asha Crasta and S. A. Khan, (2012), *Estimation of Stability derivatives of an Oscillating Hypersonic delta wings with curved leading edges*, IJMET, 3(3), pp. 483-492.

[5] Asha Crasta, M. A. Baig, S. A. Khan, (2012), *Estimation of Stability derivatives of a Delta wing in Hypersonic flow*, International Journal of Emerging trends in Engineering and Developments, 6(2), pp. 505-516.

[6] Asha Crasta and S. A. Khan, (2012*), Oscillating Supersonic delta wing with Straight Leading Edges*, International Journal of Computational Engineering Research, 2(5), pp. 1226-1233.

[7] Asha Crasta and S. A. Khan, (2014), *Effect of Angle of Incidence on Stability derivatives of a wing*, International Journal for Research in Applied Science and Engineering Technology, 2(V), pp. 411-422.

[8] Asha Crasta and S. A. Khan, (2014), *Hypersonic Similitude for Planar Wedges*, International Journal of Advanced Research in Engineering and Technology, 5(2), pp. 16-31.

[9] Asha Crasta and S. A. Khan, (2014), *Estimation of Stability Derivatives for a Planar Wedge in the Newtonian Limit*, IOSR Journal of Mathematics, 10(2), pp. 01-06.

[10] Asha Crasta and S. A. Khan, (2014), *Effect of Mach number on Stiffness and Damping Derivatives for Oscillating Hypersonic Non-Planar Wedge*,IOSR Journal of Mechanical and Civil Engineering, 11(2), pp. 04-11.

[11] Asha Crasta and S.A. Khan, (2014), *Effect of Angle of Incidence on Stiffness and Damping derivatives for Oscillating Hypersonic Non-planar Wedge*, International Journal for Research in Applied Science and Engineering Technology, 2(IV),pp. 229-242.

[12] Asha Crasta and S. A. Khan,(2014*), Supersonic Similitude for Oscillating Non-Planar wedge*,cIOSR Journal of Mathematics, 10(2), pp. 15-24.

[13] Khan S. A. and Asha Crasta, (2010), *Oscillating Supersonic delta wings with curved leading edges*, Advanced Studies in Contemporary Mathematics, 20(3), pp. 359-372.

[14] Asha Crasta and S. A. Khan (2015), "*Effect of Angle of attack on Stiffness derivative of an oscillating supersonic delta wing with curved leading edges*" IOSR-JMCE issue1, Volume12, December,pp12-25.

[15] Asha Crasta and S. A. Khan, "*Effect of Angle of attack on Damping derivative of a delta wing with full sine curved leading edges*" IJETED Journal issue5, Volume1, December- January 2015.

[16] Asha Crasta and S. A. Khan,(2015) "*Estimation of Damping derivative of a delta wing with half sine curved leading edges*" IOSR Journal of Mechanical and civil engineering, Vol.12, issue1,February ,pp40-44.

[17] Asha Crasta and S. A. Khan (2015), "*Estimation of Damping derivative in pitch of a Supersonic delta wing with curved leading edges*" IOSR Journal of Journal of Mathematics , Vol. 11,issue 1, Jan-feb ,pp.07-15.

[18] Asha Crasta and S.A.Khan (2015),"*Effect of Aspect ratio with angle of attack of an oscillating Hypersonic delta wing with Straight Leading edges*", Mathematical Sciences International research Journal",volume 4,issue2, pp.28-33., ISSN2278-8697.

[20] Aysha Shabana, Renita Monis, Asha Crasta and S.A.Khan , "*Pressure and its derivative with respect to piston Mach number for an oscillating cone* " IOSR Journal of Mechanical and Civil Engineering , volume 2 ,pp. 74-78.

[21] Renita Monis, Aysha Shabana, Asha Crasta and S.A.Khan ,"*Supersonic Flow Analysis and Evaluation of Damping Derivative*", IOSR Journal of Mechanical and Civil Engineering , volume 3 ,pp. 29- 36.

[22] Asha Crasta, S. A. Khan , " *Determination of Surface Pressure of an axisymmetric ogive in Hypersonic Flow*" in Mathematical sciences International Research Journal, Vol. 2, Issue2,August 2013,pp.333-335,ISSN:2278-8697.

[23] Asha Crasta and Khan S. A., 2014, '*Stability derivatives of a Delta Wing with Straight leading edge in the Newtonian Limit'*, International Journal of Engineering research and Applications, Vol. 4, Issue 3, Version 2, March, pp. 32-38.

[24] Asha Crasta and S. A. Khan, 2013, '*Stability Derivatives in the Newtonian Limit',* International Journal of Advanced Research in Engineering and Technology, Volume 4, Issue 7, Nov-Dec., pp. 276-289.

[25] Asha Crasta, S. A. Khan , 2014, '*Effect of angle of incidence on Roll damping derivative of a delta Wing'*, International Journal of Emerging trends in Engineering and Developments Vol. 2, Issue4, March 2014, pp. 343-356.

[26] Asha Crasta and S. A. Khan, 2014 ,'*Effect of Angle of attack on Stability derivatives of a Delta wing with Straight leading edge in Supersonic Flow'*, International Journal of Mathematics,Vol.10,issue5,Sep-oct 2014. Pp. 01-08.

[27] Asha Crasta, S. A. Khan , 2014, '*Effect of sweep angle on Roll damping derivative of a delta Wing',* International Journal of Emerging trends in Engineering and Developments Vol15, Issue4, Aug-Sep, pp. 45-55.

[28] Asha Crasta and S. A. Khan, 2014 ,'*Effect of Angle of attack on Stability derivatives of a Delta wing in Hypersonic flow with Straight leading edge'*, International Journal of Mechanical and civil engineering, Vol.11,issue5,Sep-oct (ver.2). Pp. 21-29.

[29] Asha Crasta and S.A.Khan 2014,"*Effect of sweep angle on rolling moment derivative of an oscillating supersonic/hypersonic delta wing*", International journal of modern Engg. Research, Vol.4, issue 10,ver3, pp.15-22.

[30] Asha Crasta and S.A. Khan,2016,"Effect of Aspect ratio with roll moment derivative of a delta wing in supersonic flow", *International Journal of Advances in Engineering Research, (IJAER)* 2016, Vol. No. 12, Issue No. I, July,pp.10-15, e-ISSN: 2231-5152, p-ISSN: 2454-1796.

IOP Conf. Series: Materials Science and Engineering **370** (2018) 012062 doi:10.1088/1757-899X/370/1/012062

Effect of the gap height of radial gate on the volumetric flow rate in dam

Fei Chong Ng[1], Aizat Abas[2*], Ismail Abustan[3], Z. Mohd Remy Rozainy[3], MZ Abdullah[4], Ali b. Jamaludin[5], Sharon Melissa Kon[6]

[1,2]School of Mechanical Engineering, Universiti Sains Malaysia, Engineering Campus, Nibong Tebal, 14300, Penang.
[3]School of Civil Engineering, Universiti Sains Malaysia, Engineering Campus, Nibong Tebal, 14300, Penang.
[4]School of Civil Engineering, Universiti Sains Malaysia, Engineering Campus, Nibong Tebal, 14300, Penang.
[5]TNB Research Sdn. Bhd., No. 1, Lorong Ayer Itam, Kawasan Institusi Penyelidikan, 43000 Kajang.
[6]School of Mechanical Engineering, Universiti Sains Malaysia, Engineering Campus, Nibong Tebal, 14300, Penang.

*Corresponding author: aizatabas@usm.my

Abstract. This paper investigates how the variation of gap height of radial gate in scaled down dam would affect the volumetric flow rate. Both fluid/structure interaction (FSI) numerical simulation and experimental works on a scaled-down dam model were conducted, eventually both approaches yield results with discrepancy not exceeding 9.63%. For 10 mm and 12 mm gap, lowest water flow rate is being observed in the Gate 2 compared to Gate 1 and Gate 3. Conversely, the Gate 2 registered highest water flow rate for gate with gap of 8 mm. Generally, it is found that the volumetric flow rate increases along with the increase with the gap height of radial gate. Moreover, the volumetric water flow rate in actual dam also be determined using Reynolds number.

1. Introduction

Dams are important projects for developing countries, because the population receives domestic and economic benefits from the investments [1]. However, the potential energy of the water reservoir can cause considerable damage if the dam fails [2]. During the events of dam failure, huge volume of impounding water together with immense potential energy stored in it will pound the immediate downstream areas, endangering lives, damaging properties along its path and eventually result in widespread flooding [3]. Therefore, careful planning and consistent monitoring are required for the design and construction of dams [4].

In the past, there have been various researchers that simulated the flow of water in dam, for instances MIKE 11 1-D hydrodynamic model [2, 5], smoothed particle hydrodynamics (SPH) model [6 – 9] and fluid/structure interaction (FSI) [10, 11]. Among these approaches, the FSI simulation has proven to be more reliable as it able to capture both aspects of fluid flow and structural deformation in the modelling.

To the best knowledge of the authors, research was rarely being conducted to determine the water flow front across the radial gates of dam. Essentially, the current paper would reveal the unique numerical FSI and experimental approaches that allow researchers to study the water flow rate in dam without huge expenses.

2. Simulation

The scaled-down dam model and its corresponding fluid domain after mesh are depicted in Figure 1. The dimensions of the fluid domain are 3021.4 mm (length) × 2000 mm (width) × 600 mm (height). Tetrahedral mesh with their mesh elements being optimized were adopted in present simulation. Such optimized mesh enables both low computational cost at reasonable good numerical accuracy to be achieved.

The general boundary conditions (BC) used in current simulation are as summarized in Figure 2. For the mixture phase of the inlet, velocity inlet boundary conditions are used to define the velocity and scalar properties of the flow. The velocity of the water flow into the inlet, v_{inlet} is set at 0.01 m/s; while no-slip wall condition being imposed such that the wall-fluid velocity is 0 m/s. Moreover, the pressure at the inlet and outlet equal the atmospheric pressure.

System Coupling module were used to couple both Transient Structural and Fluid Flow (FLUENT). System to achieve two-way data transfers between the structural region and fluid phase. Data transfer is defined between the fluid solid interface (FSI) of the Transient Structural and contact surface in Fluid Flow (FLUENT). The time step size used in current simulation is set at 0.1 s and the end time is 5 s. These setting will be made consistent to the similar input provided in Transient Structural. Finally, the coupled analysis is executed to solve the coupling of both structural and fluid domains.

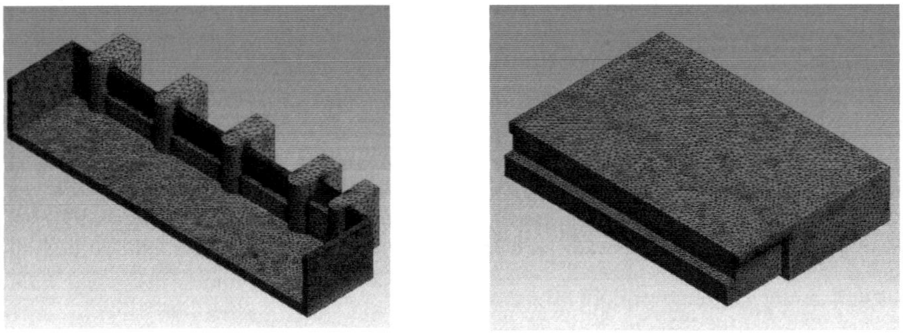

Figure 1. Meshed structural domain (left) and fluid domain (right).

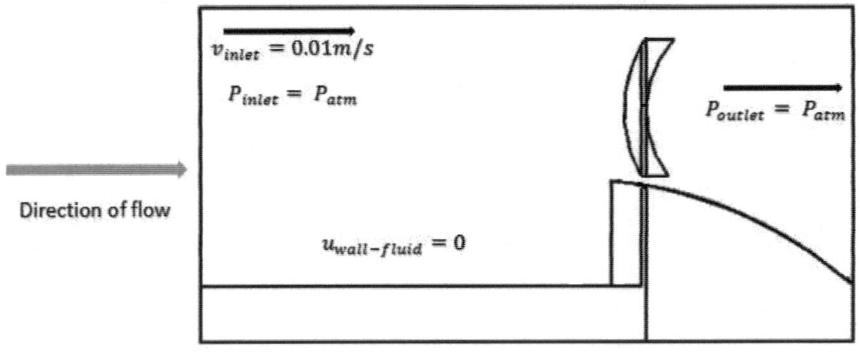

Figure 2. Summarized boundary conditions (BC) used in the numerical simulation.

3. Experiment

The experiment is conducted on the scaled-down physical dam model as shown in Figure 3 that located in the Civil School Research Lab. When water flows through the gap underneath the radial gates, force is exerted on the gate surface. The purpose of the experiment is to find out the water flow rate across the radial gate of the dam, to be subsequently compared with simulated data.

Firstly, the apparatus is set up at radial gate 1 for the following configuration as in Figure 4: (i) upstream head, h_1 = 80 mm; (ii) gap height or discharge height, h_2 = 8 mm. Then, the sensor is placed in the center of the gap beneath the radial gate and connected to the EASZ-10P Doppler Flowmeter (see Figure 5). It is used to measure the flow rate of water in a non-intrusive principle with a high speed 16-bit microprocessor unit. Once the flow rate at the radial gate 1 is taken, the measurement is continued at radial gate 2 and then 3. The experiment is repeated for gap height or discharge height, h_2 = 10 mm and 12 mm.

Figure 3. Physical scaled-down dam model.

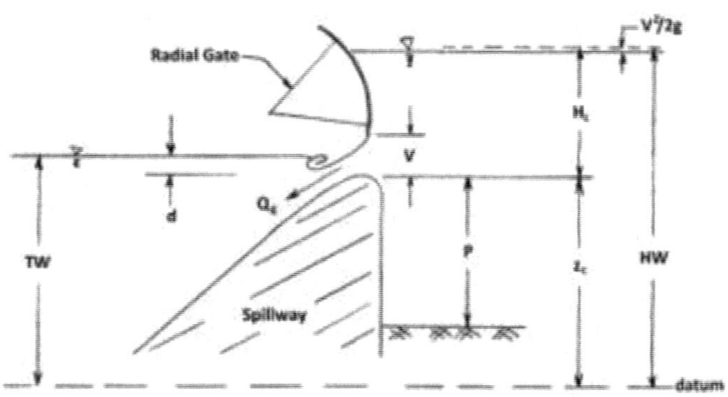

Figure 4. Position of gap or discharge height, h_2 and upstream head, h_1 for the physical model.

Figure 5. EASZ-10P Doppler Flowmeter

4. Results and Discussions

Figure 6 shows the volume fraction of water flow near the radial gate of various gap heights. Qualitatively there are more water flow front the gates of dam when the gap heights being increases. From simulation data, the water velocity at the gap of the radial gate were averaged out at three different locations to obtain its mean value near each three gates. Subsequently the flow rate of water exiting the dam is calculated. Later the simulated water flow rates for each gate at different dam cases were compared to the corresponding experimental data on water flow rates, and presented in Figure 7. The simulations also revealed that the increases in the gap height of radial gate, the volumetric flow rate of water on all three gates also increases. Based on Figure 7, it can be seen that the volumetric flow rates of water obtained from both simulation and experiment are similar. The smallest percentage difference is 1.26 % for water flowing through Gate 2 with a gap of 10 mm whereas the biggest percentage difference is 9.63 % for water flowing through Gate 3 with a gap of 8 mm. The differences are acceptable as they are less than 10 %. Several factors such as geometry, physics set up and boundary conditions are taken into consideration to ensure that the simulation model is similar to the experimental model.

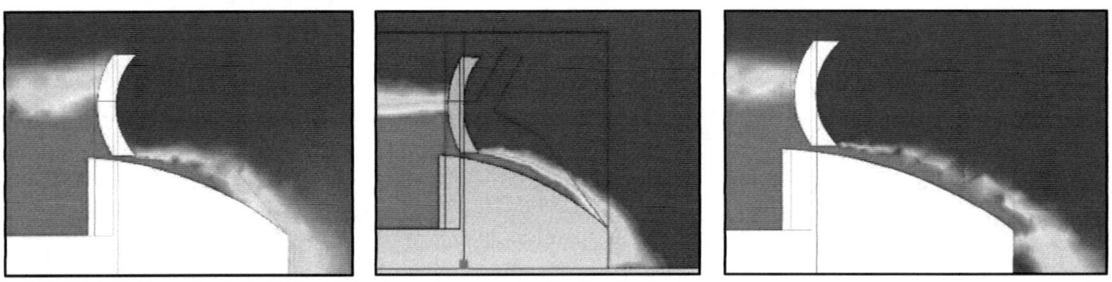

Figure 6. Water volume fraction for the flow near the radial gate of gap heights 8 mm, 10 mm and 12 mm (from left to right).

Through the dimensionless number analysis, the volumetric flow rate at actual-sized dam can be determined. Table 1 gives the dimensions of both actual dam and the scaled-down dam models used in both experimental and simulation works.

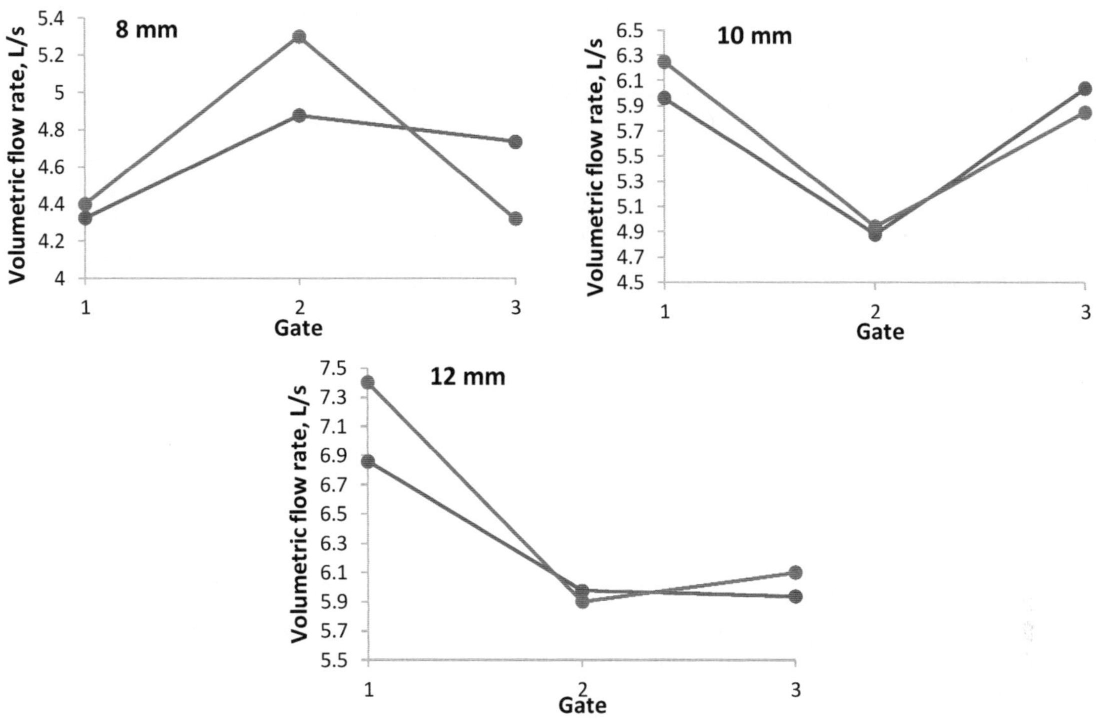

Figure 7. Plots of experimental (red) versus simulation (blue) water flow rates, across the radial gate of different gap heights (8 mm, 10 mm and 12 mm).

Table 1. Dimensions of actual and scaled-down dam models.

Dimension	Actual dam	Scaled-down 1/25 dam
Length	75535 mm	3021.4 mm
Width	24136.5 mm	965.46 mm
Height	12000 mm	480 mm
Length of one radial gate	12193 mm	487.72 mm

At ambient temperature of 28°C, the water kinematic viscosity is 0.8355 m^2/s. The dam model is assumed to be a rectangular duct with an open channel flow, thus the characteristic length, L is known as hydraulic diameter, d_h. The equation of hydraulic diameter is:

$$L = d_h = \frac{4A}{u} = \frac{4(\text{width})(\text{height})}{2(\text{width} + \text{height})}, \quad (1)$$

where A is the cross sectional area of the duct and u is the wetted length. The value of d_h for the physical model is 0.6412 m while the real dam is 16.0302 m. Using Reynolds number similarity for both scaled-up model and actual dam, the volumetric flow rate at real dam can be computed, as given in Table 2. Eventually the actual volumetric flow rate at real dam is 25 times larger than in the 1/25 scaled-down model. The Reynolds numbers of the water flow in both dams are the same. Again, it is seen that the dam with radial gate of height 0.3 m possess the highest flow rate of 156.4 L/s.

Table 2. The actual water flow rate in dam, calculated using Reynolds number.

Gap Height (m)	Average water velocity at physical model (m/s)	Average water velocity at real dam (m/s)	Volumetric flow rate at real dam (L/s)	Reynolds number
0.008	1.1906	0.0476	116.1	913727
0.010	1.1535	0.0461	140.5	885263
0.012	1.0691	0.0428	156.4	820490

5. Conclusions

This paper had presents both scaled-up experimental and numerical simulation approaches to analyze the volumetric flow rate of water across the radial gates of dam. Both methodologies eventually gave findings of relative close, with deviation not more than 10 %, affirming the results. It is reported that the water flow rate is the highest at larger gap height of the radial gate. Additionally, through Reynolds number invariant, the volumetric flow rates at real dam were determined, and the highest water flow of 156.4 L/s being recorded in 0.3 m height dam.

Acknowledgements:

The authors gratefully acknowledged the financial support provided by Institute of Postgraduate Studies (IPS), Universiti Sains Malaysia (USM), through the USM Fellowship 2017 scheme. Additional, this work was also partly supported by the Short Term Grant 60313020 from the Division of Research and Innovation, Universiti Sains Malaysia, and FRGS Grant 6071322 from the Ministry of Higher Education.

References:

[1] International Commission of Large Dams, ICOLD (2011); "Role of Dams". Available from: <www.Icold-cigb.net>. [*Extracted at* 11 March 2017]

[2] Solave and Delatte (2003), "Lessons from the Failure of the Teton Dam", Proceedings of the 3rd ASCE Forensics Congress, San Diego, California.

[3] L. M. Sidek, F. C. Ros and N. H. A. Aziz (2011), "Numerical Modelling of Dam Failure for Hydropower Development in Cameron Highlands – Batang Padang Scheme, Pahang Malaysia", Student Conference on Research and Development, Universiti Tenaga Nasional, Selangor, Malaysia.

[4] Energy Education on Dam failures. *Available from*: http://energyeducation.ca/encyclopedia/Dam_failures. [*Extracted at* 3 June 2017]

[5] Razad, A. A., & Ros, F. C. (2009). One Dimensional Dam Breach Modelling for Proposed Hydropower Development in Ulu, (October).

[6] Gomez-Gesteira, M. (2009). State-of-the-art of classical SPH for free-surface flows. Journal of Hydraulic Research, 48(extra), 0. Available from: <https://doi.org/10.3826/jhr.2010.0012>. [*Extracted at* 11 March 2017]

[7] Benedict, D., Crespo, A. C., Dominguez, J. M., Barreiro, A., & Go, M. (2011). GPUs, a New Tool of Acceleration in CFD: Efficiency and Reliability on Smoothed Particle Hydrodynamics Methods, Universidade de Vigo, Ourense, Spain and University of Manchester, Manchester, United Kingdom. Available from: <https://doi.org/10.1371/journal.pone.0020685>. [*Extracted at* 11 March 2017]

[8] Gomez-Gesteira, M. (2009). State-of-the-art of classical SPH for free-surface flows. Journal of Hydraulic Research, 48(extra), 0. Available from:<https://doi.org/10.3826/jhr.2010.0012>. [*Extracted at* 11 March 2017]

[9] Benedict, D., Crespo, A. C., Dominguez, J. M., Barreiro, A., & Go, M. (2011). GPUs, a New Tool of Acceleration in CFD: Efficiency and Reliability on Smoothed Particle

Hydrodynamics Methods, Universidade de Vigo, Ourense, Spain and University of Manchester, Manchester, United Kingdom. Available from: <https://doi.org/10.1371/journal.pone.0020685>. [*Extracted at* 11 March 2017]
[10] Gómez-Gesteira, M and Dalrymple, RA 2004 *J Waterw Port Coast Ocean Eng* **130(2)** 63
[11] Mfundo Vezi (2014). Dynamic Modelling of Arch Dams in the Ambient State, University of Cape Town, Cape Town, South Africa.

IOP Conf. Series: Materials Science and Engineering **370** (2018) 012063 doi:10.1088/1757-899X/370/1/012063

Fluid/structure interaction study on the variation of radial gate's gap height in dam

Fei Chong Ng[1], Aizat Abas[2*], Ismail Abustan[3], Z. Mohd Remy Rozainy[3], MZ Abdullah[4], Ali b. Jamaludin[5], Sharon Melissa Kon[6]

[1,2]School of Mechanical Engineering, Universiti Sains Malaysia, Engineering Campus, Nibong Tebal, 14300, Penang.
[3]School of Civil Engineering, Universiti Sains Malaysia, Engineering Campus, Nibong Tebal, 14300, Penang.
[4]School of Civil Engineering, Universiti Sains Malaysia, Engineering Campus, Nibong Tebal, 14300, Penang.
[5]TNB Research Sdn. Bhd., No. 1, Lorong Ayer Itam, Kawasan Institusi Penyelidikan, 43000 Kajang.
[6]School of Mechanical Engineering, Universiti Sains Malaysia, Engineering Campus, Nibong Tebal, 14300, Penang.

*Corresponding author: aizatabas@usm.my

Abstract. Through fluid/structure interaction (FSI) simulation and experimental work on a scaled-down dam model, the influence of radial gate height to the force exerted to the gate wall during water flow is studied. Both numerical and experimental findings gave results of great consensus, with their deviation not exceeding 14.66%. The water flow across the radial gate of the dam is visualized as modelled based on volume of fluid (VOF) scheme. It is found that the force exerted by water flow on the gate wall is the highest with intermediate gap height of 10 mm as compared to 8 mm and 10 mm.

1. Introduction

Precise planning and rigid monitoring are required for the design and construction of dams. This is because without proper handling, they can turn hazardous as they store huge volumes of water with high amount of potential energy and subsequently leads to catastrophic failures [1]. Issues that are often faced by dams here are old age, earthquakes, climate change, urban sprawl and landslides [2].

There are various works that particular research the flow and reliability in dam. Among the choice of numerical software used are: MIKE 11 1-D Hydrodynamic Model [3, 4], Smoothed Particle Hydrodynamics (SPH) model [5 – 8] and fluid/structure interaction (FSI) [9, 10]. Several researches have been done on the overall dam structure in terms of pressure and geometry. These studies are conducted using different types of models which have displayed certain inaccuracies and inflexibilities. The FSI model has proven to be one of the most reliable and commonly utilized model based on the studies done on a variety of scenarios and objects. Very little is focused on determining the flow of water as well.

To the best knowledge of the author, the research was rarely carried out to study the amount of force acting on a physical model of a dam, to ensure the structural integrity is not affected which will then be reflected on the real dam model. Therefore, the current study is a worthy research investment that can offer a new perspective to FSI simulation of concrete gravity dams.

2. Simulation

The dam model and its corresponding fluid domain after mesh are depicted in Figure 1. The length of the fluid domain is 3021.4 mm, width is 2000 mm and height is 600 mm. Meshing is done based on proximity and curvature size function with the relevance center and span angle center adjusted to the fine option. The mesh on both fluid and structural domains were optimized in terms of computational cost and numerical accuracy.

The zones of the boundary conditions in the FLUENT set-up are FSI, inlet, interior-fluid, outlet and wall-fluid. For the mixture phase of the inlet, velocity inlet boundary conditions are used to define the velocity and scalar properties of the flow. The magnitude of the velocity at inlet, v_{inlet} is 0.01 m/s so that it is nearly negligible and does not have a huge effect on the fluid flow while the wall-fluid velocity is 0. For the inlet's water phase, the volume fraction under multiphase section is set at 1. Moreover, the pressure at the inlet and outlet equal the atmospheric pressure of 0 Pa (gauge). The general boundary conditions used are summarized in Figure 2.

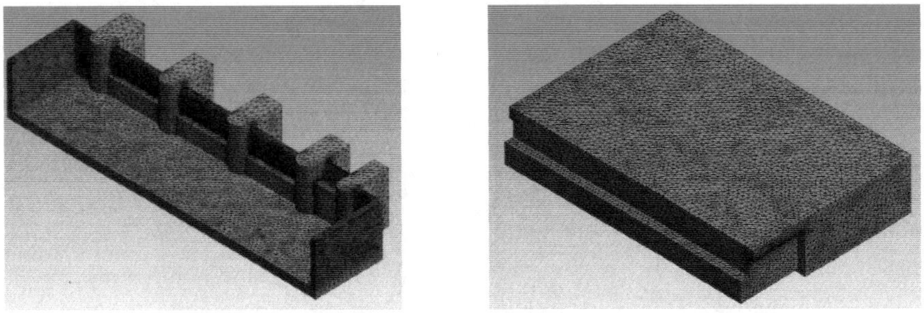

Figure 1. Meshed structural domain (left) and fluid domain (right).

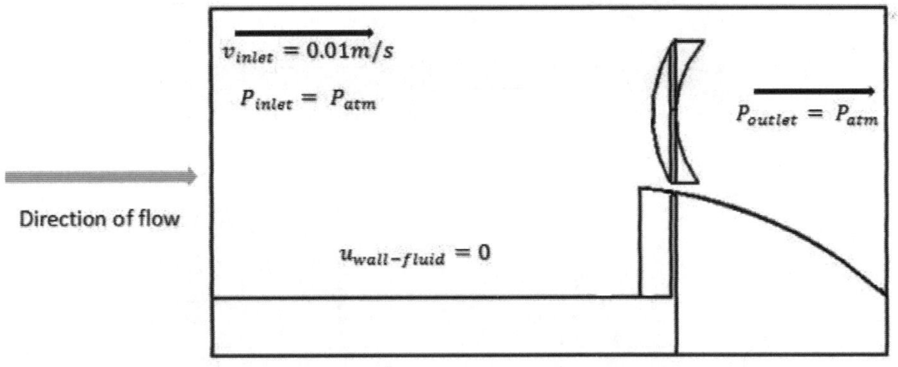

Figure 2. Schematic diagram of boundary conditions used in the numerical simulation.

The system coupling system facilitates utilizing multidisciplinary simulations between coupling participants which transfer and receive data in a coupled analysis. For this case, the analysis systems are Transient Structural and Fluid Flow (FLUENT). System coupling service manages the execution of analyses involves couplings between both the participants and two-way data transfers are conducted.

Firstly, data transfer is formed between the fluid solid interface (FSI) of the Transient Structural and Fluid Flow (FLUENT). When the data transfer happens, force from the fluid domain surface is

IOP Conf. Series: Materials Science and Engineering **370** (2018) 012063 doi:10.1088/1757-899X/370/1/012063

sent to the solid domain surface. At the same time, displacement variable is transferred from solid domain to the fluid domain surface. Next, the time step size is set at 0.1 s and the end time is 5 s consistent with settings provided in Transient Structural; while the minimum iteration is 1 and maximum iteration is 7. Finally, the coupled analysis is solved by updating the system coupling.

3. Experiment

The experiment is conducted on the physical dam model as shown in Figure 3 that located in the Civil School Research Lab to verify the simulation force data. When water flows through the gap underneath the radial gates, force is exerted on the gate surface. The purpose of the experiment is to find out the value of force acting on the surface. This is to ensure that the gate can withstand high pressure without undergoing permanent deformation.

In the experimental setup, the IEPE force sensor and accelerometer (see Figure 4) are fixed onto the steel bar by using magnets. The force sensor measures dynamic forces over a wide frequency range while the accelerometer is used to receive and monitor vibration. Both of them are connected to the LMS SCADAS Mobile (see Figure 5), a data acquisition hardware which collects signal in real time and transmits it to the Test.Xpress software platform. The software analyses and displays the vibration and force results. Then, the steel bar is held in place by a metal holder fixed on top of the radial gate. The sensor and accelerometer are pressed against the outer radial gate surface to receive incoming force and vibration from the water flowing pass the gate. All the data is sent to the software to be displayed and recorded. The experiment is carried out for dam model with varied gap heights, $h = 8, 10, 12$ mm.

Figure 3. Experimental setup for the study of dam flow.

Figure 4. Force sensor of model 1053V1 IEPE (left) and accelerometer of model 3055B2 IEPE (right).

IOP Conf. Series: Materials Science and Engineering **370** (2018) 012063 doi:10.1088/1757-899X/370/1/012063

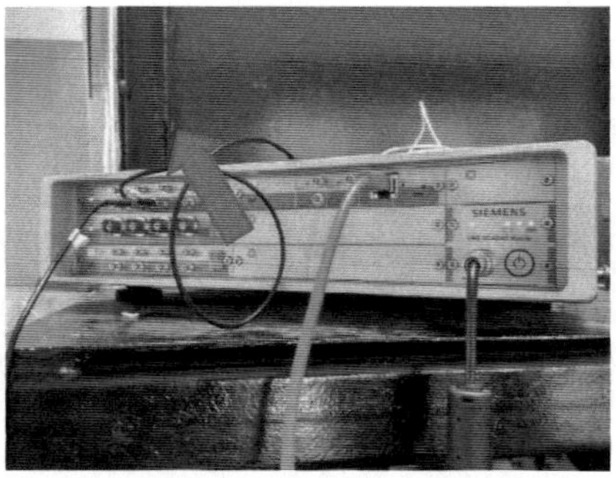

Figure 5. LMS SCADAS Mobile

4. Results and Discussions

In Table 1, the average value of the strains on the radial gate were calculated and used to determine the stress, using the Young's modulus equation. From stress, the force component is obtained through normal pressure formula. Meanwhile, Figures 6, 7 and 8 display the gate strain data taken from each gap heights at time of 5.0 s. The red points on each figure indicated the one of the three points where the strain values were taken. Hence, Figure 5 shows the differences in visual form based on the comparison made in Table 1.

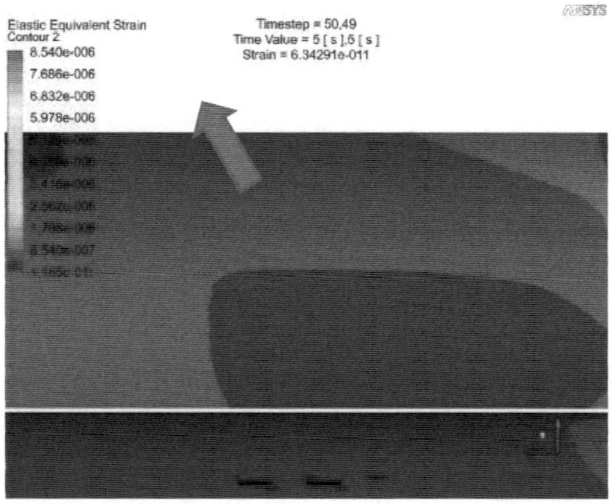

Figure 6. Strain distribution for $h = 8$ mm at simulation time of 5.0 s.

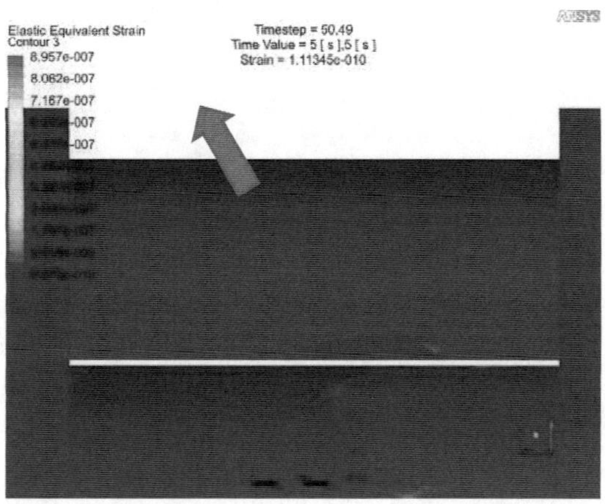

Figure 7. Strain distribution for $h = 10$ mm at simulation time of 5.0 s.

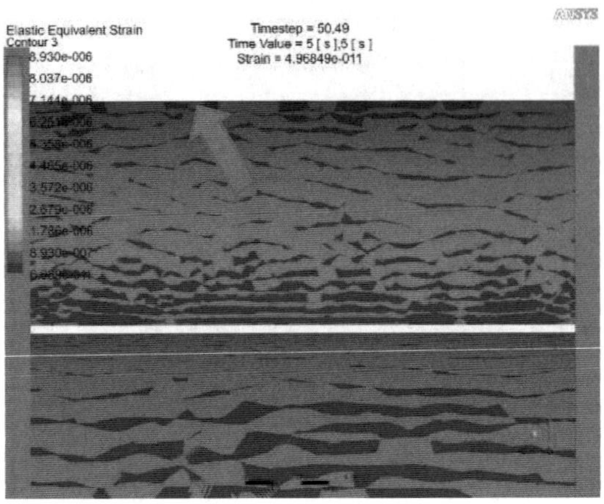

Figure 8. Strain distribution for h = 12 mm at simulation time of 5.0 s.

For gap height of 8 mm, the strain values are taken at three points on the radial gate and the mean of them are found.

$$\bar{\varepsilon} = \frac{4.56213 \times 10^{-11} + 6.34291 \times 10^{-11} + 5.55232 \times 10^{-11}}{3} = 5.48579 \times 10^{-11} \tag{1}$$

Based on the Young's modulus equation, stress is calculated when the strain is applied.

$$\bar{\sigma} = E\bar{\varepsilon} = 0.2085 \text{ Pa} \tag{2}$$

From stress, the force acting on the gate surface is obtained by dividing it with half the surface area of the gate, $A = 49940.5 \times 10^{-6}$ m^2.

$$F = \bar{\sigma}A = 0.0104 \text{ N} \tag{3}$$

The above calculations for force on radial gate were repeated for gap heights of h = 10 mm and h =12 mm. The values of force that acts on the radial gate of various gap heights are then tabulated in Table 1 and later compared with experimental data in Figure 9.

Table 1. Force obtained from simulation for different gap heights.

Gap Height (mm)	Strain ($\times 10^{-11}$)				Force (N)
	1	2	3	Average	
8	4.56213	6.34291	5.55232	5.48579	0.0104
10	10.2563	11.1345	10.3286	10.5731	0.0201
12	8.50078	4.96849	6.95983	6.8097	0.0129

Based on Figure 9, it is seen that percentage deviations between the numerical simulation and experimental results is spanning from 9.59% to 14.66%, indicating the magnitude of forces exerted on radial gate. Such varying difference between the simulated data and experimental value is due to the method of experimental set up. The force is taken from the outer surface of the radial gate at two points. When the flowing water acts against the inner surface, the force is distributed throughout the surface thus the values vary at different areas. For instance, the upper area of the gate will experience less force compared to the bottom area. It does not come in direct contact with the flowing water. The outer surface will be exerted with less force too due to energy loss in the form of sound and kinetic. Furthermore, the sensor and accelerometer may not be able to detect and transmit all the vibration motion and force as a result of lack of sensitivity. They are not calibrated and maintained consistently hence affecting the accuracy of outcomes. Whereas for the simulation part, the level of water enclosed is slightly different from the experiment. It uses a fixed and approximated value, the ability to replicate real-life fluctuating water flow scenario is limited.

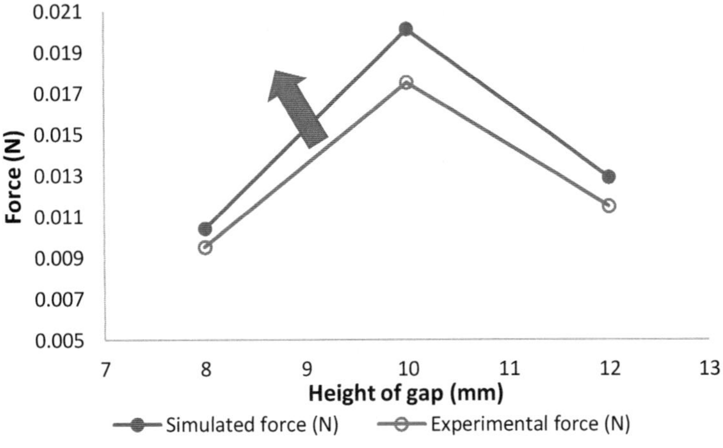

Figure 9. Comparison of simulation and experimental forces acts on the radial gates of different gap heights.

5. Conclusions

This paper enclosed the numerical and experimental methodologies in determining the forces exerted on the radial gate in a dam during water flow. Both the forces data obtained through the fluid/structure interaction (FSI) numerical scheme and experimental works agree pretty with each other, with observed discrepancy less than 14.66%. This study eventually reported that at the force exerted on the radial gate is the highest at the configuration of gap height 10 mm while compared to the 8 mm and 12 mm gates. The methodology outlined in current paper would definitely benefit to future works relating to dam design and reliability analysis.

Acknowledgements:

The authors gratefully acknowledged the financial support provided by Institute of Postgraduate Studies (IPS), Universiti Sains Malaysia (USM), through the USM Fellowship 2017 scheme. Additional, this work was also partly supported by the Short Term Grant 60313020 from the Division of Research and Innovation, Universiti Sains Malaysia, and FRGS Grant 6071322 from the Ministry of Higher Education.

References:

[1] Energy Education on Dam failures. Available from: http://energyeducation.ca/encyclopedia/Dam_failures. [Extracted at 3 June 2017]

[2] Ismail, N. (2014). Issues and Problems towards the Sustainable Dam Management System in Malaysia, Faculty of Arts and Social Science, University of Malaya, Kuala Lumpur, Malaysia.

[3] Razad, A. A., & Ros, F. C. (2009). One Dimensional Dam Breach Modelling for Proposed Hydropower Development in Ulu, (October).

[4] Sidek, L. M., Ros, F. C., & Aziz, N. H. A. (2011). Numerical Modelling of Dam Failure for Hydropower Development in Cameron Highlands-Batang Padang Scheme, Pahang Malaysia, (November).

[5] Gomez-Gesteira, M. (2009). State-of-the-art of classical SPH for free-surface flows. Journal of Hydraulic Research, 48(extra), 0. Available from: <https://doi.org/10.3826/jhr.2010.0012>. [11 March 2017]

[6] Benedict, D., Crespo, A. C., Dominguez, J. M., Barreiro, A., & Go, M. (2011). GPUs, a New Tool of Acceleration in CFD: Efficiency and Reliability on Smoothed Particle Hydrodynamics Methods, Universidade de Vigo, Ourense, Spain and University of Manchester, Manchester, United Kingdom. Available from: <https://doi.org/10.1371/journal.pone.0020685>. [11 March 2017]

[7] Gomez-Gesteira, M. (2009). State-of-the-art of classical SPH for free-surface flows. Journal of Hydraulic Research, 48(extra), 0. Available from:<https://doi.org/10.3826/jhr.2010.0012>. [11 March 2017]

[8] Benedict, D., Crespo, A. C., Dominguez, J. M., Barreiro, A., & Go, M. (2011). GPUs, a New Tool of Acceleration in CFD: Efficiency and Reliability on Smoothed Particle Hydrodynamics Methods, Universidade de Vigo, Ourense, Spain and University of Manchester, Manchester, United Kingdom. Available from: <https://doi.org/10.1371/journal.pone.0020685>. [11 March 2017]

[9] Gómez-Gesteira, M and Dalrymple, RA (2004). "Using a 3D SPH Method for Wave Impact on a Tall Structure," J Waterw Port Coast Ocean Eng, 130(2), 63-69.

[10] Mfundo Vezi (2014). Dynamic Modelling of Arch Dams in the Ambient State, University of Cape Town, Cape Town, South Africa.

IOP Conf. Series: Materials Science and Engineering **370** (2018) 012064 doi:10.1088/1757-899X/370/1/012064

Visualization of Underfill Flow in Ball Grid Array (BGA) using Particle Image Velocimetry (PIV)

Fei Chong Ng[1], Aizat Abas[2*], Ismail Abustan[3], Z. Mohd Remy Rozainy[3], MZ Abdullah[4], Ali b. Jamaludin[5], Sharon Melissa Kon[6]

[1,2]School of Mechanical Engineering, Universiti Sains Malaysia, Engineering Campus, Nibong Tebal, 14300, Penang.
[3]School of Civil Engineering, Universiti Sains Malaysia, Engineering Campus, Nibong Tebal, 14300, Penang.
[4]School of Civil Engineering, Universiti Sains Malaysia, Engineering Campus, Nibong Tebal, 14300, Penang.
[5]TNB Research Sdn. Bhd., No. 1, Lorong Ayer Itam, Kawasan Institusi Penyelidikan, 43000 Kajang.
[6]School of Mechanical Engineering, Universiti Sains Malaysia, Engineering Campus, Nibong Tebal, 14300, Penang.

*Corresponding author: aizatabas@usm.my

Abstract. This paper presents the experimental methodology using particle image velocimetry (PIV) to study the underfill process of ball grid array (BGA) chip package. PIV is a non-intrusive approach to visualize the flow behavior of underfill across the solder ball array. The BGA model of three different configurations – perimeter, middle empty and full array – were studied in current research. Through PIV experimental works, the underfill velocity distribution and vector fields for each BGA models were successfully obtained. It is found that perimeter has the shortest filling time resulting to a higher underfill velocity. Therefore, it is concluded that the flow behavior of underfill in BGA can be justified thoroughly with the aid of PIV.

1. Introduction

In electronic packaging (EP), underfill process is utmost crucial in enhancing the reliability of microelectronic device, particularly flip-chip, ball grid array (BGA) and chip device. It not only cushions the package from external contamination, but also relief the thermomechanical stress experienced by the chip and solder bumps due to the difference in the coefficient of thermal expansion (CTE) [1]. Upon realizing the importance of this underfill process and its impact to the electronics' reliability, various research on underfill process and its optimization study have been actively conducted on various approaches and point of views [2 – 7]. Various optimizing parameters and operating conditions that critically affect the underfill flow have been identified, for instance the gap height, pitch/clearance, dispensing method, bump arrangements/ball count, operating temperature and inlet pressure.

On the other hand, particle image velocimetry (PIV) is a non-intrusive optical method for flow visualization. This technique allows for the velocity of a fluid on a two-dimensional (2D) plane to be measured simultaneously. The kinematics of the flow is obtained by tracking the flow following tracing particles seeded into the fluid of interest. These particles are illuminated with the aid of high power light source. The motion of the illuminated particles is then recorded. The use of strobing laser light makes it easier to freeze the movement of the particles in view, especially in fast movement. The velocity of the flow is then calculated by knowing the distance travelled by the tracking particles between the time intervals set between camera exposures, together with the magnification of the camera.

The usage of PIV had previously reported in the underfill application of flip-chip in two literatures [8, 9]. They have successfully setting up a PIV system to study the flow in flip-chip and determine the dynamics properties of the underfill flow. However still the application of PIV in sector of electronic packaging is very limited. Therefore, this paper is devoted to outline a simple and easy-to-construct experimental setup to visualize the underfill flow based on the PIV. Furthermore, such PIV experimental approach allows researcher to identify additional data on flow vector and velocity profiles of underfill flow which unable be attained in conventional experimental in electronic packaging.

2. PIV formulation

Particle image velocimetry (PIV) is an optical method that can be used to determine the instantaneous vector measurement throughout the cross-sectional view. The motion is followed via the use of small tracer particles. Based on the positions of these tracer particle at two different instances of time can we obtained the particle displacement to infer the flow velocity field [10, 11].Velocity vector of the fluid through the tracer particles are computed based on the movement of tracer particle between two light pulses:

$$V = \frac{\Delta x}{\Delta t} \tag{1}$$

The primary source of measurement error relies on the influence of gravitational forces, g when the density of the tracer particles is different to the density, ρ of work fluid.

$$V_g = {d_p}^2 \frac{(\rho_p - \rho)}{18\mu} g \tag{2}$$

The velocity lag, V of a particle in a continuously acceleration fluid is given by:

$$\vec{V_s} = \vec{V_P} - \vec{V} = {d_p}^2 \frac{(\rho_p - \rho)}{18\mu} g \tag{3}$$

$$\vec{V_p}(t) = \vec{V}(1 - \exp(-\frac{t}{\tau_s})) \tag{4}$$

$$\tau_s = {d_p}^2 \frac{\rho_p}{18\mu} \tag{5}$$

This technique consists of the measurement of the fluid displacement, Δx over a given time interval, Δt. The LED light will illuminates the tracer particles by creating a thin laser light sheet [12]. Then subsequently, the image will be captured using a camera specifically at the target area [10].

3. Experiment

Figure 1 depicts the general experimental setup for conducting the underfill process on ball grid array. It consists of a light box where the BGA assembly is directly above the light source, forming a backlighting system. Camera video is placed perpendicularly above the BGA assembly to record the underfill flow into the gap of array. There are three different BGA models being studied of various balls arrangements, for instant full, middle empty and perimeter arrays as shown in Figure 2. Each solder ball has a diameter of 2 mm with clearance of 2 mm and gap height of 1.5 mm. The length of bump-less edge is 8 mm, resulting the overall 6 × 6 array package has a dimension of 38 × 38 mm2. For better visualization aspect, the chip/substrate is replaced by a clear smooth Perspex that exhibits the similar capillary properties as the actual surfaces. The underfill fluid that exhibits property of a Newtonian fluid with density 1042 kg/m3 and viscosity 2.2 Pa-s was carefully injected into the inlet of the BGA model, by using a syringe. All the underfill flow in the BGA will be recorded using the camera video for subsequent PIV analysis.

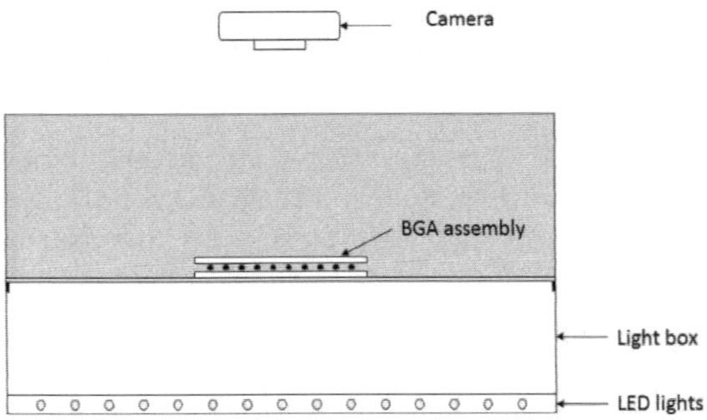

Figure 1. The experimental setup for the capillary underfill process of ball grid array (BGA).

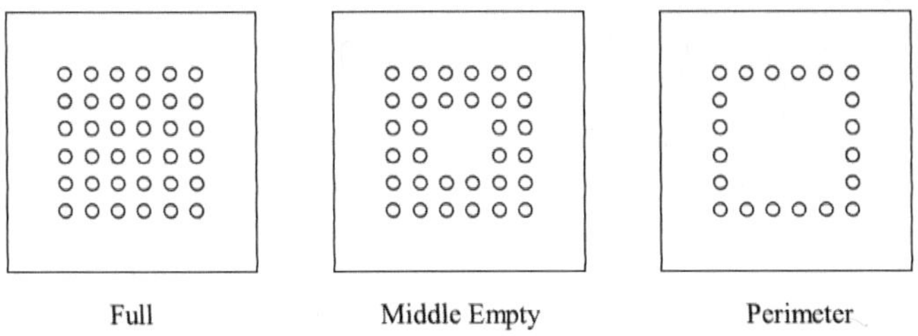

Figure 2. The three different BGA bumps arrangements being considered.

For each case, two consecutive images of interval 0.1 s at particular stage of 80% filled from the underfill video were obtained for the image pre-processing, to be ready for the PIV analysis. The original image obtained will be changed to greyscale before being pre-processed to remove as much noise as possible through high pass filter and intensity capping. Afterward, the processed images are loaded into the Matlab PIVlab software to obtain the flow vector and velocity contours in each underfill cases.

4. Results and Discussions

Figure 3 presents the unprocessed underfill flow front in three different BGA models, namely perimeter, middle empty and full array. All these underfill stages are approximately near completion, at about 80% filled. The time taken for the underfill to flow till this stage are 152 s, 147 s and 140 s, respectively for the full, middle empty and perimeter array bumps. It is cleared than the underfill flow faster in perimeter array, later follows by middle empty array and the underfill flow in full array being the slowest. This is due to the presence of large amount of solder balls in full array (36) and middle empty array (32), constitute a larger bump resistance to the underfill flow as compared to the least ball count in perimeter array (20). Such filling time trending in BGA different bumps arrangements are similar as reported in past literatures [5, 6].

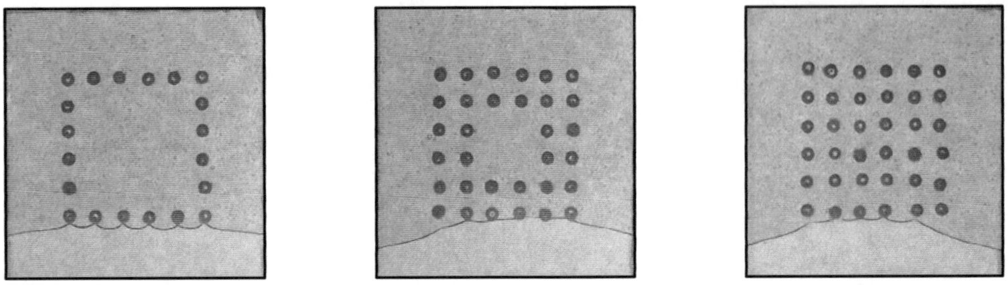

Figure 3. The underfill flow pattern across the perimeter, middle empty and full array ball grid array models (from left to right). Pinkish fluid indicated the underfill; while the white regions are unfilled.

Figure 4, Figure 5 and Figure 6 respectively show the flow vector and velocity contour profile for the underfill flow in ball grid array of different array arrangements, perimeter, middle empty and full. It is noticed that the velocity of underfill flow near the vicinity of solder ball are comparably lower than in bump-less region. As a result, for BGA with high ball counts such as the full array, there are more region of low velocity, resulting lower overall flow of underfill and causing longer filling time. This had correlated the number of solder ball present in the BGA to the flow velocity and filling time. On the side of flow vector, it is found the underfill flows directly from the inlet toward the outlet in all cases. The presence of solder balls in the array did not alter the flow direction of underfill significantly.

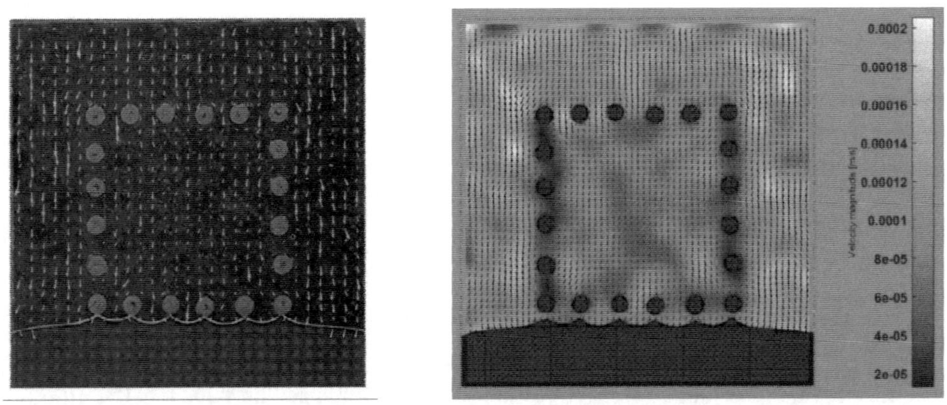

Figure 4. Flow vector and velocity contour for underfill in perimeter bump array.

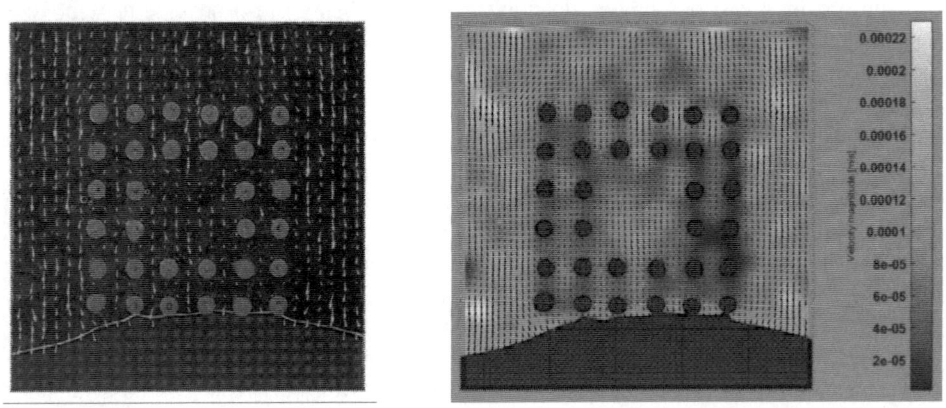

Figure 5. Flow vector and velocity contour for underfill in middle empty bump array.

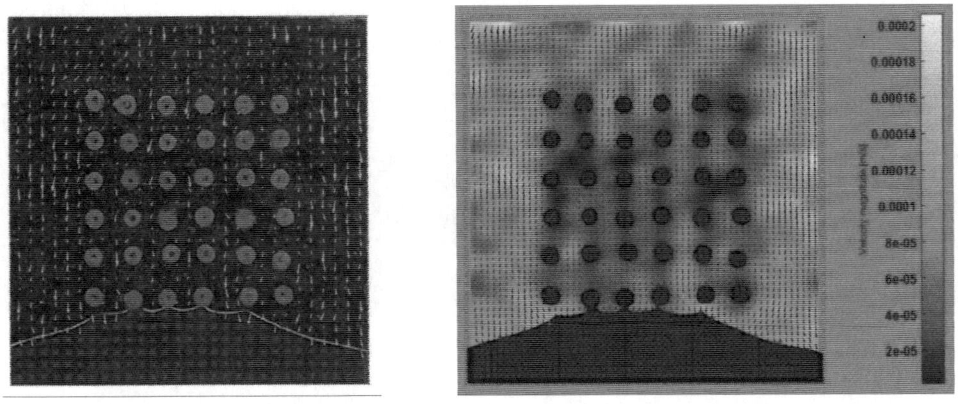

Figure 6. Flow vector and velocity contour for underfill in full bump array.

5. Conclusions

This paper had demonstrated the viability of particle image velocimetry (PIV) experimental approach in the study of underfill flow in ball grid array (BGA) packaging. Through PIV analysis, the flow vector and velocity profiles of underfill flow in the BGA can be obtained. Particularly, three BGA models of different ball arrangements and ball counts, namely perimeter, middle empty and full array are considered in this research. It is found that the trend of shortest filling time and faster underfill flow are in the ascending order of perimeter, middle empty and full array. The shortest filling time of 140 s is registered by the perimeter array; meanwhile the full array took the longest time of 152 s to reach 80% filled. However, the flow appears did not significantly affected by the presence of solder ball as justified by almost similar flow vector obtained for all three cases. Ultimately, PIV is a reliable yet simple approach to visualize the underfill flow and therefore could have its application extended in future.

Acknowledgements:

The authors gratefully acknowledged the financial support provided by Institute of Postgraduate Studies (IPS), Universiti Sains Malaysia (USM), through the USM Fellowship 2017 scheme. Additional, this work was also partly supported by the Short Term Grant 60313020 from the Division of Research and Innovation, Universiti Sains Malaysia, and FRGS Grant 6071322 from the Ministry of Higher Education.

References:

[1] H. Ardebili, M.G. Pecht 2009 Encapsulation Process Technology, Encapsulation Technologies for Electronic Applications, *Elsevier* chapter pp 129–179
[2] Fei Chong Ng, Aizat Abas, MHH Ishak, MZ Abdullah, MS Abdul Aziz 2016 *Microelectronics Reliability* **66** 143
[3] C.Y. Khor, M.Z. Abdullah, M. Abdul Mujeebu *Journal of Electronics Packaging* **134**(1) 11003.
[4] C.Y. Khor, M.Z. Abdullah, M. Abdul Mujeebu, F. Che Ani 2010 *International Communications of Heat and Mass Transfer* **37** 281
[5] A. Abas, M.S. Haslinda, M.H.H. Ishak, A.S. Nurfatin, M.Z. Abdullah, F. Che Ani 2016 *Microelectron. Eng.* **163** 83
[6] Aizat Abas, M.H.H. Ishak, M.Z. Abdullah, F. Che Ani, Soon Fuat Khor 2015 *Microelectronics Reliability*
[7] Fei Chong Ng, Aizat Abas, MHH Ishak, MZ Abdullah, MS Abdul Aziz 2016 *Microelectronics Reliability* **66** 143
[8] Young Bae Kim, Jaeyong Sung, Myeong Ho Lee 2011 *Journal of Visualization* **14** 237

[9] Seok Hwan Lee, JaeyongSung, Sarah Eunkyung Kim 2010 *Microelectronics Reliability* **50** **12** 2078
[10] N. Hafizah Mokhtar, Aizat Abas, N.A. Razak, Muhammad Najib Abdul Hamid, Soon Lay Teong 2017 *Journal of Theoretical Biology* **433** 73
[11] I. M. Richardson and C. R. Kleijn 2009 *Experimental characterization of GTA weld pool surface flow using PIV*, no. January
[12] Y. Le Sant, F. Champagnat, and G. Le Besnerais 2009 *Principles and Applications of Particle Image Velocimetry no. 1*

Simulation of Blood flow in Different Configurations Design of Bi-leaflet Mechanical Heart Valve

N. Hafizah Mokhtar[1], Aizat Abas[2]

[1, 2] School of Mechanical Engineering, Universiti Sains Malaysia, Engineering Campus, 14300 Nibong Tebal, Penang, Malaysia

Corresponding author: aizatabas@usm.my

Abstract. In this work, two different designs of artificial heart valve were devised and then compared by considering the thrombosis, wear and valve orifice to anatomical orifice ratio of each mechanical heart valve. These different design configurations of bi-leaflet mechanical heart valves model are created through the use of Computer-aided design (CAD) modelling and simulated using Computational fluid dynamic (CFD) software. Design 1 is based on existing conventional bi-leaflet valve and design 2 based on modified bi-leaflet respectively. The flow pattern, velocity, vorticity and stress analysis have been done to justify the best design. Based on results, both of the designs show a Doppler velocity index of less than the allowable standard of 2.2 which is safe to be used as replacement of the human heart valve. However, design 2 shows that it has a lower possibility of cavitation issue which will lead to lower thrombosis and provide good central flow area of blood as compared to design 1.

1. Introduction

Heart is the one most important muscular organ that pumps blood throughout the whole body. Heart contains 4 chambers which are right atrium, right ventricle, left atrium and left ventricles. The heart valves are the key components that prevent backflow of blood. Four heart valves, two on either side of the heart, make sure blood produces unidirectional flow during muscle contraction and pass through the heart valve [1]. Circulation of blood begin at the right hand side of heart, the tricuspid and pulmonic valve control the flow of blood which is returned from body to the lung to received oxygen. For the left hand side of heart, the mitral and aortic valves regulate the oxygenated blood from the lung to receive oxygen and flow to body [2]. Today, many people live long and healthy lives and never realize they have a mild valve problem. However, valve disease can seriously increase a persons' risk for sudden death or cause rapid development of problems in and around the heart that can become fatal without treatment. The introduction of artificial heart valve definitely has made an incredible impact in biomedical field due to its lifetime durability. However, there are still problems in artificial valves such as thrombosis, vegetation, cavitation and many more. Hence, a better valve design has to be created to eradicate the problems encountered.

This happened when the patient with heart valve disease that involves one or more of the four valves of the heart malfunction. Commonly, heart valve diseases are involved in the left heart valve that consist of aortic valve and mitral valve. There are two types of heart valve malfunction that include the valve did not close (regurgitation) or open (stenosis) properly causing valve insufficiency. This may cause disturbing flow of blood throughout the human body. An artificial heart valve is a

IOP Conf. Series: Materials Science and Engineering **370** (2018) 012065 doi:10.1088/1757-899X/370/1/012065

device implanted in the heart to replace the natural valve with an artificial valve [3]. There are several types for artificial heart valve that are commonly used as replacement namely tissue heart valve and mechanical heart valve [4-5]. In this project, two bi-leaflet mechanical heart valve models are designed to analyze a blood flow system through the left heart. The conceptual design on artificial mitral valve will be created using CAD software and then simulated using computational fluid dynamic (CFD) software to select the best heart valve design that meet forward and backward flow shear, reduced blood clotting, minimal leakages and carry out vorticity study of heart valve simulation

2. Methodology

2.1. *Establishment of CAD Model*

The CAD model of the bi-leaflet mechanical heart valve in this project is constructed using Solidworks software. There are two types of the bi-leaflet model 1) normal bi-leaflet 2) modified bi-leaflet. Normal bi-leaflet is flat and straight surface while modified bi-leaflet is little curvature on the leaflet as shown in Figure 1 (a) and Figure 1 (b). The model consists of housing and two leaflet. The diameter of the mechanical heart valve is 28mm. The length of the model mechanical heart valve is 14mm.

(a) (b)

Figure 1. Geometry of (a) the conventional bi-leaflet (b) modified bi-leaflet

2.2. *Simulation Using CFD Software*

In computational fluid simulation, ANSYS software is used to simulate the two different configurations of bi-leaflet mechanical heart valve.

2.2.1. Stress Analysis. As the heart valve operates under the blood pressure conditions, it might be damaged or impaired. We aimed to modify of bi-leaflet mechanical heart valve and compare with a normal mechanical heart valve using mechanical analysis. Stress and strain on mechanical mitral heart valve, during opening and closing, were determined with a finite element analysis including deformation. Support at the outer ring body is fixed with 120mmHg pressure is applied in Y-direction [6]. All the materials involved in this work is define as pyrolytic carbon which has a tensile yield strength of 120 MPa as shown in Figure 2.

Properties of Outline Row 3: Pyrolytic carbon

	A	B	C	D	E
1	Property	Value	Unit		
2	Density	2240	kg m^-3		
3	Isotropic Elasticity				
4	Derive from	Young's Mod...			
5	Young's Modulus	4800	MPa		
6	Poisson's Ratio	0.28			
7	Bulk Modulus	3.6364E+09	Pa		
8	Shear Modulus	1.875E+09	Pa		
9	Tensile Yield Strength	1.2059E+08	Pa		

Figure 2. Material property of pyrolytic carbon in CFD Software

IOP Conf. Series: Materials Science and Engineering **370** (2018) 012065 doi:10.1088/1757-899X/370/1/012065

2.2.2. Flow Analysis. The flow of blood through the valves will be simulated for different angles of leaflets and the flow pattern will be observed. Assume the flow is laminar and steady state. The density and dynamic viscosity of the blood is1060 kg/m-3 and 0.0035 Pa s respectively [7].

3. Result And Discussion

3.1. Stress Analysis Result.
From this paper, deformation analysis for conventional and modified bi-leaflet is conducted when bi-leaflet heart valve is closed and open. The result of deformation analysis for both bi-leaflet mechanical heart valve as shown in Figure 3 and Figure 4.

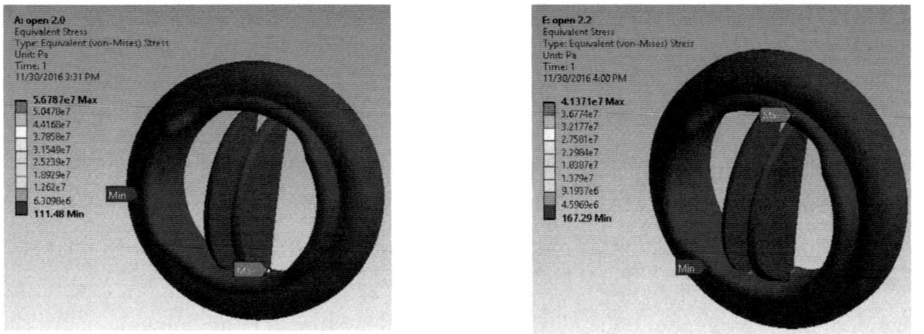

(a) (b)

Figure 3. Stress analysis for fully opening of bi-leaflet mechanical heart valve. a) Conventional bi-leaflet (b) Modified bi-leaflet

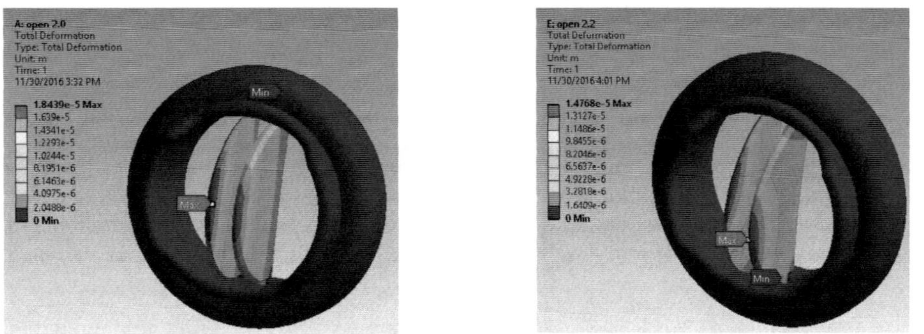

(a) (b)

Figure 4. Deformation analysis for fully opening of bi-leaflet mechanical heart valve. (a) Conventional bi-leaflet (b) Modified bi-leaflet

Table 1. Stress and deformation of convectional and modified bi-leaflet heart valve

Valve Type	Fully Opening			
	Stress (Pa)		Deformation (m)	
	Min	Max	Min	Max
Normal bi-leaflet	111.48	$5.679x10^7$	0	$1.844x10^{-5}$
Modified bi-leaflet	167.29	$4.137x10^7$	0	$1.477x10^{-5}$

From Figure 3 and Figure 4, the distribution of stress is different from tips of leaflet to the hinge of leaflet. Simulation results showed that the maximum of stress when valve is fully opened, occurred at the hinge joint of leaflet in both bi-leaflet valves. Between conventional bi-leaflet and modified bi-leaflet, stress concentration on the hinge joint in the modified bi-leaflet valve was less than that in the conventional bi-leaflet valve which is 4.137×10^7 Pa. While, for leaflet deformation in modified bi-leaflet valve was also lower compared to conventional bi-leaflet valve as shown in Table 2. This is because increasing in deformation and stress may lead to damage of mechanical heart valve [8-9]. Thus, modified bi-leaflet has shown the best structural rigidity.

3.2. *Flow Analysis of conventional and modified bi-leaflet.*

From Table 2, simulation of fully opened and closed for both bi-leaflet mechanical heart valve are done to analyse flow analysis, leaking area and vortex formation between both bi-leaflet mechanical heart valves.

Table 2. Analysis of the flow pattern of conventional and modified bi-leaflet mechanical heart valves.

Features	Conventional bi-leaflet	Modified bi-leaflet
Velocity contour (valve fully open)		
Peak velocity	$0.2329\ ms^{-1}$	$0.23\ ms^{-1}$
Leaking area (red rectangle)	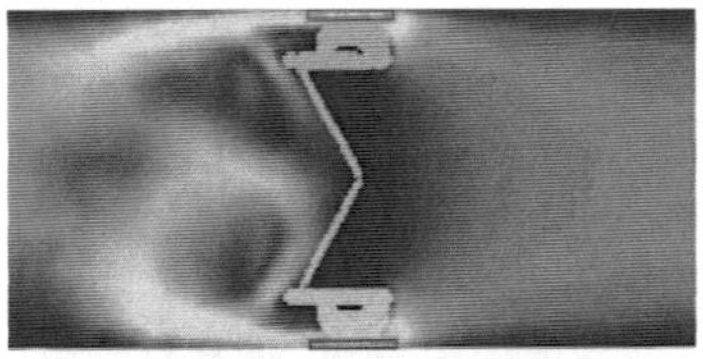	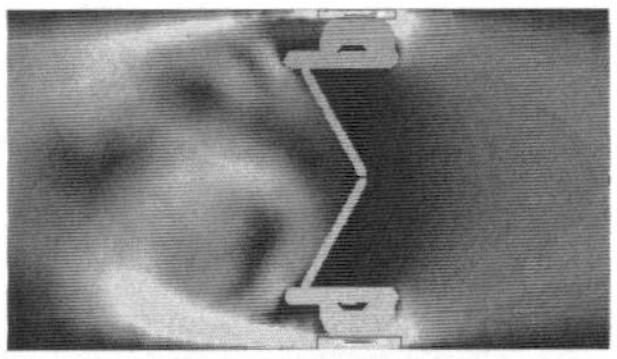
Vortex formation	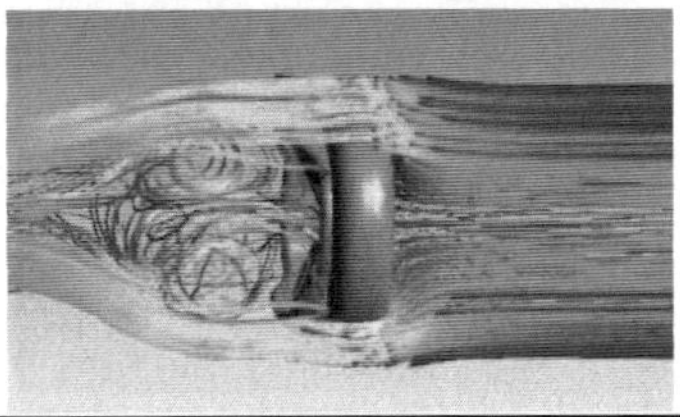	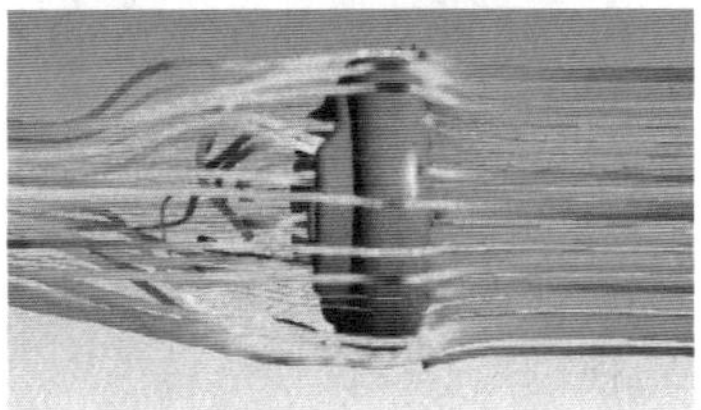

Based on the velocity contour in Table 2, we can observe modified bi-leaflet valve observe a better central flow whereas conventional of bi-leaflet valve undergo flow separation in the middle area. In addition, the maximum velocity when blood flow through is the lowest and which is ideal as the flow velocity before and after passing the mechanical valve should be close to each other. In addition, the leaking area of both bi-leaflet valves are almost similar provided the mechanical valve is still in good condition. Besides, the larger comparison between both bi-leaflet heart valves is the vortex formation during opening. The vortex formation in conventional bi-leaflet valve is greatest compared to modified bi-leaflet valve as shown in Table 2. This is because the curvature on modified bi-leaflet affects the middle section of the blood flow that reduces the vorticity during opening operation. For an artificial heart valve, vortex formation should be considered to avoid thrombosis and cavitation formation.

While, the larger vortices forms at the middle area of conventional bi-leaflet valve might induce cavitation to happen and reduce blood flow drastically. On the contrary, the vortices formed in modified bi-leaflet is less accumulated in the middle section and thus lower chance of cavitation. Smaller vortex, lesser area affected and better flow dynamic for an artificial heart valve [10-12].

3.3. *Detection and Quantification of Left heart valve Stenosis.*

One of the criteria to determine the mitral stenosis is through the Doppler velocity index (DVI). From Figure 5, the normal DVI value for mitral valve should be less than 2.2.

	Normal		Possible stenosis		Significant stenosis	
	Aortic	Mitral	Aortic	Mitral	Aortic	Mitral
Valve structure and motion						
Mechanical or bioprosthesis	Normal	Normal	Often abnormal**	Often abnormal**	Abnormal**	Abnormal**
Doppler quantitative parameters						
Peak velocity (m/s)* †	<3	<1.9	3–4	1.9–2.5	>4	>2.5
Mean gradient (mm Hg)* †	<20	≤5	20–35	6–10	>35	>10
Doppler velocity index* ¶	≥0.30	<2.2	0.25–0.29	2.2–2.5	<0.25	>2.5
Effective orifice area (cm^2)* §	>1.2	≥2	0.8–1.2	1–2	<0.8	<1
Effective orifice area versus normal reference value* §	Reference±1SD	Reference±1SD	<Reference–1SD	<Reference–1SD	<Reference–2SD	<Reference–2SD
Contour of the transprosthetic jet‡	Triangular, early peaking	–	Triangular to intermediate	–	Rounded, symmetrical	–
Acceleration time (ms)‡	<80	–	80–100	–	>100	–
Pressure half time (ms)‡ 1	–	<130	–	130–200	–	>200

Figure 5. Allowable DVI in mitral valve[13]

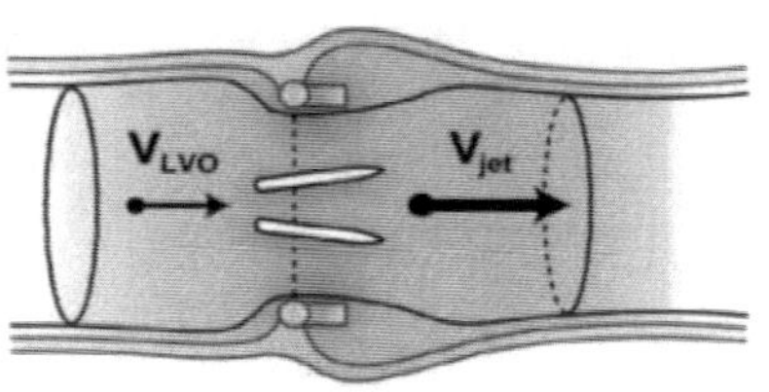

Figure 6. Parameters involved to calculate DVI

$$\text{For mitral valve: Doppler Velocity Index (DVI)} = \frac{V_{jet}}{V_{LVO}}$$

Table 3. Average velocity of blood flow depicted from computational simulation when the valve is fully open.

Valve type	Inlet velocity (m/s)						Outlet velocity (m/s)					
	1	2	3	4	5	Average	1	2	3	4	5	Average
Conventional Bileaflet	0.11	0.13	0.12	0.14	0.12	0.126	0.201	0.16	0.23	0.17	0.215	0.1952
Modified Bileaflet	0.11	0.13	0.12	0.14	0.12	0.126	0.2	0.201	0.17	0.16	0.22	0.190

Table 4. Doppler velocity index between both bi-leaflet mechanical heart valves.

Valve type	Doppler Velocity Index
Conventional bi-leaflet	1.55
Modified bi-leaflet	1.51

From Table 4, DVI value of modified bi-leaflet is the lowest compared to the others. Although conventional bi-leaflet valve has a DVI value of 1.55 but it is still within the allowable range which is the less than 2.2. Hence, all the both valves are within the permissible limit and safe to be used. However, as time goes by, the mechanical valve will consistently undergo expansion and contraction so if the valve was to subjected to sudden changes, modified bi-leaflet valve might be a better choice as it can allow at least a double increased of current peak velocity.

4. Conclusion

Two different designs for the mechanical heart valve to replace defective mitral valve have been presented and discussed in this work. Design 1 is a conventional bi-leaflet valve and Design 2 is modified bi-leaflet respectively. The stress at each hinge is much lower than conventional bi-leaflet heart valve. Both designs have a Doppler velocity index of less than the allowable 2.2 which is safe to be used as replacement of the human heart valve. However, design 2 is clearly the best design as it has lowest cavitation issue which will lead to lower thrombosis and provide good central flow area of blood.

5. Acknowledgement

The work was partly supported by the Short Term Grant 60313020 from the Division of Research and Innovation, Universiti Sains Malaysia, and FRGS Grant 6071322 from the Ministry of Higher Education.

References

[1] "How the heart work." [Online]. Available: https://www.webmd.com/heart-disease/guide/how-heart-works#1

[2] "Anatomy of Heart." [Online]. Available:https://www.nhlbi.nih.gov/health/health-topics/topics/hhw/anatomy Another reference

[3] "How artificial heart valve is made - material, making, history, used, processing, parts, components, steps, History." [Online]. Available: http://www.madehow.com/Volume-6/Artificial-Heart-Valve.html.

[4] "Prosthetic valves the past present and future i tammi raju." [Online]. Available: http://www.slideshare.net/tammirajuiragavarapu5/prosthetic-valves-the-past-present-and-future-i-tammi-raju.

[5] Dasi, L. P., Simon, H. A., Sucosky, P., & Yoganathan, A. P. (2009). *Fluid Mechanics Of Artificial Heart Valves*. Clinical and Experimental Pharmacology & Physiology, Volume 36, Issue 2, pp 225–237.

[6] Martin C., Sun W. (2011) *Bio-prosthetic Heart Valve Stress Analysis*: Impacts of Leaflet Properties and Stent Tip Deflection. In: Proulx T. (eds) Mechanics of Biological Systems and Materials, Volume 2. Conference Proceedings of the Society for Experimental Mechanics Series. Springer, New York, NY

[7] A.C. Benim, A. Nahavandi, A. Assmann, D. Schubert, P. Feindt, S.H. Suh. (2011) *Simulation of blood flow in human aorta with emphasis on outlet boundary conditions*, In Applied Mathematical Modelling, Volume 35, Issue 7, pp 3175-3188.

[8] Kwon, Y. J. 2010. *Numerical analysis for the structural strength comparison of St. Jude Medical and Edwards MIRA bileaflet mechanical heart valve prostheses*. Journal of mechanical science and technology, Volume 24(2), pp 461-469

[9] Kadhim, S.K., Nasif, M.S., Al-Kayiem, H.H. and Al Waked, R., 2006. *Using Fluid Structure Interaction Technique To Investigate The Effect Of Vibration On Bileaflet Mechanical Heart Valve Deformation*. Blood, 1060, pp.0-2.

[10] Avrahami, I., Rosenfeld, M., Einav, S. et al. Med. Biol. Eng. Comput. (2000). *Can vortices in the flow across mechanical heart valves contribute to cavitation?*. Volume 38, Issue 1, pp 93-97.

[11] Querzoli, G., Fortini, S. and Cenedese, A., 2010. *Effect of the prosthetic mitral valve on vortex dynamics and turbulence of the left ventricular flow.* Physics of fluids, 22(4), p.041901

[12] Oshkai, P., Mejia, J. and Wootton, B., 2005. Quantitative Imaging of Flow Through a Mechanical Heart Valve. In 35th AIAA Fluid Dynamics Conference and Exhibit, Toronto, ON, Canada, Jun (pp. 6-9).

[13] P. Pibarot and J. G. Dumesnil. (2012) *Doppler echocardiographic evaluation of prosthetic valve function*. Heart, vol. 98, no. 1, pp. 69–78

IOP Conf. Series: Materials Science and Engineering **370** (2018) 012066 doi:10.1088/1757-899X/370/1/012066

Simulation of Blood flow in Artificial Heart Valve Design through Left heart

N. Hafizah Mokhtar[1], Aizat Abas[2]
[1, 2] School of Mechanical Engineering, Universiti Sains Malaysia, Engineering Campus, 14300 Nibong Tebal, Penang, Malaysia

Corresponding author: aizatabas@usm.my

Abstract. In this work, an artificial heart valve is designed for use in real heart with further consideration on the effect of thrombosis, vorticity, and stress. The design of artificial heart valve model is constructed by Computer-aided design (CAD) modelling and simulated using Computational fluid dynamic (CFD) software. The effect of blood flow pattern, velocity and vorticity of the artificial heart valve design has been analysed in this research work. Based on the results, the artificial heart valve design shows that it has a Doppler velocity index that is less than the allowable standards for the left heart with values of more than 0.30 and less than 2.2. These values are safe to be used as replacement of the human heart valve.

1. Introduction

The heart is one of the most essential organs in human body. The function of human heart is to pump blood throughout the body in human circulatory system by carrying oxygen and nutrients to human tissue while removing carbon dioxide from human body [1].The heart has four main chamber which are the upper chambers and two lower chambers. The upper chamber is called atria while the lower chamber is namely ventricles. The atria is act as receiving blood while the ventricle carry blood away from the heart [2]. As human heart has four main chambers, to prevent blood from flowing back into the heart, there one-way system present in the heart namely heart valve. The heart valve has four main valves which are tricuspid valve, bicuspid valve, mitral valve and aortic valve. Tricuspid valve and aortic valve have three flaps that functioning to allow blood pass into heart and block the blood from flowing backward into ventricle. Besides, bicuspid valve and mitral valve have two flaps which are allow blood from the body system into ventricles and block blood from flowing backward to human body. Consequently, heart valve diseases are frequently affecting human health which is disturbing the flow of blood throughout the human body. There are two critical diseases conditions that may affect flow of the blood which are called stenosis and regurgitation. A stenosis is known as valve narrowing due to valve does not open fully that will obstruct flow of blood into human body [3]. Furthermore, regurgitation knowns as valve that does not close properly and causing blood flow to backwards. These diseases might harm human body due to heart need to pump more volume of blood that required for human boy. The heart valve that commonly effected from these diseases are aortic and mitral valve.

For severe heart valve diseases, a surgery is needed to replace malfunctions heart valve with an artificial heart valve. There are several types of artificial heart valves, commonly are bio-prosthesis and mechanical heart valve. In few decades, mechanical heart valve is become favourable for the replacement of heart valve due to bio-prosthesis heart valve need as anti-coagulant therapy after surgery [4]. There are several basic types of mechanical heart valve which are caged –balled valve,

IOP Conf. Series: Materials Science and Engineering **370** (2018) 012066 doi:10.1088/1757-899X/370/1/012066

mono-leaflet valve, bi-leaflet valve and tri-leaflet valve. While for bio-prosthesis, the commonly types involved in surgery are stented bio-prosthesis, stent-less bio-prosthesis and percutaneous bio-prosthesis. Haemodynamic, durability and thrombosis are main factor that need to be consider for an artificial heart valves. Therefore, in this research work, a bi-leaflet heart valve is choose to analyse the blood flow of artificial heart valve in human heart. The 3D model of bi-leaflet is constructed using computer –aided design (CAD) software and simulated by computational fluid dynamic (CFD) software to carry out stress analysis, flow pattern and vorticity of an artificial heart valve.

2. Methodology

2.1. Artificial heart valve design

The CAD model of the artificial heart valve in this project is constructed using Computer-aided design (CAD) software. The artificial heart valve consist of housing, two leaflets, flat and straight surface as shown in Figure 1 (a) and Figure 1 (b). The diameter of the mechanical heart valve is 28mm. The length of the model mechanical heart valve is 14mm.

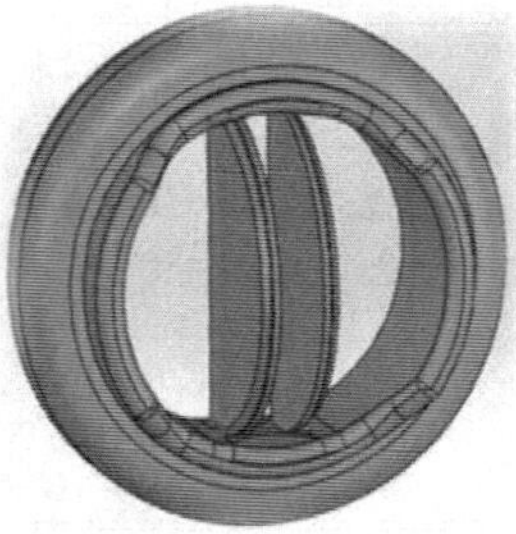

(a)

(b)

Figure 1. Geometry of artificial heart valve (a) Isometric view (b) Front view

2.2. Simulation Using CFD Software

2.2.1. FVM Formulation Prior to the commencement of the CFD simulation, it is important to look at the governing equations underlying the physics taken place. Although there are additional complex conditions such as pulsatile flow and non-Newtonian fluids properties, the governing equations are the same as other fluids problem. The fundamental governing equations are the continuity equation and the Navier-Stokes equations.

(a) Continuity equation

$$\frac{\partial \rho}{\partial t} + \nabla \cdot (\rho v) = 0 \tag{1}$$

Since blood is considered as incompressible fluid, the rate of change of density will be zero. Hence, the continuity equation will be simplified as follow:

$$\nabla \cdot v = 0 \tag{2}$$

(b) Navier-Stokes equations

$$\rho\left(\frac{dv}{dt} + v \cdot \nabla v\right) = -\nabla p + \mu \nabla^2 v + f \tag{3}$$

In Navier-Stokes equations, the viscosity coefficient, μ is not a constant.

2.2.2. Stress Analysis. As the heart valve operates under the blood pressure conditions, it might be damaged or impaired. For this works, artificial heart valve is analysed using mechanical analysis. Stress and strain on mechanical mitral heart valve, during opening and closing, were determined with a finite element analysis including deformation. Support at the outer ring body is fixed with 120mmHg pressure is applied in Y-direction [5]. All the materials involved in this work is define as pyrolytic carbon which has a tensile yield strength of 120 MPa.

2.2.3. Flow Analysis. The flow of blood through the valves will be simulated for different angles of leaflets and the flow pattern will be observed. Assume the flow is laminar and steady state. The density and dynamic viscosity of the blood is 1060 kg/m-3 and 0.0035 Pa s respectively [6].

3. Result And Discussion

3.1. Stress Analysis Result.

From this paper, stress analysis for artificial heart valve is conducted when artificial heart valve is closed and open. The result of stress analysis for artificial heart valve as shown in Figure 3 and 4.

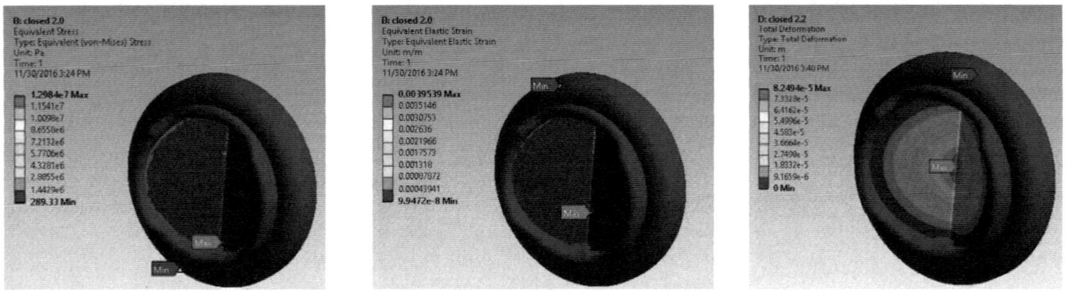

(a) (b) (c)

Figure 2. Stress analysis for artificial heart valve when case is closed. a) Stress (b) Strain (c) Deformation

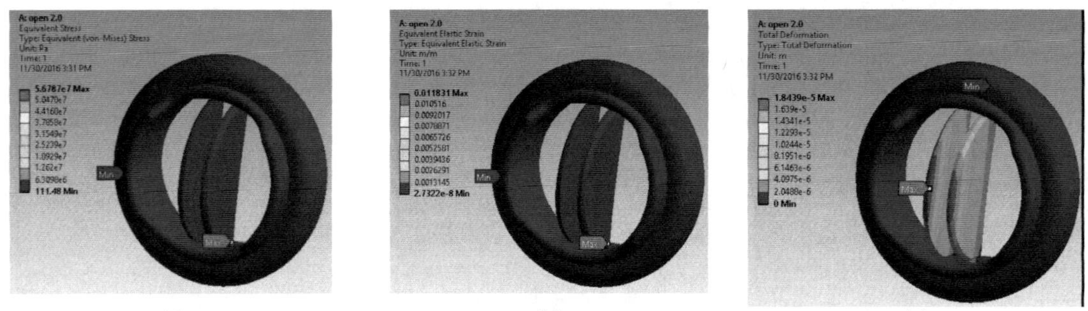

(a) (b) (c)

Figure 3. Stress analysis for artificial heart valve when case is opened. a) Stress (b) Strain (c) Deformation

Table 1. Stress and deformation of conventional and modified bi-leaflet heart valve during closed and opened.

Valve Type	Stress (Pa)		Strain		Deformation (m)	
	Min	Max	Min	Max	Min	Max
Closed	289.33	$1.298x10^7$	$9.947x10^{-8}$	0.00395	0	$1.036x10^{-5}$
Opened	111.48	5.679x107	2.732x10-8	0.0118	0	1.844x10-5

From Table 1, simulation results showed that the maximum of stress, strain and deformation when the artificial heart valve is fully opened. It is shows in Figure 3 and Figure 4, the maximum of stress and strain is occurred at the hinge joint of leaflets. . Deformation are observed in the middle of the leaflets as well as near the corners where two leaflets meet. This shows that, inertia of flow effect stress analysis result due pressure loads act into the leaflets [7]. This stress analysis is done to ensure the structural rigidity for an artificial heart valve during closed and opened.

3.2. *Flow Analysis of Artificial Heart Valve*

From Figure 5, simulation of fully opened and closed for artificial heart valve are done to analyse flow pattern analysis and vortex formation.

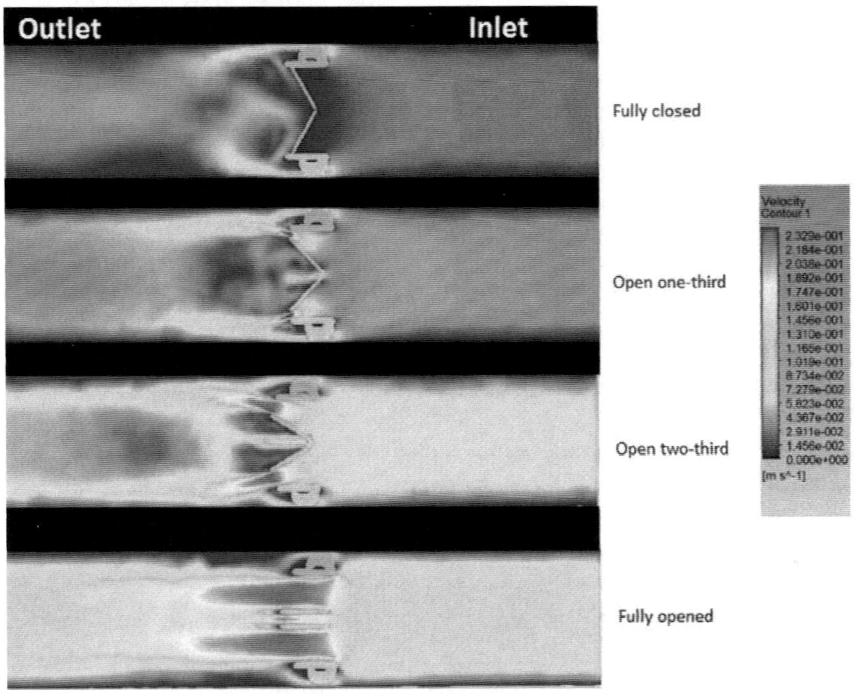

Figure 4. Flow velocity of artificial heart valve

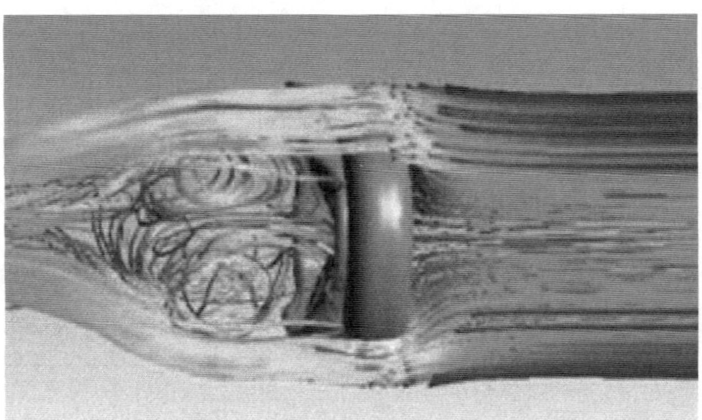

Figure 5. Vortex formation from artificial heart valve

Based on the velocity contour in Figure 5, we can observe flow pattern of artificial heart valve during is closed, open one third, open two-third and fully opened. As shown in Figure 5, in fully open condition, there are flow separation at the middle area central of artificial heart valve. The peak velocity occurring from flow analysis is 0.2329m/s. When it is fully close, there are some leakage at the hinge area of both leaflets. Besides, the vortex formation at middle section of artificial heart valve is quite larger for an artificial heart valve design as shown in Figure 6. Formation of vortex is not favourable for an artificial heart valve that may lead to cavitation thereby might damaging the heart valve [8].

3.3. Doppler Velocity Index of Left heart valve Stenosis.

One of the criteria to determine the mitral stenosis is through the Doppler velocity index (DVI). From Figure 7, the normal DVI value for aortic valve and mitral valve should be more than 0.30 and less than 2.2.

	Normal		Possible stenosis		Significant stenosis	
	Aortic	Mitral	Aortic	Mitral	Aortic	Mitral
Valve structure and motion						
Mechanical or bioprosthesis	Normal	Normal	Often abnormal**	Often abnormal**	Abnormal**	Abnormal**
Doppler quantitative parameters						
Peak velocity (m/s)* †	<3	<1.9	3–4	1.9–2.5	>4	>2.5
Mean gradient (mm Hg)* †	<20	≤5	20–35	6–10	>35	>10
Doppler velocity index* ¶	≥0.30	≤2.2	0.25–0.29	2.2–2.5	≤0.25	>2.5
Effective orifice area (cm^2)* §	>1.2	≥2	0.8–1.2	1–2	<0.8	<1
Effective orifice area versus normal reference value* §	Reference±1SD	Reference±1SD	<Reference–1SD	<Reference–1SD	<Reference–2SD	<Reference–2SD
Contour of the transprosthetic jet‡	Triangular, early peaking	–	Triangular to intermediate	–	Rounded, symmetrical	–
Acceleration time (ms)‡	<80	–	80–100	–	>100	–
Pressure half time (ms)‡ 1	–	<130	–	130–200	–	>200

Figure 6. Allowable DVI in mitral valve[9]

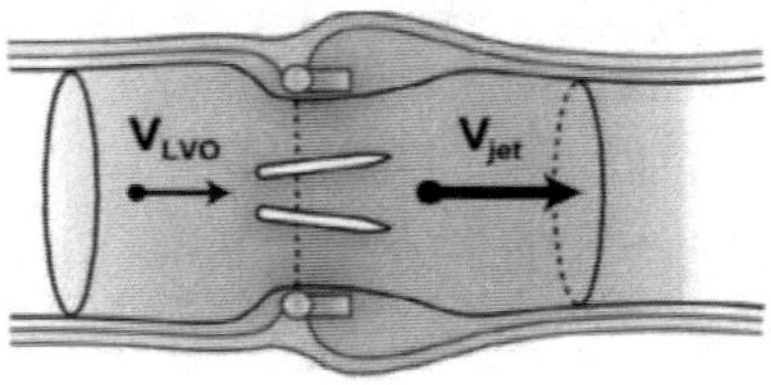

Figure 7. Parameters involved to calculate DVI

$$\text{For mitral valve: Doppler Velocity Index (DVI)} = \frac{V_{jet}}{V_{LVO}}$$

Table 2. Average velocity of blood flow and Doppler velocity index depicted from computational simulation when the valve is fully open.

Inlet velocity (m/s)						Outlet velocity (m/s)						DVI
1	2	3	4	5	Average	1	2	3	4	5	Average	
0.11	0.13	0.12	0.14	0.12	0.126	0.201	0.16	0.23	0.17	0.215	0.1952	1.55

From Table 3, DVI value of artificial heart valve is the 1.55 which is still within allowable range for left heart valve as shown in Fig. 7. Hence, the artificial heart valve is within the permissible limit and the artificial heart valve for this work is safe to be used.

4. Conclusion

For this research work, the effect of blood flow pattern, velocity and vorticity of the artificial heart valve design has been analyzed. The peak velocity shows in flow analysis is 0.2329 m/s. The stress analysis shown that higher stress, strain and deformation is occurring at the hinge of the both leaflets. Besides, the artificial heart valve design shows that it has a Doppler velocity index that is 1.55 which is less than the allowable standards for the left heart with values of more than 0.30 and less than 2.2. Therefore, from this value of DVI, the artificial heart valve in this research work is safe to be used as replacement of the human heart valve.

5. Acknowledgement

The work was partly supported by the Short Term Grant 60313020 from the Division of Research and Innovation, Universiti Sains Malaysia, and FRGS Grant 6071322 from the Ministry of Higher Education

References

[1] "Heart." [Online]. Available: https://www.nationalgeographic.com/science/health-and-human-body/human-body/heart/

[2] "Anatomy of Heart." [Online]. Available:https://www.nhlbi.nih.gov/health/health-topics/topics/hhw/anatomy Another reference

[3] "Heart valve disease." [Online]. Available:https://www.bhf.org.uk/heart-health/conditions/heart-valve-disease

[4] Dasi, L. P., Simon, H. A., Sucosky, P., & Yoganathan, A. P. (2009*). Fluid Mechanics Of Artificial Heart Valves*. Clinical and Experimental Pharmacology & Physiology, Volume 36 Issue 2, pp 225–237.

[5] Martin C., Sun W. (2011) *Bio-prosthetic Heart Valve Stress Analysis*: Impacts of Leaflet Properties and Stent Tip Deflection. In: Proulx T. (eds) Mechanics of Biological Systems and Materials, Volume 2. Conference Proceedings of the Society for Experimental Mechanics Series. Springer, New York, NY

[6] A.C. Benim, A. Nahavandi, A. Assmann, D. Schubert, P. Feindt, S.H. Suh. (2011) *Simulation of blood flow in human aorta with emphasis on outlet boundary conditions*, In Applied Mathematical Modelling, Volume 35, Issue 7, pp 3175-3188.

[7] H. Kim, J. Lu, M.S. Sacks, and K.B. Chandran (2008). *Dynamic Simulation of Bioprosthetic Heart Valves Using a Stress Resultant Shell Model.* Annals of Biomedical Engineering, Vol. 36, No. 2.

[8] Avrahami, I., Rosenfeld, M., Einav, S. et al. Med. Biol. Eng. Comput. (2000). *Can vortices in the flow across mechanical heart valves contribute to cavitation?.* Volume 38, Issue 1, pp 93-97.

[9] P. Pibarot and J. G. Dumesnil. (2012) *Doppler echocardiographic evaluation of prosthetic valve function*. Heart, vol. 98, no. 1, pp. 69–78.

[10] M. Engineering, (2012) . *Prediction of Blood Flow Velocity and Leaflet.* Volume 2, No. June, pp. 217–225.

Discrete Phase Model (DPM) study of nano-reinforced Lead Free Solder Sn-3.0Ag-0.5Cu (SAC305)

M.A.Fatah M.Mukhtar[1], Aizat Abas[2*], M.S Haslinda[2], F. Che Ani[3,4], M. Z. Abdullah[5], A.Jalar[6], R. Ismail[7]

[1]Faculty of Engineering, DRB-Hicom University of Automotive Malaysia, 26607 Pekan, Pahang, Malaysia.
[2]School of Mechanical Engineering, Engineering Campus, Universiti Sains Malaysia, Nibong Tebal, 14300, Penang, Malaysia.
[3,6,7]Institute of Microengineering and Nanoelectronics, Universiti Kebangsaan Malaysia, 43600 Bangi, Selangor, Malaysia.
[4]Jabil Circuits, 56, Hilir Sungai Kluang 1, Bayan Lepas, 11900 Penang, Malaysia.
[5]School of Aerospace Engineering, Engineering Campus, Universiti Sains Malaysia, Nibong Tebal, 14300, Penang, Malaysia.

*Corresponding author: aizatabas@usm.my

Abstract. This paper presents a preliminary study of the interaction between two models of numerical simulation namely volume of fluid (VOF) and discrete phase model (DPM). The multiphase-DPM model is capable of tracking the trajectory of the TiO2 nanoparticles that has been doped in the lead free solder Sn-3.0Ag-0.5Cu (SAC305). The current model was developed based on passive surface mount component 01005 capacitor that undergoes conventional reflow process. The trajectory of the nanoparticles at 0.01, 0.05 and 0.15wt% in the molten solder is shown in the flow front of the wetted molten solder. At 0.05wt% of nanoparticles, good dispersion of nanoparticles, pressure distribution and wetting time was found. The difference in wetting time are about 3.76% and 0.46% for 0.05wt% and 0.15wt% compared to 0.01wt% nanoparticle. The trajectory of the nanoparticles are shown to move along with the wetting formation of the molten solder. The pressure distribution and velocity vector of the different weight percentage of nanoparticles are also be studied. Higher pressure distribution is found at the capacitor area and intermetallic compound (IMC) region. The higher pressure distribution projected to reduce the micro void formation and optimize the reliability of the solder joint.

1. Introduction

Surface mount technology have allowed diversification in optimizing and sustaining the reliability of the integrated circuit (IC) package and miniaturization of passive and active components. In order to improve the reliability and performance of the electronic assembly, various types of nanoparticles have been used as the reinforced material doped in the molten solder [1]. The presence of foreign element in lead free solder such as aluminum oxide (Al2O3), nickle oxide (NiO), and titanium dioxide (TiO2) have changed the mechanical properties and microstructure of the lead free solder [2,3,4]. Y. Tang et al found that the hardness value of reinforced solder with TiO2 nanoparticles increases by 34% [5] and wettability of the solder material was improved at 0.05-0.1 wt% nanoparticles [6].

In the fluid modelling, finite volume (FVM) based model has been shown to have vast capability of modelling fluid with high accuracy [7]. This model only have limitation of solving continuous flow and limited to tracking nanoparticle movement. From literature, various study on the DPM has been introduced mainly in fluidized beds that computes gas-solid fluidization as a mean to track the particles trajectory [8]. Additionally, the ability of DPM model to simulate and track particles has also been study by S. Rashidi. et al in convective AL2O3–water nanofluid [9], N. Kharoua, et al in prediction of Black Powder distribution [10] and S. Lin et al in particle gasification [11].

Previous studies that have been conducted in the use of DPM model show the capability of particle tracking but have limitation in the study of coupling between both multiphase and discrete phase method. In this study, a two way interaction of multiphase and discrete phase model (DPM) is developed to track and analyse the trajectory of the nanoparticles that is being doped in the lead free solder Sn-3.0Ag-0.5Cu (SAC305).

2. Method

The numerical model used in this study is based on Navier-Stokes equation that involve conservation of mass and momentum:

Continuity equation:

$$\frac{\partial \rho}{\partial t} + \nabla(\rho u) = 0 \tag{1}$$

Energy equation:

$$\rho C_P(\frac{\partial T}{\partial t} + u\nabla T) = \nabla(k\nabla T) + \Phi \tag{2}$$

Momentum equation:

$$\frac{\partial(\rho u)}{\partial t} + \nabla(\rho u u) = -\nabla P + \nabla \tau + \rho g \tag{3}$$

in which, ρ denotes the density of the fluid, u is the velocity, τ is the shear stress and gravity, g = -9.81 m/s.

The model is based on two phase interactions of SAC305 and air. Both phases are distinguished by the volume of fraction (VOF) in the range 0 to 1 with f = 1 indicating the presence of SAC 305 solder as described below,

VOF transport equation:

$$\frac{\partial \rho}{\partial t} f + u\nabla f = 0 \tag{4}$$

The trajectory and tracking of the nanoparticles in the fluid are associated with the mass and other forces exerted on each of the nanoparticles as given by:

Particle force balance:

$$\frac{\partial u_P}{\partial t} = F_D(u - u_P) + g_x\frac{(\rho_P - \rho)}{\rho_P} + F_x \tag{5}$$

where FD (u -up) is the drag force associated with the velocity of fluid and particles, μ is the molecular viscosity of the fluid, ρp is particle density and dp is the particle diameter.

Additional force, Fx is related to the mass, acceleration and pressure gradient in the fluid based on the consideration that ρ> ρp.

$$F_x = \frac{\rho}{\rho_P} u_P(\frac{\partial u}{\partial x}) \tag{6}$$

The dispersion and movement of the particle in a fluid is described by the Brownian motion of the particles. The interaction of the particle with the fluid medium is associated with the Brownian force, FB as:

$$F_B = \xi\sqrt{\frac{\pi S_o}{\Delta t}} \tag{7}$$

Additionally, the force that takes into consideration the effect nanoparticle sizes in terms of the interaction of lift, shear and drag exerted on each of the particle is the Saffman's lift force. Due to the low particle Reynolds number, this force always acts upward.

Saffman's lift force:

$$F_L = \frac{2Kv^{1/2}\rho d_{ij}}{\rho_p d_p (d_{ik} d_{ki})^{1/4}}(u - u_p) \tag{8}$$

where, v = kinematic viscosity, $K = 2.594$, constant, and

d_{ij}, d_{ik}, d_{ki} = deformation tensor.

The nanoparticle is expected to move along with the formation of the wetting which show two way interaction of the multiphase-DPM.

3. Numerical Simulation Setup

The environment of conventional reflow process is replicated in Ansys FLUENT using two way interactions of multiphase VOF and DPM model. A 3-dimensional model of a 01005 capacitor was constructed and will undergo a two-way interactions with the mixed solder paste. The discrete particle phase will enable precise prediction of the trajectory nanoparticles. The schematic diagram and mesh model of the printed circuit board (PCB) and 01005 capacitor are represented in Figure 1. The mesh developed for the current model is based on tetrahedrons mesh. A 3-dimensional model is developed to mimic real PCB and 01005 capacitor as shown in Table 2.

The Sn-3.0Ag-0.5Cu (SAC305) is used as the molten solder and nanoparticles doped with TiO2 nanoparticles and the properties are tabulated in Table 2. The SAC305 is initially patched into the mesh grid with the volume fraction of 1 to simulate real industry environment. The nanoparticles with weight percentages of 0.01, 0.05, and 0.15wt% are doped in the molten solder.

The solver used to solve coupling of both pressure and velocity in this model is Semi-Implicit Method (SIMPLE) algorithm scheme. This numerical solver is used to obtain high accuracy result.

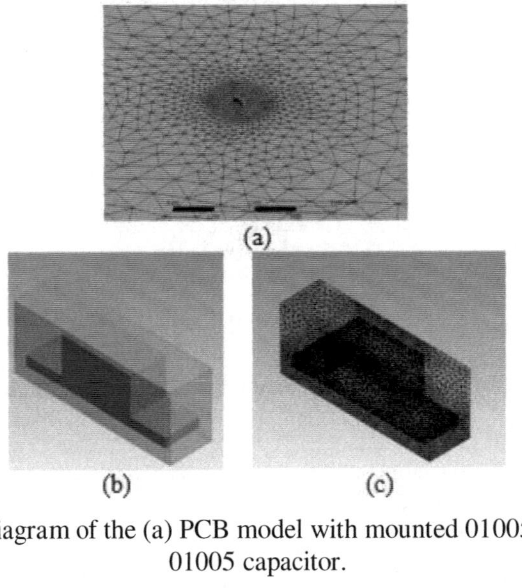

Figure 1. Schematic diagram of the (a) PCB model with mounted 01005 capacitor, and (b) (c) 01005 capacitor.

Table 1. Dimension of the FR4-PCB and 01005 capacitor.

DIMENSIONS (MM)	
FR4-PCB	255.0 x 178.0 x 2.0
01005 Capacitor	0.4 x 0.2 x 0.2

Table 2. SAC305 solder and TiO2 nanoparticle properties

Properties SAC305	
Density, p	7380 kg/m^3
Specific heat capacity, Cp	230 J kg ^{o} K
Thermal Conductivity	58 Wm/K
Viscosity	0.002 kg/ms
Standard state enthalpy	0.04 J/mol
Properties TiO_2	
Density, p	4250 kg/m^3
Specific heat capacity, Cp	686 J kg ^{o} K
Thermal Conductivity	8.95 Wm/K
Diameter, d	≈20nm

4. Result and Discussion

4.1. Flow Front Pattern

Figure 2 presents the flow front of the wetting molten solder with the trajectory of the nanoparticles at different weight percentages of 0.01, 0.05, and 0.15wt%. From the flow front, it can be shown that the nanoparticles trajectory are dispersed according to the motion of the wetting molten solder. The dispersion of the nanoparticles at some extent varies due to different concentration of nanoparticles present in the molten solder. At 0.01wt% of nanoparticles, the lowest trajectory and dispersion of the nanoparticles were found. This is due to the small amount of nanoparticles being injected in the molten solder. Better dispersion of the nanoparticles in the wetted molten solder is given by 0.05wt% of nanoparticles compared to 0.01wt% and 0.15wt%. At 0.15wt% of nanoparticles, the trajectory of the nanoparticles tends to accumulate at the upper layer of the molten solder. The dispersion of the nanoparticles at the upper layer was influenced by the buoyancy effect and density due to the density of the nanoparticles are much smaller compared to SAC305. The randomness and instability of the dispersion of nanoparticle will reduce the strength and the reliability of the solder joint. As studied previously by T. Laurila et al, failure result was found showing in drop test analysis for the case of more than 0.1wt% addition of reinforcement material into the SAC solder [12].

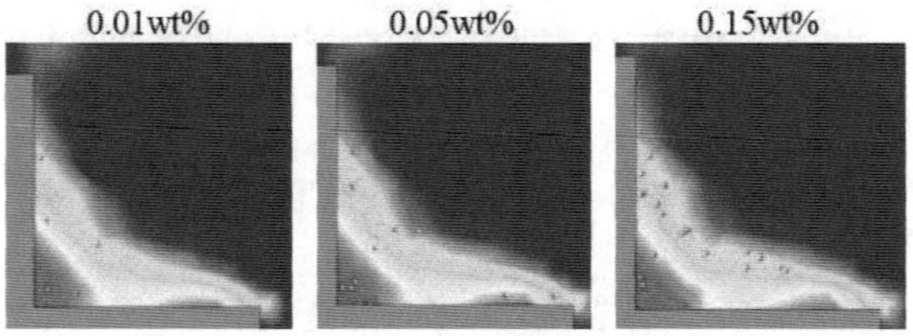

Figure 2. The flow front of the different weight percentage of nanoparticles.

The distribution of pressure and velocity in the wetted molten solder is depicted in Figure 3.The trajectory of the velocity vector for 0.15wt% is higher compared to 0.05wt% and 0.01wt% due to the higher concentration of the nanoparticle presents in the molten solder as in Figure 3. From Figure 3, the highest distribution of pressure is shown at the bottom of the PCB pad and side terminal of the 01005 capacitor. This due to the mechanism of the formation wetted solder which associates with surface tension, since higher surface tension means more wetting and more exerted force, then more pressure. At 0.05wt% show higher pressure distribution compared to 0.01wt% and 0.05wt%. During the instantaneous nanoparticles dispersion in the molten solder, the nanoparticles that are present at the high pressure contour have the possibility to inhibit the formation of micro-voids. The trajectory of the nanoparticles typically at the higher pressure region have the lower possibility of the micro-voids formation. This was proven in a research by A. Abas et al [13] and M. H.H Ishak et. As the size of the particles in the nano-scale, the nanoparticles are exerted velocity same as the al [14] which found that the possibility of void formation is lower at higher pressure distribution.

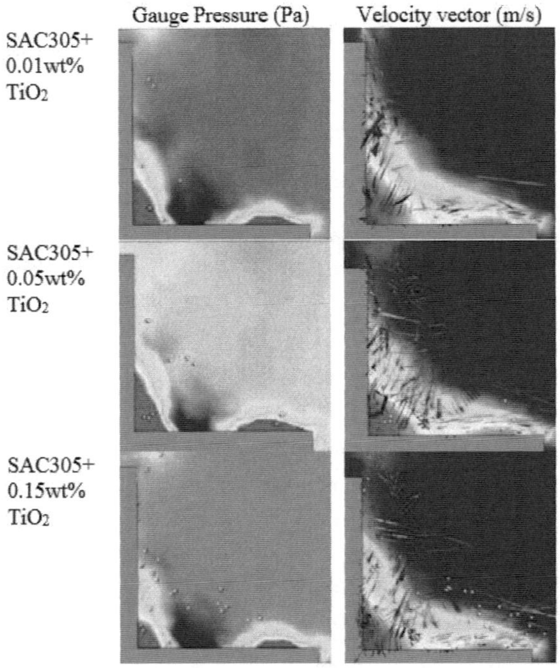

Figure 3. Pressure distribution and velocity vector at different weight percentage of nanoparticles.

4.2. Wetting Time

Figure 4 depicted the plot of wetting time against wetting percentage for different weight percentage of the nanoparticles. At higher weight percentage of the nanoparticles, the wetting time increases. The addition of more particles seems to increase the surface that subsequently promotes the wettability of the mixed SAC solder. At 20% of wetting percentage, the time taken for each case is 1.823s, 1.712s, and 1.515s for 0.01, 0.05 and 0.15wt% respectively. Additionally, at 80% wetting percentage, the wetting time for 0.01, 0.05 and 0.15wt% are 2.820s, 2.926s, and 2.833s. There occurs a slight difference about 3.76% and 0.46% of wetting time for 0.05wt% and 0.15wt% compared to 0.01wt% nanoparticles. This as previously mentioned is due to the alteration of the current surface leading to an improved wetting capability.

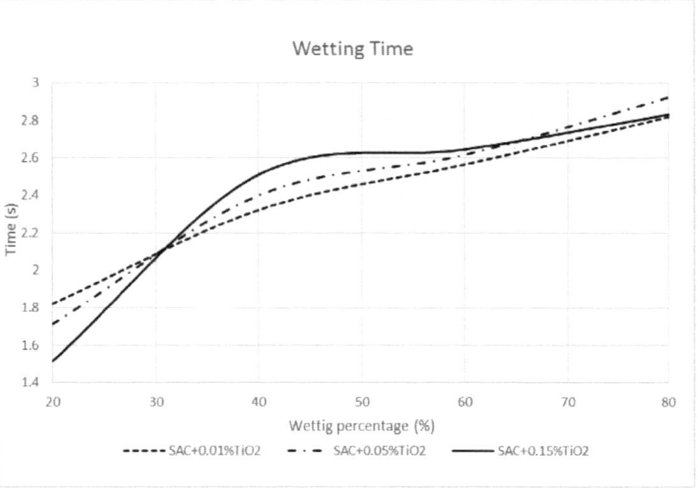

Figure 4. Graph of wetting time versus wetting percentage (%).

5. Conclusion

This paper studies the dispersion of the discrete nanoparticles doped in SAC305 molten solder. The model was developed based on two way interactions of the multiphase-DPM model finite volume based simulation. The trajectory of the nanoparticles in the wetted molten solder is being tracked based on the flow front pattern of the simulation at different weight percentage of nanoparticles at 0.01, 0.05 and 0.15wt%. The nanoparticles trajectory for 0.05wt% is shown to have the best dispersion in comparison other nanoparticles percentages. The current model also shows the pressure distribution and velocity vector of the nanoparticles due to its direct effect on the reliability and strength of the solder joint. The study also found high pressure distribution at the vicinity of the component and intermetallic compound (IMC) region. The high pressure distribution is expected to reduce micro void formation. The addition of nanoparticles found slight increment in the wetting time as weight percentages of nanoparticles are increased.

Acknowledgement

This research work was partly supported by FRGS grant, FRGS/1/2015/TK03/USM/03/2, Short Term Grant, 60313020 from the Division of Research and Innovation, Universiti Sains Malaysia and Universiti Kebangsaan Malaysia (Research grant–DIP-2014-012). The authors wish to extend their appreciation to Mr. M.Y. Tura and Mr. A. Marzukhi (Jabil Circuit Sdn Bhd) for their technical support.

References

[1] E.M.N. Ervina & A. Singh 2014 *Soldering & Surface Mount Technology*, **26**, 147
[2] L. C. Tsao, S. Y. Chang, C. I. Lee, W. H. Sun, & C. H. Huang 2010 *Materials & Design* **31** 4831
[3] S. Chellvarajoo & M.Z. Abdullah 2016 *Materials & Design* **90** 499
[4] Y. Tang, G. Y. Li, D. Q. Chen, & Y. C. Pan 2014 *Journal of Materials Science: Materials in Electronics* **25**, 981-991
[5] Y. Tang, G.Y. Li, & Y.C. Pan 2014 *Materials & Design* **55** 574
[6] Y. Tang, Y. C. Pan, & G. Y. Li 2013 *J Mater Sci: Mater Electron* **24** 1587
[7] F. C. Ng, A. Abas, Z. L. Gan, M. Z., Abdullah, F. C. Ani, & M. Y. T. Ali 2017 *Microelectronics Reliability* **72** 45
[8] L. Lu, J. Xu, W. Ge, G. Gao, Y. Jiang, M. Zhao, & J. Li 2016 *Chemical Engineering Science* **155** 314

[9] S. Rashidi, M. Bovand, J. A. Esfahani, & G. Ahmadi 2016 *Applied Thermal Engineering* **100** 39
[10] N. Kharoua, M. AlShehhi, & L. Khezzar 2015 *Powder Technology* **286** 202
[11] S. Lin, L. Ding, Z. Zhou, & G. Yu 2016 *Fuel* **186**, 656
[12] T. Laurila, V. Vuorinen, & M. Paulasto-Kröckel 2010 *Mater. Sci. Eng. R* **68** 1
[13] A. Abas, M. H. H. Ishak, M. Z. Abdullah, F. C. Ani, & S. F. Khor 2016 *Microelectronics Reliability* **56** 170
[14] M. H. H. Ishak, M. Z. Abdullah, & A. Abas 2016 *Microelectronics Reliability* **65** 205